Analysis of Variance, Design, and Regression

Linear Modeling for Unbalanced Data

Second Edition

CHAPMAN & HALL/CRC
Texts in Statistical Science Series

Series Editors

Francesca Dominici, *Harvard School of Public Health, USA*
Julian J. Faraway, *University of Bath, UK*
Martin Tanner, *Northwestern University, USA*
Jim Zidek, *University of British Columbia, Canada*

Statistical Inference: An Integrated Approach,
Second Edition
H. S. Migon, D. Gamerman, and
F. Louzada

Beyond ANOVA: Basics of Applied Statistics
R.G. Miller, Jr.

A Primer on Linear Models
J.F. Monahan

Applied Stochastic Modelling, Second Edition
B.J.T. Morgan

Elements of Simulation
B.J.T. Morgan

Probability: Methods and Measurement
A. O'Hagan

Introduction to Statistical Limit Theory
A.M. Polansky

Applied Bayesian Forecasting and Time Series
Analysis
A. Pole, M. West, and J. Harrison

Statistics in Research and Development,
Time Series: Modeling, Computation, and
Inference
R. Prado and M. West

Introduction to Statistical Process Control
P. Qiu

Sampling Methodologies with Applications
P.S.R.S. Rao

A First Course in Linear Model Theory
N. Ravishanker and D.K. Dey

Essential Statistics, Fourth Edition
D.A.G. Rees

Stochastic Modeling and Mathematical
Statistics: A Text for Statisticians and
Quantitative Scientists
F.J. Samaniego

Statistical Methods for Spatial Data Analysis
O. Schabenberger and C.A. Gotway

Bayesian Networks: With Examples in R
M. Scutari and J.-B. Denis

Large Sample Methods in Statistics
P.K. Sen and J. da Motta Singer

Spatio-Temporal Methods in Environmental
Epidemiology
G. Shaddick and J.V. Zidek

Decision Analysis: A Bayesian Approach
J.Q. Smith

Analysis of Failure and Survival Data
P. J. Smith

Applied Statistics: Handbook of GENSTAT
Analyses
E.J. Snell and H. Simpson

Applied Nonparametric Statistical Methods,
Fourth Edition
P. Sprent and N.C. Smeeton

Data Driven Statistical Methods
P. Sprent

Generalized Linear Mixed Models:
Modern Concepts, Methods and Applications
W. W. Stroup

Survival Analysis Using S: Analysis of
Time-to-Event Data
M. Tableman and J.S. Kim

Applied Categorical and Count Data Analysis
W. Tang, H. He, and X.M. Tu

Elementary Applications of Probability Theory,
Second Edition
H.C. Tuckwell

Introduction to Statistical Inference and Its
Applications with R
M.W. Trosset

Understanding Advanced Statistical Methods
P.H. Westfall and K.S.S. Henning

Statistical Process Control: Theory and
Practice, Third Edition
G.B. Wetherill and D.W. Brown

Generalized Additive Models:
An Introduction with R
S. Wood

Epidemiology: Study Design and
Data Analysis, Third Edition
M. Woodward

Practical Data Analysis for Designed
Experiments
B.S. Yandell

Texts in Statistical Science

Analysis of Variance, Design, and Regression

Linear Modeling for Unbalanced Data

Second Edition

Ronald Christensen

University of New Mexico
Albuquerque, USA

CRC Press
Taylor & Francis Group
Boca Raton London New York

CRC Press is an imprint of the
Taylor & Francis Group, an **informa** business

A CHAPMAN & HALL BOOK

CRC Press
Taylor & Francis Group
6000 Broken Sound Parkway NW, Suite 300
Boca Raton, FL 33487-2742

First issued in paperback 2020

© 2016 by Taylor & Francis Group, LLC
CRC Press is an imprint of Taylor & Francis Group, an Informa business

No claim to original U.S. Government works

ISBN-13: 978-1-4987-3014-3 (hbk)
ISBN-13: 978-0-367-73740-5 (pbk)

Visit the Taylor & Francis Web site at
http://www.taylorandfrancis.com

and the CRC Press Web site at
http://www.crcpress.com

To Mark, Karl, and John

It was great fun.

Contents

CONTENTS

Preface

Background

Big Data are the future of Statistics. The electronic revolution has increased exponentially our ability to measure things. A century ago, data were hard to come by. Statisticians put a premium on extracting every bit of information that the data contained. Now data are easy to collect; the problem is sorting through them to find meaning. To a large extent, this happens in two ways: doing a crude analysis on a massive amount of data or doing a careful analysis on the moderate amount of data that were isolated from the massive data as being meaningful. It is quite literally impossible to analyze a million data points as carefully as one can analyze a hundred data points, so "crude" is not a pejorative term but rather a fact of life.

The fundamental tools used in analyzing data have been around a long time. It is the emphases and the opportunities that have changed. With thousands of observations, we don't need a perfect statistical analysis to detect a large effect. But with thousands of observations, we might look for subtle effects that we never bothered looking for before, and such an analysis must be done carefully—as must any analysis in which only a small part of the massive data are relevant to the problem at hand. The electronic revolution has also provided us with the opportunity to perform data analysis procedures that were not practical before, but in my experience, the new procedures (often called *machine learning*), are sophisticated applications of fundamental tools.

This book explains some of the fundamental tools and the ideas needed to adapt them to big data. It is not a book that analyzes big data. The book analyzes small data sets carefully but by using tools that 1) can easily be scaled to large data sets or 2) apply to the haphazard way in which small relevant data sets are now constructed. Personally, I believe that it is not safe to apply models to large data sets until you understand their implications for small data. There is also a major emphasis on tools that look for subtle effects (interactions, homologous effects) that are hard to identify.

The fundamental tools examined here are linear structures for modeling data; specifically, how to incorporate specific ideas about the structure of the data into the model for the data. Most of the book is devoted to adapting linear structures (regression, analysis of variance, analysis of covariance) to examine measurement (continuous) data. But the exact same methods apply to either-or (Yes/No, binomial) data, count (Poisson, multinomial) data, and time-to-event (survival analysis, reliability) data. The book also places strong emphasis on foundational issues, e.g., the meaning of significance tests and the interval estimates associated with them; the difference between prediction and causation; and the role of randomization.

The platform for this presentation is the revision of a book I published in 1996, *Analysis of Variance, Design, and Regression: Applied Statistical Methods*. Within a year, I knew that the book was not what I thought needed to be taught in the 21st century, cf., Christensen (2000). This book, *Analysis of Variance, Design, and Regression: Linear Modeling of Unbalanced Data*, shares with the earlier book lots of the title, much of the data, and even some of the text, but the book is radically different. The original book focused greatly on balanced analysis of variance. This book focuses on modeling unbalanced data. As such, it generalizes much of the work in the previous book. The more general methods presented here agree with the earlier methods for balanced data. Another advantage of taking a modeling approach to unbalanced data is that by making the effort to treat unbalanced analysis of variance, one can easily handle a wide range of models for nonnormal data, because the same fundamental methods apply. To that end, I have included new chapters on logistic regression, log-linear models, and time-to-event data. These are placed near the end of the book, not because they are less important, but because the real subject of the book is modeling with linear structures and the methods for measurement data carry over almost immediately.

In early versions of this edition I made extensive comparisons between the methods used here and the balanced ANOVA methods used in the 1996 book. In particular, I emphasized how the newer methods continue to give the same results as the earlier methods when applied to balanced data. While I have toned that down, comparisons still exist. In such comparisons, I do not repeat the details of the balanced analysis given in the earlier book. CRC Press/Chapman & Hall have been kind enough to let me place a version of the 1996 book on my website so that readers can explore the comparisons in detail. Another good thing about having the old book up is that it contains a chapter on confounding and fractional replications in 2^n factorials. I regret having to drop that chapter, but the discussion is based on contrasts for balanced ANOVA and did not really fit the theme of the current edition.

When I was in high school, my two favorite subjects were math and history. On a whim, I made the good choice to major in Math for my BA. I mention my interest in history to apologize (primarily in the same sense that C.S. Lewis was a Christian "apologist") for using so much old data. Unless you are trying to convince 18-year-olds that Statistics is sexy, I don't think the age of the data should matter.

I need to thank Adam Branscum, my coauthor on Christensen et al. (2010). Adam wrote the first drafts of Chapter 7 and Appendix C of that book. Adam's work on Chapter 7 definitely influenced this work and Adam's work on Appendix C is what got me programming in R. This is also a good time to thank the people who have most influenced my career: Wes Johnson, Ed Bedrick, Don Berry, Frank Martin, and the late, great Seymour Geisser. My colleague Yan Lu taught out of a prepublication version of the book, and, with her students, pointed out a number of issues. Generally, the first person whose opinions and help I sought was my son Fletcher.

After the effort to complete this book, I'm feeling as unbalanced as the data being analyzed.

Specifics

I think of the book as something to use in the traditional Master's level year-long course on regression and analysis of variance. If one needed to actually separate the material into a regression course and an ANOVA course, the regression material is in Chapters 6–11 and 20–23. Chapters 12–19 are traditionally viewed as ANOVA. But I much prefer to use both regression and ANOVA ideas when examining the generalized linear models of Chapters 20–22. Well-prepared students could begin with Chapter 3 and skip to Chapter 6. By well-prepared, I tautologically mean students who are already familiar with Chapters 1, 2, 4, and 5.

For less well-prepared students, obviously I would start at the beginning and deemphasize the more difficult topics. This is what I have done when teaching data analysis to upper division Statistics students and graduate students from other fields. I have tried to isolate more difficult material into clearly delineated (sub)sections. In the first semester of such a course, I would skip the end of Chapter 8, include the beginning of Chapter 12, and let time and student interest determine how much of Chapters 9, 10, and 13 to cover. But the book wasn't written to be a text for such a course; it is written to address unbalanced multi-factor ANOVA.

The book requires very little pre-knowledge of math, just algebra, but does require that one not be afraid of math. It does not perform calculus, but it discusses that integrals provide areas under curves and, in an appendix, gives the integral formulae for means and variances. It largely avoids matrix algebra but presents enough of it to enable the matrix approach to linear models to be introduced. For a regression-ANOVA course, I would supplement the material after Chapter 11 with occasional matrix arguments. Any material described as a regression approach to an ANOVA problem lends itself to matrix discussion.

Although the book starts at the beginning mathematically, it is not for the intellectually unsophisticated. By Chapter 2 it discusses the impreciseness of our concepts of populations and how the deletion of outliers must change those concepts. Chapter 2 also discusses the "murky" transformation from a probability interval to a confidence interval and the differences between significance testing, Neyman–Pearson hypothesis testing, and Bayesian methods. Because a lot of these ideas are subtle, and because people learn best from specifics to generalities rather than the other way around, Chapter 3 reiterates much of Chapter 2 but for general linear models. Most of the remainder of the book can be viewed as the application of Chapter 3 to specific data structures. Well-prepared students could start with Chapter 3 despite occasional references made to results in the first two chapters.

Chapter 4 considers two-sample data. Perhaps its most unique feature is, contrary to what seems popular in introductory Statistics these days, the argument that testing equality of means for two independent samples provides much less information when the variances are different than when they are the same.

Chapter 5 exists because I believe that if you teach one- and two-sample continuous data problems, you have a duty to present their discrete data analogs. Having gone that far, it seemed silly to avoid analogs to one-way ANOVA. I do not find the one-way ANOVA F test for equal group means to be all that useful. Contrasts contain more interesting information. The last two sections of Chapter 5 contain, respectively, discrete data analogs to one-way ANOVA and a method of extracting information similar to contrasts.

Chapters 6, 7, and 8 provide tools for exploring the relationship between a single dependent variable and a single measurement (continuous) predictor. A key aspect of the discussion is that the methods in Chapters 7 and 8 extend readily to more general linear models, i.e., those involving categorial and/or multiple predictors. The title of Chapter 8 arises from my personal research interest in testing lack of fit for linear models and the recognition of its relationship to nonparametric regression.

Chapters 9, 10, and 11 examine features associated with multiple regression. Of particular note are new sections on modeling interaction through generalized additive models and on lasso regression. I consider these important concepts for serious students of Statistics. The last of these chapters is where the book's use of matrices is focused. The discussion of principal component regression is located here, not because the discussion uses matrices, but because the discussion requires matrix knowledge to understand.

The rest of the book involves categorical predictor variables. In particular, *the material after Chapter 13 is the primary reason for writing this edition*. The first edition focused on multifactor balanced data and looking at contrasts, not only in main effects but contrasts within two- and three-factor interactions. This edition covers the same material for unbalanced data.

Chapters 12 and 13 cover one-way analysis of variance (ANOVA) models and multiple comparisons but with an emphasis on the ideas needed when examining multiple categorical predictors. Chapter 12 involves one categorical predictor much like Chapter 6 involved one continuous predictor.

Chapter 14 examines the use of two categorial predictors, i.e., two-way ANOVA. It also introduces the concept of homologous factors. Chapter 15 looks at models with one continuous and one categorical factor, analysis of covariance. Chapter 16 considers models with three categorical predictors.

Chapters 17 and 18 introduce the main ideas of experimental design. Chapter 17 introduces a wide variety of standard designs and concepts of design. Chapter 18 introduces the key idea of defining treatments with factorial structure. The unusual aspect of these chapters is that the analyses presented apply when data are missing from the original design.

Chapter 19 introduces the analysis of dependent data. The primary emphasis is on the analysis of split-plot models. A short discussion is also given of multivariate analysis. Both of these methods require groups of observations that are independent of other groups but that are dependent within the groups. Both methods require balance within the groups but the groups themselves can be unbalanced. Subsection 19.2.1 even introduces a method for dealing with unbalance within groups.

It seems to have become popular to treat fixed and random effects models as merely two options for analyzing data. I think these are very different models with very different properties; random effects being far more sophisticated. As a result, I have chosen to introduce random effects as a special case of split-plot models in Subsection 19.4.2. Subsampling models can also be viewed as special cases of split-plot models and are treated in Subsection 19.4.1.

Chapters 20, 21, and 22 illustrate that the modeling ideas from the previous chapters continue to apply to generalized linear models. In addition, Chapter 20 spends a lot of time pointing out potholes that I see in standard programs for performing logistic regression.

Chapter 23 is a brief introduction to nonlinear regression. It is the only chapter, other than Chapter 11, that makes extensive use of matrices and the only one that requires knowledge of calculus. Nonlinear regression is a subject that I think deserves more attention than it gets. I think it is the form of regression that we should aspire to, in the sense that we should aspire to having science that is sophisticated enough to posit such models.

Ronald Christensen
Albuquerque, New Mexico
February 2015

Edited Preface to First Edition

This book examines the application of basic statistical methods: primarily analysis of variance and regression but with some discussion of count data. It is directed primarily towards Master's degree students in Statistics studying analysis of variance, design of experiments, and regression analysis. I have found that the Master's level regression course is often popular with students outside of Statistics. These students are often weaker mathematically and the book caters to that fact while continuing to give a complete matrix formulation of regression.

The book is complete enough to be used as a second course for upper division and beginning graduate students in Statistics and for graduate students in other disciplines. To do this, one must be selective in the material covered, but the more theoretical material appropriate only for Statistics Master's students is generally isolated in separate subsections and, less often, in separate sections.

I think the book is reasonably encyclopedic. It really contains everything I would like my students to know about Applied Statistics prior to them taking courses in linear model theory or log-linear models.

I believe that beginning students (even Statistics Master's students) often find statistical procedures to be a morass of vaguely related special techniques. As a result, this book focuses on four connecting themes.

1. Most inferential procedures are based on identifying a (scalar) parameter of interest, estimating that parameter, obtaining the standard error of the estimate, and identifying the appropriate reference distribution. Given these items, the inferential procedures are identical for various parameters.

2. Balanced one-way analysis of variance has a simple, intuitive interpretation in terms of comparing the sample variance of the group means with the mean of the sample variances for each group. All balanced analysis of variance problems can be considered in terms of computing sample variances for various group means. *These concepts exist in the new edition but are de-emphasized as are balanced data.*

3. Comparing different models provides a structure for examining both balanced and unbalanced analysis of variance problems and for examining regression problems. In some problems the most reasonable analysis is simply to find a succinct model that fits the data well. *This is the core of the new edition.*

4. Checking assumptions is a crucial part of every statistical analysis.

The object of statistical data analysis is to reveal useful structure within the data. In a model-based setting, I know of two ways to do this. One way is to find a *succinct* model for the data. In such a case, the structure revealed is simply the model. The model selection approach is particularly appropriate when the ultimate goal of the analysis is making predictions. This book uses the model selection approach for multiple regression and for general unbalanced multifactor analysis of variance. The other approach to revealing structure is to start with a general model, identify interesting one-dimensional parameters, and perform statistical inferences on these parameters. This parametric approach requires that the general model involve parameters that are easily interpretable. We exploit the parametric approach for one-way analysis of variance and simple linear regression.

All statistical models involve assumptions. Checking the validity of these assumptions is crucial because *the models we use are never correct. We hope that our models are good approximations of the true condition of the data and experience indicates that our models often work very well.* Nonetheless, to have faith in our analyses, we need to check the modeling assumptions as best we can. Some assumptions are very difficult to evaluate, e.g., the assumption that observations are statistically independent. For checking other assumptions, a variety of standard tools has been developed. Using these tools is as integral to a proper statistical analysis as is performing an appropriate confidence interval or test. For the most part, using model-checking tools without the aid of a computer is more trouble than most people are willing to tolerate.

My experience indicates that students gain a great deal of insight into balanced analysis of variance by actually doing the computations. The computation of the mean square for treatments in a balanced one-way analysis of variance is trivial on any hand calculator with a variance or standard deviation key. More importantly, the calculation reinforces the fundamental and intuitive idea behind the balanced analysis of variance test, i.e., that a mean square for treatments is just a multiple of the sample variance of the corresponding treatment means. I believe that as long as students find the balanced analysis of variance computations challenging, they should continue to do them by hand (calculator). I think that automated computation should be motivated by boredom rather than bafflement. *While I still believe this is true, it too is deemphasized in this edition.*

In addition to the four primary themes discussed above, there are several other characteristics that I have tried to incorporate into this book.

I have tried to use examples to motivate theory rather than to illustrate theory. Most chapters begin with data and an initial analysis of that data. After illustrating results for the particular data, we go back and examine general models and procedures. I have done this to make the book more palatable to two groups of people: those who only care about theory after seeing that it is useful and those unfortunates who can never bring themselves to care about theory. (The older I get, the more I identify with the first group. As for the other group, I find myself agreeing with W. Edwards Deming that experience without theory teaches nothing.) As mentioned earlier, the theoretical material is generally confined to separate subsections or, less often, separate sections, so it is easy to ignore.

I believe that the *ultimate* goal of all statistical analysis is prediction of observable quantities. I have incorporated predictive inferential procedures where they seemed natural.

The object of most Statistics books is to illustrate techniques rather than to analyze data; this book is no exception. Nonetheless, I think we do students a disservice by not showing them a substantial portion of the work necessary to analyze even 'nice' data. To this end, I have tried to consistently examine residual plots, to present alternative analyses using different transformations and case deletions, and to give some final answers in plain English. I have also tried to introduce such material as early as possible. I have included reasonably detailed examinations of a three-factor analysis of variance and of a split-plot design with four factors. I have included some examples in which, like real life, the final answers are not 'neat.' While I have tried to introduce statistical ideas as soon as possible, I have tried to keep the mathematics as simple as possible for as long as possible. For example, matrix formulations are postponed to the last chapter on multiple regression.

I never use side conditions or normal equations in analysis of variance. *But computer programs use side conditions and I discuss how they affect model interpretations.*

In multiple comparison methods, (weakly) controlling the experimentwise error rate is discussed in terms of first performing an omnibus test for no treatment effects and then choosing a criterion for evaluating individual hypotheses. Most methods considered divide into those that use the omnibus F test, those that use the Studentized range test, and the Bonferroni method, which does not use any omnibus test. *In the current edition I have focused primarily on multiple comparison methods that work for unbalanced data.*

I have tried to be very clear about the fact that experimental designs are set up for arbitrary groups of treatments and that factorial treatment structures are simply an efficient way of defining the treatments in some problems. Thus, the nature of a randomized complete block design does not depend on how the treatments happen to be defined. The analysis always begins with a breakdown of the sum of squares into blocks, treatments, and error. Further analysis of the treatments then focuses on whatever structure happens to be present.

The analysis of covariance chapter *no longer* includes an extensive discussion of how the covariates must be chosen to maintain a valid experiment. *That discussion has been moved to the chapter Basic Experimental Designs.* Tukey's one degree of freedom test for nonadditivity is presented as a test for the need to perform a power transformation rather than as a test for a particular type of interaction. *Tukey's test is now part of the Model Checking chapter, not the ACOVA chapter.*

The chapter on confounding and fractional replication has more discussion of analyzing such data than many other books contain.

Acknowledgements

Many people provided comments that helped in writing this book. My colleagues Ed Bedrick, Aparna Huzurbazar, Wes Johnson, Bert Koopmans, Frank Martin, Tim O'Brien, and Cliff Qualls helped a lot. I got numerous valuable comments from my students at the University of New Mexico. Marjorie Bond, Matt Cooney, Jeff S. Davis, Barbara Evans, Mike Fugate, Jan Mines, and Jim Shields stand out in this regard. The book had several anonymous reviewers, some of whom made excellent suggestions.

I would like to thank Martin Gilchrist and Springer-Verlag for permission to reproduce Example 7.6.1 from *Plane Answers to Complex Questions: The Theory of Linear Models.* I also thank the Biometrika Trustees for permission to use the tables in Appendix B.5. Professor John Deely and the University of Canterbury in New Zealand were kind enough to support completion of the book during my sabbatical there.

Now my only question is what to do with the chapters on quality control, p^n factorials, and response surfaces that ended up on the cutting room floor. I have pretty much given up on publishing the quality control material. Response surfaces got into *Advanced Linear Modeling (ALM)* and I'm hoping to get p^n factorials into a new edition of *ALM*.

Ronald Christensen
Albuquerque, New Mexico
February 1996
Edited, October 2014

Computing

There are two aspects to computing: generating output and interpreting output. We cannot always control the generation of output, so we need to be able to interpret a variety of outputs. The book places great emphasis on interpreting the range of output that one might encounter when dealing with the data structures in the book. This comes up most forcefully when dealing with multiple categorical predictors because arbitrary choices must be made by computer programmers to produce some output, e.g., parameter estimates. The book deals with the arbitrary choices that are most commonly made. Methods for generating output have, for the most part, been removed from the book and placed on my website.

R has taken over the Statistics computing world. While R code is in the book, **illustrations of all the analyses and all of the graphics have been performed in R and are available on my website**: www.stat.unm.edu/~fletcher. Also, substantial bodies of Minitab and SAS code (particularly for SAS's GENMOD and LOGISTIC procedures) are available on my website. While Minitab and many versions of SAS are now menu driven, the menus essentially write the code for running a procedure. Presenting the code provides the information needed by the programs and, implicitly, the information needed in the menus. That information is largely the same regardless of the program. The choices of R, Minitab, and SAS are not meant to denigrate any other software. They are merely what I am most familiar with.

The online computing aids are chapter for chapter (and for the most part, section for section) images of the book. Thus, if you want help computing something from Section 2.5 of the book, look in Section 2.5 of the online material.

My strong personal preference is for doing whatever I can in Minitab. That is largely because Minitab forces me to remember fewer arcane commands than any other system (that I am familiar with). Data analysis output from Minitab is discussed in the book because it differs from the output provided by R and SAS. For fitting large tables of counts, as discussed in Chapter 21, I highly recommend the program BMDP 4F. Fortunately, this can now be accessed through some batch versions of SAS. **My website contains files for virtually all the data.** But you need to compare each file to the tabled data and not just assume that the file looks exactly like the table.

Finally, I would like to point out a notational issue. In both Minitab and SAS, "glm" refers to fitting *general linear models*. In R, "glm" refers to fitting *generalized linear models*, which are something different. Generalized linear models contain general linear models as a special case. The models in Chapters 20, 21, and 22 are different special cases of generalized linear models. (I am not convinced that generalized linear models are anything more than a series of special cases connected by a remarkable computing trick, cf. Christensen, 1997, Chapter 9.)

BMDP Statistical Software was located at 1440 Sepulveda Boulevard, Los Angeles, CA 90025.

MINITAB is a registered trademark of Minitab, Inc., 3081 Enterprise Drive, State College, PA 16801, telephone: (814) 238-3280, telex: 881612.

Chapter 1

Introduction

Statistics has two roles in society. First, Statistics is in the business of creating stereotypes. Think of any stereotype you like, but to keep me out of trouble let's consider something innocuous, like the hypothesis that Italians talk with their hands more than Scandinavians. To establish the stereotype, you need to collect data and use it to draw a conclusion. Often the conclusion is that either the data suggest a difference or that they do not. The conclusion is (almost) never whether a difference actually exists, only whether or not the data suggest a difference and how strongly they suggest it. Statistics has been filling this role in society for at least 100 years.

Statistics' less recognized second role in society is debunking stereotypes. Statistics is about appreciating variability. It is about understanding variability, explaining it, and controlling it. I expect that with enough data, one could show that, on average, Italians really do talk with their hands more than Scandinavians. Collecting a lot of data helps control the relevant variability and allows us to draw a conclusion. But I also expect that we will never be able to predict *accurately* whether a random Italian will talk with their hands more than a random Scandinavian. There is too much variability among humans. Even when differences among groups exist, those differences often pale in comparison to the variability displayed by individuals within the groups—to the point where group differences are often meaningless when dealing with individuals. For statements about individuals, collecting a lot of data only helps us to more accurately state the limits of our (very considerable) uncertainty.

Ultimately, Statistics is about what you can conclude and, equally, what you cannot conclude from analyzing data that are subject to variability, as all data are. Statisticians use ideas from probability to quantify variability. They typically analyze data by creating probability models for the data.

In this chapter we introduce basic ideas of probability and some related mathematical concepts that are used in Statistics. Values to be analyzed statistically are generally thought of as random variables; these are numbers that result from random events. The mean (average) value of a population is defined in terms of the expected value of a random variable. The variance is introduced as a measure of the variability in a random variable (population). We also introduce some special distributions (populations) that are useful in modeling statistical data. The purpose of this chapter is to introduce these ideas, so they can be used in analyzing data and in discussing statistical models.

In writing statistical models, we often use symbols from the Greek alphabet. A table of these symbols is provided in Appendix B.6.

Rumor has it that there are some students studying Statistics who have an aversion to mathematics. Such people might be wise to focus on the concepts of this chapter and not let themselves get bogged down in the details. The details are given to provide a more complete introduction for those students who are not math averse.

1.1 Probability

Probabilities are numbers between zero and one that are used to explain random phenomena. We are all familiar with simple probability models. Flip a standard coin; the probability of heads is 1/2. Roll a die; the probability of getting a three is 1/6. Select a card from a well-shuffled deck; the probability of getting the queen of spades is 1/52 (assuming there are no jokers). One way to view probability models that many people find intuitive is in terms of random sampling from a fixed population. For example, the 52 cards form a fixed population and picking a card from a well-shuffled deck is a means of randomly selecting one element of the population. While we will exploit this idea of sampling from fixed populations, we should also note its limitations. For example, blood pressure is a very useful medical indicator, but even with a fixed population of people it would be very difficult to define a useful population of blood pressures. Blood pressure depends on the time of day, recent diet, current emotional state, the technique of the person taking the reading, and many other factors. Thinking about populations is very useful, but the concept can be very limiting both practically and mathematically. For measurements such as blood pressures and heights, there are difficulties in even specifying populations mathematically.

For mathematical reasons, probabilities are defined not on particular outcomes but on sets of outcomes (events). This is done so that continuous measurements can be dealt with. It seems much more natural to define probabilities on outcomes as we did in the previous paragraph, but consider some of the problems with doing that. For example, consider the problem of measuring the height of a corpse being kept in a morgue under controlled conditions. The only reason for getting morbid here is to have some hope of defining what the height is. Living people, to some extent, stretch and contract, so a height is a nebulous thing. But even given that someone has a fixed height, we can never know what it is. When someone's height is measured as 177.8 centimeters (5 feet 10 inches), their height is not really 177.8 centimeters, but (hopefully) somewhere between 177.75 and 177.85 centimeters. There is really no chance that anyone's height is *exactly* 177.8 cm, or exactly 177.8001 cm, or exactly 177.800000001 cm, or exactly 56.5955π cm, or exactly $(76\sqrt{5} + 4.5\sqrt{3})$ cm. In any neighborhood of 177.8, there are more numerical values than one could even imagine counting. The height should be somewhere in the neighborhood, but it won't be the particular value 177.8. The point is simply that trying to specify all the possible heights and their probabilities is a hopeless exercise. It simply cannot be done.

Even though individual heights cannot be measured exactly, when looking at a population of heights they follow certain patterns. There are not too many people over 8 feet (244 cm) tall. There are lots of males between 175.3 cm and 177.8 cm (5′9″ and 5′10″). With continuous values, each possible outcome has no chance of occurring, but outcomes do occur and occur with regularity. If probabilities are defined for sets instead of outcomes, these regularities can be reproduced mathematically. Nonetheless, initially the best way to learn about probabilities is to think about outcomes and their probabilities.

There are five key facts about probabilities:

1. Probabilities are between 0 and 1.

2. Something that happens with probability 1 is a sure thing.

3. If something has no chance of occurring, it has probability 0.

4. If something occurs with probability, say, .25, the probability that it will not occur is $1 - .25 = .75$.

5. If two events are mutually exclusive, i.e., if they cannot possibly happen at the same time, then the probability that either of them occurs is just the sum of their individual probabilities.

Individual outcomes are always mutually exclusive, e.g., you cannot flip a coin and get both heads and tails, so probabilities for outcomes can always be added together. Just to be totally correct, I should mention one other point. It may sound silly, but we need to assume that *something* occurring is always a sure thing. If we flip a coin, we must get either heads or tails with probability 1. We could even allow for the coin landing on its edge as long as the probabilities for all the outcomes add up to 1.

EXAMPLE 1.1.1.   Consider the nine outcomes that are all combinations of three heights, tall (T),

Table 1.1: *Height—eye color probabilities*.

		Eye color		
		Blue	Brown	Green
	Tall	.12	.15	.03
Height	Medium	.22	.34	.04
	Short	.06	.01	.03

medium (M), short (S), and three eye colors, blue (Bl), brown (Br) and green (G). The combinations are displayed below.

Height—eye color combinations

		Eye color		
		Blue	Brown	Green
	Tall	T, Bl	T, Br	T, G
Height	Medium	M, Bl	M, Br	M, G
	Short	S, Bl	S, Br	S, G

The set of all outcomes is

$$\{(T, Bl), (T, Br), (T, G), (M, Bl), (M, Br), (M, G), (S, Bl), (S, Br), (S, G)\}.$$

The event that someone is tall consists of the three pairs in the first row of the table, i.e.,

$$\{T\} = \{(T, Bl), (T, Br), (T, G)\}.$$

This is the union of the three outcomes (T, Bl), (T, Br), and (T, G). Similarly, the set of people with blue eyes is obtained from the first column of the table; it is the union of (T, Bl), (M, Bl), and (S, Bl) and can be written

$$\{Bl\} = \{(T, Bl), (M, Bl), (S, Bl)\}.$$

If we know that $\{T\}$ *and* $\{Bl\}$ both occur, there is only one possible outcome, (T, Bl).

The event that $\{T\}$ *or* $\{Bl\}$ occurs consists of all outcomes in either the first row or the first column of the table, i.e.,

$$\{(T, Bl), (T, Br), (T, G), (M, Bl), (S, Bl)\}. \qquad \Box$$

EXAMPLE 1.1.2. Table 1.1 contains probabilities for the nine outcomes that are combinations of height and eye color from Example 1.1.1. Note that each of the nine numbers is between 0 and 1 and that the sum of all nine equals 1. The probability of blue eyes is

$$\begin{aligned} Pr(Bl) &= Pr[(T, Bl), (M, Bl), (S, Bl)] \\ &= Pr(T, Bl) + Pr(M, Bl) + Pr(S, Bl) \\ &= .12 + .22 + .06 \\ &= .4. \end{aligned}$$

Similarly, $Pr(Br) = .5$ and $Pr(G) = .1$. The probability of not having blue eyes is

$$\begin{aligned} Pr(\text{not } Bl) &= 1 - Pr(Bl) \\ &= 1 - .4 \\ &= .6. \end{aligned}$$

Note also that $Pr(\text{not } Bl) = Pr(Br) + Pr(G)$.

The (*marginal*) probabilities for the various heights are:

$$\Pr(T) = .3, \quad \Pr(M) = .6, \quad \Pr(S) = .1.$$ □

Even if there are a countable (but infinite) number of possible outcomes, one can still define a probability by defining the probabilities for each outcome. It is only for measurement data that one really needs to define probabilities on sets.

Two random events are said to be independent if knowing that one of them occurs provides no information about the probability that the other event will occur. Formally, two events A and B are *independent* if

$$\Pr(A \text{ and } B) = \Pr(A)\Pr(B).$$

Thus the probability that *both* events A and B occur is just the product of the individual probabilities that A occurs and that B occurs. As we will begin to see in the next section, independence plays an important role in Statistics.

EXAMPLE 1.1.3.   Using the probabilities of Table 1.1 and the computations of Example 1.1.2, the events tall and brown eyes are independent because

$$\Pr(\text{tall and brown}) = \Pr(T, Br) = .15 = (.3)(.5) = \Pr(T) \times \Pr(Br).$$

On the other hand, medium height and blue eyes are *not* independent because

$$\Pr(\text{medium and blue}) = \Pr(M, Bl) = .22 \neq (.6)(.4) = \Pr(M) \times \Pr(Bl).$$ □

1.2   Random variables and expectations

A *random variable* is simply a function that relates outcomes with numbers. The key point is that any probability associated with the outcomes induces a probability on the numbers. The numbers and their associated probabilities can then be manipulated mathematically. Perhaps the most common and intuitive example of a random variable is rolling a die. The outcome is that a face of the die with a certain number of spots ends up on top. These can be pictured as

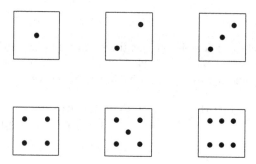

Without even thinking about it, we define a random variable that transforms these six faces into the numbers $1, 2, 3, 4, 5, 6$.

In Statistics we think of observations as random variables. These are often some number associated with a randomly selected member of a population. For example, one random variable is the height of a person who is to be randomly selected from among University of New Mexico students. (A random selection gives the same probability to every individual in the population. This random variable presumes that we have well-defined methods for measuring height and defining UNM students.) Rather than measuring height, we could define a different random variable by giving the person a score of 1 if that person is female and 0 if the person is male. We can also perform mathematical operations on random variables to yield new random variables. Suppose we plan to select a random sample of 10 students; then we would have 10 random variables with female and male scores. The sum of these random variables is another random variable that tells us the (random) number of females in the sample. Similarly, we would have 10 random variables for heights and we can define a new random variable consisting of the average of the 10 individual height random variables. Some random variables are related in obvious ways. In our example we measure both a height and a sex score on each person. If the sex score variable is a 1 (telling us that the person is female), it suggests that the height may be smaller than we would otherwise suspect. Obviously some female students are taller than some male students, but knowing a person's sex definitely changes our knowledge about their probable height. .

We do similar things in tossing a coin.

EXAMPLE 1.2.1. Consider tossing a coin twice. The four outcomes are ordered pairs of heads (H) and tails (T). The outcomes can be denoted as

$$(H,H) \qquad (H,T) \qquad (T,H) \qquad (T,T)$$

where the outcome of the first toss is the first element of the ordered pair.

The standard probability model has the four outcomes equally probable, i.e., $1/4 = \Pr(H,H) = \Pr(H,T) = \Pr(T,H) = \Pr(T,T)$. Equivalently

		Second toss		
		Heads	Tails	Total
First toss	Heads	1/4	1/4	1/2
	Tails	1/4	1/4	1/2
	Total	1/2	1/2	1

The probability of heads on each toss is 1/2. The probability of tails is 1/2. We will define two random variables:

$$y_1(r,s) = \begin{cases} 1 & \text{if } r = H \\ 0 & \text{if } r = T \end{cases}$$

$$y_2(r,s) = \begin{cases} 1 & \text{if } s = H \\ 0 & \text{if } s = T. \end{cases}$$

Thus, y_1 is 1 if the first toss is heads and 0 otherwise. Similarly, y_2 is 1 if the second toss is heads and 0 otherwise.

The event $y_1 = 1$ occurs if and only if we get heads on the first toss. We get heads on the first toss by getting either of the outcome pairs (H,H) or (H,T). In other words, the event $y_1 = 1$ is equivalent to the event $\{(H,H),(H,T)\}$. The probability of $y_1 = 1$ is just the sum of the probabilities of the outcomes in $\{(H,H),(H,T)\}$.

$$\begin{aligned} \Pr(y_1 = 1) &= \Pr(H,H) + \Pr(H,T) \\ &= 1/4 + 1/4 = 1/2. \end{aligned}$$

Similarly,

$$\begin{aligned} \Pr(y_1 = 0) &= \Pr(T,H) + \Pr(T,T) \\ &= 1/2 \\ \Pr(y_2 = 1) &= 1/2 \\ \Pr(y_2 = 0) &= 1/2. \end{aligned}$$

Now define another random variable,

$$W(r,s) = y_1(r,s) + y_2(r,s).$$

The random variable W is the total number of heads in two tosses:

$$\begin{aligned}
W(H,H) &= 2 \\
W(H,T) &= W(T,H) = 1 \\
W(T,T) &= 0.
\end{aligned}$$

Moreover,

$$\begin{aligned}
\Pr(W = 2) &= \Pr(H,H) = 1/4 \\
\Pr(W = 1) &= \Pr(H,T) + \Pr(T,H) = 1/2 \\
\Pr(W = 0) &= \Pr(T,T) = 1/4.
\end{aligned}$$

These three equalities define a probability on the outcomes 0, 1, 2. In working with W, we can ignore the original outcomes of head–tail pairs and work only with the new outcomes 0, 1, 2 and their associated probabilities. We can do the same thing for y_1 and y_2. The probability table given earlier can be rewritten in terms of y_1 and y_2.

		y_2		
		1	0	y_1 totals
y_1	1	1/4	1/4	1/2
	0	1/4	1/4	1/2
y_2 totals		1/2	1/2	1

Note that, for example, $\Pr[(y_1,y_2) = (1,0)] = 1/4$ and $\Pr(y_1 = 1) = 1/2$. This table shows the *distribution* of the probabilities for y_1 and y_2 both separately (marginally) and jointly.　　　　□

For any random variable, *a statement of the possible outcomes and their associated probabilities is referred to as the (marginal)* probability distribution *of the random variable. For two or more random variables, a table or other statement of the possible joint outcomes and their associated probabilities is referred to as the* joint probability distribution *of the random variables.*

All of the entries in the center of the distribution table given above for y_1 and y_2 are independent. For example,

$$\Pr[(y_1,y_2) = (1,0)] \equiv \Pr(y_1 = 1 \text{ and } y_2 = 0) = \Pr(y_1 = 1)\Pr(y_2 = 0).$$

We therefore say that y_1 and y_2 are independent. In general, *two random variables y_1 and y_2 are independent if any event involving only y_1 is independent of any event involving only y_2.*

Independence is an extremely important concept in Statistics. Observations to be analyzed are commonly assumed to be independent. This means that *the random aspect of one observation contains no information about the random aspect of any other observation.* (However, every observation tells us about fixed aspects of the underlying population such as the population center.) *For most purposes in Applied Statistics, this intuitive understanding of independence is sufficient.*

1.2.1　*Expected values and variances*

The *expected value (population mean)* of a random variable is a number characterizing the middle of the distribution. For a random variable y with a discrete distribution (i.e., one having a finite or countable number of outcomes), the expected value is

$$\mathrm{E}(y) \equiv \sum_{\text{all } r} r\Pr(y = r).$$

EXAMPLE 1.2.2. Let y be the result of picking one of the numbers $2, 4, 6, 8$ at random. Because the numbers are chosen at random,

$$1/4 = \Pr(y = 2) = \Pr(y = 4) = \Pr(y = 6) = \Pr(y = 8).$$

The expected value in this simple example is just the mean (average) of the four possible outcomes.

$$\begin{aligned} E(y) &= 2\left(\frac{1}{4}\right) + 4\left(\frac{1}{4}\right) + 6\left(\frac{1}{4}\right) + 8\left(\frac{1}{4}\right) \\ &= (2+4+6+8)/4 \\ &= 5. \end{aligned}$$

□

EXAMPLE 1.2.3. Five pieces of paper are placed in a hat. The papers have the numbers $2, 4, 6,$ $6,$ and 8 written on them. A piece of paper is picked at random. The expected value of the number drawn is the mean of the numbers on the five pieces of paper. Let y be the random variable that relates a piece of paper to the number on that paper. Each piece of paper has the same probability of being chosen, so, because the number 6 appears twice, the distribution of the random variable y is

$$\frac{1}{5} = \Pr(y = 2) = \Pr(y = 4) = \Pr(y = 8)$$

$$\frac{2}{5} = \Pr(y = 6).$$

The expected value is

$$\begin{aligned} E(y) &= 2\left(\frac{1}{5}\right) + 4\left(\frac{1}{5}\right) + 6\left(\frac{2}{5}\right) + 8\left(\frac{1}{5}\right) \\ &= (2+4+6+6+8)/5 \\ &= 5.2. \end{aligned}$$

□

EXAMPLE 1.2.4. Consider the coin tossing random variables $y_1, y_2,$ and W from Example 1.2.1. Recalling that y_1 and y_2 have the same distribution,

$$\begin{aligned} E(y_1) &= 1\left(\frac{1}{2}\right) + 0\left(\frac{1}{2}\right) = \frac{1}{2} \\ E(y_2) &= \frac{1}{2} \\ E(W) &= 2\left(\frac{1}{4}\right) + 1\left(\frac{1}{2}\right) + 0\left(\frac{1}{4}\right) = 1. \end{aligned}$$

The variable y_1 is the number of heads in the first toss of the coin. The two possible values 0 and 1 are equally probable, so the middle of the distribution is $1/2$. W is the number of heads in two tosses; the expected number of heads in two tosses is 1.

□

The expected value indicates the middle of a distribution, but does not indicate how spread out (dispersed) a distribution is.

EXAMPLE 1.2.5. Consider three gambles that I will allow you to take. In game z_1 you have equal chances of winning $12, 14, 16,$ or 18 dollars. In game z_2 you can again win $12, 14, 16,$ or 18 dollars, but now the probabilities are $.1$ that you will win either \$14 or \$16 and $.4$ that you will win \$12 or \$18. The third game I call z_3 and you can win $5, 10, 20,$ or 25 dollars with equal chances. Being no fool, I require you to pay me \$16 for the privilege of playing any of these games. We can write each game as a random variable.

z_1	outcome	12	14	16	18
	probability	.25	.25	.25	.25

z_2	outcome	12	14	16	18
	probability	.4	.1	.1	.4

z_3	outcome	5	10	20	25
	probability	.25	.25	.25	.25

I try to be a good casino operator, so none of these games is fair. You have to pay $16 to play, but you only expect to win $15. It is easy to see that

$$\mathrm{E}(z_1) = \mathrm{E}(z_2) = \mathrm{E}(z_3) = 15.$$

But don't forget that I'm taking a loss on the ice-water I serve to players and I also have to pay for the pictures of my extended family that I've decorated my office with.

Although the games z_1, z_2, and z_3 have the same expected value, the games (random variables) are very different. Game z_2 has the same outcomes as z_1, but much more of its probability is placed farther from the middle value, 15. The extreme observations 12 and 18 are much more probable under z_2 than z_1. If you currently have $16, need $18 for your grandmother's bunion removal, and anything less than $18 has no value to you, then z_2 is obviously a better game for you than z_1.

Both z_1 and z_2 are much more tightly packed around 15 than is z_3. If you needed $25 for the bunion removal, z_3 is the game to play because you can win it all in one play with probability .25. In either of the other games you would have to win at least five times to get $25, a much less likely occurrence. Of course you should realize that the most probable result is that Grandma will have to live with her bunion. You are unlikely to win either $18 or $25. While the ethical moral of this example is that a fool and his money are soon parted, the statistical point is that there is more to a random variable than its mean. The variability of random variables is also important.   □

The *(population) variance* is a measure of how spread out a distribution is from its expected value. Let y be a random variable having a discrete distribution with $\mathrm{E}(y) = \mu$, then the variance of y is

$$\mathrm{Var}(y) \equiv \sum_{\mathrm{all}\ r} (r - \mu)^2 \mathrm{Pr}(y = r).$$

This is the average squared distance of the outcomes from the center of the population. More technically, it is the expected squared distance between the outcomes and the mean of the distribution.

EXAMPLE 1.2.6. Using the random variables of Example 1.2.5,

$$\begin{aligned}
\mathrm{Var}(z_1) &= (12-15)^2(.25) + (14-15)^2(.25) \\
&\quad + (16-15)^2(.25) + (18-15)^2(.25) \\
&= 5 \\
\mathrm{Var}(z_2) &= (12-15)^2(.4) + (14-15)^2(.1) \\
&\quad + (16-15)^2(.1) + (18-15)^2(.4) \\
&= 7.4 \\
\mathrm{Var}(z_3) &= (5-15)^2(.25) + (10-15)^2(.25) \\
&\quad + (20-15)^2(.25) + (25-15)^2(.25) \\
&= 62.5.
\end{aligned}$$

The increasing variances from z_1 through z_3 indicate that the random variables are increasingly spread out. However, the value $\text{Var}(z_3) = 62.5$ seems too large to measure the relative variabilities of the three random variables. More on this later. □

EXAMPLE 1.2.7. Consider the coin tossing random variables of Examples 1.2.1 and 1.2.4.

$$\text{Var}(y_1) \;=\; \left(1 - \frac{1}{2}\right)^2 \frac{1}{2} + \left(0 - \frac{1}{2}\right)^2 \frac{1}{2} = \frac{1}{4}$$

$$\text{Var}(y_2) \;=\; \frac{1}{4}$$

$$\text{Var}(W) \;=\; (2-1)^2\left(\frac{1}{4}\right) + (1-1)^2\left(\frac{1}{2}\right) + (0-1)^2\left(\frac{1}{4}\right) = \frac{1}{2}.$$ □

A problem with the variance is that it is measured on the wrong scale. If y is measured in meters, $\text{Var}(y)$ involves the terms $(r - \mu)^2$; hence it is measured in meters squared. To get things back on the original scale, we consider the *standard deviation* of y

$$\text{Std. dev. } (y) \equiv \sqrt{\text{Var}(y)}.$$

EXAMPLE 1.2.8. Consider the random variables of Examples 1.2.5 and 1.2.6. The variances of the games are measured in dollars squared while the standard deviations are measured in dollars.

$$
\begin{aligned}
\text{Std. dev. } (z_1) &= \sqrt{5} &\doteq\; 2.236 \\
\text{Std. dev. } (z_2) &= \sqrt{7.4} &\doteq\; 2.720 \\
\text{Std. dev. } (z_3) &\equiv \sqrt{62.5} &\doteq\; 7.906
\end{aligned}
$$

The standard deviation of z_3 is 3 to 4 times larger than the others. From examining the distributions, the standard deviations seem to be more intuitive measures of relative variability than the variances. The variance of z_3 is 8.5 to 12.5 times larger than the other variances; these values seem unreasonably inflated. □

Standard deviations and variances are useful as measures of the relative dispersions of different random variables. The actual numbers themselves do not mean much. Moreover, there are other equally good measures of dispersion that can give results that are somewhat inconsistent with these. One reason standard deviations and variances are so widely used is because they are convenient mathematically. In addition, normal (Gaussian) distributions are widely used in Applied Statistics and are completely characterized by their expected values (means) and variances (or standard deviations). Knowing these two numbers, the mean and variance, one knows everything about a normal distribution.

1.2.2 Chebyshev's inequality

Another place in which the numerical values of standard deviations are useful is in applications of Chebyshev's inequality. Chebyshev's inequality gives a lower bound on the probability that a random variable is within an interval. Chebyshev's inequality is important in quality control work (control charts) and in evaluating prediction intervals.

Let y be a random variable with $\text{E}(y) = \mu$ and $\text{Var}(y) = \sigma^2$. Chebyshev's inequality states that for any number $k > 1$,

$$\Pr[\mu - k\sigma < y < \mu + k\sigma] \geq 1 - \frac{1}{k^2}.$$

Thus the probability that y will fall within k standard deviations of μ is at least $1 - (1/k^2)$.

The beauty of Chebyshev's inequality is that it holds for absolutely any random variable y. Thus we can always make some statement about the probability that y is in a symmetric interval about μ. In many cases, for particular choices of y, the probability of being in the interval can be much greater than $1 - k^{-2}$. For example, if $k = 3$ and y has a normal distribution as discussed in the next section, the probability of being in the interval is actually .997, whereas Chebyshev's inequality only assures us that the probability is no less than $1 - 3^{-2} = .889$. However, we know the lower bound of .889 applies regardless of whether y has a normal distribution.

1.2.3   Covariances and correlations

Often we take two (or more) observations on the same member of a population. We might observe the height and weight of a person. We might observe the IQs of a wife and husband. (Here the population consists of married couples.) In such cases we may want a numerical measure of the relationship between the pairs of observations. Data analysis related to these concepts is known as regression analysis and is introduced in Chapter 6. These ideas are also briefly used for testing normality in Section 2.5.

The *covariance* is a measure of the linear relationship between two random variables. Suppose y_1 and y_2 are discrete random variables. Let $E(y_1) = \mu_1$ and $E(y_2) = \mu_2$. The covariance between y_1 and y_2 is

$$\text{Cov}(y_1, y_2) \equiv \sum_{\text{all } (r,s)} (r - \mu_1)(s - \mu_2)\text{Pr}(y_1 = r, y_2 = s).$$

Positive covariances arise when relatively large values of y_1 tend to occur with relatively large values y_2 and small values of y_1 tend to occur with small values of y_2. On the other hand, negative covariances arise when relatively large values of y_1 tend to occur with relatively small values of y_2 and small values of y_1 tend to occur with large values of y_2. It is simple to see from the definition that, for example,

$$\text{Var}(y_1) = \text{Cov}(y_1, y_1).$$

In an attempt to get a handle on what the numerical value of the covariance means, it is often rescaled into a *correlation coefficient*.

$$\text{Corr}(y_1, y_2) \equiv \text{Cov}(y_1, y_2) / \sqrt{\text{Var}(y_1)\text{Var}(y_2)}.$$

Positive values of the correlation have the same qualitative meaning as positive values of the covariance, but now a *perfect* increasing linear relationship is indicated by a correlation of 1. Similarly, negative correlations and covariances mean similar things, but a perfect decreasing linear relationship gives a correlation of -1. The absence of any linear relationship is indicated by a value of 0.

A perfect linear relationship between y_1 and y_2 means that an increase of one unit in, say, y_1 dictates an exactly proportional change in y_2. For example, if we make a series of very accurate temperature measurements on something and simultaneously use one device calibrated in Fahrenheit and one calibrated in Celsius, the pairs of numbers should have an essentially perfect linear relationship. Estimates of covariances and correlations are called sample covariances and sample correlations. They will be considered in Section 6.7 although mention of the sample correlation is made in Section 2.5 and Section 3.9.

EXAMPLE 1.2.9.   Let z_1 and z_2 be two random variables defined by the following probability table:

		z_2			z_1 totals
		0	1	2	
	6	0	1/3	0	1/3
z_1	4	1/3	0	0	1/3
	2	0	0	1/3	1/3
z_2 totals		1/3	1/3	1/3	1

Then

$$E(z_1) = 6\left(\frac{1}{3}\right) + 4\left(\frac{1}{3}\right) + 2\left(\frac{1}{3}\right) = 4,$$

$$E(z_2) = 0\left(\frac{1}{3}\right) + 1\left(\frac{1}{3}\right) + 2\left(\frac{1}{3}\right) = 1,$$

$$\text{Var}(z_1) = (2-4)^2\left(\frac{1}{3}\right) + (4-4)^2\left(\frac{1}{3}\right) + (6-4)^2\left(\frac{1}{3}\right)$$
$$= 8/3,$$

$$\text{Var}(z_2) = (0-1)^2\left(\frac{1}{3}\right) + (1-1)^2\left(\frac{1}{3}\right) + (2-1)^2\left(\frac{1}{3}\right)$$
$$= 2/3,$$

$$\text{Cov}(z_1, z_2) = (2-4)(0-1)(0) + (2-4)(1-1)(0) + (2-4)(2-1)\left(\frac{1}{3}\right)$$
$$+ (4-4)(0-1)\left(\frac{1}{3}\right) + (4-4)(1-1)(0) + (4-4)(2-1)(0)$$
$$+ (6-4)(0-1)(0) + (6-4)(1-1)\left(\frac{1}{3}\right) + (6-4)(2-1)(0)$$
$$= -2/3,$$

$$\text{Corr}(z_1, z_2) = (-2/3)\Big/\sqrt{(8/3)(2/3)}$$
$$= -1/2.$$

This correlation indicates that relatively large z_1 values tend to occur with relatively small z_2 values. However, the correlation is considerably greater than -1, so the linear relationship is less than perfect. Moreover, the correlation measures the linear relationship and *fails to identify the perfect nonlinear relationship* between z_1 and z_2. If $z_1 = 2$, then $z_2 = 2$. If $z_1 = 4$, then $z_2 = 0$. If $z_1 = 6$, then $z_2 = 1$. If you know one random variable, you know the other, but because the relationship is nonlinear, the correlation is not ± 1. □

EXAMPLE 1.2.10. Consider the coin toss random variables y_1 and y_2 from Example 1.2.1. We earlier observed that these two random variables are independent. If so, there should be no relationship between them (linear or otherwise). We now show that their covariance is 0.

$$\text{Cov}(y_1, y_2) = \left(0 - \frac{1}{2}\right)\left(0 - \frac{1}{2}\right)\frac{1}{4} + \left(0 - \frac{1}{2}\right)\left(1 - \frac{1}{2}\right)\frac{1}{4}$$
$$+ \left(1 - \frac{1}{2}\right)\left(0 - \frac{1}{2}\right)\frac{1}{4} + \left(1 - \frac{1}{2}\right)\left(1 - \frac{1}{2}\right)\frac{1}{4}$$
$$= \frac{1}{16} - \frac{1}{16} - \frac{1}{16} + \frac{1}{16} = 0.$$ □

In general, whenever two random variables are independent, their covariance (and thus their correlation) is 0. However, just because two random variables have 0 covariance does not imply that they are independent. Independence has to do with not having any kind of relationship; covariance examines only linear relationships. Random variables with nonlinear relationships can have zero covariance but not be independent.

1.2.4   Rules for expected values and variances

We now present some extremely useful results that allow us to show that statistical estimates are reasonable and to establish the variability associated with statistical estimates. These results relate to the expected values, variances, and covariances of linear combinations of random variables. A *linear combination* of random variables is something that only involves multiplying random variables by fixed constants, adding such terms together, and adding a constant.

Proposition 1.2.11.   Let y_1, y_2, y_3, and y_4 be random variables and let a_1, a_2, a_3, and a_4 be real numbers.

1. $E(a_1 y_1 + a_2 y_2 + a_3) = a_1 E(y_1) + a_2 E(y_2) + a_3$.
2. If y_1 and y_2 are independent, $Var(a_1 y_1 + a_2 y_2 + a_3) = a_1^2 Var(y_1) + a_2^2 Var(y_2)$.
3. $Var(a_1 y_1 + a_2 y_2 + a_3) = a_1^2 Var(y_1) + 2 a_1 a_2 Cov(y_1, y_2) + a_2^2 Var(y_2)$.
4. $Cov(a_1 y_1 + a_2 y_2, a_3 y_3 + a_4 y_4) = a_1 a_3 Cov(y_1, y_3) + a_1 a_4 Cov(y_1, y_4) + a_2 a_3 Cov(y_2, y_3) + a_2 a_4 Cov(y_2, y_4)$.

All of these results generalize to linear combinations involving more than two random variables.

EXAMPLE 1.2.12.   Recall that when independently tossing a coin twice, the total number of heads, W, is the sum of y_1 and y_2, the number of heads on the first and second tosses, respectively. We have already seen that $E(y_1) = E(y_2) = .5$ and that $E(W) = 1$. We now illustrate item 1 of the proposition by finding $E(W)$ again. Since $W = y_1 + y_2$,

$$E(W) = E(y_1 + y_2) = E(y_1) + E(y_2) = .5 + .5 = 1.$$

We have also seen that $Var(y_1) = Var(y_2) = .25$ and that $Var(W) = .5$. Since the coin tosses are independent, item 2 above gives

$$Var(W) = Var(y_1 + y_2) = Var(y_1) + Var(y_2) = .25 + .25 = .5.$$

The key point is that this is an easier way of finding the expected value and variance of W than using the original definitions.   □

We now illustrate the generalizations referred to in Proposition 1.2.11. We begin by looking at the problem of estimating the mean of a population.

EXAMPLE 1.2.13.   Let y_1, y_2, y_3, and y_4 be four random variables each with the same (population) mean μ, i.e., $E(y_i) = \mu$ for $i = 1, 2, 3, 4$. We can compute the *sample mean* (average) of these, defining

$$
\begin{aligned}
\bar{y}. &\equiv \frac{y_1 + y_2 + y_3 + y_4}{4} \\
&= \frac{1}{4} y_1 + \frac{1}{4} y_2 + \frac{1}{4} y_3 + \frac{1}{4} y_4.
\end{aligned}
$$

The · in the subscript of $\bar{y}.$ indicates that the sample mean is obtained by summing over the subscripts of the y_is. The · notation is not necessary for this problem but becomes useful in dealing with the analysis of variance problems treated later in the book.

Using item 1 of Proposition 1.2.11 we find that

$$
\begin{aligned}
E(\bar{y}.) &= E\left(\frac{1}{4} y_1 + \frac{1}{4} y_2 + \frac{1}{4} y_3 + \frac{1}{4} y_4 \right) \\
&= \frac{1}{4} E(y_1) + \frac{1}{4} E(y_2) + \frac{1}{4} E(y_3) + \frac{1}{4} E(y_4) \\
&= \frac{1}{4} \mu + \frac{1}{4} \mu + \frac{1}{4} \mu + \frac{1}{4} \mu \\
&= \mu.
\end{aligned}
$$

Thus one observation on $\bar{y}.$ would make a reasonable estimate of μ.

If we also assume that the y_is are independent with the same variance, say, σ^2, then from item 2 of Proposition 1.2.11

$$
\begin{aligned}
\text{Var}(\bar{y}.) &= \text{Var}\left(\frac{1}{4}y_1 + \frac{1}{4}y_2 + \frac{1}{4}y_3 + \frac{1}{4}y_4\right) \\
&= \left(\frac{1}{4}\right)^2 \text{Var}(y_1) + \left(\frac{1}{4}\right)^2 \text{Var}(y_2) \\
&\quad + \left(\frac{1}{4}\right)^2 \text{Var}(y_3) + \left(\frac{1}{4}\right)^2 \text{Var}(y_4) \\
&= \left(\frac{1}{4}\right)^2 \sigma^2 + \left(\frac{1}{4}\right)^2 \sigma^2 + \left(\frac{1}{4}\right)^2 \sigma^2 + \left(\frac{1}{4}\right)^2 \sigma^2 \\
&= \frac{\sigma^2}{4}.
\end{aligned}
$$

The variance of $\bar{y}.$ is only one fourth of the variance of an individual observation. Thus the $\bar{y}.$ observations are more tightly packed around their mean μ than the y_is are. This indicates that one observation on $\bar{y}.$ is more likely to be close to μ than an individual y_i. □

These results for $\bar{y}.$ hold quite generally; they are not restricted to the average of four random variables. *If $\bar{y}. = (1/n)(y_1 + \cdots + y_n) = \sum_{i=1}^{n} y_i/n$ is the sample mean of n independent random variables all with the same population mean μ and population variance σ^2,*

$$
E(\bar{y}.) = \mu
$$

and

$$
\text{Var}(\bar{y}.) = \frac{\sigma^2}{n}.
$$

Proving these general results uses exactly the same ideas as the proofs for a sample of size 4.

As with a sample of size 4, the general results on $\bar{y}.$ are very important in statistical inference. If we are interested in determining the population mean μ from future data, the obvious estimate is the average of the individual observations, $\bar{y}.$. The observations are random, so the estimate $\bar{y}.$ is also a random variable and the middle of its distribution is $E(\bar{y}.) = \mu$, the original population mean. Thus $\bar{y}.$ is a reasonable estimate of μ. Moreover, $\bar{y}.$ is a better estimate than any particular observation y_i because $\bar{y}.$ has a smaller variance, σ^2/n, as opposed to σ^2 for y_i. With less variability in the estimate, any one observation of $\bar{y}.$ is more likely to be near its mean μ than a single observation y_i. In practice, we obtain data and compute a sample mean. This constitutes one observation on the random variable $\bar{y}.$. If our sample mean is to be a good estimate of μ, our one look at $\bar{y}.$ had better have a good chance of being close to μ. This occurs when the variance of $\bar{y}.$ is small. Note that the larger the sample size n, the smaller is σ^2/n, the variance of $\bar{y}.$. We will return to these ideas later.

Generally, we will use item 1 of Proposition 1.2.11 to show that estimates are *unbiased*. In other words, we will show that the expected value of an estimate is what we are trying to estimate. In estimating μ, we have $E(\bar{y}.) = \mu$, so $\bar{y}.$ is an unbiased estimate of μ. All this really does is show that $\bar{y}.$ is a reasonable estimate of μ. More important than showing unbiasedness is using item 2 to find variances of estimates. Statistical inference depends crucially on having some idea of the variability of an estimate. Item 2 is the primary tool in finding the appropriate variance for different estimators.

1.3   Continuous distributions

As discussed in Section 1.1, many things that we would like to measure are, in the strictest sense, not measurable. We cannot find a building's exact height even though we can approximate it *extremely*

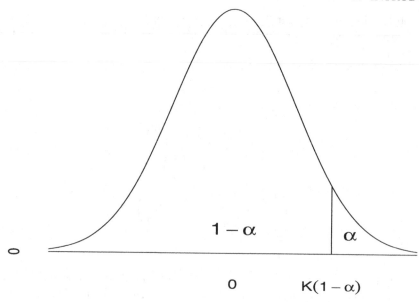

Figure 1.1: *A continuous probability density.*

accurately. This theoretical inability to measure things exactly has little impact on our practical world, but it has a substantial impact on the theory of Statistics.

The data in most statistical applications can be viewed either as counts of how often some event has occurred or as measurements. Probabilities associated with count data are easy to describe. We discuss some probability models for count data in Sections 1.4 and 1.5. With measurement data, we can never obtain an exact value, so we don't even try. With measurement data, we assign probabilities to intervals. Thus we do not discuss the probability that a person has the height 177.8 cm or 177.8001 cm or 56.5955π cm, but we do discuss the probability that someone has a height *between* 177.75 cm and 177.85 cm. Typically, we think of doing this in terms of pictures. We associate probabilities with areas under curves. (Mathematically, this involves integral calculus and is discussed in a brief appendix at the end of the chapter.) Figure 1.1 contains a picture of a continuous probability distribution (*a density*). Probabilities must be between 0 and 1, so the curve must always be nonnegative (to make all areas nonnegative) and the area under the entire curve must be 1.

Figure 1.1 also shows a point $K(1-\alpha)$. This point divides the area under the curve into two parts. The probability of obtaining a number less than $K(1-\alpha)$ is $1-\alpha$, i.e., the area under the curve to the left of $K(1-\alpha)$ is $1-\alpha$. The probability of obtaining a number greater than $K(1-\alpha)$ is α, i.e., the area under the curve to the right of $K(1-\alpha)$. $K(1-\alpha)$ is a particular number, so the probability is 0 that $K(1-\alpha)$ will actually occur. There is no area under a curve associated with any particular point.

Pictures such as Figure 1.1 are often used as models for populations of measurements. With a fixed population of measurements, it is natural to form a histogram, i.e., a bar chart that plots intervals for the measurement against the proportion of individuals that fall into a particular interval. Pictures such as Figure 1.1 can be viewed as approximations to such histograms. The probabilities described by pictures such as Figure 1.1 are those associated with randomly picking an individual from the population. Thus, randomly picking an individual from the population modeled by Figure 1.1 yields a measurement less than $K(1-\alpha)$ with probability $1-\alpha$.

Ideas similar to those discussed in Section 1.2 can be used to define expected values, variances, and covariances for continuous distributions. These extensions involve integral calculus and are discussed in the appendix. In any case, Proposition 1.2.11 continues to apply.

The most commonly used distributional model for measurement data is the *normal* distribution (also called the *Gaussian* distribution). The bell-shaped curve in Figure 1.1 is referred to as the standard normal curve. The formula for writing the curve is not too ugly; it is

$$f(x) = \frac{1}{\sqrt{2\pi}} e^{-x^2/2}.$$

Here e is the base of natural logarithms. Unfortunately, even with calculus it is very difficult to compute areas under this curve. Finding standard normal probabilities requires a table or a computer routine.

By itself, the standard normal curve has little value in modeling measurements. For one thing, the curve is centered about 0. I don't take many measurements where I think the central value should be 0. To make the normal distribution a useful model, we need to expand the standard normal into a family of distributions with different centers (expected values) μ and different spreads (standard deviations) σ. By appropriate recentering and rescaling of the plot, all of these curves will have the same shape as Figure 1.1. Another important fact that allows us to combine data into estimates is that *linear combinations of independent normally distributed observations are again normally distributed.*

The standard normal distribution is the special case of a normal with $\mu = 0$ and $\sigma = 1$. The standard normal plays an important role because it is the only normal distribution for which we actually compute probabilities. (Areas under the curve are hard to compute so we rely on computers or, heaven forbid, tables.) Suppose a measurement y has a normal distribution with mean μ, standard deviation σ, and variance σ^2. We write this as

$$y \sim N(\mu, \sigma^2).$$

Normal distributions have the property that

$$\frac{y - \mu}{\sigma} \sim N(0, 1),$$

cf. Exercise 1.6.2. This standardization process allows us to find probabilities for all normal distributions using only one difficult computational routine.

The standard normal distribution is sometimes used in constructing statistical inferences but more often a similar distribution is used. When data are normally distributed, statistical inferences often require something called Student's t distribution. (Student was the pen name of the Guinness brewmaster W. S. Gosset.) The t distribution is a family of distributions all of which look roughly like Figure 1.1. They are all symmetric about 0, but they have slightly different amounts of dispersion (spread). The amount of variability in each distribution is determined by a positive integer parameter called the *degrees of freedom*. With only 1 degree of freedom, the mathematical properties of a t distribution are fairly bizarre. (This special case is called a Cauchy distribution.) As the number of degrees of freedom get larger, the t distributions get better behaved and have less variability. As the degrees of freedom gets arbitrarily large, the t distribution approximates the standard normal distribution; see Figure 1.2.

Two other distributions that come up later are the chi-squared distribution (χ^2) and the F distribution. These arise naturally when drawing conclusions about the population variance from data that are normally distributed. Both distributions differ from those just discussed in that both are asymmetric and both are restricted to positive numbers. However, the basic idea of probabilities being areas under curves remains unchanged. The shape of a chi-squared distribution depends on one parameter called its degrees of freedom. An F depends on two parameters, its numerator and denominator degrees of freedom. Figure 1.3 illustrates a $\chi^2(8)$ and an $F(3, 18)$ distribution along with illustrating the notation for an α percentile. With three or more degrees of freedom for a χ^2 and three or more numerator degrees of freedom for an F, the distributions are shaped roughly like those in Figure 1.3, i.e., they are positive skewed distributions with densities that start at 0, increase, and

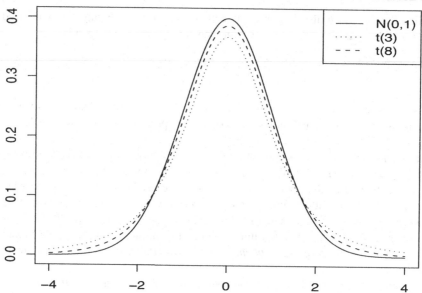

Figure 1.2: *Three t distributions.*

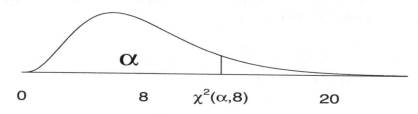

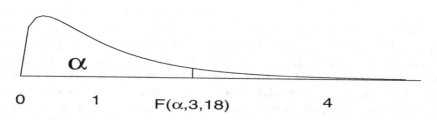

Figure 1.3 *Top: A $\chi^2(8)$ distribution with the α percentile. Bottom: An $F(3,18)$ distribution the α percentile.*

then decrease. With fewer than three degrees of freedom, the densities take on their largest values near 0.

In Section 1.2, we introduced Chebyshev's inequality. Shewhart (1931, p. 177) discusses work by Camp and Meidell that allows us to improve on Chebyshev's inequality for continuous distributions. Once again let $E(y) = \mu$ and $Var(y) = \sigma^2$. If the density, i.e., the function that defines the curve, is symmetric, unimodal (has only one peak), and always decreases as one moves farther away from the mode, then the inequality can be sharpened to

$$\Pr[\mu - k\sigma < y < \mu + k\sigma] \geq 1 - \frac{1}{(2.25)k^2}.$$

As discussed in the previous section, with y normal and $k = 3$, the true probability is .997, Chebyshev's inequality gives a lower bound of .889, and the new improved Chebyshev inequality gives a lower bound of .951. By making some relatively innocuous assumptions, we get a substantial improvement in the lower bound.

1.4 The binomial distribution

There are a few distributions that are used in the vast majority of statistical applications. The reason for this is that they tend to occur naturally. The normal distribution is one. As discussed in the next chapter, the normal distribution occurs in practice because a result called The Central Limit Theorem dictates that many distributions can be approximated by the normal. Two other distributions, the binomial and the multinomial, occur in practice because they are very simple. In this section we discuss the binomial. The next section introduces the multinomial distribution. The results of this section are used in Chapters 5, 20, and 21 and in discussions of transformations.

If you have independent identical random trials and count how often something (anything) occurs, the appropriate distribution is the binomial. What could be simpler?

EXAMPLE 1.4.1. Being somewhat lonely in my misspent youth, I decided to use the computer dating service aTonal.com. The service was to provide me with five matches. Being a very open-minded soul, I convinced myself that the results of one match would not influence my opinion about other matches. From my limited experience with the opposite sex, I have found that I enjoy about 40% of such brief encounters. I decided that my money would be well spent if I enjoyed two or more of the five matches. Unfortunately, my loan shark repossessed my 1954 Studebaker before I could indulge in this taste of nirvana. Back in those days, we chauvinists believed: no wheels—no women. Nevertheless, let us compute the probability that I would have been satisfied with the dating service. Let W be the number of matches I would have enjoyed. The simplest way to find the probability of satisfaction is

$$
\begin{aligned}
\Pr(W \geq 2) &= 1 - \Pr(W < 2) \\
&= 1 - \Pr(W = 0) - \Pr(W = 1),
\end{aligned}
$$

but that is much too easy. Let's compute

$$
\Pr(W \geq 2) = \Pr(W = 2) + \Pr(W = 3) + \Pr(W = 4) + \Pr(W = 5).
$$

In particular, we compute each term on the right-hand side.

Write the outcome of the five matches as an ordered collection of Ls and Ds. For example, (L, D, L, D, D) indicates that I like the first and third matches, but dislike the second, fourth, and fifth.

To like five matches, I must like every one of them.

$$
\Pr(W = 5) = \Pr(L, L, L, L, L).
$$

Remember, I assumed that the matches were independent and that the probability of my liking any one is .4. Thus,

$$
\begin{aligned}
\Pr(W = 5) &= \Pr(L)\Pr(L)\Pr(L)\Pr(L)\Pr(L) \\
&= (.4)^5.
\end{aligned}
$$

The probability of liking four matches is a bit more complicated. I could only dislike one match, but there are five different choices for the match that I could dislike. It could be the fifth, the fourth, the third, the second, or the first. Any pattern of 4 Ls and a D excludes the other patterns from occurring, e.g., if the only match I dislike is the fourth, then the only match I dislike cannot be the second. Since the patterns are mutually exclusive (disjoint), the probability of disliking one match is the sum of the probabilities of the individual patterns.

$$\Pr(W = 4) \;=\; \Pr(L,L,L,L,D) \qquad\qquad (1.4.1)$$
$$+\Pr(L,L,L,D,L)$$
$$+\Pr(L,L,D,L,L)$$
$$+\Pr(L,D,L,L,L)$$
$$+\Pr(D,L,L,L,L).$$

By assumption $\Pr(L) = .4$, so $\Pr(D) = 1 - \Pr(L) = 1 - .4 = .6$. The matches are independent, so

$$\Pr(L,L,L,L,D) \;=\; \Pr(L)\Pr(L)\Pr(L)\Pr(L)\Pr(D)$$
$$=\; (.4)^4 .6 .$$

Similarly,

$$\Pr(L,L,L,D,L) \;=\; \Pr(L,L,D,L,L)$$
$$=\; \Pr(L,D,L,L,L)$$
$$=\; \Pr(D,L,L,L,L)$$
$$=\; (.4)^4 .6 .$$

Summing up the values in Equation (1.4.1),

$$\Pr(W = 4) = 5(.4)^4(.6) .$$

Computing the probability of liking three matches is even worse.

$$\Pr(W = 3) \;=\; \Pr(L,L,L,D,D)$$
$$+\Pr(L,L,D,L,D)$$
$$+\Pr(L,D,L,L,D)$$
$$+\Pr(D,L,L,L,D)$$
$$+\Pr(L,L,D,D,L)$$
$$+\Pr(L,D,L,D,L)$$
$$+\Pr(D,L,L,D,L)$$
$$+\Pr(L,D,D,L,L)$$
$$+\Pr(D,L,D,L,L)$$
$$+\Pr(D,D,L,L,L).$$

Again all of these patterns have exactly the same probability. For example, using independence

$$\Pr(D,L,D,L,L) = (.4)^3(.6)^2 .$$

Adding up all of the patterns

$$\Pr(W = 3) = 10(.4)^3(.6)^2 .$$

By now it should be clear that

$$\Pr(W = 2) = (\text{number of patterns with 2 Ls and 3 Ds})(.4)^2(.6)^3 .$$

The number of patterns can be computed as

$$\binom{5}{2} \equiv \frac{5!}{2!(5-2)!} \equiv \frac{5 \cdot 4 \cdot 3 \cdot 2 \cdot 1}{(2 \cdot 1)(3 \cdot 2 \cdot 1)} = 10 .$$

The probability that I would be satisfied with the dating service is

$$\Pr(W \geq 2) = 10(.4)^2(.6)^3 + 10(.4)^3(.6)^2 + 5(.4)^4.6 + (.4)^5$$
$$= .663.$$
□

Binomial random variables can also be generated by sampling from a fixed population. If we were going to make 20 random selections from the UNM student body, the number of females would have a binomial distribution. Given a set of procedures for defining and sampling the student body, there would be some fixed number of students of which a given number would be females. Under random sampling, the probability of selecting a female on any of the 20 trials would be simply the proportion of females in the population. Although it is very unlikely to occur in this example, the sampling scheme must allow the possibility of students being selected more than once in the sample. If people were not allowed to be chosen more than once, each successive selection would change the proportion of females available for the subsequent selection. Of course, when making 20 selections out of a population of over 20,000 UNM students, even if you did not allow people to be reselected, the changes in the proportions of females are insubstantial and the binomial distribution makes a good approximation to the true distribution. On the other hand, if the entire student population was 40 rather than 20,000+, it might not be wise to use the binomial approximation when people are not allowed to be reselected.

Typically, the outcome of interest in a binomial is referred to as a success. If the probability of a success is p for each of N independent identical trials, then the number of successes y has a binomial distribution with parameters N and p. Write

$$y \sim \text{Bin}(N, p).$$

The distribution of y is

$$\Pr(y = r) = \binom{N}{r} p^r (1 - p)^{N-r}$$

for $r = 0, 1, \ldots, N$. Here

$$\binom{N}{r} \equiv \frac{N!}{r!(N-r)!}$$

where for any positive integer m, $m! \equiv m(m-1)(m-2)\cdots(2)(1)$ and $0! \equiv 1$. The notation $\binom{N}{r}$ is read "N choose r" because it is the number of distinct ways of choosing r individuals out of a collection containing N individuals.

EXAMPLE 1.4.2. The random variables in Example 1.2.1 were y_1, the number of heads on the first toss of a coin, y_2, the number of heads on the second toss of a coin, and W, the combined number of heads from the two tosses. These have the following distributions:

$$y_1 \sim \text{Bin}\left(1, \frac{1}{2}\right)$$

$$y_2 \sim \text{Bin}\left(1, \frac{1}{2}\right)$$

$$W \sim \text{Bin}\left(2, \frac{1}{2}\right).$$

Note that W, the $\text{Bin}\left(2, \frac{1}{2}\right)$, was obtained by adding together the two independent $\text{Bin}\left(1, \frac{1}{2}\right)$ random variables y_1 and y_2. This result is quite general. Any $\text{Bin}(N, p)$ random variable can be written as the sum of N independent $\text{Bin}(1, p)$ random variables. □

Given the probability distribution of a binomial, we can find the mean (expected value) and variance. By definition, if $y \sim \text{Bin}(N, p)$, the mean is

$$E(y) = \sum_{r=0}^{N} r \binom{N}{r} p^r (1-p)^{N-r}.$$

This is difficult to evaluate directly, but by writing y as the sum of N independent $\text{Bin}(1, p)$ random variables and using Exercise 1.6.1 and Proposition 1.2.11, it is easily seen that

$$E(y) = Np.$$

Similarly, the variance of y is

$$\text{Var}(y) = \sum_{r=0}^{N} (r - Np)^2 \binom{N}{r} p^r (1-p)^{N-r}$$

but by again writing y as the sum of N independent $\text{Bin}(1, p)$ random variables and using Exercise 1.6.1 and Proposition 1.2.11, it is easily seen that

$$\text{Var}(y) = Np(1-p).$$

Exercise 1.6.8 consists of proving these mean and variance formulae.

On occasion we will need to look at both the number of successes from a group of N trials and the number of failures at the same time. If the number of successes is y_1 and the number of failures is y_2, then

$$\begin{aligned} y_2 &= N - y_1 \\ y_1 &\sim \text{Bin}(N, p) \end{aligned}$$

and

$$y_2 \sim \text{Bin}(N, 1-p).$$

The last result holds because, with independent identical trials, the number of outcomes that we call failures must also have a binomial distribution. If p is the probability of success, the probability of failure is $1 - p$. Of course,

$$\begin{aligned} E(y_2) &= N(1-p) \\ \text{Var}(y_2) &= N(1-p)p. \end{aligned}$$

Note that $\text{Var}(y_1) = \text{Var}(y_2)$ regardless of the value of p. Finally,

$$\text{Cov}(y_1, y_2) = -Np(1-p)$$

and

$$\text{Corr}(y_1, y_2) = -1.$$

There is a perfect linear relationship between y_1 and y_2. If y_1 goes up one count, y_2 goes down one count. When we look at both successes and failures write

$$(y_1, y_2) \sim \text{Bin}(N, p, (1-p)).$$

This is the simplest case of the multinomial distribution discussed in the next section. But first we look at a special case of Binomial sampling.

1.4.1 Poisson sampling

The *Poisson distribution* might be used to model the number of flaws on a dvd. There is no obvious upper bound on the number of flaws. If we put a grid over the (square?) dvd, we could count whether every grid square contains a flaw. The number of grid squares with a flaw has a binomial distribution. As we make the grid finer and finer, the number of grid squares that contain flaws will become the actual number of flaws. Also, for finer grids, the probability of a flaw decreases as the size of each square decreases but the number of grid squares increases correspondingly while the expected number of squares with flaws remains the same. After all, the number of flaws we expect on the dvd has nothing to do with the grid that we decide to put over it. If we let λ be the expected number of flaws, $\lambda = Np$ where N is the number of grid squares and p is the probability of a flaw in the square.

The Poisson distribution is an approximation used for binomials with a very large number of trials, each having a very small probability of success. Under these conditions, if $Np \doteq \lambda$ we write

$$y \sim \text{Pois}(\lambda).$$

For an infinitely large number of trials, the distribution of y is

$$\Pr(y = r) = \lambda^r e^\lambda / r!,$$

$r = 0, 1, 2, \ldots$. These probabilities are just the limits of the binomial probabilities under the conditions described. The mean and variance of a $\text{Pois}(\lambda)$ are

$$E(y) = \lambda$$

and

$$\text{Var}(y) = \lambda.$$

1.5 The multinomial distribution

The *multinomial distribution* is a generalization of the binomial allowing more than two categories. The results in this section are used in Chapters 5 and 21.

EXAMPLE 1.5.1.   Consider the probabilities for the nine height and eye color categories given in Example 1.1.2. The probabilities are repeated below.

Height—eye color probabilities

		Blue	Brown	Green
	Tall	.12	.15	.03
Height	Medium	.22	.34	.04
	Short	.06	.01	.03

Suppose a random sample of 50 individuals was obtained with these probabilities. For example, one might have a population of 100 people in which 12 were tall with blue eyes, 15 were tall with brown eyes, 3 were short with green eyes, etc. We could randomly select one of the 100 people as the first individual in the sample. Then, returning that individual to the population, take another random selection from the 100 to be the second individual. We are to proceed in this way until 50 people are selected. Note that with a population of 100 and a sample of 50 there is a substantial chance that some people would be selected more than once. The numbers of selections falling into each of the nine categories has a multinomial distribution with $N = 50$ and these probabilities.

It is unlikely that one would actually perform sampling from a population of 100 people as described above. Typically, one would not allow the same person to be chosen more than once.

However, if we had a population of 10,000 people where 1200 were tall with blue eyes, 1500 were tall with brown eyes, 300 were short with green eyes, etc., with a sample size of 50 we might be willing to allow the possibility of selecting the same person more than once simply because it is extremely unlikely to happen. Technically, to obtain the multinomial distribution with $N = 50$ and these probabilities, when sampling from a fixed population we need to allow individuals to appear more than once. However, when taking a small sample from a large population, it does not matter much whether or not you allow people to be chosen more than once, so the multinomial often provides a good approximation even when individuals are excluded from reappearing in the sample.

□

Consider a group of N independent identical trials in which each trial results in the occurrence of one of q events. Let $y_i, i = 1, \ldots, q$ be the number of times that the ith event occurs and let p_i be the probability that the ith event occurs on any trial. The p_is must satisfy $p_1 + p_2 + \cdots + p_q = 1$. We say that $(y_1, \ldots, y_q)$ has a multinomial distribution with parameters $N, p_1, \ldots, p_q$. Write

$$(y_1, \ldots, y_q) \sim \text{Mult}(N, p_1, \ldots, p_q).$$

The distribution is given by the probabilities

$$\Pr(y_1 = r_1, \ldots, y_q = r_q) = \frac{N!}{r_1! \cdots r_q!} p_1^{r_1} \cdots p_q^{r_q}$$

$$= \left(N! \Big/ \prod_{i=1}^{q} r_i! \right) \prod_{i=1}^{q} p_i^{r_i}.$$

Here the r_is are allowed to be any whole numbers with each $r_i \geq 0$ and $r_1 + \cdots + r_q = N$. Note that if $q = 2$, this is just a binomial distribution. In general, each individual component y_i of a multinomial consists of N trials in which category i either occurs or does not occur, so individual components have the marginal distributions

$$y_i \sim \text{Bin}(N, p_i).$$

It follows that

$$E(y_i) = N p_i$$

and

$$\text{Var}(y_i) = N p_i (1 - p_i).$$

It can also be shown that

$$\text{Cov}(y_i, y_j) = -N p_i p_j \qquad \text{for} \qquad i \neq j.$$

EXAMPLE 1.5.2.   Suppose that the 50 individuals from Example 1.5.1 fall into the categories as listed below.

Height—eye color observations

		Eye color		
		Blue	Brown	Green
	Tall	5	8	2
Height	Medium	10	18	2
	Short	3	1	1

The probability of getting this particular table is

$$\frac{50!}{5!8!2!10!18!2!3!1!1!}(.12)^5(.15)^8(.03)^2(.22)^{10}(.34)^{18}(.04)^2(.06)^3(.01)^1(.03)^1.$$

This number is zero to over 5 decimal places. The fact that this is a very small number is not surprising. There are a lot of possible tables, so the probability of getting any particular table is very small. In fact, many of the possible tables are *much* less likely to occur than this table.

Let's return to thinking about the observations as random. The expected number of observations for each category is given by Np_i. It is easily seen that the expected counts for the cells are as follows.

Height—eye color expected values

		Eye color		
		Blue	Brown	Green
	Tall	6.0	7.5	1.5
Height	Medium	11.0	17.0	2.0
	Short	3.0	0.5	1.5

Note that the expected counts need not be integers.

The variance for, say, the number of tall blue-eyed people in this sample is $50(.12)(1 - .12) = 5.28$. The variance of the number of short green-eyed people is $50(.03)(1 - .03) = 1.455$. The covariance between the number of tall blue-eyed people and the number of short green-eyed people is $-50(.12)(.03) = -.18$. The correlation between the numbers of tall blue-eyed people and short green-eyed people is $-.18/\sqrt{(5.28)(1.455)} = -0.065$.   □

1.5.1   Independent Poissons and multinomials

Suppose that instead of sampling 50 people and cross classifying them into the height–eye color categories, we spend an hour at a shopping mall looking at people. Suppose during that time we saw 5 tall, blue-eyed people as well as the other numbers given earlier. Note that there is no obvious maximum number of tall blue-eyed people that we can see in an hour, nor obvious maximum numbers for the other categories. The Poisson distribution is a reasonable model for the count in each category and the counts could well be independent. If we happen to see 50 observations in the table, we can think about the distribution of the counts given that there is a total of 50 observations. It turns out that this conditional distribution is the multinomial distribution.

Later, in Chapter 21, we will look at methods for analyzing independent Poisson observations. Because of the relationship between independent Poissons and multinomials, the methods for independent Poisson data can also be used to analyze multinomial data.

Appendix: probability for continuous distributions

As stated in Section 1.3, probabilities are sometimes defined as areas under a curve. The curve, called a probability density function or just a density, must be defined by some nonnegative function $f(\cdot)$. (Nonnegative to ensure that probabilities are always positive.) Thus the probability that a random observation y is between two numbers, say a and b, is the area under the curve measured between a and b. Using calculus, this is

$$\Pr[a < y < b] = \int_a^b f(y)\, dy.$$

Because we are measuring areas under curves, there is no area associated with any one point, so $\Pr[a < y < b] = \Pr[a \leq y < b] = \Pr[a < y \leq b] = \Pr[a \leq y \leq b]$. The area under the entire curve must be 1, i.e.,

$$1 = \Pr[-\infty < y < \infty] = \int_{-\infty}^{\infty} f(y)\, dy.$$

Figure 1.1 indicates that the probability below $K(1 - \alpha)$ is $1 - \alpha$, i.e.,

$$1 - \alpha = \Pr[y < K(1 - \alpha)] = \int_{-\infty}^{K(1-\alpha)} f(y)\, dy$$

and that the probability above $K(1-\alpha)$ is α, i.e.,

$$\alpha = \Pr[y > K(1-\alpha)] = \int_{K(1-\alpha)}^{\infty} f(y)\,dy.$$

The expected value of y is defined as

$$E(y) = \int_{-\infty}^{\infty} yf(y)\,dy.$$

For any function $g(y)$, the expected value is

$$E[g(y)] = \int_{-\infty}^{\infty} g(y)f(y)\,dy.$$

In particular, if we let $E(y) = \mu$ and $g(y) = (y-\mu)^2$, we define the variance as

$$\text{Var}(y) = E[(y-\mu)^2] = \int_{-\infty}^{\infty} (y-\mu)^2 f(y)\,dy.$$

To define the covariance between two random variables, say y_1 and y_2, we need a joint density $f(y_1, y_2)$. We can find the density for y_1 alone as

$$f_1(y_1) = \int_{-\infty}^{\infty} f(y_1, y_2)\,dy_2$$

and we can write $E(y_1)$ in two equivalent ways,

$$E(y_1) = \int_{-\infty}^{\infty} \int_{-\infty}^{\infty} y_1 f(y_1, y_2)\,dy_1\,dy_2 = \int_{-\infty}^{\infty} y_1 f_1(y_1)\,dy_1.$$

Writing $E(y_1) = \mu_1$ and $E(y_2) = \mu_2$, we can now define the covariance between y_1 and y_2 as

$$\text{Cov}(y_1, y_2) = \int_{-\infty}^{\infty} \int_{-\infty}^{\infty} (y_1 - \mu_1)(y_2 - \mu_2) f(y_1, y_2)\,dy_1\,dy_2.$$

1.6 Exercises

EXERCISE 1.6.1. Use the definitions to find the expected value and variance of a $\text{Bin}(1, p)$ distribution.

EXERCISE 1.6.2. Let y be a random variable with $E(y) = \mu$ and $\text{Var}(y) = \sigma^2$. Show that

$$E\left(\frac{y-\mu}{\sigma}\right) = 0$$

and

$$\text{Var}\left(\frac{y-\mu}{\sigma}\right) = 1.$$

Let $\bar{y}_{\cdot}$ be the sample mean of n independent observations y_i with $E(y_i) = \mu$ and $\text{Var}(y_i) = \sigma^2$. What is the expected value and variance of

$$\frac{\bar{y}_{\cdot} - \mu}{\sigma/\sqrt{n}}?$$

Hint: For the first part, write

$$\frac{y-\mu}{\sigma} \quad \text{as} \quad \frac{1}{\sigma}y - \frac{\mu}{\sigma}$$

and use Proposition 1.2.11.

EXERCISE 1.6.3.   Let y be the random variable consisting of the number of spots that face up upon rolling a die. Give the distribution of y. Find the expected value, variance, and standard deviation of y.

EXERCISE 1.6.4.   Consider your letter grade for this course. Obviously, it is a random phenomenon. Define the 'grade point' random variable: $y(A) = 4$, $y(B) = 3$, $y(C) = 2$, $y(D) = 1$, $y(F) = 0$. If you were lucky enough to be taking the course from me, you would find that I am an easy grader. I give 5% As, 10% Bs, 35% Cs, 30% Ds, and 20% Fs. I also assign grades at random, that is to say, my tests generate random scores. Give the distribution of y. Find the expected value, variance, and standard deviation of the grade points a student would earn in my class. (Just in case you hadn't noticed, I'm being sarcastic.)

EXERCISE 1.6.5.   Referring to Exercise 1.6.4, supposing I have a class of 40 students, what is the joint distribution for the numbers of students who get each of the five grades? Note that we are no longer looking at how many grade points an individual student might get, we are now counting how many occurrences we observe of various events. What is the distribution for the number of students who get Bs? What is the expected value of the number of students who get Cs? What is the variance and standard deviation of the number of students who get Cs? What is the probability that in a class of 5 students, 1 gets an A, 2 get Cs, 1 gets a D, and 1 fails?

EXERCISE 1.6.6.   Graph the function $f(x) = 1$ if $0 < x < 1$ and $f(x) = 0$ otherwise. This is known as the uniform density on $(0,1)$. If we use this curve to define a probability function, what is the probability of getting an observation larger than $1/4$? Smaller than $2/3$? Between $1/3$ and $7/9$?

EXERCISE 1.6.7.   Arthritic ex-football players prefer their laudanum made with Old Pain-Killer Scotch by two to one. If we take a random sample of 5 arthritic ex-football players, what is the distribution of the number who will prefer Old Pain-Killer? What is the probability that only 2 of the ex-players will prefer Old Pain-Killer? What is the expected number who will prefer Old Pain-Killer? What are the variance and standard deviation of the number who will prefer Old Pain-Killer?

EXERCISE 1.6.8.   Let $W \sim \text{Bin}(N, p)$ and for $i = 1, \ldots, N$ take independent y_is that are $\text{Bin}(1, p)$. Argue that W has the same distribution as $y_1 + \cdots + y_N$. Use this fact, along with Exercise 1.6.1 and Proposition 1.2.11, to find $E(W)$ and $\text{Var}(W)$.

EXERCISE 1.6.9.   Appendix B.1 gives probabilities for a family of distributions that all look roughly like Figure 1.1. All members of the family are symmetric about zero and the members are distinguished by having different numbers of degrees of freedom (df). They are called t distributions. For $0 \le \alpha \le 1$, the α percentile of a t distribution with df degrees of freedom is the point x such that $\Pr[t(df) \le x] = \alpha$. For example, from Table B.1 the row corresponding to $df = 10$ and the column for the .90 percentile tells us that $\Pr[t(10) \le 1.372] = .90$.

(a) Find the .99 percentile of a $t(7)$ distribution.

(b) Find the .975 percentile of a $t(50)$ distribution.

(c) Find the probability that a $t(25)$ is less than or equal to 3.450.

(d) Find the probability that a $t(100)$ is less than or equal to 2.626.

(e) Find the probability that a $t(16)$ is greater than 2.92.

(f) Find the probability that a $t(40)$ is greater than 1.684.

(g) Recalling that t distributions are symmetric about zero, what is the probability that a $t(40)$ distribution is less than -1.684?

(h) What is the probability that a $t(40)$ distribution is between -1.684 and 1.684?

(i) What is the probability that a $t(25)$ distribution is less than -3.450?

(j) What is the probability that a $t(25)$ distribution is between -3.450 and 3.450?

EXERCISE 1.6.10.   Consider a random variable that takes on the values $25, 30, 45$, and 50 with probabilities $.15, .25, .35$, and $.25$, respectively. Find the expected value, variance, and standard deviation of this random variable.

EXERCISE 1.6.11.   Consider three independent random variables X, Y, and Z. Suppose $E(X) = 25, E(Y) = 40$, and $E(Z) = 55$ with $Var(X) = 4, Var(Y) = 9$, and $Var(Z) = 25$.

(a) Find $E(2X + 3Y + 10)$ and $Var(2X + 3Y + 10)$.

(b) Find $E(2X + 3Y + Z + 10)$ and $Var(2X + 3Y + Z + 10)$.

EXERCISE 1.6.12.   As of 1994, Duke University had been in the final four of the NCAA's national basketball championship tournament seven times in nine years. Suppose their appearances were independent and that they had a probability of .25 for winning the tournament in each of those years.

(a) What is the probability that Duke would win two national championships in those seven appearances?

(b) What is the probability that Duke would win three national championships in those seven appearances?

(c) What is the expected number of Duke championships in those seven appearances?

(d) What is the variance of the number of Duke championships in those seven appearances?

EXERCISE 1.6.13.   Graph the function $f(x) = 2x$ if $0 < x < 1$ and $f(x) = 0$ otherwise. If we use this curve to define a probability function, what is the probability of getting an observation larger than $1/4$? Smaller than $2/3$? Between $1/3$ and $7/9$?

EXERCISE 1.6.14.   A pizza parlor makes small, medium, and large pizzas. Over the years they make 20% small pizzas, 35% medium pizzas, and 45% large pizzas. On a given Tuesday night they were asked to make only 10 pizzas. If the orders were independent and representative of the long-term percentages, what is the probability that the orders would be for four small, three medium, and three large pizzas? On such a night, what is the expected number of large pizzas to be ordered and what is the expected number of small pizzas to be ordered? What is the variance of the number of large pizzas to be ordered and what is the variance of the number of medium pizzas to be ordered?

EXERCISE 1.6.15.   When I order a limo, 65% of the time the driver is male. Assuming independence, what is the probability that 6 of my next 8 drivers are male? What is the expected number of male drivers among my next eight? What is the variance of the number of male drivers among my next eight?

EXERCISE 1.6.16.   When I order a limo, 65% of the time the driver is clearly male, 30% of the time the driver is clearly female, and 5% of the time the gender of the driver is indeterminant. Assuming independence, what is the probability that among my next 8 drivers 5 are clearly male and 3 are clearly female? What is the expected number of indeterminant drivers among my next eight? What is the variance of the number of clearly female drivers among my next eight?

Chapter 2

One Sample

In this chapter we examine the analysis of a single *random* sample consisting of n independent observations from some population.

2.1 Example and introduction

EXAMPLE 2.1.1.   Consider the dropout rate from a sample of math classes at the University of New Mexico as reported by Koopmans (1987). The data are

$$5,22,10,12,8,17,2,25,10,10,7,7,40,7,9,17,12,12,1,$$

$$13,10,13,16,3,14,17,10,10,13,59,11,13,5,12,14,3,14,15.$$

This list of $n = 38$ observations is not very illuminating. A graphical display of the numbers is more informative. Figure 2.1 plots the data above a single axis. This is often called a *dot plot*. From Figure 2.1, we see that most of the observations are between 0 and 18. There are two conspicuously large observations. Going back to the original data we identify these as the values 40 and 59. In particular, these two *outlying* values strongly suggest that the data do not follow a bell-shaped curve and thus that the data do not follow a normal distribution.   □

Typically, for one sample of data we assume that the n observations are

Data		Distribution
$y_1, y_2, \ldots, y_n$	independent	$N(\mu, \sigma^2)$

The key assumptions are that the observations are independent and have the same distribution. In particular, we assume they have the same (unknown) mean μ and the same (unknown) variance σ^2.

These assumptions of independence and a constant distribution should be viewed as only useful approximations to actual conditions. Often the most valuable approach to evaluating these assumptions is simply to think hard about whether they are reasonable. In any case, the conclusions we reach are only as good as the assumptions we have made. The only way to be positive that these assumptions are true is if we arrange for them to be true. If we have a fixed finite population and take a random sample from the population allowing elements of the population to be observed more than once, then the assumptions (other than normality) are true. In Example 2.1.1, if we had the dropout

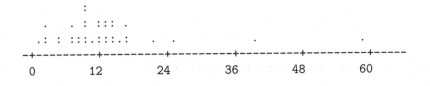

Figure 2.1: *Dot plot for drop rate percentage data.*

rates for all math classes in the year and randomly selected these 38 while allowing for classes to appear more than once in the sample, the assumptions of independence with the same distribution are satisfied.

The ideal conditions of independent sampling from a fixed population are difficult to achieve. Many populations refuse to hold still while we sample them. For example, the population of students at a large university changes almost continuously (during working hours). To my way of thinking, the populations associated with most interesting data are virtually impossible to define unambiguously. Who really cares about the dropout rates? As such, they can only be used to fix blame. Our real interest is in what the data can tell us about current and future dropout rates. If the data are representative of current or future conditions, the data can be used to fix problems. For example, one might find out whether certain instructors generate huge dropout rates, and avoid taking classes from them. Perhaps the large dropout rates are because the instructor is more demanding. You might want to seek out such a class. It is difficult to decide whether these or any data are representative of current or future conditions because we cannot possibly know the future population and we cannot practically know the current population. As mentioned earlier, often our best hope is to think hard about whether these data approximate independent observations from the population of interest.

Even when sampling from a fixed population, we use approximations. In practice we rarely allow elements of a fixed population to be observed more than once in a sample. This invalidates the assumptions. If the first sampled element is eliminated, the second element is actually being sampled from a different population than the first. (One element has been eliminated.) Fortunately, when the sample contains a small proportion of the fixed population, the standard assumptions make a good approximation. Moreover, the normal distribution is never more than an approximation to a fixed population. The normal distribution has an infinite number of possible outcomes, while fixed populations are finite. Often, the normal distribution makes a good approximation, especially if we do our best to validate it. In addition, the assumption of a normal distribution is only used when drawing conclusions from small samples. For large samples we can get by without the assumption of normality.

Our primary objective is to draw conclusions about the mean μ. We condense the data into summary statistics. These are the sample mean, the sample variance, and the sample standard deviation. The *sample mean* has the algebraic formula

$$\bar{y}. \equiv \frac{1}{n} \sum_{i=1}^{n} y_i = \frac{1}{n} [y_1 + y_2 + \cdots + y_n]$$

where the $\cdot$ in $\bar{y}.$ indicates that the mean is obtained by averaging the y_is over the subscript i. The sample mean $\bar{y}.$ estimates the population mean μ. The *sample variance* is an estimate of the population variance σ^2. The sample variance is *essentially* the average squared distance of the observations from the sample mean,

$$s^2 \equiv \frac{1}{n-1} \sum_{i=1}^{n} (y_i - \bar{y}.)^2 \tag{2.1.1}$$

$$= \frac{1}{n-1} \left[(y_1 - \bar{y}.)^2 + (y_2 - \bar{y}.)^2 + \cdots + (y_n - \bar{y}.)^2 \right].$$

The *sample standard deviation* is just the square root of the sample variance,

$$s \equiv \sqrt{s^2}.$$

EXAMPLE 2.1.2.   The sample mean of the dropout rate data is

$$\bar{y}. = \frac{5 + 22 + 10 + 12 + 8 + \cdots + 3 + 14 + 15}{38} = 13.105.$$

If we think of these data as a sample from the fixed population of math dropout rates, $\bar{y}.$ is obviously an estimate of the simple average of all the dropout rates of all the classes in that academic year. Equivalently, $\bar{y}.$ is an estimate of the expected value for the random variable defined as the dropout rate obtained when we randomly select one class from the fixed population. Alternatively, we may interpret $\bar{y}.$ as an estimate of the mean of some population that is more interesting but less well defined than the fixed population of math dropout rates.

The sample variance is

$$s^2 = \frac{\left[(5-13.105)^2 + (22-13.105)^2 + \cdots + (14-13.105)^2 + (15-13.105)^2\right]}{38-1}$$

$$= 106.42.$$

This estimates the variance of the random variable obtained when randomly selecting one class from the fixed population. The sample standard deviation is

$$s = \sqrt{106.42} = 10.32. \qquad \square$$

The only reason s^2 is *not* the average squared distance of the observations from the sample mean is that the denominator in (2.1.1) is $n-1$ instead of n. If μ were known, a better estimate of the population variance σ^2 would be

$$\hat{\sigma}^2 \equiv \sum_{i=1}^{n} (y_i - \mu)^2 / n. \qquad (2.1.2)$$

In s^2, we have used $\bar{y}.$ to estimate μ. Not knowing μ, we know less about the population, so s^2 cannot be as good an estimate as $\hat{\sigma}^2$. The quality of a variance estimate can be measured by the number of observations on which it is based; $\hat{\sigma}^2$ makes full use of all n observations for estimating σ^2. In using s^2, we lose the functional equivalent of one observation for having estimated the parameter μ. Thus s^2 has $n-1$ in the denominator of (2.1.1) and is said to have $n-1$ *degrees of freedom*. In nearly all problems that we will discuss, there is one degree of freedom available for every observation. The degrees of freedom are assigned to various estimates and we will need to keep track of them.

The statistics $\bar{y}.$ and s^2 are estimates of μ and σ^2, respectively. The *Law of Large Numbers* is a mathematical result implying that for large sample sizes n, $\bar{y}.$ gets arbitrarily close to μ and s^2 gets arbitrarily close to σ^2.

Both $\bar{y}.$ and s^2 are computed from the random observations y_i. The summary statistics are functions of random variables, so they must also be random. Each has a distribution and to draw conclusions about the unknown parameters μ and σ^2 we need to know the distributions. In particular, if the original data are normally distributed, the sample mean has the distribution

$$\bar{y}. \sim N\left(\mu, \frac{\sigma^2}{n}\right)$$

or equivalently,

$$\frac{\bar{y}. - \mu}{\sqrt{\sigma^2/n}} \sim N(0,1); \qquad (2.1.3)$$

see Exercise 1.6.2. In Subsection 1.2.4 we established that $E(\bar{y}.) = \mu$ and $Var(\bar{y}.) = \sigma^2/n$, so the only new claim made here is that the sample mean computed from *independent, identically distributed (iid)* normal random variables is again normally distributed. Actually, this is a special case of the earlier claim that any linear combinations of independent normals is again normal. Moreover, the *Central Limit Theorem* is a mathematical result stating that the normal distribution for $\bar{y}.$ is approximately true for 'large' samples n, regardless of whether the original data are normally distributed.

As we will see below, the distributions given earlier are only useful in drawing conclusions

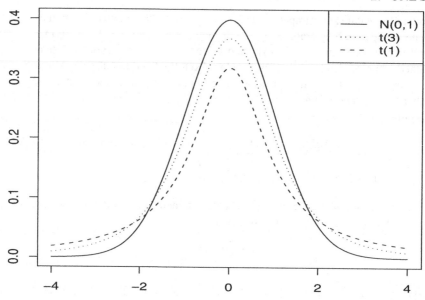

Figure 2.2: *Three distributions: solid, $N(0,1)$; long dashes, $t(1)$; short dashes, $t(3)$.*

about data when σ^2 is known. Generally, we will need to estimate σ^2 with s^2 and proceed as best we can. By the law of large numbers, s^2 becomes arbitrarily close to σ^2, so for large samples we can substitute s^2 for σ^2 in the distributions above. In other words, for large samples the *approximation*

$$\frac{\bar{y}. - \mu}{\sqrt{s^2/n}} \sim N(0,1) \qquad (2.1.4)$$

holds regardless of whether the data were originally normal.

For small samples we cannot rely on s^2 being close to σ^2, so we fall back on the assumption that the original data are normally distributed. For normally distributed data, the appropriate distribution is called a t distribution with $n-1$ degrees of freedom. In particular,

$$\frac{\bar{y}. - \mu}{\sqrt{s^2/n}} \sim t(n-1). \qquad (2.1.5)$$

The t distribution is similar to the standard normal but more spread out; see Figure 2.2. It only makes sense that if we need to estimate σ^2 rather than knowing it, our conclusions will be less exact. This is reflected in the fact that the t distribution is more spread out than the $N(0,1)$. In the previous paragraph we argued that for large n the appropriate distribution is

$$\frac{\bar{y}. - \mu}{\sqrt{s^2/n}} \sim N(0,1).$$

We are now arguing that for normal data the appropriate distribution is $t(n-1)$. It had better be the case (and is) that for large n the $N(0,1)$ distribution is approximately the same as the $t(n-1)$ distribution. In fact, we define $t(\infty)$ to be a $N(0,1)$ distribution where ∞ indicates an infinitely large number.

Formal distribution theory

By definition, the t distribution is obtained as the ratio of two things related to the sample mean and variance. We now present this general definition.

First, for normally distributed data, the sample variance s^2 has a known distribution that depends on σ^2. It is related to a distribution called the *chi-squared* (χ^2) distribution with $n-1$ degrees of freedom. In particular,

$$\frac{(n-1)s^2}{\sigma^2} \sim \chi^2(n-1).$$ (2.1.6)

Moreover, for normal data, $\bar{y}.$ and s^2 are independent.

Definition 2.1.3. A t distribution is the distribution obtained when a random variable with a $N(0,1)$ distribution is divided by an independent random variable that is the square root of a χ^2 random variable over its degrees of freedom. The t distribution has the same degrees of freedom as the chi-square.

In particular, $[\bar{y}. - \mu]/\sqrt{\sigma^2/n}$ is $N(0,1)$, $\sqrt{[(n-1)s^2/\sigma^2]/(n-1)}$ is the square root of a chi-squared random variable over its degrees of freedom, and the two are independent because $\bar{y}.$ and s^2 are independent, so

$$\frac{\bar{y}. - \mu}{\sqrt{s^2/n}} = \frac{[\bar{y}. - \mu]/\sqrt{\sigma^2/n}}{\sqrt{[(n-1)s^2/\sigma^2]/(n-1)}} \sim t(n-1).$$

The t distribution has the same degrees of freedom as the estimate of σ^2; this is typically the case in other applications.

2.2 Parametric inference about μ

Most statistical tests and confidence intervals are applications of a single theory that focuses on a single parameter. While we will make use of this parametric theory when necessary, and while people educated in Statistics are expected to know this parametric approach to inference, the current book focuses on a model-based approach to statistical inference that will be introduced in Section 2.4.

To use the parametric theory in question, we need to know four things. In the one-sample problem the four things are

1. the parameter of interest, μ,

2. the estimate of the parameter, $\bar{y}.$,

3. the standard error of the estimate, $\mathrm{SE}(\bar{y}.) \equiv \sqrt{s^2/n} = s/\sqrt{n}$, and

4. the appropriate distribution for $[\bar{y}. - \mu]/\sqrt{s^2/n}$.

In practice the appropriate distribution can always be thought of as a t distribution with some number of degrees of freedom, df. The t distribution is denoted $t(df)$. When the original observations are assumed to be independent $N(\mu, \sigma^2)$, the appropriate distribution is $t(n-1)$, that is, the degrees of freedom are $n-1$. Regardless of the original distribution, if the observations are independent with a common distribution having mean μ and variance σ^2 and if the sample size n is large, the central limit theorem and the law of large numbers suggest that the appropriate distribution is a $N(0,1)$, which is the same as a $t(\infty)$ distribution, that is, a t with an infinite number of degrees of freedom. In practice, I suspect that a $t(n-1)$ will almost always be a better approximation to the true distribution than a $t(\infty)$.

Specifically, we need a known (tabled or programmed) distribution for $[\bar{y}. - \mu]/\sqrt{s^2/n}$ that is symmetric about zero and continuous. The standard error, $\mathrm{SE}(\bar{y}.)$, is the estimated standard deviation of $\bar{y}.$. Recall that the variance of $\bar{y}.$ is σ^2/n, so its standard deviation is $\sqrt{\sigma^2/n}$ and estimating σ^2 by s^2 gives the standard error $\sqrt{s^2/n}$.

The appropriate distribution for $[\bar{y}. - \mu]/\sqrt{s^2/n}$ when the data are normally distributed is the $t(n-1)$ as in (2.1.4). For large samples, an approximate distribution is the $N(0,1)$ as in (2.1.3).

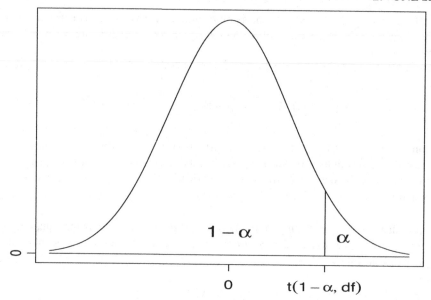

Figure 2.3: $1 - \alpha$ *percentile of the distribution of* $[\bar{y}. - \mu]/\mathrm{SE}(\bar{y}.)$.

Recall that for large samples from a normal population, it is largely irrelevant whether we use the standard normal or the t distribution because they are essentially the same. In the unrealistic case where σ^2 is known we do not need to estimate it, so we use $\sqrt{\sigma^2/n}$ instead of $\sqrt{s^2/n}$ for the standard error. In this case, the appropriate distribution is $N(0,1)$ as in (2.1.2) if either the original data are normal or the sample size is large.

We need notation for the percentage points of the known distribution and we need a name for the point that cuts off the top α of the distribution. Typically, we need to find points that cut off the top $5\%, 2.5\%, 1\%,$ or 0.5% of the distribution, so α is $0.05, 0.025, 0.01,$ or 0.005. As discussed in the previous paragraph, the appropriate distribution depends on various circumstances of the problem, so we begin by discussing percentage points with a generic notation. We use the notation $t(1 - \alpha, df)$ for the point that cuts off the top α of the distribution. Figure 2.3 displays this idea graphically for a value of α between 0 and 0.5. The distribution is described by the curve, which is symmetric about 0. $t(1 - \alpha, df)$ is indicated along with the fact that the area under the curve to the right of $t(1 - \alpha, df)$ is α. Formally the point that cuts off the top α of the distribution is $t(1 - \alpha, df)$ where

$$\Pr\left[\frac{\bar{y}. - \mu}{\mathrm{SE}(\bar{y}.)} > t(1 - \alpha, df)\right] = \alpha.$$

Note that the same point $t(1 - \alpha, df)$ also cuts off the bottom $1 - \alpha$ of the distribution, i.e.,

$$\Pr\left[\frac{\bar{y}. - \mu}{\mathrm{SE}(\bar{y}.)} < t(1 - \alpha, df)\right] = 1 - \alpha.$$

This is illustrated in Figure 2.3 by the fact that the area under the curve to the left of $t(1 - \alpha, df)$ is $1 - \alpha$. The reason the point is labeled $t(1 - \alpha, df)$ is because it cuts off the bottom $1 - \alpha$ of the distribution. The labeling depends on the percentage to the left even though our interest is in the percentage to the right.

There are at least three different ways to label these percentage points; I have simply used the one I feel is most consistent with general usage in Probability and Statistics. The key point however is to be familiar with Figure 2.3. We need to find points that cut off a fixed percentage of the area under the curve. As long as we can find such points, what we call them is irrelevant. Ultimately,

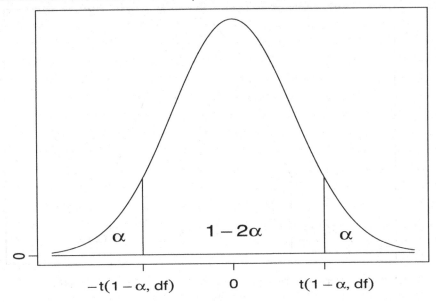

Figure 2.4: *Symmetry about 0 in the distribution of* $[\bar{y}. - \mu]/\mathrm{SE}(\bar{y}.)$.

anyone doing Statistics will need to be familiar with all three methods of labeling. One method of labeling is in terms of the area to the left of the point; this is the one we will use. A second method is labeling in terms of the area to the right of the point; thus the point we call $t(1 - \alpha, df)$ could be labeled, say, $Q(\alpha, df)$. The third method is to call this number, say, $W(2\alpha, df)$, where the area to the right of the point is doubled in the label. For example, if the distribution is a $N(0,1) = t(\infty)$, the point that cuts off the bottom 97.5% of the distribution is 1.96. This point also cuts off the top 2.5% of the area. It makes no difference if we refer to 1.96 as the number that cuts off the bottom 97.5%, $t(0.975, \infty)$, or as the number that cuts off the top 2.5%, $Q(0.025, \infty)$, or as the number $W(0.05, \infty)$ where the label involves 2×0.025; the important point is being able to identify 1.96 as the appropriate number. Henceforth, we will always refer to points in terms of $t(1 - \alpha, df)$, the point that cuts off the bottom $1 - \alpha$ of the distributions. No further reference to the alternative labelings will be made but all three labels are used in Appendix B.1. There $t(1 - \alpha, df)$s are labeled as percentiles and, for reasons related to statistical tests, $Q(\alpha, df)$s and $W(2\alpha, df)$s are labeled as *one-sided* and *two-sided* α levels, respectively.

A fundamental assumption in our inference about μ is that the distribution of $[\bar{y}. - \mu]/\mathrm{SE}(\bar{y}.)$ is symmetric about 0. By the symmetry around zero, if $t(1 - \alpha, df)$ cuts off the top α of the distribution, $-t(1 - \alpha, df)$ must cut off the bottom α of the distribution. Thus for distributions that are symmetric about 0 we have $t(\alpha, df)$, the point that cuts off the bottom α of the distribution, equal to $-t(1 - \alpha, df)$. This fact is illustrated in Figure 2.4. Algebraically, we write

$$\Pr\left[\frac{\bar{y}. - \mu}{\mathrm{SE}(\bar{y}.)} < -t(1 - \alpha, df)\right] = \Pr\left[\frac{\bar{y}. - \mu}{\mathrm{SE}(\bar{y}.)} < t(\alpha, df)\right] = \alpha.$$

Frequently, we want to create a central interval that contains a specified probability, say $1 - \alpha$. Figure 2.5 illustrates the construction of such an interval. Algebraically, the middle interval with probability $1 - \alpha$ is obtained by

$$\Pr\left[-t\left(1 - \frac{\alpha}{2}, df\right) < \frac{\bar{y}. - \mu}{\mathrm{SE}(\bar{y}.)} < t\left(1 - \frac{\alpha}{2}, df\right)\right] = 1 - \alpha.$$

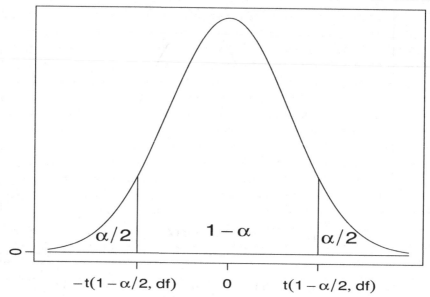

Figure 2.5: $1 - \alpha$ *central interval for the distribution of* $[\bar{y}. - \mu]/\mathrm{SE}(\bar{y}.)$.

The probability of getting something outside of this interval is

$$\alpha = \frac{\alpha}{2} + \frac{\alpha}{2} = \Pr\left[\frac{\bar{y}. - \mu}{\mathrm{SE}(\bar{y}.)} < -t\left(1 - \frac{\alpha}{2}, df\right)\right] + \Pr\left[\frac{\bar{y}. - \mu}{\mathrm{SE}(\bar{y}.)} > t\left(1 - \frac{\alpha}{2}, df\right)\right].$$

Percentiles of the t distribution are given in Appendix B.1 with the ∞ row giving percentiles of the $N(0,1)$ distribution.

2.2.1   Significance tests

A significance test is a procedure for checking the validity of a (null) model for the data. A model involves a number of assumptions; often one of those assumptions is identified as being of special importance and is called the *null hypothesis*. We wish to test whether or not the (null) model is true. If relevant data are available, we can test the model, but we cannot really test whether it is true or false; we can merely test whether the data are consistent or inconsistent with the model. Data that are inconsistent with the model suggest that the model is false. Data that are consistent with the model are just that, consistent with the model; they do not imply that the model is true because other models could equally well have generated the data.

In a one-sample problem, for some fixed known number m_0 we may want to test the *null hypothesis*

$$H_0 : \mu = m_0.$$

The number m_0 must be known; it is some number that is of interest for the specific data being analyzed. It is not just an unspecified symbol. The null model involves not only the assumption that $\mu = m_0$ but also the assumptions that the data are independent with common mean μ and common variance σ^2, and perhaps also that the data are normally distributed.

EXAMPLE 2.2.1.   For the dropout rate data, we might be interested in the hypothesis that the true dropout rate is 10%. Thus the null hypothesis is $H_0 : \mu = 10$. The other assumptions were discussed at the beginning of the chapter. They include such things as independence, normality, and all observations having the same mean and variance. While we can never confirm that these other assumptions are absolutely valid, it is a key aspect of modern statistical practice to validate the assumptions as far as is reasonably possible. When we are convinced that the other assumptions are reasonably valid, data that contradict the assumptions can be reasonably interpreted as contradicting the specific assumption H_0.   □

The test is based on all the assumptions including H_0 being true and we check to see if the data are inconsistent with those assumptions. The idea is much like the idea of a proof by contradiction. We assume a model that includes the assumption H_0. If the data contradict that model, we can conclude that something is wrong with the model. If we can satisfy ourselves that all of the assumptions other than the assumption H_0 are true, and we have data that are inconsistent with the model, then H_0 must be false. If the data do not contradict the H_0 model, we can only conclude that the data are consistent with the assumptions. *We can never conclude that the assumptions are true.* Unfortunately, *data almost never yield an absolute contradiction to the null model. We need to quantify the extent to which the data are inconsistent with the null model.*

We need to be able to identify data that are inconsistent with the null model. Under the assumptions that the data are independent with common mean and variance, with either normal distributions or a large sample and with $\mu = m_0$, the distribution of $(\bar{y}. - m_0)/\sqrt{s^2/n}$ has an approximate $t(n-1)$ distribution with density as illustrated in Figures 2.2–2.5. From those illustrations, the least likely observations to occur under a $t(n-1)$ distribtion are those that are far from 0. Thus, values of $\bar{y}.$ far from m_0 make us question the validity of the null model.

We reject the null model if the *test statistic* is too far from zero, that is, if

$$\frac{\bar{y}. - m_0}{\text{SE}(\bar{y}.)}$$

is greater than some positive cutoff value or less than some negative cutoff value. Very large and very small (large negative) values of the test statistic are those that are most inconsistent with the model that includes $\mu = m_0$.

The problem is in specifying the cutoff values. For example, we do not want to reject $\mu = 10$ if the data are consistent with $\mu = 10$. One of our basic assumptions is that we know the distribution of $[\bar{y}. - \mu]/\text{SE}(\bar{y}.)$. Thus if $H_0 : \mu = 10$ is true, we know the distribution of the test statistic $[\bar{y}. - 10]/\text{SE}(\bar{y}.)$, so we know what kind of data are consistent with the $\mu = 10$ model. For instance, when $\mu = 10$, 95% of the possible values of $[\bar{y}. - 10]/\text{SE}(\bar{y}.)$ are between $-t(0.975, n-1)$ and $t(0.975, n-1)$. Any values of $[\bar{y}. - 10]/\text{SE}(\bar{y}.)$ that fall between these numbers are reasonably consistent with $\mu = 10$ and values outside the interval are defined as being inconsistent with $\mu = 10$. Thus values of $[\bar{y}. - 10]/\text{SE}(\bar{y}.)$ greater than $t(0.975, n-1)$ or less than $-t(0.975, n-1)$ cause us to reject the null model. Note that we arbitrarily specified the central 95% of the distribution as being consistent with the $\mu = 10$ model, as opposed to the central 99% or central 90%. We get to pick our criterion for what is consistent with the null model.

EXAMPLE 2.2.2.   For the dropout rate data, consider the null hypothesis $H_0 : \mu = 10$, i.e., that the mean dropout rate is 10%. These data are not normal, so we must hope that the sample size is large enough to justify use of the t distribution. Mathematically, large n suggest a $t(\infty) = N(0,1)$ distribution, but we consider the $t(n-1)$ to be a better approximate distribution. If we choose a central 90% interval, then the probability of being outside the central interval is $\alpha = 0.10$, and the upper cutoff value is $t\left(1 - \frac{\alpha}{2}, 37\right) = t(0.95, 37) = 1.687$.

The $\alpha = 0.10$ level test for the model incorporating $H_0 : \mu = 10$ is to reject the null model if

$$\frac{\bar{y}. - 10}{s/\sqrt{38}} > 1.687,$$

or if

$$\frac{\bar{y}. - 10}{s/\sqrt{38}} < -1.687.$$

The estimate of μ is $\bar{y}. = 13.105$ and the observed standard error is $s/\sqrt{n} = 10.32/\sqrt{38} = 1.673$, so *the observed value of the test statistic* is

$$t_{obs} \equiv \frac{13.105 - 10}{1.673} = 1.856.$$

Comparing this to the cutoff value of 1.687 we have $1.856 > 1.687$, so the null model is rejected. There is evidence at the $\alpha = 0.10$ level that the model with mean dropout rate of 10% is incorrect. In fact, since $\bar{y}. = 13.105 > 10$, if the assumptions other than H_0 are correct, there is the suggestion that the dropout rate is greater than 10%.

This conclusion depends on the choice of the α level. If we choose $\alpha = 0.05$, then the appropriate cutoff value is $t(0.975, 37) = 2.026$. Since the observed value of the test statistic is 1.856, which is neither greater than 2.026 nor less than -2.026, we do not reject the null model. When we do not reject the H_0 model, we cannot say that the true mean dropout rate is 10%, but we can say that, at the $\alpha = 0.05$ level, the data are consistent with the (null) model that has a true mean dropout rate of 10%. □

Generally, a test of significance is based on an α level that indicates how unusual the data are relative to the assumptions of the null model. The α-level test for the model that incorporates $H_0 : \mu = m_0$ is to reject the null model if

$$\frac{\bar{y}. - m_0}{SE(\bar{y}.)} > t\left(1 - \frac{\alpha}{2}, n - 1\right)$$

or if

$$\frac{\bar{y}. - m_0}{SE(\bar{y}.)} < -t\left(1 - \frac{\alpha}{2}, n - 1\right).$$

This is equivalent to saying, reject H_0 if

$$\frac{|\bar{y}. - m_0|}{SE(\bar{y}.)} > t\left(1 - \frac{\alpha}{2}, n - 1\right).$$

Also note that we are rejecting the H_0 model for those values of $[\bar{y}. - m_0]/SE(\bar{y}.)$ that are most inconsistent with the $t(n-1)$ distribution, those being the values of the test statistic with large absolute values.

In significance testing, a null model should never be accepted; it is either rejected or not rejected. A better way to think of a significance test is that one concludes that the data are either consistent or inconsistent with the null model. The statement that the data are inconsistent with the H_0 model is a strong statement. It suggests in some specified degree that something is wrong with the H_0 model. The statement that the data are consistent with H_0 is not a strong statement; it does not suggest the H_0 model is true. For example, the dropout data happen to be consistent with $H_0 : \mu = 12$; the test statistic

$$\frac{\bar{y}. - 12}{SE(\bar{y}.)} = \frac{13.105 - 12}{1.673} = 0.66$$

is quite small. However, the data are equally consistent with $\mu = 12.00001$. These data cannot possibly indicate that $\mu = 12$ rather than $\mu = 12.00001$. In fact, we established earlier that based on an $\alpha = 0.05$ test, these data are even consistent with $\mu = 10$. *Data that are consistent with the H_0 model do not imply that the null model is correct.*

With these data there is very little hope of distinguishing between $\mu = 12$ and $\mu = 12.00001$. The probability of getting data that lead to rejecting $H_0 : \mu = 12$ when $\mu = 12.00001$ is only just slightly more than the probability of getting data that lead to rejecting H_0 when $\mu = 12$. The probability of getting data that lead to rejecting $H_0 : \mu = 12$ when $\mu = 12.00001$ is called the *power* of the test when $\mu = 12.00001$. *The power is the probability of appropriately rejecting H_0 and depends on the* *particular value of μ ($\neq 12$).* The fact that the power is very small for detecting $\mu = 12.00001$ is not much of a problem because no one would really care about the difference between a dropout rate of 12 and a dropout rate of 12.00001. However, a small power for a difference that one cares about is a major concern. The power is directly related to the standard error and can be increased by reducing the standard error. One natural way to reduce the standard error $s/\sqrt{n}$ is by increasing the sample size n. *Of course this discussion of power presupposes that all assumptions in the model other than H_0 are correct.*

One of the difficulties in a general discussion of significance testing is that the actual null hypothesis is always context specific. You cannot give general rules for what to use as a null hypothesis because the null hypothesis needs to be some interesting claim about the population mean μ. When you sample different populations, the population mean differs, and interesting claims about the population mean depend on the exact nature of the population. The best practice for setting up null hypotheses is simply to look at lots of problems and ask yourself what claims about the population mean are of interest to you. As we examine more sophisticated data structures, some interesting hypotheses will arise from the structures themselves. For example, if we have two samples of similar measurements we might be interested in testing the null hypothesis that they have the same population means. Note that there are lots of ways in which the means could be different, but only one way in which they can be the same. Of course if the specific context suggests that one mean should be, say, 25 units greater than the other, we can use that as the null hypothesis. Similarly, if we have a sample of objects and two different measurements on each object, we might be interested in whether or not the measurements are related. In that case, an interesting null hypothesis is that the measurements are *not* related. Again, there is only one way in which measurements can be unrelated (independent), but there are many ways for measurements to display a relationship.

In practice, nobody actually uses the procedures just presented. These procedures require us to pick specific values for m_0 in $H_0 : \mu = m_0$ and for α. In practice, one either picks an α level and presents results for all values of m_0 by giving a confidence interval, or one picks a value m_0 and presents results for all α levels by giving a P value.

2.2.2   *Confidence intervals*

A $(1 - \alpha)$ confidence interval for μ consists of all the values m_0 that would not be rejected by an α-level test of $H_0 : \mu = m_0$. Confidence intervals are commonly viewed as the most useful single procedure in statistical inference but it should be pointed out that they require the validity of all the model assumptions other than $H_0 : \mu = m_0$. A 95% confidence interval for μ is based on the fact that an $\alpha = 0.05$ level test will not be rejected when

$$-t(0.975, n-1) < \frac{\bar{y}. - m_0}{\text{SE}(\bar{y}.)} < t(0.975, n-1).$$

Some algebra shows that these inequalities are equivalent to

$$\bar{y}. - t(0.975, n-1)\text{SE}(\bar{y}.) < m_0 < \bar{y}. + t(0.975, n-1)\text{SE}(\bar{y}.).$$

Thus, the value m_0 is not rejected by an α-level test if and only if m_0 is within the interval having endpoints $\bar{y}. \pm t(0.975, n-1)\text{SE}(\bar{y}.)$.

More generally, a $(1 - \alpha)100\%$ confidence interval for μ is based on observing that an α-level test of $H_0 : \mu = m_0$ does not reject when

$$-t\left(1 - \frac{\alpha}{2}, n-1\right) < \frac{\bar{y}. - m_0}{\text{SE}(\bar{y}.)} < t\left(1 - \frac{\alpha}{2}, n-1\right)$$

which is algebraically equivalent to

$$\bar{y}. - t\left(1 - \frac{\alpha}{2}, n-1\right)\text{SE}(\bar{y}.) < m_0 < \bar{y}. + t\left(1 - \frac{\alpha}{2}, n-1\right)\text{SE}(\bar{y}.).$$

A proof of the algebraic equivalence is given in the appendix to the next chapter. The endpoints of the interval can be written

$$\bar{y}_{\cdot} \pm t\left(1 - \frac{\alpha}{2}, n-1\right) SE(\bar{y}_{\cdot}),$$

or, substituting the form of the standard error,

$$\bar{y}_{\cdot} \pm t\left(1 - \frac{\alpha}{2}, n-1\right) \frac{s}{\sqrt{n}}.$$

The $1 - \alpha$ confidence interval contains all the values of μ that are consistent with both the data and the model as determined by an α-level test. Note that increasing the sample size n decreases the standard error and thus makes the confidence interval narrower. Narrower confidence intervals give more precise information about μ. In fact, by taking n large enough, we can make the confidence interval arbitrarily narrow.

EXAMPLE 2.2.3. For the dropout rate data presented at the beginning of the chapter, the parameter is the mean dropout rate for math classes, the estimate is $\bar{y}_{\cdot} = 13.105$, and the standard error is $s/\sqrt{n} = 10.32/\sqrt{38} = 1.673$. As seen in the dot plot, the original data are not normally distributed. The plot looks nothing at all like the bell-shaped curve in Figure 1.1, which is a picture of a normal distribution. Thus we hope that a sample of size 38 is sufficiently large to justify use of the central limit theorem and the law of large numbers. We use the $t(37)$ distribution as a small sample approximation to the $t(\infty) = N(0,1)$ distribution that is suggested by the mathematical results. For a 95% confidence interval, $95 = (1 - \alpha)100$, $.95 = (1 - \alpha)$, $\alpha = 1 - 0.95 = 0.05$, and $1 - \alpha/2 = 0.975$, so the number we need from the t table is $t(0.975, 37) = 2.026$. The endpoints of the confidence interval are

$$13.105 \pm 2.026(1.673)$$

giving an interval of

$$(9.71, 16.50).$$

Rounding to simple numbers, we are 95% confident that the true dropout rate is between 10% and 16.5%, but only in the sense that these are the parameter values that are consistent with the data and the model based on a $\alpha = 0.05$ test. □

Many people think that a 95% confidence interval for μ has a 95% probability of containing the parameter μ. The definition of the confidence interval just given does not lend itself towards that misinterpretation. There is another method of developing confidence intervals, one that has never made any sense to me. This alternative development does lends itself to being misinterpreted as a statement about the probability that the parameter is contained in the interval. Traditionally, statisticians have worked very hard to correct this misinterpretation. Personally, I do not think the misinterpretation does any real harm since it can be justified using arguments from Bayesian Statistics.

2.2.3 P values

Rather than having formal rules for when to reject the null model, one can report the evidence against the null model. This is done by reporting the *significance level* of the test, also known as the *P value*. The *P* value is computed assuming that the null model including $\mu = m_0$ is true and the *P* value *is the probability of seeing data that are as weird or more weird than those that were actually observed.* In other words, it is the α level at which the test would just barely not be rejected. Remember, based on Figures 2.2 through 2.5, weird data are those that lead to t values that are far from 0.

EXAMPLE 2.2.4. For $H_0: \mu = 10$ the observed value of the test statistic is 1.856. Clearly, data that give values of the test statistic that are greater than 1.856 are more weird than the actual data. Also, by symmetry, data that give a test statistic of -1.856 are just as weird as data that yield a 1.856. Finally, data that give values smaller than -1.856 are more weird than data yielding a statistic of 1.856. As before, we use the $t(37)$ distribution. From an appropriate computer program,

$$
\begin{aligned}
P &= \Pr[t(37) \geq 1.856] + \Pr[t(37) \leq -1.856] \\
&= 0.0357 + 0.0357 \\
&= 0.0715.
\end{aligned}
$$

Thus the approximate P value is 0.07. The P value is approximate because the use of the $t(37)$ distribution is an approximation based on large samples. Algebraically,

$$
P = \Pr[t(37) \geq 1.856] + \Pr[t(37) \leq -1.856] = \Pr[|t(37)| \geq |1.856|].
$$

We can see from this that the P value corresponds to the α level of a test where $H_0 : \mu = 10$ would just barely not be rejected. Thus, with a P value of 0.07, any test of $H_0 : \mu = 10$ with $\alpha > 0.07$ will be rejected while any test with $\alpha \leq 0.07$ will not be rejected. In this case, 0.07 is less than 0.10, so an $\alpha = 0.10$ level test of the null model with $H_0 : \mu = 10$ will reject H_0. On the other hand, 0.07 is greater than 0.05, so an $\alpha = 0.05$ test does not reject the null model.

If you do not have access to a computer, rough P values can be determined from a t table. Comparing $|1.856|$ to the t tables of Appendix B.1, we see that

$$
t(0.95, 37) = 1.687 < |1.856| < 2.026 = t(0.975, 37),
$$

so the P value satisfies

$$
2(1 - 0.95) = 0.10 > P > 0.05 = 2(1 - 0.975).
$$

In other words, $t(0.95, 37)$ is the cutoff value for an $\alpha = 0.10$ test and $t(0.975, 37)$ is the cutoff value for an $\alpha = 0.05$ test; $|1.856|$ falls between these values, so the P value is between 0.10 and 0.05. When only a t table is available, P values are most simply specified in terms of bounds such as these.   □

The P value is a measure of the evidence against the null hypothesis in which the smaller the P value the more evidence against H_0. The P value can be used to perform various α-level tests.

2.3 Prediction intervals

In many situations, rather than trying to learn about μ, it is more important to obtain information about future observations from the same process. A $1 - \alpha$ prediction interval will consist of all future observations that are consistent with the current observations and the model as determined by an α-level test. With independent observations, the natural point prediction for a future observation is just the estimate of μ. Unfortunately, we do not know μ, so our point prediction is our estimate of μ, the sample mean $\bar{y}.$. Our ideas about where future observations will lie involve two sources of variability. First, there is the variability that a new observation y_0 displays about its mean value μ. Second, we need to deal with the fact that we do not know μ, so there is variability associated with $\bar{y}.$, our estimate of μ. In the dropout rate example, $\bar{y}. = 13.105$ and $s^2 = 106.42$. If we could assume that the observations are normally distributed (which is a poor assumption), we could create a 99% prediction interval. The theory for constructing prediction intervals is discussed in the next subsection. The interval for the new observation is centered about $\bar{y}.$, our best point predictor, and is similar to a confidence interval but uses a standard error that is appropriate for prediction. The actual interval has endpoints

$$
\bar{y}. \pm t\left(1 - \frac{\alpha}{2}, n - 1\right) \sqrt{s^2 + \frac{s^2}{n}}.
$$

In our example of a 99% interval, $.99 = 1 - \alpha$, so $\alpha = 0.01$ and with $n = 38$ we use $t(0.995, 37) = 2.715$. The endpoints of the interval become

$$13.105 \pm 2.715 \sqrt{106.42 + \frac{106.42}{38}}$$

or

$$13.105 \pm 28.374$$

for an interval of $(-15.27, 41.48)$. In practice, dropout percentages cannot be less than 0, so a more practical interval is $(0, 41.44)$. To the limits of our assumptions, a math class will be consistent with the model and past data if its dropout rate falls between 0 and 41.5%. It is impossible to validate assumptions about future observations (as long as they remain in the future), thus the exact confidence levels of prediction intervals are always suspect.

The key difference between the 99% prediction interval and a 99% confidence interval is the standard error. In a confidence interval, the standard error is $\sqrt{s^2/n}$. In a prediction interval, we mentioned the need to account for two sources of variability and the corresponding standard error is $\sqrt{s^2 + s^2/n}$. The first term in this square root estimates the variance of the new observation, while the second term in the square root estimates the variance of $\bar{y}.$, the point predictor.

As mentioned earlier and as will be shown in Section 2.5, the assumption of normality is pretty poor for the 38 observations on dropout rates. Even without the assumption of normality we can get an approximate evaluation of the interval. The interval uses the value $t(0.995, 37) = 2.71$, and we will see that even without the assumption of normality, the approximate confidence level of this prediction interval is at least

$$100 \left(1 - \frac{1}{(2.71)^2} \right) \% = 86\%.$$

Theory

In this chapter we assume that the observations y_i are independent from a population with mean μ and variance σ^2. We have assumed that all our previous observations on the process have been independent, so it is reasonable to assume that the future observation y_0 is independent of the previous observations with the same mean and variance. The prediction interval is actually based on the difference $y_0 - \bar{y}.$, i.e., we examine how far a new observation may reasonably be from our point predictor. Note that

$$E(y_0 - \bar{y}.) = \mu - \mu = 0.$$

To proceed we need a standard error for $y_0 - \bar{y}.$ and a distribution that is symmetric about 0. The standard error of $y_0 - \bar{y}.$ is just the standard deviation of $y_0 - \bar{y}.$ when available or, more often, an estimate of the standard deviation. First we need to find the variance. As $\bar{y}.$ is computed from the previous observations, it is independent of y_0 and, using Proposition 1.2.11,

$$\text{Var}(y_0 - \bar{y}.) = \text{Var}(y_0) + \text{Var}(\bar{y}.) = \sigma^2 + \frac{\sigma^2}{n} = \sigma^2 \left[1 + \frac{1}{n} \right].$$

The standard deviation is the square root of the variance. Typically, σ^2 is unknown, so we estimate it with s^2 and our standard error becomes

$$\text{SE}(y_0 - \bar{y}.) = \sqrt{s^2 + \frac{s^2}{n}} = \sqrt{s^2 \left[1 + \frac{1}{n} \right]} = s\sqrt{1 + \frac{1}{n}}.$$

For future reference, note that the first equality in this equation can be rewritten as

$$\text{SE}(y_0 - \bar{y}.) = \sqrt{s^2 + \text{SE}(\bar{y}.)^2}.$$

To get an appropriate distribution, we assume that all the observations are normally distributed. In this case,

$$\frac{y_0 - \bar{y}.}{\text{SE}(y_0 - \bar{y}.)} \sim t(n-1).$$

The validity of the $t(n-1)$ distribution is established in Exercise 2.8.10.

Using the distribution based on normal observations, a 99% prediction interval is obtained from the following inequalities:

$$-t(0.995, n-1) < \frac{y_0 - \bar{y}.}{\text{SE}(y_0 - \bar{y}.)} < t(0.995, n-1)$$

which occurs if and only if

$$\bar{y}. - t(0.995, n-1)\text{SE}(y_0 - \bar{y}.) < y_0 < \bar{y}. + t(0.995, n-1)\text{SE}(y_0 - \bar{y}.).$$

The key point is that the two sets of inequalities are algebraically equivalent. A 99% prediction interval has endpoints

$$\bar{y}. \pm t(0.995, n-1)\text{SE}(y_0 - \bar{y}.).$$

This looks similar to a 99% confidence interval for μ but the standard error is very different. In the prediction interval, the endpoints are

$$\bar{y}. \pm t(0.995, n-1) s \sqrt{\left[1 + \frac{1}{n}\right]},$$

while in a confidence interval the endpoints are

$$\bar{y}. \pm t(0.995, n-1) s \sqrt{\frac{1}{n}}.$$

The standard error for the prediction interval is typically much larger than the standard error for the confidence interval. Moreover, unlike the confidence interval, the prediction interval cannot be made arbitrarily small by taking larger and larger sample sizes n. Of course, to compute an arbitrary $(1-\alpha)100\%$ prediction interval, simply replace the value $t(0.995, n-1)$ with $t(1-\alpha/2, n-1)$.

Even when the data are not normally distributed, we can obtain an approximate worst-case confidence coefficient or α level for large samples. In other words, if the data are not normal but we still use the cutoff values from the $t(n-1)$ distribution, what can we say about how weird it is to see something outside the cutoff values?

The approximation comes from using the *Law of Large Numbers* to justify treating s as if it were the actual population standard deviation σ. With this approximation, Chebyshev's inequality states that

$$\frac{1}{t(0.995, n-1)^2} \geq \Pr\left[\frac{|y_0 - \bar{y}.|}{\text{SE}(y_0 - \bar{y}.)} > t(0.995, n-1)\right],$$

cf. Subsection 1.2.2. The 99% prediction interval based on 38 observations and cutoff values from the $t(37)$ distribution corresponds to not rejecting an α-level test where α is somewhere below

$$\frac{1}{(2.71)^2} = 0.14,$$

for an approximate confidence coefficient above $0.86 = 1 - 0.14$. This assumes that the past observations and the future observation form a random sample from the same population and assumes that 38 observations is large enough to justify using the Law of Large Numbers. Similarly, if we can apply the improved version of Chebyshev's inequality from Section 1.3, we get an upper bound on α of $1/2.25(2.71)^2 = 0.061$ for an approximate confidence coefficient above $0.94 = 1 - 0.06$.

2.4 Model testing

We return to the subject of testing hypotheses about μ but now we use model-based tests. If we were only going to perform tests on the mean of one sample, there would be little point in introducing this alternative test procedure, but testing models works in many situations where testing a single parameter is difficult. Moreover, model testing can provide tests of more than one parameter. The focus of this section is to introduce model-based tests and to show the relationship between parametric tests and model-based tests for hypotheses about the mean μ. Throughout, we have assumed that the process of generating the data yields independent observations from some population. In quality control circles this is referred to as having a process that is under *statistical control*.

Model-based tests depend on measures of how well different models explain the data. For many problems, we use variance estimates to quantify how well a model explains that data. A better explanation will lead to a smaller variance estimate. For one-sample problems, the variance estimates we will use are s^2 as defined in Equation (2.1.1) and $\hat{\sigma}^2$ as defined in (2.1.2). Recall that $\hat{\sigma}^2$ is the variance estimate used when μ is known.

Under the one-sample model with μ unknown, our variance estimate is s^2. Under the one-sample model with the null hypothesis $H_0 : \mu = m_0$ assumed to be true, the variance estimate is

$$\hat{\sigma}_0^2 \equiv \frac{1}{n} \sum_{i=1}^{n} (y_i - m_0)^2 .$$

If the null model is true, the two variance estimates should be about the same. If the two variance estimates are different, it suggests that something is wrong with the null model. One way to evaluate whether the estimates are about the same is to evaluate whether $\hat{\sigma}^2 / s^2$ is about 1.

Actually, it is not common practice to compare the two variance estimates $\hat{\sigma}_0^2$ and s^2 directly. Typically, one rewrites the variance estimate from the null model $\hat{\sigma}_0^2$ as a weighted average of the more general estimate s^2 and *something else*. This something else will also be an estimate of σ^2 when the null model is true; see Chapter 3. It turns out that in a one-sample problem, the something else has a particularly nice form. The formula for the weighted average turns out to be

$$\hat{\sigma}_0^2 = \frac{n-1}{n} s^2 + \frac{1}{n} n (\bar{y}. - m_0)^2 .$$

This estimate has weight $(n-1)/n$ on s^2 and weight $1/n$ on the something else, $n(\bar{y}. - m_0)^2$. When the null model is true, $n(\bar{y}. - m_0)^2$ is an estimate of σ^2 with 1 degree of freedom. The estimate $\hat{\sigma}_0^2$ has n degrees of freedom and it is being split into s^2 with $n-1$ degrees of freedom and $n(\bar{y}. - m_0)^2$, so there is only 1 degree of freedom left for $n(\bar{y}. - m_0)^2$.

The test is based on looking at whether $n(\bar{y}. - m_0)^2 / s^2$ is close to 1 or not. Under the null model including normality, this ratio has a distribution called an F distribution. The variance estimate in the numerator has 1 degree of freedom and the variance estimate in the denominator has $n-1$ degrees of freedom. The degrees of freedom identify a particular member of the family of F distributions. Thus we write,

$$\frac{n(\bar{y}. - m_0)^2}{s^2} \sim F(1, n-1).$$

To compare this test to the parameter-based test, note that

$$\frac{n(\bar{y}. - m_0)^2}{s^2} = \left[\frac{|\bar{y}. - m_0|}{\sqrt{s^2/n}} \right]^2 ,$$

with the right-hand side being the square of the t statistic for testing the null model with $H_0 : \mu = m_0$.

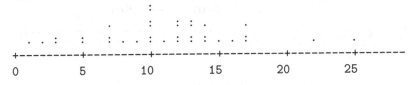

Figure 2.6: *Dot plot for drop rate percentage data: outliers deleted.*

2.5 Checking normality

From Figure 2.1, we identified two *outliers* in the dropout rate data, the 40% and the 59% dropout rates. If we delete these two points from the data, the remaining data may have a more nearly normal distribution. The dot plot with the two cases deleted is given in Figure 2.6. This is much more nearly normally distributed, i.e., looks much more like a bell-shaped curve, than the complete data.

Dot plots and other versions of histograms are not effective in evaluating normality. Very large amounts of data are needed before one can evaluate normality from a histogram. A more useful technique for evaluating the normality of small and moderate-size samples is the construction of a *normal probability plot*, also known as a *normal plot*, or a *rankit plot*, or a *normal quantile plot*, or a *normal q–q plot*. The idea is to order the data from smallest to largest and then to compare the ordered values to what one would expect the ordered values to be if they were truly a random sample from a normal distribution. These pairs of values should be roughly equal, so if we plot the pairs we would expect to see a line with a slope of about 1 that goes through the origin.

The problem with this procedure is that finding the expected ordered values requires us to know the mean μ and standard deviation σ of the appropriate population. These are generally not available. To avoid this problem, the expectations of the ordered values are computed assuming $\mu = 0$ and $\sigma = 1$. The expected ordered values from this standard normal distribution are called *normal scores* or *rankits*, or *(theoretical) normal quantiles*. Computing the expected values this way, we no longer anticipate a line with slope 1 and intercept 0. We now anticipate a line with slope σ and intercept μ. While it is possible to obtain estimates of the mean and standard deviation from a normal plot, our primary interest is in whether the plot looks like a line. A linear plot is consistent with normal data; a nonlinear plot is inconsistent with normal data. Christensen (2011, Section 13.2) gives a more detailed motivation for normal plots.

The normal scores are difficult to compute, so we generally get a computer program to do the work. In fact, just creating a plot is considerable work without a computer.

EXAMPLE 2.5.1. Consider the dropout rate data. Figure 2.7 contains the normal plot for the complete data. The two outliers cause the plot to be severely nonlinear. Figure 2.8 contains the normal plot for the dropout rate data with the two outliers deleted. It is certainly not horribly nonlinear. There is a little shoulder at the bottom end and some wiggling in the middle.

We can eliminate the shoulder in this plot by transforming the original data. Figure 2.9 contains a normal plot for the square roots of the data with the outliers deleted. While the plot no longer has a shoulder on the lower end, it seems to be a bit less well behaved in the middle.

We might now repeat our tests and confidence intervals for the 36 observations left when the outliers are deleted. We can do this for either the original data or the square roots of the original data. In either case, it now seems reasonable to treat the data as normal, so we can more confidently use a $t(36 - 1)$ distribution instead of hoping that the sample is large enough to justify use of the $t(37)$ distribution. We will consider these tests and confidence intervals in the next chapter.

It is important to remember that *if outliers are deleted, the conclusions reached are not valid for data containing outliers.* For example, a confidence interval will be for the mean dropout rate excluding the occasional classes with extremely large dropout rates. If we are confident that any deleted outliers are not really part of the population of interest, this causes no problem. Thus, if we were sure that the large dropout rates were the result of clerical errors and did not provide any

Normal Q–Q Plot

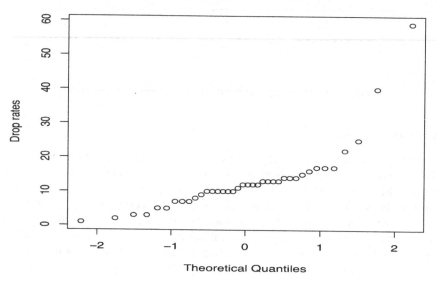

Figure 2.7: *Normal plot for drop rate percentage data: full data.*

Normal Q–Q Plot

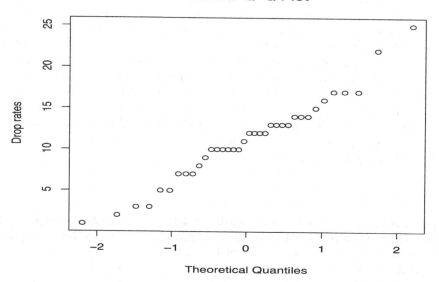

Figure 2.8: *Normal plot for drop rate percentage data: outliers deleted.*

information about true dropout rates, our conclusions about the population should be based on the data excluding the outliers. More often, though, we do not know that outliers are simple mistakes. *Often, outliers are true observations and often they are the most interesting and useful observations in the data.* If the outliers are true observations, systematically deleting them changes both the sample and the population of interest. In this case, the confidence interval is for the mean of a population implicitly defined by the process of deleting outliers. Admittedly, the idea of the mean dropout rate excluding the occasional outliers is not very clearly defined, but remember that the real population of interest is not too clearly defined either. We do not really want to learn about the clearly defined population of dropout rates; we really want to treat the dropout rate data as a sample

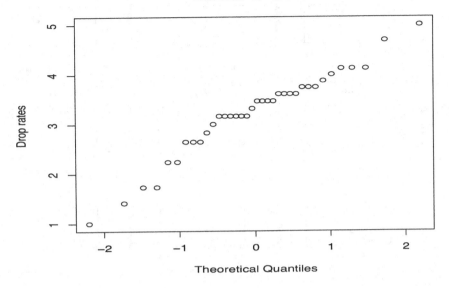

Figure 2.9: *Normal plot for square roots of drop rate percentage data: outliers deleted.*

from a population that allows us to draw useful inferences about current and future dropout rates. If we really cared about the fixed population, we could specify exactly what kinds of observations we would exclude and what we meant by the population mean of the observations that would be included. Given the nature of the true population of interest, I think that such technicalities are more trouble than they are worth at this point. □

Normal plots are subject to random variation because the data used in them are subject to random variation. Typically, normal plots are not perfectly straight. Figures 2.10 through 2.13 each present nine normal plots for which the data are in fact normally distributed. The figures differ by the number of observations in each plot, which are $10, 25, 50, 100$, respectively. By comparison to these, Figures 2.8 and 2.9, the normal plots for the dropout rate data and the square root of the dropout rates both with outliers deleted, look reasonably normal. Of course, if the dropout rate data are truly normal, the square root of these data cannot be truly normal and vice versa. However, both are reasonably close to normal distributions.

From Figures 2.10 through 2.13 we see that as the sample size n gets bigger, the plots get straighter. Normal plots based on even larger normal samples tend to appear straighter than these. Normal plots based on smaller normal samples can look much more crooked.

Testing normality

In an attempt to quantify the straightness of a normal plot, Shapiro and Francia (1972) proposed the summary statistic W', which is the squared sample correlation between the pairs of points in the plots. The population correlation coefficient was introduced in Subsection 1.2.3. The sample correlation coefficient is introduced in Chapter 6. At this point, it is sufficient to know that sample correlation coefficients near 0 indicate very little linear relationship between two variables and sample correlation coefficients near 1 or -1 indicate a very strong linear relationship. Since you need a computer to get the normal scores (rankits) anyway, just rely on the computer to give you the squared sample correlation coefficient.

A sample correlation coefficient near 1 indicates a strong tendency of one variable to increase (linearly) as the other variable increases, and sample correlation coefficients near -1 indicate a

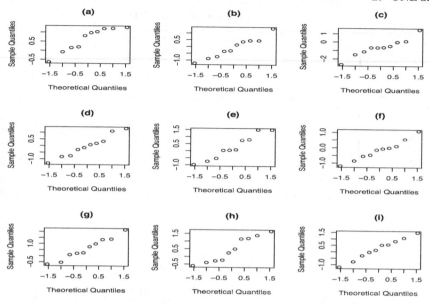

Figure 2.10: *Normal plots for normal data, n = 10.*

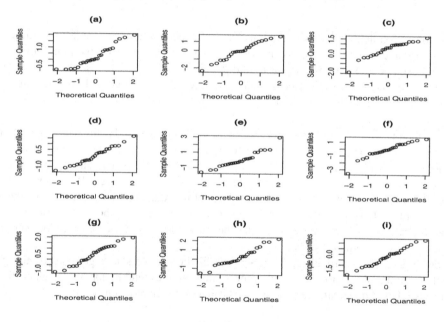

Figure 2.11: *Normal plots for normal data, n = 25.*

strong tendency for one variable to decrease (linearly) as the other variable increases. In normal plots we are looking for a strong tendency for one variable, the ordered data, to increase as the other variable, the rankits, increases, so normal data should display a sample correlation coefficient near 1 and thus the square of the sample correlation, W', should be near 1. If W' is too small, it indicates that the data are inconsistent with the assumption of normality. If W' is smaller than, say, 95% of the values one would see from normally distributed data, it is substantial evidence that the data are not normally distributed. If W' is smaller than, say, 99% of the values one would see from normally distributed data, it is strong evidence that the data are not normally distributed. Appendix B.3 presents tables of the values $W'(0.05, n)$ and $W'(0.01, n)$. These are the points above

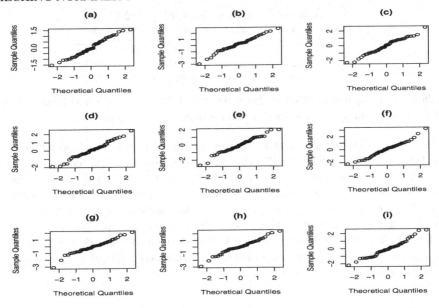

Figure 2.12: *Normal plots for normal data, n* = 50.

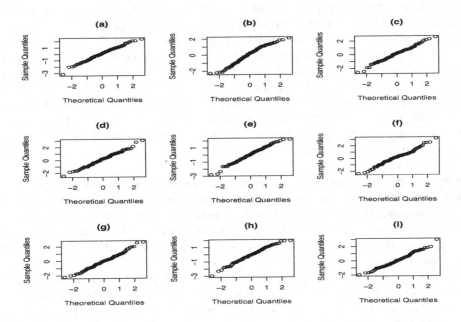

Figure 2.13: *Normal plots for normal data, n* = 100.

which fall, respectively, 95% and 99% of the W' values one would see from normally distributed data. Of course the W' percentiles are computed using not only the assumption of normality, but also the assumptions that the observations are independent with the same mean and variance. Note also that the values of these percentiles depend on the sample size n. The tabled values are consistent with our earlier observation that the plots are more crooked for smaller numbers of observations and straighter for larger numbers of observations in that the tabled values get larger with n. For comparison, we give the observed W' values for the data used in Figure 2.11.

Shapiro–Francia statistics for Figure 2.11

Plot	W'	Plot	W'	Plot	W'
(a)	0.940	(d)	0.982	(g)	0.977
(b)	0.976	(e)	0.931	(h)	0.965
(c)	0.915	(f)	0.915	(i)	0.987

These should be compared to $W'(0.05, 25) \doteq 0.918$ and $W'(0.01, 25) \doteq 0.88$ from Appendix B.3. Two of these nine values are below the 5% point, which is quite strange.

EXAMPLE 2.5.2.   For the dropout rate data we have three normal plots. The complete, untransformed data yield a W' value of 0.697. This value is inconsistent with the assumption that the dropout rate data has a normal distribution. Deleting the two outliers, W' is 0.978 for the untransformed data and 0.960 for the square roots of the data. The tabled percentiles are $W'(0.05, 36) = 0.940$ and $W'(0.01, 36) = 0.91$, so the untransformed data and the square root data look alright. In addition, W' was computed for the square roots of the complete data. Its value, 0.887, is still significantly low, but is a vast improvement over the untransformed complete data. The outliers are not nearly as strange when the square roots of the data are considered. Sometimes it is possible to find a transformation that eliminates outliers.   □

2.6   Transformations

In analyzing a collection of numbers, we assume that the observations are a random sample from some population. Often, the population from which the observations come is not as well defined as we might like. For example, if our observations are the yields of corn grown on 30 one-acre plots of ground in the summer of 2013, what is the larger population from which this is a sample? Typically, we do not have a large number of one-acre plots from which we randomly select 30. Even if we had a large collection of plots, these plots are subject to different weather conditions, have different fertilities, etc. Most importantly, we are rarely interested in corn grown in 2013 for its own sake. If we are studying corn grown in 2013, we are probably interested in predicting how that same type of corn would behave if we planted it at some time in the future. No population that currently exists could be completely appropriate for drawing conclusions about plant growths in a future year. Thus *the assumption that the observations are a random sample from some population is often only a useful approximation.*

When making approximations, it is often necessary to adjust things to make the approximations more accurate. In Statistics, *two approximations we frequently make are that all the data have the same variance and that the data are normally distributed. Making numerical transformations of the data is a primary tool for improving the accuracy of these approximations.* When sampling from a fixed population, we are typically interested in transformations that improve the normality assumption because having different variances is not a problem associated with sampling from a fixed population. With a fixed population, the variance of an object is the variance of randomly choosing an object from the population. This is a constant regardless of which object we end up choosing. But data are rarely as simple as random samples from a fixed population. Once we have an object from the population, we have to obtain an observation (measurement or count) from the object. These observations on a given object are also subject to random error and the error may well depend on the specific object being observed.

We now examine the fact that observations often have different variances, depending on the object being observed. First consider taking length measurements using a 30-centimeter ruler that has millimeters marked on it. For measuring objects that are less than 30 centimeters long, like this page, we can make very accurate measurements. We should be able to measure things within half a millimeter. Now consider trying to measure the height of a doghouse that is approximately 3.5 feet tall. Using the 30-cm ruler, we measure up from the base, mark 30 cm, measure from the mark up another 30 cm, make another mark, measure from the new mark up another 30 cm, mark again, and finally we measure from the last mark to the top of the house. With all the marking and moving of the ruler, we have much more opportunity for error than we have in measuring the length of the book. Obviously, if we try to measure the height of a house containing two stories, we will have much more error. If we try to measure the height of the Burj Khalifa in Dubai using a 30 cm ruler, we will not only have a lot of error, but large psychiatric expenses as well. The moral of this tale is that, when making measurements, larger objects tend to have more variability. If the objects are about the same size, this causes little or no problem. One can probably measure female heights with approximately the same accuracy for all women in a sample. One probably cannot measure the weights of a large sample of marine animals with constant variability, especially if the sample includes both shrimp and blue whales. *When the observations are the measured* amounts *of something, often the standard deviation of an observation is proportional to its mean. When the standard deviation is proportional to the mean, analyzing the logarithms of the observations is more appropriate than analyzing the original data.*

Now consider the problem of counting up the net financial worth of a sample of people. For simplicity, let's think of just three people, me, my 10-year-old grandson (the one my son has yet to provide), and my rich uncle, Scrooge. In fact, let's just think of having a stack of one dollar bills in front of each person. My pile is of a decent size, my grandson's is small, and my uncle's is huge. When I count my pile, it is large enough that I could miscount somewhere and make a significant, but not major, error. When I count my son's pile, it is small enough that I should get it about right. When I count my uncle's pile, it is large enough that I will, almost inevitably, make several significant errors. As with measuring amounts of things, the larger the observation, the larger the potential error. However, the process of making these errors is very different than that described for measuring amounts. In such cases, the variance of the observations is often proportional to the mean of the observations. The standard corrective measure for counts is different from the standard corrective measure for amounts. *When the observations are* counts *of something, often the variance of the count is proportional to its mean. In this case, analyzing the square roots of the observations is more appropriate than analyzing the original data.*

Suppose we are looking at yearly sales for a sample of corporations. The sample may include both the corner gas (petrol) station and Exxon. It is difficult to argue that one can really *count* sales for a huge company such as Exxon. In fact, it may be difficult to count even yearly sales for a gas station. Although in theory one should be able to count sales, it may be better to think of yearly sales as measured amounts. It is not clear how to transform such data. Another example is age. We usually think of counting the years a person has been alive, but one could also argue that we are measuring the amount of time a person has been alive. *In practice, we often try both logarithmic and square root transformations and use the transformation that seems to work best,* even when the type of observation (count or amount) seems clear.

Finally, consider the proportion of times people drink a particular brand of soda pop, say, Dr. Pepper. The idea is simply that we ask a group of people what proportion of the time they drink Dr. Pepper. People who always drink Dr. Pepper are aware of that fact and should give a quite accurate proportion. Similarly, people who never drink Dr. Pepper should be able to give an accurate proportion. Moreover, people who drink Dr. Pepper about 90% of the time or about 10% of the time, can probably give a fairly accurate proportion. The people who will have a lot of variability in their replies are those who drink Dr. Pepper about half the time. They will have little idea whether they drink it 50% of the time, or 60%, or 40%, or just what. With observations that are counts or amounts, larger observations have larger variances. With observations that are proportions, observations near 0 and 1 have small variability and observations near 0.5 have large variability. Proportion data call for a completely different type of transformation. *The standard transformation for proportion data is the inverse sine (arcsine) of the square root of the proportion. When the observations are proportions, often the variance of the proportion is a constant times* $\mu(1 - \mu)/N$, *where* μ *is the mean and* N *is the number of trials. In this case, analyzing the inverse sine (arcsine) of the square root of the proportion is more appropriate than analyzing the original data.*

In practice, the square root transformation is sometimes used with proportion data. After all, many proportions are obtained as a count divided by the total number of trials. For example, the best data we could get in the Dr. Pepper drinking example would be the count of the number of Dr. Peppers consumed divided by the total number of sodas imbibed.

There is a subtle but important point that was glossed over in the previous paragraphs. If we take multiple measurements on a house, the variance depends on the true height, but the true height is the same for all observations. Such a dependence of the variance on the mean causes no problems. The problem arises when we measure a random sample of buildings, each with a variance depending on its true height.

EXAMPLE 2.6.1.   For the dropout rate data, we earlier considered the complete, untransformed data and after deleting two outliers, we looked at the untransformed data and the square roots of the data. In Examples 2.5.1 and 2.5.2 we saw that the untransformed data with the outliers deleted and the square roots of the data with the outliers deleted had approximate normal distributions. Based on the W' statistic, the untransformed data seemed to be more nearly normal. The data are proportions of people who drop from a class, so our discussion in this section suggests transforming by the inverse sine of the square roots of the proportions. Recall that proportions are values between 0 and 1, while the dropout rates were reported as values between 0 and 100, so the reported rates need to be divided by 100. For the complete data, this transformation yields a W' value of 0.85, which is much better than the untransformed value of 0.70, but worse than the value 0.89 obtained with the square root transformation. With the two outliers deleted, the inverse sine of the square roots of the proportions yields the respectable value $W' = 0.96$, but the square root transformation is simpler and gives almost the same value, while the untransformed data give a much better value of 0.98. Examination of the six normal plots (only three of which have been presented here) reinforce the conclusions given above.

With the outliers deleted, it seems reasonable to analyze the untransformed data and, to a lesser extent, the data after either transformation. *Other things being equal*, we prefer using the simplest transformation that seems to work. Simple transformations are easier to explain, justify, and interpret. The square root transformation is simpler, and thus better, than the inverse sine of the square roots of the proportions. Of course, not making a transformation seems to work best and not transforming is always the simplest transformation. Actually some people would point out, and it is undeniably true, that the act of deleting outliers is really a transformation of the data. However, we will not refer to it as such.   □

Theory

The standard transformations given above are referred to as *variance-stabilizing transformations*. The idea is that each observation is a look at something with a different mean and variance, where the variance depends on the mean. For example, when we measure the height of a house, the house has some 'true' height and we simply take a measurement of it. The variability of the measurement depends on the true height of the house. Variance-stabilizing transformations are designed to eliminate the dependence of the variance on the mean. Although variance-stabilizing transformations are used quite generally for counts, amounts, and proportions, they are derived for certain assumptions about the relationship between the mean and the variance. These relationships are tied to theoretical distributions that are appropriate for some counts, amounts, and proportions. Rao (1973, Section 6g) gives a nice discussion of the mathematical theory behind variance-stabilizing transformations.

Proportions are related to the binomial distribution for numbers of successes. We have a fixed number of trials; the proportion is the number of successes divided by the number of trials. The mean of a $\text{Bin}(N, p)$ distribution is Np and the variance is $Np(1 - p)$. This relationship between the mean and variance of a binomial leads to the inverse sine of the square root transformation.

Counts are related to the Poisson distribution. Poisson data has the property that the variance equals the mean of the observation. This relationship leads to the square root as the variance-stabilizing transformation.

For amounts, the log transformation comes from having the standard deviation proportional to the mean. The standard deviation divided by the mean is called the *coefficient of variation*, so the log transformation is appropriate for observations that have a constant coefficient of variation. (The square root transformation comes from having the variance, rather than the standard deviation, proportional to the mean.) A family of continuous distributions called the gamma distributions has a constant coefficient of variation; see Section 22.2.

The variance-stabilizing transformations are given below. In each case we assume $E(y_i) = \mu_i$ and $\mathrm{Var}(y_i) = \sigma_i^2$. The symbol $\propto$ means "proportional to."

<table>
<tr><td colspan="5" align="center">Variance-stabilizing transformations</td></tr>
<tr><td></td><td></td><td align="center">Mean, variance</td><td></td></tr>
<tr><td>Data</td><td>Distribution</td><td align="center">relationship</td><td>Transformation</td></tr>
<tr><td>Count</td><td>Poisson</td><td>$\mu_i \propto \sigma_i^2$</td><td>$\sqrt{y_i}$</td></tr>
<tr><td>Amount</td><td>Gamma</td><td>$\mu_i \propto \sigma_i$</td><td>$\log(y_i)$</td></tr>
<tr><td>Proportion</td><td>Binomial/N</td><td>$\frac{\mu_i(1-\mu_i)}{N} \propto \sigma_i^2$</td><td>$\sin^{-1}\left(\sqrt{y_i}\right)$</td></tr>
</table>

I cannot honestly recommend using variance-stabilizing transformations to analyze either binomial or Poisson data. A large body of statistical techniques has been developed specifically for analyzing binomial and Poisson data; see Chapters 5, 20, and 21. I would recommend using these alternative methods. Many people would make a similar recommendation for gamma distributed data citing the applicability of generalized linear models, cf. Chapter 22. McCullagh and Nelder (1989), Christensen (1997), and many other books provide information on generalized linear models. When applied to binomial, Poisson, or gamma distributed data, variance-stabilizing transformations provide a way to force the methods developed for normally distributed data into giving a reasonable analysis for data that are not normally distributed. If you have a clear idea about the true distribution of the data, you should use methods developed specifically for that distribution. The problem is that we often have little idea of the appropriate distribution for a set of data. For example, if we simply ask people the proportion of times they drink Dr. Pepper, we have proportion data that is not binomial. In such cases, we seek a transformation that will make a normal theory analysis approximately correct. We often pick transformations by trial and error. *The variance-stabilizing transformations provide little more than a place to start when considering transformations.*

At the beginning of this section, we mentioned two key approximations that we frequently make. These are that all the data have the same variance and that the data are normally distributed. While the rationale given above for picking transformations was based on stabilizing variances, in practice we typically choose a transformation for a single sample to attain approximate normality. To evaluate whether a transformation really stabilizes the variance, we need more information than is contained in a single sample. Control chart methods can be used to evaluate variance-stabilization for a single sample, cf. Shewhart (1931). Those methods require formation of rational subgroups and that requires additional information. We could also plot the sample against appropriately chosen variables to check variance-stabilization, but finding appropriate variables can be quite difficult and would depend on properties of the particular sampling process. Variance-stabilizing transformations are probably best suited to problems that compare samples from several populations, where the variance in each population depends on the mean of the population.

On the other hand, we already have examined methods for evaluating the normality of a single sample. Thus, since we cannot (actually, do not) evaluate variance-stabilization in a single sample, if we think that the variance of observations should increase with their mean, we might try both the log and square root transformations and pick the one for which the transformed data best approximate normality. Systematic methods for choosing a transformation are discussed in Chapter 7.

2.7 Inference about σ^2

If the data are normally distributed, we can also perform confidence intervals and tests for the population variance σ^2. While these are not typically of primary importance, they can be useful.

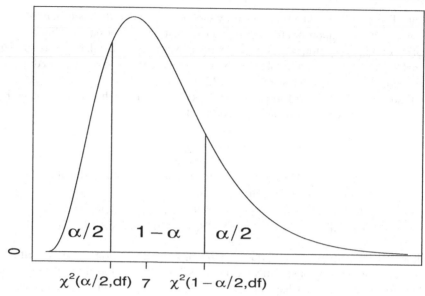

Figure 2.14: *Central χ^2 interval with probability $1 - \alpha$.*

They also tend to be sensitive to the assumption of normality. The procedures do not follow the same pattern used for most inferences that involve 1) a parameter of interest, 2) an estimate of the parameter, 3) the standard error of the estimate, and 4) a known distribution symmetric about zero; however, there are similarities. Procedures for variances typically require a parameter, an estimate, and a known distribution.

The procedures discussed in this section actually apply to all the problems in this book that involve a single variance parameter σ^2. One need only substitute the relevant estimate of σ^2 and use its degrees of freedom. Applications to the data and models considered in Chapter 19 are not quite as straightforward because there the models involve more than one variance.

In the one-sample problem, the parameter is σ^2, the estimate is s^2, and the distribution, as discussed in Equation (2.1.6), is

$$\frac{(n-1)s^2}{\sigma^2} \sim \chi^2(n-1).$$

The notation $\chi^2(1-\alpha, n-1)$ is used to denote the point that cuts off the bottom $1-\alpha$ (top α) of the χ^2 distribution with $n-1$ degrees of freedom. Note that $(n-1)s^2/\sigma^2$ is nonnegative, so the curve in Figure 2.14 illustrating the χ^2 distribution is also nonnegative. Figure 2.14 shows a central interval with probability $1-\alpha$ for a χ^2 distribution.

To test $H_0: \sigma^2 = \sigma_0^2$ the value σ_0^2 *must be known.* As usual, we assume that the null hypothesis is true, i.e., $\sigma^2 = \sigma_0^2$, so under this assumption an α-level test is based on

$$1 - \alpha = \Pr\left[\chi^2\left(\frac{\alpha}{2}, n-1\right) < \frac{(n-1)s^2}{\sigma_0^2} < \chi^2\left(1 - \frac{\alpha}{2}, n-1\right)\right];$$

see Figure 2.14. If we observe data yielding an s^2 such that $(n-1)s^2/\sigma_0^2$ is between the values $\chi^2(\frac{\alpha}{2}, n-1)$ and $\chi^2(1-\frac{\alpha}{2}, n-1)$, the data are consistent with the assumption that $\sigma^2 = \sigma_0^2$ at level α. Conversely, we reject $H_0: \sigma^2 = \sigma_0^2$ with a two-sided α-level test if

$$\frac{(n-1)s^2}{\sigma_0^2} > \chi^2\left(1 - \frac{\alpha}{2}, n-1\right)$$

or if

$$\frac{(n-1)s^2}{\sigma_0^2} < \chi^2\left(\frac{\alpha}{2}, n-1\right).$$

More specifically, we reject the null model that the data are independent, normally distributed with a constant variance σ^2, that we have the correct model for the mean structure, and that $\sigma^2 = \sigma_0^2$.

EXAMPLE 2.7.1. For the dropout rate data consider testing $H_0 : \sigma^2 = 50$ with $\alpha = 0.01$. Again, we use the data with the two outliers deleted, because they are more nearly normal. Thus, our concept of the population variance σ^2 must account for our deletion of weird cases. The deleted data contain 36 observations and s^2 for the deleted data is 27.45. The test statistic is

$$\frac{(n-1)s^2}{\sigma_0^2} = \frac{35(27.45)}{50} = 19.215.$$

The *critical region*, the region for which we reject H_0, contains all values greater than $\chi^2(0.995, 35) = 60.275$ and all values less than $\chi^2(0.005, 35) = 17.19$. The test statistic is certainly not greater than 60.275 and it is also not less than 17.19, so we have no basis for rejecting the null hypothesis at the $\alpha = 0.01$ level. At the 0.01 level, the data are consistent with the claim that $\sigma^2 = 50$. $\square$

Confidence intervals are defined in terms of testing the hypothesis $H_0 : \sigma^2 = \sigma_0^2$. A $(1-\alpha)100\%$ confidence interval for σ^2 is based on the following inequalities:

$$\chi^2\left(\frac{\alpha}{2}, n-1\right) < \frac{(n-1)s^2}{\sigma_0^2} < \chi^2\left(1-\frac{\alpha}{2}, n-1\right)$$

which occurs if and only if

$$\frac{(n-1)s^2}{\chi^2\left(1-\frac{\alpha}{2}, n-1\right)} < \sigma_0^2 < \frac{(n-1)s^2}{\chi^2\left(\frac{\alpha}{2}, n-1\right)}.$$

The first inequality corresponds to Figure 2.14 and just reflects the definition of the percentage points $\chi^2\left(\frac{\alpha}{2}, n-1\right)$ and $\chi^2\left(1-\frac{\alpha}{2}, n-1\right)$. These are defined to be the points that cut out the middle $1-\alpha$ of the chi-squared distribution and are tabled in Appendix B.2. The second inequality is based on algebraic manipulation of the terms in the first inequality. The actual derivation is given later in this section. The second inequality gives an interval that contains σ_0^2 values that are consistent with the data and the model.

$$\left(\frac{(n-1)s^2}{\chi^2\left(1-\frac{\alpha}{2}, n-1\right)}, \frac{(n-1)s^2}{\chi^2\left(\frac{\alpha}{2}, n-1\right)}\right). \tag{2.7.1}$$

The confidence interval for σ^2 requires the data to be normally distributed. This assumption is more vital for inferences about σ^2 than it is for inferences about μ. For inferences about μ, the central limit theorem indicates that the sample means are approximately normal even when the data are not normal. There is no similar result indicating that the sample variance is approximately χ^2 even when the data are not normal. (For large n, both s^2 and $\chi^2(n-1)$ approach normal distributions, but in general the approximate normal distribution for $(n-1)s^2/\sigma^2$ is not the approximately normal $\chi^2(n-1)$ distribution.)

EXAMPLE 2.7.2. Consider again the dropout rate data. We have seen that the complete data are not normal, but that after deleting the two outliers, the remaining data are reasonably normal. We find a 95% confidence interval for σ^2 from the deleted data. The percentage points for the $\chi^2(36-1)$ distribution are $\chi^2(0.025, 35) = 20.57$ and $\chi^2(0.975, 35) = 53.20$. The 95% confidence interval is

$$\left(\frac{35(27.45)}{53.20}, \frac{35(27.45)}{20.57},\right.$$

or equivalently $(18.1, 46.7)$. The interval contains all values of σ^2 that are consistent with the data and the model as determined by a two-sided $\alpha = 0.05$ level test. The interval does not contain 50, so we do have evidence against $H_0 : \sigma^2 = 50$ at the $\alpha = 0.05$ level. Remember that this is the true variance *after the deletion of outliers*. Again, when we delete outliers we are a little fuzzy about the exact definition of our parameter, but we are also being fuzzy about the exact population of interest. The exception to this is when we believe that the only outliers that exist are observations that are not really part of the population. $\square$

2.7.1 Theory

Alas, these procedures for σ^2 are merely ad hoc. They are neither appropriate significance testing results nor appropriate Neyman–Pearson theory results. Neither of those tests would use the central χ^2 interval illustrated in Figure 2.14. However, the confidence interval has a Bayesian justification.

These methods are valid Neyman–Pearson procedures but not the optimal procedures. As Neyman–Pearson procedures, the endpoints of the confidence interval (2.7.2) are random. To use the interval, we replace the random variable s^2 with the observed value of s^2 and replace the term "probability $(1 - \alpha)$" with "$(1 - \alpha)100\%$ confidence." *Once the observed value of s^2 is substituted into the interval, nothing about the interval is random any longer*, the fixed unknown value of σ^2 is either in the interval or it is not; there is no probability associated with it. The probability statement about random variables is mystically transformed into a 'confidence' statement. This is not unreasonable, but the rationale is, to say the least, murky.

A significance test would not use the cutoff values $\chi^2(\alpha/2, n - 1)$ and $\chi^2(1 - \alpha/2, n - 1)$. Let the vertical axis in Figure 2.14 be z and the horizontal axis be w. The density function being plotted is $f(w)$. Any positive value of z corresponds to two points w_1 and w_2 with

$$z = f(w_1) = f(w_2).$$

For an α-level significance test, one would find z_0 so that the corresponding points w_{01} and w_{02} have the property that

$$\alpha = \Pr[\chi^2(n - 1) \leq w_{01}] + \Pr[\chi^2(n - 1) \geq w_{02}].$$

The significance test then uses w_{01} and w_{02} as the cutoff values for an α-level test. As you can see, the method presented earlier is much simpler than this. (The optimal Neyman–Pearson test is even more complicated than the significance test.) But the significance test uses an appropriate measure of how weird any particular data value s^2 is relative to a null model based on σ_0^2. Given the cut-off values w_{01} and w_{02}, finding a confidence interval works pretty much as for the ad hoc method.

While methods for drawing inferences about variances *do not* fit our standard pattern for a single parameter of interest based on 1) a parameter of interest, 2) an estimate of the parameter, 3) the standard error of the estimate, and 4) a known distribution symmetric about zero, it should be noted that the basic logic behind these confidence intervals and tests is the same. The correspondence to model testing is strong since we are comparing the variance estimate of the original model s^2 to the variance under the null model σ_0^2. The only real difference is that the appropriate reference distribution turns out to be a $\chi^2(n - 1)$ rather than an F. In any case, significance tests are based on evaluating whether the data are consistent with the null model. Consistency is defined in terms of a known distribution that applies when the null model is true. If the data are inconsistent with the null model, the null model is rejected as being inconsistent with the observed data.

Below is a series of equalities that justify the confidence interval.

$$\chi^2\left(\frac{\alpha}{2}, n - 1\right) < \frac{(n - 1)s^2}{\sigma^2} < \chi^2\left(1 - \frac{\alpha}{2}, n - 1\right)$$

$$\frac{1}{\chi^2\left(\frac{\alpha}{2}, n - 1\right)} > \frac{\sigma^2}{(n - 1)s^2} > \frac{1}{\chi^2\left(1 - \frac{\alpha}{2}, n - 1\right)}$$

Table 2.1: *Weights of rats.*

59	54	56	59	57	52	52	61	59
53	59	51	51	56	58	46	53	57
60	52	49	56	46	51	63	49	57

$$\frac{1}{\chi^2\left(1-\frac{\alpha}{2},n-1\right)} < \frac{\sigma^2}{(n-1)s^2} < \frac{1}{\chi^2\left(\frac{\alpha}{2},n-1\right)}$$

$$\frac{(n-1)s^2}{\chi^2\left(1-\frac{\alpha}{2},n-1\right)} < \sigma^2 < \frac{(n-1)s^2}{\chi^2\left(\frac{\alpha}{2},n-1\right)}.$$

2.8 Exercises

EXERCISE 2.8.1.   Mulrow et al. (1988) presented data on the melting temperature of biphenyl as measured on a differential scanning calorimeter. The data are given below; they are the observed melting temperatures in Kelvin less 340.

$$3.02, 2.36, 3.35, 3.13, 3.33, 3.67, 3.54, 3.11, 3.31, 3.41, 3.84, 3.27, 3.28, 3.30$$

Compute the sample mean, variance, and standard deviation. Give a 99% confidence interval for the population mean melting temperature of biphenyl as measured by this machine. (Note that we don't know whether the calorimeter is accurately calibrated.)

EXERCISE 2.8.2.   Box (1950) gave data on the weights of rats that were about to be used in an experiment. The data are repeated in Table 2.1. Assuming that these are a random sample from a broader population of rats, give a 95% confidence interval for the population mean weight. Test the null hypothesis that the population mean weight is 60 using a 0.01 level test.

EXERCISE 2.8.3.   Fuchs and Kenett (1987) presented data on citrus juice for fruits grown during a specific season at a specific location. The sample size was 80 but many variables were measured on each sample. Sample statistics for some of these variables are given below.

Variable	BX	AC	SUG	K	FORM	PECT
Mean	10.4	1.3	7.7	1180.0	22.2	451.0
Variance	0.38	0.036	0.260	43590.364	6.529	16553.996

The variables are BX—total soluble solids produced at 20°C, AC—acidity as citric acid unhydrons, SUG—total sugars after inversion, K—potassium, FORM—formol number, PECT—total pectin. Give a 99% confidence interval for the population mean of each variable. Give a 99% prediction interval for each variable. Test whether the mean of BX equals 10. Test whether the mean of SUG is equal to 7.5. Use $\alpha = 0.01$ for each test.

EXERCISE 2.8.4.   Jolicoeur and Mosimann (1960) gave data on female painted turtle shell lengths. The data are presented in Table 2.2. Give a 95% confidence interval for the population mean length. Give a 99% prediction interval for the shell length of a new female.

EXERCISE 2.8.5.   Mosteller and Tukey (1977) extracted data from the *Coleman Report*. Among the variables considered was the percentage of sixth-graders whose fathers were employed in white-collar jobs. Data for 20 New England schools are given in Table 2.3. Are the data reasonably normal? Do any of the standard transformations improve the normality? After finding an appropriate transformation (if necessary), test the null hypothesis that the percentage of white-collar fathers is 50%.

Table 2.2: *Female painted turtle shell lengths.*

98	138	123	155	105	147	133	159
103	138	133	155	109	149	134	162
103	141	133	158	123	153	136	177

Table 2.3: *Percentage of fathers with white-collar jobs.*

28.87	20.10	69.05	65.40	29.59
44.82	77.37	24.67	65.01	9.99
12.20	22.55	14.30	31.79	11.60
68.47	42.64	16.70	86.27	76.73

Use a 0.05 level test. Give a 99% confidence interval for the percentage of fathers with white-collar jobs. If a transformation was needed, relate your conclusions back to the original measurement scale.

EXERCISE 2.8.6.   Give a 95% confidence interval for the population variance associated with the data of Exercise 2.8.5. Remember that inferences about variances require the assumption of normality. Could the variance reasonably be 10?

EXERCISE 2.8.7.   Give a 95% confidence interval for the population variance associated with the data of Exercise 2.8.4. Remember that the inferences about variances require the assumption of normality.

EXERCISE 2.8.8.   Give 99% confidence intervals for the population variances of all the variables in Exercise 2.8.3. Assume that the original data were normally distributed. Using $\alpha = 0.01$, test whether the potassium variance could reasonably be 45,000. Could the formol number variance be 8?

EXERCISE 2.8.9.   Shewhart (1931, p. 62) reproduces Millikan's data on the charge of an election. These are repeated in Table 2.4. Check for outliers and nonnormality. Adjust the data appropriately if there are any problems. Give a 98% confidence interval for the population mean value. Give a 98% prediction interval for a new measurement. (Millikan argued that some adjustments were needed before these data could be used in an optimal fashion but we will ignore his suggestions.)

EXERCISE 2.8.10.   Let $y_0, y_1, \ldots, y_n$ be independent $N(\mu, \sigma^2)$ random variables and compute $\bar{y}.$, and s^2 from observations 1 through n. Show that $(y_0 - \bar{y}.)/\sqrt{\sigma^2 + \sigma^2/n} \sim N(0,1)$ using results from Chapter 1 and the fact that linear combinations of independent normals are normal. Recalling that $y_0, \bar{y}.$, and s^2 are independent and that $(n-1)s^2/\sigma^2 \sim \chi^2(n-1)$, use Definition 2.1.3 to show that $(y - \bar{y}.)/\sqrt{s^2 + s^2/n} \sim t(n-1)$.

Table 2.4: *Observations on the charge of an electron.*

4.781	4.764	4.777	4.809	4.761	4.769	4.795	4.776
4.765	4.790	4.792	4.806	4.769	4.771	4.785	4.779
4.758	4.779	4.792	4.789	4.805	4.788	4.764	4.785
4.779	4.772	4.768	4.772	4.810	4.790	4.775	4.789
4.801	4.791	4.799	4.777	4.772	4.764	4.785	4.788
4.779	4.749	4.791	4.774	4.783	4.783	4.797	4.781
4.782	4.778	4.808	4.740	4.790	4.767	4.791	4.771
4.775	4.747						

Chapter 3

General Statistical Inference

Before we can perform a statistical analysis on data, we need to make assumptions about the data. A *model* for the data is simply a statement of those assumptions. Typical assumptions are that the observations are independent, have equal variances, and that either the observations are normally distributed or involve large sample sizes. (We don't really know what "large" means, so large samples is an assumption.) Typically, models also say something about the expected values of the observations. In fact, it is the expected values that generally receive most of the attention when discussing models. Most statistical procedures, e.g., confidence intervals, prediction intervals, and tests of a null hypothesis, rely on the validity of the model for the validity of the procedure. As such, it is vitally important that we do what we can to establish the validity of the model. Sections 2.5 and 2.6 contained our first steps in that direction.

This chapter focuses on significance testing as a fundamental procedure in statistical inference. Confidence intervals and P values are presented as extensions of a basic testing procedure. The approach is very much in the spirit of the traditional approach used by R.A. Fisher as opposed to a later approach to testing and confidence intervals introduced by Jerzy Neyman and E.S. Pearson. As such, we do our best to avoid the artifacts of the Neyman–Pearson approach including alternative hypotheses, one-sided testing, and the concept of the probability of Type I error. Although I am a strong proponent of the use of Bayesian procedures—see Christensen et al. (2010)—they receive little attention in this book.

The basic idea of significance testing is that one has a model for the data and seeks to determine whether the data are consistent with that model or whether they are inconsistent with the model. Determining that the data are inconsistent with the model is a strong statement. It suggests that the model is wrong. It is a characteristic of statistical analysis that data rarely give an absolute contradiction to a model, so we need to measure the extent to which the data are inconsistent with the model. On the other hand, observing that the data are consistent with the model is a weak statement. Although the data may be consistent with the current model, we could always construct other models for which the data would also be consistent.

Frequently, when constructing tests, we have an underlying model for the data to which we add some additional assumption, and then we want to test whether this new model is consistent with the data. There are two terminologies for this procedure. First, the additional assumption is often referred to as a *null hypothesis*, so the original model along with the additional assumption is called the *null model*. Alternatively, the original model is often called the *full model* and the null model is called the *reduced model*. The null model is a reduced model in the sense that it is a special case of the full model, that is, it consists of the full model with the added restriction of the null hypothesis. When discussing full and reduced models, we might not bother to specify the null hypothesis, but every reduced model corresponds to some null hypothesis.

The most commonly used statistical tests and confidence intervals derive from a theory based on a single parameter of interest, i.e., the null hypothesis is a specific assumption about a single parameter. While we use this single parameter theory when convenient, the focus of this book is on models rather than parameters. We begin with a general statement of our model-based approach to testing and then turn to an examination of the single parameter approach. A key aspect of the model-based approach is that it easily allows for testing many parameters at once. The basic ideas of both theories were illustrated in Chapter 2. The point of the current chapter is to present the theories in general form and to reemphasize fundamental techniques. The general theories will then be used throughout the book. Because the theories are stated in quite general terms, some prior familiarity with the ideas as discussed in Chapter 2 is highly recommended.

3.1 Model-based testing

Our goal in data analysis is frequently to find the simplest model that provides an adequate explanation of the data. A fundamental tool in that process is testing a given model, the *full model*, against a special case of the full model, the *reduced model*. The tests are based on measures of how well the models explain the data. We begin with a discussion of measuring how well a model fits the data.

Suppose we have a model that involves independent data y_i, $i = 1,\ldots,n$, with $E(y_i) = \mu_i$ and some common variance, $Var(y_i) = \sigma^2$. This model is not very interesting because the only thing we could do with it would be to use y_i to estimate μ_i. A model becomes more interesting if we develop some relationships between the μ_is. The simplest model is that the μ_is all equal some common value μ. That is the one-sample model of the previous chapter. Other interesting models divide the data into two groups with a common mean within each group (the two-sample problems of Chapter 4), or divide the data into multiple groups with a common mean within each group (one-way analysis of variance, Chapter 12), or use some other observations x_i and assume a linear relationship, e.g., $\mu_i = \beta_0 + \beta_1 x_i$ (simple linear regression, Chapter 6). The general point is that a model allows us to estimate the expected values of the y_is. The estimate of the expected value μ_i might well be denoted $\hat{\mu}_i$ but more commonly it is known as a *fitted value* and denoted $\hat{y}_i$. To measure the error in modeling the mean values, compute the sum of the squared differences between the actual data y_i and the fitted values $\hat{y}_i$. This *sum of squares for error* (*SSE*) is defined as

$$SSE = \sum_{i=1}^{n}(y_i - \hat{y}_i)^2.$$

The values $\hat{\varepsilon}_i \equiv y_i - \hat{y}_i$ are called *residuals* and are also used to evaluate model assumptions like independence, equal variances, and normality.

Typically the model involves parameters that describe the interrelations between the μ_is. If there are r (functionally distinct) parameters for the mean values, we define the *degrees of freedom for error* (*dfE*) as

$$dfE = n - r.$$

The degrees of freedom can be thought of as the effective number of observations that are available for estimating the variance σ^2 after using the model to estimate the means. Finally, our estimate of σ^2 is the *mean squared error* (*MSE*) defined as

$$MSE = \frac{SSE}{dfE}.$$

To test two models, identify the full model (Full) and compute $SSE(Full)$, $dfE(Full)$, and $MSE(Full)$. Similarly, for the reduced model (Red.), compute $SSE(Red.)$, $dfE(Red.)$, and $MSE(Red.)$. The identification of a full model and a reduced model serves to suggest a test statistic, i.e., something on which to base a test. The test is a test of whether the reduced model is correct. We establish the behavior (distribution) of the test statistic when the reduced model is correct, and if the observed value of the test statistic looks unusual relative to this *reference distribution*, we conclude that something is wrong with the reduced model. (Or that we got unlucky in collecting our data—always a possibility.)

Although the test assumes that the reduced model is correct and checks whether the data tend to contradict that assumption, for the purpose of *developing* a test we often act as if the full model is true, regardless of whether the reduced model is true. This focuses the search for abnormal behavior in a certain direction. Nonetheless, concluding that the reduced model is wrong does not imply that the full model is correct.

Since the reduced model is a special case of the full model, the full model must always explain the data at least as well as the reduced model. In other words, the error from Model (Red.) must be as large as the error from Model (Full), i.e., $SSE(Red.) \geq SSE(Full)$. The reduced model being smaller than the full, it also has more degrees of freedom for error, i.e., $dfE(Red.) \geq dfE(Full)$.

If the reduced model is true, we will show later that the statistic

$$MSTest \equiv \frac{SSE(Red.) - SSE(Full)}{dfE(Red.) - dfE(Full)}$$

is an estimate of the variance, σ^2, with degrees of freedom $dfE(Red.) - dfE(Full)$. Since the reduced model is a special case of the full model, whenever the reduced model is true, the full model is also true. Thus, if the reduced model is true, $MSE(Full)$ is also an estimate of σ^2, and the ratio

$$F \equiv MSTest/MSE(Full)$$

should be about 1, since it is the ratio of two estimates of σ^2. This ratio is called the *F statistic* in honor of R.A. Fisher.

Everybody's favorite reduced model takes

$$E(y_i) = \mu$$

so that every observation has the same mean. This is the reduced model being tested in nearly all of the three-line ANOVA tables given by computer programs, but we have much more flexibility than that.

The F statistic is an actual number that we can compute from the data, so we eventually have an actual observed value for the F statistic, say F_{obs}. If F_{obs} is far from 1, it suggests that something may be wrong with the assumptions in the reduced model, i.e., either the full model is wrong or the null hypothesis is wrong. The question becomes, "What constitutes an F_{obs} far from 1?" Even when Model (Red.) is absolutely correct, the variability in the data causes variability in the F statistic. Since $MSTest$ and $MSE(Full)$ are always nonnegative, the F statistic is nonnegative. Huge values of F_{obs} are clearly far from 1. But we will see that sometimes values of F_{obs} very near 0 are also far from 1. By quantifying the variability in the F statistic when Model (Red.) is correct, we get an idea of what F statistics are consistent with Model (Red.) and what F values are inconsistent with Model (Red.).

When, in addition to the assumption of independent observations with common variance σ^2 and the assumption that the reduced model for the means is correct, we also assume that the data are normally distributed and that both the full and reduced models are "linear" so that they have nice mathematical properties, the randomness in the F statistic is described by an F distribution. Properties of the F distribution can be tabled, or more commonly, determined by computer programs. The F distribution depends on two parameters, the degrees of freedom for $MSTest$ and the degrees of freedom for $MSE(Full)$; thus we write

$$F = \frac{MSTest}{MSE(Full)} \sim F[dfE(Red.) - dfE(Full), dfE(Full)].$$

The shape (density) of the $F[dfE(Red.) - dfE(Full), dfE(Full)]$ distribution determines which values of the F statistic are inconsistent with the null model. A typical F density is shown in Figure 3.1. F values for which the curve takes on small values are F values that are unusual under the null model. Thus, in Figure 3.1, unusual values of F occur when F is either very much larger than 1 or very close to 0. Generally, when $dfE(Red.) - dfE(Full) \geq 3$ both large values of F and values of

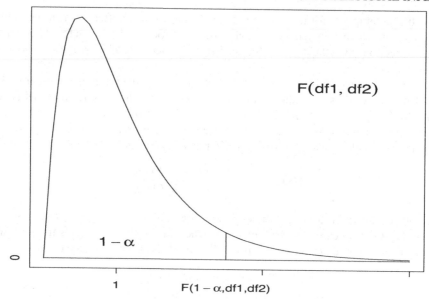

Figure 3.1: *Percentiles of F(df1, df2) distributions; df1 ≥ 3.*

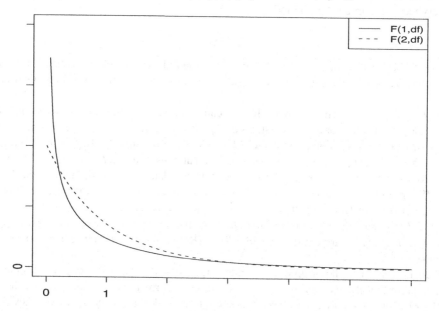

Figure 3.2: *F(1, df) and F(2, df) densities.*

F near 0 are inconsistent with the null model. As shown in Figure 3.2, when $dfE(Red.) - dfE(Full)$ is one or two, only very large values of the F statistic are inconsistent with the null model because in those cases the density is large for F values near 0.

It can be shown that when the full model is wrong (which implies that the reduced model is also wrong), it is possible for the F statistic to get either much larger than 1 or much smaller than 1. Either case calls the reduced model in question.

Traditionally, people and computer programs have concerned themselves only with values of the F statistic that are much larger than 1. If the full model is true but the reduced model is not true, for linear models it can be shown that *MSTest* estimates $\sigma^2 + \delta$ where δ is some positive number.

Since the full model is true, $MSE(Full)$ still estimates σ^2, so $MSTest/MSE(Full)$ estimates $(\sigma^2 + \delta)/\sigma^2 = 1 + (\delta/\sigma^2) > 1$. Thus, if the full model is true but the reduced model is not true, the F statistic tends to get larger than 1 (and not close to 0). Commonly, an α-level test of whether Model (Red.) is an adequate substitute for Model (Full) has been rejected when

$$\frac{[SSE(Red.) - SSE(Full)] / [dfE(Red.) - dfE(Full)]}{MSE(Full)}$$

$$> F[1 - \alpha, dfE(Red.) - dfE(Full), dfE(Full)]. \quad (3.1.1)$$

The argument that the F statistic tends to be larger than 1 when the reduced model is false depends on the validity of the assumptions made about the full model. These include the data being independent with equal variances and involve some structure on the mean values. As discussed earlier, comparing the F statistic to the F distribution additionally presumes that the full and reduced models are linear and that the data are normally distributed.

Usually, the probability that an $F[dfE(Red.) - dfE(Full), dfE(Full)]$ distribution is larger than the observed value of the F statistic is reported as something called a P *value*. For three or more numerator degrees of freedom, I do not think this usual computation of a P value is really a P value at all. It is slightly too small. A P value is supposed to be the probability of seeing data as weird or weirder than you actually observed. With three or more numerator degrees of freedom, F statistics near 0 can be just as weird as F statistics much larger than 1. Weird values should be determined as values that have a low probability of occurring or, in continuous cases like these, have a low probability density function, i.e., the curve plotted in Figure 3.1. The probability density for an F distribution with three or more degrees of freedom in the numerator gets small both for values much larger than 1 and for values near 0. To illustrate, consider an $F(5,20)$ distribution and an observed test statistic of $F_{obs} = 2.8$. The usual reported P value would be 0.0448, the probability of being at least 2.8. By our two-sided definition, the actual P value should be 0.0456. The computations depend on the fact that, from the shape of the $F(5,20)$ distribution; seeing an F statistic of 0.036 is just as weird as seeing the 2.8 and the probability of seeing something smaller than 0.036 is 0.0008. Technically, $F_{obs} = 2.8$ and $F_{obs} = 0.036$ have the same density, and the densities get smaller as the F values get closer to infinity and to zero, respectively. The P value should be the probability of being below 0.036 and above 2.8, not just the probability of being above 2.8.

For this little example, the difference between our two-sided and the usual one-sided P values is 0.0008, so as commonly reported, a one-sided P value of $0.9992 = 1 - 0.0008$, which would be reported for $F_{obs} = 0.036$, would make us just as suspicious of the null model as the P value 0.0448, which would be reported when seeing $F_{obs} = 2.8$.

Alas, I suspect that the "one-sided" P values will be with us for quite a while. I doubt that many software packages are going to change how they compute the P values for F tests simply because I disapprove of their current practice. Besides, as a practical matter, checking whether the one-sided P values are very close to 1 works reasonably well.

We now establish that $MSTest$ is a reasonable estimate of σ^2 when the reduced model (Red.) holds. The basic idea is this: If we have three items where the first is an average of the other two, and if the first item and one of the other two both estimate σ^2, then the third item must also be an estimate of σ^2; see Exercise 3.10.10. Write

$$
\begin{aligned}
MSE(Red.) &= \frac{1}{dfE(Red.)}[SSE(Red.) - SSE(Full) + SSE(Full)] \\
&= \frac{dfE(Red.) - dfE(Full)}{dfE(Red.)}\left(\frac{SSE(Red.) - SSE(Full)}{dfE(Red.) - dfE(Full)}\right) + \frac{dfE(Full)}{dfE(Red.)}MSE(Full) \\
&= \frac{dfE(Red.) - dfE(Full)}{dfE(Red.)}MSTest + \frac{dfE(Full)}{dfE(Red.)}MSE(Full).
\end{aligned}
$$

This displays $MSE(Red.)$ as a weighted average of $MSTest$ and $MSE(Full)$ because the multipliers

$$\frac{dfE(Red.) - dfE(Full)}{dfE(Red.)} \quad \text{and} \quad \frac{dfE(Full)}{dfE(Red.)}$$

are both between 0 and 1 and they add to 1. Since the reduced model is a special case of the full model, when the reduced model is true, both $MSE(Red.)$ and $MSE(Full)$ are reasonable estimates of σ^2. Since one estimate of σ^2, the $MSE(Red.)$, has been written as a weighted average of another estimate of σ^2, the $MSE(Full)$, and something else, $MSTest$, it follows that the something else must also be an estimate of σ^2.

In data analysis, we are looking for a (relatively) succinct way of summarizing the data. The smaller the model, the more succinct the summarization. However, we do not want to eliminate useful aspects of a model, so we test the smaller (more succinct) model against the larger model to see if the smaller model gives up significant explanatory power. Note that the larger model always has at least as much explanatory power as the smaller model because the larger model includes everything in the smaller model plus more. Although a reduced model may be an adequate substitute for a full model on a particular set of data, it does *not* follow that the reduced model will be an adequate substitute for the full model with any data collected on the variables in the full model. Our models are really approximations and a good approximate model for some data might not be a good approximation for data on the same variables collected differently.

Finally, we mention an alternative way of specifying models. Here we supposed that the model involves independent data y_i, $i = 1, \ldots, n$, with $E(y_i) = \mu_i$ and some common variance, $\text{Var}(y_i) = \sigma^2$. We generally impose some structure on the μ_is and sometimes we assume that the y_is are normally distributed. An equivalent way of specifying the model is to write

$$y_i = \mu_i + \varepsilon_i$$

and make the assumptions that the ε_is are independent with $E(\varepsilon_i) = 0$, $\text{Var}(\varepsilon_i) = \sigma^2$, and are normally distributed. Using the rules for means and variances, it is easy to see that once again,

$$E(y_i) = E(\mu_i + \varepsilon_i) = \mu_i + E(\varepsilon_i) = \mu_i + 0 = \mu_i$$

and

$$\text{Var}(y_i) = \text{Var}(\mu_i + \varepsilon_i) = \text{Var}(\varepsilon_i) = \sigma^2.$$

It also follows that if the ε_is are independent, the y_is are independent, and if the ε_is are normally distributed, the y_is are normally distributed. The ε_is are called *errors* and the residuals $\hat{\varepsilon}_i = y_i - \hat{y}_i$ are estimates (actually predictors) of the errors.

Typically, the full model specifies a relationship among the μ_is that depends on some parameters, say, $\theta_1, \ldots, \theta_r$. Typically, a reduced model specifies some additional relationship among the θ_js that is called a null hypothesis (H_0), for example, $\theta_1 = \theta_2$. As indicated earlier, everybody's favorite reduced model has a common mean for all observations, hence

$$y_i = \mu + \varepsilon_i.$$

We now apply this theory to the one-sample problems of Chapter 2. The full model is simply the one-sample model, thus the variance estimate is $MSE(Full) = s^2$, which we know has $dfE(Full) = n - 1$. A little algebra gives $SSE(Full) = dfE(Full) \times MSE(Full) = (n-1)s^2$. For testing the null model with $H_0 : \mu = m_0$, the variance estimate for the reduced model is

$$MSE(Red.) = \hat{\sigma}_0^2 \equiv \frac{1}{n} \sum_{i=1}^{n} (y_i - m_0)^2$$

with $dfE(Red.) = n$ and $SSE(Red.) = n\hat{\sigma}_0^2$. We also discussed in Chapter 2 that

$$SSE(Red.) - SSE(Full) = n(\bar{y}_. - m_0)^2$$

so that

$$MSTest = [SSE(Red.) - SSE(Full)]/[dfE(Red.) - dfE(Full)] = n(\bar{y}. - m_0)^2/[n - (n-1)]$$

and

$$F = \frac{MSTest}{MSE(Full)} = \frac{n(\bar{y}. - m_0)^2}{s^2} = \left[\frac{\bar{y}. - m_0}{s/\sqrt{n}}\right]^2.$$

The F statistic should be close to 1 if the null model is correct. If the data are normally distributed under the null model, the F statistic should be one observation from an $F(1, n-1)$ distribution, which allows us more precise determinations of the extent to which an F statistic far from 1 contradicts the null model. Recall that with one or two degrees of freedom in the numerator of the F test, values close to 0 are the values most consistent with the reduced model, cf. Figure 3.2.

EXAMPLE 3.1.1.   Years ago, 16 people were independently abducted by S.P.E.C.T.R.E after a Bretagne Swords concert and forced to submit to psychological testing. Among the tests was a measure of audio acuity. From many past abductions in other circumstances, S.P.E.C.T.R.E knows that such observations form a normal population. The observed values of $\bar{y}$. and s^2 were 22 and 0.25, respectively, for the audio acuity scores. Now the purpose of all this is that S.P.E.C.T.R.E. had a long-standing plot that required the use of a loud rock band. They had been planning to use the famous oriental singer Perry Cathay but Bretagne Swords' fans offered certain properties they preferred, provided that those fans' audio acuity scores were satisfactory. From extremely long experience with abducting Perry Cathay fans, S.P.E.C.T.R.E. knows that they have a population mean of 20 on the audio acuity test. S.P.E.C.T.R.E. wishes to know whether Bretagne Swords fans differ from this value. Naturally, they tested $H_0 : \mu = 20$.

The test is to reject the null model if

$$F = \frac{16(\bar{y}. - 20)^2}{s^2}$$

is far from 1 or, if the data are normally distributed, if the F statistic looks unusual relative to an $F(1, 15)$ distribution. Using the observed data,

$$F_{obs} = \frac{16(22 - 20)^2}{0.25} = 256$$

which is very far from 1.   □

EXAMPLE 3.1.2.   The National Association for the Abuse of Student Yahoos (also known as NAASTY) has established guidelines indicating that university dropout rates for math classes should be 15%. In Chapter 2 we considered data from the University of New Mexico's 1984–85 academic year on dropout rates for math classes. We found that the 38 observations on dropout rates were not normally distributed; they contained two outliers. Based on an $\alpha = .05$ test, we wish to know if the University of New Mexico (UNM) meets the NAASTY guidelines when treating the 1984–85 academic year data as a random sample. As is typical in such cases, NAASTY has specified that the central value of the distribution of dropout rates should be 15% but it has not stated a specific definition of the central value. We interpret the central value to be the population mean of the dropout rates and test the null hypothesis $H_0 : \mu = 15\%$.

The complete data consist of 38 observations from which we compute $\bar{y}. = 13.11$ and $s^2 = 106.421$. The data are nonnormal so, although the F statistic is reasonable, we have little to justify comparing the F statistic to the $F(1, 37)$ distribution. Substituting the observed values for $\bar{y}$. and s^2 into the F statistic gives the observed value of the test statistic

$$F_{obs} = \frac{38(13.11 - 15)^2}{106.421} = 1.275,$$

which is not far from 1. The 1984–85 data provide no evidence that UNM violates the NAASTY guidelines.

If we delete the two outliers, the analysis changes. The summary statistics become $\bar{y}_d = 11.083$ and $s_d^2 = 27.45$. Here the subscript d is used as a reminder that the outliers have been deleted. Without the outliers, the data are approximately normal and we can more confidently use the $F(1,35)$ reference distribution,

$$F_{obs,d} = \frac{36(11.083 - 15)}{27.45} = 20.2.$$

This is far from 1. In fact, the 0.999 percentile of an $F(1,35)$ is $F(0.999,1,35) \doteq 12.9$, so an observed F_d of 20.2 constitutes very unusual data relative to the null model. Now we have evidence that dropout rates differ from 15% (or that something else is wrong with the model) but only for a population that no longer includes "outliers." $\qquad\square$

3.1.1   An alternative F test

Not infrequently, when testing models, both the full model (Full) and the reduced model (Red.) are special cases of a biggest model (Big.). In these situations, typically we have fitted a model, the biggest model, and are exploring various submodels that may adequately fit the data. Testing full versus reduced models provides a tool in evaluating their relative merits. In cases with multiple tests and a biggest model, the process of choosing the full model tends to bias $MSE(Full)$ as an estimate of σ^2, so the best practice is to replace $MSE(Full)$ in the denominator of the test by the mean squared error from the biggest model, $MSE(Big.)$. In such cases we prefer to reject the null model at the α level when

$$\frac{[SSE(Red.) - SSE(Full)] / [dfE(Red.) - dfE(Full)]}{MSE(Big.)}$$

$$> F[1 - \alpha, dfE(Red.) - dfE(Full), dfE(Big.)] \quad (3.1.2)$$

rather than using the critical region defined by (3.1.1).

3.2   Inference on single parameters: assumptions

A commonly used alternative to model testing is to focus attention on a single parameter that is important in modeling the data. Most statistical inference on a single parameter devolves from one general theory of inference. To use the general theory of inference on a single parameter, we need to know four things:

1. the parameter of interest, Par,
2. the estimate of the parameter, Est,
3. the standard error of the estimate, $SE(Est)$, and
4. the appropriate reference distribution.

Specifically, what we need to know about the reference distribution is that

$$\frac{Est - Par}{SE(Est)}$$

has a distribution that is some member of the family of t distributions, say $t(df)$, where df specifies the degrees of freedom. The estimate Est is taken to be a random variable. The standard error, $SE(Est)$, is the standard deviation of the estimate if that is known, but more commonly it is an estimate of the standard deviation. If the $SE(Est)$ is estimated, it typically involves an estimate of σ^2 and the estimate of σ^2 determines the degrees of freedom for the t distribution. If the $SE(Est)$ is known, then typically σ^2 is known, and the distribution is usually the standard normal distribution,

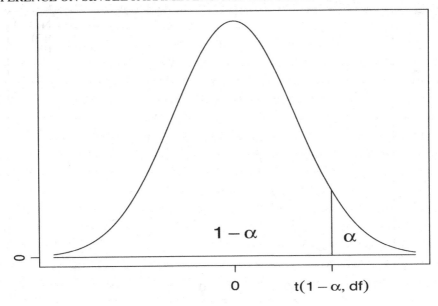

Figure 3.3: *Percentiles of* $t(df)$ *distributions.*

i.e., $t(\infty)$. In some problems, e.g., problems involving the binomial distribution, the central limit theorem is used to get an approximate distribution and inferences proceed as if that distribution were correct. Although appealing to the central limit theorem, so the known distribution is the standard normal, we generally use a t with finite degrees of freedom hoping that it provides a better approximation to the true reference distribution than a standard normal.

Identifying a parameter of interest and an estimate of that parameter is relatively easy. The more complicated part of the procedure is obtaining the standard error. To do that, one typically derives the variance of Est, estimates it (if necessary), and takes the square root. Obviously, rules for deriving variances play an important role in finding standard errors.

These four items—Par, Est, SE(Est), reference distribution—depend crucially on the assumptions made in modeling the data. They depend on assumptions made about the expected values of the observations but also on assumptions of independence, equal variances (homoscedasticity), and normality or large sample sizes. For the purposes of this discussion, we refer to the assumptions made to obtain the four items as the (full) model.

We need notation for the percentage points of the t distribution. In particular, we need a name for the point that cuts off the top α of the distribution. The point that cuts off the top α of the distribution also cuts off the bottom $1 - \alpha$ of the distribution. These ideas are illustrated in Figure 3.3. The notation $t(1 - \alpha, df)$ is used for the point that cuts off the top α.

The illustration in Figure 3.3 is written formally as

$$\Pr\left[\frac{Est - Par}{SE(Est)} > t(1 - \alpha, df)\right] = \alpha.$$

By symmetry about zero we also have

$$\Pr\left[\frac{Est - Par}{SE(Est)} < -t(1 - \alpha, df)\right] = \alpha.$$

The value $t(1 - \alpha, df)$ is called a percentile or percentage point. It is most often found from a computer program but can also be found from a t table or, in the case of $t(\infty)$, from a standard normal table. One can get a feeling for how similar a $t(df)$ distribution is to a standard normal simply

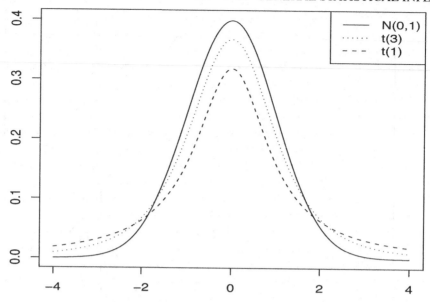

Figure 3.4: $t(df)$ densities for $df = 1, 3, \infty$.

by examining the t tables in Appendix B.1 and noting how quickly the t percentiles approach the values given for infinite degrees of freedom. Alternatively, Figure 3.4 shows that $t(df)$ distributions are centered around 0 and that a $t(1)$ distribution is more spread out than a $t(3)$ distribution, which is more spread out than a $N(0, 1) \equiv t(\infty)$ distribution.

Although we have advertised the methods to be developed in the next sections as being based on parameters rather than models, our discussion of parametric testing will continue to be based on the models assumed for the data and the more specific null models determined by specifying a particular value for the parameter.

3.3 Parametric tests

Tests are often used to check whether *Par* has some specified value. For some fixed known number m_0, we may want to test the *null hypothesis*

$$H_0 : Par = m_0.$$

In this context, the null (reduced) model consists of the assumptions made to obtain the four elements discussed in the previous section together with H_0.

The number m_0 must be known; it is some number that is of interest for the specific data being analyzed. It is impossible to give general rules for picking m_0 because the choice must depend on the context of the data. As mentioned in the previous chapter, *the structure of the data (but not the actual values of the data) sometimes suggests interesting hypotheses* such as testing whether two populations have the same mean or testing whether there is a relationship between two variables. Ultimately the researcher must determine what hypotheses are of interest and these hypotheses determine both *Par* and m_0. In any case, m_0 is never just an unspecified symbol; it must have meaning within the context of the problem.

The test of the null model involving $H_0 : Par = m_0$ is based on the four elements discussed in the previous section and therefore relies on all of the assumptions of the basic model for the data. In addition, the test assumes H_0 is true, so the test is performed assuming the validity of the null model. The idea of the test is to check whether the data seem to be consistent with the null model.

When the (full) model is true, *Est* provides an estimate of *Par*, regardless of the value of

Par. Under the null model, $Par = m_0$, so Est should be close to m_0, and thus the t *statistic* $[Est - m_0]/\text{SE}(Est)$ should be close to 0. Large positive and large negative values of the t statistic indicate data that are inconsistent with the null model. The problem is in specifying what we mean by "large." We will conclude that the data contradict the null model if we observe a value of $[Est - m_0]/\text{SE}(Est)$ that is farther from 0 than some cutoff values.

The problem is then to make intelligent choices for the cutoff values. The solution is based on the fact that if the null model is true,

$$\frac{Est - m_0}{\text{SE}(Est)} \sim t(df).$$

In other words, the t statistic, computed from the data and H_0, has a $t(df)$ distribution. From Figure 3.3, values of the $t(df)$ distribution close to 0 are common and values far from 0 are unusual. We use the $t(df)$ distribution to quantify how unusual values of the t statistic are.

When we substitute the observed values of Est and $\text{SE}(Est)$ into the t statistic we get one observation on the random t statistic, say t_{obs}. When the null model is true, this observation comes from the reference distribution $t(df)$. The question is whether it is reasonable to believe that this one observation came from the $t(df)$ distribution. If so, the data are consistent with the null model. If the observation could not reasonably have come from the reference distribution, the data contradict the null model. Contradicting the null model is a strong inference; it implies that something about the null model is false. (Either there is something wrong with the basic model or with the assumption that $Par = m_0$.) On the other hand, inferring that the data are consistent with the null model does not suggest that it is true. Such data can also be consistent with models other than the null model.

The cutoff values for testing are determined by choosing an α level. The α-level test for $H_0 : Par = m_0$ is to reject the null model if

$$\frac{Est - m_0}{\text{SE}(Est)} > t\left(1 - \frac{\alpha}{2}, df\right)$$

or if

$$\frac{Est - m_0}{\text{SE}(Est)} < -t\left(1 - \frac{\alpha}{2}, df\right).$$

This is equivalent to rejecting H_0 if

$$\frac{|Est - m_0|}{\text{SE}(Est)} > t\left(1 - \frac{\alpha}{2}, df\right).$$

We are rejecting H_0 for those values of $[Est - m_0]/\text{SE}(Est)$ that are most inconsistent with the $t(df)$ distribution, those being the values far from zero. The α level is just a measure of how weird we require the data to be before we reject the null model.

EXAMPLE 3.3.1.   Consider again the 16 people who were independently abducted by S.P.E.C.T.R.E after a Bretagne Swords concert and forced to submit to audio acuity testing. S.P.E.C.T.R.E knows that the observations are normal and observed $\bar{y}. = 22$ and $s^2 = .25$. S.P.E.C.T.R.E. wishes to know whether Bretagne Swords fans differ from the population mean of 20 that Perry Cathay fans display. Naturally, they tested $H_0 : \mu = 20$. They chose an α level of 0.01.

1) $Par = \mu$

2) $Est = \bar{y}.$

3) $\text{SE}(Est) = s/\sqrt{16}$. In this case the $\text{SE}(Est)$ is estimated.

4) $[Est - Par]/\text{SE}(Est) = [\bar{y}. - \mu]/[s/\sqrt{16}]$ has a $t(15)$ distribution. This follows because the data are normally distributed and the standard error is estimated using s.

The $\alpha = 0.01$ test is to reject H_0 if

$$\frac{|\bar{y}. - 20|}{s/\sqrt{16}} > 2.947 = t(0.995, 15).$$

To find the appropriate cutoff value, note that $1 - \alpha/2 = 1 - 0.01/2 = 0.995$, so $t(1 - \alpha/2, 15) = t(0.995, 15)$. With $\bar{y}. = 22$ and $s^2 = 0.25$, we reject H_0 if

$$|t_{obs}| \equiv \frac{|22 - 20|}{\sqrt{0.25/16}} > 2.947.$$

Since $|22 - 20|/\sqrt{0.25/16} = 16$ is greater than 2.947, we reject the null model at the $\alpha = 0.01$ level. If the assumptions of the basic model are correct, there is clear (indeed, overwhelming) evidence that the Bretagne Swords fans have higher scores. (Unfortunately, my masters will not let me inform you whether high scores mean better hearing or worse.) □

EXAMPLE 3.3.2.   We again consider data from the University of New Mexico's 1984–85 academic year on dropout rates for math classes and compare them to the NAASTY guidelines of 15% dropout rates. Based on an $\alpha = .05$ test, we wish to know if the University of New Mexico meets the NAASTY guidelines of 15% dropout rates when treating the 1984–85 academic year data as a random sample. We test the null hypothesis $H_0 : \mu = 15\%$. The 38 observations on dropout rates were not normally distributed; they contained two outliers.

From the complete data of 38 observations we compute $\bar{y}. = 13.11$ and $s^2 = 106.421$. The data are nonnormal, so we have little choice but to hope that 38 observations constitute a sufficiently large sample to justify the use of a t approximation, i.e.,

$$\frac{\bar{y}. - \mu}{\sqrt{s^2/38}} \sim t(37).$$

With an α level of 0.05 and the $t(37)$ distribution, the test rejects H_0 if

$$\frac{\bar{y}. - 15}{\sqrt{s^2/38}} > 2.026 = t(0.975, 37) = t\left(1 - \frac{\alpha}{2}, 37\right)$$

or if

$$\frac{\bar{y}. - 15}{\sqrt{s^2/38}} < -2.026.$$

Substituting the observed values for $\bar{y}.$ and s^2 gives the observed value of the test statistic

$$t_{obs} = \frac{13.11 - 15}{\sqrt{106.421/38}} = -1.13.$$

The value of -1.13 is neither greater than 2.026 nor less than -2.026, so the null hypothesis cannot be rejected at the 0.05 level. The 1984–85 data provide no evidence that UNM violates the NAASTY guidelines (or that anything else is wrong with the null model). Many people would use a $t(\infty)$ distribution in this example based on the hope that $n = 38$ qualifies as a large sample size, but the $t(\infty)$ seems too optimistic to me.

If we delete the two outliers, the analysis changes. Again, the subscript d is used as a reminder that the outliers have been deleted. Without the outliers, the data are approximately normal and we can more confidently use the reference distribution

$$\frac{\bar{y}_d - \mu_d}{\sqrt{s_d^2/36}} \sim t(35).$$

For this reference distribution the $\alpha = 0.05$ test rejects $H_0 : \mu_d = 15$ if

$$\frac{\bar{y}_d - 15}{\sqrt{s_d^2/36}} > 2.030 = t(0.975, 35)$$

or if

$$\frac{\bar{y}_d - 15}{\sqrt{s_d^2/36}} < -2.030 = -t(0.975, 35).$$

With $\bar{y}_d = 11.083$ and $s_d^2 = 27.45$ from the data without the outliers, the observed value of the t statistic is

$$t_{obs,d} = \frac{11.083 - 15}{\sqrt{27.45/36}} = -4.49.$$

The absolute value of -4.49 is greater than 2.030, i.e., $-4.49 < -2.030$, so we reject the null model with $H_0 : \mu_d = 15\%$ at the 0.05 level. When we exclude the two extremely high observations, we have evidence that the typical dropout rate was different from 15%, *provided the other assumptions are true*. In particular, *since the test statistic is negative*, we have evidence that the population mean dropout rate with outliers deleted was actually *less than* 15%. Obviously, most of the UNM math faculty during 1984–85 were not sufficiently nasty.

Finally, we consider the role of transformations in testing. As in Chapter 2, we again consider the square roots of the dropout rates with the two outliers deleted. As discussed earlier, NAASTY has specified that the central value of the distribution of dropout rates should be 15% but has not stated a specific definition of the central value. We are reasonably free to interpret their guideline and we now interpret it as though the population mean of the square roots of the dropout rates should be $\sqrt{15}$. This interpretation leads us to the null hypothesis $H_0 : \mu_{rd} = \sqrt{15}$. Here the subscript r reminds us that square *roots* have been taken and the subscript d reminds us that outliers have been deleted. As discussed earlier, a reasonable appropriate reference distribution is

$$\frac{\bar{y}_{rd} - \mu_{rd}}{\sqrt{s_{rd}^2/36}} \sim t(35),$$

so the test rejects H_0 if

$$\frac{|\bar{y}_{rd} - \sqrt{15}|}{\sqrt{s_{rd}^2/36}} > 2.030 = t(0.975, 35).$$

The sample mean and variance of the transformed, deleted data are $\bar{y}_{rd} = 3.218$ and $s_{rd}^2 = 0.749574$, so the observed value of the t statistic is

$$t_{obs,rd} = \frac{3.218 - 3.873}{\sqrt{0.749574/36}} = -4.54.$$

The test statistic is similar to that in the previous paragraph. The null hypothesis is again rejected and all conclusions drawn from the rejection are essentially the same. I believe that when two analyses both appear to be valid, either the practical conclusions agree or neither analysis should be trusted. □

In practice, people rarely use the procedures presented in this section. These procedures require one to pick specific values for m_0 in $H_0 : Par = m_0$ and for α. In practice, one either picks an α level and presents results for all values of m_0 or one picks a value m_0 and presents results for all α levels. The first of these options is discussed in the next section.

3.4 Confidence intervals

Confidence intervals are commonly viewed as the single most useful procedure in statistical inference. (I don't think I agree with that view.) A $(1-\alpha)$ confidence interval for *Par* consists of all the values m_0 that would not be rejected by an α-level test of $H_0 : Par = m_0$. In other words, the confidence interval consists of all the parameter values that are consistent with both the data and the model as determined by an α-level test. (Since the parameter is part of the model, it seems *a little* redundant to specify that these are parameter values that are consistent with the model. One might take that to be understood.)

A 95% confidence interval for *Par* is based on the fact that an $\alpha = .05$ level test of $H_0 : Par = m_0$ will not be rejected when

$$-t(0.975, df) < \frac{Est - m_0}{SE(Est)} < t(0.975, df).$$

Some algebra (given in the appendix to the chapter) shows that the test will not be rejected when

$$Est - t(0.975, df)\, SE(Est) < m_0 < Est + t(.975, df)\, SE(Est).$$

Thus, the value m_0 is not rejected by a 0.05 level test if and only if m_0 is within the interval having endpoints $Est \pm t(0.975, df)\, SE(Est)$.

EXAMPLE 3.4.1.   In Example 3.3.1 we considered past data on audio acuity in a post-rock environment. Those data were collected on fans of Bretagne Swords from her days of playing Statler Brothers Solitaire. The nefarious organization responsible for this study found it necessary to update their findings after she found her missing card. This time they abducted for themselves 10 independent observations and they were positive that the data would follow a normal distribution with variance 6. (Such arrogance is probably responsible for the failure of S.P.E.C.T.R.E.'s plans of world domination. In any case, their resident statistician was in no position to question these assumptions.) S.P.E.C.T.R.E. found that $\bar{y}.$ was 17. They seek a 95% confidence interval for μ, the mean of the population.

1) $Par = \mu$,

2) $Est = \bar{y}.$,

3) $SE(Est) = \sqrt{6/10}$, in this case $SE(Est)$ is known and not estimated.

4) $[Est - Par]/SE(Est) = [\bar{y}. - \mu]/\sqrt{6/10}$ has a $t(\infty)$ distribution.

For a 95% confidence interval, observe that $1 - \alpha = 95\% = 0.95$ and $\alpha = 0.05$. It follows that $t\left(1 - \frac{\alpha}{2}, \infty\right) = t(0.975, \infty) = 1.96$. The limits of the 95% confidence interval are

$$\bar{y}. \pm 1.96\sqrt{6/10}$$

or, since $\bar{y}. = 17$,

$$17 \pm 1.96\sqrt{6/10}.$$

S.P.E.C.T.R.E. concluded that *for this model* the data were consistent with a mean hearing score between 15.5 and 18.5 for people at this concert (or at least for the population they were considering for abduction) based on a 0.05 level test.   □

EXAMPLE 3.4.2.   In Chapter 2 we considered data on dropout rates for math classes. The 38 observations contained two outliers. Our parameter for these data is μ, the population mean dropout rate for math classes, the estimate is the sample mean $\bar{y}.$, and the standard error is $\sqrt{s^2/38}$ where s^2 is the sample variance. Based on the central limit theorem and the law of large numbers, we used the *approximate* reference distribution

$$\frac{\bar{y}. - \mu}{\sqrt{s^2/38}} \sim t(37).$$

From the 38 observations, we computed $\bar{y}_\cdot = 13.11$ and $s^2 = 106.421$ and found a 95% confidence interval for the dropout rate of $(9.7, 16.5)$. The endpoints of the confidence interval are computed as

$$13.11 \pm 2.026(\sqrt{106.421/38}).$$

If we drop the two outliers, the remaining data seem to be normally distributed. Recomputing the sample mean and sample variance with the outliers deleted, we get $\bar{y}_d = 11.083$ and $s_d^2 = 27.45$. Without the outliers, we can use the reference distribution

$$\frac{\bar{y}_d - \mu_d}{\sqrt{s_d^2/36}} \sim t(35).$$

This $t(35)$ distribution relies on the assumption of normality (which we have validated) rather than relying on the unvalidated large sample approximations from the central limit theorem and law of large numbers. Philosophically, the $t(35)$ distribution should give more accurate results, but we have no way to establish whether that is actually true for these data. To compute a 95% confidence interval based on the data without the outliers, we need to find the appropriate tabled values. Observe once again that $1 - \alpha = 95\% = 0.95$ and $\alpha = 0.05$. It follows that $t\left(1 - \frac{\alpha}{2}, df\right) = t(0.975, 35) = 2.030$, and, substituting the observed values of $\bar{y}_d$ and s_d^2, the confidence interval has endpoints

$$11.083 \pm 2.030(\sqrt{27.45/36}).$$

The actual interval is $(9.3, 12.9)$. *Excluding the extremely high values that occasionally occur*, the model and data are consistent with a mean dropout rate between 9.3 and 12.9 percent based on a 0.05 test. Remember, this is a confidence interval for the mean of math classes; it *does not* indicate that you can be 95% confident that your next math class will have a dropout rate between 9.3 and 12.9 percent. Such an inference requires a prediction interval, cf. Section 3.7.

The interval $(9.3, 12.9)$ is much narrower than the one based on all 38 observations, largely because our estimate of the variance is much smaller when the outliers have been deleted. Note also that with the outliers deleted, we are drawing inferences about a different parameter than when they are present. With the outliers deleted, our conclusions are only valid for the bulk of the observations. While occasional weird observations can be eliminated from our analysis, we cannot stop them from occurring.

We have also looked at the square roots of the dropout rate data. *We now consider the effect on confidence intervals of transforming the data*. With the two outliers deleted and taking square roots of the observations, we found earlier that the data are reasonably normal. The sample mean and variance of the transformed, deleted data are $\bar{y}_{rd} = 3.218$ and $s_{rd}^2 = 0.749574$. Using the reference distribution

$$\frac{\bar{y}_{rd} - \mu_{rd}}{\sqrt{s_{rd}^2/36}} \sim t(35),$$

we obtain a 95% confidence interval with endpoints

$$3.218 \pm 2.030 \left(\sqrt{\frac{0.749574}{36}} \right).$$

The confidence interval reduces to $(2.925, 3.511)$. This is a 95% confidence interval for the population mean of the square roots of the dropout rate percentages with 'outliers' removed from the population.

The confidence interval $(2.925, 3.511)$ does not really address the issue that we set out to investigate. We wanted some idea of the value of the population mean dropout rate. We have obtained a 95% confidence interval for the population mean of the square roots of the dropout rate percentages (with outliers removed from the population). There is no simple, direct relationship between the population mean dropout rate and the population mean of the square roots of the dropout rate percentages, but a simple device can be used to draw conclusions about typical values for dropout rates when the analysis is performed on the square roots of the dropout rates.

If the square root data are normal, the mean is the same as the median. *The median is a value with 50% of observations falling at or below it and 50% falling at or above it.* Although the mean on the square root scale does not transform back to the mean on the original scale, the median does. Since $(2.925, 3.511)$ provides a 95% confidence interval for the median from the *square roots* of the dropout rate percentages, we simply *square* all the values in the interval to draw conclusions about the median dropout rate percentages. Squaring the endpoints of the interval gives the new interval $(2.925^2, 3.511^2) = (8.6, 12.3)$. We are now 95% confident that the median of the population of dropout rates is between 8.6 and 12.3. Interestingly, we will see in Section 3.7 that prediction intervals do not share these difficulties in interpretation associated with transforming the data.

Note that the back transformed interval $(8.6, 12.3)$ for the median obtained from the transformed, deleted data is similar to the interval $(9.3, 12.9)$ for the mean (which is also the median of the assumed model) obtained earlier from the untransformed data with the outliers deleted. Again, when two distinct analyses both seem reasonably valid, I would be very hesitant about drawing practical conclusions that could not be justified from *both* analyses.   □

The confidence intervals obtained from this theory can frequently be obtained by another approach to statistical inference using 'Bayesian' arguments; see Berry (1996). In the Bayesian justification, the *correct* interpretation of a 95% confidence interval is that *the probability is 95% that the parameter is in the interval*.

Rather than the testing interpretation or the Bayesian interpretation, most statisticians seem to favor the Neyman–Pearson definition for confidence intervals based on the idea that in a long run of performing 95% confidence intervals, about 95% will contain the true parameter. Of course this does not actually tell you anything about the confidence interval at hand. It also assumes that all the models are correct in the long run of confidence intervals. It is difficult to get students to accept this definition as anything other than a memorized fact. Students frequently misinterpret this definition as the Bayesian interpretation.

The long run interpretation of confidence intervals tempts people to make a mistake in interpretation. If I am about to flip a coin, we can agree that the physical mechanism involved gives probability $1/2$ to both heads and tails. If I flip the coin but don't show it to you, you still *feel* like the probabilities are both $1/2$. But I know the result! Therefore, the probabilities based on the physical mechanism no longer apply, and your feeling that probability $1/2$ is appropriate is entirely in your head. It feels good, but what is the justification? Bayesian Statistics involves developing justifications for such probabilities.

The long run interpretation of confidence intervals is exactly the same as flipping a coin that turns up heads, say, 95% of the time. The parameter being in the interval is analogous to the coin being heads. Maybe it is; maybe it isn't. How the number 0.95 applies to a particular interval or flip, after it has been determined, is a mystery. Of course many statisticians simply recite the correct probability statement and ignore its uselessness. The significance testing and Bayesian interpretations of the intervals both seem reasonable to me.

Confidence intervals give all the possible parameter values that seem to be consistent with the data and the model. In particular, they give the results of testing $H_0 : Par = m_0$ for a *fixed α but every choice of m_0.* In the next section we discuss P values that give the results of testing $H_0 : Par = m_0$ for a *fixed m_0 but every choice of α.*

3.5   *P* values

Rather than having formal rules for when to reject the null model, one can report the evidence against the null model. That is done by reporting the P value. The P value is computed under the null model. It is the probability of seeing data that are as weird or more weird than those that were actually observed. Formally, with $H_0 : Par = m_0$ we write t_{obs} for the observed value of the t statistic as computed from the *observed* values of Est and $SE(Est)$. Thus t_{obs} is our summary of the data that were actually observed. Recalling our earlier discussion that the most unusual values of t_{obs} are those far from 0, the probability under the null model of seeing something as or more weird than we actually saw is the probability that a $t(df)$ distribution is farther from 0 than $|t_{obs}|$. Formally, we can write this as

$$P = Pr\left[\left|\frac{Est - m_0}{SE(Est)}\right| \geq |t_{obs}|\right].$$

Here Est (and usually $SE(Est)$) are viewed as random and it is assumed that $Par = m_0$ so that $(Est - m_0)/SE(Est)$ has the known reference distribution $t(df)$. The value of t_{obs} is a fixed known number, so we can actually compute P. Using the symmetry of the $t(df)$ distribution, the basic idea is that for, say, t_{obs} positive, any value of $(Est - m_0)/SE(Est)$ greater than t_{obs} is more weird than t_{obs}. Any data that yield $(Est - m_0)/SE(Est) = -t_{obs}$ are just as weird as t_{obs} and values of $(Est - m_0)/SE(Est)$ less than $-t_{obs}$ are more weird than observing t_{obs}.

EXAMPLE 3.5.1.   Again consider the Bretagne Swords data. We have 16 observations taken from a normal population and we wish to test $H_0 : \mu = 20$. As before, 1) $Par = \mu$, 2) $Est = \bar{y}.$, 3) $SE(Est) = s/\sqrt{16}$, and 4) $[Est - Par]/SE(Est) = [\bar{y}. - \mu]/[s/\sqrt{16}]$ has a $t(15)$ distribution. This time we take $\bar{y}. = 19.78$ and $s^2 = .25$, so the observed test statistic is

$$t_{obs} = \frac{19.78 - 20}{\sqrt{0.25/16}} = -1.76.$$

From a t table, $t(0.95, 15) = 1.75$, so

$$P = Pr[|t(15)| \geq |-1.76|] \doteq Pr[|t(15)| \geq 1.75] = 0.10.$$

Alternatively, $t(0.95, 15) \doteq |1.76|$, so $P \doteq 2(1 - .95)$.   □

The P value is the smallest α level for which the test would be rejected. Thus, if we perform an α-level test where α *is less than the P value, we can conclude immediately that the null model is not rejected. If we perform an α-level test where α is greater than the P value, we know immediately that the null model is rejected.* Thus computing a P value eliminates the need to go through the formal testing procedures described in Section 3.3. Knowing the P value immediately gives the test results for any choice of α. The P value is a measure of how consistent the data are with the null model. Large values (near 1) indicate great consistency. Small values (near 0) indicate data that are inconsistent with the null model.

EXAMPLE 3.5.2.   In Example 3.3.2 we considered tests for the drop rate data. Using the complete untransformed data and the null hypothesis $H_0 : \mu = 15$, we observed the test statistic

$$t_{obs} = \frac{13.11 - 15}{\sqrt{106.421/38}} = -1.13.$$

Using a computer program, we can compute

$$P = Pr[|t(37)| \geq |-1.13|] = 0.26.$$

An $\alpha = 0.26$ test would be just barely rejected by these data. Any test with an α level smaller than 0.26 is more stringent (the cutoff values are farther from 0 than 1.13) and would not be rejected. Thus the commonly used $\alpha = 0.05$ and $\alpha = 0.01$ tests would not be rejected. Similarly, any test with an α level greater than 0.26 is less stringent and would be rejected. Of course, it is extremely rare that one would use a test with an α level greater than 0.26. Recall that the P value of 0.26 is a highly questionable number because it was based on a highly questionable reference distribution, the $t(37)$.

Using the untransformed data with outliers deleted and the null hypothesis $H_0 : \mu_d = 15$, we observed the test statistic

$$t_{obs,d} = \frac{11.083 - 15}{\sqrt{27.45/36}} = -4.49.$$

We compute

$$P = \Pr[|t(35)| \geq |-4.49|] = 0.000.$$

This P value is not really zero; it is a number that is so small that when we round it off to three decimal places the number is zero. In any case, the test is rejected for any reasonable choice of α. In other words, the test is rejected for any choice of α that is greater than 0.000. (Actually for any α greater than 0.0005 because of the round-off issue.)

Using the square roots of the data with outliers deleted and the null hypothesis $H_0 : \mu_{rd} = \sqrt{15}$, the observed value of the test statistic is

$$t_{obs,rd} = \frac{3.218 - 3.873}{\sqrt{0.749574/36}} = -4.54.$$

We compute

$$P = \Pr[|t(35)| \geq |-4.54|] = 0.000.$$

Once again, the test result is highly significant. But remember, unless you are reasonably sure that the model is right, you cannot be reasonably sure that H_0 is wrong.   □

EXAMPLE 3.5.3.   In Example 3.3.1 we considered audio acuity data for Bretagne Swords fans and tested whether their mean score differed from fans of Perry Cathay. In this example we test whether their mean score differs from that of Tangled Female Sibling fans. Recall that the observed values of n, $\bar{y}.$, and s^2 for Bretagne Swords fans were 16, 22, and 0.25, respectively and that the data were normal. Tangled Female Sibling fans have a population mean score of 22.325, so we test $H_0 : \mu = 22.325$. The test statistic is $(22 - 22.325)/\sqrt{0.25/16} = -2.6$. If we do an $\alpha = 0.05$ test, $|-2.6| > 2.13 = t(0.975, 15)$, so we reject H_0, but if we do an $\alpha = 0.01$ test, $|-2.6| < 2.95 = t(0.995, 15)$, so we do not reject H_0. In fact, $|-2.6| \doteq t(0.99, 15)$, so the P value is essentially $.02 = 2(1 - .99)$. The P value is larger than 0.01, so the 0.01 test does not reject H_0; the P value is less than 0.05, so the test rejects H_0 at the 0.05 level.

If we consider confidence intervals, the 99% interval has endpoints $22 \pm 2.95\sqrt{0.25/16}$ for an interval of $(21.631, 22.369)$ and the 95% interval has endpoints $22 \pm 2.13\sqrt{0.25/16}$ for an interval of $(21.734, 22.266)$. Notice that the hypothesized value of 22.325 is inside the 99% interval, so it is not rejected by a 0.01 level test, but 22.325 is outside the 95% interval, so a 0.05 test rejects $H_0 : \mu = 22.325$. The 98% interval has endpoints $22 \pm 2.60\sqrt{0.25/16}$ for an interval of $(21.675, 22.325)$ and the hypothesized value is on the edge of the interval.   □

In the absence of other assumptions, a large P value does not constitute evidence in support of the null model. A large P value indicates that the data are consistent with the null model but, typically, it is easy to find other models even more consistent with the data. In Example 3.5.1, the data are even more consistent with $\mu = 19.78$.

Philosophically, it would be more proper to define the P value prior to defining an α-level test, defining an α-level test as one that rejects when the P value is less than or equal to α. One would then define confidence intervals relative to α-level tests. I changed the order because I caved to my perception that people are more interested in confidence intervals.

3.6 Validity of tests and confidence intervals

In significance testing, we make an assumption, namely the null model, and check whether the data are consistent with the null model or inconsistent with it. If the data are consistent with the null model, that is all that we can say. If the data are inconsistent with the null model, it suggests that the null model is somehow wrong. (This is very similar to the mathematical idea of a proof by contradiction.)

Often people want a test of the null hypothesis $H_0 : Par = m_0$ rather than the null model. The null model involves a series of assumptions in addition to $H_0 : Par = m_0$. Typically we assume that observations are independent and have equal variances. In most tests that we will consider, we assume that the data have normal distributions. As we consider more complicated data structures, we will need to make more assumptions. The proper conclusion from a test is that either the data are consistent with our assumptions or the data are inconsistent with our assumptions. If the data are inconsistent with the assumptions, it suggests that at least one of them is invalid. In particular, if the data are inconsistent with the assumptions, it does not necessarily imply that the particular assumption embodied in the null hypothesis is the one that is invalid. Before we can reasonably conclude that the null hypothesis is untrue, we need to ensure that the other assumptions are reasonable. Thus it is crucial to check our assumptions as fully as we can. Plotting the data, or more often plotting the residuals, plays a vital role in checking assumptions. Plots are used throughout the book, but special emphasis on plotting is given in Chapter 7.

In Section 3.2 it is typically quite easy to define parameters Par and estimates Est. The role of the assumptions is crucial in obtaining a valid $SE(Est)$ and an appropriate reference distribution. If our assumptions are reasonably valid, our $SE(Est)$ and reference distribution will be reasonably valid and the procedures outlined here lead to conclusions about Par with reasonable validity. Of course the assumptions that need to be checked depend on the precise nature of the analysis being performed, i.e., the precise model that has been assumed.

3.7 Theory of prediction intervals

Some slight modifications of the general theory allow us to construct prediction intervals for future observations from the model. Many of us would argue that the fundamental purpose of science is making accurate predictions of things that can be observed in the future. As with estimation, predicting the occurrence of a particular value (point prediction) is less valuable than interval prediction because a point prediction gives no idea of the variability associated with the prediction.

In constructing prediction intervals for a new observation y_0, we make a number of assumptions. First, we assume that we will obtain data $y_1, \ldots, y_n$ that are independent with common variance σ^2 and normally distributed. The random observation to be predicted is y_0. It is assumed that y_0 is independent of $y_1, \ldots, y_n$ with variance σ^2 and normal. Our parameter is $Par = E(y_0)$ and Est uses $y_1, \ldots, y_n$ to estimate $E(y_0)$, i.e., our point prediction for y_0. We also assume that σ^2 has an estimate $\hat{\sigma}^2$ computed from $y_1, \ldots, y_n$, that $SE(Est) = \hat{\sigma}A$ for some known constant A, and that $(Est - Par)/SE(Est)$ has a t distribution with, say, df degrees of freedom. (Technically, we need Est to have a normal distribution, $df \times (\hat{\sigma}^2/\sigma^2)$, to have a $\chi^2(df)$ distribution, and independence of Est and $\hat{\sigma}^2$.)

A prediction interval for y_0 is based on the distribution of $y_0 - Est$ because we need to evaluate how far y_0 can reasonably be from our point prediction of y_0. The value of the future observation y_0 is independent of the past observations and thus of Est. It follows that the variance of $y_0 - Est$ is

$$\text{Var}(y_0 - Est) = \text{Var}(y_0) + \text{Var}(Est) = \sigma^2 + \text{Var}(Est) = \sigma^2 + \sigma^2 A^2$$

and that the standard error of $y_0 - Est$ is

$$SE(y_0 - Est) = \sqrt{\hat{\sigma}^2 + [SE(Est)]^2} = \sqrt{\hat{\sigma}^2[1 + A^2]}. \qquad (3.7.1)$$

$SE(y_0 - Est)$ is called the standard error of prediction and is sometimes written $SE(Prediction)$. Using the standard error of prediction, one can show that

$$\frac{y_0 - Est}{SE(y_0 - Est)} \sim t(df).$$

A $(1 - \alpha)$ prediction interval is based on testing whether a particular future y_0 value would be consistent with the assumptions we have made (our model) and the other data. An α-level test for y_0 would not be rejected if

$$-t\left(1 - \frac{\alpha}{2}, df\right) < \frac{y_0 - Est}{SE(y_0 - Est)} < t\left(1 - \frac{\alpha}{2}, df\right).$$

Rearranging the terms leads to the inequalities

$$Est - t\left(1 - \frac{\alpha}{2}, df\right) SE(y_0 - Est) < y_0 < Est + t\left(1 - \frac{\alpha}{2}, df\right) SE(y_0 - Est).$$

The prediction interval consists of all y_0 values that fall between these two observable limits. These are the y_0 values that are consistent with our model and data. The endpoints of the interval can be written

$$Est \pm t\left(1 - \frac{\alpha}{2}, df\right) \sqrt{\hat{\sigma}^2 + [SE(Est)]^2}. \tag{3.7.2}$$

Unfortunately, it is impossible to even attempt to validate assumptions about observations to be taken in the future. How could we possibly validate that a future observation is going to be independent of previous observations? Thus, the validity of prediction intervals is always suspect.

The prediction interval determined by (3.7.2) is similar to, but wider than, the confidence interval for $Par = E(y_0)$, which is

$$Est \pm t\left(1 - \frac{\alpha}{2}, df\right) SE(Est).$$

From the form of $SE(y_0 - Est)$ given in (3.7.1), we see that

$$SE(y_0 - Est) = \sqrt{\hat{\sigma}^2 + [SE(Est)]^2} \geq SE(Est).$$

Typically, the prediction standard error is much larger than the standard error of the estimate, so prediction intervals are much wider than confidence intervals for $E(y_0)$. In particular, increasing the number of observations typically decreases the standard error of the estimate but has a *relatively* minor effect on the standard error of prediction. Increasing the sample size is not intended to make $\hat{\sigma}^2$ smaller, it only makes $\hat{\sigma}^2$ a more accurate estimate of σ^2.

EXAMPLE 3.7.1.   As in Example 3.3.2, we eliminate the two outliers from the dropout rate data. The 36 remaining observations are approximately normal. A 95% confidence interval for the mean had endpoints

$$11.083 \pm 2.030\sqrt{27.45/36}.$$

A 95% prediction interval has endpoints

$$11.083 \pm 2.030\sqrt{27.45 + \frac{27.45}{36}}$$

or

$$11.083 \pm 10.782.$$

The prediction interval is $(0.301, 21.865)$, which is much wider than the confidence interval of $(9.3, 12.9)$. Dropout rates for a new math class between 0.3% and 21.9% are consistent with the data and the model based on a 0.05 level test. Population mean dropout rates for math classes between

9% and 13% are consistent with the data and the model. Of course the prediction interval assumes that the new class is from a population similar to the 1984–85 math classes with huge dropout rates deleted. Such assumptions are almost impossible to validate. Moreover, there is some chance that the new observation will be one with a huge dropout rate and this interval says nothing about such observations.

In Example 3.3.2 we also considered the square roots of the dropout rate data with the two outliers eliminated. To predict the square root of a new observation, we use the 95% interval

$$3.218 \pm 2.030 \left(\sqrt{0.749574 + \frac{0.749574}{36}} \right),$$

which reduces to $(1.436, 5.000)$. This is a prediction interval for the square root of a new observation, so actual values of the new observation between $(1.436^2, 5.000^2)$, i.e., $(2.1, 25)$ are consistent with the data and model based on a 0.05 level test. Retransforming a prediction interval back into the original scale causes no problems of interpretation whatsoever. This prediction interval and the one in the previous paragraph are comparable. Both include values from near 0 up to the low to mid twenties. $\qquad\square$

Lower bounds on prediction confidence

If the normal and χ^2 distributional assumptions stated at the beginning of the section break down, our measure of how unusual a future data point might be is invalid. The cut-off value for our test is based on rejecting y_0 values that are unusual relative to the t distribution. If we use the cut-off values from the t distribution even when the distribution is not valid, what can we say about the weirdness of data that exceed the cut-off values?

Relying primarily on the independence assumptions and having sufficient data to use $\hat{\sigma}^2$ as an approximation to σ^2, we can find an approximate lower bound for the confidence that a new observation is in the prediction interval. Chebyshev's inequality from Subsection 1.2.2 gives

$$1 - t\left(1 - \frac{\alpha}{2}, df\right)^{-2} \le \Pr\left[-t\left(1 - \frac{\alpha}{2}, df\right) < \frac{y_0 - Est}{SE(y_0 - Est)} < t\left(1 - \frac{\alpha}{2}, df\right)\right],$$

or equivalently,

$$1 - t\left(1 - \frac{\alpha}{2}, df\right)^{-2} \le \Pr\left[Est - t\left(1 - \frac{\alpha}{2}, df\right) SE(y_0 - Est) < y_0 \right.$$
$$\left. < Est + t\left(1 - \frac{\alpha}{2}, df\right) SE(y_0 - Est)\right].$$

This indicates that the confidence coefficient for the prediction interval given by

$$Est \pm t\left(1 - \frac{\alpha}{2}, df\right) SE(y_0 - Est)$$

is (approximately) at least

$$\left[1 - t\left(1 - \frac{\alpha}{2}, df\right)^{-2}\right] 100\%.$$

In other words, the probability of seeing data as weird or weirder than $t\left(1 - \frac{\alpha}{2}, df\right)$ is no more than

$$t\left(1 - \frac{\alpha}{2}, df\right)^{-2}.$$

If we can use the improved version of Chebyshev's inequality from Section 1.3, we can raise the confidence coefficient to

$$\left[1 - (2.25)^{-1} t\left(1 - \frac{\alpha}{2}, df\right)^{-2}\right] 100\%$$

or lower the α level to

$$(2.25)^{-1}t\left(1 - \frac{\alpha}{2}, df\right)^{-2}.$$

EXAMPLE 3.7.2.   Assuming that a sample of 36 observations is enough to ensure that s^2 is essentially equal to σ^2, the nominal 95% prediction interval given in Example 3.7.1 for dropout rates has a confidence level, regardless of the distribution of the data, that is at least

$$\left(1 - \frac{1}{2.030^2}\right) = 76\% \quad \text{or even} \quad \left(1 - \frac{1}{2.25(2.030)^2}\right) = 89\%.$$

The true α level for the corresponding test is no more than 0.24, or, if the improved version of Chebyshev applies, 0.11.

3.8   Sample size determination and power

Suppose we wish to estimate the mean height of the men officially enrolled in Statistics classes at the University of New Mexico on April 5, 2010 at 3 pm. How many observations should we take? The answer to that question depends on how accurate our estimate needs to be and on our having some idea of the variability in the population.

To get a rough indication of the variability we argue as follows. Generally, men have a mean height of about 69 inches and I would guess that about 95% of them are between 63 inches and 75 inches. The probability that a $N(\mu, \sigma^2)$ random variable is between $\mu \pm 2\sigma$ is approximately 0.95, which suggests that $\sigma = [(\mu + 2\sigma) - (\mu - 2\sigma)]/4$ may be about $(75 - 63)/4 = 3$ for a typical population of men.

Before proceeding with sample size determination, observe that sample sizes have a real effect on the usefulness of confidence intervals. Suppose $\bar{y}. = 72$ and $n = 9$, so the 95% confidence interval for mean height has endpoints of roughly $72 \pm 2(3/\sqrt{9})$, or 72 ± 2, with an interval of $(70, 74)$. Here we use 3 as a rough indication of σ in the standard error and 2 as a rough indication of the tabled value for a 95% interval. If having an estimate that is off by 1 inch is a big deal, the confidence interval is totally inadequate. There is little point in collecting the data, because regardless of the value of $\bar{y}.$, we do not have enough accuracy to draw interesting conclusions. For example, if I claimed that the true mean height for this population was 71 inches and I cared whether my claim was off by an inch, the data are not only consistent with my claim but also with the claims that the true mean height is 70 inches and 72 inches and even 74 inches. The data are inadequate for my purposes. Now suppose $\bar{y}. = 72$ and $n = 3600$, the confidence interval has endpoints $72 \pm 2(3/\sqrt{3600})$ or 72 ± 0.1 with an interval of $(71.9, 72.1)$. We can tell that the population mean may be 72 inches but we are quite confident that it is not 72.11 inches. Would anyone really care about the difference between a mean height of 72 inches and a mean height of 72.11 inches? Three thousand six hundred observations gives us more information than we really need. We would like to find a middle ground.

Now suppose we wish to learn the mean height to within 1 inch with 95% confidence. From a sample of size n, a 95% confidence interval for the mean has endpoints that are roughly $\bar{y}. \pm 2(3/\sqrt{n})$. With 95% confidence, the mean height could be as high as $\bar{y}. + 2(3/\sqrt{n})$ or as low as $\bar{y}. - 2(3/\sqrt{n})$. We want the difference between these numbers to be no more than 1 inch. The difference between the two numbers is $12/\sqrt{n}$, so to obtain the required difference of 1 inch, set $1 = 12/\sqrt{n}$, so that $\sqrt{n} = 12/1$ or $n = 144$.

The semantics of these problems can be a bit tricky. We asked for an interval that would tell us the mean height to within 1 inch with 95% confidence. If instead we specified that we wanted our estimate to be off by no more than 1 inch, the estimate is in the middle of the interval, so the distance from the middle to the endpoint needs to be 1 inch. In other words, $1 = 2(3/\sqrt{n})$, so $\sqrt{n} = 6/1$ or $n = 36$. Note that learning the parameter to within 1 inch is the same as having an estimate that is off by no more than $1/2$ inch.

The concepts illustrated above work quite generally. Typically an observation y has $\text{Var}(y) = \sigma^2$ and Est has $\text{SE}(Est) = \sigma A$. The constant A in $\text{SE}(Est)$ is a known function of the sample size (or sample sizes in situations involving more than one sample). In inference problems we replace σ in the standard error with an estimate of σ obtained from the data. In determining sample sizes, the data have not yet been observed, so σ has to be approximated from previous data or knowledge. The length of a $(1 - \alpha)100\%$ confidence interval is

$$[Est + t(1 - \alpha/2, df)\text{SE}(Est)] - [Est - t(1 - \alpha/2, df)\text{SE}(Est)]$$
$$= 2t(1 - \alpha/2, df)\text{SE}(Est) = 2t(1 - \alpha/2, df)\sigma A.$$

The tabled value $t(1 - \alpha/2, df)$ can be approximated by $t(1 - \alpha/2, \infty)$. If we specify that the confidence interval is to be w units wide, set

$$w = 2t(1 - \alpha/2, \infty)\sigma A \tag{3.8.1}$$

and solve for the (approximate) appropriate sample size. In Equation (3.8.1), w, $t(1 - \alpha/2, \infty)$, and σ are all known and A is a known function of the sample size.

Unfortunately it is not possible to take Equation (3.8.1) any further and show directly how it determines the sample size. The discussion given here is general and thus the ultimate solution depends on the type of data being examined. In the only case we have examined as yet, there is one-sample, $Par = \mu$, $Est = \bar{y}_\cdot$, and $\text{SE}(Est) = \sigma/\sqrt{n}$. Thus, $A = 1/\sqrt{n}$ and Equation (3.8.1) becomes

$$w = 2t(1 - \alpha/2, \infty)\sigma/\sqrt{n}.$$

Rearranging this gives

$$\sqrt{n} = 2t(1 - \alpha/2, \infty)\sigma/w$$

and

$$n = [2t(1 - \alpha/2, \infty)\sigma/w]^2.$$

But this formula only applies to one-sample problems. For other problems considered in this book, e.g., comparing two independent samples, comparing more than two independent samples, and simple linear regression, Equation (3.8.1) continues to apply but the constant A becomes more complicated. In cases where there is more than one sample involved, the various sample sizes are typically assumed to all be the same, and in general their relative sizes need to be specified, e.g., we could specify that the first sample will have 10 more observations than the second or that the first sample will have twice as many observations as the second.

Another approach to determining approximate sample sizes is based on the power of an α-level test. (Here we are sinking, or at least wading, into the morass of Neyman–Pearson testing.) If the model is correct but the null hypothesis is noticeably wrong, we want a sample size that gives us a decent chance of recognizing that fact. Tests are set up assuming that, say, $H_0 : Par = m_0$ is true. Power is computed assuming that $Par \neq m_0$. Suppose that $Par = m_A \neq m_0$, then *the power when $Par = m_A$ is the probability that the $(1 - \alpha)100\%$ confidence interval will not contain m_0.* Another way of saying that the confidence interval does not contain m_0 is saying that an α-level two-sided test of $H_0 : Par = m_0$ rejects H_0. In determining sample sizes, you need to pick m_A as some value you care about. You need to care about it in the sense that if $Par = m_A$ rather than $Par = m_0$, you would like to have a reasonably good chance of rejecting $H_0 : Par = m_0$.

Cox (1958, p. 176) points out that it often works well to choose the sample size so that

$$|m_A - m_0| \doteq 3\text{SE}(Est). \tag{3.8.2}$$

Cox shows that this procedure gives reasonable powers for common choices of α. Here m_A and m_0 are known and $\text{SE}(Est) = \sigma A$, where σ is known and A is a known function of sample size. Also note that this suggestion does not depend explicitly on the α level of the test. As with Equation

(3.8.1), Equation (3.8.2) can be solved to give n in particular cases, but a general solution for n is not possible because it depends on the exact nature of the value A.

Consider again the problem of determining the mean height. If my null hypothesis is $H_0 : \mu = 72$ and I want a reasonable chance of rejecting H_0 when $\mu = 73$, Cox's rule suggests that I should have $1 = |73 - 72| \doteq 3\,(3/\sqrt{n})$ so that $\sqrt{n} \doteq 9$ or $n \doteq 81$.

It is important to remember that these are only rough guides for sample sizes. They involve several approximations, the most important of which is approximating σ. If there is more than one parameter of interest in a study, sample size computations can be performed for each and a compromise sample size can be selected.

In the early years of my career, I was amazed at my own lack of interest in teaching students about statistical power until Cox (1958, p. 161) finally explained it for me. He points out that power is very important in planning investigations but it is not very important in analyzing them. I might even go so far as to say that once the data have been collected, a power analysis can at best tell you whether you have been wasting your time. In other words, a power analysis will only tell you how likely you were to find differences given the design of your experiment and the choice of test.

Although the simple act of rejecting a null model does nothing to suggest what models might be correct, it can still be interesting to see whether we have a reasonable chance of rejecting the null model when some alternative model is true. Hence our discussion. However, the theory of testing presented here is not an appropriate theory for making choices between a null model and some alternative. Our theory is a procedure for (possibly) falsifying a null model.

3.9   The shape of things to come

To keep the discussion in this chapter as simple as possible, the examples have thus far been restricted to one-sample problems. However, the results of this chapter apply to more complicated problems such as two-sample problems, regression, and analysis of variance. For these different problems, the only thing that changes is how we model the means of the observations.

Through the vast majority of this book, we will assume that a model exists to predict a measurement random variable y based on a (nonrandom) predictor x. The predictor x can be a single measurement (continuous) variable or a single categorical (factor or classification) variable. A categorical variable is one that defines groups of observations. A categorical variable can identify which observations are male and which are female. It can identify racial groups or socio-economic groups or age groups. (Although age could also be a measurement.) The predictor x can be either a single variable or x can be a vector of measurement and categorical variables.

Our models are written

$$y = m(x) + \varepsilon,$$

where $m(\cdot)$ is some fixed function that determines the mean of y for a given x and ε is some unobservable error term with mean 0. Thus

$$E(y) = E[m(x) + \varepsilon] = m(x) + E(\varepsilon) = m(x) + 0 = m(x).$$

With n observations on this model, write

$$y_h = m(x_h) + \varepsilon_h, \qquad h = 1, \dots, n. \tag{3.9.1}$$

We typically assume

$$\varepsilon_h \text{s} \quad \text{independent} \quad N(0, \sigma^2). \tag{3.9.2}$$

Here σ^2 is an unknown parameter that we must estimate. Together (3.9.1) and (3.9.2) constitute our model for the observations. The function $m(\cdot)$ is our model for the mean of y. We make assumptions about the form of $m(\cdot)$ that typically include unknown (mean) parameters that we must estimate.

Frequently, we find it more convenient to express the model in terms of the observations. These are independent, normally distributed, and have the same variance σ^2, i.e.,

$$y_h\text{s} \quad \text{independent} \quad N[m(x_h), \sigma^2]. \tag{3.9.3}$$

If x is a single variable that only ever takes on one value, say, $x \equiv 1$, then we have the model for a one-sample problem as discussed in Chapter 2. In particular, Model (3.9.1) becomes

$$y_h = m(1) + \varepsilon_h, \qquad h = 1, \ldots, n.$$

If we make the identification

$$\mu \equiv m(1),$$

we get a one-sample model with one mean parameter to estimate,

$$y_h = \mu + \varepsilon_h, \qquad h = 1, \ldots, n,$$

or, we more often write it in terms of Model (3.9.3),

$$y_h\text{s} \quad \text{independent} \quad N(\mu, \sigma^2).$$

In Chapter 6 we deal with a model that involves a single measurement predictor. In particular, we discuss verbal abilities y in a school and relate them to a measurement of socio-economic status x for the school. In *simple linear regression* we assume that

$$m(x) = \beta_0 + \beta_1 x,$$

so our model incorporates a linear relationship between x and the expected value of y. For a set of n observations, write

$$y_h = \beta_0 + \beta_1 x_h + \varepsilon_h, \qquad h = 1, \ldots, n.$$

Here x is a known value but β_0 and β_1 are unknown, uniquely defined mean parameters that we must estimate.

In Chapter 8 we introduce models with more complicated functions of a single predictor x. These include polynomial models. A third-degree *polynomial regression model* has

$$m(x) = \beta_0 + \beta_1 x + \beta_2 x^2 + \beta_3 x^3.$$

Again the βs are unknown, uniquely defined mean parameters and x is treated as fixed. If the relationship between x and $\mathrm{E}(y)$ is nonlinear, polynomials provide one method of modeling the nonlinear relationship.

In Section 6.9 we introduce, and in Chapters 9, 10, and 11 we consider in detail, models for measurement variables with a vector of predictors $x = (x_1, \ldots, x_p)'$. With $p = 3$, a typical *multiple regression model* incorporates

$$m(x) = \beta_0 + \beta_1 x_1 + \beta_2 x_2 + \beta_3 x_3.$$

When written for all n observations, the model becomes

$$y_h = \beta_0 + \beta_1 x_{h1} + \beta_2 x_{h2} + \beta_3 x_{h3} + \varepsilon_h. \tag{3.9.4}$$

Here the β_js are unknown parameters and the x_{hj} values are all treated as fixed.

The predictors x_{hj} used in (3.9.4) are necessarily numerical. Typically, they are either measurements of some sort or 0-1 indicators of group membership. Categorical variables do not have to be numerical (Sex, Race) but categories are often coded as numbers, e.g., Female = 1, Male = 2. It would be *inappropriate* to use a (non-binary) categorical variable taking numerical values in (3.9.4).

A categorical variable with, say, five categories should be incorporated into a multiple regression by incorporating four predictor variables that take on 0-1 values. More on this in Sections 6.8 and 12.3.

Chapter 4 deals with two-sample problems, so it deals with a single categorical predictor that only takes on two values. Suppose x takes on just the two values 1 and 2 for, say, females and males. Then our model for the mean of y reduces to the *two-sample model*

$$m(x) = \begin{cases} \mu_1 \equiv m(1), & \text{if } x = 1 \\ \mu_2 \equiv m(2), & \text{if } x = 2. \end{cases}$$

We have only two uniquely defined mean parameters to estimate: μ_1 and μ_2. This $m(\cdot)$ gives the model used in Section 4.2.

Unlike simple, polynomial, and multiple regression, there is no convenient way to write the specific two-sample model in the general form (3.9.1). Although the two-sample model clearly fits the general form, to deal with categorical variables it is convenient to play games with our subscripts. We replace the single subscript h that indicates all n of the observations in the data with a pair of subscripts: i that identifies the group and j that identifies observations within the group. If we have N_i observations in group i, the total number of observations n must equal $N_1 + N_2$. Now we can rewrite Model (3.9.1), when x is a two-group categorical predictor, as

$$y_{ij} = m(i) + \varepsilon_{ij}, \qquad i = 1, 2, \ j = 1, \dots, N_i.$$

Identifying $\mu_i \equiv m(i)$ gives

$$y_{ij} = \mu_i + \varepsilon_{ij}, \qquad i = 1, 2, \ j = 1, \dots, N_i.$$

A single categorical predictor variable with more than two groups works pretty much the same way. If there are a groups and the categorical predictor variable takes on the values $1, \dots, a$, the model has

$$m(x) = \begin{cases} \mu_1 \equiv m(1), & \text{if } x = 1 \\ \mu_2 \equiv m(2), & \text{if } x = 2 \\ \quad \vdots & \qquad \vdots \\ \mu_a \equiv m(a), & \text{if } x = a, \end{cases}$$

with a uniquely defined mean parameters to estimate. We can rewrite Model (3.9.1) when x is an a group categorical predictor as

$$y_{ij} = \mu_i + \varepsilon_{ij}, \qquad i = 1, \dots, a, \ j = 1, \dots, N_i.$$

These *one-way analysis of variance (ANOVA)* models are examined in Chapter 12.

It really does not matter what values the categorical predictor actually takes as long as there are only a distinct values. Thus, x can take on any a numbers or it can take on any a letter values or a symbols of any kind, as long as they constitute distinct group identifiers. If the category is sex, the values may be the words "male" and "female."

Sometimes group identifiers can simultaneously be meaningful measurement variables. In Chapter 12 we examine data on the strength of trusses built with metal plates of different lengths. The metal plates are 4, 6, 8, 10, or 12 inches long. There are 7 observations for each length of plate, so we create a predictor variable x with $n = 35$ that takes on these five numerical values. We now have two options. We can treat x as a categorical variable with five groups, or we can treat x as a measurement predictor variable and fit a linear or other polynomial regression model. We will see in Chapter 12 that fitting a polynomial of degree four (one less than the number of categories) is equivalent to treating the variable as a five-category predictor.

If we have two categorical predictors, say, x_1 a type of drug and x_2 a racial group, we have considerable variety in the models we can build. Suppose x_1 takes on the values $1, \dots, a$, x_2 takes on the values $1,\ldots,b$, and $x = (x_1, x_2)$. Perhaps the simplest two-category predictor model to state is the *interaction model*

$$m(x) = m(i, j) \equiv \mu_{ij}, \quad \text{if } x_1 = i \text{ and } x_2 = j,$$

with ab uniquely defined mean parameters. Using alternative subscripts we can write this model as

$$y_{ijk} = \mu_{ij} + \varepsilon_{ijk}, \quad i = 1,\ldots,a, \; j = 1,\ldots,b, \; k = 1,\ldots,N_{ij},$$

where N_{ij} is the number of observations that have both $x_1 = i$ and $x_2 = j$. For the interaction model, we could replace the two categorical variables having a and b groups, respectively, with a single categorical variable that takes on ab distinct categories.

Two categorical variables naturally allow some useful flexibility. We can write an *additive-effects model*, also called a *no-interaction model*, as

$$m(x) = m(i, j) \equiv \mu + \alpha_i + \eta_j, \quad \text{if } x_1 = i \text{ and } x_2 = j,$$

or

$$y_{ijk} = \mu + \alpha_i + \eta_j + \varepsilon_{ijk}, \quad i = 1,\ldots,a, \; j = 1,\ldots,b, \; k = 1,\ldots,N_{ij}.$$

Here there are $1 + a + b$ parameters, μ, the α_is, and the η_js. Two of the parameters (not just any two) are redundant, so there are $(1 + a + b) - 2$ functionally distinct parameters. Models with two categorical predictors are discussed in Chapter 14. Models with one categorical predictor and one continuous predictor are discussed in Chapter 15 along with instances when the continuous predictor can also be viewed as a categorical predictor. Models for three categorical predictors are discussed in Chapter 16.

Models based on two categorical predictors are called *two-factor ANOVA* models. A model based on two or more categorical predictors is called a *multifactor ANOVA* model. Models with three or more categorical predictors may also be called *higher-order ANOVAs*. An ANOVA model is considered *balanced* if the number of observations on each group or combination of groups is the same. For a one-way ANOVA that means $N_1 = \cdots = N_a \equiv N$ and for a two-factor ANOVA it means $N_{ij} \equiv N$ for all i and j. Computations for ANOVA models are much simpler when they are balanced.

Analysis of covariance (ACOVA or ANCOVA) consists of situations in which we have both types of predictors (measurement and categorical) in the same model. ACOVA is primarily introduced in Chapter 15. Some use of it is also made in Section 8.4. When group identifiers are simultaneously meaningful measurements, we have the option of performing ACOVA, multifactor ANOVA, or multiple regression, depending on whether we view the predictors as a mix of categorical and measurement, all categorical, or all measurement.

The models $m(\cdot)$ all involve some unknown parameters that we must estimate, although some of the parameters may be redundant. Call the number of nonredundant, i.e., functionally distinct, mean parameters r. Upon estimating the mean parameters, we get an estimated model $\hat{m}(\cdot)$. Applying this estimated model to the predictor variables in our data gives the *fitted values*, also called the *predicted values*,

$$\hat{y}_h \equiv \hat{m}(x_h), \quad h = 1,\ldots,n.$$

From the fitted values, we compute the *residuals*,

$$\hat{\varepsilon}_h \equiv y_h - \hat{y}_h, \quad h = 1,\ldots,n.$$

As discussed in Chapter 7, we use the residuals to check the assumptions made in (3.9.2).

We also use the residuals to estimate the unknown variance, σ^2, in (3.9.2). The *degrees of freedom for error* is defined as the number of observations minus the number of functionally distinct mean parameters, i.e.,

$$dfE = n - r.$$

The *sum of squares error* is defined as the sum of the squared residuals, i.e.,

$$SSE = \hat{\varepsilon}_1^2 + \cdots + \hat{\varepsilon}_n^2 = \sum_{h=1}^{n} \hat{\varepsilon}_h^2.$$

Finally, our estimate of the variance of an observation, σ^2, is the *mean squared error* defined by

$$MSE = \frac{SSE}{dfE}.$$

Two models are considered to be equivalent if they give exactly the same fitted values for any set of observations. In such cases the number of functionally distinct mean parameters will be the same, as will the residuals, SSE, dfE, and MSE.

It is possible to put all of the models we have discussed here in the form of a multiple regression model by properly selecting or constructing the predictor variables. Such models are called *linear models*. *Unless otherwise stated, we will assume that all of our measurement data models are linear models*. Linear models are "linear" in the parameters, not the predictor variables x. For example, polynomial regression models are linear in the regression parameters β_j but they are not linear in the predictor variable x. The models used for count data at the end of the book differ somewhat from the linear models for continuous measurements but all use similar mean structures $m(x)$ that allow us to exploit the tools developed in earlier chapters.

A valuable measure of the predictive ability of a model is R^2, the squared sample correlation coefficient between the pairs $(\hat{y}_h, y_h)$, cf. Section 6.7. Values near 0 indicate little predictive ability while values near 1 indicate great predictive ability. (Actually, it is *possible* to get a high R^2 with lousy predictions but it is then easy to turn those lousy predictions into very good predictions.) R^2 measures predictive ability, not the correctness of the model. Incorrect models can be very good predictors and have very high R^2s while perfect models can be poor predictors and have very low R^2s. Models with more parameters in them tend to have higher values of R^2 because the larger models can do a better job of approximating the y_h values in the fitted data. Unfortunately, this can happen when the bigger models actually do a worse job of predicting y values that are outside the fitted data.

On occasion, better to satisfy the assumptions (3.9.2), we might transform the original data y into y_*, for example $y_* = \log(y)$, cf. Section 7.3. If we then fit a model $y_{*h} = m(x_h) + \varepsilon_h$ and get fitted values $\hat{y}_{*h}$, these can be back transformed to the original scale giving $\hat{y}_h$, say, $\hat{y}_h = e^{\hat{y}_{*h}}$. R^2 values computed in this way between $(\hat{y}_h, y_h)$ are comparable regardless of any transformations involved.

Appendix: derivation of confidence intervals

We wish to establish the equivalence of the inequalities

$$-t\left(1 - \frac{\alpha}{2}, df\right) < \frac{Est - Par}{SE(Est)} < t\left(1 - \frac{\alpha}{2}, df\right)$$

and

$$Est - t\left(1 - \frac{\alpha}{2}, df\right) SE(Est) < Par < Est + t\left(1 - \frac{\alpha}{2}, df\right) SE(Est).$$

We do this by establishing a series of equivalences. The justifications for the equivalences are given at the end:

$$-t\left(1 - \frac{\alpha}{2}, df\right) < \frac{Est - Par}{SE(Est)} < t\left(1 - \frac{\alpha}{2}, df\right) \tag{1}$$

if and only if

$$-t\left(1 - \frac{\alpha}{2}, df\right) SE(Est) < Est - Par < t\left(1 - \frac{\alpha}{2}, df\right) SE(Est) \tag{2}$$

if and only if

$$t\left(1 - \frac{\alpha}{2}, df\right) SE(Est) > -Est + Par > -t\left(1 - \frac{\alpha}{2}, df\right) SE(Est) \tag{3}$$

if and only if

$$Est + t\left(1 - \frac{\alpha}{2}, df\right) SE(Est) > Par > Est - t\left(1 - \frac{\alpha}{2}, df\right) SE(Est) \tag{4}$$

if and only if

$$Est - t\left(1 - \frac{\alpha}{2}, df\right) SE(Est) < Par < Est + t\left(1 - \frac{\alpha}{2}, df\right) SE(Est). \tag{5}$$

JUSTIFICATION OF STEPS.

For (1) iff (2):   if $c > 0$, then $a < b$ if and only if $ac < bc$.
For (2) iff (3): $a < b$ if and only if $-a > -b$.
For (3) iff (4): $a < b$ if and only if $a + c < b + c$.
For (4) iff (5): $a > b$ if and only if $b < a$.

3.10   Exercises

EXERCISE 3.10.1.   Identify the parameter, estimate, standard error of the estimate, and reference distribution for Exercise 2.8.1.

EXERCISE 3.10.2.   Identify the parameter, estimate, standard error of the estimate, and reference distribution for Exercise 2.8.2.

EXERCISE 3.10.3.   Identify the parameter, estimate, standard error of the estimate, and reference distribution for Exercise 2.8.4.

EXERCISE 3.10.4.   Consider that I am collecting (normally distributed) data with a variance of 4 and I want to test a null hypothesis of $H_0 : \mu = 10$. What sample size should I take according to Cox's rule if I want a reasonable chance of rejecting H_0 when $\mu = 13$? What if I want a reasonable chance of rejecting H_0 when $\mu = 12$? What sample size should I take if I want a 95% confidence interval that is no more than 2 units long? What if I want a 99% confidence interval that is no more than 2 units long?

EXERCISE 3.10.5.   The turtle shell data of Jolicoeur and Mosimann (1960) given in Exercise 2.7.4 has a standard deviation of about 21.25. If we were to collect a new sample, how large should the sample size be in order to have a 95% confidence interval with a length of (about) four units? According to Cox's rule, what sample size should I take if I want a reasonable chance of rejecting $H_0 : \mu = 130$ when $\mu = 140$?

EXERCISE 3.10.6.   With reference to Exercise 2.8.3, give the approximate number of observations necessary to estimate the mean of BX to within 0.01 units with 99% confidence. How large a sample is needed to get a reasonable test of $H_0 : \mu = 10$ when $\mu = 11$ using Cox's rule?

EXERCISE 3.10.7.   With reference to Exercise 2.8.3, give the approximate number of observations necessary to get a 99% confidence for the mean of K that has a length of 60. How large a sample is needed to get a reasonable test of $H_0 : \mu = 1200$ when $\mu = 1190$ using Cox's rule? What is the number when $\mu = 1150$?

EXERCISE 3.10.8.   With reference to Exercise 2.8.3, give the approximate number of observations necessary to estimate the mean of FORM to within 0.5 units with 95% confidence. How large a sample is needed to get a reasonable test of $H_0 : \mu = 20$ when $\mu = 20.2$ using Cox's rule?

EXERCISE 3.10.9.   With reference to Exercise 2.8.2, give the approximate number of observations necessary to estimate the mean rat weight to within 1 unit with 95% confidence. How large a sample is needed to get a reasonable test of $H_0 : \mu = 55$ when $\mu = 54$ using Cox's rule?

EXERCISE 3.10.10.   Suppose we have three random variables y, y_1, and y_2 and let α be a number between 0 and 1. Show that if $y = \alpha y_1 + (1 - \alpha) y_2$ and if $E(y) = E(y_2) = \theta$ then $E(y_1) = \theta$.

EXERCISE 3.10.11.   Given that $y_1, \ldots, y_n$ are independent with $E(y_i) = \mu_i$ and $\sigma^2 = \mathrm{Var}(y_i) = E[y_i - \mu_i]^2$, give intuitive justifications for why both $\hat{\sigma}^2 \equiv \sum_{i=1}^{n}(y_i - \hat{y}_i)^2/n$ and $MSE \equiv \sum_{i=1}^{n}(y_i - \hat{y}_i)^2/dfE$ are reasonable estimates of σ^2. Recall that $\hat{y}_i$ is an estimate of μ_i.

Chapter 4

Two Samples

In this chapter we consider several situations where it is of interest to compare two samples. First we consider two samples of correlated data. These are data that consist of pairs of observations measuring comparable quantities. Next we consider two independent samples from populations with the same variance. This data form is generalized to several independent samples with a common variance in Chapter 12, a problem that is known as *analysis of variance* or more commonly as ANOVA. We then examine two independent samples from populations with different variances. Finally we consider the problem of testing whether the variances of two populations are equal.

4.1 Two correlated samples: Paired comparisons

Paired comparisons involve pairs of observations on similar variables. Often these are two observations taken on the same object under different circumstances or two observations taken on related objects. No new statistical methods are needed for analyzing such data.

EXAMPLE 4.1.1. Shewhart (1931, p. 324) presents data on the hardness of an item produced by welding two parts together. Table 4.1 gives the hardness measurements for each of the two parts. The hardness of part 1 is denoted y_1 and the hardness of part 2 is denoted y_2. For $i = 1, 2$, the data for part i are denoted y_{ij}, $j = 1, \ldots, 27$. These data are actually a subset of the data presented by Shewhart.

We are interested in the difference between μ_1, the population mean for part one, and μ_2, the population mean for part two. In other words, the parameter of interest is $Par = \mu_1 - \mu_2$. Note that if there is no difference between the population means, $\mu_1 - \mu_2 = 0$. The natural estimate of this parameter is the difference between the sample means, i.e., $Est = \bar{y}_1. - \bar{y}_2.$. Here we use the bar

Table 4.1: *Shewhart's hardness data.*

Case	y_1	y_2	$d = y_1 - y_2$	Case	y_1	y_2	$d = y_1 - y_2$
1	50.9	44.3	6.6	15	46.6	31.5	15.1
2	44.8	25.7	19.1	16	50.4	38.1	12.3
3	51.6	39.5	12.1	17	45.9	35.2	10.7
4	43.8	19.3	24.5	18	47.3	33.4	13.9
5	49.0	43.2	5.8	19	46.6	30.7	15.9
6	45.4	26.9	18.5	20	47.3	36.8	10.5
7	44.9	34.5	10.4	21	48.7	36.8	11.9
8	49.0	37.4	11.6	22	44.9	36.7	8.2
9	53.4	38.1	15.3	23	46.8	37.1	9.7
10	48.5	33.0	15.5	24	49.6	37.8	11.8
11	46.0	32.6	13.4	25	51.4	33.5	17.9
12	49.0	35.4	13.6	26	45.8	37.5	8.3
13	43.4	36.2	7.2	27	48.5	38.3	10.2
14	44.4	32.5	11.9				

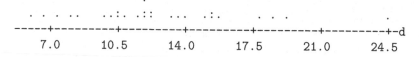

Figure 4.1: *Dot plot of differences.*

and the dot ($\cdot$) in place of the second subscript to indicate averaging over the second subscript, i.e.,
$\bar{y}_{i\cdot} = (y_{i1} + \cdots + y_{i27})/27$.

To perform parametric statistical inferences, we need the standard error of the estimate, i.e.,
$SE(\bar{y}_{1\cdot} - \bar{y}_{2\cdot})$. As indicated earlier, finding an appropriate standard error is often the most difficult aspect of statistical inference. In problems such as this, where the data are paired, finding the standard error is complicated by the fact that the two observations in each pair are not independent. In data such as these, *different pairs are often independent but observations within a pair are not*.

In paired comparisons, we use a trick to reduce the problem to consideration of only one sample. It is a simple algebraic fact that the difference of the sample means, $\bar{y}_{1\cdot} - \bar{y}_{2\cdot}$ is the same as the sample mean of the differences $d_j = y_{1j} - y_{2j}$, i.e., $\bar{d} = \bar{y}_{1\cdot} - \bar{y}_{2\cdot}$. Thus $\bar{d}$ is an estimate of the parameter of interest $\mu_1 - \mu_2$. The differences are given in Table 4.1 along with the data. Summary statistics are listed below for each variable and the differences. Note that for the hardness data, $\bar{d} = 12.663 = 47.552 - 34.889 = \bar{y}_{1\cdot} - \bar{y}_{2\cdot}$. In particular, if the positive value for $\bar{d}$ means anything (other than random variation), it indicates that part one is harder than part two.

Sample statistics				
Variable	N_i	Mean	Variance	Std. dev.
y_1	27	47.552	6.79028	2.606
y_2	27	34.889	26.51641	5.149
$d = y_1 - y_2$	27	12.663	17.77165	4.216

Given that $\bar{d}$ is an estimate of $\mu_1 - \mu_2$, we can base the entire analysis on the differences. The differences constitute a single sample of data, so the standard error of $\bar{d}$ is simply the usual one-sample standard error,

$$SE(\bar{d}) = s_d/\sqrt{27},$$

where s_d is the sample standard deviation as computed from the 27 differences. The differences are plotted in Figure 4.1. Note that there is one potential outlier. We leave it as an exercise to reanalyze the data with the possible outlier removed.

We now have *Par*, *Est*, and SE(*Est*); it remains to find the appropriate distribution. Figure 4.2 gives a normal plot for the differences. While there is an upward curve at the top due to the possible outlier, the curve is otherwise reasonably straight. The Wilk–Francia statistic of $W' = 0.955$ is above the fifth percentile of the null distribution. With normal data we use the reference distribution

$$\frac{\bar{d} - (\mu_1 - \mu_2)}{s_d/\sqrt{27}} \sim t(27 - 1)$$

and we are now in a position to perform parametric statistical inferences.

Our observed values of the mean and standard error are $\bar{d} = 12.663$ and $SE(\bar{d}) = 4.216/\sqrt{27} = 0.811$. From a $t(26)$ distribution, we find $t(0.995, 26) = 2.78$. A 99% confidence interval for the difference in hardness has endpoints

$$12.663 \pm 2.78(0.811),$$

which gives an interval of, roughly, $(10.41, 14.92)$. Based on a .01 level test, the data and the model are consistent with the population mean hardness for part 1 being between 10.41 and 14.92 units harder than that for part 2.

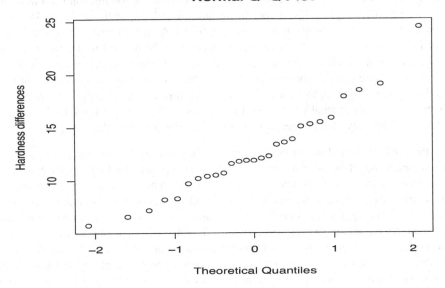

Figure 4.2: *Normal plot of differences, $W' = 0.955$.*

We can also get a 99% prediction interval for the difference in hardness to be observed on a new welded piece. The prediction interval has endpoints of

$$12.663 \pm 2.78 \sqrt{4.216^2 + 0.811^2}$$

for an interval of $(0.73, 24.60)$.

To test the hypothesis that the two parts have the same hardness, we set up the hypothesis $H_0 : \mu_1 = \mu_2$, or equivalently, $H_0 : \mu_1 - \mu_2 = 0$. The test statistic is

$$\frac{12.663 - 0}{0.811} = 15.61.$$

This is far from zero, so the data are inconsistent with the null model. Thus, if the other assumptions in the model are true, there is strong evidence that the hardness of part 1 is different than the hardness of part 2. Since the test statistic is positive, we conclude that $\mu_1 - \mu_2 > 0$ and that part 1 is harder than part 2. Note that this is consistent with our 99% confidence interval $(10.41, 14.92)$, which contains only positive values for $\mu_1 - \mu_2$.

Inferences and predictions for an individual population are made ignoring the other population, i.e., they are made using methods for one sample. For example, using the sample statistics for y_1 gives a 99% confidence interval for μ_1, the population mean hardness for part 1, with endpoints

$$47.552 \pm 2.78 \sqrt{\frac{6.79028}{27}}$$

and a 99% prediction interval for the hardness of part 1 in a new piece has endpoints

$$47.552 \pm 2.78 \sqrt{6.79028 + \frac{6.79028}{27}}$$

and interval $(40.175, 59.929)$. Of course, the use of the $t(26)$ distribution requires that we validate the assumption that the observations on part 1 are a random sample from a normal distribution.

When finding a prediction interval for y_1, we can typically improve the interval if we know the corresponding value of y_2. As we saw earlier, the 99% prediction interval for a new difference $d = y_1 - y_2$ has $0.73 < y_1 - y_2 < 24.60$. If we happen to know that, say, $y_2 = 35$, the interval becomes $0.73 < y_1 - 35 < 24.60$ or $35.73 < y_1 < 59.60$. As it turns out, with these data the new 99% prediction interval for y_1 is not an improvement over the interval in the previous paragraph. The new interval is noticeably wider. However, these data are somewhat atypical. Typically in paired data, the two measurements are highly correlated, so that the sample variances of the differences is substantially less than the sample variance of the individual measurements. In such situations, the new interval will be substantially narrower. In these data, the sample variance for the differences is 17.77165 and is actually much larger than the sample variance of 6.79028 for y_1.   □

The trick of looking at differences between pairs is necessary because the two observations in a pair are not independent. While different pairs of welded parts are assumed to behave independently, it seems unreasonable to *assume* that two hardness measurements on a single item that has been welded together would behave independently. This lack of independence makes it difficult to find a standard error for comparing the sample means unless we look at the differences. In the remainder of this chapter, we consider two-sample problems in which all of the observations are assumed to be independent. The observations in each sample are independent of each other and independent of all the observations in the other sample. Paired comparison problems almost fit those assumptions but they break down at one key point. In a paired comparison, we assume that every observation is independent of the other observations in the same sample and that each observation is independent of all the observations in the other sample *except* for the observation in the other sample that it is paired with. When analyzing two samples, if we can find any reason to identify individuals as being part of a pair, that fact is sufficient to make us treat the data as a paired comparison.

Since paired comparisons reduce to one-sample procedures, the model-based procedures of Chapter 2 apply.

The method of paired comparisons is also the name of a totally different statistical procedure. Suppose one wishes to compare five brands of chocolate chip cookies: A, B, C, D, E. It would be difficult to taste all five and order them appropriately. As an alternative, one can taste test pairs of cookies, e.g., $(A,B), (A,C), (A,D), (A,E), (B,C), (B,D)$, etc. and identify the better of the two. The benefit of this procedure is that it is much easier to rate two cookies than to rate five. See David (1988) for a survey and discussion of procedures developed to analyze such data.

4.2 Two independent samples with equal variances

The most commonly used two-sample technique consists of comparing independent samples from two populations with the same variance. The sample sizes for the two groups are possibly different, say, N_1 and N_2, and we write the common variance as σ^2.

EXAMPLE 4.2.1.   The data in Table 4.2 are final point totals for an introductory Statistics class. The data are divided by the sex of the student. We investigate whether the data display sex differences. The data are plotted in Figure 4.3. Figures 4.4 and 4.5 contain normal plots for the two sets of data. Figure 4.4 is quite straight but Figure 4.5 looks curved. Our analysis is not particularly sensitive to nonnormality and the W' statistic for Figure 4.5 is 0.937, which is well above the fifth percentile, so we proceed under the assumption that both samples are normal. We also assume that all of the observations are independent. This assumption may be questionable because some students probably studied together; nonetheless, independence seems like a reasonable working assumption.
                                                                                           □

The methods in this section rely on the assumption that the two populations are normally distributed and have the same variance. In particular, we assume two independent samples

Sample	Data		Distribution
1	$y_{11}, y_{12}, \ldots, y_{1N_1}$	iid	$N(\mu_1, \sigma^2)$
2	$y_{21}, y_{22}, \ldots, y_{2N_2}$	iid	$N(\mu_2, \sigma^2)$

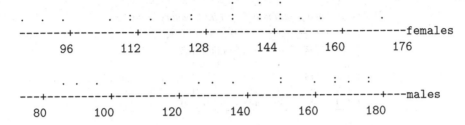

Figure 4.3: *Dot plots for final point totals.*

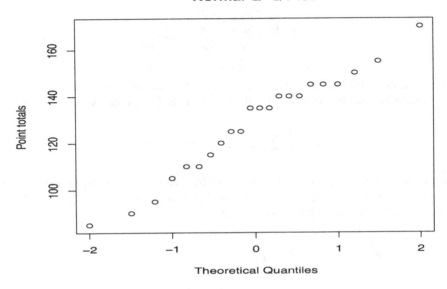

Figure 4.4: *Normal plot for females,* $W' = 0.974$.

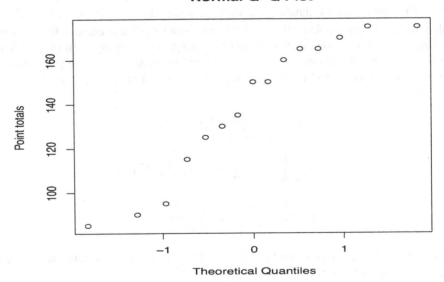

Figure 4.5: *Normal plot for males,* $W' = 0.937$.

Table 4.2: *Final point totals for an introductory Statistics class.*

Females					Males		
140	125	90	105	145	165	175	135
135	155	170	140	85	175	160	165
150	115	125	95		170	115	150
135	145	110	135		150	85	130
110	120	140	145		90	95	125

and compute summary statistics from the samples. The summary statistics are just the sample mean and the sample variance for each individual sample.

Sample statistics

Sample	Size	Mean	Variance
1	N_1	$\bar{y}_{1\cdot}$	s_1^2
2	N_2	$\bar{y}_{2\cdot}$	s_2^2

Except for checking the validity of our assumptions, these summary statistics are more than sufficient for the entire analysis. Algebraically, the sample mean for population $i, i = 1, 2$, is

$$\bar{y}_{i\cdot} \equiv \frac{1}{N_i} \sum_{j=1}^{N_i} y_{ij} = \frac{1}{N_i} [y_{i1} + y_{i2} + \cdots + y_{iN_i}]$$

where the $\cdot$ in $\bar{y}_{i\cdot}$ indicates that the mean is obtained by averaging over j, the second subscript in the y_{ij}s. The sample means, $\bar{y}_{1\cdot}$ and $\bar{y}_{2\cdot}$, are estimates of μ_1 and μ_2.

The sample variance for population $i, i = 1, 2$, is

$$
\begin{aligned}
s_i^2 &= \frac{1}{N_i - 1} \sum_{j=1}^{N_i} (y_{ij} - \bar{y}_{i\cdot})^2 \\
&= \frac{1}{N_i - 1} \left[(y_{i1} - \bar{y}_{i\cdot})^2 + (y_{i2} - \bar{y}_{i\cdot})^2 + \cdots + (y_{iN_i} - \bar{y}_{i\cdot})^2 \right].
\end{aligned}
$$

The s_i^2s both estimate σ^2. Combining the s_i^2s can yield a better estimate of σ^2 than either individual estimate. We form a pooled estimate of the variance, say s_p^2, by averaging s_1^2 and s_2^2. With unequal sample sizes an efficient pooled estimate of σ^2 must be a weighted average of the s_i^2s. Obviously, if we have $N_1 = 100,000$ observations in the first sample and only $N_2 = 10$ observations in the second sample, the variance estimate s_1^2 is much better than s_2^2 and we want to give it more weight. The weights are the degrees of freedom associated with the estimates. The pooled estimate of the variance is

$$
\begin{aligned}
s_p^2 &\equiv \frac{(N_1 - 1)s_1^2 + (N_2 - 1)s_2^2}{(N_1 - 1) + (N_2 - 1)} \\
&= \frac{1}{N_1 + N_2 - 2} \left[\sum_{j=1}^{N_1} (\bar{y}_{1j} - \bar{y}_{1\cdot})^2 + \sum_{j=1}^{N_2} (\bar{y}_{2j} - \bar{y}_{2\cdot})^2 \right] \\
&= \frac{1}{N_1 + N_2 - 2} \sum_{i=1}^{2} \sum_{j=1}^{N_i} (\bar{y}_{ij} - \bar{y}_{i\cdot})^2.
\end{aligned}
$$

The degrees of freedom for s_p^2 are $N_1 + N_2 - 2 = (N_1 - 1) + (N_2 - 1)$, i.e., the sum of the degrees of freedom for the individual estimates s_i^2.

EXAMPLE 4.2.2.   For the data on final point totals, the sample statistics follow.

Sample Statistics

Sample	N_i	$\bar{y}_{i\cdot}$	s_i^2	s_i
females	22	127.954545	487.2835498	22.07
males	15	139.000000	979.2857143	31.29

From these values, we obtain the pooled estimate of the variance,

$$s_p^2 = \frac{(N_1-1)s_1^2 + (N_2-1)s_2^2}{N_1+N_2-2} = \frac{(21)487.28 + (14)979.29}{35} = 684.08. \qquad \square$$

We are now in a position to draw statistical inferences about the μ_is. The main problem in obtaining tests and confidence intervals is in finding appropriate standard errors. The crucial fact is that the samples are independent so that the $\bar{y}_{i\cdot}$s are independent.

For inferences about the difference between the two means, say, $\mu_1 - \mu_2$, use the general procedure of Chapter 3 with

$$Par = \mu_1 - \mu_2$$

and

$$Est = \bar{y}_{1\cdot} - \bar{y}_{2\cdot}.$$

Note that $\bar{y}_{1\cdot} - \bar{y}_{2\cdot}$ is unbiased for estimating $\mu_1 - \mu_2$ because

$$E(\bar{y}_{1\cdot} - \bar{y}_{2\cdot}) = E(\bar{y}_{1\cdot}) - E(\bar{y}_{2\cdot}) = \mu_1 - \mu_2.$$

The two means are independent, so the variance of $\bar{y}_{1\cdot} - \bar{y}_{2\cdot}$ is the variance of $\bar{y}_{1\cdot}$ plus the variance of $\bar{y}_{2\cdot}$, i.e.,

$$\text{Var}(\bar{y}_{1\cdot} - \bar{y}_{2\cdot}) = \text{Var}(\bar{y}_{1\cdot}) + \text{Var}(\bar{y}_{2\cdot}) = \frac{\sigma^2}{N_1} + \frac{\sigma^2}{N_2} = \sigma^2\left[\frac{1}{N_1} + \frac{1}{N_2}\right].$$

The standard error of $\bar{y}_{1\cdot} - \bar{y}_{2\cdot}$ is the estimated standard deviation of $\bar{y}_{1\cdot} - \bar{y}_{2\cdot}$,

$$SE(\bar{y}_{1\cdot} - \bar{y}_{2\cdot}) = \sqrt{s_p^2\left[\frac{1}{N_1} + \frac{1}{N_2}\right]}.$$

Under our assumption that the original data are normal, the reference distribution is

$$\frac{(\bar{y}_{1\cdot} - \bar{y}_{2\cdot}) - (\mu_1 - \mu_2)}{\sqrt{s_p^2\left[\frac{1}{N_1} + \frac{1}{N_2}\right]}} \sim t(N_1+N_2-2).$$

The degrees of freedom for the t distribution are the degrees of freedom for s_p^2.

Having identified the parameter, estimate, standard error, and distribution, inferences follow the usual pattern. A 95% confidence interval for $\mu_1 - \mu_2$ is

$$(\bar{y}_{1\cdot} - \bar{y}_{2\cdot}) \pm t(0.975, N_1+N_2-2)\sqrt{s_p^2\left[\frac{1}{N_1} + \frac{1}{N_2}\right]}.$$

A test of hypothesis that the means are equal, $H_0 : \mu_1 = \mu_2$, can be converted into the equivalent hypothesis involving $Par = \mu_1 - \mu_2$, namely $H_0 : \mu_1 - \mu_2 = 0$. The test is handled in the usual way. An $\alpha = 0.01$ test rejects H_0 if

$$\frac{|(\bar{y}_{1\cdot} - \bar{y}_{2\cdot}) - 0|}{\sqrt{s_p^2\left[\frac{1}{N_1} + \frac{1}{N_2}\right]}} > t(0.995, N_1+N_2-2).$$

As discussed in Chapter 3, what we are really doing is testing the validity of the null model that incorporates all of the assumptions, including the assumption of the null hypothesis $H_0 : \mu_1 - \mu_2 = 0$. Fisher (1925) quite rightly argues that the appropriate test is often a test of whether the two samples come from the same normal population, rather than a test of whether the means are equal given that the variances are (or are not) equal.

In our discussion of comparing differences, we have defined the parameter as $\mu_1 - \mu_2$. We could just as well have defined the parameter as $\mu_2 - \mu_1$. This would have given an entirely equivalent analysis.

Inferences about a single mean, say, μ_2, use the general procedures with $Par = \mu_2$ and $Est = \bar{y}_2.$. The variance of $\bar{y}_2.$ is σ^2/N_2, so $SE(\bar{y}_2.) = \sqrt{s_p^2/N_2}$. Note the use of s_p^2 rather than s_2^2. The reference distribution is $[\bar{y}_2. - \mu_2]/SE(\bar{y}_2.) \sim t(N_1 + N_2 - 2)$. A 95% confidence interval for μ_2 is

$$\bar{y}_2. \pm t(0.975, N_1 + N_2 - 2)\sqrt{s_p^2/N_2}.$$

A 95% prediction interval for a new observation on variable y_2 is

$$\bar{y}_2. \pm t(0.975, N_1 + N_2 - 2)\sqrt{s_p^2 + \frac{s_p^2}{N_2}}.$$

An $\alpha = 0.01$ test of the hypothesis, say $H_0 : \mu_2 = 5$, rejects H_0 if

$$\frac{|\bar{y}_2. - 5|}{\sqrt{s_p^2/N_2}} > t(0.995, N_1 + N_2 - 2).$$

EXAMPLE 4.2.3.   For comparing females and males on final point totals, the parameter of interest is

$$Par = \mu_1 - \mu_2$$

where μ_1 indicates the population mean final point total for females and μ_2 indicates the population mean final point total for males. The estimate of the parameter is

$$Est = \bar{y}_1. - \bar{y}_2. = 127.95 - 139.00 = -11.05.$$

The pooled estimate of the variance is $s_p^2 = 684.08$, so the standard error is

$$SE(\bar{y}_1. - \bar{y}_2.) = \sqrt{s_p^2\left(\frac{1}{N_1} + \frac{1}{N_2}\right)} = \sqrt{684.08\left(\frac{1}{22} + \frac{1}{15}\right)} = 8.7578.$$

The data have reasonably normal distributions and the variances are not too different (more on this later), so the reference distribution is taken as

$$\frac{(\bar{y}_1. - \bar{y}_2.) - (\mu_1 - \mu_2)}{\sqrt{s_p^2\left(\frac{1}{22} + \frac{1}{15}\right)}} \sim t(35)$$

where $35 = N_1 + N_2 - 2$. The tabled value for finding 95% confidence intervals and $\alpha = 0.05$ tests is

$$t(0.975, 35) = 2.030.$$

A 95% confidence interval for $\mu_1 - \mu_2$ has endpoints

$$-11.05 \pm (2.030)8.7578$$

which yields an interval $(-28.8, 6.7)$. Population mean scores between, roughly, 29 points *less* for females and 7 points *more* for females are consistent with the data and the model based on a 0.05 test.

An $\alpha = 0.05$ test of $H_0 : \mu_1 - \mu_2 = 0$ is not rejected because 0, the hypothesized value of $\mu_1 - \mu_2$, is contained in the 95% confidence interval for $\mu_1 - \mu_2$. The P value for the test is based on the observed value of the test statistic

$$t_{obs} = \frac{(\bar{y}_{1.} - \bar{y}_{2.}) - 0}{\sqrt{s_p^2 \left(\frac{1}{22} + \frac{1}{15}\right)}} = \frac{-11.05 - 0}{8.7578} = -1.26.$$

The probability of obtaining an observation from a $t(35)$ distribution that is as extreme or more extreme than $|-1.26|$ is 0.216. There is very little evidence that the population mean final point total for females is different (smaller) than the population mean final point total for males. The P value is greater than 0.2, so, as we established earlier, neither an $\alpha = 0.05$ nor an $\alpha = 0.01$ test is rejected. If we were silly enough to do an $\alpha = 0.25$ test, we would then reject the null hypothesis.

A 95% confidence interval for μ_1, the mean of the females, has endpoints

$$127.95 \pm (2.030)\sqrt{684.08/22},$$

which gives the interval $(116.6, 139.3)$. A mean final point total for females between 117 and 139 is consistent with the data and the model. A 95% prediction interval for a new observation on a female has endpoints

$$127.95 \pm (2.030)\sqrt{684.08 + \frac{684.08}{22}},$$

which gives the interval $(73.7, 182.2)$. A new observation on a female between 74 and 182 is consistent with the data and the model. This assumes that the new observation is randomly sampled from the same population as the previous data.

A test of the assumption of equal variances is left for the final section but we will see in the next section that the results for these data do not depend substantially on the equality of the variances.

□

4.2.1 Model testing

The full model has been described in great detail: one need only make the identification $MSE(Full) = s_p^2, dfE(Full) = N_1 + N_2 - 2$, and $SSE(Full) = (N_1 + N_2 - 2)s_p^2$.

The usual reduced model has $\mu_1 = \mu_2$. In that case, *all* of the observations are independent with the same mean and variance. Denote the sample variance computed from all $N_1 + N_2$ observations as s_y^2. It has $N_1 + N_2 - 1$ degrees of freedom. Identify $MSE(Red.) = s_y^2, dfE(Red.) = N_1 + N_2 - 1$, and $SSE(Red.) = (N_1 + N_2 - 1)s_y^2$.

EXAMPLE 4.2.4.   Using earlier computations,

$$MSE(Full) = 684.08, \quad dfE(Full) = 22 + 15 - 2 = 35,$$

$$SSE(Full) = (35)684.08 = 23942.8.$$

Computing the sample variance while treating all observations as one sample gives

$$MSE(Red.) = 695.31, \quad dfE(Red.) = 22 + 15 - 1 = 36,$$

$$SSE(Red.) = (36)695.31 = 25031.16.$$

It follows that

$$MSTest = \frac{25031.16 - 23942.8}{36 - 35} = 1088.36$$

and the F statistic is

$$F = \frac{1088.36}{684.08} = 1.59 = (-1.26)^2.$$

Note that the F statistic is simply the square of the corresponding t statistic for testing $H_0 : \mu_1 = \mu_2$.

□

Algebraic formulae are available for $SSE(Full)$ and $SSE(Red.)$:

$$SSE(Full) = \sum_{i=1}^{2} \sum_{j=1}^{N_i} (y_{ij} - \bar{y}_{i\cdot})^2,$$

and

$$SSE(Red.) = \sum_{i=1}^{2} \sum_{j=1}^{N_i} (y_{ij} - \bar{y}_{\cdot\cdot})^2,$$

where

$$\bar{y}_{i\cdot} = \frac{1}{N_i} \sum_{j=1}^{N_i} y_{ij} \quad \text{and} \quad \bar{y}_{\cdot\cdot} = \frac{1}{N_1 + N_2} \sum_{i=1}^{2} \sum_{j=1}^{N_i} y_{ij}.$$

Some additional algebra establishes that for this two-sample problem

$$SSE(Red.) - SSE(Full) = \sum_{i=1}^{2} N_i (\bar{y}_{i\cdot} - \bar{y}_{\cdot\cdot})^2 = \frac{N_1 N_2}{N_1 + N_2} (\bar{y}_{1\cdot} - \bar{y}_{2\cdot})^2 = \frac{(\bar{y}_{1\cdot} - \bar{y}_{2\cdot})^2}{\frac{1}{N_1} + \frac{1}{N_2}}.$$

From the last relationship, it is easy to see that the model-based F statistic is simply the square of the parameter-based t statistic.

4.3 Two independent samples with unequal variances

We now consider two independent samples with unequal variances σ_1^2 and σ_2^2. In this section we examine inferences about the means of the two populations. While inferences about means can be valuable, great care is required when drawing practical conclusions about populations with unequal variances. For example, if you want to produce gasoline with an octane of at least 87, you may have a choice between two processes. One process y_1 gives octanes distributed as $N(89,4)$ and the other y_2 gives $N(90,4)$. The two processes have the same variance, so the process with the higher mean gives more gas with an octane of at least 87. On the other hand, if y_1 gives $N(89,4)$ and y_2 gives $N(90,16)$, the y_1 process with mean 89 has a higher probability (0.84) of achieving an octane of 87 than the y_2 process with mean 90 (probability 0.77); see Figure 4.6 and Exercise 4.5.10. This is a direct result of the y_2 process having more variability.

   We have illustrated that for two normal distributions with different variances, the difference in the means may not be a very interesting parameter. More generally, anytime the distributions of the raw data have different shapes for the two groups, statements about the difference in the means may be uninteresting. Nonetheless, having given this warning that mean differences may not be the thing to look at, we proceed with our discussion on drawing statistical inferences for the means of two groups. The first thing to note in doing this is that our model testing procedures in Section 3.1 assumed that all the data had the same variance, so they do not apply.

EXAMPLE 4.3.1.   Jolicoeur and Mosimann (1960) present data on the sizes of turtle shells (carapaces). Table 4.3 presents data on the shell heights for 24 females and 24 males. These data are not paired; it is simply a caprice that 24 carapaces were measured for each sex. Our interest centers on estimating the population means for female and male heights, estimating the difference between the heights, and testing whether the difference is zero.

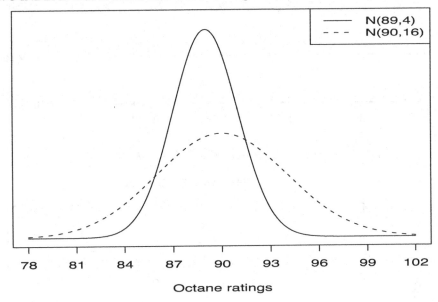

Figure 4.6: *Octane ratings*.

Table 4.3: *Turtle shell heights*.

Female				Male			
51	38	63	46	39	42	37	43
51	38	60	51	39	45	35	41
53	42	62	51	38	45	35	41
57	42	63	51	40	45	39	41
55	44	61	48	40	46	38	40
56	50	67	49	40	47	37	44

Following Christensen (2001) and others, we take natural logarithms of the data, i.e.,

$$y_1 = \log(\text{female height}) \qquad y_2 = \log(\text{male height}).$$

(*All logarithms in this book are natural logarithms.*) The log data are plotted in Figure 4.7. The female heights give the impression of being both larger and more spread out. Figures 4.8 and 4.9 contain normal plots for the females and males, respectively. Neither is exceptionally straight but they do not seem too bad. Summary statistics follow; they are consistent with the visual impressions given by Figure 4.7. The summary statistics will be used later to illustrate our statistical inferences.

Group	Size	Mean	Variance	Standard deviation
Females	24	3.9403	0.02493979	0.1579
Males	24	3.7032	0.00677276	0.0823

In general we assume two independent samples

Sample	Data		Distribution
1	$y_{11}, y_{12}, \ldots, y_{1N_1}$	iid	$N(\mu_1, \sigma_1^2)$
2	$y_{21}, y_{22}, \ldots, y_{2N_2}$	iid	$N(\mu_2, \sigma_2^2)$

and compute summary statistics from the samples.

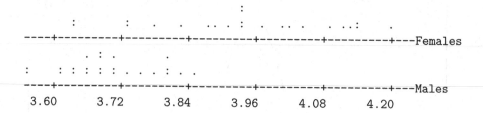

Figure 4.7: *Dot plots of turtle shell log heights.*

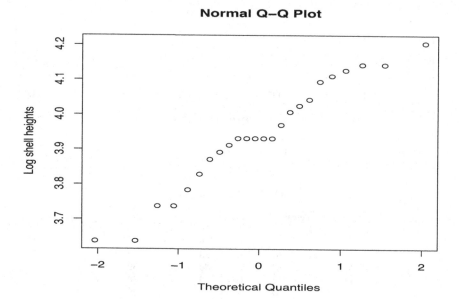

Figure 4.8: *Normal plot for female turtle shell log heights.*

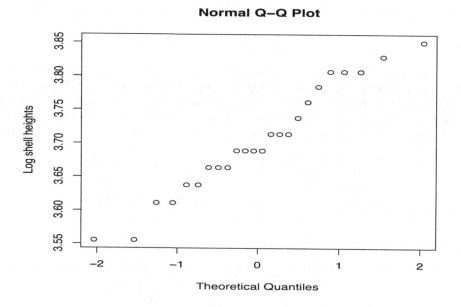

Figure 4.9: *Normal plot for male turtle shell log heights.*

Sample	Size	Mean	Variance
1	N_1	$\bar{y}_{1.}$	s_1^2
2	N_2	$\bar{y}_{2.}$	s_2^2

Again, the sample means, $\bar{y}_{1.}$ and $\bar{y}_{2.}$, are estimates of μ_1 and μ_2, but now s_1^2 and s_2^2 estimate σ_1^2 and σ_2^2. We have two different variances, so it is inappropriate to pool the variance estimates. Once again, the crucial fact in obtaining a standard error is that the samples are independent.

For inferences about the difference between the two means, say, $\mu_1 - \mu_2$, again use the general procedure with

$$Par = \mu_1 - \mu_2$$

and

$$Est = \bar{y}_{1.} - \bar{y}_{2.}.$$

Just as before, $\bar{y}_{1.} - \bar{y}_{2.}$ is unbiased for estimating $\mu_1 - \mu_2$. The two sample means are independent, so

$$\text{Var}(\bar{y}_{1.} - \bar{y}_{2.}) = \text{Var}(\bar{y}_{1.}) + \text{Var}(\bar{y}_{2.}) = \frac{\sigma_1^2}{N_1} + \frac{\sigma_2^2}{N_2}.$$

The standard error of $\bar{y}_{1.} - \bar{y}_{2.}$ is

$$\text{SE}(\bar{y}_{1.} - \bar{y}_{2.}) = \sqrt{\frac{s_1^2}{N_1} + \frac{s_2^2}{N_2}}.$$

Even when the original data are normal, the appropriate reference distribution is *not* a t distribution. As a matter of fact, the appropriate reference distribution is not known. However, a good approximate distribution is

$$\frac{(\bar{y}_{1.} - \bar{y}_{2.}) - (\mu_1 - \mu_2)}{\sqrt{s_1^2/N_1 + s_2^2/N_2}} \sim t(\nu)$$

where

$$\nu \equiv \frac{\left(s_1^2/N_1 + s_2^2/N_2\right)^2}{\left(s_1^2/N_1\right)^2/(N_1 - 1) + \left(s_2^2/N_2\right)^2/(N_2 - 1)} \tag{4.3.1}$$

is an approximate number of degrees of freedom. This approximate distribution was proposed by Satterthwaite (1946) and was discussed by Snedecor and Cochran (1980). Having identified the parameter, estimate, standard error, and reference distribution, inferences follow the usual pattern.

EXAMPLE 4.3.2. Consider the turtle data. Recall that

Group	Size	Mean	Variance	Standard deviation
Females	24	3.9403	0.02493979	0.1579
Males	24	3.7032	0.00677276	0.0823

We begin by considering a test of $H_0 : \mu_1 = \mu_2$ or equivalently $H_0 : \mu_1 - \mu_2 = 0$. As before, $Par = \mu_1 - \mu_2$ and $Est = 3.9403 - 3.7032 = 0.2371$. The standard error is now

$$\text{SE}(\bar{y}_{1.} - \bar{y}_{2.}) = \sqrt{\frac{0.02493979}{24} + \frac{0.00677276}{24}} = 0.03635.$$

Using $s_1^2/N_1 = 0.02493979/24 = 0.001039158$ and $s_2^2/N_2 = 0.00677276/24 = 0.000282198$ in Equation (4.3.1), the approximate degrees of freedom are

$$\nu = \frac{(0.001039158 + 0.000282198)^2}{(0.001039158)^2/23 + (0.000282198)^2/23} = 34.6.$$

An $\alpha = 0.01$ test is rejected if the observed value of the test statistic is farther from zero than the cutoff value $t(0.995, 34.6) \doteq t(0.995, 35) = 2.72$. The observed value of the test statistic is

$$t_{obs} = \frac{0.2371 - 0}{0.03635} = 6.523,$$

which is greater than the cutoff value, so the test is rejected. There is evidence at the 0.01 level that the mean shell height for females is different from the mean shell height for males. Obviously, since $\bar{y}_{1.} - \bar{y}_{2.} = 0.2371$ is positive, there is evidence that the females have shells of greater height. As always, such a conclusion relies on the other assumptions being true. With these sample sizes, the main thing that could invalidate the conclusion would be a lack of independence among the observations. Actually, the conclusion is that the means of the log(heights) are different, but if these are different we conclude that the mean heights are different, or more particularly that the median heights are different.

The 95% confidence interval for the difference between mean log shell heights for females and males, i.e., $\mu_1 - \mu_2$, uses $t(0.975, 34.6) \doteq t(0.975, 35) = 2.03$. The endpoints are

$$0.2371 \pm 2.03 (0.03635),$$

and the interval is $(0.163, 0.311)$. We took logs of the data, so for transformed data that are normal (or any other symmetric distribution), e^{μ_i} is the median height for group i even though e^{μ_i} is not the mean height for group i. Thus, $e^{\mu_1 - \mu_2}$ is the ratio of the median of the female heights to the median of the male heights. If we transform back to the original scale the interval is $(e^{0.163}, e^{0.311})$ or $(1.18, 1.36)$. The data are consistent with the population median for females being, roughly, between one and a sixth and one and a third *times* the median shell heights for males. Note that a difference between 0.163 and 0.311 on the log scale transforms into a *multiplicative effect* between 1.18 and 1.36 on the original scale. This idea is discussed in more detail in Example 12.1.1.

It is inappropriate to pool the variance estimates, so inferences about μ_1 and μ_2 are performed just as for one sample. The 95% confidence interval for the mean log shell height for females, μ_1, uses the estimate $\bar{y}_{1.}$, the standard error $s_1/\sqrt{24}$, and the tabled value $t(0.975, 24 - 1) = 2.069$. It has endpoints

$$3.9403 \pm 2.069 \left(0.1579/\sqrt{24}\right)$$

which gives the interval $(3.87, 4.01)$. Transforming to the original scale gives the interval $(47.9, 55.1)$. The data are consistent with a median shell height for females between, roughly, 48 and 55 millimeters based on a 0.05 level test. Males also have 24 observations, so the interval for μ_2 also uses $t(0.975, 24 - 1)$, has endpoints

$$3.7032 \pm 2.069 \left(0.0823/\sqrt{24}\right),$$

and an interval $(3.67, 3.74)$. Transforming the interval back to the original scale gives $(39.3, 42.1)$. The data are consistent with a median shell height for males between, roughly, 39 and 42 millimeters. The 95% prediction interval for the transformed shell height of a future male has endpoints

$$3.7032 \pm 2.069 \left(0.0823\sqrt{1 + \frac{1}{24}}\right),$$

which gives the interval $(3.529, 3.877)$. Transforming the prediction interval back to the original scale gives $(34.1, 48.3)$. Transforming a prediction interval back to the original scale creates no problems of interpretation. $\qquad\qquad\square$

EXAMPLE 4.3.3.   Reconsider the final point totals data of Section 4.2. Without the assumption of equal variances, the standard error is

$$SE(\bar{y}_{1.} - \bar{y}_{2.}) = \sqrt{\frac{487.28}{22} + \frac{979.29}{15}} = 9.3507.$$

From Equation (4.3.1), the degrees of freedom for the approximate t distribution are 23. A 95% confidence interval for the difference is $(-30.4, 8.3)$ and the observed value of the statistic for testing equal means is $t_{obs} = -1.18$. This gives a P value of 0.22. These values are all quite close to those obtained using the equal variance assumption. □

It is an algebraic fact that if $N_1 = N_2$, the observed value of the test statistic for $H_0 : \mu_1 = \mu_2$ based on unequal variances is the same as that based on equal variances. In the turtle example, the sample sizes are both 24 and the test statistic of 6.523 is the same as the equal variances test statistic. The algebraic equivalence occurs because with equal sample sizes, the standard errors from the two procedures are the same. With equal sample sizes, the only practical difference between the two procedures for examining $Par = \mu_1 - \mu_2$ is in the choice of degrees of freedom for the t distribution. In the turtle example above, the unequal variances procedure had approximately 35 degrees of freedom, while the equal variance procedure has 46 degrees of freedom. The degrees of freedom are sufficiently close that the substantive results of the turtle analysis are essentially the same, regardless of method. The other fact that should be recalled is that the reference distribution associated with $\mu_1 - \mu_2$ for the equal variance method is exactly correct for data that satisfy the model assumptions. Even for data that satisfy the unequal variance method assumptions, the reference distribution is just an approximation.

4.4 Testing equality of the variances

We assume that the original data are independent random samples. Our goal is to test the hypothesis that the variances are equal, i.e.,

$$H_0 : \sigma_2^2 = \sigma_1^2.$$

The hypothesis can be converted into an equivalent hypothesis,

$$H_0 : \frac{\sigma_2^2}{\sigma_1^2} = 1.$$

An obvious test statistic is

$$\frac{s_2^2}{s_1^2}.$$

We will reject the hypothesis of equal variances if the test statistic is too much greater than 1 or too much less than 1. As always, the problem is in identifying a precise meaning for "too much." To do this, we need to know that both samples have normal distributions, so that the distribution of the test statistic can be found when the variances are equal. The distribution turns out to be an F distribution, i.e., if H_0 is true

$$\frac{s_2^2}{s_1^2} \sim F(N_2 - 1, N_1 - 1).$$

The distribution depends on the degrees of freedom for the two estimates. The first parameter in $F(N_2 - 1, N_1 - 1)$ is $N_2 - 1$, the degrees of freedom for the variance estimate in the numerator of s_2^2/s_1^2, and the second parameter is $N_1 - 1$, the degrees of freedom for the variance estimate in the denominator. The test statistic s_2^2/s_1^2 is nonnegative, so our reference distribution $F(N_2 - 1, N_1 - 1)$ is nonnegative. Tables are given in Appendix B.

In some sense, the F distribution is 'centered' around one and we reject H_0 if s_2^2/s_1^2 is too large or too small to have reasonably come from an $F(N_2 - 1, N_1 - 1)$ distribution. A commonly used $\alpha = 0.01$ level test is rejected, i.e., we conclude that either $\sigma_2^2 \neq \sigma_1^2$ or some other of our assumptions is wrong, if

$$\frac{s_2^2}{s_1^2} > F(0.995, N_2 - 1, N_1 - 1)$$

or if

$$\frac{s_2^2}{s_1^2} < F(0.005, N_2 - 1, N_1 - 1)$$

where $F(0.995, N_2 - 1, N_1 - 1)$ cuts off the top 0.005 of the distribution and $F(0.005, N_2 - 1, N_1 - 1)$ cuts off the bottom 0.005 of the distribution. It is rare that one finds the bottom percentiles of an F distribution tabled but they can be obtained from the top percentiles. In particular,

$$F(0.005, N_2 - 1, N_1 - 1) = \frac{1}{F(0.995, N_1 - 1, N_2 - 1)}.$$

Note that the degrees of freedom have been reversed in the right-hand side of the equality.

While this is a convenient way to construct a test, it is neither a significance test nor the optimal Neyman–Pearson test. In fact, this problem raises serious issues about significance testing for continuous distributions.

EXAMPLE 4.4.1.   We again consider the log turtle height data. The sample variance of log female heights is $s_1^2 = 0.02493979$ and the sample variance of log male heights is $s_2^2 = 0.00677276$. The $\alpha = 0.01$ level test is rejected, i.e., we conclude that the null model with $\sigma_2^2 = \sigma_1^2$ is wrong if

$$0.2716 = \frac{0.00677276}{0.02493979} = \frac{s_2^2}{s_1^2} > F(0.995, 23, 23) = 3.04$$

or if

$$0.2716 < F(0.005, 23, 23) = \frac{1}{F(0.995, 23, 23)} = \frac{1}{3.04} = 0.33.$$

The second of these inequalities is true, so the null model with equal variances is rejected at the 0.01 level. We have evidence that $\sigma_2^2 \neq \sigma_1^2$ if the model is true and, since the statistic is less than one, evidence that $\sigma_2^2 < \sigma_1^2$.   □

EXAMPLE 4.4.2.   Consider again the final point total data. The sample variance for females is $s_1^2 = 487.28$ and the sample variance for males is $s_2^2 = 979.29$. The test statistic is

$$\frac{s_1^2}{s_2^2} = \frac{487.28}{979.29} = 0.498.$$

For the tests being used, it does not matter which variance estimate we put in the numerator as long as we keep the degrees of freedom straight. The observed test statistic is not less than $1/F(0.95, 14, 21) = 1/2.197 = 0.455$ nor greater than $F(0.95, 21, 14) = 2.377$, so the null model is not rejected at the $\alpha = 0.10$ level.   □

In practice, *tests for the equality of variances are rarely performed*. As misguided as it may be, typically, the main emphasis is on drawing conclusions about the μ_is. The motivation for testing equality of variances is frequently to justify the use of the pooled estimate of the variance. The test assumes that the null hypothesis of equal variances is true and data that are inconsistent with the assumptions indicate that the assumptions are false. We generally hope that this indicates that the assumption about the null hypothesis is false, but, in fact, unusual data may be obtained if any of the assumptions are invalid. The equal variances test assumes that the data are independent and normal and that the variances are equal. Minor deviations from normality may cause the test to be rejected. While procedures for comparing μ_is based on the pooled estimate of the variance are sensitive to unequal variances, they are not particularly sensitive to nonnormality. The test for equality of variances is so sensitive to nonnormality that when rejecting this test one has little idea if the problem is really unequal variances or if it is nonnormality. Thus one has little idea whether there is a problem with the pooled estimate procedures or not. Since the test is not very informative, it is rarely performed. Moreover, as discussed at the beginning of this section, if the variances of the two groups are substantially different, inferences about the means may be irrelevant to the underlying practical issues.

Theory

The F distribution used here is related to the fact that for independent random samples of normal data,

$$\frac{(N_i - 1)s_i^2}{\sigma_i^2} \sim \chi^2(N_i - 1).$$

Definition 4.4.3. An F distribution is the ratio of two independent chi-squared random variables divided by their degrees of freedom. The numerator and denominator degrees of freedom for the F distribution are the degrees of freedom for the respective chi-squareds.

In this problem, the two chi-squared random variables divided by their degrees of freedom are

$$\frac{(N_i - 1)s_i^2/\sigma_i^2}{N_i - 1} = \frac{s_i^2}{\sigma_i^2},$$

$i = 1, 2$. They are independent because they are taken from independent samples and their ratio is

$$\frac{s_2^2}{\sigma_2^2} \Big/ \frac{s_1^2}{\sigma_1^2} = \frac{s_2^2}{s_1^2}\frac{\sigma_1^2}{\sigma_2^2}.$$

When the null hypothesis is true, i.e., $\sigma_2^2/\sigma_1^2 = 1$, by definition, we get

$$\frac{s_2^2}{s_1^2} \sim F(N_2 - 1, N_1 - 1),$$

so the test statistic has an F distribution under the null hypothesis and the normal sampling model. Note that we could equally well have reversed the roles of the two groups and set the test up as

$$H_0 : \frac{\sigma_1^2}{\sigma_2^2} = 1$$

with the test statistic

$$\frac{s_1^2}{s_2^2}.$$

A non-optimal Neyman–Pearson α level test is rejected if

$$\frac{s_1^2}{s_2^2} > F\left(1 - \frac{\alpha}{2}, N_1 - 1, N_2 - 1\right)$$

or if

$$\frac{s_1^2}{s_2^2} < F\left(\frac{\alpha}{2}, N_1 - 1, N_2 - 1\right).$$

Using the fact that for any α between zero and one and any degrees of freedom r and s,

$$F(\alpha, r, s) = \frac{1}{F(1 - \alpha, s, r)}, \tag{4.4.1}$$

it is easily seen that this test is equivalent to the one we constructed. Relation (4.4.1) is a result of the fact that with equal variances both s_2^2/s_1^2 and s_1^2/s_2^2 have F distributions. Clearly, the smallest, say, 5% of values from s_2^2/s_1^2 are also the largest 5% of the values of s_1^2/s_2^2.

Table 4.4: *Weights of rats on thiouracil.*

Rat	Start	Finish	Rat	Start	Finish
1	61	129	6	51	119
2	59	122	7	56	108
3	53	133	8	58	138
4	59	122	9	46	107
5	51	140	10	53	122

Table 4.5: *Weight gain comparison.*

Control		Thyroxin	
115	107	132	88
117	90	84	119
133	91	133	
115	91	118	
95	112	87	

4.5 Exercises

EXERCISE 4.5.1. Box (1950) gave data on the weights of rats that were given the drug Thiouracil. The rats were measured at the start of the experiment and at the end of the experiment. The data are given in Table 4.4. Give a 99% confidence interval for the difference in weights between the finish and the start. Test the null hypothesis that the population mean weight gain was less than or equal to 50 with $\alpha = 0.02$.

EXERCISE 4.5.2. Box (1950) also considered data on rats given Thyroxin and a control group of rats. The weight *gains* are given in Table 4.5. Give a 95% confidence interval for the difference in weight gains between the Thyroxin group and the control group. Give the P value for a test of whether the control group has weight gains different than the Thyroxin group.

EXERCISE 4.5.3. Conover (1971, p. 226) considered data on the physical fitness of male seniors in a particular high school. The seniors were divided into two groups based on whether they lived on a farm or in town. The results in Table 4.6 are from a physical fitness test administered to the students. High scores indicate that an individual is physically fit. Give a 95% confidence interval for the difference in mean fitness scores between the town and farm students. Test the hypothesis of no difference at the $\alpha = 0.10$ level. Give a 99% confidence interval for the mean fitness of town boys. Give a 99% prediction interval for a future fitness score for a farm boy.

EXERCISE 4.5.4. Use the data of Exercise 4.5.3 to test whether the fitness scores for farm boys are more or less variable than fitness scores for town boys.

Table 4.6: *Physical fitness of male high school seniors.*

Town	12.7	16.9	7.6	2.4	6.2	9.9
Boys	14.2	7.9	11.3	6.4	6.1	10.6
	12.6	16.0	8.3	9.1	15.3	14.8
	2.1	10.6	6.7	6.7	10.6	5.0
	17.7	5.6	3.6	18.6	1.8	2.6
	11.8	5.6	1.0	3.2	5.9	4.0
Farm	14.8	7.3	5.6	6.3	9.0	4.2
Boys	10.6	12.5	12.9	16.1	11.4	2.7

Table 4.7: *Turtle lengths.*

Females			Males			
98	138	123	155	121	104	116	93
103	138	133	155	125	106	117	94
103	141	133	158	127	107	117	96
105	147	133	159	128	112	119	101
109	149	134	162	131	113	120	102
123	153	136	177	135	114	120	103

Table 4.8: *Verbal ability test scores.*

8 yr. olds			10 yr. olds		
324	344	448	428	399	414
366	390	372	366	412	396
322	434	364	386	436	
398	350		404	452	

EXERCISE 4.5.5.   Jolicoeur and Mosimann (1960) gave data on turtle shell *lengths*. The data for females and males are given in Table 4.7. Explore the need for a transformation. Test whether there is a difference in lengths using $\alpha = 0.01$. Give a 95% confidence interval for the difference in lengths.

EXERCISE 4.5.6.   Koopmans (1987) gave the data in Table 4.8 on verbal ability test scores for 8 year-olds and 10 year-olds. Test whether the two groups have the same mean with $\alpha = 0.01$ and give a 95% confidence interval for the difference in means. Give a 95% prediction interval for a new 10 year old. Check your assumptions.

EXERCISE 4.5.7.   Burt (1966) and Weisberg (1985) presented data on IQ scores for identical twins that were raised apart, one by foster parents and one by the genetic parents. Variable y_1 is the IQ score for a twin raised by foster parents, while y_2 is the corresponding IQ score for the twin raised by the genetic parents. The data are given in Table 4.9.

We are interested in the difference between μ_1, the population mean for twins raised by foster parents, and μ_2, the population mean for twins raised by genetic parents. Analyze the data. Check your assumptions.

EXERCISE 4.5.8.   Table 4.10 presents data given by Shewhart (1939, p. 118) on various atomic weights as reported in 1931 and again in 1936. Analyze the data. Check your assumptions.

Table 4.9: *Burt's IQ data.*

Case	y_1	y_2	Case	y_1	y_2	Case	y_1	y_2
1	82	82	10	93	82	19	97	87
2	80	90	11	95	97	20	87	93
3	88	91	12	88	100	21	94	94
4	108	115	13	111	107	22	96	95
5	116	115	14	63	68	23	112	97
6	117	129	15	77	73	24	113	97
7	132	131	16	86	81	25	106	103
8	71	78	17	83	85	26	107	106
9	75	79	18	93	87	27	98	111

Table 4.10: *Atomic weights in 1931 and 1936.*

Compound	1931	1936	Compound	1931	1936
Arsenic	74.93	74.91	Lanthanum	138.90	138.92
Caesium	132.81	132.91	Osmium	190.8	191.5
Columbium	93.3	92.91	Potassium	39.10	39.096
Iodine	126.932	126.92	Radium	225.97	226.05
Krypton	82.9	83.7	Ytterbium	173.5	173.04

Table 4.11: *Peel-strengths.*

Adhesive	Observations					
A	60	63	57	53	56	57
B	52	53	44	48	48	53

EXERCISE 4.5.9.   Reanalyze the data of Example 4.1.1 after deleting the one possible outlier. Does the analysis change much? If so, how?

EXERCISE 4.5.10.   Let $y_1 \sim N(89, 4)$ and $y_2 \sim N(90, 16)$. Show that $\Pr[y_1 \geq 87] > \Pr[y_2 \geq 87]$, so that the population with the lower mean has a higher probability of exceeding 87. Recall that $(y_1 - 89)/\sqrt{4} \sim N(0, 1)$ with a similar result for y_2 so that both probabilities can be rewritten in terms of a $N(0, 1)$.

EXERCISE 4.5.11.   Mandel (1972) reported stress test data on elongation for a certain type of rubber. Four pieces of rubber sent to one laboratory yielded a sample mean and variance of 56.50 and 5.66, respectively. Four different pieces of rubber sent to another laboratory yielded a sample mean and variance of 52.50 and 6.33, respectively. Are the data two independent samples or a paired comparison? Is the assumption of equal variances reasonable? Give a 99% confidence interval for the difference in population means and give an approximate P value for testing that there is no difference between population means.

EXERCISE 4.5.12.   Bethea et al. (1985) reported data on the peel-strengths of adhesives. Some of the data are presented in Table 4.11. Give an approximate P value for testing no difference between adhesives, a 95% confidence interval for the difference between mean peel-strengths, and a 95% prediction interval for a new observation on Adhesive A.

EXERCISE 4.5.13.   Garner (1956) presented data on the tensile strength of fabrics. Here we consider a subset of the data. The complete data and a more extensive discussion of the experimental procedure are given in Exercise 11.5.2. The experiment involved testing fabric strengths on different machines. Eight homogeneous strips of cloth were divided into samples and each machine was used on a sample from each strip. The data are given in Table 4.12. Are the data two independent samples or a paired comparison? Give a 98% confidence interval for the difference in population means. Give an approximate P value for testing that there is no difference between population means. What is the result of an $\alpha = 0.05$ test?

Table 4.12: *Tensile strength.*

Strip	1	2	3	4	5	6	7	8
m_1	18	9	7	6	10	7	13	1
m_2	7	11	11	4	8	12	5	11

Table 4.13: *Acreage in corn for different farm acreages.*

Size	Corn acreage				
240	65	80	65	85	30
400	75	35	140	90	110

Table 4.14: *Cutting dates.*

Year	29	30	31	32	33	34	35	36	37	38
June 1	201	230	324	512	399	891	449	595	632	527
June 15	301	296	543	778	644	1147	585	807	804	749

EXERCISE 4.5.14.   Snedecor and Cochran (1967) presented data on the number of acres planted in corn for two sizes of farms. Size was measured in acres. Some of the data are given in Table 4.13. Are the data two independent samples or a paired comparison? Is the assumption of equal variances reasonable? Test for differences between the farms of different sizes. Clearly state your α level. Give a 98% confidence interval for the mean difference between different farms.

EXERCISE 4.5.15.   Snedecor and Haber (1946) presented data on cutting dates of asparagus. On two plots of land, asparagus was grown every year from 1929 to 1938. On the first plot the asparagus was cut on June 1, while on the second plot the asparagus was cut on June 15. Note that growing conditions will vary considerably from year to year. Also note that the data presented have cutting dates confounded with the plots of land. If one plot of land is intrinsically better for growing asparagus than the other, there will be no way of separating that effect from the effect of cutting dates. Are the data two independent samples or a paired comparison? Give a 95% confidence interval for the difference in population means and give an approximate P value for testing that there is no difference between population means. Give a 95% prediction interval for the difference in a new year. The data are given in Table 4.14.

EXERCISE 4.5.16.   Snedecor (1945b) presented data on a pesticide spray. The treatments were the number of units of active ingredient contained in the spray. Several different sources for breeding mediums were used and each spray was applied on each distinct breeding medium. The data consisted of numbers of dead adult flies found in cages that were set over the breeding medium containers. Some of the data are presented in Table 4.15. Give a 95% confidence interval for the difference in population means. Give an approximate P value for testing that there is no difference between population means and an $\alpha = 0.05$ test. Give a 95% prediction interval for a new observation with 8 units. Give a 95% prediction interval for a new observation with 8 units when the corresponding 0 unit value is 300.

EXERCISE 4.5.17.   Using the data of Example 4.2.1 give a 95% prediction interval for the difference in total points between a new female and a new male. This was not discussed earlier so it requires a deeper understanding of Section 3.5.

Table 4.15: *Dead adult flies.*

Medium	A	B	C	D	E	F	G
0 units	423	326	246	141	208	303	256
8 units	414	127	206	78	172	45	103

Chapter 5

Contingency Tables

In this chapter we consider data that consist of counts. We begin in Section 5.1 by examining a set of data on the number of females admitted into graduate school at the University of California, Berkeley. A key feature of these data is that only two outcomes are possible: admittance or rejection. Data with only two outcomes are referred to as *binary (or dichotomous) data*. Often the two outcomes are referred to generically as success and failure. In Section 5.2, we expand our discussion by comparing two sets of dichotomous data; we compare Berkeley graduate admission rates for females and males. Section 5.3 examines *polytomous data*, i.e., count data in which there are more than two possible outcomes. For example, numbers of Swedish females born in the various months of the year involve counts for 12 possible outcomes. Section 5.4 examines comparisons between two samples of polytomous data, e.g., comparing the numbers of females and males that are born in the different months of the year. Section 5.5 looks at comparisons among more than two samples of polytomous data. The last section considers a method of reducing large tables of counts that involve several samples of polytomous data into smaller more interpretable tables.

Sections 5.1 and 5.2 involve analogues of Chapters 2 and 4 that are appropriate for dichotomous data. The basic analyses in these sections simply involve new applications of the ideas in Chapter 3. Sections 5.3, 5.4, and 5.5 are polytomous data analogues of Chapters 2, 4, and 12. Everitt (1977) and Fienberg (1980) give more detailed introductions to the analysis of count data. Sophisticated analyses of count data frequently use analogues of ANOVA and regression called logistic regression and log-linear models. These are discussed in Chapters 20 and 21, respectively.

5.1 One binomial sample

The few distributions that are most commonly used in Statistics arise naturally. The normal distribution arises for measurement data because the variability in the data often results from the mean of a large number of small errors and the central limit theorem indicates that such means tend to be normally distributed.

The binomial distribution arises naturally with count data because of its simplicity. Consider a number of trials, say n, each a success or failure. If each trial is independent of the other trials and if the probability of obtaining a success is the same for every trial, then the random number of successes has a binomial distribution. *The beauty of discrete data is that the probability models can often be justified solely by how the data were collected. This does not happen with measurement data.* The binomial distribution depends on two parameters, n, the number of independent trials, and the constant probability of success, say p. Typically, we know the value of n, while p is the unknown parameter of interest. Binomial distributions were examined in Section 1.4.

Bickel et al. (1975) report data on admissions to graduate school at the University of California, Berkeley. The numbers of females that were admitted and rejected are given below along with the total number of applicants.

Graduate admissions at Berkeley

	Admitted	Rejected	Total
Female	557	1278	1835

It seems reasonable to view the 1835 females as a random sample from a population of potential female applicants. We are interested in the probability p that a female applicant is admitted to graduate school. A natural estimate of the parameter p is the proportion of females that were actually admitted, thus our estimate of the parameter is

$$\hat{p} = \frac{557}{1835} = 0.30354.$$

We have a parameter of interest, p, and an estimate of that parameter, $\hat{p}$; if we can identify a standard error and an appropriate distribution, we can use methods from Chapter 3 to perform statistical inferences.

The key to finding a standard error is to find the variance of the estimate. As we will see later,

$$\text{Var}(\hat{p}) = \frac{p(1-p)}{n}. \tag{5.1.1}$$

To estimate the standard deviation of $\hat{p}$, we simply use $\hat{p}$ to estimate p in (5.1.1) and take the square root. Thus the standard error is

$$\text{SE}(\hat{p}) = \sqrt{\frac{\hat{p}(1-\hat{p})}{n}} = \sqrt{\frac{0.30354(1-0.30354)}{1835}} = 0.01073.$$

The final requirement for using the results of Chapter 3 is to find an appropriate reference distribution for

$$\frac{\hat{p} - p}{\text{SE}(\hat{p})}.$$

We can think of each trial as scoring either a 1, if the trial is a success, or a 0, if the trial is a failure. With this convention $\hat{p}$, the proportion of successes, is really the average of the 0-1 scores and since $\hat{p}$ is an average we can apply the central limit theorem. (In fact, $\text{SE}(\hat{p})$ is very nearly $s/\sqrt{n}$, where s is computed from the 0-1 scores.) The central limit theorem simply states that for a large number of trials n, the distribution of $\hat{p}$ is approximately normal with a population mean that is the population mean of $\hat{p}$ and a population variance that is the population variance of $\hat{p}$. We have already given the variance of $\hat{p}$ and we will see later that $\text{E}(\hat{p}) = p$, thus for large n we have the approximation

$$\hat{p} \sim N\left(p, \frac{p(1-p)}{n}\right).$$

The variance is unknown but by the law of large numbers it is approximately equal to our estimate of it, $\hat{p}(1-\hat{p})/n$. Standardizing the normal distribution (cf. Exercise 1.6.2) gives the approximation

$$\frac{\hat{p} - p}{\text{SE}(\hat{p})} \sim N(0,1) \equiv t(\infty). \tag{5.1.2}$$

This distribution requires a sample size that is large enough for both the central limit theorem approximation and the law of large numbers approximation to be reasonably valid. For values of p that are not too close to 0 or 1, the approximation works reasonably well with sample sizes as small as 20. However, the normal distribution is unrealistically precise, since it is based on both a normal approximation and a law of large numbers approximation. We use the $t(n-1)$ distribution instead, hoping that it provides a more realistic view of the reference distribution.

We now have $Par = p$, $Est = \hat{p}$, $\text{SE}(\hat{p}) = \sqrt{\hat{p}(1-\hat{p})/n}$, and the distribution in (5.1.2) or $t(n-1)$. As in Chapter 3, a 95% confidence interval for p has limits

$$\hat{p} \pm 1.96\sqrt{\frac{\hat{p}(1-\hat{p})}{n}}.$$

Here $1.96 = t(0.975, \infty) \doteq t(0.975, 1834)$. Recall that a $(1 - \alpha)100\%$ confidence interval requires the $(1 - \alpha/2)$ percentile of the distribution. For the female admissions data, the limits are

$$0.30354 \pm 1.96(0.01073),$$

which gives the interval $(0.28, 0.32)$. The data are consistent with a population proportion of females admitted to Berkeley's graduate school between 0.28 and 0.32. (As is often the case, it is not exactly clear what population these data relate to.) Agresti and Coull (1998) discuss alternative methods of constructing confidence intervals, some of which have better Neyman–Pearson coverage rates.

We can also perform, say, an $\alpha = 0.01$ test of the null hypothesis $H_0 : p = 1/3$. The test rejects H_0 if

$$\frac{\hat{p} - 1/3}{SE(\hat{p})} > 2.58$$

or if

$$\frac{\hat{p} - 1/3}{SE(\hat{p})} < -2.58.$$

Here $2.58 = t(0.995, \infty) \doteq t(0.995, 1834)$. An α-level test requires the $(1 - \frac{\alpha}{2})100\%$ point of the distribution. The Berkeley data yield the test statistic

$$\frac{0.30354 - 0.33333}{0.01073} = -2.78,$$

which is smaller than -2.58, so we reject the null model with $p = 1/3$ at $\alpha = 0.01$. In other words, we can reject, with strong assurance, the claim that one third of female applicants are admitted to graduate school at Berkeley, provided the data really are binomial. Since the test statistic is negative, we have evidence that the true proportion is less than one third. The test as constructed here is equivalent to checking whether $p = 1/3$ is within a 99% confidence interval.

There is an alternative, slightly different, way of performing tests such as $H_0 : p = 1/3$. The difference involves using a different standard error. The variance of the estimate $\hat{p}$ is $p(1 - p)/n$. In obtaining a standard error, we estimated p with $\hat{p}$ and took the square root of the estimated variance. Recalling that tests are performed *assuming that the null hypothesis is true*, it makes sense in the testing problem to use the assumption $p = 1/3$ in computing a standard error for $\hat{p}$. Thus an *alternative standard error* for $\hat{p}$ in this testing problem is

$$\sqrt{\frac{1}{3}\left(1 - \frac{1}{3}\right) \Big/ 1835} = 0.01100.$$

The test statistic now becomes

$$\frac{0.30354 - 0.33333}{0.01100} = -2.71.$$

Obviously, since the test statistic is slightly different, one could get slightly different answers for tests using the two different standard errors. Moreover, the results of this test will not always agree with a corresponding confidence interval for p because this test uses a different standard error than the confidence interval. (It hardly seems worth the trouble to compute a confidence interval using these standard errors, although it could be done; see Lindgren, 1968, Sec. 5.3.)

The difference between the two standard errors is often minor compared to the level of approximation inherent in using either the standard normal or the $t(n - 1)$ as a reference distribution. In any case, whether we ascribe the differences to the standard errors or to the quality of the normal approximations, the exact behavior of the two test statistics can be quite different when the sample size is small. Moreover, *when p is near 0 or 1, the sample sizes must be quite large to get a good normal approximation.*

The theoretical results needed for analyzing a single binomial sample are establishing that $\hat{p}$ is a reasonable estimate of p and that the variance formula given earlier is correct. The data are $y \sim \text{Bin}(n, p)$. As seen in Section 1.4, $\text{E}(y) = np$ and $\text{Var}(y) = np(1 - p)$. The estimate of p is $\hat{p} = y/n$. The estimate is unbiased because

$$\text{E}(\hat{p}) = \text{E}(y/n) = \text{E}(y)/n = np/n = p.$$

The variance of the estimate is

$$\text{Var}(\hat{p}) = \text{Var}(y/n) = \text{Var}(y)/n^2 = np(1 - p)/n^2 = p(1 - p)/n.$$

5.1.1   The sign test

We now consider an alternative analysis for paired comparisons based on the binomial distribution. Consider Burt's data on IQs of identical twins raised apart from Exercise 4.5.7 and Table 4.9. The earlier discussion of paired comparisons involved assuming and validating the normal distribution for the differences in IQs between twins. In the current discussion, we make the same assumptions as before except we replace the normality assumption with the weaker assumption that the distribution of the differences is symmetric. In the earlier discussion, we would test $H_0 : \mu_1 - \mu_2 = 0$. In the current discussion, we test whether there is a $50 : 50$ chance that y_1, the IQ for the foster-parent-raised twin, is larger than y_2, the IQ for the genetic-parent-raised twin. In other words, we test whether $\Pr(y_1 - y_2 > 0) = 0.5$. We have a sample of $n = 27$ pairs of twins. If $\Pr(y_1 - y_2 > 0) = 0.5$, the number of pairs with $y_1 - y_2 > 0$ has a $\text{Bin}(27, 0.5)$ distribution. From Table 4.9, 13 of the 27 pairs have larger IQs for the foster-parent-raised child. (These are the differences with a positive sign, hence the name sign test.) The proportion is $\hat{p} = 13/27 = 0.481$. The test statistic is

$$\frac{0.481 - 0.5}{\sqrt{0.5(1 - 0.5)/27}} = -0.20,$$

which is clearly consistent with the null model.

A similar method could be used to test, say, whether there is a $50 : 50$ chance that y_1 is at least 3 IQ points greater than y_2. This hypothesis translates into $\Pr(y_1 - y_2 \geq 3) = 0.5$. The test is then based on the number of differences that are 3 or more.

The point of the sign test is the weakening of the assumption of normality. If the normality assumption is appropriate, the t test of Section 4.1 is more appropriate. When the normality assumption is not appropriate, some modification like the sign test should be used. In this book, the usual approach is to check the normality assumption and, if necessary, to transform the data to make the normality assumption reasonable. For a more detailed introduction to *nonparametric* methods such as the sign test; see, for example, Conover (1971).

5.2   Two independent binomial samples

In this section we compare two independent binomial samples. Consider again the Berkeley admissions data. Table 5.1 contains data on admissions and rejections for the 1835 females considered in Section 5.1 along with data on 2691 males. We assume that the sample of females is independent of the sample of males. Throughout, we refer to the females as the first sample and the males as the second sample.

We consider being admitted to graduate school a "success." Assuming that the females are a binomial sample, they have a sample size of $n_1 = 1835$ and some probability of success, say, p_1. The observed proportion of female successes is

$$\hat{p}_1 = \frac{557}{1835} = 0.30354.$$

Table 5.1: *Graduate admissions at Berkeley.*

	Admitted	Rejected	Total
Females	557	1278	1835
Males	1198	1493	2691

Treating the males as a binomial sample, the sample size is $n_2 = 2691$ and the probability of success is, say, p_2. The observed proportion of male successes is

$$\hat{p}_2 = \frac{1198}{2691} = 0.44519.$$

Our interest is in comparing the success rate of females and males. The appropriate parameter is the difference in proportions,

$$Par = p_1 - p_2.$$

The natural estimate of this parameter is

$$Est = \hat{p}_1 - \hat{p}_2 = 0.30354 - 0.44519 = -0.14165.$$

With independent samples, we can find the variance of the estimate and thus the standard error. Since the females are independent of the males,

$$\text{Var}(\hat{p}_1 - \hat{p}_2) = \text{Var}(\hat{p}_1) + \text{Var}(\hat{p}_2).$$

Using the variance formula in Equation (5.1.1),

$$\text{Var}(\hat{p}_1 - \hat{p}_2) = \frac{p_1(1 - p_1)}{n_1} + \frac{p_2(1 - p_2)}{n_2}. \tag{5.2.1}$$

Estimating p_1 and p_2 and taking the square root gives the standard error,

$$
\begin{aligned}
\text{SE}(\hat{p}_1 - \hat{p}_2) &= \sqrt{\frac{\hat{p}_1(1 - \hat{p}_1)}{n_1} + \frac{\hat{p}_2(1 - \hat{p}_2)}{n_2}} \\
&= \sqrt{\frac{0.30354(1 - 0.30354)}{1835} + \frac{0.44519(1 - 0.44519)}{2691}} \\
&= 0.01439.
\end{aligned}
$$

For large sample sizes n_1 and n_2, both $\hat{p}_1$ and $\hat{p}_2$ have approximate normal distributions and they are independent, so $\hat{p}_1 - \hat{p}_2$ has an approximate normal distribution and the appropriate reference distribution is approximately

$$\frac{(\hat{p}_1 - \hat{p}_2) - (p_1 - p_2)}{\text{SE}(\hat{p}_1 - \hat{p}_2)} \sim N(0, 1).$$

Alternatively, we could use the method of Section 4.3 to determine approximate degrees of freedom for a t distribution.

We now have all the requirements for applying the results of Chapter 3. A 95% confidence interval for $p_1 - p_2$ has endpoints

$$(\hat{p}_1 - \hat{p}_2) \pm 1.96 \sqrt{\frac{\hat{p}_1(1 - \hat{p}_1)}{n_1} + \frac{\hat{p}_2(1 - \hat{p}_2)}{n_2}},$$

where the value $1.96 = t(0.975, \infty)$ seems reasonable given the large sample sizes involved. For comparing the female and male admissions, the 95% confidence interval for the population difference in proportions has endpoints

$$-0.14165 \pm 1.96(0.01439).$$

The interval is $(-0.17, -0.11)$. Proportions of women being admitted to graduate school at Berkeley between 0.11 and 0.17 less than that for men are consistent with the data and the model at $\alpha = 0.05$.

To test $H_0 : p_1 = p_2$, or equivalently $H_0 : p_1 - p_2 = 0$, reject an $\alpha = 0.10$ test if

$$\frac{(\hat{p}_1 - \hat{p}_2) - 0}{SE(\hat{p}_1 - \hat{p}_2)} > 1.645$$

or if

$$\frac{(\hat{p}_1 - \hat{p}_2) - 0}{SE(\hat{p}_1 - \hat{p}_2)} < -1.645.$$

Again, the value 1.645 is obtained from the $t(\infty)$ distribution and presumes very large samples. With the Berkeley data, the observed value of the test statistic is

$$\frac{-0.14165 - 0}{0.01439} = -9.84.$$

This is far smaller than -1.645, so the test rejects the null hypothesis of equal proportions at the 0.10 level pretty much regardless of how we determine degrees of freedom. The test statistic is negative, so there is evidence that the proportion of women admitted to graduate school is lower than the proportion of men.

Once again, an alternative standard error is often used in testing problems. The test assumes that the null hypothesis is true, i.e. $p_1 = p_2$, so in constructing a standard error for the test statistic it makes sense to pool the data into one estimate of this common proportion. The pooled estimate is a weighted average of the individual estimates,

$$
\begin{aligned}
\hat{p}_* &= \frac{n_1 \hat{p}_1 + n_2 \hat{p}_2}{n_1 + n_2} \\
&= \frac{1835(0.30354) + 2691(0.44519)}{1835 + 2691} \\
&= \frac{557 + 1198}{1835 + 2691} \\
&= 0.38776.
\end{aligned}
$$

Using $\hat{p}_*$ to estimate both p_1 and p_2 in Equation (5.2.1) and taking the square root gives the alternative standard error

$$
\begin{aligned}
SE(\hat{p}_1 - \hat{p}_2) &= \sqrt{\frac{\hat{p}_*(1 - \hat{p}_*)}{n_1} + \frac{\hat{p}_*(1 - \hat{p}_*)}{n_2}} \\
&= \sqrt{\hat{p}_*(1 - \hat{p}_*)\left[\frac{1}{n_1} + \frac{1}{n_2}\right]} \\
&= \sqrt{0.38776(1 - 0.38776)\left[\frac{1}{1835} + \frac{1}{2691}\right]} \\
&= 0.01475.
\end{aligned}
$$

The alternative test statistic is

$$\frac{-0.14165 - 0}{0.01475} = -9.60.$$

Table 5.2: *Swedish female births by month.*

Month	Females	$\hat{p}$	Probability	E	$(O-E)/\sqrt{E}$
January	3537	0.083	1/12	3549.25	−0.20562
February	3407	0.080	1/12	3549.25	−2.38772
March	3866	0.091	1/12	3549.25	5.31678
April	3711	0.087	1/12	3549.25	2.71504
May	3775	0.087	1/12	3549.25	3.78930
June	3665	0.086	1/12	3549.25	1.94291
July	3621	0.085	1/12	3549.25	1.20435
August	3596	0.084	1/12	3549.25	0.78472
September	3491	0.082	1/12	3549.25	−0.97775
October	3391	0.080	1/12	3549.25	−2.65629
November	3160	0.074	1/12	3549.25	−6.53372
December	3371	0.079	1/12	3549.25	−2.99200
Total	42591	1	1	42591.00	

Again, the two test statistics are slightly different but the difference should be minor compared to the level of approximation involved in using the normal distribution.

A final note. Before you conclude that the data in Table 5.1 provide evidence of sex discrimination, you should realize that females tend to apply to different graduate programs than males. A more careful examination of the complete Berkeley data shows that the difference observed here largely results from females applying more frequently than males to highly restrictive programs, cf. Christensen (1997, p. 114). Rejecting the test suggests that something is wrong with the null model. In this case, the assumption of binomial sampling is wrong. Some people have different probabilities of being admitted than other people, depending on what department they applied to.

5.3 One multinomial sample

In this section we investigate the analysis a single polytomous variable, i.e., a count variable with more than two possible outcomes. In particular, we assume that the data are a sample from a *multinomial* distribution, cf. Section 1.5. The multinomial distribution is a generalization of the binomial that allows more than two outcomes. We assume that each trial gives one of, say, q possible outcomes. Each trial must be independent and the probability of each outcome must be the same for every trial. The multinomial distribution gives probabilities for the number of trials that fall into each of the possible outcome categories. The binomial distribution is a special case of the multinomial distribution in which $q = 2$.

The first two columns of Table 5.2 give months and numbers of Swedish females born in each month. The data are from Cramér (1946) who did not name the months. We assume that the data begin in January.

With polytomous data such as those listed in Table 5.2, there is no one parameter of primary interest. One might be concerned with the proportions of births in January, or December, or in any of the twelve months. With no one parameter of interest, the one-parameter methods of Chapter 3 do not apply. Column 3 of Table 5.2 gives the observed proportions of births for each month. These are simply the monthly births divided by the total births for the year. Note that the proportion of births in March seems high and the proportion of births in November seems low.

A simplistic, yet interesting, hypothesis is that the proportion of births is the same for every month. In this case the model is multinomial sampling, the null hypothesis is equal probabilities, and they are combined into a null model. To test this null model, we compare the number of observed births to the number of births we would expect to see if the hypothesis were true. The number of births we expect to see in any month is just the probability of having a birth in that month times the total number of births. The equal probabilities are given in column 4 of Table 5.2 and the expected values are given in column 5. The entries in column 5 are labeled E for expected value and are computed as $(1/12)42591 = 3549.25$. *It cannot be overemphasized that the expectations are computed under the assumption that the null model is true.*

Comparing observed values with expected values can be tricky. Suppose an observed value is 2145 and the expected value is 2149. The two numbers are off by 4; the observed value is pretty close to the expected. Now suppose the observed value is 1 and the expected value is 5. Again the two numbers are off by 4 but now the difference between observed and expected seems quite substantial. A difference of 4 means something very different depending on how large both numbers are. To account for this phenomenon, we standardized the difference between observed and expected counts. We do this by dividing the difference by the square root of the expected count. Thus, when we compare observed counts with expected counts we look at

$$\frac{O-E}{\sqrt{E}} \tag{5.3.1}$$

where O stands for the observed count and E stands for the expected count. The values in (5.3.1) are called *Pearson residuals*, after Karl Pearson.

The Pearson residuals for the Swedish female births are given in column 6 of Table 5.2. As noted earlier, the two largest deviations from the assumption of equal probabilities occur for March and November. Reasonably large deviations also occur for May and to a lesser extent December, April, October, and February. In general, *the Pearson residuals can be compared to observations from a $N(0,1)$ distribution to evaluate whether a residual is large.* For example, the residuals for March and November are 5.3 and -6.5. These are not values one is likely to observe from a $N(0,1)$ distribution; they provide strong evidence that the birth rate in March is really larger than $1/12$ and that the birth rate in November is really smaller than $1/12$.

Births seem to peak in March and they, more or less, gradually decline until November. After November, birth rates are still low but gradually increase until February. In March birth rates increase markedly. Birth rates are low in the fall and lower in the winter; they jump in March and remain relatively high, though decreasing, until September. This analysis could be performed using the monthly proportions of column 2 but the results are clearer using the residuals.

A statistic for testing whether the null model of equal proportions is reasonable can be obtained by squaring the residuals and adding them together. This statistic is known as *Pearson's χ^2* (chi-squared) statistic and is computed as

$$X^2 = \sum_{\text{all cells}} \frac{(O-E)^2}{E}.$$

For the female Swedish births,

$$X^2 = 121.24.$$

Note that small values of X^2 indicate observed values that are similar to the expected values, so small values of X^2 are consistent with the null model. (However, with 3 or more degrees of freedom, values that are too small can indicate that the multinomial sampling assumption is suspect.) Large values of X^2 occur whenever one or more observed values are far from the expected values. To perform a test, we need some idea of how large X^2 could reasonably be when the null model is true. It can be shown that for a problem such as this with 1) a fixed number of cells q, here $q = 12$, with 2) a null model consisting of known probabilities such as those given in column 4 of Table 5.2, and with 3) large sample sizes for each cell, the null distribution of X^2 is approximately

$$X^2 \sim \chi^2(q-1).$$

The degrees of freedom are only $q-1$ because the $\hat{p}$s *must* add up to 1. Thus, if we know $q-1 = 11$ of the proportions, we can figure out the last one. Only $q-1$ of the cells are really free to vary. From Appendix B.2, the 99.5th percentile of a $\chi^2(11)$ distribution is $\chi^2(0.995, 11) = 26.76$. The observed X^2 value of 121.24 is much larger than this, so the observed value of X^2 could not reasonably come

Table 5.3: *Swedish births: monthly observations ($O_{ij}s$) and monthly proportions by sex.*

Month	Observations			Proportions	
	Female	Male	Total	Female	Male
January	3537	3743	7280	0.083	0.082
February	3407	3550	6957	0.080	0.078
March	3866	4017	7883	0.091	0.088
April	3711	4173	7884	0.087	0.091
May	3775	4117	7892	0.089	0.090
June	3665	3944	7609	0.086	0.086
July	3621	3964	7585	0.085	0.087
August	3596	3797	7393	0.084	0.083
September	3491	3712	7203	0.082	0.081
October	3391	3512	6903	0.080	0.077
November	3160	3392	6552	0.074	0.074
December	3371	3761	7132	0.079	0.082
Total	42591	45682	88273	1.000	1.000

from a $\chi^2(11)$ distribution. Tests based on X^2, like F tests, are commonly viewed as being rejected only for large values of the test statistics and P values are computed correspondingly. However, X^2 values that are too small also suggest that something is awry with the null model. In any case, there is overwhelming evidence that monthly female Swedish births are not multinomial with constant probabilities.

In this example, our null hypothesis was that the probability of a female birth was the same in every month. A more reasonable hypothesis might be that the probability of a female birth is the same on every day. The months have different numbers of days so under this null model they have different probabilities. For example, assuming a 365-day year, the probability of a female birth in January is $31/365$, which is somewhat larger than $1/12$. Exercise 5.8.4 involves testing the corresponding null model.

We can use results from Section 5.1 to help in the analysis of multinomial data. If we consider only the month of December, we can view each trial as a success if the birth is in December and a failure otherwise. Writing the probability of a birth in December as p_{12}, from Table 5.2 the estimate of p_{12} is

$$\hat{p}_{12} = \frac{3371}{42591} = 0.07915$$

with standard error

$$SE(\hat{p}_{12}) = \sqrt{\frac{0.07915(1 - 0.07915)}{42591}} = 0.00131$$

and a 95% confidence interval has endpoints

$$0.07915 \pm 1.96(0.00131).$$

The interval reduces to $(0.077, 0.082)$. Tests for monthly proportions can be performed in a similar fashion.

5.4 Two independent multinomial samples

Table 5.3 gives monthly births for Swedish females and males along with various marginal totals. We wish to determine whether monthly birth rates differ for females and males. Denote the females as population 1 and the males as population 2. Thus we have a sample of 42,591 females and, by assumption, an independent sample of 45,682 males.

In fact, it is more likely that there is actually only one sample here, one consisting of 88,273 births. It is more likely that the births have been divided into 24 categories depending on sex and birth month. Such data can be treated as two independent samples with (virtually) no loss of generality. *The interpretation of results for two independent samples is considerably simpler than the interpretation necessary for one sample cross-classified by both sex and month, thus we discuss such data as though they are independent samples.* The alternative interpretation involves a multinomial sample with the probabilities for each month and sex being independent.

The number of births in month i for sex j is denoted O_{ij}, where $i = 1, \ldots, 12$ and $j = 1, 2$. Thus, for example, the number of males born in December is $O_{12,2} = 3761$. Let $O_{i\cdot}$ be the total for month i, $O_{\cdot j}$ be the total for sex j, and $O_{\cdot\cdot}$ be the total over all months and sexes. For example, May has $O_{5\cdot} = 7892$, males have $O_{\cdot 2} = 45{,}682$, and the grand total is $O_{\cdot\cdot} = 88{,}273$.

Our interest now is in whether the population proportion of births for each month is the same for females as for males. We no longer make any assumption about the numerical values of these proportions; our null hypothesis is simply that whatever the proportions are, they are the same for females and males in each month. Again, we wish to compare the observed values, the O_{ij}s with expected values, but now, since we do not have specific hypothesized proportions for any month, we must estimate the expected values.

Under the null hypothesis that the proportions are the same for females and males, it makes sense to pool the male and female data to get an estimate of the proportion of births in each month. Using the column of monthly totals in Table 5.3, the estimated proportion for January is the January total divided by the total for the year, i.e.,

$$\hat{p}_1^0 = \frac{7280}{88273} = 0.0824714.$$

In general, for month i we have

$$\hat{p}_i^0 = \frac{O_{i\cdot}}{O_{\cdot\cdot}}$$

where the superscript of 0 is used to indicate that these proportions are estimated under the null hypothesis of identical monthly rates for males and females. The estimate of the expected number of females born in January is just the number of females born in the year times the estimated probability of a birth in January,

$$\hat{E}_{11} = 42591(0.0824714) = 3512.54.$$

The expected number of males born in January is the number of males born in the year times the estimated probability of a birth in January,

$$\hat{E}_{12} = 45682(0.0824714) = 3767.46.$$

In general,

$$\hat{E}_{ij} = O_{\cdot j}\hat{p}_i^0 = O_{\cdot j}\frac{O_{i\cdot}}{O_{\cdot\cdot}} = \frac{O_{i\cdot}O_{\cdot j}}{O_{\cdot\cdot}}.$$

Again, *the estimated expected values are computed assuming that the proportions of births are the same for females and males in every month, i.e., assuming that the null model is true.* The estimated expected values under the null model are given in Table 5.4. Note that the totals for each month and for each sex remain unchanged.

The estimated expected values are compared to the observations using Pearson residuals, just as in Section 5.3. The Pearson residuals are

$$\tilde{r}_{ij} \equiv \frac{O_{ij} - \hat{E}_{ij}}{\sqrt{\hat{E}_{ij}}}.$$

A more apt name for the Pearson residuals in this context may be *crude standardized residuals*. It is the standardization here that is crude and not the residuals. The standardization in the Pearson residuals ignores the fact that $\hat{E}$ is itself an estimate. Better, but considerably more complicated,

Table 5.4: *Estimated expected Swedish births by month ($\hat{E}_{ij}$s) and pooled proportions.*

| Month | Expectations | | Total | Pooled proportions |
	Female	Male		
January	3512.54	3767.46	7280	0.082
February	3356.70	3600.30	6957	0.079
March	3803.48	4079.52	7883	0.089
April	3803.97	4080.03	7884	0.089
May	3807.83	4084.17	7892	0.089
June	3671.28	3937.72	7609	0.086
July	3659.70	3925.30	7585	0.086
August	3567.06	3825.94	7393	0.084
September	3475.39	3727.61	7203	0.082
October	3330.64	3572.36	6903	0.078
November	3161.29	3390.71	6552	0.074
December	3441.13	3690.87	7132	0.081
Total	42591.00	45682.00	88273	1.000

Table 5.5: *Pearson residuals for Swedish birth months ($\tilde{r}_{ij}$s).*

Month	Female	Male
January	0.41271	−0.39849
February	0.86826	−0.83837
March	1.01369	−0.97880
April	−1.50731	1.45542
May	−0.53195	0.51364
June	−0.10365	0.10008
July	−0.63972	0.61770
August	0.48452	−0.46785
September	0.26481	−0.25570
October	1.04587	−1.00987
November	−0.02288	0.02209
December	−1.19554	1.15438

standardized residuals can be defined for count data, cf. Christensen (1997, Section 6.7) and Chapters 20 and 21. For the Swedish birth data, the Pearson residuals are given in Table 5.5. Note that when compared to a $N(0,1)$ distribution, none of the residuals is very large; all are smaller than 1.51 in absolute value.

As in Section 5.3, the sum of the squared Pearson residuals gives Pearson's χ^2 statistic for testing the null model of no differences between females and males. Pearson's test statistic is

$$X^2 = \sum_{ij} \frac{(O_{ij} - \hat{E}_{ij})^2}{\hat{E}_{ij}}.$$

For the Swedish birth data, computing the statistic from the 24 cells in Table 5.5 gives

$$X^2 = 14.9858.$$

For a formal test, X^2 is compared to a χ^2 distribution. The appropriate number of degrees of freedom for the χ^2 test is the number of cells in the table adjusted to account for all the parameters we have estimated as well as the constraint that the sex totals sum to the grand total. There are 12×2 cells but only $12 - 1$ free months and only $2 - 1$ free sex totals. The appropriate distribution is $\chi^2((12 - 1)(2 - 1)) = \chi^2(11)$. *The degrees of freedom are the number of data rows in Table 5.3 minus 1 times the number of data columns in Table 5.3 minus 1.* The 90th percentile of a $\chi^2(11)$ distribution is $\chi^2(0.9, 11) = 17.28$, so the observed test statistic $X^2 = 14.9858$ could reasonably come from a $\chi^2(11)$ distribution. Moreover, $\chi^2(0.75, 11) = 13.70$, so a one-sided P value is between 0.25

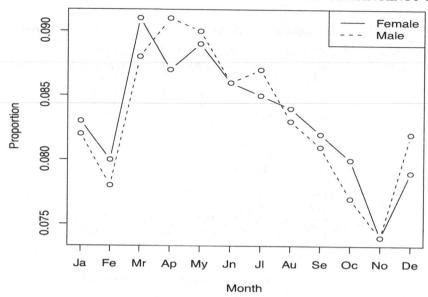

Figure 5.1: *Monthly Swedish birth proportions by sex.*

and 0.10, and the more appropriate two-sided P value would be even bigger. There is no evidence of any differences in the monthly birth rates for males and females.

Another way to evaluate the null model is by comparing the observed monthly birth proportions by sex. These observed proportions are given in Table 5.3. If the populations of females and males have the same proportions of births in each month, the observed proportions of births in each month should be similar (except for sampling variation). One can compare the numbers directly in Table 5.3 or one can make a visual display of the observed proportions as in Figure 5.1.

The methods just discussed apply equally well to the binomial data of Table 5.1. Applying the X^2 test given here to the data of Table 5.1 gives

$$X^2 = 92.2.$$

The statistic X^2 is equivalent to the test statistic given in Section 5.2 using the pooled estimate $\hat{p}_*$ to compute the alternative standard error. The test statistic in Section 5.2 is -9.60, and if we square that we get

$$(-9.60)^2 = 92.2 = X^2.$$

The -9.60 is compared to a $N(0,1)$, while the 92.2 is compared to a $\chi^2(1)$ because Table 5.1 has 2 rows and 2 columns. A $\chi^2(1)$ distribution is obtained by squaring a $N(0,1)$ distribution; P values are identical and critical values are equivalent.

R and Minitab code for fitting contingency tables is given on my website.

5.5 Several independent multinomial samples

The methods of Section 5.4 extend easily to dealing with more than two samples. Consider the data in Table 5.6 that were extracted from Lazerwitz (1961). The data involve samples from three religious groups and consist of numbers of people in various occupational groups. The occupations are labeled A, professions; B, owners, managers, and officials; C, clerical and sales; and D, skilled. The three religious groups are Protestant, Roman Catholic, and Jewish. This is a subset of a larger collection of data that includes many more religious and occupational groups. The fact that we are restricting ourselves to a subset of a larger data set has no effect on the analysis. As discussed in Section 5.4, the analysis of these data is essentially identical regardless of whether the data come from

Table 5.6: *Religion and occupations.*

	Occupation				
Religion	A	B	C	D	Total
Protestant	210	277	254	394	1135
Roman Catholic	102	140	127	279	648
Jewish	36	60	30	17	143
Total	348	477	411	690	1926

one sample of 1926 individuals cross-classified by religion and occupation, or four independent samples of sizes 348, 477, 411, and 690 taken from the occupational groups, or three independent samples of sizes 1135, 648, and 143 taken from the religious groups. We choose to view the data as independent samples from the three religious groups. The data in Table 5.6 constitutes a 3×4 table because, excluding the totals, the table has 3 rows and 4 columns.

We again test whether the populations are the same. In other words, the null hypothesis is that the probability of falling into any occupational group is identical for members of the various religions. Under this null hypothesis, it makes sense to pool the data from the three religions to obtain estimates of the common probabilities. For example, under the null hypothesis of identical populations, the estimate of the probability that a person is a professional is

$$\hat{p}_1^0 = \frac{348}{1926} = 0.180685.$$

For skilled workers the estimated probability is

$$\hat{p}_4^0 = \frac{690}{1926} = 0.358255.$$

Denote the observations as O_{ij} with i identifying a religious group and j indicating occupation. We use a dot to signify summing over a subscript. Thus the total for religious group i is

$$O_{i.} = \sum_j O_{ij},$$

the total for occupational group j is

$$O_{.j} = \sum_i O_{ij},$$

and

$$O_{..} = \sum_{ij} O_{ij}$$

is the grand total. Recall that the null hypothesis is that the probability of being in an occupation group is the same for each of the three populations. Pooling information over religions, we have

$$\hat{p}_j^0 = \frac{O_{.j}}{O_{..}}$$

as the estimate of the probability that someone in the study is in occupational group j. *This estimate is only appropriate when the null model is true.*

The estimated expected count under the null model for a particular occupation and religion is obtained by multiplying the number of people sampled in that religion by the probability of the occupation. For example, the estimated expected count under the null model for Jewish professionals is

$$\hat{E}_{31} = 143(0.180685) = 25.84.$$

Table 5.7: *Estimated expected counts* $(\hat{E}_{ij}s)$.

Religion	A	B	C	D	Total
Protestant	205.08	281.10	242.20	406.62	1135
Roman Catholic	117.08	160.49	138.28	232.15	648
Jewish	25.84	35.42	30.52	51.23	143
Total	348.00	477.00	411.00	690.00	1926

Table 5.8: *Residuals* $(\tilde{r}_{ij}s)$.

Religion	A	B	C	D
Protestant	0.34	−0.24	0.76	−0.63
Roman Catholic	−1.39	−1.62	−0.96	3.07
Jewish	2.00	4.13	−0.09	−4.78

Similarly, the estimated expected count for Roman Catholic skilled workers is

$$\hat{E}_{24} = 648(0.358255) = 232.15.$$

In general,

$$\hat{E}_{ij} = O_{i\cdot}\hat{p}_j^0 = O_{i\cdot}\frac{O_{\cdot j}}{O_{\cdot\cdot}} = \frac{O_{i\cdot}O_{\cdot j}}{O_{\cdot\cdot}}.$$

Again, *the estimated expected values are computed assuming that the null model is true.* The expected values for all occupations and religions are given in Table 5.7.

The estimated expected values are compared to the observations using Pearson residuals. The Pearson residuals are

$$\tilde{r}_{ij} = \frac{O_{ij} - \hat{E}_{ij}}{\sqrt{\hat{E}_{ij}}}.$$

These crude standardized residuals are given in Table 5.8 for all occupations and religions. The largest negative residual is −4.78 for Jewish people with occupation D. This indicates that Jewish people were substantially underrepresented among skilled workers relative to the other two religious groups. On the other hand, Roman Catholics were substantially overrepresented among skilled workers, with a positive residual of 3.07. The other large residual in the table is 4.13 for Jewish people in group B. Thus Jewish people were more highly represented among owners, managers, and officials than the other religious groups. Only one other residual is even moderately large, the 2.00 indicating a high level of Jewish people in the professions. The main feature of these data seems to be that the Jewish group was different from the other two. A substantial difference appears in every occupational group except clerical and sales.

As in Sections 5.3 and 5.4, the sum of the squared Pearson residuals gives Pearson's χ^2 statistic for testing the null model that the three populations are the same and the samples are independent multinomials. Pearson's test statistic is

$$X^2 = \sum_{ij} \frac{(O_{ij} - \hat{E}_{ij})^2}{\hat{E}_{ij}}.$$

Summing the squares of the values in Table 5.8 gives

$$X^2 = 60.0.$$

The appropriate number of degrees of freedom for the χ^2 test is the number of data rows in Table 5.6 minus 1 times the number of data columns in Table 5.6 minus 1. Thus the appropriate

Table 5.9: *Observed proportions by religion.*

Religion	Occupation				Total
	A	B	C	D	
Protestant	0.185	0.244	0.224	0.347	1.00
Roman Catholic	0.157	0.216	0.196	0.431	1.00
Jewish	0.252	0.420	0.210	0.119	1.00

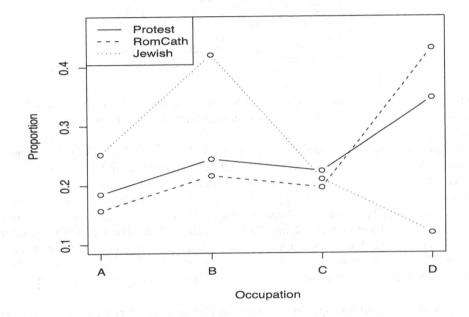

Figure 5.2: *Occupational proportions by religion.*

reference distribution is $\chi^2((3-1)(4-1)) = \chi^2(6)$. The 99.5th percentile of a $\chi^2(6)$ distribution is $\chi^2(0.995,6) = 18.55$ so the observed statistic $X^2 = 60.0$ could not reasonably come from a $\chi^2(6)$ distribution. In particular, for multinomial sampling the test clearly indicates that the proportions of people in the different occupation groups differ with religious category.

As in the previous section, we can informally evaluate the null model by examining the observed proportions for each religious group. The observed proportions are given in Table 5.9. Under the null model, the observed proportions in each occupation category should be the same for all the religions (up to sampling variability). Figure 5.2 displays the observed proportions graphically. The Jewish group is obviously very different from the other two groups in occupations B and D and is very similar in occupation C. The Jewish proportion seems somewhat different for occupation A. The Protestant and Roman Catholic groups seem similar except that the Protestants are a bit underrepresented in occupation D and therefore are overrepresented in the other three categories. (Remember that the four proportions for each religion must add up to one, so being underrepresented in one category forces an overrepresentation in one or more other categories.)

5.6 Lancaster–Irwin partitioning

Lancaster–Irwin partitioning is a method for breaking a table of count data into smaller tables. (For readers with prior knowledge of analysis of variance, when used to its maximum extent, partitioning is similar in spirit to looking at orthogonal contrasts in analysis of variance.) The basic idea is that a table of counts can be broken into two component tables, a reduced table and a collapsed table. Table 5.10 illustrates such a partition for the data of Table 5.6. In the reduced table, the row for the

Table 5.10: *A Lancaster–Irwin partition of Table 5.6.*

Reduced table					
Religion	A	B	C	D	Total
Protestant	210	277	254	394	1135
Roman Catholic	102	140	127	279	648
Total	312	417	381	673	1783

Collapsed table					
Religion	A	B	C	D	Total
Prot. & R.C.	312	417	381	673	1783
Jewish	36	60	30	17	143
Total	348	477	411	690	1926

Jewish group has been eliminated, leaving a subset of the original table. In the collapsed table, the two rows in the reduced table, Protestant and Roman Catholic, have been collapsed into a single row.

In Lancaster–Irwin partitioning, we pick a group of either rows or columns, say rows. The reduced table involves all of the columns but only the chosen subgroup of rows. The collapsed table involves all of the columns and all of the rows *not* in the chosen subgroup, along with a row that combines (collapses) all of the subgroup rows into a single row. In Table 5.10 the chosen subgroup of rows contains the Protestants and Roman Catholics. The reduced table involves all occupational groups but only the Protestants and Roman Catholics. In the collapsed table the occupational groups are unaffected but the Protestants and Roman Catholics are combined into a single row. The other rows remain the same; in this case the other rows consist only of the Jewish row. As alluded to above, rather than picking a group of rows to form the partitioning, we could select a group of columns.

Lancaster–Irwin partitioning is by no means a unique process. There are as many ways to partition a table as there are ways to pick a group of rows or columns. In Table 5.10 we made a particular selection based on the residual analysis of these data from the previous section. The main feature we discovered in the residual analysis was that the Jewish group seemed to be different from the other two groups. Thus it seemed to be of interest to compare the Jewish group with a combination of the others and then to investigate what differences there might be among the other religious groups. The partitioning of Table 5.10 addresses precisely these questions. In the remainder of our discussion we assume that multinomial sampling is valid so that we have tests of the null hypothesis and not just the null model.

Tables 5.11 and 5.12 provide statistics for the analysis of the reduced table and collapsed table. The reduced table simply reconfirms our previous conclusions. The X^2 value of 12.3 indicates substantial evidence of a difference between Protestants and Roman Catholics. The percentage point $\chi^2(0.995, 3) = 12.84$ indicates that the one-sided P value for the test is a bit greater than 0.005. The residuals indicate that the difference was due almost entirely to the fact that Roman Catholics have relatively higher representation among skilled workers. (Or equivalently, that Protestants have relatively lower representation among skilled workers.) Overrepresentation of Roman Catholics among skilled workers forces their underrepresentation among other occupational groups but the level of underrepresentation in the other groups was approximately constant as indicated by the approximately equal residuals for Roman Catholics in the other three occupation groups. We will see later that for Roman Catholics in the other three occupation groups, their distribution among those groups was almost the same as those for Protestants. This reinforces the interpretation that the difference was due almost entirely to the difference in the skilled group.

The conclusions that can be reached from the collapsed table are also similar to those drawn in the previous section. The X^2 value of 47.5 on 3 degrees of freedom indicates overwhelming evidence that the Jewish group was different from the combined Protestant–Roman Catholic group. The residuals can be used to isolate the sources of the differences. The two groups differed in

Table 5.11: *Reduced table.*

Religion	Observations				Total
	A	B	C	D	
Protestant	210	277	254	394	1135
Roman Catholic	102	140	127	279	648
Total	312	417	381	673	1783
Religion	Estimated expected counts				Total
	A	B	C	D	
Protestant	198.61	265.45	242.53	428.41	1135
Roman Catholic	113.39	151.55	138.47	244.59	648
Total	312.00	417.00	381.00	673.00	1783
Religion	Pearson residuals				
	A	B	C	D	
Protestant	0.81	0.71	0.74	−1.66	
Roman Catholic	−1.07	−0.94	−0.97	2.20	

$X^2 = 12.3, df = 3$

Table 5.12: *Collapsed table.*

Religion	Observations				Total
	A	B	C	D	
Prot. & R.C.	312	417	381	673	1783
Jewish	36	60	30	17	143
Total	348	477	411	690	1926
Religion	Estimated expected counts				Total
	A	B	C	D	
Prot. & R.C.	322.16	441.58	380.48	638.77	1783
Jewish	25.84	35.42	30.52	51.23	143
Total	348.00	477.00	411.00	690.00	1926
Religion	Pearson residuals				
	A	B	C	D	
Prot. & R.C.	−0.57	−1.17	0.03	1.35	
Jewish	2.00	4.13	−0.09	−4.78	

$X^2 = 47.5, df = 3$

proportions of skilled workers and proportions of owners, managers, and officials. There was a substantial difference in the proportions of professionals. There was almost no difference in the proportion of clerical and sales workers between the Jewish group and the others.

The X^2 value computed for Table 5.6 was 60.0. The X^2 value for the collapsed table is 47.5 and the X^2 value for the reduced table is 12.3. Note that $60.0 \doteq 59.8 = 47.5 + 12.3$. It is not by chance that the sum of the X^2 values for the collapsed and reduced tables is approximately equal to the X^2 value for the original table. In fact, this relationship is a primary reason for using the Lancaster–Irwin partitioning method. The approximate equality $60.0 \doteq 59.8 = 47.5 + 12.3$ indicates that the vast bulk of the differences between the three religious groups is due to the collapsed table, i.e., the difference between the Jewish group and the other two. Roughly 80% (47.5/60) of the original X^2 value is due to the difference between the Jewish group and the others. Of course the X^2 value 12.2 for the reduced table is still large enough to strongly suggest differences between Protestants and Roman Catholics.

Not all data will yield an approximation as close as $60.0 \doteq 59.8 = 47.5 + 12.3$ for the partitioning. The fact that we have an approximate equality rather than an exact equality is due to our choice of the test statistic X^2. Pearson's statistic is simple and intuitive; it compares observed values with expected values and standardizes by the size of the expected value. An alternative test statistic also

Table 5.13: *Reduced table.*

Religion	Observations			Total
	A	B	C	
Protestant	210	277	254	741
Roman Catholic	102	140	127	369
Total	312	417	381	1110

Religion	Estimated expected counts			Total
	A	B	C	
Protestant	208.28	278.38	254.34	741
Roman Catholic	103.72	138.62	126.66	369
Total	312.00	417.00	381.00	1110

Religion	Pearson residuals			
	A	B	C	
Protestant	0.12	−0.08	0.00	
Roman Catholic	−0.17	0.12	0.03	

$X^2 = 0.065, df = 2$

exists called the *likelihood ratio test statistic*,

$$G^2 = 2 \sum_{\text{all cells}} O_{ij} \log \left(O_{ij} / \hat{E}_{ij} \right).$$

The motivation behind the likelihood ratio test statistic is not as transparent as that behind Pearson's statistic, so we will not discuss the likelihood ratio test statistic in any detail until later. However, one advantage of the likelihood ratio test statistic is that the sum of its values for the reduced table and collapsed table gives *exactly* the likelihood ratio test statistic for the original table. Likelihood ratio statistics will be used extensively in Chapters 20 and 21, and in Chapter 21, Lancaster-Irwin partitioning will be revisited.

Further partitioning

We began this section with the 3×4 data of Table 5.6 that has 6 degrees of freedom for its X^2 test. We partitioned the data into two 2×4 tables, each with 3 degrees of freedom. We can continue to use the Lancaster–Irwin method to partition the reduced and collapsed tables given in Table 5.10. The process of partitioning previously partitioned tables can be continued until the original table is broken into a collection of 2×2 tables. Each 2×2 table has one degree of freedom for its chi-squared test, so partitioning provides a way of breaking a large table into one degree of freedom components.

What we have been calling the reduced table involves all four occupational groups along with the two religious groups Protestant and Roman Catholic. The table was given in both Table 5.10 and Table 5.11. We now consider this table further. It was discussed earlier that the difference between Protestants and Roman Catholics can be ascribed almost entirely to the difference in the proportion of skilled workers in the two groups. To explore this we choose a new partition based on a group of *columns* that includes all occupations other than the skilled workers. Thus we get the 'reduced' table in Table 5.13 with occupations A, B, and C and the 'collapsed' table in Table 5.14 with occupation D compared to the accumulation of the other three.

Table 5.13 allows us to examine the proportions of Protestants and Catholics in the occupational groups A, B, and C. We are not investigating whether Catholics were more or less likely than Protestants to enter these occupational groups; we are examining their distribution *within* the groups. The analysis is based only on those individuals *that were in this collection of three occupational groups*. The X^2 value is exceptionally small, only 0.065. There is no evidence of any difference between Protestants and Catholics for these three occupational groups.

Table 5.14: *Collapsed table.*

Religion	Observations		Total
	A & B & C	D	
Protestant	741	394	1135
Roman Catholic	369	279	648
Total	1110	673	1783

Religion	Estimated expected counts		Total
	A & B & C	D	
Protestant	706.59	428.41	1135
Roman Catholic	403.41	244.59	648
Total	1110.00	673.00	1783

Religion	Pearson residuals	
	A & B & C	D
Protestant	1.29	−1.66
Roman Catholic	−1.71	2.20

$X^2 = 12.2, df = 1$

Table 5.15: *Collapsed table.*

Religion	Observations		Total
	A & B & D	C	
Prot. & R.C.	1402	381	1783
Jewish	113	30	143
Total	1515	411	1926

Religion	Estimated expected counts		Total
	A & B & D	C	
Prot. & R.C.	1402.52	380.48	1783
Jewish	112.48	30.52	143
Total	1515.00	411.00	1926

Religion	Pearson residuals	
	A & B & D	C
Prot. & R.C.	−0.00	0.03
Jewish	0.04	−0.09

$X^2 = 0.01, df = 1$

Table 5.13 is a 2×3 table. We could partition it again into two 2×2 tables but there is little point in doing so. We have already established that there is no evidence of differences.

Table 5.14 has the three occupational groups A, B, and C collapsed into a single group. This table allows us to investigate whether Catholics were more or less likely than Protestants to enter this group of three occupations. The X^2 value is a substantial 12.2 on one degree of freedom, so we can tentatively conclude that there was a difference between Protestants and Catholics. From the residuals, we see that *among people in the four occupational groups*, Catholics were more likely than Protestants to be in the skilled group and less likely to be in the other three.

Table 5.14 is a 2×2 table so no further partitioning is possible. Note again that the X^2 of 12.3 from Table 5.11 is approximately equal to the sum of the 0.065 from Table 5.13 and the 12.2 from Table 5.14.

Finally, we consider additional partitioning of the collapsed table given in Tables 5.10 and 5.12. It was noticed earlier that the Jewish group seemed to differ from Protestants and Catholics in every occupational group except C, clerical and sales. Thus we choose a partitioning that isolates group C. Table 5.15 gives a collapsed table that compares C to the combination of groups A, B, and D. Table 5.16 gives a reduced table that involves only occupational groups A, B, and D.

Table 5.15 demonstrates no difference between the Jewish group and the combined Protestant–Catholic group. Thus the proportion of people in clerical and sales was about the same for the

Table 5.16: *Reduced table.*

Religion	Observations			Total
	A	B	D	
Prot. & R.C.	312	417	673	1402
Jewish	36	60	17	113
Total	348	477	690	1515

Religion	Estimated expected counts			Total
	A	B	D	
Prot. & R.C.	322.04	441.42	638.53	1402
Jewish	25.96	35.58	51.47	113
Total	348.00	477.00	690.00	1515

Religion	Pearson residuals		
	A	B	D
Prot. & R.C.	−0.59	−1.16	1.36
Jewish	1.97	4.09	−4.80

$X^2 = 47.2, df = 2$

Table 5.17: *Reduced table.*

Religion	Observations		Total
	B	D	
Prot. & R.C.	417	673	1090
Jewish	60	17	77
Total	477	690	1167

Religion	Estimated expected counts		Total
	B	D	
Prot. & R.C.	445.53	644.47	1090
Jewish	31.47	45.53	77
Total	477.00	690.00	1167

Religion	Pearson residuals	
	B	D
Prot. & R.C.	−1.35	1.12
Jewish	5.08	−4.23

$X^2 = 46.8, df = 1$

Jewish group as for the combined Protestant and Roman Catholic group. Any differences between the Jewish and Protestant–Catholic groups must be in the proportions of people *within* the three occupational groups A, B, and D.

Table 5.16 demonstrates major differences between occupations A, B, and D for the Jewish group and the combined Protestant–Catholic group. As seen earlier and reconfirmed here, skilled workers had much lower representation among the Jewish group, while professionals and especially owners, managers, and officials had much higher representation among the Jewish group.

Table 5.16 can be further partitioned into Tables 5.17 and 5.18. Table 5.17 is a reduced 2 × 2 table that considers the difference between the Jewish group and others with respect to occupational groups B and D. Table 5.18 is a 2 × 2 collapsed table that compares occupational group A with the combination of groups B and D.

Table 5.17 shows a major difference between occupational groups B and D. Table 5.18 may or may not show a difference between group A and the combination of groups B and D. The X^2 values are 46.8 and 5.45, respectively. The question is whether an X^2 value of 5.45 is suggestive of a difference between religious groups when we have examined the data in order to choose the partitions of Table 5.6. Note that the two X^2 values sum to 52.25, whereas the X^2 value for Table 5.16, from which they were constructed, is only 47.2. The approximate equality is a very rough approximation. Nonetheless, we see from the relative sizes of the two X^2 values that the majority of the difference

Table 5.18: *Collapsed table.*

Religion	Observations		Total
	A	B & D	
Prot. & R.C.	312	1090	1402
Jewish	36	77	113
Total	348	1167	1515

Religion	Estimated expected counts		Total
	A	B & D	
Prot. & R.C.	322.04	1079.96	1402
Jewish	25.96	87.04	113
Total	348.00	1167.00	1515

Religion	Pearson residuals	
	A	B & D
Prot. & R.C.	−0.56	0.30
Jewish	1.97	−1.08

$X^2 = 5.45, df = 1$

between the Jewish group and the other religious groups was in the proportion of owners, managers, and officials as compared to the proportion of skilled workers.

Ultimately, we have partitioned Table 5.6 into Tables 5.13, 5.14, 5.15, 5.17, and 5.18. These are all 2×2 tables except for Table 5.13. We could also have partitioned Table 5.13 into two 2×2 tables but we chose to leave it because it showed so little evidence of any difference between Protestants and Roman Catholics for the three occupational groups considered. The X^2 value of 60.0 for Table 5.6 was approximately partitioned into X^2 values of 0.065, 12.2, 0.01, 46.8, and 5.45, respectively. Except for the 0.065 from Table 5.13, each of these values is computed from a 2×2 table, so each has 1 degree of freedom. The 0.065 is computed from a 2×3 table, so it has 2 degrees of freedom. The sum of the five X^2 values is 64.5, which is roughly equal to the 60.0 from Table 5.6.

The five X^2 values can all be used in testing but we let the data suggest the partitions. It is inappropriate to compare these X^2 values to their usual χ^2 percentage points to obtain tests. A simple way to adjust for both the multiple testing and the data dredging (letting the data suggest partitions) is to compare all X^2 values to the percentage points appropriate for Table 5.6. For example, the $\alpha = 0.05$ test for Table 5.6 uses the critical value $\chi^2(0.95, 6) = 12.58$. By this standard, Table 5.17 with $X^2 = 46.8$ shows a significant difference between religious groups and Table 5.14 with $X^2 = 12.2$ nearly shows a significant difference between religious groups. The value of $X^2 = 5.45$ for Table 5.18 gives no evidence of a difference based on this criterion even though such a value would be highly suggestive if we could compare it to a $\chi^2(1)$ distribution. This method is similar in spirit to Scheffé's method to be considered in Section 13.4 and suffers from the same extreme conservatism.

5.7 Exercises

EXERCISE 5.7.1. Reiss et al. (1975) and Fienberg (1980) reported that 29 of 52 virgin female undergraduate university students who used a birth control clinic thought that extramarital sex is not always wrong. Give a 99% confidence interval for the population proportion of virgin undergraduate university females who use a birth control clinic and think that extramarital sex is not always wrong.

In addition, 67 of 90 virgin females who did not use the clinic thought that extramarital sex is not always wrong. Give a 99% confidence interval for the difference in proportions between the two groups and give a 0.05 level test that there is no difference.

EXERCISE 5.7.2. In France in 1827, 6929 people were accused in the courts of assize and 4236 were convicted. In 1828, 7396 people were accused and 4551 were convicted. Give a 95% confidence interval for the proportion of people convicted in 1827. At the 0.01 level, test the null model

Table 5.19: *French convictions.*

Year	Convictions	Accusations
1825	4594	7234
1826	4348	6988
1827	4236	6929
1828	4551	7396
1829	4475	7373
1830	4130	6962

Table 5.20: *Occupation and religion.*

Religion	A	B	C	D	E	F	G	H
White Baptist	43	78	64	135	135	57	86	114
Black Baptist	9	2	9	23	47	77	18	41
Methodist	73	80	80	117	102	58	66	153
Lutheran	23	36	43	59	46	26	49	46
Presbyterian	35	54	38	46	19	22	11	46
Episcopalian	27	27	20	14	7	5	2	15

that the conviction rate in 1827 was different than $2/3$. Does the result of the test depend on the choice of standard error? Give a 95% confidence interval for the difference in conviction rates between the two years. Test the hypothesis of no difference in conviction rates using $\alpha = 0.05$ and both standard errors.

EXERCISE 5.7.3. Pauling (1971) reports data on the incidence of colds among French skiers who where given either ascorbic acid or a placebo. Of 139 people given ascorbic acid, 17 developed colds. Of 140 people given the placebo, 31 developed colds. Do these data suggest that the proportion of people who get colds differs depending on whether they are given ascorbic acid?

EXERCISE 5.7.4. Use the data in Table 5.2 to test whether the probability of a birth in each month is the number of days in the month divided by 365. Thus the null probability for January is $31/365$ and the null probability for February is $28/365$.

EXERCISE 5.7.5. Snedecor and Cochran (1967) report data from an unpublished report by E. W. Lindstrom. The data concern the results of cross-breeding two types of corn (maize). In 1301 crosses of two types of plants, 773 green, 231 golden, 238 green-golden, and 59 golden-green-striped plants were obtained. If the inheritance of these properties is particularly simple, Mendelian genetics suggests that the probabilities for the four types of corn may be $9/16, 3/16, 3/16$, and $1/16$, respectively. Test whether these probabilities are appropriate. If they are inappropriate, identify the problem.

EXERCISE 5.7.6. Quetelet (1842) and Stigler (1986, p. 175) report data on conviction rates in the French Courts of Assize (Law Courts) from 1825 to 1830. The data are given in Table 5.19. Test whether the conviction rate is the same for each year. Use $\alpha = 0.05$. (Hint: Table 5.19 is written in a nonstandard form. You need to modify it before applying the methods of this chapter.) If there are differences in conviction rates, use residuals to explore these differences.

EXERCISE 5.7.7. Table 5.20 contains additional data from Lazerwitz (1961). These consist of a breakdown of the Protestants in Table 5.6 but with the addition of four more occupational categories. The additional categories are E, semiskilled; F, unskilled; G, farmers; H, no occupation. Analyze the data with an emphasis on partitioning the table.

Table 5.21: *Heights and chest circumferences.*

Chest	Heights					Total
	64–65	66–67	68–69	70–71	71–73	
39	142	442	341	117	20	1062
40	118	337	436	153	38	1082
Total	260	779	777	270	58	2144

EXERCISE 5.7.8.   Stigler (1986, p. 208) reports data from the *Edinburgh Medical and Surgical Journal* (1817) on the relationship between heights and chest circumferences for Scottish militia men. Measurements were made in inches. We concern ourselves with two groups of men, those with 39-inch chests and those with 40-inch chests. The data are given in Table 5.21. Test whether the distribution of heights is the same for these two groups.

Chapter 6

Simple Linear Regression

This chapter examines data that come as pairs of numbers, say (x,y), and the problem of fitting a line to them. More generally, it examines the problem of predicting one variable (y) from values of another variable (x). Consider for the moment the popular wisdom that people who read a lot tend to have large vocabularies and poor eyes. Thus, reading causes both conditions: large vocabularies and poor eyes. If this is true, it may be possible to predict the size of someone's vocabulary from the condition of their eyes. Of course this does not mean that having poor eyes causes large vocabularies. Quite the contrary, if anything, poor eyes probably keep people from reading and thus cause small vocabularies. Regression analysis is concerned with predictive ability, not with causation.

Section 6.1 of this chapter introduces an example along with many of the basic ideas and methods of simple linear regression (SLR). The rest of the chapter goes into the details of simple linear regression. Section 6.7 deals with an idea closely related to simple linear regression: the correlation between two variables. Section 6.9 provides an initial introduction to multiple regression, i.e., regression with more than one predictor variable.

6.1 An example

Data from *The Coleman Report* were reproduced in Mosteller and Tukey (1977). The data were collected from schools in the New England and Mid-Atlantic states of the USA. For now we consider only two variables: y—the mean verbal test score for sixth graders, and x—a composite measure of socioeconomic status. The data are presented in Table 6.1.

Figure 6.1 contains a scatter plot of the data. Note the rough linear relationship. The higher the composite socioeconomic status variable, the higher the mean verbal test score. However, there is considerable error in the relationship. By no means do the points lie exactly on a straight line.

We assume a basic linear relationship between the ys and xs, something like $y = \beta_0 + \beta_1 x$. Here β_1 is the slope of the line and β_0 is the intercept. Unfortunately, the observed y values do not fit exactly on a line, so $y = \beta_0 + \beta_1 x$ is only an approximation. We need to modify this equation to allow for the variability of the observations about the line. We do this by building a random error

Table 6.1: *Coleman Report data.*

School	y	x	School	y	x
1	37.01	7.20	11	23.30	−12.86
2	26.51	−11.71	12	35.20	0.92
3	36.51	12.32	13	34.90	4.77
4	40.70	14.28	14	33.10	−0.96
5	37.10	6.31	15	22.70	−16.04
6	33.90	6.16	16	39.70	10.62
7	41.80	12.70	17	31.80	2.66
8	33.40	−0.17	18	31.70	−10.99
9	41.01	9.85	19	43.10	15.03
10	37.20	−0.05	20	41.01	12.77

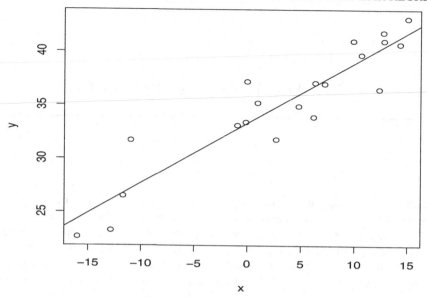

Figure 6.1: *Coleman Report data.*

term into the linear relationship. Write the relationship as $y = \beta_0 + \beta_1 x + \varepsilon$, where ε indicates the random error. In this model for the behavior of the data, ε accounts for the deviations between the y values we actually observe and the line $\beta_0 + \beta_1 x$ where we expect to observe any y value that corresponds to x. As we are interested in predicting y from known x values, *we treat x as a known (nonrandom) variable.*

We assume that the relationship $y = \beta_0 + \beta_1 x + \varepsilon$ applies to all of our observations. For the current data, that means we assume the relationship holds for all of the 20 pairs of values in Table 6.1. This assumption is stated as *the simple linear regression model* for these data,

$$y_i = \beta_0 + \beta_1 x_i + \varepsilon_i, \tag{6.1.1}$$

$i = 1, \ldots, 20$. For this model to be useful, we need to make some assumptions about the errors, the ε_is. The standard assumptions are that the

$$\varepsilon_i \text{s are independent } N(0, \sigma^2).$$

Given data for which these assumptions are reasonable, we can estimate the unknown parameters. Although we assume a linear relationship between the ys and xs, the model does not assume that we know the slope β_1 or the intercept β_0. Together, these unknown parameters would tell us the exact nature of the linear relationship but both need to be estimated. We use the notation $\hat{\beta}_1$ and $\hat{\beta}_0$ to denote estimates of β_1 and β_0, respectively. To perform statistical inferences we also need to estimate the variance of the errors, σ^2. Note that σ^2 is also the variance of the y observations because none of β_0, β_1, and x are random.

The simple linear regression model involves many assumptions. It assumes that the relationship between y and x is linear, it assumes that the errors are normally distributed, it assumes that the errors all have the same variance, it assumes that the errors are all independent, and it assumes that the errors all have mean 0. This last assumption is redundant. It turns out that the errors all have mean 0 if and only if the relationship between y and x is linear. As far as possible, we will want to verify (validate) that these assumptions are reasonable before we put much faith in the estimates and statistical inferences that can be obtained from simple linear regression. Chapters 7 and 8 deal with checking these assumptions.

Before getting into a detailed discussion of simple linear regression, we illustrate some highlights using the *Coleman Report* data. We need to fit Model (6.1.1) to the data. A computer program typically yields parameter estimates, standard errors for the estimates, t ratios for testing whether the parameters are zero, P values for the tests, and an analysis of variance table. These results are often displayed as illustrated below. We refer to them as the *table of coefficients* and the *analysis of variance (ANOVA) table*, respectively.

Table of Coefficients

Predictor	$\hat{\beta}_k$	SE($\hat{\beta}_k$)	t	P
Constant	33.3228	0.5280	63.11	0.000
x	0.56033	0.05337	10.50	0.000

Analysis of Variance

Source	df	SS	MS	F	P
Regression	1	552.68	552.68	110.23	0.000
Error	18	90.25	5.01		
Total	19	642.92			

Much can be learned from these two tables of statistics. The estimated regression equation is

$$\hat{y} = 33.3 + 0.560x.$$

This equation allows us to predict a value for y when the value of x is given. In particular, for these data an increase of one unit in socioeconomic status increases our prediction of mean verbal test scores by about 0.56 units. This is not to say that some program to increase socioeconomic statuses by one unit will increase mean verbal test scores by about 0.56 unit. The 0.56 *describes* the *current* data, *it does not imply a causal relationship*. If we want to predict the mean verbal test score for a school that is very similar to the ones in this study, this equation should give good predictions. If we want to predict the mean verbal test score for a school that is very different from the ones in this study, this equation is likely to give poor predictions. In fact, if we collect new data from schools with very different socioeconomic statuses, the data are not similar to these, so this fitted model would be highly questionable if applied in the new situation. Nevertheless, a simple linear regression model with a different intercept and slope might fit the new data well. Similarly, data collected after a successful program to raise socioeconomic statuses are unlikely to be similar to the data collected before such a program. The relationship between socioeconomic status and mean verbal test scores may be changed by such a program. In particular, the things causing both socioeconomic status and mean verbal test score may be changed in unknown ways by such a program. These are crucial points and bear repeating. *The regression equation describes an observed relationship between mean verbal test scores and socioeconomic status. It can be used to predict mean verbal test scores from socioeconomic status in similar situations. It does not imply that changing the socioeconomic status a fixed amount will cause the mean verbal test scores to change by a proportional amount.*

The table of coefficients

The table of coefficients allows us to perform a variety of inferences on single parameters. In simple linear regression, the reference distribution for statistical inferences is almost invariably $t(dfE)$ where dfE is the degrees of freedom for error from the analysis of variance table. For these data, $dfE = n - 2 = 18$ because we are estimating 2 regression parameters β_0 and β_1. We now consider some illustrations of statistical inferences.

The t statistics for testing $H_0 : \beta_k = 0$ are reported in the table of coefficients. For example, the test of $H_0 : \beta_1 = 0$ has

$$t_{obs} = \frac{0.56033}{0.05337} = 10.50.$$

The significance level of the test is the P value,

$$P = \Pr[|t(18)| > 10.50] = 0.000.$$

The value 0.000 indicates a large amount of evidence against the null model. If we are convinced that all the assumptions of the simple linear regression model are correct, then the only thing that could be wrong with the null model is that $\beta_1 \neq 0$. Note that if $\beta_1 = 0$, the linear relationship becomes $y = \beta_0 + \varepsilon$, so there is no relationship between y and x, i.e., y does not depend on x. The small P value indicates that the slope is not zero and thus the variable x helps to explain the variable y.

The table of coefficients also allows us to compute a variety of other t statistics. For example, if we wanted to test $H_0 : \beta_1 = 1$

$$t_{obs} = \frac{0.56033 - 1}{0.05337} = -8.24.$$

The significance level of the test is the P value,

$$P = \Pr[|t(18)| > |-8.24|] = 0.000.$$

Alternatively, we could compute the 95% confidence interval for β_1, which has endpoints

$$\hat{\beta}_1 \pm t(0.975, dfE) \, \text{SE}\left(\hat{\beta}_1\right).$$

From a t table, $t(0.975, 18) = 2.101$, so, using the tabled statistics, the endpoints are

$$0.56033 \pm 2.101(0.05337).$$

The confidence interval is $(0.448, 0.672)$, so values of the slope β_1 between 0.448 and 0.672 are consistent with the data and the model based on a 0.05 level test.

Consider the problem of estimating the value of the line at $x = -16.04$. This value of x is the minimum observed value for socioeconomic status, so it is somewhat dissimilar to the other x values in the data. Its dissimilarity causes there to be substantial variability in estimating the regression line (expected value of y) at this point. The point on the line is $\beta_0 + \beta_1(-16.04)$ and the estimator is

$$\hat{\beta}_0 + \hat{\beta}_1(-16.04) = 33.32 + 0.560(-16.04) = 24.34.$$

For constructing 95% t intervals, the percentile needed is $t(0.975, 18) = 2.101$. The standard error for the estimate of the point on the line is usually available from computer programs (cf. Section 6.10.); in this example it is $\text{SE}(Line) = 1.140$. The 95% confidence interval for the point on the line $\beta_0 + \beta_1(-16.04)$ has endpoints

$$24.34 \pm 2.101(1.140),$$

which gives the interval $(21.9, 26.7)$. Values of the population mean of the schoolwise mean verbal test scores for New England and Mid-Atlantic sixth graders with a school socioeconomic measure of -16.04 that are consistent with the data are those between 21.9 and 26.7.

The prediction $\hat{y}$ for a new observation with $x = -16.04$ is simply the estimated point on the line

$$\hat{y} = \hat{\beta}_0 + \hat{\beta}_1(-16.04) = 24.34.$$

Prediction of a new observation is subject to more error than estimation of a point on the line. A new observation has the same variance as all other observations, so the prediction interval must account for this variance as well as for the variance of estimating the point on the line. The standard error for the prediction interval is computed as

$$\text{SE}(Prediction) = \sqrt{MSE + \text{SE}(Line)^2}. \tag{6.1.2}$$

In this example,

$$SE(Prediction) = \sqrt{5.01 + (1.140)^2} = 2.512.$$

The prediction interval endpoints are

$$24.34 \pm 2.101(2.512),$$

and the 95% prediction interval is $(19.1, 29.6)$. Future values of sixth graders's mean verbal test scores that are consistent with the data and model are those between 19.1 and 29.6. These are for a *different* New England or Mid-Atlantic school with a socioeconomic measure of -16.04. Note that the prediction interval is considerably wider than the corresponding confidence interval. Note also that this is just another special case of the prediction theory in Section 3.7. As such, these results are analogous to those obtained for the one-sample and two-sample data structures.

The ANOVA Table

The primary value of the analysis of variance table is that it gives the degrees of freedom, the sum of squares, and the mean square for error. The mean squared error is the estimate of σ^2 and the sum of squares error and degrees of freedom for error are vital for comparing different regression models. Note that the sums of squares for regression and error add up to the sum of squares total and that the degrees of freedom for regression and error also add up to the degrees of freedom total.

The analysis of variance table gives an alternative but equivalent test for whether the x variable helps to explain y. The alternative test of $H_0 : \beta_1 = 0$ is based on

$$F = \frac{MSReg}{MSE} = \frac{552.68}{5.01} = 110.23.$$

Note that the value of this statistic is $110.23 = (10.50)^2$; the F statistic is just the square of the corresponding t statistic for testing $H_0 : \beta_1 = 0$. The F and t tests are equivalent. In particular, the P values are identical. In this case, both are infinitesimal, zero to three decimal places. Our conclusion that $\beta_1 \neq 0$ means that the x variable helps to explain the variation in the y variable. In other words, it is possible to predict the mean verbal test scores for a school's sixth grade classes from the socioeconomic measure. Of course, the fact that some predictive ability exists does not mean that the predictive ability is sufficient to be useful.

6.1.1 Computer commands

Minitab, SAS, and R commands are all given on the website. However, since R is both free and more complicated than the other options, we now present a reasonably complete set of R commands for performing an analysis. The commands are also available on the website.

The following R script will give you most of what you need. When you open R, go to the File menu and open a new script window. Copy this script into the new window. To run part of the script, highlight the part you want to run, right click your mouse, and choose "Run line or selection." The last part of the script produces items discussed in Chapter 7. The material is integral to a good data analysis, so I consider it is more useful to keep all of the commands together here. The Chapter 7 material includes the production of four plots. As far as I know, only one graph will show up at a time, so graphs need to be run sequentially. If you run them all, you will only see the last one. (There are ways to produce all four at once but the graphs are smaller; see the website.)

```
coleman.slr <- read.table("C:\\E-drive\\Books\\ANREG2\\NewData\\tab6-1.dat",
          sep="",col.names=c("School","x","y"))

attach(coleman.slr)
coleman.slr
```

```
#Summary tables
cr <- lm(y ~ x)
crp=summary(cr)
crp
anova(cr)
```

```
#compute confidence intervals
confint(cr, level=0.95)
# or do it from scratch
R=crp$cov.unscaled
se <- sqrt(diag(R)) * crp$sigma
ci=c(cr$coef-qt(.975,crp$df[2])*se, cr$coef+qt(.975,crp$df[2])*se)
CI95 = matrix(ci,crp$df[1],2)
CI95
```

```
#prediction
new = data.frame(x=c(-16.04))
predict(lm(y~x),new,se.fit=T,interval="confidence")
predict(lm(y~x),new,interval="prediction")
```

```
#ploting data with fitted line
plot(x,y)
abline(cr)
```

The rest of the script gives procedures discussed in Chapter 7. First we create a table of diagnostic values. Then we perform residual plots.

```
infv = c(y,cr$fit,hatvalues(cr),rstandard(cr),
          rstudent(cr),cooks.distance(cr))
inf=matrix(infv,I(crp$df[1]+crp$df[2]),6,dimnames = list(NULL,
                   c("y", "yhat", "lev","r","t","C")))
inf
# Note: delete y from table if it contains missing observations
```

```
#Normal and two residual plots:  Do one plot at a time!
qqnorm(rstandard(cr),ylab="Standardized residuals")

plot(cr$fit,rstandard(cr),xlab="Fitted",
        ylab="Standardized residuals",main="Residual-Fitted plot")

plot(x,rstandard(cr),xlab="x",ylab="Standardized residuals",
        main="Residual-Socio plot")
```

```
#leverage plot
Leverage=hatvalues(cr)
plot(School,Leverage,main="School-Leverage plot")
```

```
# Wilk-Francia Statistic
rankit=qnorm(ppoints(rstandard(cr),a=I(3/8)))
ys=sort(rstandard(cr))
Wprime=(cor(rankit,ys))^2
```

`Wprime`

The vast majority of the analyses we will run can be computed by changing the (two lines of the) `read.table` command to enter the appropriate data, and changing the `cr <- lm(y ~ x)` command to allow for fitting an appropriate model.

6.2 The simple linear regression model

In general, simple linear regression seeks to fit a line to pairs of numbers (x, y) that are subject to error. These pairs of numbers may arise when there is a perfect linear relationship between x and a variable y_* but where y_* cannot be measured without error. Our actual observations y are then the sum of y_* and the measurement error. Alternatively, we may sample a population of objects and take two measurements on each object. In this case, both elements of the pair (x, y) are random. In simple linear regression we think of using the x measurement to predict the y measurement. While x is actually random in this scenario, we use it as if it were fixed because we cannot predict y until we have actually observed the x value. We want to use the particular observed value of x to predict y, so for our purposes x is a fixed number. In any case, *the xs are always treated as fixed numbers in simple linear regression.*

The model for simple linear regression is a line with the addition of errors

$$y_i = \beta_0 + \beta_1 x_i + \varepsilon_i, \quad i = 1, \ldots, n$$

where y is the variable of primary interest and x is the predictor variable. Both the y_is and the x_is are observable, the y_is are assumed to be random, and the x_is are assumed to be known fixed constants. The unknown coefficients (regression parameters) β_0 and β_1 are the intercept and the slope of the line, respectively. The ε_is are unobservable errors that are assumed to be independent of each other with mean zero and the same variance, i.e.,

$$E(\varepsilon_i) = 0, \quad \text{Var}(\varepsilon_i) = \sigma^2.$$

Typically the errors are also assumed to have normal distributions, i.e.,

$$\varepsilon_i \text{s independent } N(0, \sigma^2).$$

Sometimes the assumption of independence is replaced by the weaker assumption that $\text{Cov}(\varepsilon_i, \varepsilon_j) = 0$ for $i \neq j$.

Note that since β_0, β_1, and the x_is are all assumed to be fixed constants,

$$E(y_i) = E(\beta_0 + \beta_1 x_i + \varepsilon_i) = \beta_0 + \beta_1 x_i + E(\varepsilon_i) = \beta_0 + \beta_1 x_i,$$

$$\text{Var}(y_i) = \text{Var}(\beta_0 + \beta_1 x_i + \varepsilon_i) = \text{Var}(\varepsilon_i) = \sigma^2.$$

If the ε_is are independent, the y_is are independent, and if the ε_is are normally distributed, so are the y_is. When making assumptions about the errors, these facts about the y_is are derived from the assumptions. Alternatively, we could just specify our model assumptions in terms of the y_is. That is, we could just assume that the y_is are independent with $E(y_i) = \beta_0 + \beta_1 x_i$, $\text{Var}(y_i) = \sigma^2$, and that the y_is are normally distributed.

The regression parameter estimates $\hat{\beta}_1$ and $\hat{\beta}_0$ are *least squares estimates*. Least squares estimates are choices of β_0 and β_1 that minimize

$$\sum_{i=1}^{n} (y_i - \beta_0 - \beta_1 x_i)^2.$$

Formulae for the estimates are given in Section 6.10. These estimates provide *fitted (predicted) values* $\hat{y}_i = \hat{\beta}_0 + \hat{\beta}_1 x_i$ and *residuals* $\hat{\varepsilon}_i = y_i - \hat{y}_i, i = 1, \ldots, n$. The sum of squares error is

$$SSE = \sum_{i=1}^{n} \hat{\varepsilon}_i^2.$$

The model involves two parameters for the mean values $E(y_i)$, namely β_0 and β_1, so the degrees of freedom for error are

$$dfE = n - 2.$$

Our estimate of the variance σ^2 is the mean squared error defined as

$$MSE = \frac{SSE}{dfE}.$$

Formulae for computing the least squares estimates are given in Section 6.10. The least squares estimates and the mean squared error are unbiased in the sense that their expected values are the parameters they estimate.

Proposition 6.2.1. $E\left(\hat{\beta}_1\right) = \beta_1, E\left(\hat{\beta}_0\right) = \beta_0,$ and $E(MSE) = \sigma^2$.

Proofs of the unbiasedness of the slope and intercept estimates are given in the appendix that appears at the end of Section 6.10.

We briefly mention the standard optimality properties of least squares estimates but for a detailed discussion see Christensen (2011, Chapter 2). Assuming that the errors have independent normal distributions, the estimates $\hat{\beta}_0$, $\hat{\beta}_1$, and MSE have the smallest variance of any unbiased estimates. The least squares estimates $\hat{\beta}_0$ and $\hat{\beta}_1$ are also maximum likelihood estimates. Maximum likelihood estimates are those values of the parameters that are most likely to generate the data that were actually observed. Without assuming that the errors are normally distributed, $\hat{\beta}_0$ and $\hat{\beta}_1$ have the smallest variance of any unbiased estimates that are linear functions of the y observations. (Linear functions allow multiplying the y_is by constants and adding terms together. Remember, the x_is are constants, as are any functions of the x_is.) Note that with this weaker assumption, i.e., giving up normality, we get a weaker result, minimum variance among only linear unbiased estimates instead of all unbiased estimates. To summarize, *under the standard assumptions, least squares estimates of the regression parameters are best (minimum variance) linear unbiased estimates (BLUEs), and for normally distributed data they are minimum variance unbiased estimates and maximum likelihood estimates*.

The *coefficient of determination*, R^2, measures the predictive ability of the model. When we discuss sample correlations in Section 6.7, we will define R^2 as the squared correlation between the pairs $(\hat{y}_i, y_i)$. Alternatively, R^2 can be computed from the ANOVA table as

$$R^2 \equiv \frac{SSReg}{SSTot}.$$

As such, it measures the percentage of the total variability in y that is explained by the x variable. In our example,

$$R^2 = \frac{552.68}{642.92} = 86.0\%,$$

so 86.0% of the total variability is explained by the regression model. This is a large percentage, so it appears that the x variable has substantial predictive power. However, *a large R^2 does not imply that the model is good in absolute terms*. It may be possible to show that this model does not fit the data adequately. In other words, while this model is explaining much of the variability, we may be able to establish that it is not explaining as much of the variability as it ought. (Example 7.2.2 involves a model with a high R^2 that is demonstrably inadequate.) Conversely, a model with a low R^2 may be the perfect model but the data may simply have a great deal of variability. For example, if you have temperature measurements obtained by having someone walk outdoors and guess the Celsius temperature and then use the true Fahrenheit temperatures as a predictor, the exact linear relationship between Celsius and Fahrenheit temperatures may make a line the ideal model. Nonetheless, the obvious inaccuracy involved in people guessing Celsius temperatures may cause a

Table 6.2: *Analysis of Variance.*

Source	df	SS	MS	F
Intercept(β_0)	1	$n\bar{y}_.^2 \equiv C$	$n\bar{y}_.^2$	
Regression(β_1)	1	$\sum_{i=1}^{n} (\hat{y}_i - \bar{y}_.)^2$	$SSReg$	$\frac{MSReg}{MSE}$
Error	$n-2$	$\sum_{i=1}^{n} (y_i - \hat{y}_i)^2$	$SSE/(n-2)$	
Total	n	$\sum_{i=1}^{n} y_i^2$		

Table 6.3: *Analysis of Variance.*

Source	df	SS	MS	F
Regression(β_1)	1	$\sum_{i=1}^{n} (\hat{y}_i - \bar{y}_.)^2$	$SSReg$	$\frac{MSReg}{MSE}$
Error	$n-2$	$\sum_{i=1}^{n} (y_i - \hat{y}_i)^2$	$SSE/(n-2)$	
Total	$n-1$	$\sum_{i=1}^{n} (y_i - \bar{y}_.)^2$		

low R^2. Moreover, even a high R^2 of 86% may provide inadequate predictions for the purposes of the study, while in other situations an R^2 of, say, 14% may be perfectly adequate. It depends on the purpose of the study. Finally, it must be recognized that a large R^2 may be an unrepeatable artifact of a particular data set. *The coefficient of determination is a useful tool but it must be used with care. In particular, it is a much better measure of the predictive ability of a model than of the correctness of a model.*

6.3 The analysis of variance table

A standard tool in regression analysis is the construction of an analysis of variance table. Tables 6.2 and 6.3 give alternative forms, both based on the sample mean, the fitted values and the data.

The best form is given in Table 6.2. In this form there is one degree of freedom for every observation, cf. the total line, and the sum of squares total is the sum of all of the squared observations. The degrees of freedom and sums of squares for intercept, regression, and error can be added to obtain the degrees of freedom and sums of squares total. We see that one degree of freedom is used to estimate the intercept, one is used for the slope, and the rest are used to estimate the variance.

The more commonly used form for the analysis of variance table is given as Table 6.3. It eliminates the line for the intercept and corrects the total line so that the degrees of freedom and sums of squares still add up.

Note that

$$\sum_{i=1}^{n} (y_i - \bar{y}_.)^2 = \sum_{i=1}^{n} y_i^2 - C = SSTot - C.$$

6.4 Model-based inference

We now repeat the testing procedures of Section 6.1 using the model-based approach of Section 3.1. The simple linear regression model is

$$y_i = \beta_0 + \beta_1 x_i + \varepsilon_i. \tag{6.4.1}$$

This will be the full model for our tests, thus $SSE(Full) = 90.25$, $dfE(Full) = 18$, and $MSE(Full) = 5.01$ are all as reported in the "Error" line of the simple linear regression ANOVA table.

The model-based test of $H_0 : \beta_1 = 0$ is precisely the F test provided by the ANOVA table. To find the reduced model, we need to incorporate $\beta_1 = 0$ into Model (6.4.1). The reduced model is $y_i = \beta_0 + 0x_i + \varepsilon_i$ or

$$y_i = \beta_0 + \varepsilon_i.$$

This is just the model for a one-sample problem, so the $MSE(Red.) = s_y^2$, the sample variance of the y_is, $dfE(Red.) = n - 1 = dfTot - C$, and $SSE(Red.) = (n-1)s_y^2 = SSTot - C$ from the ANOVA table. To obtain the F statistic, compute

$$MSTest \equiv \frac{SSE(Red.) - SSE(Full)}{dfE(Red.) - dfE(Full)} = \frac{642.92 - 90.25}{19 - 18} = 552.68$$

and

$$F = \frac{MSTest}{MSE(Full)} = \frac{552.68}{5.01} = 110.23$$

as discussed earlier. The value 110.23 is compared to an $F(1, 18)$ distribution.

To test $H_0 : \beta_1 = 1$, we incorporate the null hypothesis into Model (6.4.1) to obtain $y_i = \beta_0 + 1x_i + \varepsilon_i$. We now move the completely known term $1x_i$, known as an *offset*, to the left side of the equality and write the reduced model as

$$y_i - x_i = \beta_0 + \varepsilon_i. \tag{6.4.2}$$

Again, this is just the model for a one-sample problem, but now the dependent variable is $y_i - x_i$. It follows that $MSE(Red.) = s_{y-x}^2 = 22.66$, the sample variance of the numbers $y_i - x_i$, $dfE(Red.) = n - 1 = 19$, and $SSE(Red.) = 19s_{y-x}^2 = 430.54$. To obtain the F statistic, compute

$$MSTest \equiv \frac{SSE(Red.) - SSE(Full)}{dfE(Red.) - dfE(Full)} = \frac{430.54 - 90.25}{19 - 18} = 340.29$$

and

$$F = \frac{340.29}{5.01} = 67.87 = (-8.24)^2.$$

Note that the F statistic is the square of the t statistic from Section 6.1. The F value 67.87 is much larger than one but could be compared to an $F(1, 18)$ distribution.

Testing Model (6.4.1) against Model (6.4.2) may seem unusual because we are comparing models that have different dependent variables. The reason this works is because Model (6.4.1) is equivalent to

$$y_i - x_i = \beta_0 + \beta_{1*}x_i + \varepsilon_i.$$

In particular, this model gives the same $SSE(Full)$, $dfE(Full)$, and $MSE(Full)$ as Model (6.4.1). Parameter estimates are a little different but in an appropriate way. For example, $\hat{\beta}_{1*}$ from this model equals $\hat{\beta}_1 - 1$ from Model (6.4.1).

To test $H_0 : \beta_0 = 0$, we fit the reduced model

$$y_i = \beta_1 x_i + \varepsilon_i. \tag{6.4.3}$$

This is known as a *simple linear regression through the origin*. Most computer programs have fitting an intercept as the default option but one can choose to fit the model without an intercept. Fitting this model gives a new table of coefficients and ANOVA table.

Table of Coefficients

Predictor	$\hat{\beta}_k$	SE($\hat{\beta}_k$)	t	P
x	1.6295	0.7344	2.22	0.039

Analysis of Variance

Source	df	SS	MS	F	P
Regression	1	5198	5198	4.92	0.039
Error	19	20061	1056		
Total	20	25259			

Note that the ANOVA table for regression through the origin is similar to Table 6.2 in that it does not correct the total line for the intercept. As before, the table of coefficients and the ANOVA table provide equivalent tests of $H_0 : \beta_1 = 0$, e.g., $4.92 = (2.22)^2$, but the tests are now based on the new model, (6.4.3), so the tests are quite different from those discussed earlier.

For our purpose of testing $H_0 : \beta_0 = 0$, we only need $dfE(Red.) = 19$ and $SSE(Red.) = 20061$ from the Error line of this ANOVA table as well as the results from the full model. To obtain the F statistic, compute

$$MSTest \equiv \frac{SSE(Red.) - SSE(Full)}{dfE(Red.) - dfE(Full)} = \frac{20061 - 90.25}{19 - 18} = 19970.75$$

and

$$F = \frac{19970.75}{5.01} = 3983 = (63.11)^2.$$

If you check these computations, you will notice some round-off error. $MSE(Full)$ is closer to 5.0139. The value 3983 could be compared to an $F(1, 18)$ distribution but that is hardly necessary since it is huge. The value 63.11 was reported as the t statistic in Section 6.1.

6.5 Parametric inferential procedures

The general theory for a single parameter from Chapter 3 applies to inferences about regression parameters. The theory requires 1) a parameter (Par), 2) an estimate (Est) of the parameter, 3) the standard error of the estimate $(SE(Est))$, and 4) a known (tabled) distribution for

$$\frac{Est - Par}{SE(Est)}$$

that is symmetric about 0. The computations for most of the applications considered in this section were illustrated in Section 6.1 for the *Coleman Report* data.

Consider inferences about the slope parameter β_1. Formulae for the estimate $\hat{\beta}_1$ and the standard error of $\hat{\beta}_1$ are as given in Section 6.10. The appropriate reference distribution is

$$\frac{\hat{\beta}_1 - \beta_1}{SE(\hat{\beta}_1)} \sim t(n-2).$$

Using standard methods, the 99% confidence interval for β_1 has endpoints

$$\hat{\beta}_1 \pm t(0.995, n-2) \, SE(\hat{\beta}_1).$$

An $\alpha = .05$ test of, say, $H_0 : \beta_1 = 0$ rejects H_0 if

$$\frac{|\hat{\beta}_1 - 0|}{SE(\hat{\beta}_1)} > t(0.975, n-2).$$

An $\alpha = .05$ test of $H_0 : \beta_1 = 1$ rejects H_0 if

$$\frac{|\hat{\beta}_1 - 1|}{SE(\hat{\beta}_1)} > t(0.975, n-2).$$

For inferences about the intercept parameter β_0, formulae for the estimate $\hat{\beta}_0$ and the standard error of $\hat{\beta}_0$ are as given in Section 6.10. The appropriate reference distribution is

$$\frac{\hat{\beta}_0 - \beta_0}{\text{SE}\left(\hat{\beta}_0\right)} \sim t(n-2).$$

A 95% confidence interval for β_0 has endpoints

$$\hat{\beta}_0 \pm t(0.975, n-2)\,\text{SE}\left(\hat{\beta}_0\right).$$

An $\alpha = .01$ test of $H_0 : \beta_0 = 0$ rejects H_0 if

$$\frac{|\hat{\beta}_0 - 0|}{\text{SE}\left(\hat{\beta}_0\right)} > t(0.995, n-2).$$

Typically, inferences about β_0 are not of substantial interest. β_0 is the intercept; it is the value of the line when $x = 0$. Typically, the line is only an approximation to the behavior of the (x,y) pairs in the neighborhood of the observed data. This approximation is only valid in the neighborhood of the observed data. If we have not collected data near $x = 0$, the intercept is describing behavior of the line outside the range of valid approximation.

We can also draw inferences about a point on the line $y = \beta_0 + \beta_1 x$. For any fixed point x, $\beta_0 + \beta_1 x$ has an estimate

$$\hat{y} \equiv \hat{\beta}_0 + \hat{\beta}_1 x.$$

To get a standard error for $\hat{y}$, we first need its variance. As shown in the appendix at the end of Section 6.10, the variance of $\hat{y}$ is

$$\text{Var}\left(\hat{\beta}_0 + \hat{\beta}_1 x\right) = \sigma^2 \left[\frac{1}{n} + \frac{(x - \bar{x}.)^2}{(n-1)s_x^2}\right], \tag{6.5.1}$$

so the standard error of $\hat{y}$ is

$$\text{SE}\left(\hat{\beta}_0 + \hat{\beta}_1 x\right) = \sqrt{MSE\left[\frac{1}{n} + \frac{(x - \bar{x}.)^2}{(n-1)s_x^2}\right]}. \tag{6.5.2}$$

The appropriate distribution for inferences about the point $\beta_0 + \beta_1 x$ is

$$\frac{\left(\hat{\beta}_0 + \hat{\beta}_1 x\right) - (\beta_0 + \beta_1 x)}{\text{SE}\left(\hat{\beta}_0 + \hat{\beta}_1 x\right)} \sim t(n-2).$$

Using standard methods, the 99% confidence interval for $(\beta_0 + \beta_1 x)$ has endpoints

$$\left(\hat{\beta}_0 + \hat{\beta}_1 x\right) \pm t(0.995, n-2)\,\text{SE}\left(\hat{\beta}_0 + \hat{\beta}_1 x\right).$$

We typically prefer to have small standard errors. Even when σ^2, and thus MSE, is large, using Equation (6.5.2) we see that the standard error of $\hat{y}$ will be small when the number of observations n is large, when the x_i values are well spread out, i.e., s_x^2 is large, and when x is close to $\bar{x}.$. In other words, the line can be estimated efficiently in the neighborhood of $\bar{x}.$ by collecting a lot of data. Unfortunately, if we try to estimate the line far from where we collected the data, the standard error of the estimate gets large. The standard error gets larger as x gets farther away from the center of the data, $\bar{x}.$, because the term $(x-\bar{x}.)^2$ gets larger. This effect is standardized by the original observations; the term in question is $(x-\bar{x}.)^2/(n-1)s_x^2$, so $(x-\bar{x}.)^2$ must be large relative to $(n-1)s_x^2$ before a problem develops. In other words, the distance between x and $\bar{x}.$ must be several times the standard deviation s_x before a problem develops. Nonetheless, large standard errors occur when we try to estimate the line far from where we collected the data. Moreover, the regression line is typically just an approximation that holds in the neighborhood of where the data were collected. This approximation is likely to break down for data points far from the original data. So, in addition to the problem of having large standard errors, estimates far from the neighborhood of the original data may be totally invalid.

Estimating a point on the line is distinct from prediction of a new observation for a given x value. Ideally, the prediction would be the true point on the line for the value x. However, the true line is an unknown quantity, so our prediction is the estimated point on the line at x. The distinction between prediction and estimating a point on the line arises because a new observation is subject to variability about the line. In making a prediction we must account for the variability of the new observation even when the line is known, as well as account for the variability associated with our need to estimate the line. The new observation is assumed to be independent of the past data, so the variance of the prediction is σ^2 (the variance of the new observation) plus the variance of the estimate of the line as given in (6.5.1). The standard error replaces σ^2 with MSE and takes the square root, i.e.,

$$SE(Prediction) = \sqrt{MSE\left[1+\frac{1}{n}+\frac{(x-\bar{x}.)^2}{(n-1)s_x^2}\right]}.$$

Note that this is the same as the formula given in Equation (6.1.2). Prediction intervals follow in the usual way. For example, the 99% prediction interval associated with x has endpoints

$$\hat{y} \pm t(0.995,n-2)\,SE(Prediction).$$

As discussed earlier, estimation of points on the line should be restricted to x values in the neighborhood of the original data. For similar reasons, predictions should also be made only in the neighborhood of the original data. While it is possible, by collecting a lot of data, to estimate the line well even when the variance σ^2 is large, it is not always possible to get good prediction intervals. Prediction intervals are subject to the variability of both the observations and the estimate of the line. The variability of the observations cannot be eliminated or reduced. If this variability is too large, we may get prediction intervals that are too large to be useful. If the simple linear regression model is the "truth," there is nothing to be done, i.e., no way to improve the prediction intervals. If the simple linear regression model is only an approximation to the true process, a more sophisticated model may give a better approximation and produce better prediction intervals.

6.6 An alternative model

For some purposes, it is more convenient to work with an alternative to the model $y_i = \beta_0 + \beta_1 x_i + \varepsilon_i$. The alternative model is

$$y_i = \beta_{*0} + \beta_1(x_i - \bar{x}.) + \varepsilon_i$$

where we have adjusted the predictor variable for its mean. The key difference between the parameters in the two models is that

$$\beta_0 = \beta_{*0} - \beta_1 \bar{x}..$$

In fact, this is the basis for our formula for estimating β_0 in Section 6.10. The new parameter β_{*0} has a very simple estimate, $\hat{\beta}_{*0} \equiv \bar{y}..$ It then follows that

$$\hat{\beta}_0 = \bar{y}. - \hat{\beta}_1 \bar{x}..$$

The reason that this model is useful is because the predictor variable $x_i - \bar{x}.$ has the property $\sum_{i=1}^{n} (x_i - \bar{x}.) = 0$. This property leads to the simple estimate of β_{*0} but also to the fact that $\bar{y}.$ and $\hat{\beta}_1$ are independent. Independence simplifies the computation of variances for regression line estimates. We will not go further into these claims at this point but the results follow trivially from the matrix approach to regression that will be treated in Chapter 11.

The key point about the alternative model is that it is equivalent to the original model. The β_1 parameters are the same, as are their estimates and standard errors. The models give the same predictions, the same ANOVA table F test, and the same R^2. Even the intercept parameters are equivalent, i.e., they are related in a precise fashion so that knowing about the intercept in either model yields equivalent information about the intercept in the other model.

6.7 Correlation

The correlation coefficient is a measure of the linear relationship between two variables. The population correlation coefficient, usually denoted ρ, was discussed in Chapter 1 along with the population covariance. The sample covariance between x and y is defined as

$$s_{xy} = \frac{1}{n-1} \sum_{i=1}^{n} (x_i - \bar{x}.)(y_i - \bar{y}.)$$

and the sample correlation is defined as

$$r = \frac{s_{xy}}{s_x s_y} = \frac{\sum_{i=1}^{n} (x_i - \bar{x}.)(y_i - \bar{y}.)}{\sqrt{\sum_{i=1}^{n} (x_i - \bar{x}.)^2 \sum_{i=1}^{n} (y_i - \bar{y}.)^2}}.$$

As with the population correlation, a sample correlation of 0 indicates no linear relationship between the (x_i, y_i) pairs. A sample correlation of 1 indicates a perfect increasing linear relationship. A sample correlation of -1 indicates a perfect decreasing linear relationship.

The sample correlation coefficient is related to the estimated slope. From Equation (6.10.1) it will be easily seen that

$$r = \hat{\beta}_1 \frac{s_x}{s_y}.$$

EXAMPLE 6.7.1.   *Simulated data with various correlations*
Figure 6.2 contains plots of 25 pairs of observations with four different correlations. These are presented so the reader can get some feeling for the meaning of various correlation values. The caption gives the sample correlation r corresponding to each population correlation ρ. The population correlation is useful in that it provides some feeling for the amount of sampling variation to be found in r based on samples of 25 from (jointly) normally distributed data. Chapter 7 provides plots of uncorrelated data.   □

A commonly used statistic in regression analysis is the coefficient of determination, R^2. The best definition of R^2 is as the square of the sample correlation between the pairs $(\hat{y}_i, y_i)$. *This applies to virtually any predictive model.* Equivalently, for linear regression, R^2 can be computed from the ANOVA table as

$$R^2 \equiv \frac{SSReg}{SSTot}.$$

This is the percentage of the total variation in the dependent variable that is explained by the regression. For simple linear regression (and only for simple linear regression), using formulae in Section 6.10,

$$R^2 = r^2,$$

where r is the sample correlation between x and y.

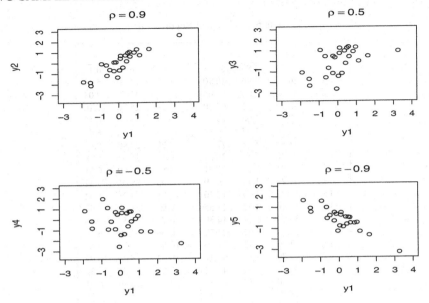

Figure 6.2 *Correlated data. The actual sample correlations are* $r = 0.889, 0.522, -0.388, -0.868$, *respectively.*

6.8 Two-sample problems

The problem of two independent samples with equal variances can be treated as a regression problem. For the data of Table 4.2, let y be the total point score and let x be a 0-1 indicator variable for whether someone is female. In Chapter 4 we wrote the observations as y_{ij}, $i = 1, 2$, $j = 1, \ldots, N_i$ with $i = 1$ indicating females, so $x_{ij} = 1$ if $i = 1$ and $x_{ij} = 0$ if $i = 2$. Alternatively, we could list all the data in one string, say, y_k, $k = 1, \ldots, n$ where $n = N_1 + N_2$ and use $x_k = 1$ to indicate females, so that $x_k = 0$ indicates males.

If we fit the simple linear regression

$$y_{ij} = \beta_0 + \beta_1 x_{ij} + \varepsilon_{ij}$$

or equivalently

$$y_k = \beta_0 + \beta_1 x_k + \varepsilon_k,$$

then for males

$$\mu_2 \equiv E(y_{2j}) = \beta_0 + \beta_1 x_{2j} = \beta_0 + \beta_1 \times 0 = \beta_0,$$

and for females

$$\mu_1 \equiv E(y_{1j}) = \beta_0 + \beta_1 x_{1j} = \beta_0 + \beta_1 \times 1 = \beta_0 + \beta_1.$$

It follows that

$$\mu_1 - \mu_2 = \beta_1.$$

Fitting the simple linear regression model gives

Table of Coefficients

Predictor	$\hat{\beta}_k$	SE($\hat{\beta}_k$)	t	P
Constant	139.000	6.753	20.58	0.000
x	−11.045	8.758	−1.26	0.216

Note that $\hat{\beta}_0$ is the sample mean for the females and the standard error is the same as that reported in Section 4.2. The test of $H_0 : \hat{\beta}_0 = 0$ is not terribly interesting.

Table 6.4: Coleman Report *data*.

School	y	x_1	x_2	x_3	x_4	x_5
1	37.01	3.83	28.87	7.20	26.60	6.19
2	26.51	2.89	20.10	−11.71	24.40	5.17
3	36.51	2.86	69.05	12.32	25.70	7.04
4	40.70	2.92	65.40	14.28	25.70	7.10
5	37.10	3.06	29.59	6.31	25.40	6.15
6	33.90	2.07	44.82	6.16	21.60	6.41
7	41.80	2.52	77.37	12.70	24.90	6.86
8	33.40	2.45	24.67	−0.17	25.01	5.78
9	41.01	3.13	65.01	9.85	26.60	6.51
10	37.20	2.44	9.99	−0.05	28.01	5.57
11	23.30	2.09	12.20	−12.86	23.51	5.62
12	35.20	2.52	22.55	0.92	23.60	5.34
13	34.90	2.22	14.30	4.77	24.51	5.80
14	33.10	2.67	31.79	−0.96	25.80	6.19
15	22.70	2.71	11.60	−16.04	25.20	5.62
16	39.70	3.14	68.47	10.62	25.01	6.94
17	31.80	3.54	42.64	2.66	25.01	6.33
18	31.70	2.52	16.70	−10.99	24.80	6.01
19	43.10	2.68	86.27	15.03	25.51	7.51
20	41.01	2.37	76.73	12.77	24.51	6.96

The estimate of β_1 is just the difference in the means between the females and the males; up to round-off error, the standard error and the t statistic are exactly as reported in Section 4.2 for inferences related to $Par = \mu_1 - \mu_2$. Moreover, the MSE as reported in the ANOVA table is precisely the pooled estimate of the variance with the appropriate degrees of freedom.

Analysis of Variance

Source	df	SS	MS	F	P
Regression	1	1088.1	1088.1	1.59	0.216
Error	35	23943.0	684.1		
Total	36	25031.1			

6.9 A multiple regression

In our discussion of simple linear regression, we considered data from *The Coleman Report*. The data given were only two of six variables reported in Mosteller and Tukey (1977). We now begin our consideration of the entire collection of variables. (Chapter 9 provides a detailed examination of the data.) Recall that the data are from schools in the New England and Mid-Atlantic states. The variables are y, the mean verbal test score for sixth graders; x_1, staff salaries per pupil; x_2, percentage of sixth graders whose fathers have white-collar jobs; x_3, a composite measure of socioeconomic status; x_4, the mean score of a verbal test given to the teachers; and x_5, the mean educational level of the sixth graders' mothers (one unit equals two school years). The dependent variable y is the same as in the simple linear regression example and the variable x_3 was used as the sole predictor variable in the earlier analysis. The data are given in Table 6.4.

We assume the data satisfy the *multiple regression model*

$$y_i = \beta_0 + \beta_1 x_{i1} + \beta_2 x_{i2} + \beta_3 x_{i3} + \beta_4 x_{i4} + \beta_5 x_{i5} + \varepsilon_i, \tag{6.9.1}$$

$i = 1, \ldots, 20$, where the ε_is are unobservable independent $N(0, \sigma^2)$ random variables and the βs are fixed unknown parameters. Fitting Model (6.9.1) with a computer program typically yields a table of coefficients and an analysis of variance table.

Table of Coefficients

Predictor	$\hat{\beta}_k$	$SE(\hat{\beta}_k)$	t	P
Constant	19.95	13.63	1.46	0.165
x_1	−1.793	1.233	−1.45	0.168
x_2	0.04360	0.05326	0.82	0.427
x_3	0.55576	0.09296	5.98	0.000
x_4	1.1102	0.4338	2.56	0.023
x_5	−1.811	2.027	−0.89	0.387

Analysis of Variance

Source	df	SS	MS	F	P
Regression	5	582.69	116.54	27.08	0.000
Error	14	60.24	4.30		
Total	19	642.92			

From just these two tables of statistics much can be learned. As will be illustrated in Chapter 9, using the single parameter methods of Chapter 3 we can produce a variety of inferential results for the β_j coefficients from the table of coefficients. Moreover, the estimated regression equation is

$$\hat{y} = 19.9 - 1.79x_1 + 0.0436x_2 + 0.556x_3 + 1.11x_4 - 1.81x_5,$$

which provides us with both our fitted values and our residuals. The analysis of variance table F test is a test of the full model (6.9.1) versus the reduced model $y_i = \beta_0 + \varepsilon_i$. In Chapter 9 we will use the Error line from the ANOVA table to construct other model tests. Similar ideas will be exploited in the next two chapters for special cases of multiple regression.

6.10 Estimation formulae for simple linear regression

In this age of computing, most people are content to have a computer program give them the estimates and standard errors needed to analyze a simple linear regression model. However, some people might still be interested in the process.

The unknown parameters in the simple linear regression model are the slope, β_1, the intercept, β_0, and the variance, σ^2. All of the estimates $\hat{\beta}_1$, $\hat{\beta}_0$, and MSE, can be computed from just six summary statistics

$$n, \quad \bar{x}., \quad s_x^2, \quad \bar{y}., \quad s_y^2, \quad \sum_{i=1}^{n} x_i y_i,$$

i.e., the sample size, the sample mean and variance of the x_is, the sample mean and variance of the y_is, and $\sum_{i=1}^{n} x_i y_i$. The only one of these that is any real work to obtain on a decent hand calculator is $\sum_{i=1}^{n} x_i y_i$. The standard estimates of the parameters are, respectively,

$$\hat{\beta}_1 = \frac{\sum_{i=1}^{n} (x_i - \bar{x}.) y_i}{\sum_{i=1}^{n} (x_i - \bar{x}.)^2},$$

$$\hat{\beta}_0 = \bar{y}. - \hat{\beta}_1 \bar{x}.,$$

and the *mean squared error*

$$MSE = \frac{\sum_{i=1}^{n} \left(y_i - \hat{\beta}_0 - \hat{\beta}_1 x_i \right)^2}{n-2}$$

$$= \frac{1}{n-2} \left[\sum_{i=1}^{n} (y_i - \bar{y}.)^2 - \hat{\beta}_1^2 \sum_{i=1}^{n} (x_i - \bar{x}.)^2 \right]$$

$$= \frac{1}{n-2} \left[(n-1)s_y^2 - \hat{\beta}_1^2 (n-1)s_x^2 \right].$$

The slope estimate $\hat{\beta}_1$ given above is the form that is most convenient for deriving its statistical properties. In this form it is just a linear combination of the y_is. However, $\hat{\beta}_1$ is commonly written in a variety of ways to simplify various computations and, unfortunately for students, they are expected to recognize all of them. Observing that $0 = \sum_{i=1}^{n}(x_i - \bar{x}.)$ so that $0 = \sum_{i=1}^{n}(x_i - \bar{x}.)\bar{y}.$, we can also write

$$\hat{\beta}_1 = \frac{\sum_{i=1}^{n}(x_i - \bar{x}.)(y_i - \bar{y}.)}{\sum_{i=1}^{n}(x_i - \bar{x}.)^2} = \frac{s_{xy}}{s_x^2} = \frac{(\sum_{i=1}^{n}x_iy_i) - n\bar{x}.\bar{y}.}{(n-1)s_x^2}, \tag{6.10.1}$$

where $s_{xy} = \sum_{i=1}^{n}(x_i - \bar{x}.)(y_i - \bar{y}.)/(n-1)$ is the sample covariance between x and y. The last equality on the right of

EXAMPLE 6.10.1.   For the simple linear regression on the *Coleman Report* data,

$$n = 20, \quad \bar{x}. = 3.1405, \quad s_x^2 = 92.64798395,$$

$$\bar{y}. = 35.0825, \quad s_y^2 = 33.838125, \quad \sum_{i=1}^{n} x_iy_i = 3189.8793.$$

The estimates are

$$\hat{\beta}_1 = \frac{3189.8793 - 20(3.1405)(35.0825)}{(20-1)92.64798395} = 0.560325468,$$

$$\hat{\beta}_0 = 35.0825 - 0.560325468(3.1405) = 33.32279787$$

and

$$\begin{aligned}
MSE &= \frac{1}{20-2}\left[(20-1)33.838125 - (0.560325468)^2(20-1)92.64798395\right] \\
&= \frac{1}{18}\left[642.924375 - 552.6756109\right] \tag{6.10.2} \\
&= \frac{90.2487641}{18} = 5.01382.
\end{aligned}$$

Up to round-off error, these are the same results as tabled in Section 6.1.   □

It is not clear that these estimates of β_0, β_1, and σ^2 are even reasonable. The estimate of the slope β_1 seems particularly unintuitive. However, from Proposition 6.2.1, the estimates are unbiased, so they are at least estimating what we claim that they estimate.

The parameter estimates are unbiased but that alone does not ensure that they are good estimates. These estimates are the best estimates available in several senses as discussed in Section 6.2.

To draw statistical inferences about the regression parameters, we need standard errors for the estimates. To find the standard errors we need to know the variance of each estimate.

Proposition 6.10.1.

$$Var\left(\hat{\beta}_1\right) = \frac{\sigma^2}{\sum_{i=1}^{n}(x_i - \bar{x}.)^2} = \frac{\sigma^2}{(n-1)s_x^2}$$

and

$$Var\left(\hat{\beta}_0\right) = \sigma^2\left[\frac{1}{n} + \frac{\bar{x}.^2}{\sum_{i=1}^{n}(x_i - \bar{x}.)^2}\right] = \sigma^2\left[\frac{1}{n} + \frac{\bar{x}.^2}{(n-1)s_x^2}\right].$$

The proof is given in the appendix at the end of the section. Note that, except for the unknown

Table 6.5: *Analysis of Variance.*

Source	df	SS	MS	F
Intercept(β_0)	1	$n\bar{y}_.^2 \equiv C$	$n\bar{y}_.^2$	
Regression(β_1)	1	$\hat{\beta}_1^2(n-1)s_x^2$	SSReg	$\frac{MSReg}{MSE}$
Error	$n-2$	$\sum_{i=1}^n \left(y_i - \hat{\beta}_0 - \hat{\beta}_1 x_i\right)^2$	$SSE/(n-2)$	
Total	n	$\sum_{i=1}^n y_i^2$		

parameter σ^2, the variances can be computed using the same six numbers we used to compute $\hat{\beta}_0$, $\hat{\beta}_1$, and MSE. Using MSE to estimate σ^2 and taking square roots, we get the standard errors,

$$SE\left(\hat{\beta}_1\right) = \sqrt{\frac{MSE}{(n-1)s_x^2}}$$

and

$$SE\left(\hat{\beta}_0\right) = \sqrt{MSE\left[\frac{1}{n} + \frac{\bar{x}_.^2}{(n-1)s_x^2}\right]}.$$

EXAMPLE 6.10.2.   For the *Coleman Report* data, using the numbers $n, \bar{x}_.$, and s_x^2,

$$Var\left(\hat{\beta}_1\right) = \frac{\sigma^2}{(20-1)92.64798395} = \frac{\sigma^2}{1760.311695}$$

and

$$Var\left(\hat{\beta}_0\right) = \sigma^2\left[\frac{1}{20} + \frac{3.1405^2}{(20-1)92.64798395}\right] = \sigma^2\left[0.055602837\right].$$

The MSE is 5.014, so the standard errors are

$$SE\left(\hat{\beta}_1\right) = \sqrt{\frac{5.014}{1760.311695}} = 0.05337$$

and

$$SE\left(\hat{\beta}_0\right) = \sqrt{5.014\left[0.055602837\right]} = 0.5280.$$

□

   We always like to have estimates with small variances. The forms of the variances show how to achieve this. For example, the variance of $\hat{\beta}_1$ gets smaller when n or s_x^2 gets larger. Thus, more observations (larger n) result in a smaller slope variance and more dispersed x_i values (larger s_x^2) also result in a smaller slope variance. Of course all of this assumes that the simple linear regression model is correct.
   The ANOVA tables Table 6.2 and Table 6.3 can be rewritten now in terms of the parameter estimates as Tables 6.5 and 6.6, respectively. The more commonly used form for the analysis of variance table is given as Table 6.6. It eliminates the line for the intercept and corrects the total line so that the degrees of freedom and sums of squares still add up.

EXAMPLE 6.10.3.   Consider again simple linear regression on the *Coleman Report* data. The analysis of variance table was given in Section 6.1; Table 6.7 illustrates the necessary computations.

Table 6.6: *Analysis of Variance.*

Source	df	SS	MS	F
Regression(β_1)	1	$\hat{\beta}_1^2(n-1)s_x^2$	SSReg	$\frac{MSReg}{MSE}$
Error	$n-2$	$\sum_{i=1}^{n}\left(y_i - \hat{\beta}_0 - \hat{\beta}_1 x_i\right)^2$	$SSE/(n-2)$	
Total	$n-1$	$(n-1)s_y^2$		

Table 6.7: *Analysis of Variance.*

Source	df	SS	MS	F
Regression(β_1)	1	$0.560325^2(20-1)92.64798$	552.6756109	$\frac{552.68}{5.014}$
Error	$20-2$	90.2487641	90.2487641/18	
Total	$20-1$	$(20-1)33.838125$		

Most of the computations were made earlier in Equation (6.10.2) during the process of obtaining the *MSE* and all are based on the usual six numbers, $n, \bar{x}., s_x^2, \bar{y}., s_y^2$, and $\sum x_i y_i$. More directly, the computations depend on $n, \hat{\beta}_1, s_x^2$, and s_y^2. Note that the *SSE* is obtained as $SSTot - SSReg$. The correction factor C in Tables 6.2 and 6.5 is $20(35.0825)^2$ but it is not used in these computations for Table 6.7. $\square$

Finally, from Table 6.6 and (6.10.1)

$$R^2 = \frac{\hat{\beta}_1^2(n-1)s_x^2}{(n-1)s_y^2} = \hat{\beta}_1^2 \frac{s_x^2}{s_y^2} = r^2.$$

Appendix: simple linear regression proofs

PROOF OF UNBIASEDNESS FOR THE REGRESSION ESTIMATES.
As established earlier, the βs and x_is are all fixed numbers, so

$$E(y_i) = E(\beta_0 + \beta_1 x_i + \varepsilon_i) = \beta_0 + \beta_1 x_i + E(\varepsilon_i) = \beta_0 + \beta_1 x_i.$$

Also note that $\sum_{i=1}^{n}(x_i - \bar{x}.) = 0$, so $\sum_{i=1}^{n}(x_i - \bar{x}.)\bar{x}. = 0$. It follows that

$$\sum_{i=1}^{n}(x_i - \bar{x}.)^2 = \sum_{i=1}^{n}(x_i - \bar{x}.)x_i - \sum_{i=1}^{n}(x_i - \bar{x}.)\bar{x}. = \sum_{i=1}^{n}(x_i - \bar{x}.)x_i.$$

Now consider the slope estimate.

$$
\begin{aligned}
E\left(\hat{\beta}_1\right) &= E\left(\frac{\sum_{i=1}^{n}(x_i - \bar{x}.)y_i}{\sum_{i=1}^{n}(x_i - \bar{x}.)^2}\right) \\
&= \frac{\sum_{i=1}^{n}(x_i - \bar{x}.)E(y_i)}{\sum_{i=1}^{n}(x_i - \bar{x}.)^2} \\
&= \frac{\sum_{i=1}^{n}(x_i - \bar{x}.)(\beta_0 + \beta_1 x_i)}{\sum_{i=1}^{n}(x_i - \bar{x}.)^2}
\end{aligned}
$$

$$= \beta_0 \frac{\sum_{i=1}^{n}(x_i - \bar{x}.)}{\sum_{i=1}^{n}(x_i - \bar{x}.)^2} + \beta_1 \frac{\sum_{i=1}^{n}(x_i - \bar{x}.)x_i}{\sum_{i=1}^{n}(x_i - \bar{x}.)^2}$$

$$= \beta_0 \frac{0}{\sum_{i=1}^{n}(x_i - \bar{x}.)^2} + \beta_1 \frac{\sum_{i=1}^{n}(x_i - \bar{x}.)^2}{\sum_{i=1}^{n}(x_i - \bar{x}.)^2}$$

$$= \beta_1.$$

The proof for the intercept goes as follows:

$$
\begin{aligned}
\mathrm{E}\left(\hat{\beta}_0\right) &= \mathrm{E}\left(\bar{y}. - \hat{\beta}_1\bar{x}.\right) \\
&= \mathrm{E}\left(\frac{1}{n}\sum_{i=1}^{n} y_i\right) - \mathrm{E}\left(\hat{\beta}_1\right)\bar{x}. \\
&= \frac{1}{n}\sum_{i=1}^{n}\mathrm{E}(y_i) - \beta_1\bar{x}. \\
&= \frac{1}{n}\sum_{i=1}^{n}(\beta_0 + \beta_1 x_i) - \beta_1\bar{x}. \\
&= \beta_0 + \beta_1\frac{1}{n}\sum_{i=1}^{n}(x_i) - \beta_1\bar{x}. \\
&= \beta_0 + \beta_1\bar{x}. - \beta_1\bar{x}. \\
&= \beta_0.
\end{aligned}
$$

PROOF OF VARIANCE FORMULAE.   As established earlier,

$$\mathrm{Var}(y_i) = \mathrm{Var}(\beta_0 + \beta_1 x_i + \varepsilon_i) = \mathrm{Var}(\varepsilon_i) = \sigma^2.$$

Now consider the slope estimate. Recall that the y_is are independent.

$$
\begin{aligned}
\mathrm{Var}\left(\hat{\beta}_1\right) &= \mathrm{Var}\left(\frac{\sum_{i=1}^{n}(x_i - \bar{x}.)y_i}{\sum_{i=1}^{n}(x_i - \bar{x}.)^2}\right) \\
&= \frac{1}{\left[\sum_{i=1}^{n}(x_i - \bar{x}.)^2\right]^2}\mathrm{Var}\left(\sum_{i=1}^{n}(x_i - \bar{x}.)y_i\right) \\
&= \frac{1}{\left[\sum_{i=1}^{n}(x_i - \bar{x}.)^2\right]^2}\sum_{i=1}^{n}(x_i - \bar{x}.)^2\mathrm{Var}(y_i) \\
&= \frac{1}{\left[\sum_{i=1}^{n}(x_i - \bar{x}.)^2\right]^2}\sum_{i=1}^{n}(x_i - \bar{x}.)^2\sigma^2 \\
&= \frac{\sigma^2}{\sum_{i=1}^{n}(x_i - \bar{x}.)^2} \\
&= \frac{\sigma^2}{(n-1)s_x^2}.
\end{aligned}
$$

Rather than establishing the variance of $\hat{\beta}_0$ directly, we find $\mathrm{Var}(\hat{\beta}_0 + \hat{\beta}_1 x)$ for an arbitrary value x. The variance of $\hat{\beta}_0$ is the special case with $x = 0$. A key result is that $\bar{y}.$ and $\hat{\beta}_1$ are independent. This was discussed in relation to the alternative regression model of Section 6.6. The independence of these estimates is based on the errors having independent normal distributions with the same

Table 6.8: *Age and maintenance costs of truck tractors.*

Age	Cost	Age	Cost	Age	Cost
0.5	163	4.0	495	5.0	890
0.5	182	4.0	723	5.0	1522
1.0	978	4.0	681	5.0	1194
1.0	466	4.5	619	5.5	987
1.0	549	4.5	1049	6.0	764
		4.5	1033	6.0	1373

variance. More generally, if the errors have the same variance and zero covariance, we still get $\text{Cov}(\bar{y}_., \hat{\beta}_1) = 0$; see Exercise 6.11.6.

$$
\begin{aligned}
\text{Var}\left(\hat{\beta}_0 + \hat{\beta}_1 x\right) &= \text{Var}\left(\bar{y}_. - \hat{\beta}_1 \bar{x}_. + \hat{\beta}_1 x\right) \\
&= \text{Var}\left(\bar{y}_. + \hat{\beta}_1 (x - \bar{x}_.)\right) \\
&= \text{Var}(\bar{y}_.) + \text{Var}\left(\hat{\beta}_1\right)(x - \bar{x}_.)^2 - 2(x - \bar{x}_.)\text{Cov}\left(\bar{y}_., \hat{\beta}_1\right) \\
&= \frac{1}{n^2}\sum_{i=1}^{n}\text{Var}(y_i) + \text{Var}\left(\hat{\beta}_1\right)(x - \bar{x}_.)^2 \\
&= \frac{1}{n^2}\sum_{i=1}^{n}\sigma^2 + \frac{\sigma^2(x - \bar{x}_.)^2}{(n-1)s_x^2} \\
&= \sigma^2\left[\frac{1}{n} + \frac{(x - \bar{x}_.)^2}{(n-1)s_x^2}\right].
\end{aligned}
$$

In particular, when $x = 0$ we get

$$
\text{Var}\left(\hat{\beta}_0\right) = \sigma^2\left[\frac{1}{n} + \frac{\bar{x}_.^2}{(n-1)s_x^2}\right].
$$

6.11   Exercises

EXERCISE 6.11.1.   Draper and Smith (1966, p. 41) considered data on the relationship between the age of truck tractors (in years) and the cost (in dollars) of maintaining them over a six-month period. The data are given in Table 6.8. Plot cost versus age and fit a regression of cost on age. Give 95% confidence intervals for the slope and intercept. Give a 99% confidence interval for the mean cost of maintaining tractors that are 2.5 years old. Give a 99% prediction interval for the cost of maintaining a particular tractor that is 2.5 years old.

Reviewing the plot of the data, how much faith should be placed in these estimates for tractors that are 2.5 years old?

EXERCISE 6.11.2.   Stigler (1986, p. 6) reported data from Cassini (1740) on the angle between the plane of the equator and the plane of the Earth's revolution about the Sun. The data are given in Table 6.9. The years -229 and -139 indicate 230 B.C. and 140 B.C. respectively. The angles are listed as the minutes above 23 degrees.

Plot the data. Are there any obvious outliers (weird data)? If outliers exist, compare the fit of the line with and without the outliers. In particular, compare the different 95% confidence intervals for the slope and intercept.

EXERCISE 6.11.3.   Mulrow et al. (1988) presented data on the calibration of a differential scanning calorimeter. The melting temperatures of mercury and naphthalene are known to be 234.16

Table 6.9: *Angle between the plane of the equator and the plane of rotation about the Sun.*

Year	Angle	Year	Angle	Year	Angle	Year	Angle
−229	51.33$\overline{3}$	880	35.000	1500	28.400	1600	31.000
−139	51.33$\overline{3}$	1070	34.000	1500	29.26$\overline{6}$	1656	29.03$\overline{3}$
140	51.16$\overline{6}$	1300	32.000	1570	29.91$\overline{6}$	1672	28.900
390	30.000	1460	30.000	1570	31.500	1738	28.33$\overline{3}$

Table 6.10: *Melting temperatures.*

Chemical	x	y
Naphthalene	353.24	354.62
	353.24	354.26
	353.24	354.29
	353.24	354.38
Mercury	234.16	234.45
	234.16	234.06
	234.16	234.61
	234.16	234.48

and 353.24 Kelvin, respectively. The data are given in Table 6.10. Plot the data. Fit a simple linear regression $y = \beta_0 + \beta_1 x + \varepsilon$ to the data. Under ideal conditions, the simple linear regression should have $\beta_0 = 0$ and $\beta_1 = 1$; test whether these hypotheses are true using $\alpha = 0.05$. Give a 95% confidence interval for the population mean of observations taken on this calorimeter for which the true melting point is 250. Give a 95% prediction interval for a new observation taken on this calorimeter for which the true melting point is 250.

Is there any way to check whether it is appropriate to use a line in modeling the relationship between x and y? If so, do so.

EXERCISE 6.11.4.   Exercise 6.11.3 involves the calibration of a measuring instrument. Often, calibration curves are used in reverse, i.e., we would use the calorimeter to measure a melting point y and use the regression equation to give a point estimate of x. If a new substance has a measured melting point of 300 Kelvin, using the simple linear regression model, what is the estimate of the true melting point? Use a prediction interval to determine whether the measured melting point of $y = 300$ is consistent with the true melting point being $x = 300$. Is an observed value of 300 consistent with a true value of 310?

EXERCISE 6.11.5.   Working–Hotelling confidence bands are a method for getting confidence intervals for every point on a line with a guaranteed simultaneous coverage. The method is essentially the same as Scheffé's method for simultaneous confidence intervals discussed in Section 13.3. For estimating the point on the line at a value x, the endpoints of the $(1 - \alpha)100\%$ simultaneous confidence intervals are

$$(\hat{\beta}_0 + \hat{\beta}_1 x) \pm \sqrt{2F(1 - \alpha, 2, dfE)}\, \mathrm{SE}(\hat{\beta}_0 + \hat{\beta}_1 x).$$

Using the *Coleman Report* data of Table 6.1, find 95% simultaneous confidence intervals for the values $x = -17, -6, 0, 6, 17$. Plot the estimated regression line and the Working–Hotelling confidence bands. We are 95% confident that the entire line $\beta_0 + \beta_1 x$ lies between the confidence bands. Compute the regular confidence intervals for $x = -17, -6, 0, 6, 17$ and compare them to the results of the Working–Hotelling procedure.

EXERCISE 6.11.6.   Use part (4) of Proposition 1.2.11 to show that $\mathrm{Cov}(\bar{y}_\cdot, \hat{\beta}_1) = 0$ whenever $\mathrm{Var}(\varepsilon_i) = \sigma^2$ for all i and $\mathrm{Cov}(\varepsilon_i, \varepsilon_j) = 0$ for all $i \neq j$. Hint: write out $\bar{y}_\cdot$ and $\hat{\beta}_1$ in terms of the y_is.

Chapter 7

Model Checking

In this chapter we consider methods for checking model assumptions and the use of transformations to correct problems with the assumptions. The primary method for checking model assumptions is the use of residual plots. If the model assumptions are valid, residual plots should display no detectable patterns. We begin in Section 7.1 by familiarizing readers with the look of plots that display no detectable patterns. Section 7.2 deals with methods for checking the assumptions made in simple linear regression (SLR). If the assumptions are violated, we need alternative methods of analysis. Section 7.3 presents methods for transforming the original data so that the assumptions become reasonable on the transformed data. Chapter 8 deals with tests for lack of fit. These are methods of constructing more general models that may fit the data better. *Chapters 7 and 8 apply quite generally to regression, analysis of variance, and analysis of covariance models.* They are not restricted to simple linear regression.

7.1 Recognizing randomness: Simulated data with zero correlation

Just as it is important to be able to look at a plot and tell when the x and y variables are related, it is important to be able to look at a plot and tell that two variables are unrelated. In other words, we need to be able to identify plots that only display random variation. This skill is of particular importance in Section 7.2 where we use plots to evaluate the assumptions made in simple linear regression. To check the assumptions of the regression model, we use plots that should display only random variation when the assumptions are true. Any systematic pattern in the model checking plots indicates a problem with our assumed regression model.

EXAMPLE 7.1.1.   *Simulated data with zero correlation*
We now examine data on five uncorrelated variables, y_1 through y_5. Figures 7.1 and 7.2 contain various plots of the variables. Since all the variable pairs have zero correlation, i.e., $\rho = 0$, any 'patterns' that are recognizable in these plots are due entirely to random variation. In particular, note that there is no real pattern in the y_2-y_3 plot.

The point of this example is to familiarize the reader with the appearance of random plots. The reader should try to identify systematic patterns in these plots, remembering that there are none. This suggests that in the model checking plots that appear later, any systematic pattern of interest should be more pronounced than anything that can be detected in Figures 7.1 and 7.2.

Below are the sample correlations r for each pair of variables. Although $\rho = 0$, none of the r values is zero and some of them are reasonably far from 0.

Sample correlations

	y_1	y_2	y_3	y_4	y_5
y_1	1.000				
y_2	−0.248	1.000			
y_3	−0.178	0.367	1.000		
y_4	−0.163	0.130	0.373	1.000	
y_5	0.071	0.279	0.293	0.054	1.000

□

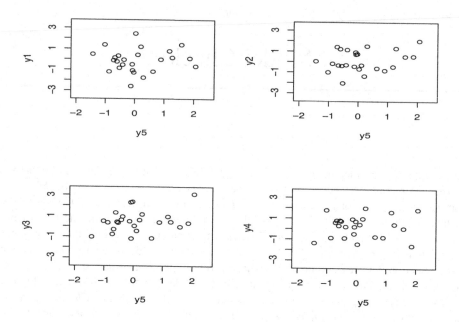

Figure 7.1: *Plots of data with $\rho = 0$.*

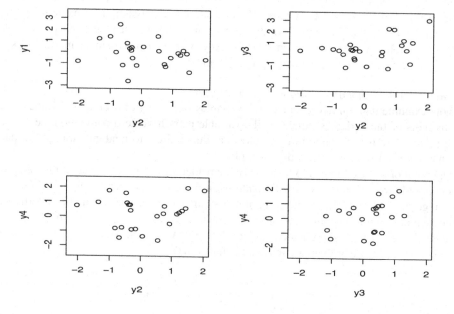

Figure 7.2: *Plots of data with $\rho = 0$.*

7.2 Checking assumptions: Residual analysis

As in Chapter 3 our standard model is

$$y_h = m(x_h) + \varepsilon_h, \qquad h = 1, \ldots, n,$$

$$\varepsilon_h \text{s independent } N(0, \sigma^2).$$

For example, the simple linear regression model posits

$$m(x_h) = \beta_0 + \beta_1 x_h.$$

The assumptions involved can all be thought of in terms of the errors. The assumptions are that

1. the ε_hs are independent,
2. $E(\varepsilon_h) = 0$ for all h,
3. $\text{Var}(\varepsilon_h) = \sigma^2$ for all h,
4. the ε_hs are normally distributed.

To have faith in our analysis, we need to validate these assumptions as far as possible. These are *assumptions* and cannot be validated completely, but we can try to detect gross violations of the assumptions.

The first assumption, that the ε_hs are independent, is the most difficult to validate. If the observations are taken at regular time intervals, they may lack independence and standard time series methods may be useful in the analysis. We will not consider this further; the interested reader can consult the time series literature, e.g., Shumway and Stoffer (2000). More general methods for checking independence were developed by Christensen and Bedrick (1997) and are reviewed in Christensen (2011). In general, we rely on the data analyst to think hard about whether there are reasons for the data to lack independence.

The second assumption is that $E(\varepsilon_h) = 0$. This is violated when we have the wrong model. The simple linear regression model with $E(\varepsilon_h) = 0$ specifies that

$$E(y_h) = \beta_0 + \beta_1 x_h.$$

If we fit this model when it is incorrect, we will not have errors with $E(\varepsilon_h) = 0$. More generally, if we fit a mean model $m(x_h)$, then

$$E(y_h) = m(x_h),$$

and if the model is incorrect, we will not have errors with $E(\varepsilon_h) = 0$. Having the wrong model for the means is called *lack of fit*.

The last two assumptions are that the errors all have some common variance σ^2 and that they are normally distributed. The term *homoscedasticity* refers to having a constant (homogeneous) variance. The term *heteroscedasticity* refers to having nonconstant (heterogeneous) variances.

In checking the error assumptions, we are hampered by the fact that the errors are not observable; we must estimate them. The SLR model involves

$$y_h = \beta_0 + \beta_1 x_h + \varepsilon_h$$

or equivalently,

$$y_h - \beta_0 - \beta_1 x_h = \varepsilon_h.$$

Given the fitted values $\hat{y}_h = \hat{\beta}_0 + \hat{\beta}_1 x_h$, we estimate ε_h with the *residual*

$$\hat{\varepsilon}_h = y_h - \hat{y}_h.$$

Similarly, in Chapter 3 we defined fitted values and residuals for general models. I actually prefer referring to predicting the error rather than estimating it. *One estimates fixed unknown parameters and predicts unobserved random variables*. Our discussion depends only on having fitted values and residuals; it does not depend specifically on the SLR model.

Two of the error assumptions are independence and homoscedasticity of the variances. Unfortunately, even when these assumptions are true, the residuals are neither independent nor do they have the same variance. For example, the SLR residuals all involve the random variables $\hat{\beta}_0$ and $\hat{\beta}_1$, so they are not independent. Moreover, the ith residual involves $\hat{\beta}_0 + \hat{\beta}_1 x_h$, the variance of which depends on $(x_h - \bar{x}.)$. Thus the variance of $\hat{\varepsilon}_h$ depends on x_h. There is little we can do about the lack of independence except hope that it does not cause severe problems. On the other hand, we can adjust for the differences in variances. In linear models the variance of a residual is

$$\text{Var}(\hat{\varepsilon}_i) = \sigma^2(1 - h_i)$$

where h_i is the *leverage* of the ith case. Leverages are discussed a bit later in this section and more extensively in relation to multiple regression. (In discussing leverages I have temporarily switched from using the meaningless subscript h to identify individual cases to using the equally meaningless subscript i. There are two reasons. First, many people use the notation h_i for leverages, to the point of writing it as "HI." Second, h_h looks funny. My preference would be to denote the leverages m_{hh}, cf. Chapter 11.)

Given the variance of a residual, we can obtain a standard error for it,

$$\text{SE}(\hat{\varepsilon}_i) = \sqrt{MSE(1 - h_i)}.$$

We can now adjust the residuals so they all have a variance of about 1; these *standardized residuals* are

$$r_i = \frac{\hat{\varepsilon}_i}{\sqrt{MSE(1 - h_i)}}.$$

The main tool used in checking assumptions is plotting the residuals or, more commonly, the standardized residuals. Normality is checked by performing a normal plot on the standardized residuals. If the assumptions (other than normality) are correct, plots of the standardized residuals versus any variable should look random. If the variable plotted against the r_is is continuous with no major gaps, the plots should look similar to the plots given in the previous section. In problems where the predictor variables are just group indicators (e.g., two-sample problems like Section 6.8 or the analysis of variance problems of Chapter 12), we often plot the residuals against identifiers of the groups, so the discrete nature of the number of groups keeps the plots from looking like those of the previous section. The single most popular diagnostic plot is probably the plot of the standardized residuals against the predicted (fitted) values $\hat{y}_i$, however the r_is can be plotted against any variable that provides a value associated with each case.

Violations of the error assumptions are indicated by any systematic pattern in the residuals. This could be, for example, a pattern of increased variability as the predicted values increase, or some curved pattern in the residuals, or any change in the variability of the residuals.

A residual plot that displays an increasing variance looks roughly like a horn opening to the right.

A residual plot indicating a decreasing variance is a horn opening to the left.

Residual–Fitted plot

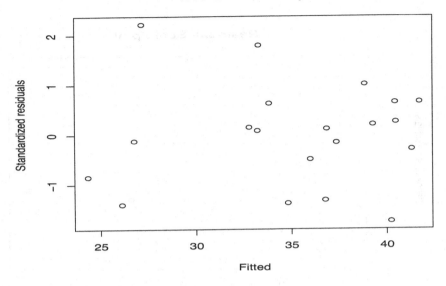

Figure 7.3: *Plot of the standardized residuals r versus ŷ*, Coleman Report.

Plots that display curved shapes typically indicate lack of fit. One example of a curve is given below.

EXAMPLE 7.2.1.   *Coleman Report data*
Figures 7.3 through 7.5 contain standardized residual plots for the simple linear regression on the *Coleman Report* data. Figure 7.3 is a plot against the predicted values; Figure 7.4 is a plot against the sole predictor variable x. The shapes of these two plots are identical. This always occurs in simple linear regression because the predictions ŷ are a linear function of the one predictor x. The one caveat to the claim of identical shapes is that the plots may be reversed. If the estimated slope is negative, the largest x values correspond to the smallest ŷ values. Figures 7.3 and 7.4 look like random patterns but it should be noted that if the smallest standardized residual were dropped (the small one on the right), the plot might suggest decreasing variability. The normal plot of the standardized residuals in Figure 7.5 does not look too bad.   □

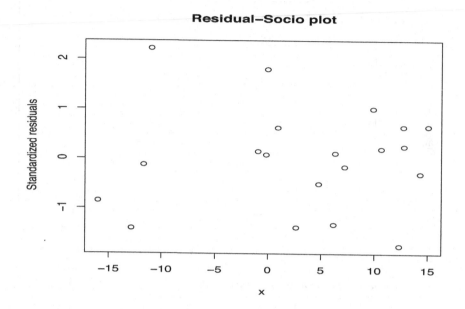

Figure 7.4: *Plot of the standardized residuals r versus x,* Coleman Report.

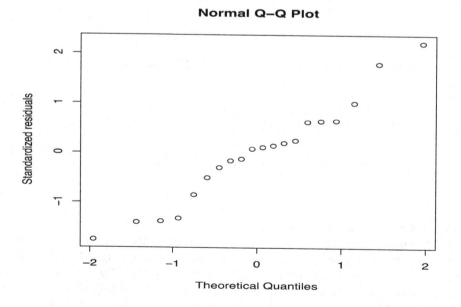

Figure 7.5: *Normal plot,* Coleman Report, $W' = 0.966$.

Table 7.1: *Hooker data.*

Case	Temperature	Pressure	Case	Temperature	Pressure
1	180.6	15.376	17	191.1	19.490
2	181.0	15.919	18	191.4	19.758
3	181.9	16.106	19	193.4	20.480
4	181.9	15.928	20	193.6	20.212
5	182.4	16.235	21	195.6	21.605
6	183.2	16.385	22	196.3	21.654
7	184.1	16.959	23	196.4	21.928
8	184.1	16.817	24	197.0	21.892
9	184.6	16.881	25	199.5	23.030
10	185.6	17.062	26	200.1	23.369
11	185.7	17.267	27	200.6	23.726
12	186.0	17.221	28	202.5	24.697
13	188.5	18.507	29	208.4	27.972
14	188.8	18.356	30	210.2	28.559
15	189.5	18.869	31	210.8	29.211
16	190.6	19.386			

7.2.1 Another example

EXAMPLE 7.2.2. *Hooker data.*
Forbes (1857) reported data on the relationship between atmospheric pressure and the boiling point of water that were collected in the Himalaya mountains by Joseph Hooker. Weisberg (1985, p. 28) presented a subset of 31 observations that are reproduced in Table 7.1.

A scatter plot of the data is given in Figure 7.6. The data appear to fit a line very closely. The usual summary tables follow for regressing pressure on temperature.

Table of Coefficients: Hooker data.

Predictor	$\hat{\beta}_k$	$SE(\hat{\beta}_k)$	t	P
Constant	−64.413	1.429	−45.07	0.000
Temperature	0.440282	0.007444	59.14	0.000

Analysis of Variance: Hooker data.

Source	df	SS	MS	F	P
Regression	1	444.17	444.17	3497.89	0.000
Error	29	3.68	0.13		
Total	30	447.85			

The coefficient of determination is exceptionally large:

$$R^2 = \frac{444.17}{447.85} = 99.2\%.$$

The plot of residuals versus predicted values is given in Figure 7.7. A pattern is very clear; the residuals form something like a parabola. In spite of a very large R^2 and a scatter plot that looks quite linear, the residual plot shows that a lack of fit obviously exists. After seeing the residual plot, you can go back to the scatter plot and detect suggestions of nonlinearity. The simple linear regression model is clearly inadequate, so we do not bother presenting a normal plot. In the next two sections, we will examine ways to deal with this lack of fit. □

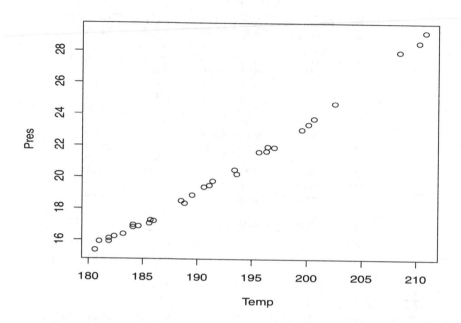

Figure 7.6: *Scatter plot of Hooker data.*

Residual–Fitted plot

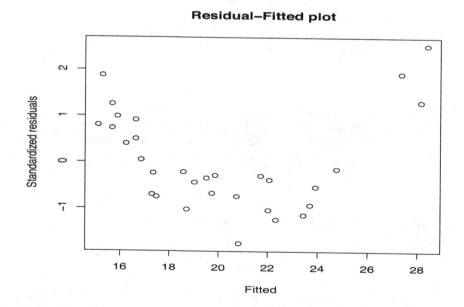

Figure 7.7: *Standardized residuals versus predicted values for Hooker data.*

7.2.2   Outliers

Outliers are bizarre data points. They are points that do not seem to fit with the other observations in a data set. We can characterize bizarre points as having either bizarre x values or bizarre y values. There are two valuable tools for identifying outliers.

Leverages are values between 0 and 1 that measure how bizarre an x value is relative to the other x values in the data. *A leverage near 1 is a very bizarre point.* Leverages that are small are similar to the other data. The sum of all the leverages in a simple linear regression is always 2, thus the average leverage is $2/n$. Points with leverages larger than $4/n$ cause concern and leverages above $6/n$ cause considerable concern.

For the more general models of Section 3.9 we used r to denote the number of uniquely defined parameters in a model $m(\cdot)$. For general linear models the average leverage is r/n. Points with leverages larger than $2r/n$ or $3r/n$ are often considered high-leverage points. The concept of leverage will be discussed in more detail when we discuss multiple regression.

Outliers in the y values can be detected from the *standardized deleted residuals*. Standardized deleted residuals are also called t *residuals* (in Minitab) and *studentized residuals* (in the R language). Standardized deleted residuals are just standardized residuals, but the residual for the hth case is computed from a regression that does not include the hth case. For example, in SLR the third deleted residual is

$$\hat{\varepsilon}_{[3]} = y_3 - \hat{\beta}_{0[3]} - \hat{\beta}_{1[3]}x_3$$

where the estimates $\hat{\beta}_{0[3]}$ and $\hat{\beta}_{1[3]}$ are computed from a regression in which case 3 has been dropped from the data. More generally, we could write this as

$$\hat{\varepsilon}_{[3]} = y_3 - \hat{m}_{[3]}(x_3)$$

where $\hat{m}_{[3]}(\cdot)$ is the estimate of the mean model based on all data except the third observation. The third standardized deleted residual is simply the third deleted residual divided by its standard error. The standardized deleted residuals, denoted t_h, really contain the same information as the standardized residuals r_h; the largest standardized deleted residuals are also the largest standardized residuals. The main virtue of the standardized deleted residuals is that they can be compared to a $t(dfE - 1)$ distribution to test whether they could reasonably have occurred when the model is true. The degrees of freedom in the SLR test are $n - 3$ because the simple linear regression model was fitted without the ith case, so there are only $n - 1$ data points in the fit and $(n - 1) - 2$ degrees of freedom for error. If we had reason to suspect that, say, case 3 might be an outlier, we would reject it being consistent with the model and the other data if for $h = 3$

$$|t_h| \geq t\left(1 - \frac{\alpha}{2}, dfE - 1\right).$$

If one examines the *largest* absolute standardized deleted residual, the appropriate α-level test rejects if

$$\max_h |t_h| \geq t\left(1 - \frac{\alpha}{2n}, dfE - 1\right).$$

An unadjusted $t(dfE - 1)$ test is no longer appropriate. The distribution of the maximum of a group of identically distributed random variables is not the same as the original distribution. For n variables, the true P value is no greater than nP^* where P^* is the "P value" computed by comparing the maximum to the original distribution. This is known as a *Bonferroni adjustment* and is discussed in more detail in Chapter 13.

EXAMPLE 7.2.3.   The leverages and standardized deleted residuals are given in Table 7.2 for the *Coleman Report* data with one predictor. Compared to the leverage rule of thumb $4/n = 4/20 = 0.2$, only case 15 has a noticeably large leverage. None of the cases is above the $6/n$ rule of thumb. In simple linear regression, one does not really need to evaluate the leverages directly because the

Table 7.2: *Outlier diagnostics for the* Coleman Report *data.*

Case	Leverages	Std. del. residuals	Case	Leverages	Std. del. residuals
1	0.059362	−0.15546	11	0.195438	−1.44426
2	0.175283	−0.12019	12	0.052801	0.61394
3	0.097868	−1.86339	13	0.051508	−0.49168
4	0.120492	−0.28961	14	0.059552	0.14111
5	0.055707	0.10792	15	0.258992	−0.84143
6	0.055179	−1.35054	16	0.081780	0.19341
7	0.101914	0.63059	17	0.050131	−1.41912
8	0.056226	0.07706	18	0.163429	2.52294
9	0.075574	1.00744	19	0.130304	0.63836
10	0.055783	1.92501	20	0.102677	0.24410

necessary information about bizarre x values is readily available from the x,y plot of the data. In multiple regression with three or more predictor variables, leverages are vital because no one scatter plot can give the entire information on bizarre x values. In the scatter plot of the *Coleman Report* data, Figure 6.1, there are no outrageous x values, although there is a noticeable gap between the smallest four values and the rest. From Table 6.1 we see that the cases with the smallest x values are 2, 11, 15, and 18. These cases also have the highest leverages reported in Table 7.2. The next two highest leverages are for cases 4 and 19; these have the largest x values.

For an overall $\alpha = 0.05$ level test of the deleted residuals, the tabled value needed is

$$t\left(1 - \frac{0.05}{2(20)}, 17\right) = 3.54.$$

None of the standardized deleted residuals (t_hs) approach this, so there is no evidence of any unaccountably bizarre y values.

A handy way to identify cases with large leverages, residuals, standardized residuals, or standardized deleted residuals is with an index plot. This is simply a plot of the value against the case number as in Figure 7.8 for leverages. □

7.2.3 Effects of high leverage

EXAMPLE 7.2.4. Figure 7.9 contains some data along with their least squares estimated line. The four points on the left form a perfect line with slope 1 and intercept 0. There is one high-leverage point far away to the right. The actual data are given below along with their leverages.

Case	1	2	3	4	5
y	1	2	3	4	−3
x	1	2	3	4	20
Leverage	0.30	0.26	0.24	0.22	0.98

The case with $x = 20$ is an extremely high-leverage point; it has a leverage of nearly 1. The estimated regression line is forced to go very nearly through this high-leverage point. In fact, this plot has two clusters of points that are very far apart, so a rough approximation to the estimated line is the line that goes through the mean x and y values for each of the two clusters. This example has one cluster of four cases on the left of the plot and another cluster consisting solely of the one case on the right. The average values for the four cases on the left give the point $(\bar{x}, \bar{y}) = (2.5, 2.5)$. The one case on the right is $(20, -3)$. A little algebra shows the line through these two points to be $\hat{y} = 3.286 - 0.314x$. The estimated line using least squares turns out to be $\hat{y} = 3.128 - 0.288x$, which is not too different. The least squares line goes through the two points $(2.5, 2.408)$ and $(20, -2.632)$, so the least squares line is a little lower at $x = 2.5$ and a little higher at $x = 20$.

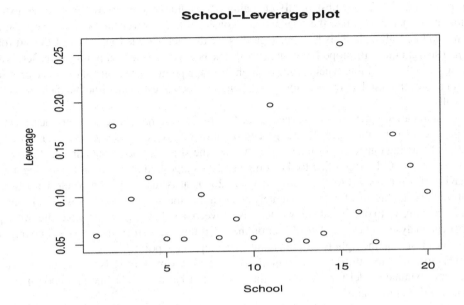

Figure 7.8: *Index plot of leverages for* Coleman Report *data.*

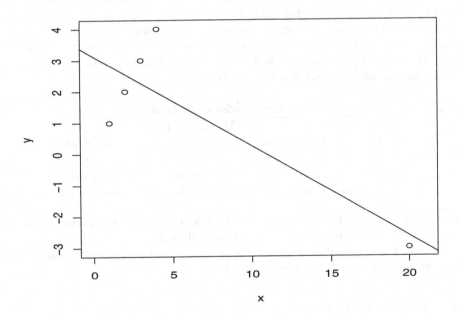

Figure 7.9: *Plot of y versus x.*

Obviously, the single point on the right of Figure 7.9 dominates the estimated straight line. For example, if the point on the right was $(20, 15)$, the estimated line would go roughly through that point and $(2.5, 2.5)$. Substantially changing the y value at $x = 20$ always gives an extremely different estimated line than the ones we just considered. Wherever the point on the right is, the estimated line follows it. This happens regardless of the fact that the four cases on the left follow a *perfect* straight line with slope 1 and intercept 0. The behavior of the four points on the left is almost irrelevant to the fitted line when there is a high-leverage point on the right. They have an effect on the quality of the rough two-point approximation to the actual estimated line but their overall effect is small.

To summarize what can be learned from Figure 7.9, we have a reasonable idea about what happens to y for x values near the range 1 to 4, and we have some idea of what happens when x is 20, but, barring outside information, we have not the slightest idea what happens to y when x is between 4 and 20. Fitting a line to the complete data suggests that we know something about the behavior of y for any value of x between 1 and 20. That is just silly! We would be better off to analyze the two clusters of points separately and to admit that learning about y when x is between 4 and 20 requires us to obtain data on y when x is between 4 and 20. In this example, the two separate statistical analyses are trivial. The cluster on the left follows a perfect line so we simply report that line. The cluster on the right is a single point so we report the point.

This example also illustrates a point about good approximate models. A straight line makes a great approximate model for the four points on the left but a straight line is a poor approximate model for the entire data.

□

7.3   Transformations

If the residuals show a problem with lack of fit, heteroscedasticity, or nonnormality, one way to deal with the problem is to try transforming the y_hs. Typically, this only works well when y_{max}/y_{min} is reasonably large. The use of transformations is often a matter of trial and error. Various transformations are tried and the one that gives the best-fitting model is used. In this context, the best-fitting model should have residual plots indicating that the model assumptions are reasonably valid. The first approach to transforming the data should be to consider transformations that are suggested by any theory associated with the data collection. Another approach to choosing a transformation is to try a variance-stabilizing transformation. These were discussed in Section 2.5 and are repeated below for data y_h with $E(y_h) = \mu_h$ and $Var(y_h) = \sigma_h^2$.

Variance-stabilizing transformations			
Data	Distribution	Mean, variance relationship	Transformation
Count	Poisson	$\mu_h \propto \sigma_h^2$	$\sqrt{y_h}$
Amount	Gamma	$\mu_h \propto \sigma_h$	$\log(y_h)$
Proportion	Binomial/N	$\mu_h(1-\mu_h)/N \propto \sigma_h^2$	$\sin^{-1}\left(\sqrt{y_h}\right)$

Whenever the data have the indicated mean-variance relationship, the corresponding variance-stabilizing transformation is supposed to work reasonably well.

Personally, I usually start by trying the log or square root transformations and, if they do not work, then I worry about how to find something better.

7.3.1   Circle of transformations

For a simple linear regression that displays lack of fit, the curved shape of an x, y plot suggests possible transformations to straighten it out. We consider power transformations of both y and x, thus y is transformed into, say, y^λ and x is transformed into x^γ. Note that $\lambda = 1$ and $\gamma = 1$ indicate no transformation. As we will justify later, we treat $\lambda = 0$ and $\gamma = 0$ as log transformations.

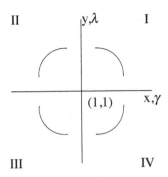

Figure 7.10: *The circle of x, y transformations.*

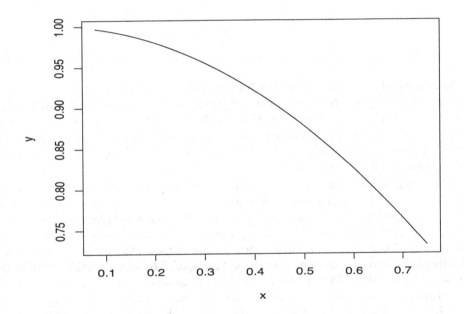

Figure 7.11: *Curved x, y plot.*

Figure 7.10 indicates the kinds of transformations appropriate for some different shapes of x, y curves. For example, if the x, y curve is similar to that in quadrant I, i.e., the y values decrease as x increases and the curve opens to the lower left, appropriate transformations involve increasing λ or increasing γ or both. Here we refer to increasing λ and γ relative to the no transformation values of $\lambda = 1$ and $\gamma = 1$. In particular, Figure 7.11 gives an x, y plot for part of a cosine curve that is shaped like the curve in quadrant I. Figure 7.12 is a plot of the numbers after x has been transformed into $x^{1.5}$ and y has been transformed into $y^{1.5}$. Note that the curve in Figure 7.12 is much straighter than the curve in Figure 7.11. If the x, y curve increases and opens to the lower right, such as those in quadrant II, appropriate transformations involve increasing λ or decreasing γ or both. An x, y curve similar to that in quadrant III suggests decreasing λ or decreasing γ or both. The graph given in Figure 7.10 is often referred to as *the circle of x, y transformations.*

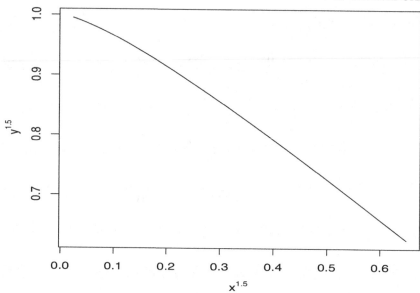

Figure 7.12: *Plot of $x^{1.5}, y^{1.5}$.*

We established in the previous section that the Hooker data does not fit a straight line and that the scatter plot in Figure 7.6 increases with a slight tendency to open to the upper left. This is the same shaped curve as in quadrant IV of Figure 7.10. The circle of x, y transformations suggests that to straighten the curve, we should try transformations with decreased values of λ or increased values of γ or both. Thus we might try transforming y into $y^{1/2}, y^{1/4}, \log(y),$ or y^{-1}. Similarly, we might try transforming x into $x^{1.5}$ or x^2.

To get a preliminary idea of how well various transformations work, we should do a series of plots. We might begin by examining the four plots in which $y^{1/2}, y^{1/4}, \log(y),$ and y^{-1} are plotted against x. We might then plot y against both $x^{1.5}$ and x^2. We should also plot all possibilities involving one of $y^{1/2}, y^{1/4}, \log(y),$ and y^{-1} plotted against one of $x^{1.5}$ and x^2 and we may need to consider other choices of λ and γ. For the Hooker data, looking at these plots would probably only allow us to eliminate the worst transformations. Recall that Figure 7.6 looks remarkably straight and it is only after fitting a simple linear regression model and examining residuals that the lack of fit (the curvature of the x, y plot) becomes apparent. Evaluating the transformations would require fitting a simple linear regression for every pair of transformed variables that has a plot that looks reasonably straight.

Observe that many of the power transformations considered here break down with values of x or y that are negative. For example, it is difficult to take square roots and logs of negative numbers. Fortunately, data are often positive or at least nonnegative. Measured amounts, counts, and proportions are almost always nonnegative. When problems arise, a small constant is often added to all cases so that they all become positive. Of course, it is unclear what constant should be added.

Obviously, the circle of transformations, just like the variance-stabilizing transformations, provides only suggestions on how to transform the data. The process of choosing a particular transformation remains one of trial and error. We begin with reasonable candidates and examine how well these transformations agree with the simple linear regression model. When we find a transformation that agrees well with the assumptions of simple linear regression, we proceed to analyze the data. Obviously, an alternative to transforming the data is to change the model. In Chapter 8 we consider a new class of models that incorporate transformations of the x variable. In the remainder of this section, we focus on a systematic method for choosing a transformation of y.

7.3.2 Box–Cox transformations

We now consider a systematic method, introduced by Box and Cox (1964), for choosing a power transformation for general models. Consider the family of power transformations, say, y_h^λ. This includes the square root transformation as the special case $\lambda = 1/2$ and other interesting transformations such as the reciprocal transformation y_h^{-1}. By making a minor adjustment, we can bring log transformations into the power family. Consider the transformations

$$y_h^{(\lambda)} = \begin{cases} (y_h^\lambda - 1)/\lambda & \lambda \neq 0 \\ \log(y_h) & \lambda = 0 \end{cases}.$$

For any fixed $\lambda \neq 0$, the transformation $y_h^{(\lambda)}$ is equivalent to y_h^λ, because the difference between the two transformations consists of subtracting a constant and dividing by a constant. In other words, for $\lambda \neq 0$, fitting the model

$$y_h^\lambda = m(x_h) + \varepsilon_h$$

is equivalent to fitting the model

$$y_h^{(\lambda)} = m(x_h) + \varepsilon_h,$$

in that fitted values $\hat{y}_h^\lambda$ satisfy $\hat{y}_h^{(\lambda)} = \left(\hat{y}_h^\lambda - 1\right)/\lambda$. (This happens whenever $m(x)$ is a linear model that includes an intercept or terms equivalent to fitting an intercept.) Parameters in the two models have slightly different meanings. While the transformation $\left(y_h^\lambda - 1\right)/\lambda$ is undefined for $\lambda = 0$, as λ approaches 0, $\left(y_h^\lambda - 1\right)/\lambda$ approaches $\log(y_h)$, so the log transformation fits in naturally.

Unfortunately, the results of fitting models to $y_h^{(\lambda)}$ with different values of λ are not directly comparable. Thus it is difficult to decide which transformation in the family to use. This problem is easily evaded (cf. Cook and Weisberg, 1982) by further modifying the family of transformations so that the results of fitting with different λs are comparable. Let $\tilde{y}$ be the geometric mean of the y_hs, i.e.,

$$\tilde{y} = \left[\prod_{i=1}^n y_h\right]^{1/n} = \exp\left[\frac{1}{n}\sum_{i=1}^n \log(y_h)\right]$$

and define the family of transformations

$$z_h^{(\lambda)} = \begin{cases} \left[y_h^\lambda - 1\right] / \left[\lambda\, \tilde{y}^{\lambda-1}\right] & \lambda \neq 0 \\ \tilde{y}\log(y_h) & \lambda = 0 \end{cases}.$$

The results of fitting the model

$$z_h^{(\lambda)} = m(x_h) + \varepsilon_h$$

can be summarized via $SSE(\lambda)$. These values are directly comparable for different values of λ. The choice of λ that yields the smallest $SSE(\lambda)$ is the best-fitting model. (It maximizes the likelihood with respect to λ.)

Box and Draper (1987, p. 290) discuss finding a confidence interval for the transformation parameter λ. An approximate $(1-\alpha)100\%$ confidence interval consists of all λ values that satisfy

$$\log SSE(\lambda) - \log SSE(\hat{\lambda}) \leq \chi^2(1-\alpha, 1)/dfE$$

where $\hat{\lambda}$ is the value of λ that minimizes $SSE(\lambda)$. When y_{max}/y_{min} is not large, the interval tends to be wide.

EXAMPLE 7.3.1. *Hooker data*
In the previous section, we found that Hooker's data on atmospheric pressure and boiling points displayed a lack of fit when regressing pressure on temperature. We now consider using power transformations to eliminate the lack of fit.

Table 7.3: *Choice of power transformation.*

λ	1/2	1/3	1/4	0	$-1/4$	$-1/2$
$SSE(\lambda)$	1.21	0.87	0.78	0.79	1.21	1.98

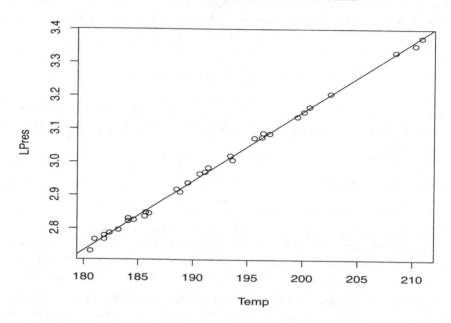

Figure 7.13: *Plot of log(Pres) versus Temp.*

Table 7.3 contains $SSE(\lambda)$ values for some reasonable choices of λ. Assuming that $SSE(\lambda)$ is a very smooth (convex) function of λ, the best λ value is probably between 0 and 1/4. If the curve being minimized is very flat between 0 and 1/4, there is a possibility that the minimizing value is between 1/4 and 1/3. One could pick more λ values and compute more $SSE(\lambda)$s but I have a bias towards simple transformations. (They are easier to sell to clients.)

The log transformation of $\lambda = 0$ is simple (certainly simpler than the fourth root) and $\lambda = 0$ is near the optimum, so we will consider it further. *We now use the simple log transformation, rather than adjusting for the geometric mean.* The data are displayed in Figure 7.13. The usual summary tables follow.

Table of Coefficients: Log Hooker data.

Predictor	$\hat{\beta}_k$	$SE(\hat{\beta}_k)$	t	P
Constant	-1.02214	0.03365	-30.38	0.000
Temp.	0.0208698	0.0001753	119.08	0.000

Analysis of Variance: Log Hooker data.

Source	df	SS	MS	F	P
Regression	1	0.99798	0.99798	14180.91	0.000
Error	29	0.00204	0.00007		
Total	30	1.00002			

The plot of the standardized residuals versus the predicted values is given in Figure 7.14. There is no obvious lack of fit or inconstancy of variances. Figure 7.15 contains a normal plot of the

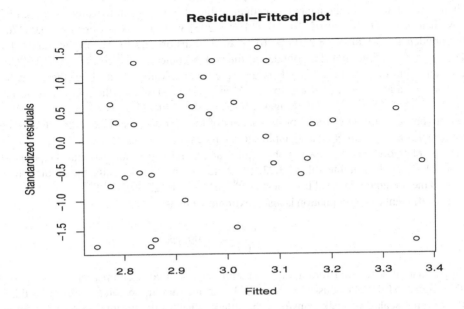

Figure 7.14: *Standardized residuals versus predicted values, logs of Hooker data.*

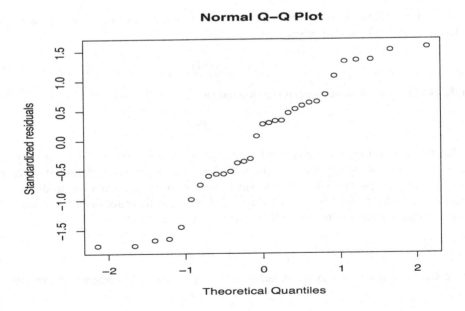

Figure 7.15: *Normal plot for logs of Hooker data, $W' = 0.961$.*

standardized residuals. The normal plot is not horrible but it is not wonderful either. There is a pronounced shoulder at the bottom and perhaps even an S shape.

If we are interested in the mean (or median) value of log pressure for a temperature of 205°F, the estimate is $3.2562 = -1.02214 + 0.0208698(205)$ with a standard error of 0.00276 and a 95% confidence interval of $(3.2505, 3.2618)$. Back transforming to the original units, the estimate is $e^{3.2562} = 25.95$ and the confidence interval becomes $(e^{3.2505}, e^{3.2618})$ or $(25.80, 26.10)$. The mean value of the line at 205°F is also the median value of the line, the point that has 50% of the data above it and 50% below it. After back transforming, the regression surface, say, $\hat{m}_*(x) = e^{-1.02214 + 0.0208698x}$, is no longer a line and the value 25.95 is no longer an estimate of the mean, but it is still an estimate of the median of the regression surface at 205°F and the back transformed confidence interval also applies to the median.

The point prediction for a new log observation at 205°F has the same value as the point estimate and has a 95% prediction interval of $(3.2381, 3.2742)$. In the original units, the prediction is again 25.95 and the prediction interval becomes $(e^{3.2381}, e^{3.2742})$ or $(25.49, 26.42)$.

The coefficient of determination is again extremely high,

$$R^2 = \frac{0.99798}{1.00002} = 99.8\%,$$

although because of the transformation this number is not directly comparable to the R^2 of 0.992 for the original SLR. As discussed in Section 3.9, to measure the predictive ability of this model on the original scale, we back transform the fitted values to the original scale and compute the squared sample correlation between the original data and these predictors on the original scale. For the Hooker data this also gives $R^2 = 0.998$, which is larger than the original SLR R^2 of 0.992. (It is a mere quirk that the R^2 on the log scale and the back transformed R^2 happen to agree to three decimal places.)   □

7.3.3   Constructed variables

One way to test whether a transformation is needed is to use a *constructed variable* as introduced by Atkinson (1973). Using the geometric mean $\tilde{y}$, let

$$w_h = y_h \left[\log(y_h/\tilde{y}) - 1 \right]. \qquad (7.3.1)$$

For the Hooker data, fit the multiple regression model

$$y_h = \beta_0 + \beta_1 x_h + \beta_2 w_h + \varepsilon_h.$$

As illustrated in Section 6.9, multiple regression gives results similar to those for simple linear regression; typical output includes a table of coefficients and an ANOVA table. A test of $H_0 : \beta_2 = 0$ from the table of coefficients gives an approximate test that no transformation is needed. The test is performed using the standard methods of Chapter 3. Details are illustrated in the following example. In addition, the estimate $\hat{\beta}_2$ provides, indirectly, an estimate of λ,

$$\hat{\lambda} = 1 - \hat{\beta}_2.$$

Frequently, this is not a very good estimate of λ but it gives an idea of where to begin a search for good λs.

EXAMPLE 7.3.2.   *Hooker data*
Performing the multiple regression of pressure on both temperature and the constructed variable w gives the following results.

Table of Coefficients

Predictor	$\hat{\beta}_k$	SE($\hat{\beta}_k$)	t	P
Constant	−43.426	2.074	−20.94	0.000
Temperature	0.411816	0.004301	95.75	0.000
w	0.80252	0.07534	10.65	0.000

The t statistic is $10.65 = 0.80252/.07534$ for testing that the regression coefficient of the constructed variable is 0. The P value of 0.000 strongly indicates the need for a transformation. The estimate of λ is

$$\hat{\lambda} = 1 - \hat{\beta}_2 = 1 - 0.80 = 0.2,$$

which is consistent with what we learned from Table 7.3. From Table 7.3 we suspected that the best transformation would be between 0 and 0.25. Of course this estimate of λ is quite crude; finding the 'best' transformation requires a more extensive version of Table 7.3. I limited the choices of λ in Table 7.3 because I was unwilling to consider transformations that I did not consider simple. □

In general, to test the need for a transformation in a linear model

$$y_h = m(x_h) + \varepsilon_h, \tag{7.3.2}$$

we add the constructed variable w_h from (7.3.1) to get the model

$$y_h = m(x_h) + \gamma w_h + \varepsilon_h, \tag{7.3.3}$$

and test $H_0 : \gamma = 0$. This gives only an *approximate* test of whether a power transformation is needed. The usual t distribution is not really appropriate. The problem is that the constructed variable w involves the ys, so the ys appear on both sides of the equality in Model (7.3.3). This is enough to invalidate the theory behind the usual test.

It turns out that this difficulty can be avoided by using the predicted values from Model (7.3.2). We write these as $\hat{y}_{h(2)}$s, where the subscript (2) is a reminder that the predicted values come from Model (7.3.2). We can now define a new constructed variable,

$$\tilde{w}_h = \hat{y}_{h(2)} \log(\hat{y}_{h(2)}),$$

and fit

$$y_h = m(x_h) + \gamma \tilde{w}_h + \varepsilon_h. \tag{7.3.4}$$

The new constructed variable $\tilde{w}_h$ simply replaces y_h with $\hat{y}_{h(2)}$ in the definition of w_h and deletes some terms made redundant by using the $\hat{y}_{h(2)}$s. If Model (7.3.2) is valid, the usual test of $H_0 : \gamma = 0$ from Model (7.3.4) has the standard t distribution in spite of the fact that the $\tilde{w}_h$s depend on the y_hs. By basing the constructed variable on the $\hat{y}_{h(2)}$s, we are able to get an exact t test for $\gamma = 0$ and restrict the weird behavior of the test statistic to situations in which $\gamma \neq 0$.

Tukey (1949) uses neither the constructed variable w_h nor $\tilde{w}_h$ but a third constructed variable that is an approximation to $\tilde{w}_h$. Using a method from calculus known as Taylor's approximation (expanding about $\bar{y}$.) and simplifying the approximation by eliminating terms that have no effect on the test of $H_0 : \gamma = 0$, we get $\hat{y}^2_{h(2)}$ as a new constructed variable. This leads to fitting the model

$$y_h = m(x_h) + \gamma \hat{y}^2_{h(2)} + \varepsilon_h, \tag{7.3.5}$$

and testing the need for a transformation by testing $H_0 : \gamma = 0$. When applied to an additive two-way model as discussed in Chapter 14 (without replication), this is Tukey's famous one degree of freedom test for nonadditivity. Recall that t tests are equivalent to F tests with one degree of freedom in the numerator, hence the reference to one degree of freedom in the name of Tukey's test.

Models (7.3.3), (7.3.4), and (7.3.5) all provide *rough* estimates of the appropriate power transformation. From models (7.3.3) and (7.3.4), the appropriate power is estimated by $\hat{\lambda} = 1 - \hat{\gamma}$. In Model (7.3.5), because of the simplification employed after the approximation, the estimate is $\hat{\lambda} = 1 - 2\bar{y} \cdot \hat{\gamma}$.

Atkinson (1985, Section 8.1) gives an extensive discussion of various constructed variables for testing power transformations. In particular, he suggests (on page 158) that while the tests based on $\tilde{w}_h$ and $\hat{y}_{h(2)}^2$ have the advantage of giving exact t tests and being easier to compute, the test using w_h may be more sensitive in detecting the need for a transformation, i.e., may be more powerful.

The tests used with models (7.3.4) and (7.3.5) are special cases of a general procedure introduced by Rao (1965) and Milliken and Graybill (1970); see also Christensen (2011, Section 9.5). In addition, Cook and Weisberg (1982), and Emerson (1983) contain useful discussions of constructed variables and methods related to Tukey's test.

EXAMPLE 7.3.3.   *Hooker data*
Fitting the simple linear regression while adding the alternative constructed variables gives

Table of Coefficients

Predictor	$\hat{\beta}_k$	SE($\hat{\beta}_k$)	t	P
Constant	184.67	22.84	8.09	0.000
temp	-1.1373	0.1446	-7.86	0.000
$\tilde{w}$	0.88557	0.08115	10.91	0.000

and

Table of Coefficients

Predictor	$\hat{\beta}_k$	SE($\hat{\beta}_k$)	t	P
Constant	1.696	6.069	0.28	0.782
temp	0.05052	0.03574	1.41	0.169
$\hat{y}^2$	0.020805	0.001899	10.95	0.000

Not surprisingly, the t statistics for both constructed variables are huge.   □

Transforming a predictor variable

Weisberg (1985, p. 156) suggests applying a log transformation to the predictor variable x in simple linear regression whenever x_{max}/x_{min} is larger than 10 or so. There is also a procedure, originally due to Box and Tidwell (1962), that is akin to the constructed variable test but that is used for checking the need to transform x. As presented by Weisberg, this procedure consists of fitting the original model

$$y_h = \beta_0 + \beta_1 x_h + \varepsilon_h$$

to obtain $\hat{\beta}_1$ and then fitting the model

$$y_h = \eta_0 + \eta_1 x_h + \eta_2 x_h \log(x_h) + \varepsilon_h.$$

Here, $x_h \log(x_h)$ is just an additional predictor variable that we compute from the values of x_h. The test of $H_0 : \eta_2 = 0$ is a test for whether a transformation of x is needed. If $\eta_2 \neq 0$, transforming x into x^γ is suggested where a rough estimate of γ is

$$\hat{\gamma} = \frac{\hat{\eta}_2}{\hat{\beta}_1} + 1$$

and $\gamma = 0$ is viewed as the log transformation. Typically, only γ values between about -2 and 2 are considered useable. Of course, none of this is going to make any sense if x takes on negative values, and if x_{max}/x_{min} is not large, computational problems may occur when trying to fit a model that contains both x_h and $x_h \log(x_h)$.

In multiple regression, to test the need for transforming any particular predictor variable, just add another variable that is the original variable times its log.

Table 7.4: *Acreage in corn for different farm acreages.*

Farm	Corn	Farm	Corn	Farm	Corn
x	y	x	y	x	y
80	25	160	45	320	110
80	10	160	40	320	30
80	20	240	65	320	55
80	32	240	80	320	60
80	20	240	65	400	75
160	60	240	85	400	35
160	35	240	30	400	140
160	20	320	70	400	90
				400	110

7.4 Exercises

EXERCISE 7.4.1.   Using the complete data of Exercise 6.11.2, test the need for a transformation of the simple linear regression model. Repeat the test after eliminating any outliers. Compare the results.

EXERCISE 7.4.2.   Snedecor and Cochran (1967, Section 6.18) presented data obtained in 1942 from South Dakota on the relationship between the size of farms (in acres) and the number of acres planted in corn. The data are given in Table 7.4.

Plot the data. Fit a simple linear regression to the data. Examine the residuals and discuss what you find. Test the need for a power transformation. Is it reasonable to examine the square root or log transformations? If so, do so.

EXERCISE 7.4.3.   Repeat Exercise 7.4.2 but instead of using the number of acres of corn as the dependent variable, use the proportion of acreage in corn as the dependent variable. Compare the results to those given earlier.

## Chapter 8

Lack of Fit and Nonparametric Regression

In analyzing data we often start with an initial model that is relatively complicated, that we hope fits reasonably well, and look for simpler versions that still fit the data adequately. *Lack of fit* involves an initial model that does not fit the data adequately. Most often, we start with a full model and look at reduced models. When dealing with lack of fit, our initial model is the reduced model, and we look for models that fit significantly better than the reduced model. In this chapter, we introduce methods for testing lack of fit for a simple linear regression model. As with the chapter on model checking, these ideas translate with (relatively) minor modifications to testing lack of fit for other models. The issue of testing lack of fit will arise again in later chapters.

The full models that we create in order to test lack of fit are all models that involve fitting more than one predictor variable. These are multiple regression models. Multiple regression was introduced in Section 6.9 and special cases were applied in Section 7.3. This chapter makes extensive use of special cases of multiple regression. The general topic, however, is considered in Chapter 9.

We illustrate lack-of-fit test testing methods by testing for lack of fit in the simple linear regression on the Hooker data of Table 7.1 and Example 7.2.2. Figure 8.1 displays the data with the fitted line and we again provide the ANOVA table for this (reduced) model.

Analysis of Variance: Hooker data SLR.

Source	df	SS	MS	F	P
Regression	1	444.17	444.17	3497.89	0.000
Error	29	3.68	0.13		
Total	30	447.85			

Section 8.1 considers extending the simple linear regression model by fitting a polynomial in the predictor x. Section 2 considers some strange things that can happen when fitting high-order polynomials. Section 3 introduced the idea of extending the model by using functions of x other than polynomials. Section 4 looks at fitting the model to disjoint subsets of the data. Section 5 examines how the partitioning ideas of Section 4 lead naturally to the idea of fitting "splines." Finally, Section 6 gives a brief introduction of Fisher's famous lack-of-fit test. The ideas of fitting models based on various functions of x and fitting models on subsets of the data (and then recombining the results) are fundamental in the field of *nonparametric regression*.

8.1 Polynomial regression

With Hooker's data, the simple linear regression of pressure on temperature shows a lack of fit. The residual plot in Figure 7.7 clearly shows nonrandom structure. In Section 7.3, we used a power transformation to eliminate the lack of fit. In this section we introduce an alternative method called *polynomial regression*. Polynomial regression is a special case of the multiple regression model that was introduced in Section 6.9 and is discussed more fully in Chapter 9.

With a single predictor variable x, we can try to eliminate lack of fit in the simple linear regression $y_i = \beta_0 + \beta_2 x_i + \varepsilon_i$ by fitting larger models. In particular, we can fit the *quadratic* (parabolic)

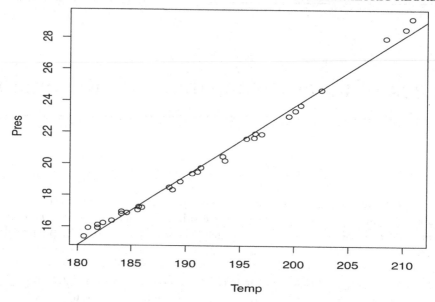

Figure 8.1: *Hooker data, linear fit.*

model

$$y_i = \beta_0 + \beta_1 x_i + \beta_2 x_i^2 + \varepsilon_i.$$

We could also try a *cubic* model

$$y_i = \beta_0 + \beta_1 x_i + \beta_2 x_i^2 + \beta_3 x_i^3 + \varepsilon_i,$$

the *quartic* model

$$y_i = \beta_0 + \beta_1 x_i + \beta_2 x_i^2 + \beta_3 x_i^3 + \beta_4 x_i^4 + \varepsilon_i,$$

or higher-degree polynomials. If we view our purpose as finding good, easily interpretable approximate models for the data, *high-degree polynomials can behave poorly*. As we will see later, the process of fitting the observed data can cause high-degree polynomials to give very erratic results in areas very near the observed data. A good approximate model should work well, not only at the observed data, but also near it. Thus, we focus on low-degree polynomials. The problem of erratic fits is addressed in the next section. We now examine issues related to fitting polynomials.

EXAMPLE 8.1.1.   Computer programs give output for polynomial regression that is very similar to that for simple linear regression. We fit a fifth-degree (quintic) polynomial to Hooker's data,

$$y_i = \gamma_0 + \gamma_1 x_i + \gamma_2 x_i^2 + \gamma_3 x_i^3 + \gamma_4 x_i^4 + \gamma_5 x_i^5 + \varepsilon_i. \tag{8.1.1}$$

Actually, we tried fitting a cubic model to these data and encountered numerical instability. (Some computer programs object to fitting it.) This may be related to the fact that the R^2 is so high. To help with the numerical instability of the procedure, before computing the powers of the x variable we subtracted the mean $\bar{x}_. = 191.787$. Thus, we actually fit,

$$y_i = \beta_0 + \beta_1 (x_i - \bar{x}_.) + \beta_2 (x_i - \bar{x}_.)^2 + \beta_3 (x_i - \bar{x}_.)^3 + \beta_4 (x_i - \bar{x}_.)^4 + \beta_5 (x_i - \bar{x}_.)^5 + \varepsilon_i. \tag{8.1.2}$$

These two models are equivalent in that they always give the same fitted values, residuals, and degrees of freedom. Moreover, $\gamma_5 \equiv \beta_5$ but none of the other γ_js are equivalent to the corresponding β_js. (The equivalences are obtained by the rather ugly process of actually multiplying out the powers of $(x_i - \bar{x}.)$ in Model (8.1.2) so that the model can be rewritten in the form of Model (8.1.1).) The fitted model, (8.1.2), is summarized by the table of coefficients and the ANOVA table.

Table of Coefficients: Model (8.1.2).

Predictor	$\hat{\beta}_k$	$SE(\hat{\beta}_k)$	t	P
Constant	19.7576	0.0581	340.19	0.000
$(x - \bar{x}.)$	0.41540	0.01216	34.17	0.000
$(x - \bar{x}.)^2$	0.002179	0.002260	0.96	0.344
$(x - \bar{x}.)^3$	0.0000942	0.0001950	0.48	0.633
$(x - \bar{x}.)^4$	0.00001522	0.00001686	0.90	0.375
$(x - \bar{x}.)^5$	−0.00000080	0.00000095	−0.84	0.409

Analysis of Variance: Model (8.1.2).

Source	df	SS	MS	F	P
Regression	5	447.175	89.435	3315.48	0.000
Error	25	0.674	0.027		
Total	30	447.850			

The most important things here are that we now know the SSE, dfE, and MSE from the fifth-degree polynomial. The ANOVA table also provides an F test for comparing the fifth-degree polynomial against the reduced model $y_i = \beta_0 + \varepsilon_i$, not a terribly interesting test.

Usually, *the only interesting t test for a regression coefficient in polynomial regression is the one for the highest term in the polynomial.* In this case the t statistic for the fifth-degree term is −0.84 with a P value of 0.409, so there is little evidence that we need the fifth-degree term in the polynomial. All the t statistics are computed as if the variable in question was the only variable being dropped from the fifth-degree polynomial. For example, it usually makes little sense to have a quintic model that does not include a quadratic term, so there is little point in examining the t statistic for testing $\beta_2 = 0$. One reason for this is that simple linear transformations of the predictor variable change the roles of lower-order terms. For example, something as simple as subtracting $\bar{x}.$ completely changes the meaning of γ_2 from Model (8.1.1) to β_2 in Model (8.1.2). Another way to think about this is that the Hooker data uses temperature measured in Fahrenheit as a predictor variable. The quintic model, (8.1.2), for the Hooker data is consistent with $\beta_2 = 0$ with a P value of 0.344. If we changed to measuring temperature in Celsius, there is no reason to believe that the new quintic model would still be consistent with $\beta_2 = 0$. When there is a quintic term in the model, a quadratic term based on Fahrenheit measurements has a completely different meaning than a quadratic term based on Celsius measurements. The same is true for all the other terms except the highest-order term, here the quintic term. On the other hand, the Fahrenheit and Celsius quintic models that include all lower-order terms are equivalent, just as the simple linear regressions based on Fahrenheit and Celsius are equivalent. Of course these comments apply to all polynomial regressions. Exercise 8.7.7 explores the relationships among regression parameters for quadratic models that have and have not adjusted the predictor for its sample mean.

A lack-of-fit test is provided by testing the quintic model against the original simple linear regression model. The F statistic is

$$F_{obs} = \frac{(3.68 - 0.674)/(29 - 25)}{0.027} = 27.83$$

which is much bigger than 1 and easily significant at the 0.01 level when compared to an $F(4,25)$ distribution. The test suggests lack of fit (or some other problem with the assumptions). $\square$

8.1.1 *Picking a polynomial*

We now consider the problem of finding a small-order polynomial that fits the data well.

The table of coefficients for the quintic polynomial on the Hooker data provides a t test for whether we can drop each variable out of the model, but for the most part these tests are uninteresting. The only t statistic that is of interest is that for x^5. It makes little sense, when dealing with a fifth-degree polynomial, to worry about whether you can drop out, say, the quadratic term. The only t statistic of interest is the one that tests whether you can drop x^5 so that you could get by with a quartic polynomial. If you are then satisfied with a quartic polynomial, it makes sense to test whether you can get by with a cubic. In other words, what we would really like to do is fit the sequence of models

$$y_i = \beta_0 + \varepsilon_i, \tag{8.1.3}$$

$$y_i = \beta_0 + \beta_1 x_i + \varepsilon_i, \tag{8.1.4}$$

$$y_i = \beta_0 + \beta_1 x_i + \beta_2 x_i^2 + \varepsilon_i, \tag{8.1.5}$$

$$y_i = \beta_0 + \beta_1 x_i + \beta_2 x_i^2 + \beta_3 x_i^3 + \varepsilon_i, \tag{8.1.6}$$

$$y_i = \beta_0 + \beta_1 x_i + \beta_2 x_i^2 + \beta_3 x_i^3 + \beta_4 x_i^4 + \varepsilon_i, \tag{8.1.7}$$

$$y_i = \beta_0 + \beta_1 x_i + \beta_2 x_i^2 + \beta_3 x_i^3 + \beta_4 x_i^4 + \beta_5 x_i^5 + \varepsilon_i, \tag{8.1.8}$$

and find the smallest model that fits the data. It is equivalent to fit the sequence of polynomials with x adjusted for its mean, $\bar{x}.$. In subsequent discussion we refer to SSEs and other statistics for models (8.1.3) through (8.1.8) as $SSE(3)$ through $SSE(8)$ with other similar notations that are obvious. Recall that models (8.1.1), (8.1.2), and (8.1.8) are equivalent.

Many regression programs fit an overall model by fitting a sequence of models and provide key results from the sequence. Most often the results are the sequential sums of squares, which are simply the difference in error sums of squares for consecutive models in the sequence. Note that you must specify the variables to the computer program in the order you want them fitted. For the Hooker data, sequential fitting of models (8.1.3) through (8.1.8) gives

Source	Model Comparison	df	Seq SS	F
$(x - \bar{x}.)$	$SSE(3) - SSE(4)$	1	444.167	16465.9
$(x - \bar{x}.)^2$	$SSE(4) - SSE(5)$	1	2.986	110.7
$(x - \bar{x}.)^3$	$SSE(5) - SSE(6)$	1	0.000	0.0
$(x - \bar{x}.)^4$	$SSE(6) - SSE(7)$	1	0.003	0.1
$(x - \bar{x}.)^5$	$SSE(7) - SSE(8)$	1	0.019	0.7

Using these and statistics reported in Example 8.1.1, the F statistic for dropping the fifth-degree term from the polynomial is

$$F_{obs} = \frac{SSE(7) - SSE(8)}{MSE(8)} = \frac{0.019}{0.027} = 0.71 = (-0.84)^2.$$

The corresponding t statistic reported earlier for testing $H_0 : \beta_5 = 0$ in Model (8.1.2) was -0.84. The data are consistent with a fourth-degree polynomial.

The F test for dropping to a third-degree polynomial from a fourth-degree polynomial is

$$F_{obs} = \frac{SSE(6) - SSE(7)}{MSE(8)} = \frac{0.003}{0.027} = 0.1161.$$

In the denominator of the test we again use the MSE from the fifth-degree polynomial. *When doing a series of tests on related models one generally uses the MSE from the largest model in the denominator of all tests*, cf. Subsection 3.1.1. The t statistic corresponding to this F statistic is $\sqrt{0.1161} \doteq 0.341$, not the value 0.90 reported earlier for the fourth-degree term in the table of coefficients for the fifth-degree model, (8.1.2). The t value of 0.341 is a statistic for testing $\beta_4 = 0$ in the fourth-degree model. The value $t_{obs} = 0.341$ is not quite the t statistic (0.343) you would get in the

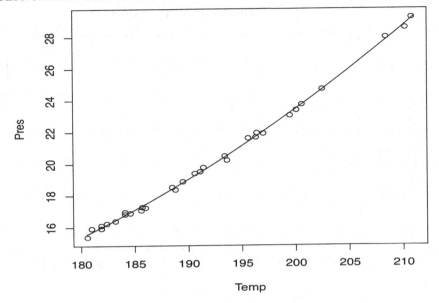

Figure 8.2: *Hooker data with quadratic fit.*

table of coefficients for fitting the fourth-degree polynomial (8.1.7) because the table of coefficients would use the *MSE* from Model (8.1.7) whereas this statistic is using the *MSE* from Model (8.1.8). Nonetheless, t_{obs} provides a test for $\beta_4 = 0$ in a model that has already specified that $\beta_5 = 0$ whereas $t = 0.90$ from the table of coefficients for the fifth-degree model, (8.1.2), is testing $\beta_4 = 0$ without specifying that $\beta_5 = 0$.

The other F statistics listed are also computed as Seq $SS/MSE(8)$. From the list of F statistics, we can clearly drop any of the polynomial terms down to the quadratic term.

8.1.2 Exploring the chosen model

We now focus on the polynomial model that fits these data well: the quadratic model

$$y_i = \beta_0 + \beta_1 x_i + \beta_2 x_i^2 + \varepsilon_i.$$

We have switched to fitting the polynomial without correcting the predictor for its mean value. Summary tables for fitting the quadratic model are

Table of Coefficients: Hooker data, quadratic model.

Predictor	$\hat{\beta}_k$	$SE(\hat{\beta}_k)$	t	P
Constant	88.02	13.93	6.32	0.000
x	-1.1295	0.1434	-7.88	0.000
x^2	0.0040330	0.0003682	10.95	0.000

Analysis of Variance: Hooker data, quadratic model.

Source	df	SS	MS	F	P
Regression	2	447.15	223.58	8984.23	0.000
Error	28	0.70	0.02		
Total	30	447.85			

The *MSE*, regression parameter estimates, and standard errors are used in the usual way. The t statistics and P values are for the tests of whether the corresponding β parameters are 0. The t

Residual–Fitted plot

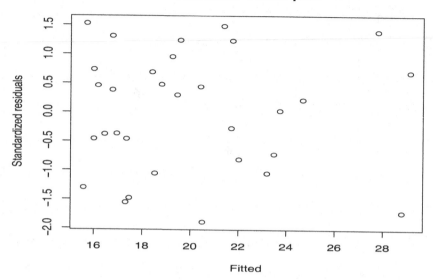

Figure 8.3: *Standardized residuals versus predicted values, quadratic model.*

statistics for β_0 and β_1 are of little interest. The t statistic for β_2 is 10.95, which is highly significant, so the quadratic model accounts for a significant amount of the lack of fit displayed by the simple linear regression model. Figure 8.2 gives the data with the fitted parabola.

We will not discuss the ANOVA table in detail, but note that with two predictors, x and x^2, there are 2 degrees of freedom for regression. In general, if we fit a polynomial of degree a, there will be a degrees of freedom for regression, one degree of freedom for every term other than the intercept. Correspondingly, when fitting a polynomial of degree a, there are $n - a - 1$ degrees of freedom for error. *The ANOVA table F statistic provides a test of whether the polynomial (in this case quadratic) model explains the data better than the model with only an intercept.*

The fitted values are obtained by substituting the x_i values into

$$\hat{y} = 88.02 - 1.1295x + 0.004033x^2.$$

The residuals are $\hat{\varepsilon}_i = y_i - \hat{y}_i$.

The coefficient of determination is computed and interpreted as before. It is the squared correlation between the pairs $(\hat{y}_i, y_i)$ and also $SSReg$ divided by the $SSTot$, so it measures the amount of the total variability that is explained by the predictor variables temperature and temperature squared. For these data, $R^2 = 99.8\%$, which is an increase from 99.2% for the simple linear regression model. It is not appropriate to compare the R^2 for this model to the R^2 from the log transformed model of Section 7.3 because they are computed from data that use different scales. However, if we back transform the fitted log values to the original scale to give $\hat{y}_{i\ell}$ values and compute R_ℓ^2 as the squared correlation between the $(\hat{y}_{i\ell}, y_i)$ values, then R_ℓ^2 and R^2 are comparable.

The standardized residual plots are given in Figures 8.3 and 8.4. The plot against the predicted values looks good, just as it did for the transformed data examined in the Section 7.3. The normal plot for this model has a shoulder at the top but it looks much better than the normal plot for the simple linear regression on the log transformed data.

If we are interested in the mean value of pressure for a temperature of 205°F, the quadratic model estimate is (up to a little round-off error)

$$\hat{y} = 25.95 = 88.02 - 1.1295\,(205) + 0.004033\,(205)^2.$$

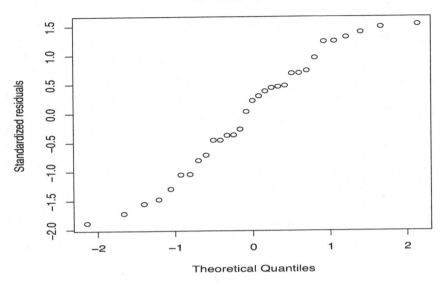

Figure 8.4: *Normal plot for quadratic model, W′ = 0.966.*

The standard error (as reported by the computer program) is 0.0528 and a 95% confidence interval is (25.84, 26.06). This compares to a point estimate of 25.95 and a 95% confidence interval of (25.80, 26.10) obtained in Section 7.3 from regressing the log of pressure on temperature and back transforming. The quadratic model *prediction* for a new observation at 205°F is again 25.95 with a 95% prediction interval of (25.61, 26.29). The corresponding back transformed prediction interval from the log transformed data is (25.49, 26.42). In this example, the results of the two methods for dealing with lack of fit are qualitatively very similar, at least at 205°F.

Finally, consider testing the quadratic model for lack of fit by comparing it to the quintic model (8.1.2). The F statistic is

$$F_{obs} = \frac{(0.70 - 0.674)/(28 - 25)}{0.027} = 0.321,$$

which is much smaller than 1 and makes no suggestion of lack of fit.

One thing we have not addressed is why we chose a fifth-degree polynomial rather than a fourth-degree or a sixth-degree or a twelfth-degree. The simplest answer is just to pick something that clearly turns out to be large enough to catch the important features of the data. If you start with too small a polynomial, go back and pick a bigger one.   □

8.2 Polynomial regression and leverages

We now present a simple example that illustrates two points: that leverages depend on the model and that high-order polynomials can fit the data in very strange ways.

EXAMPLE 8.2.1.   The data for the example follow. They were constructed to have most observations far from the middle.

Case	1	2	3	4	5	6	7
y	0.445	1.206	0.100	-2.198	0.536	0.329	-0.689
x	0.0	0.5	1.0	10.0	19.0	19.5	20.0

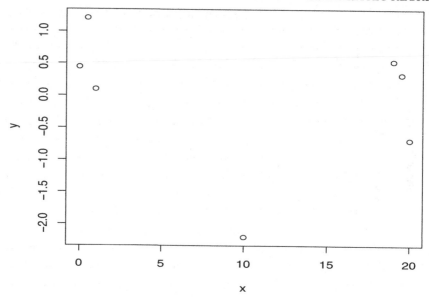

Figure 8.5: *Plot of y versus x.*

I selected the x values. The y values are a sample of size 7 from a $N(0,1)$ distribution. Note that with seven distinct x values, we can fit a polynomial of degree 6.

The data are plotted in Figure 8.5. Just by chance (honest, folks), I observed a very small y value at $x = 10$, so the data appear to follow a parabola that opens up. The small y value at $x = 10$ totally dominates the impression given by Figure 8.5. If the y value at $x = 10$ had been near 3 rather than near -2, the data would appear to be a parabola that opens down. If the y value had been between 0 and 1, the data would appear to fit a line with a slightly negative slope. When thinking about fitting a parabola, the case with $x = 10$ is an extremely high-leverage point.

Depending on the y value at $x = 10$, the data suggest a parabola opening up, a parabola opening down, or that we do not need a parabola to explain the data. Regardless of the y value observed at $x = 10$, the fitted parabola must go nearly through the point $(10,y)$. On the other hand, if we think only about fitting a line to these data, the small y value at $x = 10$ has much less effect. In fitting a line, the value $y = -2.198$ will look unusually small (it will have a very noticeable standardized residual), but it will not force the fitted line to go nearly through the point $(10, -2.198)$.

Table 8.1 gives the leverages for all of the polynomial models that can be fitted to these data. Note that there are no large leverages for the simple linear regression model (the linear polynomial). For the quadratic (parabolic) model, all of the leverages are reasonably small except the leverage of 0.96 at $x = 10$ that very nearly equals 1. Thus, in the quadratic model, the value of y at $x = 10$ dominates the fitted polynomial. The cubic model has extremely high leverage at $x = 10$, but the leverages are also beginning to get large at $x = 0, 1, 19, 20$. For the quartic model, the leverage at $x = 10$ is 1, to two decimal places; the leverages for $x = 0, 1, 19, 20$ are also nearly 1. The same pattern continues with the quintic model but the leverages at $x = 0.5, 19.5$ are also becoming large. Finally, with the sixth-degree (hexic) polynomial, all of the leverages are exactly one. This indicates that the sixth-degree polynomial has to go through every data point exactly and thus every data point is extremely influential on the estimate of the sixth-degree polynomial. (It is fortunate that there are only seven distinct x values. This discussion would really tank if we had to fit a seventh-degree polynomial. [Think about it: quartic, quintic, hexic, ... tank.])

As we fit larger polynomials, we get more high-leverage cases (and more numerical instability). Actually, as in our example, this occurs when the size of the polynomial nears one less than the num-

Table 8.1: *Leverages.*

			Model			
x	Linear	Quadratic	Cubic	Quartic	Quintic	Hexic
0.0	0.33	0.40	0.64	0.87	0.94	1.00
0.5	0.31	0.33	0.33	0.34	0.67	1.00
1.0	0.29	0.29	0.55	0.80	0.89	1.00
10.0	0.14	0.96	0.96	1.00	1.00	1.00
19.0	0.29	0.29	0.55	0.80	0.89	1.00
19.5	0.31	0.33	0.33	0.34	0.67	1.00
20.0	0.33	0.40	0.64	0.87	0.94	1.00

ber of distinct x values and nearly all data points have distinct x values. *The estimated polynomials must go very nearly through all high-leverage cases. To accomplish this the estimated polynomials may get very strange.* We now give all of the fitted polynomials for these data.

Model		Estimated polynomial
Linear	$\hat{y} =$	$0.252 - 0.029x$
Quadratic	$\hat{y} =$	$0.822 - 0.536x + 0.0253x^2$
Cubic	$\hat{y} =$	$1.188 - 1.395x + 0.1487x^2 - 0.0041x^3$
Quartic	$\hat{y} =$	$0.713 - 0.141x - 0.1540x^2 + 0.0199x^3$
		$\quad - 0.00060x^4$
Quintic	$\hat{y} =$	$0.623 + 1.144x - 1.7196x^2 + 0.3011x^3$
		$\quad - 0.01778x^4 + 0.000344x^5$
Hexic	$\hat{y} =$	$0.445 + 3.936x - 5.4316x^2 + 1.2626x^3$
		$\quad - 0.11735x^4 + 0.004876x^5$
		$\quad - 0.00007554x^6$

Figures 8.6 and 8.7 contain graphs of these estimated polynomials.

Figure 8.6 contains the estimated linear, quadratic, and cubic polynomials. The linear and quadratic curves fit about as one would expect from looking at the scatter plot Figure 8.5. For x values near the range 0 to 20, we could use these curves to predict y values and get reasonable, if not necessarily good, results. One could not say the same for the estimated cubic polynomial. The cubic curve takes on $\hat{y}$ values near -3 for some x values that are near 6. The y values in the data are between about -2 and 1.2; nothing in the data suggests that y values near -3 are likely to occur. Such predicted values are entirely the product of fitting a cubic polynomial. If we really knew that a cubic polynomial was correct for these data, the estimated polynomial would be perfectly appropriate. But most often we use polynomials to approximate the behavior of the data and for these data the cubic polynomial gives a poor approximation.

Figure 8.7 gives the estimated quartic, quintic, and hexic polynomials. Note that the scale on the y axis has changed drastically from Figure 8.6. Qualitatively, the fitted polynomials behave like the cubic except their behavior is even worse. These polynomials do very strange things everywhere except near the observed data.

It is a theoretical fact that when the degrees of freedom for error get small, the *MSE* should be an erratic estimate of σ^2. In my experience, another phenomenon that sometimes occurs when fitting large models to data is that the mean squared error gets unnaturally *small*. Table 8.2 gives the analysis of variance tables for all of the polynomial models. Our original data were a sample from a $N(0,1)$ distribution. The data were constructed with no regression structure so the best estimate of the variance comes from the total line and is $7.353/6 = 1.2255$. This value is a reasonable estimate of the true value 1. The *MSE* from the simple linear regression model also provides a reasonable estimate of $\sigma^2 = 1$. The larger models do not work as well. Most have variance estimates near 0.5, while the hexic model does not even allow an estimate of σ^2 because it fits every data point

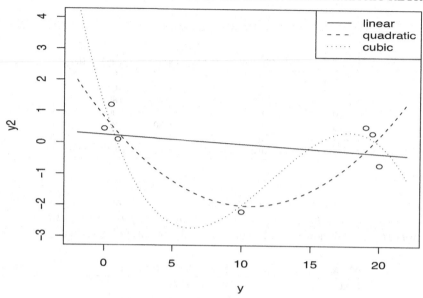

Figure 8.6: *Plots of linear (solid), quadratic (dashes), and cubic (dots) regression curves.*

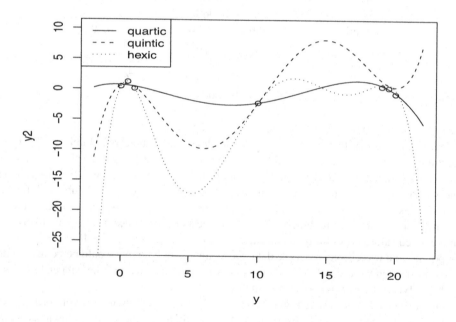

Figure 8.7: *Plots of quartic (solid), quintic (dashes), and hexic (dots) regression curves.*

perfectly. By fitting models that are too large it seems that one can often make the *MSE* artificially small. For example, the quartic model has a *MSE* of 0.306 and an *F* statistic of 5.51; if it were not for the small value of *dfE*, such an *F* value would be highly significant. *If you find a large model that has an unnaturally small MSE with a reasonable number of degrees of freedom, everything can appear to be significant even though nothing you look at is really significant.*

Just as the mean squared error often gets unnaturally small when fitting large models, R^2 gets unnaturally large. As we have seen, there can be no possible reason to use a larger model than the

Table 8.2: *Analysis of variance tables.*

Simple linear regression				
Source	df	SS	MS	F	P
Regression	1	0.457	0.457	0.33	0.59
Error	5	6.896	1.379		
Total	6	7.353			

Quadratic model				
Source	df	SS	MS	F	P
Regression	2	5.185	2.593	4.78	0.09
Error	4	2.168	0.542		
Total	6	7.353			

Cubic model				
Source	df	SS	MS	F	P
Regression	3	5.735	1.912	3.55	0.16
Error	3	1.618	0.539		
Total	6	7.353			

Quartic model				
Source	df	SS	MS	F	P
Regression	4	6.741	1.685	5.51	0.16
Error	2	0.612	0.306		
Total	6	7.353			

Quintic model				
Source	df	SS	MS	F	P
Regression	5	6.856	1.371	2.76	0.43
Error	1	0.497	0.497		
Total	6	7.353			

Hexic model				
Source	df	SS	MS	F	P
Regression	6	7.353	1.2255	—	—
Error	0	0.000	—		
Total	6	7.353			

quadratic with its R^2 of 0.71 for these 7 data points, but the cubic, quartic, quintic, and hexic models have R^2s of $0.78, 0.92, 0.93$, and 1, respectively. □

8.3 Other basis functions

In a SLR, one method for testing lack of fit was to fit a larger polynomial model. In particular, for the Hooker data we fit a fifth-degree polynomial,

$$y_i = \beta_0 + \beta_1 x_i + \beta_2 x_i^2 + \beta_3 x_i^3 + \beta_4 x_i^4 + \beta_5 x_i^5 + \varepsilon_i.$$

There was no particularly good reason to fit a fifth-degree, rather than a third-degree or seventh-degree polynomial. We just picked a polynomial that we hoped would be larger than we needed.

Rather than expanding the SLR model by adding polynomial terms, we can add other functions of x to the model. Most commonly used functions are simplified if we rescale x into a new variable taking values between 0 and 1, say, $\tilde{x}$. Commonly used functions are trig. functions, so we might fit a full model consisting of

$$y_i = \beta_0 + \beta_1 x_i + \beta_2 \cos(\pi \tilde{x}_i) + \beta_3 \sin(\pi \tilde{x}_i) + \beta_4 \cos(\pi 2 \tilde{x}_i) + \beta_5 \sin(\pi 2 \tilde{x}_i) + \varepsilon_i \qquad (8.3.1)$$

or a full model

$$y_i = \beta_0 + \beta_1 x_i + \beta_2 \cos(\pi \tilde{x}_i) + \beta_3 \cos(\pi 2 \tilde{x}_i) + \beta_4 \cos(\pi 3 \tilde{x}_i) + \beta_5 \cos(\pi 4 \tilde{x}_i) + \varepsilon_i. \qquad (8.3.2)$$

As with the polynomial models, the number of additional predictors to add depends on how complicated the data are. For the purpose of testing lack of fit, we simply need the number to be large enough to find any salient aspects of the data that are not fitted well by the SLR model.

Another approach is to add a number of indicator functions. An *indicator function* of a set A is defined as

$$I_A(\theta) = \begin{cases} 1 & \text{if } \theta \in A \\ 0 & \text{if } \theta \notin A \end{cases} \quad (8.3.3)$$

We can fit models like

$$y_i = \beta_0 + \beta_1 x_i + \beta_2 I_{[0,.25)}(\tilde{x}_i) + \beta_3 I_{[.25,.5)}(\tilde{x}_i) + \beta_4 I_{[.5,.75)}(\tilde{x}_i) + \beta_5 I_{[.75,1]}(\tilde{x}_i) + \varepsilon_i.$$

Adding indicator functions of length 2^{-j} defined on $\tilde{x}$ is equivalent to adding *Haar wavelets* to the model, cf. Christensen (2001). Unfortunately, no regression programs will fit this model because it is no longer a regression model. It is no longer a regression model because there is a redundancy in the predictor variables. The model includes an intercept, which corresponds to a predictor variable that always takes on the value 1. However, if we add together our four indicator functions, their sum is also a variable that always takes on the value 1. To evade this problem, we need either to delete one of the indicator functions (doesn't matter which one) or remove the intercept from the model. Dropping the last indicator is convenient, so we fit

$$y_i = \beta_0 + \beta_1 x_i + \beta_2 I_{[0,.25)}(\tilde{x}_i) + \beta_3 I_{[.25,.5)}(\tilde{x}_i) + \beta_4 I_{[.5,.75)}(\tilde{x}_i) + \varepsilon_i. \quad (8.3.4)$$

Any continuous function defined on an interval $[a, b]$ can be approximated arbitrarily well by a sufficiently large polynomial. Similar statements can be made about the other classes of functions introduced here. Because of this, these classes of functions are known as *basis functions*.

EXAMPLE 8.3.1.   We illustrate the methods on the Hooker data. With x the temperature, we defined $\tilde{x} = (x - 180.5)/30.5$. Fitting Model (8.3.1) gives

Analysis of Variance: Sines and Cosines.

Source	df	SS	MS	F	P
Regression	5	447.185	89.437	3364.82	0.000
Residual Error	25	0.665	0.0266		
Total	30	447.850			

A test of whether Model (8.3.1) fits significantly better than SLR has statistic

$$F_{obs} = \frac{(3.68 - 0.665)/(29 - 25)}{0.0266} = 28.4.$$

Clearly the reduced model of a simple linear regression fits worse than the model with two additional sine and cosine terms.

Fitting Model (8.3.2) gives

Analysis of Variance: Cosines.

Source	df	SS	MS	F	P
Regression	5	447.208	89.442	3486.60	0.000
Residual Error	25	0.641	0.0257		
Total	30	447.850			

A test of whether the cosine model fits significantly better than SLR has statistic

$$F_{obs} = \frac{(3.68 - 0.641)/(29 - 25)}{0.0257} = 29.6.$$

Clearly the reduced model of a simple linear regression fits worse than the model with four additional cosine terms.

Fitting Model (8.3.4) gives

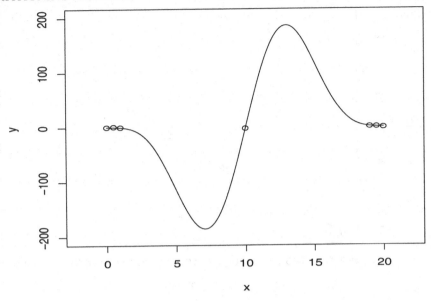

Figure 8.8: *Plot of fifth-order cosine model.*

Analysis of Variance: Haar Wavelets.

Source	df	SS	MS	F	P
Regression	4	446.77	111.69	2678.37	0.000
Residual Error	26	1.08	0.0417		
Total	30	447.85			

A test of whether this Haar wavelet model fits significantly better than SLR has statistic

$$F_{obs} = \frac{(3.68 - 1.08)/(29 - 26)}{0.0417} = 20.8.$$

Clearly the reduced model of a simple linear regression fits worse than the model with three additional indicator functions.   □

8.3.1 High-order models

For continuous basis functions like the trig functions, high-order models can behave as strangely between the data points as polynomials. For example, Figure 8.8 contains a plot of the 7 data points discussed in Section 8.2 and, using $\tilde{x} = x/20$, a fitted cosine model with 5 terms and an intercept,

$$y_i = \beta_0 + \beta_1 \cos(\pi \tilde{x}_i) + \beta_2 \cos(\pi 2\tilde{x}_i) + \beta_3 \cos(\pi 3\tilde{x}_i) + \beta_4 \cos(\pi 4\tilde{x}_i) + \beta_5 \cos(\pi 5\tilde{x}_i) + \varepsilon_i.$$

The fit away from the data is even worse than for fifth- and sixth-order polynomials.

8.4 Partitioning methods

The basic idea of the partitioning method is quite simple. Suppose we are fitting a simple linear regression but that the actual relationship between x and y is a quadratic. If you can split the x values into two parts near the maximum or minimum of the quadratic, you can get a much better approximate fit using two lines instead of one. More generally, the idea is that an approximate model should work better on a smaller set of data that has predictor variables that are more similar. Thus, if the original model is wrong, we should get a better approximation to the truth by fitting the original

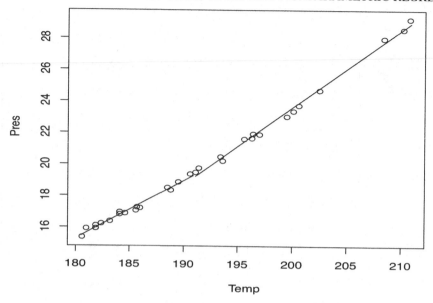

Figure 8.9: *Hooker data, partition method.*

model on a series of smaller subsets of the data. Of course if the original model is correct, it should work about the same on each subset as it does on the complete data. The statistician partitions the data into disjoint subsets, fits the original model on each subset, and compares the overall fit of the subsets to the fit of the original model on the entire data. The statistician is free to select the partitions, including the number of distinct sets, but the subsets need to be chosen based on the predictor variable(s) alone.

EXAMPLE 8.4.1.   We illustrate the partitioning method by splitting the Hooker data into two parts. Our partition sets are the data with the 16 smallest temperatures and the data with the 15 largest temperatures. We then fit a separate regression line to each partition. The two fitted lines are given in Figure 8.9. The ANOVA table is

Analysis of Variance: Partitioned Hooker data.

Source	df	SS	MS	F	P
Regression	3	446.66	148.89	3385.73	0.000
Error	27	1.19	0.04		
Total	30	447.85			

A test of whether this partitioning fits significantly better than SLR has statistic

$$F_{obs} = \frac{(3.68 - 1.19)/(29 - 27)}{0.04} = 31.125.$$

Clearly the reduced model of a simple linear regression fits worse than the model with two SLRs. Note that this is a simultaneous test of whether the slopes and intercepts are the same in each partition.   □

8.4.1   Fitting the partitioned model

We now consider three different ways to fit this partitioned model. Our computations will be subject to some round-off error. One way to fit this model is simply to divide the data into two parts and fit a simple linear regression to each one. Fitting the lowest 16 x (temperature) values gives

Table of Coefficients: Low x values.

Predictor	$\hat{\beta}_k$	SE($\hat{\beta}_k$)	t	P
Constant	−50.725	2.596	−19.54	0.000
x-low	0.36670	0.01404	26.13	0.0001

Analysis of Variance: Low x values.

Source	df	SS	MS	F	P
Regression	2	4687.1	2342.5	81269.77	0.000
Error	14	0.4	0.0		
Total	16	4687.5			

To get some extra numerical accuracy, from the F statistic we can compute $MSE = 2342.5/81269.77 = 0.028836$ so $SSE = 0.4037$. Fitting the highest 15 x values gives

Table of Coefficients: High x values.

Predictor	$\hat{\beta}_k$	SE($\hat{\beta}_k$)	t	P
Constant	−74.574	2.032	−36.70	0.000
x-high	0.49088	0.01020	48.12	0.000

Analysis of Variance: High x values.

Source	df	SS	MS	F	P
Regression	2	8193.9	4096.9	67967.66	0.000
Error	13	0.8	0.1		
Total	15	8194.7			

Again, from the F statistic $MSE = 4096.9/67967.66 = 0.060277$, so $SSE = 0.7836$. The fit of the overall model is obtained by pooling the two Error terms to give $dfE(Full) = 14 + 13 = 27$, $SSE(Full) = 0.4037 + 0.7836 = 1.1873$, with $MSE(Full) = 0.044$.

A more efficient way to proceed is to fit both simple linear regressions at once. Construct a variable h that identifies the 15 high values of x. In other words, h is 1 for the 15 highest temperature values and 0 for the 16 lowest values. Define $x_1 = h \times x$, $h_2 = 1 - h$, and $x_2 = h_2 \times x$. Fitting these four variables in a *regression through the origin*, i.e., fitting

$$y_i = \beta_1 h_{i2} + \beta_2 x_{i2} + \beta_3 h_i + \beta_4 x_{i1} + \varepsilon_i,$$

gives

Table of Coefficients: Separate lines.

Predictor	$\hat{\beta}_k$	SE($\hat{\beta}_k$)	t	P
h_2	−50.725	3.205	−15.82	0.000
x_2	0.36670	0.01733	21.16	0.000
h	−74.574	1.736	−42.97	0.000
x_1	0.490875	0.008712	56.34	0.000

Analysis of Variance: Separate lines.

Source	df	SS	MS	F	P
Regression	4	12881.0	3220.2	73229.01	0.000
Error	27	1.2	0.0		
Total	31	12882.2			

Note that these regression estimates agree with those obtained from fitting each set of data separately. The standard errors differ because here we are pooling the information in the error rather than using separate estimates of σ^2 from each subset of data. Although the ANOVA table reports $MSE = 0.0$, we can see that it actually agrees with earlier calculations by noting that $MSE = MSReg/F = 0.04397$.

The way the model was originally fitted for our discussion was regressing on x, h, and x_1, i.e., fitting

$$y_i = \beta_0 + \beta_1 x_i + \beta_2 h_i + \beta_3 x_{i1} + \varepsilon_i. \tag{8.4.1}$$

This is a model that has the low group of temperature values as a baseline and for the high group incorporates deviations from the baseline. The ANOVA table gives the same Error as the previous table and the table of regression coefficients is

Table of Coefficients: Low group baseline.

Predictor	$\hat{\beta}_k$	$SE(\hat{\beta}_k)$	t	P
Constant	−50.725	3.205	−15.82	0.000
x	0.36670	0.01733	21.16	0.000
h	−23.849	3.645	−6.54	0.000
x_1	0.12418	0.01940	6.40	0.000

The slope for the low group is 0.36670 and for the high group it is $0.36670 + 0.12418 = 0.49088$. The t test for whether the slopes are different, in a model that retains separate intercepts, is based on the x_1 row of this table and has $t = 6.40$. The intercepts also look different. The estimated intercept for the low group is −50.725 and for the high group it is $-50.725 + (-23.849) = -74.574$. The t test for whether the intercepts are different, in a model that retains separate slopes, is based on the h row and has $t = -6.54$.

8.4.2   Output for categorical predictors*

In Section 3.9 we discussed the fact that predictor variables can be of two types: continuous or categorical. Regression analysis and computer programs for regression analysis consider only continuous variables. Various programs for fitting *linear models* (as distinct from fitting regression) handle both types of variables. Of the packages discussed on the website, R's command lm and SAS's PROC GENMOD treat all (numerical) variables as continuous unless otherwise specified. In particular, if no variables are specified as categorical, both lm and GENMOD act as regression programs. Minitab's glm, on the other hand, treats all variables as categorical (factors) unless otherwise specified. Not only are the defaults different, but how the programs deal with categorical variables differs. Since partitioning the data defines categories, we have cause to introduce these issues here. Categorical variables will become ubiquitous beginning in Chapter 12.

In our partitioning example, x is continuous but h is really a categorical variable indicating which points are in the high group. When a categorical variable has only two groups, or more specifically, if it is a 0-1 indicator variable like h (or h_2), it can be treated the same way that continuous variables are treated in regression software. Indeed, we have exploited that fact up to this point. The remainder of this subsection discusses how various software treat variables that are identified as factors.

As indicated earlier, R's lm command and SAS's PROC GENMOD both have x defaulting to a continuous variable but h can be specified as a factor. Minitab's glm output has h defaulting to a factor but x must be specified as a covariate. In all of them we fit a model that specifies effects for each variable plus we fit an "interaction" between the two variables. To mimic these procedures using regression, we need to construct and use variables h_2, x_1, x_2 and two new variables h_3, x_3. One advantage of specifying h as a factor variable is that you do not have to construct any new variables.

R's lm program with h as a factor, essentially, fits Model (8.4.1), i.e., a model that uses the low temperatures as a baseline. The output is the same as the regression output that we already examined.

SAS's PROC GENMOD with h as a classification variable (factor), essentially, fits a model that uses the high group as the baseline, that is, it fits

$$y_i = \beta_0 + \beta_1 x_i + \beta_2 h_{i2} + \beta_3 x_{i2} + \varepsilon_i.$$

For the low group, the model incorporates deviations from the baseline. The three-line ANOVA table does not change from Model (8.4.1) but the table of regression coefficients is

Table of Coefficients: High group baseline.

Predictor	$\hat{\beta}_k$	SE($\hat{\beta}_k$)	t	P
Constant	−74.574	1.736	−42.97	0.000
x	0.49088	0.00871	56.34	0.000
h_2	23.849	3.645	6.54	0.000
x_2	−0.12418	0.01940	−6.40	0.000

The estimated slope for the high group is 0.49088 and for the low group it is $0.49088 + (-0.12418) = 0.36670$. The t test for whether the slopes are different, in a model that retains separate intercepts, is based on the x_2 row of this table and has $t = -6.40$. The intercepts also look different. The estimated intercept for the high group is −74.574 and for the low group it is $-74.574 + 23.849 = -50.725$. The t test for whether the intercepts are different, in a model that retains separate slopes, is based on the h_2 row and has $t = 6.54$.

The following table is how PROC GENMOD reports these results.

Table of Coefficients: SAS PROC GENMOD.

Predictor	df		$\hat{\beta}_k$	SE$_m$($\hat{\beta}_k$)	95% Conf.	Limits	t^2	P
Intercept	1		−74.5741	1.6198	−77.7489	−71.3993	2119.55	< .0001
h	0	1	23.8490	3.4019	17.1815	30.5166	49.15	< .0001
h	1	0	0.0000	0.0000	0.0000	0.0000	.	
x		1	0.4909	0.0081	0.4749	0.5068	3644.94	< .0001
x *h	0	1	−0.1242	0.0181	−0.1597	−0.0887	47.04	< .0001
x *h	1	0	0.0000	0.0000	0.0000	0.0000	.	
Scale		1	0.1957	0.0249	0.1526	0.2510		

While the parameter estimates agree in obvious ways, the standard errors are different from the regression output. The coefficients for the highest level of the factor h are forced to be zero (R does this for the lowest level of h) and the corresponding standard errors are 0 because estimates that have been forced to be zero have no variability. The nonzero standard errors are also different in GENMOD because they are not based on the MSE but rather the maximum likelihood estimate of the variance,

$$\hat{\sigma}^2 \equiv \frac{SSE}{n}.$$

We used the notation SE$_m$($\hat{\beta}_k$) with a subscript of m to indicate this difference. The relationship between the standard errors is

$$SE(\hat{\beta}_k) = \frac{\sqrt{n}}{\sqrt{dfE}} SE_m(\hat{\beta}_k).$$

Note also that GENMOD gives t^2 rather than t, provides 95% confidence intervals, and reports very small P values in a more appropriate fashion than merely reporting 0.0000. SAS also has a PROC GLM procedure that will fit the model, but it does not readily report parameter estimates.

R and SAS use variations on a theme, i.e., fix a baseline group. Minitab takes a different course. Minitab, essentially, defines variables $h_3 = h_2 - h$ and $x_3 = x \times h_3$ and fits

$$y_i = \beta_0 + \beta_1 x_i + \beta_2 h_{i3} + \beta_3 x_{i3} + \varepsilon_i.$$

This gives the regression coefficients

Table of Coefficients.

Predictor	$\hat{\beta}_k$	$SE(\hat{\beta}_k)$	t	P
Constant	−62.64962	1.82259	−34.374	0.000
x	0.42879	0.00970	44.206	0.000
h3	11.92452	1.82259	6.543	0.000
x3	−0.06209	0.00970	−6.401	0.000

Minitab's glm yields the following output for coefficients.

Table of Coefficients: Minitab glm.

Predictor	$\hat{\beta}_k$	$SE(\hat{\beta}_k)$	t	P
Constant	−62.650	1.823	−34.37	0.000
h				
0	11.925	1.823	6.54	0.000
x	0.428787	0.009700	44.21	0.000
x*h				
0	−0.062089	0.009700	−6.40	0.000

provided you ask Minitab for coefficients for all terms. (The default does not give coefficients associated with h.) The "constant" value of −62.650 is the average of the two intercept estimates that were reported earlier for the separate lines. The intercept for the low group ($h = 0$) is −62.650 + 11.925 and the intercept for the high group is −62.650 − 11.925. Note that the t test for "h 0" is the same 6.54 that was reported earlier for testing whether the intercepts were different. Minitab is fitting effects for both $h = 0$ and $h = 1$ but forcing them to sum to zero, rather than what R and SAS do, which is picking a level of h and forcing the effect of that level to be zero (hence making it the baseline). Similarly, the "x" value 0.428787 is the average of the two slope estimates reported earlier. The slope for the low group ($h = 0$) is 0.428787 + (−0.062089) and the slope for the high group is 0.428787 − (−0.062089). The t test for "x*h 0" is the same −6.40 as that reported earlier for testing whether the slopes were different. Minitab provides coefficient output that is more traditional than either R or SAS, but is often more difficult to interpret. However, given the wide variety of software and output that one may be confronted with, it is important to be able to cope with all of it.

Our discussion used the variable h that partitions the data into the smallest 16 observations and the largest 15 observations. Minitab's regression program provides a lack-of-fit test that partitions the data into the 18 observations below $\bar{x}. = 191.79$ and the 13 observations larger than the mean. Their test gets considerably more complicated when there is more than one predictor variable. They perform both this test (in more complicated situations, these tests) and a version of the test described in the next subsection, and combine the results from the various tests.

8.4.3 Utts' method

Utts (1982) proposed a lack-of-fit test based on comparing the original (reduced) model to a full model that consists of fitting the original model on a subset of the original data. In other words, you fit the model on all the data and test that against a full model that consists of fitting the model on a subset of the data. The subset is chosen to contain the points closest to $\bar{x}.$. Although it seems like fitting the model to a reduced set of points should create a reduced model, just the opposite is true. To fit a model to a reduced set of points, we can think of fitting the original model and then adding a separate parameter for every data point that we want to exclude from the fitting procedure. In fact, that is what makes this a partitioning method. There is one subset that consists of the central data and the rest of the partition has every data point in a separate set.

The central subset is chosen to be a group of points close to $\bar{x}.$. With only one predictor variable, it is easy to determine a group of central points. It turns out that for models with an intercept, the leverages are really measures of distance from $\bar{x}.$; see Christensen (2011, Section 13.1), so even with more predictor variables, one could choose a group of points that have the lowest leverages in the original model.

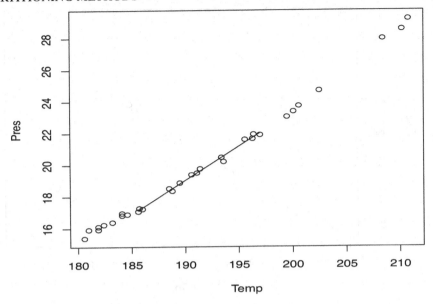

Figure 8.10: *Hooker data, Utts' method with 15 points.*

EXAMPLE 8.4.1.   We consider first the use of 15 central points with leverages below 0.05; about half the data. We then consider a group of 6 central points; about a fifth of the data.

The ANOVA table when fitting a simple linear regression to 15 central points is

Analysis of Variance: 15 central points.

Source	df	SS	MS	F	P
Regression	1	40.658	40.658	1762.20	0.000
Error	13	0.300	0.023		
Total	14	40.958			

The lack-of-fit test against a reduced model of simple linear regression on the entire data has

$$F_{obs} = \frac{(3.68 - 0.300)/(29 - 13)}{0.023} = 9.18,$$

which is highly significant. Figure 8.10 illustrates the fitting method.

When using 6 central points having leverages below 0.035, the ANOVA table is

Analysis of Variance: 6 central points.

Source	df	SS	MS	F	P
Regression	1	1.6214	1.6214	75.63	0.001
Error	4	0.0858	0.0214		
Total	5	1.7072			

and the *F* statistic is

$$F_{obs} = \frac{(3.68 - 0.0858)/(29 - 4)}{0.0214} = 6.72.$$

This is much bigger than 1 and easily significant at the 0.05 level. Both tests suggest lack of fit. Figure 8.11 illustrates the fitting method.   □

My experience is that Utt's test tends to work better with relatively small groups of central points. (Even though the *F* statistic here was smaller for the smaller group.) Minitab's regression

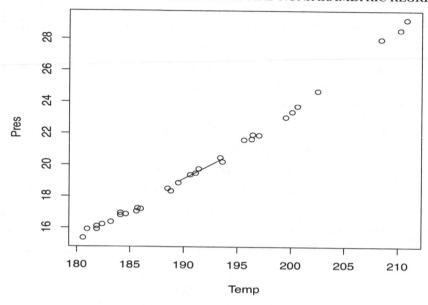

Figure 8.11: *Hooker data, Utts' method with 6 points.*

program incorporates a version of Utt's test that defines the central region as those points with leverages less than $1.1p/n$ where p is the number of regression coefficients in the model, so for a simple linear regression $p = 2$. For these data, their central region consists of the 22 observations with temperature between 183.2 and 200.6.

8.5 Splines

When fitting a polynomial to a single predictor variable, the partitioning method is extremely similar to the nonparametric regression method known as fitting *splines*. When using partitioning to test for lack of fit, our fitting of the model on each subset was merely a device to see whether the original fitted model gave better approximations on smaller subsets of the data than it did overall. The only difference when fitting splines is that we take the results obtained from fitting on the partition sets seriously as a model for the regression function $m(x)$. As such, we typically do not want to allow discontinuities in $m(x)$ at the partition points (known as "knots" in spline theory), so we include conditions that force continuity. Typically when fitting splines one uses a large number of partition sets, so there are a large number of conditions to force continuity. We illustrate the ideas on the Hooker data with only two partition sets. Generalizations are available for more than one predictor variable; see Wahba (1990).

EXAMPLE 8.5.1.   *Hooker data.*
Again, our partition sets are the data with the 16 smallest temperatures and the data with the 15 largest temperatures. Referring back to Table 7.1 we see that the partition point must be somewhere between 190.6 and 191.1. For convenience, let's set the partition point at 191. We model a separate regression line for each partition,

$$m(x) = \begin{cases} \beta_1 + \beta_2 x & \text{if } x \leq 191 \\ \beta_3 + \beta_4 x & \text{if } x > 191 \, . \end{cases}$$

Fitting two regression lines was discussed in Subsection 8.4.1 where we found the estimated lines

$$\hat{m}(x) = \begin{cases} -50.725 + 0.36670x & \text{if } x \leq 191 \\ -74.574 + 0.490875x & \text{if } x > 191 \, . \end{cases}$$

The two fitted lines were displayed in Figure 8.9.

To change this into a linear spline model, we need the two lines to match up at the knot, that is, we need to impose the continuity condition that

$$\beta_1 + \beta_2 191 = \beta_3 + \beta_4 191.$$

The condition can be rewritten in many ways but we will use

$$\beta_3 = \beta_1 + \beta_2 191 - \beta_4 191.$$

You can see from Figure 8.9 that the two separate fitted lines are already pretty close to matching up at the knot.

In Subsection 8.4.1 we fitted the partitioned model as a single linear model in two ways. The first was more transparent but the second had advantages. The same is true about the modifications needed to generate linear spline models. To begin, we constructed a variable h that identifies the 15 high values of x. In other words, h is 1 for the 15 highest temperature values and 0 for the 16 lowest values. We might now write

$$h(x) = I_{(191,\infty)}(x),$$

where we again use the indicator function introduced in Section 8.3. With slightly different notation for the predictor variables, we first fitted the two separate lines model as

$$y_i = \beta_1[1 - h(x_i)] + \beta_2 x_i[1 - h(x_i)] + \beta_3 h(x_i) + \beta_4 x_i h(x_i) + \varepsilon_i.$$

Imposing the continuity condition by substituting for β_3, the model becomes

$$y_i = \beta_1[1 - h(x_i)] + \beta_2 x_i[1 - h(x_i)] + \{\beta_1 + \beta_2 191 - \beta_4 191\} h(x_i) + \beta_4 x_i h(x_i) + \varepsilon_i$$

or

$$y_i = \beta_1 \{[1 - h(x_i)] + h(x_i)\} + \beta_2 \{x_i[1 - h(x_i)] + 191 h(x_i)\} + \beta_4 [x_i h(x_i) - 191 h(x_i)] + \varepsilon_i$$

or

$$y_i = \beta_1 + \beta_2 \{x_i[1 - h(x_i)] + 191 h(x_i)\} + \beta_4 (x_i - 191) h(x_i) + \varepsilon_i, \qquad (8.5.1)$$

where now β_1 is an overall intercept for the model.

As mentioned earlier, the two-lines model was originally fitted (with different symbols for the unknown parameters) as

$$y_i = \beta_1 + \beta_2 x_i + \gamma_1 h(x_i) + \gamma_2 x_i h(x_i) + \varepsilon_i.$$

This is a model that has the low group of temperature values as a baseline and for the high group incorporates deviations from the baseline, e.g., the slope above 191 is $\beta_2 + \gamma_2$. For this model the continuity condition is that

$$\beta_1 + \beta_2 191 = \beta_1 + \beta_2 191 + \gamma_1 + \gamma_2 191$$

or that

$$0 = \gamma_1 + \gamma_2 191$$

or that

$$\gamma_1 = -\gamma_2 191.$$

Imposing this continuity condition, the model becomes

$$y_i = \beta_1 + \beta_2 x_i - \gamma_2 191 h(x_i) + \gamma_2 x_i h(x_i) + \varepsilon_i$$

or

$$y_i = \beta_1 + \beta_2 x_i + \gamma_2 (x_i - 191) h(x_i) + \varepsilon_i. \qquad (8.5.2)$$

In discussions of splines, the function $(x_i - 191)h(x_i)$ is typically written $(x_i - 191)_+$ where for any scalar a,

$$(x - a)_+ \equiv \begin{cases} x - a & \text{if } x > a \\ 0 & \text{if } x \le a. \end{cases}$$

Fitting models (8.5.1) and (8.5.2) to the Hooker data gives

Table of Coefficients: Model (8.5.1).

Predictor	Est	SE(Est)	t	P
Constant	−48.70931	2.252956	−21.62	0.000
$x[1 - h(x)] + 191h(x)$	0.35571	0.012080	29.45	0.000
$(x - 191)_+$	0.48717	0.007619	63.95	0.000

and

Table of Coefficients: Model (8.5.2).

Predictor	Est	SE(Est)	t	P
Constant	−48.70931	2.25296	−21.620	0.000
x	0.35571	0.01208	29.447	0.000
$(x - 191)_+$	0.13147	0.01751	7.509	0.000

Notice that the slope for x values above 191, $\hat{\beta}_4 = 0.48717$, equals the slope below 191 plus the change in slopes, $\hat{\beta}_2 + \hat{\gamma}_2 = 0.35571 + 0.13147$, there being round-off error in the last digit.

Both models give $dfE = 28$, $SSE = 1.2220$, and $MSE = 0.04364$. We can even use the linear spline model as the basis for a lack-of-fit test of the simple linear regression on the Hooker data,

$$F_{obs} = \frac{(3.6825 - 1.2220)/(29 - 28)}{0.04364} = 56.38.$$

Obviously, fitting different lines on each partition set is a more general model than fitting the same line on each partition set. But since fitting a single line to all the data gives continuity at each knot, fitting different lines on each partition set and forcing them to be continuous is still a more general model than fitting the same line on all the data.   □

In general, to fit a linear spline model, you need to decide on a group of knots at which the slope will change. Call these $\tilde{x}_j$, $j = 1, \ldots, r$. The linear spline model then becomes

$$y_i = \beta_0 + \beta_1 x_i + \sum_{j=1}^{r} \gamma_j (x_i - \tilde{x}_j)_+ + \varepsilon_i.$$

Similar ideas work with higher-degree polynomials. The most popular polynomial to use is cubic; see Exercise 8.7.8. The general cubic spline model is

$$y_i = \beta_0 + \beta_1 x_i + \beta_2 x_i^2 + \beta_3 x^3 + \sum_{j=1}^{r} \gamma_j [(x_i - \tilde{x}_j)_+]^3 + \varepsilon_i.$$

See Christensen (2001, Section 7.6) for more discussion in a similar vein.

8.6   Fisher's lack-of-fit test

We now introduce Fisher's lack-of-fit test for the Hooker data. The test is discussed in much more detail in Chapter 12 and extended in Chapter 15. For now, notice that the predictor variable includes two replicate temperatures: $x = 181.9$ with y values 15.106 and 15.928, and $x = 184.1$ with y values 16.959 and 16.817. In this case, the computation for Fisher's lack-of-fit test is quite simple. We use the replicated x values to obtain a measure of pure error. First, compute the sample variance of the

Table 8.3: *IQs and achievement scores.*

IQ	Achiev.	IQ	Achiev.	IQ	Achiev.	IQ	Achiev.	IQ	Achiev.
100	49	105	50	134	78	107	43	122	66
117	47	89	72	125	39	121	75	130	63
98	69	96	45	140	66	90	40	116	43
87	47	105	47	137	69	132	80	101	44
106	45	95	46	142	68	116	55	92	50
134	55	126	67	130	71	137	73	120	60
77	72	111	66	92	31	113	48	80	31
107	59	121	59	125	53	110	41	117	55
125	27	106	49	120	64	114	29	93	50

y_is at each replicated x value. There are 2 observations at each replicated x, so the sample variance computed at each x has 1 degree of freedom. Since there are two replicated xs each with one degree of freedom for the variance estimate, the pure error has $1 + 1 = 2$ degrees of freedom. To compute the sum of squares for pure error, observe that when $x = 181.9$, the mean y is 15.517. The contribution to the sum of squares pure error from this x value is $(15.106 - 15.517)^2 + (15.928 - 15.517)^2$. A similar contribution is computed for $x = 184.1$ and they are added to get the sum of squares pure error. The degrees of freedom and sum of squares for lack of fit are found by taking the values from the original error and subtracting the values for the pure error. The F test for lack of fit examines the mean square lack of fit divided by the mean square pure error.

Analysis of Variance.

Source	df	SS	MS	F	P
Regression	1	444.17	444.17	3497.89	0.000
Error	29	3.68	0.13		
(Lack of Fit)	27	3.66	0.14	10.45	0.091
(Pure Error)	2	0.03	0.01		
Total	30	447.85			

The F statistic for lack of fit, 10.45, seems substantially larger than 1, but because there are only 2 degrees of freedom in the denominator, the P value is a relatively large 0.09. This method is closely related to one-way analysis of variance as discussed in Chapter 12.

8.7 Exercises

EXERCISE 8.7.1.   Dixon and Massey (1969) presented data on the relationship between IQ scores and results on an achievement test in a general science course. Table 8.3 contains a subset of the data. Fit the simple linear regression model of achievement on IQ and the quadratic model of achievement on IQ and IQ squared. Evaluate both models and decide which is the best.

EXERCISE 8.7.2.   In Exercise 7.4.2 we considered data on the relationship between farm sizes and the acreage in corn. Fit the linear, quadratic, cubic, and quartic polynomial models to the logs of the acreages in corn. Find the model that fits best. Check the assumptions for this model.

EXERCISE 8.7.3.   Use two methods other than fitting polynomial models to test for lack of fit in Exercise 8.7.1

EXERCISE 8.7.4.   Based on the height and weight data given in Table 8.4, fit a simple linear regression of weight on height for these data and check the assumptions. Give a 99% confidence interval for the mean weight of people with a 72-inch height. Test for lack of fit of the simple linear regression model.

Table 8.4: *Weights for various heights.*

Ht.	Wt.	Ht.	Wt.
65	120	63	110
65	140	63	135
65	130	63	120
65	135	72	170
66	150	72	185
66	135	72	160

Table 8.5: *Jensen's crank pin data.*

Day	Diameter	Day	Diameter	Day	Diameter	Day	Diameter
4	93	10	93	16	82	22	90
4	100	10	88	16	72	22	92
4	88	10	87	16	80	22	82
4	85	10	87	16	72	22	77
4	89	10	87	16	89	22	89

EXERCISE 8.7.5.   Jensen (1977) and Weisberg (1985, p. 101) considered data on the outside diameter of crank pins that were produced in an industrial process. The diameters of batches of crank pins were measured on various days; if the industrial process is "under control" the diameters should not depend on the day they were measured. A subset of the data is given in Table 8.5 in a format consistent with performing a regression analysis on the data. The diameters of the crank pins are actually $.742 + y_{ij}10^{-5}$ inches, where the y_{ij}s are reported in Table 8.5. Perform polynomial regressions on the data. Give two lack-of-fit tests for the simple linear regression not based on polynomial regression.

EXERCISE 8.7.6.   Beineke and Suddarth (1979) and Devore (1991, p. 380) consider data on roof supports involving trusses that use light-gauge metal connector plates. Their dependent variable is an axial stiffness index (ASI) measured in kips per inch. The predictor variable is the length of the light-gauge metal connector plates. The data are given in Table 8.6.
    Fit linear, quadratic, cubic, and quartic polynomial regression models using powers of x, the plate length, and using powers of $x - \bar{x}.$, the plate length minus the average plate length. Compare the results of the two procedures. If your computer program will not fit some of the models, report on that in addition to comparing results for the models you could fit.

EXERCISE 8.7.7.   Consider fitting quadratic models $y_i = \gamma_0 + \gamma_1 x_i + \gamma_2 x_i^2 + \varepsilon_i$ and $y_i = \beta_0 + \beta_1 (x_i - \bar{x}.) + \beta_2 (x_i - \bar{x}.)^2 + \varepsilon_i$. Show that $\gamma_2 = \beta_2$, $\gamma_1 = \beta_1 - 2\beta_2\bar{x}.$, and $\gamma_0 = \beta_0 - \beta_1\bar{x}. + \beta_2\bar{x}.^2$.

EXERCISE 8.7.8.   *Cubic Splines.*

Table 8.6: *Axial stiffness index data.*

Plate	ASI	Plate	ASI	Plate	ASI	Plate	ASI	Plate	ASI
4	309.2	6	402.1	8	392.4	10	346.7	12	407.4
4	409.5	6	347.2	8	366.2	10	452.9	12	441.8
4	311.0	6	361.0	8	351.0	10	461.4	12	419.9
4	326.5	6	404.5	8	357.1	10	433.1	12	410.7
4	316.8	6	331.0	8	409.9	10	410.6	12	473.4
4	349.8	6	348.9	8	367.3	10	384.2	12	441.2
4	309.7	6	381.7	8	382.0	10	362.6	12	465.8

To fit two cubic polynomials on the Hooker partition sets, we can fit the regression function

$$
\begin{aligned}
m(x) &= \beta_0 + \beta_1 x + \beta_2 x^2 + \beta_3 x^3 + \gamma_0 h(x) + \gamma_1 x h(x) + \gamma_2 x^2 h(x) + \gamma_3 x^3 h(x) \\
&= (\beta_0 + \beta_1 x + \beta_2 x^2 + \beta_3 x^3) + h(x)(\gamma_0 + \gamma_1 x + \gamma_2 x^2 + \gamma_3 x^3),
\end{aligned}
$$

where the polynomial coefficients below the knot are the β_js and above the knot are the $(\beta_j + \gamma_j)$s. Define the change polynomial as

$$
C(x) \equiv \gamma_0 + \gamma_1 x + \gamma_2 x^2 + \gamma_3 x^3.
$$

To turn the two polynomials into cubic splines, we require that the two cubic polynomials be equal at the knot but also that their first and second derivatives be equal at the knot. It is not hard to see that this is equivalent to requiring that the change polynomial have

$$
0 = C(191) = \left.\frac{dC(x)}{dx}\right|_{x=191} = \left.\frac{d^2 C(x)}{dx^2}\right|_{x=191},
$$

where our one knot for the Hooker data is at $x = 191$. Show that imposing these three conditions leads to the model

$$
\begin{aligned}
m(x) &= \beta_0 + \beta_1 x + \beta_2 x^2 + \beta_3 x^3 + \gamma_3 (x - 191)^3 h(x) \\
&= \beta_0 + \beta_1 x + \beta_2 x^2 + \beta_3 x^3 + \gamma_3 [(x - 191)_+]^3.
\end{aligned}
$$

(It is easy to show that $C(x) = \gamma_3 (x - 191)^3$ satisfies the three conditions. It is a little harder to show that satisfying the three conditions implies that $C(x) = \gamma_3 (x - 191)^3$.)

Chapter 9

Multiple Regression: Introduction

Multiple (linear) regression involves predicting values of a dependent variable from the values on a collection of other (predictor) variables. In particular, linear combinations of the predictor variables are used in modeling the dependent variable. For the most part, the use of categorical predictors in multiple regression is inappropriate. To incorporate categorical predictors, they need to be replaced by 0-1 indicators for the various categories.

9.1 Example of inferential procedures

In Section 6.9 we introduced Mosteller and Tukey's *Coleman Report* data; see Table 6.4. The variables are y, the mean verbal test score for sixth graders; x_1, staff salaries per pupil; x_2, percentage of sixth graders whose fathers have white-collar jobs; x_3, a composite measure of socioeconomic status; x_4, the mean of verbal test scores given to the teachers; and x_5, the mean educational level of the sixth grader's mothers (one unit equals two school years). Figures 9.1 through 9.4 plot all of the variables.

It is of interest to examine the correlations between y and the predictor variables.

	x_1	x_2	x_3	x_4	x_5
Correlation with y	0.192	0.753	0.927	0.334	0.733

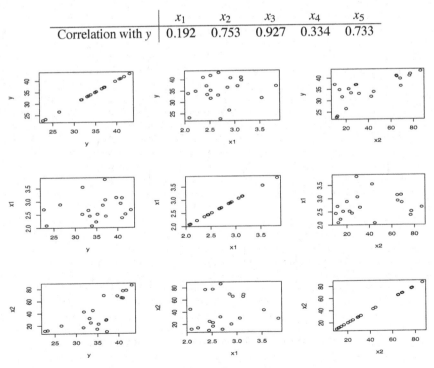

Figure 9.1: *Scatterplot matrix for* Coleman Report *data.*

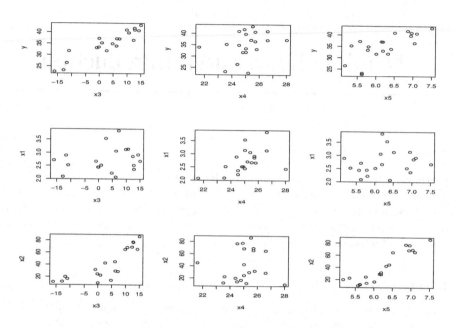

Figure 9.2: *Scatterplot matrix for* Coleman Report *data*.

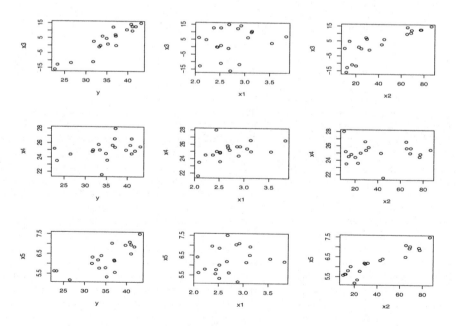

Figure 9.3: *Scatterplot matrix for* Coleman Report *data*.

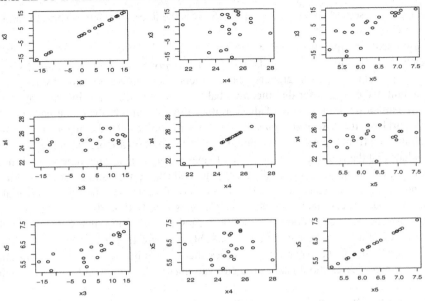

Figure 9.4: *Scatterplot matrix for* Coleman Report *data.*

Of the five variables, x_3, the one used in the simple linear regression, has the highest correlation. Thus it explains more of the y variability than any other single variable. Variables x_2 and x_5 also have reasonably high correlations with y. Low correlations exist between y and both x_1 and x_4. Interestingly, x_1 and x_4 turn out to be more important in explaining y than either x_2 or x_5. However, the explanatory power of x_1 and x_4 only manifests itself after x_3 has been fitted to the data.

The model is

$$y_i = \beta_0 + \beta_1 x_{i1} + \beta_2 x_{i2} + \beta_3 x_{i3} + \beta_4 x_{i4} + \beta_5 x_{i5} + \varepsilon_i, \qquad (9.1.1)$$

$i = 1, \ldots, 20$, where the ε_is are unobservable independent $N(0, \sigma^2)$ random variables and the βs are fixed unknown parameters. Fitting Model (9.1.1) with a computer program typically yields a table of coefficients with parameter estimates, standard errors for the estimates, t ratios for testing whether the parameters are zero, P values, and an analysis of variance table.

Table of Coefficients: Model (9.1.1)

Predictor	$\hat{\beta}_k$	$SE(\hat{\beta}_k)$	t	P
Constant	19.95	13.63	1.46	0.165
x_1	−1.793	1.233	−1.45	0.168
x_2	0.04360	0.05326	0.82	0.427
x_3	0.55576	0.09296	5.98	0.000
x_4	1.1102	0.4338	2.56	0.023
x_5	−1.811	2.027	−0.89	0.387

Analysis of Variance: Model (9.1.1)

Source	df	SS	MS	F	P
Regression	5	582.69	116.54	27.08	0.000
Error	14	60.24	4.30		
Total	19	642.92			

From just these two tables of statistics much can be learned. In particular, the estimated regression equation is

$$\hat{y} = 19.9 - 1.79x_1 + 0.0436x_2 + 0.556x_3 + 1.11x_4 - 1.81x_5.$$

Substituting the observed values x_{ij}, $j = 1,\ldots,5$ gives the fitted (predicted) values $\hat{y}_i$ and the residuals $\hat{\varepsilon}_i = y_i - \hat{y}_i$.

As discussed in simple linear regression, this equation *describes* the relationship between y and the predictor variables for the *current* data; *it does not imply a causal relationship*. If we go out and increase the percentage of sixth graders whose fathers have white-collar jobs by 1%, i.e., increase x_2 by one unit, we cannot infer that mean verbal test scores will tend to increase by 0.0436 units. In fact, we cannot think about any of the variables in a vacuum. No variable has an effect in the equation apart from the *observed values* of all the other variables. If we conclude that some variable can be eliminated from the model, we cannot conclude that the variable has no effect on y, we can only conclude that the variable is not necessary to explain *these* data. The same variable may be very important in explaining other, rather different, data collected on the same variables. All too often, people choose to interpret the estimated regression coefficients as if the predictor variables cause the value of y but the estimated regression coefficients simply describe an observed relationship. Frankly, since the coefficients do not describe a causal relationship, many people, including the author, find regression coefficients to be remarkably uninteresting quantities. What this model is good at is predicting values of y for new cases that are similar to those in the current data. In particular, such new cases should have predictor variables with values similar to those in the current data.

The t statistics for testing $H_0 : \beta_k = 0$ were reported in the table of coefficients. For example, the test of $H_0 : \beta_4 = 0$ has

$$t_{obs} = \frac{1.1102}{.4338} = 2.56.$$

The P value is

$$P = \Pr[|t(dfE)| \geq 2.56] = 0.023.$$

The value 0.023 indicates a reasonable amount of evidence that variable x_4 is needed in the model. We can be reasonably sure that dropping x_4 from the model harms the explanatory (predictive) power of the model. In particular, with a P value of 0.023, the test of the null model with $H_0 : \beta_4 = 0$ is rejected at the $\alpha = 0.05$ level (because $0.05 > 0.023$), but the test is not rejected at the $\alpha = 0.01$ level (because $0.023 > 0.01$).

A 95% confidence interval for β_3 has endpoints $\hat{\beta}_3 \pm t(0.975, dfE)\,\mathrm{SE}(\hat{\beta}_3)$. From a t table, $t(0.975, 14) = 2.145$ and from the table of coefficients the endpoints are

$$0.55576 \pm 2.145(0.09296).$$

The confidence interval is $(0.356, 0.755)$, so the data are consistent with β_3 between 0.356 and 0.755.

The primary value of the analysis of variance table is that it gives the degrees of freedom, the sum of squares, and the mean square for error. The mean squared error is the estimate of σ^2, and the sum of squares error and degrees of freedom for error are vital for comparing various regression models. The degrees of freedom for error are $n - 1 - $ (the number of predictor variables). The minus 1 is an adjustment for fitting the intercept β_0.

The analysis of variance table also gives the test for whether any of the x variables help to explain y, i.e., of whether $y_i = \beta_0 + \varepsilon_i$ is an adequate model. This test is rarely of interest because it is almost always highly significant. It is a poor scholar who cannot find any predictor variables that are related to the measurement of primary interest. (Ok, I admit to being a little judgmental here.) The test of

$$H_0 : \beta_1 = \cdots = \beta_5 = 0$$

is based on

$$F_{obs} = \frac{MSReg}{MSE} = \frac{116.5}{4.303} = 27.08$$

and (typically) is rejected for large values of F. The numerator and denominator degrees of freedom come from the ANOVA table. As suggested, the corresponding P value in the ANOVA table is infinitesimal, zero to three decimal places. Thus these x variables, as a group, help to explain the variation in the y variable. In other words, it is possible to predict the mean verbal test scores for a school's sixth grade class from the five x variables measured. Of course, the fact that some predictive ability exists does not mean that the predictive ability is sufficient to be useful.

The *coefficient of determination*, R^2, measures the predictive ability of the model. It is the squared correlation between the $(\hat{y}_i, y_i)$ pairs and also is the percentage of the total variability in y that is explained by the x variables. If this number is large, it suggests a substantial predictive ability. In this example

$$R^2 \equiv \frac{SSReg}{SSTot} = \frac{582.69}{642.92} = 0.906,$$

so 90.6% of the total variability is explained by the regression model. This large percentage suggests that the five x variables have substantial predictive power. However, we saw in Section 7.1 that a large R^2 does not imply that the model is good in absolute terms. It may be possible to show that this model does not fit the data adequately. In other words, this model is explaining much of the variability but we may be able to establish that it is not explaining as much of the variability as it ought. Conversely, a model with a low R^2 value may be the perfect model but the data may simply have a great deal of variability. Moreover, even an R^2 of 0.906 may be inadequate for the predictive purposes of the researcher, while in some situations an R^2 of 0.3 may be perfectly adequate. It depends on the purpose of the research. Finally, a large R^2 may be just an unrepeatable artifact of a particular data set. The coefficient of determination is a useful tool but it must be used with care. Recall from Section 7.1 that the R^2 was 0.86 when using just x_3 to predict y.

9.1.1 Computing commands

Performing multiple regression without a computer program is impractical. Mintab's reg command is menu driven, hence very easy to use. SAS's regression procedures are a bit more complicated, but the commands listed on my website are easily followed, as are the website commands for Minitab, most of which can be avoided by using the menus. R, on the other hand, is really a programming language and much more complicated to use. Because multiple regression is, arguably, the fundamental model considered in this book, we include some R code for it.

The following R code should work for computing most of the statistics used in this chapter and the next. Of course you have to replace the location of the data file `C:\\tab9-1.dat` with the location where you stored the data.

```
coleman <- read.table("C:\\tab6-4.dat",
        sep="",col.names=c("School","x1","x2","x3","x4","x5","y"))
attach(coleman)
coleman
summary(coleman)

#Coefficient and ANOVA tables
co <- lm(y ~ x1+x2+x3+x4+x5)
cop=summary(co)
cop
anova(co)

#Confidence intervals
confint(co, level=0.95)

#Predictions
new = data.frame(x1=2.07, x2=9.99,x3=-16.04,x4= 21.6, x5=5.17)
```

```
predict(co,new,se.fit=T,interval="confidence")
predict(co,new,interval="prediction")

# Diagnostics table
infv = c(y,co$fit,hatvalues(co),rstandard(co),rstudent(co),cooks.distance(co))
inf=matrix(infv,I(cop$d \! f[1]+cop$d \! f[2]),6,dimnames = list(NULL,
                    c("y", "yhat", "lev","r","t","C")))
inf

# Normal and fitted values plots
qqnorm(rstandard(co),ylab="Standardized residuals")
plot(co$fit,rstandard(co),xlab="Fitted",
        ylab="Standardized residuals",main="Residual-Fitted plot")

#Wilk-Francia Statistic
rankit=qnorm(ppoints(rstandard(co),a=I(3/8)))
ys=sort(rstandard(co))
Wprime=(cor(rankit,ys))^2
Wprime
```

9.1.2 General statement of the multiple regression model

In general we consider a dependent variable y that is a random variable of interest. We also consider $p-1$ nonrandom predictor variables $x_1,\ldots,x_{p-1}$. The general multiple (linear) regression model relates n observations on y to a linear combination of the corresponding observations on the x_js plus a random error ε. In particular, we assume

$$y_i = \beta_0 + \beta_1 x_{i1} + \cdots + \beta_{p-1} x_{i,p-1} + \varepsilon_i,$$

where the subscript $i = 1,\ldots,n$ indicates different observations and the ε_is are independent $N(0,\sigma^2)$ random variables. The β_js and σ^2 are unknown constants and are the fundamental parameters of the regression model.

Estimates of the β_js are obtained by the method of least squares. The least squares estimates are those that minimize

$$\sum_{i=1}^{n} (y_i - \beta_0 - \beta_1 x_{i1} - \beta_2 x_{i2} - \cdots - \beta_{p-1} x_{i,p-1})^2.$$

In this function the y_is and the x_{ij}s are all known quantities. Least squares estimates have a number of interesting statistical properties. If the errors are independent with mean zero, constant variance, and are normally distributed, the least squares estimates are maximum likelihood estimates (MLEs) and minimum variance unbiased estimates (MVUEs). If we keep the assumptions of mean zero and constant variance but weaken the independence assumption to that of the errors being merely uncorrelated and stop assuming normal distributions, the least squares estimates are best (minimum variance) linear unbiased estimates (BLUEs).

In checking assumptions we often use the predictions (fitted values) $\hat{y}$ corresponding to the observed values of the predictor variables, i.e.,

$$\hat{y}_i = \hat{\beta}_0 + \hat{\beta}_1 x_{i1} + \cdots + \hat{\beta}_{p-1} x_{i,p-1},$$

$i = 1,\ldots,n$. Residuals are the values

$$\hat{\varepsilon}_i = y_i - \hat{y}_i.$$

The other fundamental parameter to be estimated, besides the β_js, is the variance σ^2. The *sum of squares error* is

$$SSE = \sum_{i=1}^{n} \hat{\varepsilon}_i^2$$

and the estimate of σ^2 is the *mean squared error* (residual mean square)

$$MSE = SSE/(n-p).$$

The *MSE* is an unbiased estimate of σ^2 in that $E(MSE) = \sigma^2$. Under the standard normality assumptions, *MSE* is the minimum variance unbiased estimate of σ^2. However, the maximum likelihood estimate of σ^2 is $\hat{\sigma}^2 = MSE/n$, Unless discussing SAS's PROC GENMOD, we will never use the MLE of σ^2.

Details of the estimation procedures are given in Chapter 11.

9.2 Regression surfaces and prediction

One of the most valuable aspects of regression analysis is its ability to provide good predictions of future observations. Of course, to obtain a prediction for a new value y we need to know the corresponding values of the predictor variables, the x_js. Moreover, to obtain good predictions, the values of the x_js need to be similar to those on which the regression model was fitted. Typically, a fitted regression model is only an approximation to the true relationship between y and the predictor variables. These approximations can be very good, but, because they are only approximations, they are not valid for predictor variables that are dissimilar to those on which the approximation was based. Trying to predict for x_j values that are far from the original data is always difficult. Even if the regression model is true and not an approximation, the variance of such predictions is large. When the model is only an approximation, the approximation is typically invalid for such predictor variables and the predictions can be utter nonsense.

The regression surface for the Coleman data is the set of all values z that satisfy

$$z = \beta_0 + \beta_1 x_1 + \beta_2 x_2 + \beta_3 x_3 + \beta_4 x_4 + \beta_5 x_5$$

for some values of the predictor variables. The estimated regression surface is

$$z = \hat{\beta}_0 + \hat{\beta}_1 x_1 + \hat{\beta}_2 x_2 + \hat{\beta}_3 x_3 + \hat{\beta}_4 x_4 + \hat{\beta}_5 x_5.$$

There are two problems of interest. The first is estimating the value z on the regression surface for a fixed set of predictor variables. The second is predicting the value of a new observation to be obtained with a fixed set of predictor variables. For any set of predictor variables, the estimate of the regression surface and the prediction are identical. What differs are the standard errors associated with the different problems.

Consider estimation and prediction at

$$(x_1, x_2, x_3, x_4, x_5) = (2.07, 9.99, -16.04, 21.6, 5.17).$$

These are the minimum values for each of the variables, so there will be substantial variability in estimating the regression surface at this point. The estimator (predictor) is

$$\hat{y} = \hat{\beta}_0 + \sum_{j=1}^{5} \hat{\beta}_j x_j = 19.9 - 1.79(2.07) + 0.0436(9.99)$$

$$+ 0.556(-16.04) + 1.11(21.6) - 1.81(5.17) = 22.375.$$

For constructing 95% t intervals, the percentile needed is $t(0.975, 14) = 2.145$.

The 95% confidence interval for the point $\beta_0 + \sum_{j=1}^{5} \beta_j x_j$ on the regression surface uses the standard error for the regression surface, which is

$$SE(Surface) = 1.577.$$

The standard error is obtained from the regression program and depends on the specific value of $(x_1, x_2, x_3, x_4, x_5)$. The formula for the standard error is given in Section 11.4. This interval has endpoints

$$22.375 \pm 2.145(1.577),$$

which gives the interval

$$(18.992, 25.757).$$

The 95% prediction interval is

$$(16.785, 27.964).$$

This is about 4 units wider than the confidence interval for the regression surface. The standard error for the prediction interval can be computed from the standard error for the regression surface.

$$SE(Prediction) = \sqrt{MSE + SE(Surface)^2}.$$

In this example,

$$SE(Prediction) = \sqrt{4.303 + (1.577)^2} = 2.606,$$

and the prediction interval endpoints are

$$22.375 \pm 2.145(2.606).$$

We mentioned earlier that even if the regression model is true, the variance of predictions is large when the x_j values for the prediction are far from the original data. We can use this fact to identify situations in which the predictions are unreliable because the locations are too far away. Let $p - 1$ be the number of predictor variables so that, including the intercept, there are p regression parameters. Let n be the number of observations. A sensible rule of thumb is that we should start worrying about the validity of the prediction whenever

$$\frac{SE(Surface)}{\sqrt{MSE}} \geq \sqrt{\frac{2p}{n}}$$

and we should be very concerned about the validity of the prediction whenever

$$\frac{SE(Surface)}{\sqrt{MSE}} \geq \sqrt{\frac{3p}{n}}.$$

Recall that for simple linear regression we suggested that leverages greater than $4/n$ cause concern and those greater than $6/n$ cause considerable concern. In general, leverages greater than $2p/n$ and $3p/n$ cause these levels of concern. The simple linear regression guidelines are based on having $p = 2$. We are comparing $SE(Surface)/\sqrt{MSE}$ to the square roots of these guidelines. In our example, $p = 6$ and $n = 20$, so

$$\frac{SE(Surface)}{\sqrt{MSE}} = \frac{1.577}{\sqrt{4.303}} = 0.760 < 0.775 = \sqrt{\frac{2p}{n}}.$$

The location of this prediction is near the boundary of those locations for which we feel comfortable making predictions.

9.3 Comparing regression models

A frequent goal in regression analysis is to find the simplest model that provides an adequate explanation of the data. In examining the full model with all five x variables, there is little evidence that any of x_1, x_2, or x_5 are needed in the regression model. The t tests reported in Section 9.1 for the corresponding regression parameters gave P values of $0.168, 0.427$, and 0.387. We could drop *any* *one* of the three variables without significantly harming the model. While this does not imply that all three variables can be dropped without harming the model, dropping the three variables makes an interesting point of departure.

Fitting the reduced model

$$y_i = \beta_0 + \beta_3 x_{i3} + \beta_4 x_{i4} + \varepsilon_i$$

gives

Table of Coefficients

Predictor	$\hat{\beta}_k$	$SE(\hat{\beta}_k)$	t	P
Constant	14.583	9.175	1.59	0.130
x_3	0.54156	0.05004	10.82	0.000
x_4	0.7499	0.3666	2.05	0.057

Analysis of Variance

Source	df	SS	MS	F	P
Regression	2	570.50	285.25	66.95	0.000
Error	17	72.43	4.26		
Total	19	642.92			

We can test whether this reduced model is an adequate explanation of the data as compared to the full model. The sum of squares for error from the full model was reported in Section 9.1 as $SSE(Full) = 60.24$ with degrees of freedom $dfE(Full) = 14$ and mean squared error $MSE(Full) = 4.30$. For the reduced model we have $SSE(Red.) = 72.43$ and $dfE(Red.) = 17$. The test statistic for the adequacy of the reduced model is

$$F_{obs} = \frac{[SSE(Red.) - SSE(Full)]/[dfE(Red.) - dfE(Full)]}{MSE(Full)} = \frac{[72.43 - 60.24]/[17 - 14]}{4.30} = 0.94.$$

F has $[dfE(Red.) - dfE(Full)]$ and $dfE(Full)$ degrees of freedom in the numerator and denominator, respectively. Here F is about 1, so it is not significant. In particular, 0.94 is less than $F(0.95, 3, 14)$, so a formal $\alpha = .05$ level one-sided F test does not reject the adequacy of the reduced model. In other words, the .05 level one-sided test of the null model with $H_0 : \beta_1 = \beta_2 = \beta_5 = 0$ is not rejected.

This test lumps the three variables x_1, x_2, and x_5 together into one big test. It is possible that the uselessness of two of these variables could hide the fact that one of them is (marginally) significant when added to the model with x_3 and x_4. To fully examine this possibility, we need to fit three additional models. Each variable should be added, in turn, to the model with x_3 and x_4. We consider in detail only one of these three models, the model with x_1, x_3, and x_4. From fitting this model, the t statistic for testing whether x_1 is needed in the model turns out to be -1.47. This has a P value of 0.162, so there is little indication that x_1 is useful. We could also construct an F statistic as illustrated previously. The sum of squares for error in the model with x_1, x_3, and x_4 is 63.84 on 16 degrees of freedom, so

$$F_{obs} = \frac{[72.43 - 63.84]/[17 - 16]}{63.84/16} = 2.16.$$

Note that, up to round-off error, $F = t^2$. The tests are equivalent and the P value for the F statistic is also 0.162. F tests are only equivalent to a corresponding t test when the numerator of the F statistic has one degree of freedom. Methods similar to these establish that neither x_2 nor x_5 are important when added to the model that contains x_3 and x_4.

Here we are testing two models: the full model with x_1, x_3, and x_4 against a reduced model with only x_3 and x_4. Both of these models are special cases of a biggest model that contains all of x_1, x_2, x_3, x_4, and x_5. In Subsection 3.1.1, for cases like this, we recommended an alternative F statistic,

$$F_{obs} = \frac{[72.43 - 63.84]/[17 - 16]}{4.30} = 2.00,$$

where the denominator MSE of 4.30 comes from the biggest model.

In testing the reduced model with only x_3 and x_4 against the full five-variable model, we observed that one might miss recognizing a variable that was (marginally) significant. In this case we did not miss anything important. However, if we had taken the reduced model as containing only x_3 and tested it against the full five-variable model, we would have missed the importance of x_4. The F statistic for this test turns out to be only 1.74.

In the model with x_1, x_3, and x_4, the t test for x_4 turns out to have a P value of 0.021. As seen in the table given previously, if we drop x_1 and use the model with only x_3, and x_4, the P value for x_4 goes to 0.057. Thus dropping a weak variable, x_1, can make a reasonably strong variable, x_4, look weaker. There is a certain logical inconsistency here. If x_4 is important in the x_1, x_3, x_4 model or the full five-variable model (P value 0.023), it is illogical that dropping some of the other variables could make it unimportant. Even though x_1 is not particularly important by itself, it augments the evidence that x_4 is useful. The problem in these apparent inconsistencies is that the x variables are all related to each other; this is known as the problem of *collinearity*. One reason for using the alternative F tests that employ $MSE(Big.)$ in the denominator is that it ameliorates this phenomenon.

Although a reduced model may be an adequate substitute for a full model on a particular set of data, it does *not* follow that the reduced model will be an adequate substitute for the full model with any data collected on the variables in the full model.

9.3.1   *General discussion*

Suppose that we want to compare two regression models, say,

$$y_i = \beta_0 + \beta_1 x_{i1} + \cdots + \beta_{q-1} x_{i,q-1} + \cdots + \beta_{p-1} x_{i,p-1} + \varepsilon_i \tag{9.3.1}$$

and

$$y_i = \beta_0 + \beta_1 x_{i1} + \cdots + \beta_{q-1} x_{i,q-1} + \varepsilon_i. \tag{9.3.2}$$

For convenience, in this subsection we refer to equations such as (9.3.1) and (9.3.2) simply as (1) and (2). The key fact here is that all of the variables in Model (2) are also in Model (1). In this comparison, we dropped the last variables $x_{i,q}, \ldots, x_{i,p-1}$ for notational convenience only; the discussion applies to dropping any group of variables from Model (1). *Throughout, we assume that Model (1) gives an adequate fit to the data and then compare how well Model (2) fits the data with how well Model (1) fits.* Before applying the results of this subsection, the validity of the model (1) assumptions should be evaluated.

We want to know if the variables $x_{i,q}, \ldots, x_{i,p-1}$ are needed in the model, i.e., whether they are useful predictors. In other words, we want to know if Model (2) is an adequate model; whether it gives an adequate explanation of the data. The variables $x_q, \ldots, x_{p-1}$ are extraneous if and only if $\beta_q = \cdots = \beta_{p-1} = 0$. The test we develop can be considered as a test of

$$H_0: \beta_q = \cdots = \beta_{p-1} = 0.$$

Parameters are very tricky things; you never get to see the value of a parameter. I strongly prefer the interpretation of testing one model against another model rather than the interpretation of testing whether $\beta_q = \cdots = \beta_{p-1} = 0$. In practice, useful regression models are rarely correct models, although they can be *very* good approximations. Typically, we do not really care whether Model (1) is true, only whether it is useful, but dealing with parameters in an incorrect model becomes tricky.

In practice, we are looking for a (relatively) succinct way of summarizing the data. The smaller the model, the more succinct the summarization. However, we do not want to eliminate useful explanatory variables, so we test the smaller (more succinct) model against the larger model to see if the smaller model gives up significant explanatory power. Note that the larger model always has at least as much explanatory power as the smaller model because the larger model includes all the variables in the smaller model plus some more.

Applying our model testing procedures to this problem yields the following test: Reject the hypothesis

$$H_0 : \beta_q = \cdots = \beta_{p-1} = 0$$

at the α level if

$$F \equiv \frac{[SSE(Red.) - SSE(Full)]/(p-q)}{MSE(Full)} > F(1-\alpha, p-q, n-p).$$

For $p - q \geq 3$, this one-sided test is not a significance test, cf. Chapter 3.

The notation $SSE(Red.) - SSE(Full)$ focuses on the ideas of full and reduced models. Other notations that focus on variables and parameters are also commonly used. One can view the model comparison procedure as fitting Model (2) first and then seeing how much better Model (1) fits. The notation based on this refers to the (extra) *sum of squares for regressing* on $x_q, \ldots, x_{p-1}$ *after regressing* on $x_1, \ldots, x_{q-1}$ and is written

$$SSR(x_q, \ldots, x_{p-1} | x_1, \ldots, x_{q-1}) \equiv SSE(Red.) - SSE(Full).$$

This notation assumes that the model contains an intercept. Alternatively, one can think of fitting the parameters $\beta_q, \ldots, \beta_{p-1}$ after fitting the parameters $\beta_0, \ldots, \beta_{q-1}$. The relevant notation refers to the *reduction in sum of squares* (for error) due to fitting $\beta_q, \ldots, \beta_{p-1}$ after $\beta_0, \ldots, \beta_{q-1}$ and is written

$$R(\beta_q, \ldots, \beta_{p-1} | \beta_0, \ldots, \beta_{q-1}) \equiv SSE(Red.) - SSE(Full).$$

Note that it makes perfect sense to refer to $SSR(x_q, \ldots, x_{p-1} | x_1, \ldots, x_{q-1})$ as the reduction in sum of squares for fitting $x_q, \ldots, x_{p-1}$ after $x_1, \ldots, x_{q-1}$.

It was mentioned earlier that the degrees of freedom for $SSE(Red.) - SSE(Full)$ is $p - q$. Note that $p - q$ is the number of variables to the left of the vertical bar in $SSR(x_q, \ldots, x_{p-1} | x_1, \ldots, x_{q-1})$ and the number of parameters to the left of the vertical bar in $R(\beta_q, \ldots, \beta_{p-1} | \beta_0, \ldots, \beta_{q-1})$.

A point that is quite clear when thinking of model comparisons is that if you change either model, (1) or (2), the test statistic and thus the test changes. This point continues to be clear when dealing with the notations $SSR(x_q, \ldots, x_{p-1} | x_1, \ldots, x_{q-1})$ and $R(\beta_q, \ldots, \beta_{p-1} | \beta_0, \ldots, \beta_{q-1})$. If you change any variable on either side of the vertical bar, you change $SSR(x_q, \ldots, x_{p-1} | x_1, \ldots, x_{q-1})$. Similarly, the parametric notation $R(\beta_q, \ldots, \beta_{p-1} | \beta_0, \ldots, \beta_{q-1})$ is also perfectly precise, but confusion can easily arise when dealing with parameters if one is not careful. For example, when testing, say, $H_0 : \beta_1 = \beta_3 = 0$, the tests are completely different in the three models

$$y_i = \beta_0 + \beta_1 x_{i1} + \beta_3 x_{i3} + \varepsilon_i, \tag{9.3.3}$$

$$y_i = \beta_0 + \beta_1 x_{i1} + \beta_2 x_{i2} + \beta_3 x_{i3} + \varepsilon_i, \tag{9.3.4}$$

and

$$y_i = \beta_0 + \beta_1 x_{i1} + \beta_2 x_{i2} + \beta_3 x_{i3} + \beta_4 x_{i4} + \varepsilon_i. \tag{9.3.5}$$

In Model (3) the test is based on $SSR(x_1, x_3) \equiv R(\beta_1, \beta_3 | \beta_0)$, i.e., the sum of squares for regression ($SSReg$) in the model with only x_1 and x_3 as predictor variables. In Model (4) the test uses

$$SSR(x_1, x_3 | x_2) \equiv R(\beta_1, \beta_3 | \beta_0, \beta_2).$$

Model (5) uses $SSR(x_1, x_3 | x_2, x_4) \equiv R(\beta_1, \beta_3 | \beta_0, \beta_2, \beta_4)$. In all cases we are testing $\beta_1 = \beta_3 = 0$ *after* fitting all the other parameters in the model. In general, we think of testing $H_0 : \beta_q = \cdots = \beta_{p-1} = 0$ after fitting $\beta_0, \ldots, \beta_{q-1}$.

If the reduced model is obtained by dropping out only one variable, e.g., if $q - 1 = p - 2$, the parametric hypothesis is $H_0 : \beta_{p-1} = 0$. We have just developed an F test for this and we have earlier used a t test for the hypothesis. In multiple regression, just as in simple linear regression, the F test is equivalent to the t test. It follows that the t test must be considered as a test for the parameter *after fitting all* of the other parameters in the model. In particular, the t tests reported in the table of coefficients when fitting a regression tell you only whether a variable can be dropped relative to the model that contains all the other variables. These t tests cannot tell you whether more than one variable can be dropped from the fitted model. If you drop any variable from a regression model, all of the t tests change. It is only for notational convenience that we are discussing testing $\beta_{p-1} = 0$; the results hold for any β_k.

The *SSR* notation can also be used to find *SSE*s. Consider models (3), (4), and (5) and suppose we know $SSR(x_2|x_1,x_3)$, $SSR(x_4|x_1,x_2,x_3)$, and the *SSE* from Model (5). We can easily find the *SSE*s for models (3) and (4). By definition,

$$
\begin{aligned}
SSE(4) &= [SSE(4) - SSE(5)] + SSE(5) \\
&= SSR(x_4|x_1,x_2,x_3) + SSE(5).
\end{aligned}
$$

Also

$$
\begin{aligned}
SSE(3) &= [SSE(3) - SSE(4)] + SSE(4) \\
&= SSR(x_2|x_1,x_3) + \{SSR(x_4|x_1,x_2,x_3) + SSE(5)\}.
\end{aligned}
$$

Moreover, we see that

$$
\begin{aligned}
SSR(x_2,x_4|x_1,x_3) &= SSE(3) - SSE(5) \\
&= SSR(x_2|x_1,x_3) + SSR(x_4|x_1,x_2,x_3).
\end{aligned}
$$

Note also that we can change the order of the variables.

$$
SSR(x_2,x_4|x_1,x_3) = SSR(x_4|x_1,x_3) + SSR(x_2|x_1,x_3,x_4).
$$

9.4 Sequential fitting

Multiple regression analysis is largely impractical without the aid of a computer. One specifies a regression model and the computer returns the vital statistics for that model. Many computer programs actually fit a sequence of models rather than fitting the model all at once.

EXAMPLE 9.4.1.   Suppose you want to fit the model

$$
y_i = \beta_0 + \beta_1 x_{i1} + \beta_2 x_{i2} + \beta_3 x_{i3} + \beta_4 x_{i4} + \varepsilon_i.
$$

Many regression programs actually fit the sequence of models

$$
\begin{aligned}
y_i &= \beta_0 + \beta_1 x_{i1} + \varepsilon_i, \\
y_i &= \beta_0 + \beta_1 x_{i1} + \beta_2 x_{i2} + \varepsilon_i, \\
y_i &= \beta_0 + \beta_1 x_{i1} + \beta_2 x_{i2} + \beta_3 x_{i3} + \varepsilon_i, \\
y_i &= \beta_0 + \beta_1 x_{i1} + \beta_2 x_{i2} + \beta_3 x_{i3} + \beta_4 x_{i4} + \varepsilon_i.
\end{aligned}
$$

The sequence is determined by the order in which the variables are specified. If the identical model is specified in the form

$$
y_i = \beta_0 + \beta_3 x_{i3} + \beta_1 x_{i1} + \beta_4 x_{i4} + \beta_2 x_{i2} + \varepsilon_i,
$$

the end result is exactly the same but the sequence of models is

$$y_i = \beta_0 + \beta_3 x_{i3} + \varepsilon_i,$$
$$y_i = \beta_0 + \beta_3 x_{i3} + \beta_1 x_{i1} + \varepsilon_i,$$
$$y_i = \beta_0 + \beta_3 x_{i3} + \beta_1 x_{i1} + \beta_4 x_{i4} + \varepsilon_i,$$
$$y_i = \beta_0 + \beta_3 x_{i3} + \beta_1 x_{i1} + \beta_4 x_{i4} + \beta_2 x_{i2} + \varepsilon_i.$$

Frequently, programs that fit sequences of models also provide sequences of sums of squares. Thus the first sequence of models yields

$$SSR(x_1), \ SSR(x_2|x_1), \ SSR(x_3|x_1,x_2), \text{ and } SSR(x_4|x_1,x_2,x_3)$$

while the second sequence yields

$$SSR(x_3), \ SSR(x_1|x_3), \ SSR(x_4|x_3,x_1), \text{ and } SSR(x_2|x_3,x_1,x_4).$$

These can be used in a variety of ways. For example, as shown at the end of the previous section, to test

$$y_i = \beta_0 + \beta_1 x_{i1} + \beta_3 x_{i3} + \varepsilon_i$$

against

$$y_i = \beta_0 + \beta_1 x_{i1} + \beta_2 x_{i2} + \beta_3 x_{i3} + \beta_4 x_{i4} + \varepsilon_i$$

we need $SSR(x_2,x_4|x_3,x_1)$. This is easily obtained from the second sequence as

$$SSR(x_2,x_4|x_3,x_1) = SSR(x_4|x_3,x_1) + SSR(x_2|x_3,x_1,x_4). \qquad \square$$

EXAMPLE 9.4.2.   If we fit the model

$$y_i = \beta_0 + \beta_1 x_{i1} + \beta_2 x_{i2} + \beta_3 x_{i3} + \beta_4 x_{i4} + \beta_5 x_{i5} + \varepsilon_i$$

to the *Coleman Report* data, we get the sequential sums of squares listed below.

Source	df	Seq SS	Notation
x_1	1	23.77	$SSR(x_1)$
x_2	1	343.23	$SSR(x_2|x_1)$
x_3	1	186.34	$SSR(x_3|x_1,x_2)$
x_4	1	25.91	$SSR(x_4|x_1,x_2,x_3)$
x_5	1	3.43	$SSR(x_5|x_1,x_2,x_3,x_4)$

Recall that the *MSE* for the five-variable model is 4.30 on 14 degrees of freedom.

From the sequential sums of squares we can test a variety of hypotheses related to the full model. For example, we can test whether variable x_5 can be dropped from the five-variable model. The F statistic is $3.43/4.30$, which is less than 1, so the effect of x_5 is insignificant. This test is equivalent to the t test for x_5 given in Section 9.1 when fitting the five-variable model. We can also test whether we can drop both x_4 and x_5 from the full model. The F statistic is

$$F_{obs} = \frac{(25.91 + 3.43)/2}{4.30} = 3.41.$$

$F(0.95, 2, 14) = 3.74$, so this F statistic provides little evidence that the pair of variables is needed. (The relative importance of x_4 is somewhat hidden by combining it in a test with the unimportant x_5.) Similar tests can be constructed for dropping x_3, x_4, and x_5, for dropping x_2, x_3, x_4, and x_5, and for dropping x_1, x_2, x_3, x_4, and x_5 from the full model. The last of these is just the ANOVA table F test.

We can also make a variety of tests related to 'full' models that do not include all five variables. In the previous paragraph, we found little evidence that the pair x_4 and x_5 help explain the data in the five-variable model. We now test whether x_4 can be dropped when we have already dropped x_5. In other words, we test whether x_4 adds explanatory power to the model that contains x_1, x_2, and x_3. The numerator has one degree of freedom and is $SSR(x_4|x_1, x_2, x_3) = 25.91$. The usual denominator mean square for this test is the MSE from the model with x_1, x_2, x_3, and x_4, i.e., $\{14(4.303) + 3.43\}/15$. (For numerical accuracy we have added another significant digit to the MSE from the five-variable model. The SSE from the model without x_5 is just the SSE from the five-variable model plus the sequential sum of squares $SSR(x_5|x_1, x_2, x_3, x_4)$.) Our best practice would be to construct the test using the same numerator mean square but the MSE from the five-variable model in the denominator of the test. Using this second denominator, the F statistic is $25.91/4.30 = 6.03$. Corresponding F percentiles are $F(0.95, 1, 14) = 4.60$ and $F(0.99, 1, 14) = 8.86$, so x_4 may be contributing to the model. If we had used the MSE from the model with x_1, x_2, x_3, and x_4, the F statistic would be equivalent to the t statistic for dropping x_4 that is obtained when fitting this four-variable model.

If we wanted to test whether x_2 and x_3 can be dropped from the model that contains x_1, x_2, and x_3, the usual denominator is $[14(4.303) + 25.91 + 3.43]/16 = 5.60$. (The SSE for the model without x_4 or x_5 is just the SSE from the five-variable model plus the sequential sum of squares for x_4 and x_5.) Again, we would alternatively use the MSE from the five-variable model in the denominator. Using the first denominator, the test is

$$F_{obs} = \frac{(343.23 + 186.34)/2}{5.60} = 47.28,$$

which is much larger than $F(0.999, 2, 16) = 10.97$, so there is overwhelming evidence that variables x_2 and x_3 cannot be dropped from the x_1, x_2, x_3 model.

The argument for basing tests on the MSE from the five-variable model is that it is less subject to bias than the other MSEs. In the test given in the previous paragraph, the MSE from the usual 'full' model incorporates the sequential sums of squares for x_4 and x_5. A reason for doing this is that we have tested x_4 and x_5 and are not convinced that they are important. As a result, their sums of squares are incorporated into the error. Even though we may not have established an overwhelming case for the importance of either variable, there is some evidence that x_4 is a useful predictor when added to the first three variables. The sum of squares for x_4 may or may not be large enough to convince us of its importance but it is large enough to change the MSE from 4.30 in the five-variable model to 5.60 in the x_1, x_2, x_3 model. In general, if you test terms and pool them with the Error whenever the test is insignificant, you are biasing the MSE that results from this pooling.   □

In general, *when given the ANOVA table and the sequential sums of squares, we can test any model in the sequence against any reduced model that is part of the sequence. We cannot use these statistics to obtain a test involving a model that is not part of the sequence.*

9.5   Reduced models and prediction

Fitted regression models are, not surprisingly, very dependent on the observed values of the predictor variables. We have already discussed the fact that fitted regression models are particularly good for making predictions but only for making predictions on new cases with predictor variables that are similar to those used in fitting the model. Fitted models are not good at predicting observations with predictor variable values that are far from those in the observed data. We have also discussed the fact that in evaluating a reduced model we are evaluating whether the reduced model is an adequate explanation of the data. An adequate reduced model should serve well as a prediction equation but only for new cases with predictor variables similar to those in the original data. It should not be overlooked that *when using a reduced model for prediction, new cases need to be similar to the observed data on* all *predictor variables and not just on the predictor variables in the reduced model.*

Good prediction from reduced models requires that new cases be similar to observed cases on all predictor variables because of the process of selecting reduced models. Predictor variables are eliminated from a model if they are not necessary to explain the data. This can happen in two ways. If a predictor variable is truly unrelated to the dependent variable, it is both proper and beneficial to eliminate that variable. The other possibility is that a predictor variable may be related to the dependent variable but that the relationship is hidden by the nature of the observed predictor variables. In the *Coleman Report* data, suppose the true response depends on both x_3 and x_5. We know that x_3 is clearly the best single predictor but the observed values of x_5 and x_3 are closely related; the sample correlation between them is 0.819. Because of their high correlation *in these data*, much of the actual dependence of y on x_5 could be accounted for by the regression on x_3 alone. Variable x_3 acts as a surrogate for x_5. As long as we try to predict new cases that have values of x_5 and x_3 similar to those in the original data, a reduced model based on x_3 should work well. Variable x_3 should continue to act as a surrogate. On the other hand, if we tried to predict a new case that had an x_3 value similar to that in the observed data but where the pair x_3, x_5 was not similar to x_3, x_5 pairs in the observed data, the reduced model that uses x_3 as a surrogate for x_5 would be inappropriate. Predictions could be very bad and, if we thought only about the fact that the x_3 value is similar to those in the original data, we might expect the predictions to be good. Unfortunately, when we eliminate a variable from a regression model, we typically have no idea if it is eliminated because the variable really has no effect on y or because its effect is being masked by some other set of predictor variables. For further discussion of these issues see Mandel (1989a, b).

Of course there is reason to hope that predictions will typically work well for reduced models. If the data come from an observational study in which the cases are some kind of sample from a population, there is reason to expect that future cases that are *sampled in the same way* will behave similarly to those in the original study. In addition, if the data come from an experiment in which the predictor variables are under the control of the investigator, it is reasonable to expect the investigator to select values of the predictor variables that cover the full range over which predictions will be made. Nonetheless, regression models give good approximations and good predictions only within the range of the observed data and, when a reduced model is used, the definition of the range of the observed data includes the values of all predictor variables that were in the full model. In fact, *even this statement is too weak.* When using a reduced model or even when using the full model for prediction, *new cases need to be similar to the observed cases in all relevant ways.* If there is some unmeasured predictor that is related to y and if the observed predictors are highly correlated with this unmeasured variable, then for good prediction a new case needs to have a value of the unmeasured variable that is similar to those for the observed cases. In other words, *the variables in any model may be acting as surrogates for some unmeasured variables and to obtain good predictions the new cases must be similar on both the observed predictor variables and on these unmeasured variables.*

Prediction should work well whenever $(x_{i1}, x_{i2}, \ldots, x_{i,p-1}, y_i)$, $i = 1, \ldots, n$ constitutes a random sample from some population and when the point we want to predict, say y_0, corresponds to predictor variables $(x_{01}, x_{02}, \ldots, x_{0,p-1})$ that are sampled from the same population. In practice, we rarely have this ideal, but the ideal illuminates what can go wrong in practice.

9.6 Partial correlation coefficients and added variable plots

Partial correlation coefficients measure the linear relationship between two variables after adjusting for a group of other variables. The square of a partial correlation coefficient is also known as a *coefficient of partial determination.* The squared sample partial correlation coefficient between y and x_1 after adjusting for x_2, x_3, and x_4 is

$$r_{y1\cdot234}^2 = \frac{SSR(x_1|x_2,x_3,x_4)}{SSE(x_2,x_3,x_4)},$$

where $SSE(x_2, x_3, x_4)$ is the sum of squares error from a model with an intercept and the three predictors x_2, x_3, x_4. The squared sample partial correlation coefficient between y and x_2 given x_1, x_3, and x_4 is

$$r_{y2 \cdot 134}^2 = \frac{SSR(x_2 | x_1, x_3, x_4)}{SSE(x_1, x_3, x_4)}.$$

Alternatively, the sample partial correlation $r_{y2 \cdot 134}$ is precisely the ordinary sample correlation computed between the residuals from fitting

$$y_i = \beta_0 + \beta_1 x_{i1} + \beta_3 x_{i3} + \beta_4 x_{i4} + \varepsilon_i \qquad (9.6.1)$$

and the residuals from fitting

$$x_{i2} = \gamma_0 + \gamma_1 x_{i1} + \gamma_3 x_{i3} + \gamma_4 x_{i4} + \varepsilon_i. \qquad (9.6.2)$$

The information in $r_{y2 \cdot 134}^2$ is equivalent to the information in the F statistic for testing $H_0: \beta_2 = 0$ in the model

$$y_i = \beta_0 + \beta_1 x_{i1} + \beta_2 x_{i2} + \beta_3 x_{i3} + \beta_4 x_{i4} + \varepsilon_i. \qquad (9.6.3)$$

To see this, observe that

$$
\begin{aligned}
F &= \frac{SSR(x_2 | x_1, x_3, x_4)/1}{\{SSE(x_1, x_3, x_4) - [SSE(x_1, x_3, x_4) - SSR(x_1, x_2, x_3, x_4)]\}/(n-5)} \\[2mm]
&= \frac{SSR(x_2 | x_1, x_3, x_4)/1}{[SSE(x_1, x_3, x_4) - SSR(x_2 | x_1, x_3, x_4)]/(n-5)} \\[2mm]
&= (n-5)\frac{SSR(x_2 | x_1, x_3, x_4)/SSE(x_1, x_3, x_4)}{1 - SSR(x_2 | x_1, x_3, x_4)/SSE(x_1, x_3, x_4)} \\[2mm]
&= (n-5)\frac{r_{y2 \cdot 134}^2}{1 - r_{y2 \cdot 134}^2}.
\end{aligned}
$$

EXAMPLE 9.6.1.   In the school data,

$$r_{y3 \cdot 1245} = 0.8477.$$

Thus even after adjusting for all of the other predictor variables, socioeconomic status has major predictive abilities for mean verbal test scores.   □

Actually, the residuals from models (9.6.1) and (9.6.2) give the basis for the perfect plot to evaluate whether adding variable x_2 will improve Model (9.6.1). Simply plot the residuals $y_i - \hat{y}_i$ from Model (9.6.1) against the residuals $x_{i2} - \hat{x}_{i2}$ from Model (9.6.2). If there seems to be no relationship between the $y_i - \hat{y}_i$s and the $x_{i2} - \hat{x}_{i2}$s, x_2 will not be important in Model (9.6.3). If the plot looks clearly linear, x_2 will be important in Model (9.6.3). When a linear relationship exists in the plot but is due to the existence of a few points, those points are the dominant cause for x_2 being important in Model (9.6.3). The reason these *added variable* plots work is because the least squares estimate of β_2 from Model (9.6.3) is identical to the least squares estimate of β_2 from the regression through the origin

$$(y_i - \hat{y}_i) = \beta_2 (x_{i2} - \hat{x}_{i2}) + \varepsilon_i.$$

See Christensen (2011, Exercise 9.2).

EXAMPLE 9.6.2.   For the school data, Figure 9.5 gives the added variable plot to determine whether the variable x_3 adds to the model that already contains x_1, x_2, x_4, and x_5. A clear linear relationship exists, so x_3 will improve the model. Here the entire data support the linear relationship, but there are a couple of unusual cases. The second smallest x_3 residual has an awfully large y residual and the largest x_3 residual has a somewhat surprisingly small y residual.   □

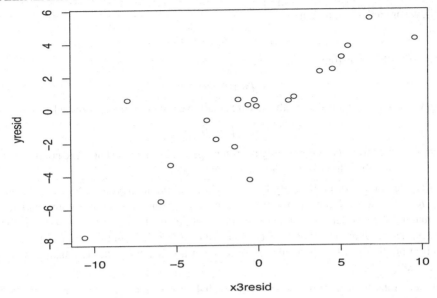

Figure 9.5: *Added variable plot: y residuals versus x_3 residuals;* Coleman Report *data.*

9.7 Collinearity

Collinearity exists when the predictor variables $x_1, \ldots, x_{p-1}$ are correlated. We have n observations on each of these variables, so we can compute the sample correlations between them. Typically, the x variables are assumed to be fixed and not random. For data like the *Coleman Report*, we have a sample of schools so the predictor variables really are random. But for the purpose of fitting the regression we treat them as fixed. (Probabilistically, we look at the conditional distribution of y given the predictor variables.) In some applications, the person collecting the data actually has control over the predictor variables so they truly are fixed. If the x variables are fixed and not random, there is some question as to what a correlation between two x variables means. Actually, we are concerned with whether the observed variables are *orthogonal*, but that turns out to be equivalent to having sample correlations of zero between the x variables. Nonzero sample correlations indicate nonorthogonality, thus we need not concern ourselves with the interpretation of sample correlations between nonrandom samples.

In regression, it is almost unheard of to have x variables that display no collinearity (correlation) [unless the variables are constructed to have no correlation]. In other words, observed x variables are almost never orthogonal. The key ideas in dealing with collinearity were previously incorporated into the discussion of comparing regression models. In fact, the methods discussed earlier were built around dealing with the collinearity of the x variables. This section merely reviews a few of the main ideas.

1. The estimate of any parameter, say $\hat{\beta}_2$, depends on *all* the variables that are included in the model.

2. The sum of squares for any variable, say x_2, depends on *all* the other variables that are included in the model. For example, none of $SSR(x_2)$, $SSR(x_2|x_1)$, and $SSR(x_2|x_3,x_4)$ would typically be equal.

3. Suppose the model

$$y_i = \beta_0 + \beta_1 x_{i1} + \beta_2 x_{i2} + \beta_3 x_{i3} + \varepsilon_i$$

is fitted and we obtain t statistics for each parameter. If the t statistic for testing $H_0 : \beta_1 = 0$ is small, we are led to the model

$$y_i = \beta_0 + \beta_2 x_{i2} + \beta_3 x_{i3} + \varepsilon_i.$$

If the t statistic for testing $H_0 : \beta_2 = 0$ is small, we are led to the model

$$y_i = \beta_0 + \beta_1 x_{i1} + \beta_3 x_{i3} + \varepsilon_i.$$

However, if the t statistics for both tests are small, we are *not* led to the model

$$y_i = \beta_0 + \beta_3 x_{i3} + \varepsilon_i.$$

To arrive at the model containing only the intercept and x_3, one must at some point use the model containing only the intercept and x_3 as a reduced model.

4. A moderate amount of collinearity has little effect on predictions and therefore little effect on SSE, R^2, and the explanatory power of the model. Collinearity increases the variance of the $\hat{\beta}_k$s, making the estimates of the parameters less reliable. (I told you not to rely on parameters anyway.) Depending on circumstances, sometimes a large amount of collinearity can have an effect on predictions. Just by chance, one may get a better fit to the data than can be justified scientifically.

The complications associated with points 1 through 4 all vanish if the sample correlations between the x variables are *all zero*.

Many computer programs will print out a matrix of correlations between the variables. One would like to think that if all the correlations between the x variables are reasonably small, say less than 0.3 or 0.4, then the problems of collinearity would not be serious. Unfortunately, that is simply not true. To avoid difficulties with collinearity, not only do all the correlations need to be small but *all of the partial correlations among the x variables must be small*. Thus, small correlations alone do not ensure small collinearity.

EXAMPLE 9.7.1. The correlations among predictors for the Coleman data are given below.

	x_1	x_2	x_3	x_4	x_5
x_1	1.000	0.181	0.230	0.503	0.197
x_2	0.181	1.000	0.827	0.051	0.927
x_3	0.230	0.827	1.000	0.183	0.819
x_4	0.503	0.051	0.183	1.000	0.124
x_5	0.197	0.927	0.819	0.124	1.000

A visual display of these relationships was provided in Figures 9.1–9.4.

Note that x_3 is highly correlated with x_2 and x_5. Since x_3 is highly correlated with y, the fact that x_2 and x_5 are also quite highly correlated with y is not surprising. Recall that the correlations with y were given at the beginning of Section 9.1. Moreover, since x_3 is highly correlated with x_2 and x_5, it is also not surprising that x_2 and x_5 have little to add to a model that already contains x_3. We have seen that it is the two variables x_1 and x_4, i.e., the variables that do not have high correlations with either x_3 or y, that have the greater impact on the regression equation.

Having regressed y on x_3, the sample correlations between y and any of the other variables are no longer important. Having done this regression, it is more germane to examine the partial correlations between y and the other variables after adjusting for x_3. However, as we will see in our discussion of variable selection in Chapter 10, even this has its drawbacks.   □

As long as points 1 through 4 are kept in mind, a moderate amount of collinearity is not a big problem. For severe collinearity, there are four common approaches: a) classical ridge regression, b) generalized inverse regression, c) principal components regression, and d) canonical regression. Classical ridge regression is probably the best known of these methods. The other three methods are closely related and seem quite reasonable. Principal components regression is discussed in Section 11.6. Another procedure, *lasso regression*, is becoming increasingly popular but it is considerably more difficult to understand how it works, cf. Section 10.5.

Table 9.1: *L. A. heart study data.*

i	x_1	x_2	x_3	x_4	x_5	y	i	x_1	x_2	x_3	x_4	x_5	y
1	44	124	80	254	70	190	31	42	136	82	383	69	187
2	35	110	70	240	73	216	32	28	124	82	360	67	148
3	41	114	80	279	68	178	33	40	120	85	369	71	180
4	31	100	80	284	68	149	34	40	150	100	333	70	172
5	61	190	110	315	68	182	35	35	100	70	253	68	141
6	61	130	88	250	70	185	36	32	120	80	268	68	176
7	44	130	94	298	68	161	37	31	110	80	257	71	154
8	58	110	74	384	67	175	38	52	130	90	474	69	145
9	52	120	80	310	66	144	39	45	110	80	391	69	159
10	52	120	80	337	67	130	40	39	106	80	248	67	181
11	52	130	80	367	69	162	41	40	130	90	520	68	169
12	40	120	90	273	68	175	42	48	110	70	285	66	160
13	49	130	75	273	66	155	43	29	110	70	352	66	149
14	34	120	80	314	74	156	44	56	141	100	428	65	171
15	37	115	70	243	65	151	45	53	90	55	334	68	166
16	63	140	90	341	74	168	46	47	90	60	278	69	121
17	28	138	80	245	70	185	47	30	114	76	264	73	178
18	40	115	82	302	69	225	48	64	140	90	243	71	171
19	51	148	110	302	69	247	49	31	130	88	348	72	181
20	33	120	70	386	66	146	50	35	120	88	290	70	162
21	37	110	70	312	71	170	51	65	130	90	370	65	153
22	33	132	90	302	69	161	52	43	122	82	363	69	164
23	41	112	80	394	69	167	53	53	120	80	343	71	159
24	38	114	70	358	69	198	54	58	138	82	305	67	152
25	52	100	78	336	70	162	55	67	168	105	365	68	190
26	31	114	80	251	71	150	56	53	120	80	307	70	200
27	44	110	80	322	68	196	57	42	134	90	243	67	147
28	31	108	70	281	67	130	58	43	115	75	266	68	125
29	40	110	74	336	68	166	59	52	110	75	341	69	163
30	36	110	80	314	73	178	60	68	110	80	268	62	138

9.8 More on model testing

In this section, we take the opportunity to introduce various methods of defining reduced models. To this end we introduce some new data, a subset of the *Chapman data*.

EXAMPLE 9.8.1. Dixon and Massey (1983) report data from the Los Angeles Heart Study supervised by J. M. Chapman. The variables are y, weight in pounds; x_1, age in years; x_2, systolic blood pressure in millimeters of mercury; x_3, diastolic blood pressure in millimeters of mercury; x_4, cholesterol in milligrams per dl; x_5, height in inches. The data from 60 men are given in Table 9.1.

For now, our interest is not in analyzing the data but in illustrating modeling techniques. We fitted the basic multiple regression model

$$y_i = \beta_0 + \beta_1 x_{i1} + \beta_2 x_{i2} + \beta_3 x_{i3} + \beta_4 x_{i4} + \beta_5 x_{i5} + \varepsilon_i. \tag{9.8.1}$$

The table of coefficients and ANOVA table follow.

Table of Coefficients: Model (9.8.1)

Predictor	$\hat{\beta}_k$	SE($\hat{\beta}_k$)	t	P
Constant	−112.50	89.56	−1.26	0.214
x_1-age	0.0291	0.2840	0.10	0.919
x_2-sbp	0.0197	0.3039	0.06	0.949
x_3-dbp	0.7274	0.4892	1.49	0.143
x_4-chol	−0.02103	0.04859	−0.43	0.667
x_5-ht	3.248	1.241	2.62	0.011

Analysis of Variance: Model (9.8.1)

Source	df	SS	MS	F	P
Regression	5	7330.4	1466.1	3.30	0.011
Residual Error	54	24009.6	444.6		
Total	59	31340.0			

One plausible reduced model is that systolic and diastolic blood pressure have the same regression coefficient, i.e., $H_0 : \beta_2 = \beta_3$. Incorporating this into Model (9.8.1) gives

$$
\begin{aligned}
y_i &= \beta_0 + \beta_1 x_{i1} + \beta_2 x_{i2} + \beta_2 x_{i3} + \beta_4 x_{i4} + \beta_5 x_{i5} + \varepsilon_i \\
&= \beta_0 + \beta_1 x_{i1} + \beta_2 (x_{i2} + x_{i3}) + \beta_4 x_{i4} + \beta_5 x_{i5} + \varepsilon_i,
\end{aligned}
\tag{9.8.2}
$$

which involves regressing y on the four variables $x_1, x_2 + x_3, x_4, x_5$. The fitted equation is

$$
\hat{y} = -113 + 0.018 x_1 + 0.283 (x_2 + x_3) - 0.0178 x_4 + 3.31 x_5.
$$

The ANOVA table

Analysis of Variance for Model (9.8.2).

Source	df	SS	MS	F	P
Regression	4	6941.9	1735.5	3.91	0.007
Residual Error	55	24398.1	443.6		
Total	59	31340.0			

leads to the test statistic for whether the reduced model fits,

$$
F_{obs} = \frac{(24398.1 - 24009.6)/(55 - 54)}{444.6} \doteq 1.
$$

The reduced model based on the sum of the blood pressures fits as well as the model with the individual blood pressures.

The table of coefficients for Model (9.8.2)

Table of Coefficients: Model (9.8.2)

Predictor	$\hat{\beta}_k$	SE($\hat{\beta}_k$)	t	P
Constant	−113.16	89.45	−1.27	0.211
x_1-age	0.0182	0.2834	0.06	0.949
$x_2 + x_3$	0.2828	0.1143	2.47	0.016
x_4-chol	−0.01784	0.04841	−0.37	0.714
x_5-ht	3.312	1.237	2.68	0.010

shows a significant effect for the sum of the blood pressures. Although neither blood pressure looked important in the table of coefficients for the full model, we find that the sum of the blood pressures is a good predictor of weight, with a positive regression coefficient. Although high blood pressure is not likely to cause high weight, there is certainly a correlation between weight and blood pressure, so it is plausible that blood pressure could be a good predictor of weight. The reader should investigate whether x_2, x_3, and $x_2 + x_3$ are all acting as surrogates for one another, i.e., whether it is sufficient to include *any one* of the three in the model, after which the others add no appreciable predictive ability.

Another plausible idea, perhaps more so for other dependent variables rather than weight, is that it could be the difference between the blood pressure readings that is important. In this case, the corresponding null hypothesis is $H_0 : \beta_2 + \beta_3 = 0$. Writing $\beta_3 = -\beta_2$, the model becomes

$$
\begin{aligned}
y_i &= \beta_0 + \beta_1 x_{i1} + \beta_2 x_{i2} - \beta_2 x_{i3} + \beta_4 x_{i4} + \beta_5 x_{i5} + \varepsilon_i \\
&= \beta_0 + \beta_1 x_{i1} + \beta_2 (x_{i2} - x_{i3}) + \beta_4 x_{i4} + \beta_5 x_{i5} + \varepsilon_i.
\end{aligned}
\tag{9.8.3}
$$

With

Analysis of Variance for Model (9.8.3)

Source	df	SS	MS	F	P
Regression	4	4575.5	1143.9	2.35	0.065
Residual Error	55	26764.5	486.6		
Total	59	31340.0			

the test statistic for whether the reduced model fits is

$$F_{obs} = \frac{(26764.5 - 24009.6)/(55 - 54)}{444.6} = 6.20.$$

The one-sided P value is 0.016, i.e., $6.20 = F(1 - .016, 1, 54)$. Clearly the reduced model fits inadequately. Replacing the blood pressures by their difference does not predict as well as having the blood pressures in the model.

It would have worked equally well to have written $\beta_3 = -\beta_2$ and fitted the reduced model

$$y_i = \beta_0 + \beta_1 x_{i1} + \beta_3(x_{i3} - x_{i2}) + \beta_4 x_{i4} + \beta_5 x_{i5} + \varepsilon_i.$$

Tests for proportional coefficients are similar to the previous illustrations. For example, we could test if the coefficient for x_2 (sbp) is 40 times smaller than for x_3 (dbp). To test $H_0 : 40\beta_2 = \beta_3$, the reduced model becomes

$$\begin{aligned} y_i &= \beta_0 + \beta_1 x_{i1} + \beta_2 x_{i2} + 40\beta_2 x_{i3} + \beta_4 x_{i4} + \beta_5 x_{i5} + \varepsilon_i \\ &= \beta_0 + \beta_1 x_{i1} + \beta_2(x_{i2} + 40x_{i3}) + \beta_4 x_{i4} + \beta_5 x_{i5} + \varepsilon_i. \end{aligned}$$

We leave it to the reader to evaluate this hypothesis.

Now let's test whether the regression coefficient for diastolic blood pressure is 0.5 units higher than for systolic. The hypothesis is $H_0 : \beta_2 + 0.5 = \beta_3$. Substitution gives

$$\begin{aligned} y_i &= \beta_0 + \beta_1 x_{i1} + \beta_2 x_{i2} + (\beta_2 + 0.5)x_{i3} + \beta_4 x_{i4} + \beta_5 x_{i5} + \varepsilon_i \\ &= 0.5x_{i3} + \beta_0 + \beta_1 x_{i1} + \beta_2(x_{i2} + x_{i3}) + \beta_4 x_{i4} + \beta_5 x_{i5} + \varepsilon_i. \end{aligned} \quad (9.8.4)$$

The term $0.5x_{i3}$ is a known constant for each observation i, often called an *offset*. Such terms are easy to handle in linear models, just take them to the other side of the equation,

$$y_i - 0.5x_{i3} = \beta_0 + \beta_1 x_{i1} + \beta_2(x_{i2} + x_{i3}) + \beta_4 x_{i4} + \beta_5 x_{i5} + \varepsilon_i, \quad (9.8.5)$$

and fit the model with the new dependent variable $y_i - 0.5x_{i3}$.

The fitted regression equation is

$$\hat{y} - 0.5x_3 = -113 + 0.026x_1 + 0.097(x_2 + x_3) - 0.0201x_4 + 3.27x_5$$

or

$$\hat{y} = -113 + 0.026x_1 + 0.097x_2 + 0.597x_3 - 0.0201x_4 + 3.27x_5.$$

The ANOVA table for the reduced model (9.8.5) is

Analysis of Variance for Model (9.8.5)

Source	df	SS	MS	F	P
Regression	4	3907.7	976.9	2.23	0.077
Residual Error	55	24043.1	437.1		
Total	59	27950.8			

It may not be obvious but Model (9.8.5) can be tested against the full model (9.8.1) in the usual way. Since x_{i3} is already included in Model (9.8.1), subtracting 0.5 times it from y_i has little effect on

Model (9.8.1): the fitted values differ only by the constant $0.5x_{i3}$ being subtracted; the residuals and degrees of freedom are identical. Performing the test of Model (9.8.5) versus Model (9.8.1) gives

$$F_{obs} = \frac{(24043.1 - 24009.6)/(55 - 54)}{444.6} = 0.075$$

for a one-sided P value of 0.79, so the equivalent reduced models (9.8.4) and (9.8.5) are consistent with the data.

We could similarly test whether the height coefficient is 3.5 in Model (9.8.1), i.e., test $H_0 : \beta_5 = 3.5$ by fitting

$$y_i = \beta_0 + \beta_1 x_{i1} + \beta_2 x_{i2} + \beta_3 x_{i3} + \beta_4 x_{i4} + 3.5x_{i5} + \varepsilon_i$$

or

$$y_i - 3.5x_{i5} = \beta_0 + \beta_1 x_{i1} + \beta_2 x_{i2} + \beta_3 x_{i3} + \beta_4 x_{i4} + \varepsilon_i. \qquad (9.8.6)$$

Fitting Model (9.8.6) gives the regression equation

$$\hat{y} - 3.5x_5 = -130 + 0.045x_1 + 0.019x_2 + 0.719x_3 - 0.0203x_4$$

or

$$\hat{y} = -130 + 0.045x_1 + 0.019x_2 + 0.719x_3 - 0.0203x_4 + 3.5x_5.$$

The ANOVA table is

Analysis of Variance for Model (9.8.6)

Source	df	SS	MS	F	P
Regression	4	3583.3	895.8	2.05	0.100
Residual Error	55	24027.9	436.9		
Total	59	27611.2			

and testing the models in the usual way gives

$$F_{obs} = \frac{(24027.9 - 24009.6)/(55 - 54)}{444.6} = 0.041$$

for a one-sided P value of 0.84. The reduced model (9.8.6) is consistent with the data.

Alternatively, we could test $H_0 : \beta_5 = 3.5$ from the original table of coefficients for Model (9.8.1) by computing

$$t_{obs} = \frac{3.248 - 3.5}{1.241} = -0.203$$

and comparing the result to a $t(54)$ distribution. The square of the t statistic equals the F statistic.

Finally, we illustrate a simultaneous test of the last two hypotheses, i.e., we test $H_0 : \beta_2 + 0.5 = \beta_3; \beta_5 = 3.5$. The reduced model is

$$\begin{aligned} y_i &= \beta_0 + \beta_1 x_{i1} + \beta_2 x_{i2} + (\beta_2 + 0.5)x_{i3} + \beta_4 x_{i4} + 3.5x_{i5} + \varepsilon_i \\ &= 0.5x_{i3} + 3.5x_{i5} + \beta_0 + \beta_1 x_{i1} + \beta_2(x_{i2} + x_{i3}) + \beta_4 x_{i4} + \varepsilon_i \end{aligned}$$

or

$$y_i - 0.5x_{i3} - 3.5x_{i5} = \beta_0 + \beta_1 x_{i1} + \beta_2(x_{i2} + x_{i3}) + \beta_4 x_{i4} + + \varepsilon_i. \qquad (9.8.7)$$

The fitted regression equation is

$$\hat{y} - .5x_3 - 3.5x_5 = -129 + 0.040x_1 + 0.094(x_2 + x_3) - 0.0195x_4$$

or

$$\hat{y} = -129 + 0.040x_1 + 0.094x_2 + 0.594x_3 - 0.0195x_4 + 3.5x_5.$$

The ANOVA table is

Analysis of Variance for Model (9.8.7)

Source	df	SS	MS	F	P
Regression	3	420.4	140.1	0.33	0.806
Residual Error	56	24058.8	429.6		
Total	59	24479.2			

and testing Model (9.8.7) against Model (9.8.1) in the usual way gives

$$F_{obs} = \frac{(24058.8 - 24009.6)/(56 - 54)}{444.6} = 0.055$$

for a one-sided P value of 0.95. In this case, the high one-sided P value is probably due less to any problems with Model (9.8.7) and due more to me looking at the table of coefficients for Model (9.8.1) and choosing a null hypothesis that seemed consistent with the data. Typically, *hypotheses should be suggested by previous theory or data, not inspection of the current data.*

9.9 Additive effects and interaction

For the *Coleman Report* data, one of the viable models had two predictors: x_3, socioeconomic status, and x_4, teacher's verbal score. If the model displayed lack of fit, there are a number of ways that we could expand the model.

In general, the simplest multiple regression model for $E(y)$ based on two predictors is

$$m(x) = \beta_0 + \beta_1 x_1 + \beta_2 x_2. \tag{9.9.1}$$

This model displays *additive effects*. The relative effect of changing x_1 into, say, $\tilde{x}_1$ is the same for any value of x_2. Specifically,

$$[\beta_0 + \beta_1 \tilde{x}_1 + \beta_2 x_2] - [\beta_0 + \beta_1 x_1 + \beta_2 x_2] = \beta_2(\tilde{x}_1 - x_1).$$

This effect does not depend on x_2, which allows us to speak about an effect for x_1. If the effect of x_1 depends on x_2, no single effect for x_1 exists and we would always need to specify the value of x_2 before discussing the effect of x_1. An exactly similar argument shows that in Model (9.9.1) the effect of changing x_2 does not depend on the value of x_1.

Generally, for any two predictors x_1 and x_2, an *additive effects (no-interaction) model* takes the form

$$m(x) = h_1(x_1) + h_2(x_2) \tag{9.9.2}$$

where $x = (x_1, x_2)$ and $h_1(\cdot)$ and $h_2(\cdot)$ are arbitrary functions. In this case, the relative effect of changing x_1 to $\tilde{x}_1$ is the same for any value of x_2 because

$$m(\tilde{x}_1, x_2) - m(x_1, x_2) = [h_1(\tilde{x}_1) + h_2(x_2)] - [h_1(x_1) + h_2(x_2)] = h_1(\tilde{x}_1) - h_1(x_1),$$

which does not depend on x_2. An exactly similar argument shows that the effect of changing x_2 does not depend on the value of x_1. In an additive model, the effect as x_1 changes can be anything at all; it can be any function h_1, and similarly for x_2. However, the combined effect must be the sum of the two individual effects. Other than Model (9.9.1), the most common no-interaction models for two measurement predictors are probably a polynomial in x_1 plus a polynomial in x_2, say,

$$m(x) = \beta_0 + \sum_{r=1}^{R} \beta_{r0} x_1^r + \sum_{s=1}^{S} \beta_{0s} x_2^s. \tag{9.9.3}$$

An *interaction model* is literally any model that does not display the additive effects structure of (9.9.2). When generalizing no-interaction polynomial models, cross-product terms are often added to model interaction. For example, Model (9.9.1) might be expanded to

$$m(x) = \beta_0 + \beta_1 x_1 + \beta_2 x_2 + \beta_3 x_1 x_2.$$

This is an interaction model because the relative effect of changing x_1 to $\tilde{x}_1$ depends on the value of x_2. Specifically,

$$[\beta_0 + \beta_1\tilde{x}_1 + \beta_2 x_2 + \beta_3\tilde{x}_1 x_2] - [\beta_0 + \beta_1 x_1 + \beta_2 x_2 + \beta_3 x_1 x_2] = \beta_2(\tilde{x}_1 - x_1) + \beta_3(\tilde{x}_1 - x_1)x_2,$$

where the second term depends on the value of x_2. To include interaction, the no-interaction polynomial model (9.9.3) might be extended to an interaction polynomial model

$$m(x) = \sum_{r=0}^{R} \sum_{s=0}^{S} \beta_{rs} x_1^r x_2^s. \tag{9.9.4}$$

These devices are easily extended to more than two predictor variables, cf. Section 10.

EXAMPLE 9.9.1. Using the *Coleman Report* data, we begin by considering

$$y_h = \beta_0 + \beta_3 x_{h3} + \beta_4 x_{h4} + \varepsilon_h$$

which was earlier examined in Section 9.3. First we fit a simple quadratic additive model

$$y_h = \beta_0 + \beta_{10} x_{h3} + \beta_{20} x_{h3}^2 + \beta_{01} x_{h4} + \beta_{02} x_{h4}^2 + \varepsilon_h.$$

From the table of coefficients ·

Table of Coefficients

Predictor	$\hat{\beta}_k$	SE($\hat{\beta}_k$)	t	P
Constant	38.0	106.5	0.36	0.726
x_3	0.54142	0.05295	10.22	0.000
x_3^2	−0.001892	0.006411	−0.30	0.772
x_4	−1.124	8.602	−0.13	0.898
x_4^2	0.0377	0.1732	0.22	0.831

we see that neither quadratic term is adding anything after the other terms because both quadratic terms have large P values. To make a simultaneous test of dropping the quadratic terms, we need to compare the error in the ANOVA table

Analysis of Variance

Source	df	SS	MS	F	P
Regression	4	571.47	142.87	29.99	0.000
Residual Error	15	71.46	4.76		
Total	19	642.92			

to the error given in Section 9.3. The F statistic becomes

$$F_{obs} = \frac{[72.43 - 71.46]/[17 - 15]}{71.46/15} = \frac{0.485}{4.76} = 0.102,$$

so together the quadratic terms are contributing virtually nothing.

The simplest interaction model is

$$y_h = \beta_0 + \beta_3 x_{h3} + \beta_4 x_{h4} + \beta_{34} x_{h3} x_{h4} + \varepsilon_h.$$

Fitting gives the table of coefficients.

Table of Coefficients

Predictor	$\hat{\beta}_k$	SE($\hat{\beta}_k$)	t	P
Constant	10.31	10.48	0.98	0.340
x_3	1.900	1.569	1.21	0.244
x_4	0.9264	0.4219	2.20	0.043
$x_3 x_4$	−0.05458	0.06304	−0.87	0.399

This shows no effect for adding the $\beta_{34}x_{h3}x_{h4}$ interaction ($P = 0.399$). Alternatively, we could compare the error from the ANOVA table

Analysis of Variance

Source	df	SS	MS	F	P
Regression	3	573.74	191.25	44.23	0.000
Residual Error	16	69.18	4.32		
Total	19	642.92			

to that given in Section 9.3 to get the F statistic

$$F_{obs} = \frac{[72.43 - 69.18]/[17 - 16]}{69.18/16} = \frac{3.25}{4.32} = 0.753 = (-0.87)^2,$$

which also gives the P value 0.399.   □

9.10   Generalized additive models

Suppose we wanted to fit a cubic interaction model to the *Coleman Report* data. With five predictor variables, the model is

$$m(x) = \sum_{r=0}^{3}\sum_{s=0}^{3}\sum_{t=0}^{3}\sum_{u=0}^{3}\sum_{v=0}^{3} \beta_{rstuv} x_1^r x_2^s x_3^t x_4^u x_5^v \tag{9.10.1}$$

and includes $5^4 = 625$ mean parameters β_{rstuv}. We might want to think twice about trying to estimate 625 parameters from just 20 schools.

This is a common problem with fitting polynomial interaction models. When we have even a moderate number of predictor variables, the number of parameters quickly becomes completely unwieldy. And it is not only a problem for polynomial interaction models. In Section 8.3 we discussed replacing polynomials with other basis functions $\phi_r(x)$. The polynomial models happen to have $\phi_r(x) = x^r$. Other choices of ϕ_r include cosines, or both cosines and sines, or indicator functions, or wavelets. Typically, $\phi_0(x) \equiv 1$. In the basis function approach, the additive polynomial model (9.9.3) generalizes to

$$m(x) = \beta_0 + \sum_{r=1}^{R} \beta_{r0}\phi_r(x_1) + \sum_{s=1}^{S} \beta_{0s}\phi_s(x_2) \tag{9.10.2}$$

and the polynomial interaction model (9.9.4) generalizes to

$$m(x) = \sum_{r=0}^{R}\sum_{s=0}^{S} \beta_{rs}\phi_r(x_1)\phi_s(x_2). \tag{9.10.3}$$

When expanding Model (9.10.3) to include more predictors, the generalized interaction model has exactly the same problem as the polynomial interaction model (9.10.1) in that it requires fitting too many parameters.

Generalized additive models provide a means for circumventing the problem. They do so by restricting the orders of the interactions. In Model (9.10.1) we have five variables, all of which can interact with one another. Instead, suppose variables x_1 and x_4 can interact with one another but with no other variables and that variables x_2, x_3, and x_5 can interact with one another but with no other variables. We can then write a generalized additive model

$$m(x) \equiv m(x_1, x_2, x_3, x_4, x_5) = h_1(x_1, x_4) + h_2(x_2, x_3, x_5). \tag{9.10.4}$$

Using the basis function approach to model each of the two terms on the right gives

$$m(x) = \sum_{r=0}^{R}\sum_{u=0}^{U} \beta_{ru}\phi_r(x_1)\phi_u(x_4) + \sum_{s=0}^{S}\sum_{t=0}^{T}\sum_{v=0}^{V} \gamma_{stu}\phi_s(x_2)\phi_t(x_3)\phi_v(x_5) - \gamma_{000}.$$

We subtracted γ_{000} from the model because both β_{00} and γ_{000} serve as intercept terms, hence they are redundant parameters. This section started by considering the cubic interaction model (9.10.1) for the *Coleman Report* data. The model has $3 = R = S = T = U = V$ and involves 625 mean parameters. Using similar cubic polynomials to model the generalized additive model (9.10.4) we need only $2^4 + 3^4 - 1 = 96$ parameters. While that is still far too many parameters to fit to the *Coleman Report* data, you can see that fitting generalized additive models are much more feasible than fitting full interaction models.

Another generalized additive model that we could propose for five variables is

$$m(x) = h_1(x_1, x_2) + h_2(x_2, x_3) + h_3(x_4, x_5).$$

A polynomial version of the model is

$$m(x) = \sum_{r=0}^{R} \sum_{s=0}^{S} \beta_{rs} x_1^r x_2^s + \sum_{s=0}^{S} \sum_{t=0}^{T} \gamma_{st} x_2^s x_3^t + \sum_{u=0}^{U} \sum_{v=0}^{V} \delta_{uv} x_4^u x_5^v. \qquad (9.10.5)$$

In this case, not only are β_{00}, γ_{00}, and δ_{00} all redundant intercept parameters, but $\sum_{s=0}^{S} \beta_{0s} x_1^0 x_2^s$ and $\sum_{s=0}^{S} \gamma_{s0} x_2^s x_3^0$ are redundant simple polynomials in x_2. In this case it is more convenient to write Model (9.10.5) as

$$m(x) = \sum_{r=0}^{R} \sum_{s=0}^{S} \beta_{rs} x_1^r x_2^s + \sum_{s=0}^{S} \sum_{t=1}^{T} \gamma_{st} x_2^s x_3^t + \sum_{u=0}^{U} \sum_{v=0}^{V} \delta_{uv} x_4^u x_5^v - \delta_{00}.$$

Of course, the catch with generalized additive models is that you need to have some idea of what variables may interact with one another. And the only obvious way to check that assumption is to test the assumed generalized additive model against the full interaction model. But this whole discussion started with the fact that fitting the full interaction model is frequently infeasible.

9.11   Final comment

The maxim for unbalanced data, and regression data are typically unbalanced, is that *if you change anything, you change everything*. If you change a predictor variable in a model, you change the meaning of the regression coefficients (to the extent that they have any meaning), you change the estimates, the fitted values, the residuals, the leverages: *everything*! If you drop out a data point, you change the meaning of the regression coefficients, the estimates, the fitted values, the residuals, the leverages: *everything*! If you change anything, you change everything. There are a few special cases where this is not true, but they are only special cases.

9.12   Exercises

EXERCISE 9.12.1.   Younger (1979, p. 533) presents data from a sample of 12 discount department stores that advertise on television, radio, and in the newspapers. The variables x_1, x_2, and x_3 represent the respective amounts of money spent on these advertising activities during a certain month while y gives the store's revenues during that month. The data are given in Table 9.2. Complete the following tasks using multiple regression.

(a) Give the theoretical model along with the relevant assumptions.

(b) Give the fitted model, i.e., repeat (a) substituting the estimates for the unknown parameters.

(c) Test $H_0 : \beta_2 = 0$ versus $H_A : \beta_2 \neq 0$ at $\alpha = 0.05$.

(d) Test the hypothesis $H_0 : \beta_1 = \beta_2 = \beta_3 = 0$.

(e) Give a 99% confidence interval for β_2.

Table 9.2: *Younger's advertising data.*

Obs.	y	x_1	x_2	x_3	Obs.	y	x_1	x_2	x_3
1	84	13	5	2	7	34	12	7	2
2	84	13	7	1	8	30	10	3	2
3	80	8	6	3	9	54	8	5	2
4	50	9	5	3	10	40	10	5	3
5	20	9	3	1	11	57	5	6	2
6	68	13	5	1	12	46	5	7	2

(f) Test whether the reduced model $y_i = \beta_0 + \beta_1 x_{i1} + \varepsilon_i$ is an adequate explanation of the data as compared to the full model.

(g) Test whether the reduced model $y_i = \beta_0 + \beta_1 x_{i1} + \varepsilon_i$ is an adequate explanation of the data as compared to the model $y_i = \beta_0 + \beta_1 x_{i1} + \beta_2 x_{i2} + \varepsilon_i$.

(h) Write down the ANOVA table for the 'full' model used in (g).

(i) Construct an added variable plot for adding variable x_3 to a model that already contains variables x_1 and x_2. Interpret the plot.

(j) Compute the sample partial correlation $r_{y3 \cdot 12}$. What does this value tell you?

EXERCISE 9.12.2.   The information below relates y, a second measurement on wood volume, to x_1, a first measurement on wood volume, x_2, the number of trees, x_3, the average age of trees, and x_4, the average volume per tree. Note that $x_4 = x_1/x_2$. Some of the information has not been reported, so that you can figure it out on your own.

Table of Coefficients

Predictor	$\hat{\beta}_k$	SE($\hat{\beta}_k$)	t	P
Constant	23.45	14.90		0.122
x_1	0.93209	0.08602		0.000
x_2		0.4721	1.5554	0.126
x_3	−0.4982	0.1520		0.002
x_4	3.486	2.274		0.132

Analysis of Variance

Source	df	SS	MS	F	P
Regression	4	887994			0.000
Error					
Total	54	902773			

Source	df	Sequential SS
x_1	1	883880
x_2	1	183
x_3	1	3237
x_4	1	694

(a) How many observations are in the data?

(b) What is R^2 for this model?

(c) What is the mean squared error?

(d) Give a 95% confidence interval for β_2.

(e) Test the null hypothesis $\beta_3 = 0$ with $\alpha = 0.05$.

(f) Test the null hypothesis $\beta_1 = 1$ with $\alpha = 0.05$.

Table 9.3: *Prater's gasoline–crude oil data.*

y	x_1	x_2	x_3	x_4	y	x_1	x_2	x_3	x_4
6.9	38.4	6.1	220	235	24.8	32.2	5.2	236	360
14.4	40.3	4.8	231	307	26.0	38.4	6.1	220	365
7.4	40.0	6.1	217	212	34.9	40.3	4.8	231	395
8.5	31.8	0.2	316	365	18.2	40.0	6.1	217	272
8.0	40.8	3.5	210	218	23.2	32.2	2.4	284	424
2.8	41.3	1.8	267	235	18.0	31.8	0.2	316	428
5.0	38.1	1.2	274	285	13.1	40.8	3.5	210	273
12.2	50.8	8.6	190	205	16.1	41.3	1.8	267	358
10.0	32.2	5.2	236	267	32.1	38.1	1.2	274	444
15.2	38.4	6.1	220	300	34.7	50.8	8.6	190	345
26.8	40.3	4.8	231	367	31.7	32.2	5.2	236	402
14.0	32.2	2.4	284	351	33.6	38.4	6.1	220	410
14.7	31.8	0.2	316	379	30.4	40.0	6.1	217	340
6.4	41.3	1.8	267	275	26.6	40.8	3.5	210	347
17.6	38.1	1.2	274	365	27.8	41.3	1.8	267	416
22.3	50.8	8.6	190	275	45.7	50.8	8.6	190	407

(g) Give the F statistic for testing the null hypothesis $\beta_3 = 0$.

(h) Give $SSR(x_3|x_1,x_2)$ and find $SSR(x_3|x_1,x_2,x_4)$.

(i) Test the model with only variables x_1 and x_2 against the model with all of variables x_1, x_2, x_3, and x_4.

(j) Test the model with only variables x_1 and x_2 against the model with variables x_1, x_2, and x_3.

(k) Should the test in part (g) be the same as the test in part (j)? Why or why not?

(l) For estimating the point on the regression surface at $(x_1,x_2,x_3,x_4) = (100,25,50,4)$, the standard error of the estimate for the point on the surface is 2.62. Give the estimated point on the surface, a 95% confidence interval for the point on the surface, and a 95% prediction interval for a new point with these x values.

(m) Test the null hypothesis $\beta_1 = \beta_2 = \beta_3 = \beta_4 = 0$ with $\alpha = 0.05$.

EXERCISE 9.12.3. Atkinson (1985) and Hader and Grandage (1958) have presented Prater's data on gasoline. The variables are y, the percentage of gasoline obtained from crude oil; x_1, the crude oil gravity °API; x_2, crude oil vapor pressure measured in lbs/in^2; x_3, the temperature, in °F, at which 10% of the crude oil is vaporized; and x_4, the temperature, in °F, at which all of the crude oil is vaporized. The data are given in Table 9.3. Find a good model for predicting gasoline yield from the other four variables.

EXERCISE 9.12.4. Analyze the Chapman data of Example 9.8.1. Find a good model for predicting weight from the other variables.

EXERCISE 9.12.5. Table 9.4 contains a subset of the pollution data analyzed by McDonald and Schwing (1973). The data are from various years in the early 1960s. They relate air pollution to mortality rates for various standard metropolitan statistical areas in the United States. The dependent variable y is the total age-adjusted mortality rate per 100,000 as computed for different metropolitan areas. The predictor variables are, in order, mean annual precipitation in inches, mean January temperature in degrees F, mean July temperature in degrees F, population per household, median school years completed by those over 25, percent of housing units that are sound and with all facilities, population per sq. mile in urbanized areas, percent non-white population in urbanized areas, relative pollution potential of sulphur dioxide, annual average of percent relative humidity at 1 pm. Find a good predictive model for mortality.

Table 9.4: *Pollution data.*

x_1	x_2	x_3	x_4	x_5	x_6	x_7	x_8	x_9	x_{10}	y
36	27	71	3.34	11.4	81.5	3243	8.8	42.6	59	921.870
35	23	72	3.14	11.0	78.8	4281	3.5	50.7	57	997.875
44	29	74	3.21	9.8	81.6	4260	0.8	39.4	54	962.354
47	45	79	3.41	11.1	77.5	3125	27.1	50.2	56	982.291
43	35	77	3.44	9.6	84.6	6441	24.4	43.7	55	1071.289
53	45	80	3.45	10.2	66.8	3325	38.5	43.1	54	1030.380
43	30	74	3.23	12.1	83.9	4679	3.5	49.2	56	934.700
45	30	73	3.29	10.6	86.0	2140	5.3	40.4	56	899.529
36	24	70	3.31	10.5	83.2	6582	8.1	42.5	61	1001.902
36	27	72	3.36	10.7	79.3	4213	6.7	41.0	59	912.347
52	42	79	3.39	9.6	69.2	2302	22.2	41.3	56	1017.613
33	26	76	3.20	10.9	83.4	6122	16.3	44.9	58	1024.885
40	34	77	3.21	10.2	77.0	4101	13.0	45.7	57	970.467
35	28	71	3.29	11.1	86.3	3042	14.7	44.6	60	985.950
37	31	75	3.26	11.9	78.4	4259	13.1	49.6	58	958.839
35	46	85	3.22	11.8	79.9	1441	14.8	51.2	54	860.101
36	30	75	3.35	11.4	81.9	4029	12.4	44.0	58	936.234
15	30	73	3.15	12.2	84.2	4824	4.7	53.1	38	871.766
31	27	74	3.44	10.8	87.0	4834	15.8	43.5	59	959.221
30	24	72	3.53	10.8	79.5	3694	13.1	33.8	61	941.181
31	45	85	3.22	11.4	80.7	1844	11.5	48.1	53	891.708
31	24	72	3.37	10.9	82.8	3226	5.1	45.2	61	871.338
42	40	77	3.45	10.4	71.8	2269	22.7	41.4	53	971.122
43	27	72	3.25	11.5	87.1	2909	7.2	51.6	56	887.466
46	55	84	3.35	11.4	79.7	2647	21.0	46.9	59	952.529
39	29	75	3.23	11.4	78.6	4412	15.6	46.6	60	968.665
35	31	81	3.10	12.0	78.3	3262	12.6	48.6	55	919.729
43	32	74	3.38	9.5	79.2	3214	2.9	43.7	54	844.053
11	53	68	2.99	12.1	90.6	4700	7.8	48.9	47	861.833
30	35	71	3.37	9.9	77.4	4474	13.1	42.6	57	989.265
50	42	82	3.49	10.4	72.5	3497	36.7	43.3	59	1006.490
60	67	82	2.98	11.5	88.6	4657	13.5	47.3	60	861.439
30	20	69	3.26	11.1	85.4	2934	5.8	44.0	64	929.150
25	12	73	3.28	12.1	83.1	2095	2.0	51.9	58	857.622
45	40	80	3.32	10.1	70.3	2682	21.0	46.1	56	961.009
46	30	72	3.16	11.3	83.2	3327	8.8	45.3	58	923.234
54	54	81	3.36	9.7	72.8	3172	31.4	45.5	62	1113.156
42	33	77	3.03	10.7	83.5	7462	11.3	48.7	58	994.648
42	32	76	3.32	10.5	87.5	6092	17.5	45.3	54	1015.023
36	29	72	3.32	10.6	77.6	3437	8.1	45.5	56	991.290
37	38	67	2.99	12.0	81.5	3387	3.6	50.3	73	893.991
42	29	72	3.19	10.1	79.5	3508	2.2	38.8	56	938.500
41	33	77	3.08	9.6	79.9	4843	2.7	38.6	54	946.185
44	39	78	3.32	11.0	79.9	3768	28.6	49.5	53	1025.502
32	25	72	3.21	11.1	82.5	4355	5.0	46.4	60	874.281

Alternatively, you can obtain the complete data from the Internet statistical service STATLIB by going to http://lib.stat.cmu.edu/datasets/ and clicking on "pollution." The data consist of 16 variables on 60 cases.

EXERCISE 9.12.6. Go to http://lib.stat.cmu.edu/datasets/ and click on "bodyfat." There are data for 15 variables along with a description of the data.

(a) Using the body density measurements as a dependent variable, perform a multiple regression using all of the other variables except body fat as predictor variables. What variables can be safely eliminated from the analysis? Discuss any surprising or expected results in terms of the variables that seem to be most important.

(b) Using the body fat measurements as a dependent variable, perform a multiple regression using all of the other variables except density as predictor variables. What variables can be safely eliminated from the analysis? Discuss any surprising or expected results in terms of the variables that seem to be most important.

Chapter 10

Diagnostics and Variable Selection

In this chapter we continue our discussion of multiple regression. In particular, we focus on checking the assumptions of regression models by looking at diagnostic statistics. If problems with assumptions become apparent, one way to deal with them is to try transformations. The discussion of transformations in Section 7.3 continues to apply. Among the methods discussed there, only the circle of transformations depends on having a simple linear regression model. The other methods apply with multiple regression as well as the analysis of variance models introduced in Chapter 12 and later. In particular, the discussion of transforming x at the end of Section 7.3 takes on new importance in multiple regression because multiple regression involves several predictor variables, each of which is a candidate for transformation. Incidentally, the modified Box–Tidwell procedure evaluates each predictor variable separately, so it involves adding only one predictor variable $x_{ij} \log(x_{ij})$ to the multiple regression model at a time.

This chapter also examines methods for choosing good reduced models. Variable selection methods fall into two categories: best subset selection methods and stepwise regression methods. Both are discussed. In Section 4 we examine the interplay between influential cases and model selection techniques. Finally, Section 5 gives a brief introduction to lasso regression. We continue to illustrate techniques on the data from the *Coleman Report* given in Section 6.9 (Table 6.4) and discussed in Chapter 9.

10.1 Diagnostics

Table 10.1 contains a variety of measures for checking the assumptions of the multiple regression model with five predictor variables that was fitted in Section 6.9 and Chapter 9 to the *Coleman Report* data. The table includes case indicators, the data y, the predicted values $\hat{y}$, the leverages, the standardized residuals r, the standardized deleted residuals t, and Cook's distances C. All of these, except for *Cook's distance*, were introduced in Section 7.2. Recall that leverages measure the distance between the predictor variables of a particular case and the overall center of those data. Cases with leverages near 1 dominate any fitted regression. As a rule of thumb, leverages greater than $2p/n$ cause concern and leverages greater than $3p/n$ cause (at least mild) consternation. Here n is the number of observations in the data and p is the number of regression coefficients, including the intercept. The standardized deleted residuals t contain essentially the same information as the standardized residuals r but t values can be compared to a $t(dfE - 1)$ distribution to obtain a formal test of whether a case is consistent with the other data. (A formal test based on the r values requires a more exotic distribution than the $t(dfE - 1)$.) Cook's distance for case i is defined as

$$C_i = \frac{\sum_{h=1}^{n} \left(\hat{y}_h - \hat{y}_{h[i]} \right)^2}{pMSE}, \tag{10.1.1}$$

where $\hat{y}_h$ is the predictor of the hth case and $\hat{y}_{h[i]}$ is the predictor of the hth case when case i has been removed from the data. Cook's distance measures the effect of deleting case i on the prediction of all of the original observations.

Figures 10.1 and 10.2 are plots of the standardized residuals versus normal scores and against

Table 10.1: *Diagnostics:* Coleman Report, *full data.*

Case	y	ŷ	Leverage	r	t	C
1	37.01	36.66	0.482	0.23	0.23	0.008
2	26.51	26.86	0.486	−0.24	−0.23	0.009
3	36.51	40.46	0.133	−2.05	−2.35	0.107
4	40.70	41.17	0.171	−0.25	−0.24	0.002
5	37.10	36.32	0.178	0.42	0.40	0.006
6	33.90	33.99	0.500	−0.06	−0.06	0.001
7	41.80	41.08	0.239	0.40	0.38	0.008
8	33.40	33.83	0.107	−0.22	−0.21	0.001
9	41.01	40.39	0.285	0.36	0.34	0.008
10	37.20	36.99	0.618	0.16	0.16	0.007
11	23.30	25.51	0.291	−1.26	−1.29	0.110
12	35.20	33.45	0.403	1.09	1.10	0.133
13	34.90	35.95	0.369	−0.64	−0.62	0.040
14	33.10	33.45	0.109	−0.18	−0.17	0.001
15	22.70	24.48	0.346	−1.06	−1.07	0.099
16	39.70	38.40	0.157	0.68	0.67	0.014
17	31.80	33.24	0.291	−0.82	−0.81	0.046
18	31.70	26.70	0.326	2.94	4.56	0.694
19	43.10	41.98	0.285	0.64	0.63	0.027
20	41.01	40.75	0.223	0.14	0.14	0.001

the predicted values. The largest standardized residual, that for case 18, appears to be somewhat unusually large. To test whether the data from case 18 are consistent with the other data, we can compare the standardized deleted residual to a $t(dfE - 1)$ distribution. From Table 10.1, the t residual is 4.56. The corresponding P value is 0.0006. Actually, we chose to perform the test on the t residual for case 18 only because it was the largest of the 20 t residuals. Because the test is based on the largest of the t values, it is appropriate to multiply the P value by the number of t statistics considered. This gives $20 \times 0.0006 = 0.012$, which is still a very small P value. There is considerable evidence that the data of case 18 are inconsistent, for whatever reason, with the other data. This fact cannot be discovered from a casual inspection of the raw data.

The only point of any concern with respect to the leverages is case 10. Its leverage is 0.618, while $2p/n = 0.6$. This is only a mildly high leverage and case 10 seems well behaved in all other respects; in particular, C_{10} is small, so deleting case 10 has very little effect on predictions.

We now reconsider the analysis with case 18 deleted. The regression equation is

$$y = 34.3 - 1.62x_1 + 0.0854x_2 + 0.674x_3 + 1.11x_4 - 4.57x_5$$

and $R^2 = 0.963$. Table 10.2 contains the table of coefficients. Table 10.3 contains the analysis of variance. Table 10.4 contains diagnostics. Note that the MSE is less than half of its previous value when case 18 was included in the analysis. It is no surprise that the MSE is smaller, since the case being deleted is often the single largest contributor to the SSE. Correspondingly, the regression parameter t statistics in Table 10.2 are all much more significant. The actual regression coefficient estimates have changed a bit but not greatly. Predictions have not changed radically either, as can be seen by comparing the predictions given in Tables 10.1 and 10.4. Although the predictions have not changed radically, they have changed more than they would have if we deleted any observation other than case 18. From the definition of Cook's distance given in Equation (10.1.1), C_{18} is precisely the sum of the squared differences between the predictions in Tables 10.1 and 10.4 divided by 6 times the MSE from the full data. From Table 10.1, Cook's distance when dropping case 18 is much larger than Cook's distance from dropping any other case.

Consider again Table 10.4 containing the diagnostic statistics when case 18 has been deleted. Case 10 has moderately high leverage but seems to be no real problem. Figures 10.3 and 10.4 give the normal plot and the standardized residual versus predicted value plot, respectively, with case

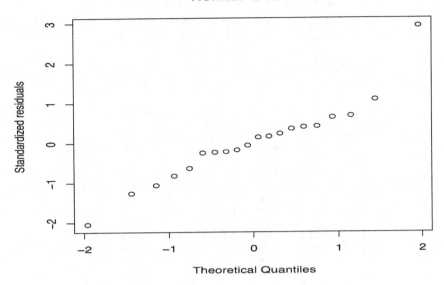

Figure 10.1: *Normal plot, full data,* $W' = 0.903$.

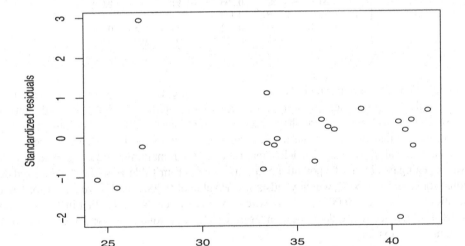

Figure 10.2: *Standardized residuals versus predicted values, full data.*

Table 10.2: *Table of Coefficients: Case 18 deleted.*

Predictor	$\hat{\beta}$	$SE(\hat{\beta})$	t	P
Constant	34.287	9.312	3.68	0.003
x_1	−1.6173	0.7943	−2.04	0.063
x_2	0.08544	0.03546	2.41	0.032
x_3	0.67393	0.06516	10.34	0.000
x_4	1.1098	0.2790	3.98	0.002
x_5	−4.571	1.437	−3.18	0.007

Table 10.3: *Analysis of Variance: Case 18 deleted.*

Source	df	SS	MS	F	P
Regression	5	607.74	121.55	68.27	0.000
Error	13	23.14	1.78		
Total	18	630.88			

Table 10.4: *Diagnostics: Case 18 deleted.*

Case	y	ŷ	Leverage	r	t	C
1	37.01	36.64	0.483	0.39	0.37	0.023
2	26.51	26.89	0.486	−0.39	−0.38	0.024
3	36.51	40.21	0.135	−2.98	−5.08	0.230
4	40.70	40.84	0.174	−0.12	−0.11	0.001
5	37.10	36.20	0.179	0.75	0.73	0.020
6	33.90	33.59	0.504	0.33	0.32	0.018
7	41.80	41.66	0.248	0.12	0.12	0.001
8	33.40	33.65	0.108	−0.20	−0.19	0.001
9	41.01	41.18	0.302	−0.15	−0.15	0.002
10	37.20	36.79	0.619	0.50	0.49	0.068
11	23.30	23.69	0.381	−0.37	−0.35	0.014
12	35.20	34.54	0.435	0.66	0.64	0.055
13	34.90	35.82	0.370	−0.87	−0.86	0.074
14	33.10	32.38	0.140	0.58	0.57	0.009
15	22.70	22.36	0.467	0.35	0.33	0.017
16	39.70	38.25	0.158	1.18	1.20	0.044
17	31.80	32.82	0.295	−0.91	−0.90	0.058
18		24.28	0.483			
19	43.10	41.44	0.292	1.48	1.56	0.151
20	41.01	41.00	0.224	0.00	0.00	0.000

18 deleted. Figure 10.4 is particularly interesting. At first glance, it appears to have a horn shape opening to the right. But there are only three observations on the left of the plot and many on the right, so one would *expect* a horn shape because of the data distribution. Looking at the right of the plot, we see that in spite of the data distribution, much of the horn shape is due to a single very small residual. If we mentally delete that residual, the remaining residuals contain a hint of an upward opening parabola. The potential outlier is case 3. From Table 10.4, the standardized deleted residual for case 3 is −5.08, which yields a raw P value of 0.0001, and if we adjust for having 19 t statistics, the P value is 0.0019, still an extremely small value. Note also that in Table 10.1, when case 18 was included in the data, the standardized deleted residual for case 3 was somewhat large but not nearly so extreme.

With cases 3 and 18 deleted, the regression equation becomes

$$y = 29.8 - 1.70x_1 + 0.0851x_2 + 0.666x_3 + 1.18x_4 - 4.07x_5.$$

The R^2 for these data is 0.988. The table of coefficients is in Table 10.5, the analysis of variance is in Table 10.6, and the diagnostics are in Table 10.7.

Deleting the outlier, case 3, again causes a drop in the *MSE*, from 1.78 with only case 18 deleted to 0.61 with both cases 3 and 18 deleted. This creates a corresponding drop in the standard errors for all regression coefficients and makes them all appear to be more significant. The actual estimates of the regression coefficients do not change much from Table 10.2 to Table 10.5. The largest changes seem to be in the constant and in the coefficient for x_5.

From Table 10.7, the leverages, t statistics, and Cook's distances seem reasonable. Figures 10.5 and 10.6 contain a normal plot and a plot of standardized residuals versus predicted values. Both plots look good. In particular, the suggestion of lack of fit in Figure 10.4 appears to be unfounded.

Normal Q–Q Plot

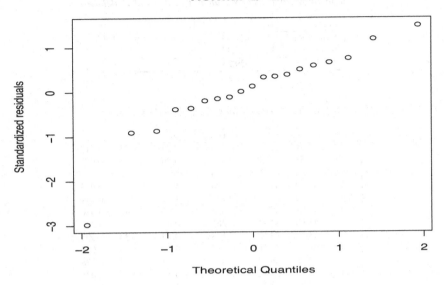

Figure 10.3: *Normal plot, case 18 deleted, $W' = 0.852$.*

Residual–Fitted plot

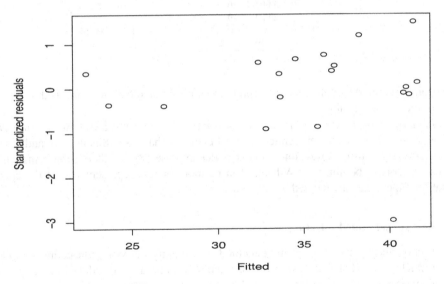

Figure 10.4: *Standardized residuals versus predicted values, case 18 deleted.*

Table 10.5: *Table of Coefficients: Cases 3 and 18 deleted.*

Predictor	$\hat{\beta}$	SE($\hat{\beta}$)	t	P
Constant	29.758	5.532	5.38	0.000
x_1	−1.6985	0.4660	−3.64	0.003
x_2	0.08512	0.02079	4.09	0.001
x_3	0.66617	0.03824	17.42	0.000
x_4	1.1840	0.1643	7.21	0.000
x_5	−4.0668	0.8487	−4.79	0.000

Table 10.6: *Analysis of Variance: Cases 3 and 18 deleted.*

Source	df	SS	MS	F	P
Regression	5	621.89	124.38	203.20	0.000
Error	12	7.34	0.61		
Total	17	629.23			

Table 10.7: *Diagnostics: Cases 3 and 18 deleted.*

Case	y	ŷ	Leverage	r	t	C
1	37.01	36.83	0.485	0.33	0.31	0.017
2	26.51	26.62	0.491	−0.20	−0.19	0.007
3		40.78	0.156			
4	40.70	41.43	0.196	−1.04	−1.05	0.044
5	37.10	36.35	0.180	1.07	1.07	0.041
6	33.90	33.67	0.504	0.42	0.41	0.030
7	41.80	42.11	0.261	−0.46	−0.44	0.012
8	33.40	33.69	0.108	−0.39	−0.38	0.003
9	41.01	41.56	0.311	−0.84	−0.83	0.053
10	37.20	36.94	0.621	0.54	0.52	0.078
11	23.30	23.66	0.381	−0.58	−0.57	0.035
12	35.20	34.24	0.440	1.65	1.79	0.356
13	34.90	35.81	0.370	−1.47	−1.56	0.212
14	33.10	32.66	0.145	0.60	0.59	0.010
15	22.70	22.44	0.467	0.46	0.44	0.031
16	39.70	38.72	0.171	1.38	1.44	0.066
17	31.80	33.02	0.298	−1.85	−2.10	0.243
18		24.50	0.486			
19	43.10	42.22	0.332	1.37	1.43	0.155
20	41.01	41.49	0.239	−0.70	−0.68	0.025

Once again, Figure 10.6 could be misinterpreted as a horn shape but the 'horn' is due to the distribution of the predicted values.

Ultimately, someone must decide whether or not to delete unusual cases based on subject matter considerations. There is only moderate statistical evidence that case 18 is unusual and case 3 does not look severely unusual unless one previously deletes case 18. Are there subject matter reasons for these schools to be unusual? Will the data be more or less representative of the appropriate population if these data are deleted?

10.2   Best subset model selection

In this section and the next, we examine methods for identifying good reduced models *relative to a given (full) model*. Reduced models are of interest because a good reduced model provides an adequate explanation of the current data and, typically, the reduced model is more understandable because it is more succinct. Even more importantly, *for data collected in a similar fashion, a good reduced model often provides* better *predictions and parameter estimates than the full model*, cf. the subsection below on Mallows's C_p statistic and Christensen (2011, Section 14.7). Of course, difficulties with predictions arise when a good reduced model is used with new cases that are not similar to those on which the reduced model was fitted and evaluated. In particular, a good fitted reduced model should not be used for prediction of a new case unless *all* of the predictor variables in the new case are similar to those in the original data. *It is not enough that new cases be similar on just the variables in the reduced model.* In fact it is not sufficient that they be similar on all of the variables in the full model because some important variable may not have been measured for the full model, yet a new case with a very different value of this unmeasured variable can act very differently.

This section presents three methods for examining all possible reduced models. These methods

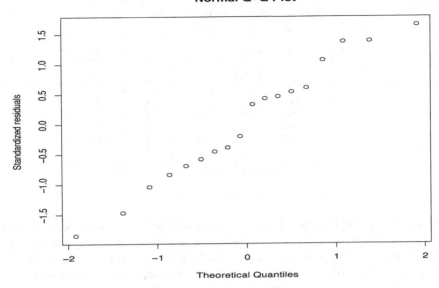

Figure 10.5: *Normal plot, cases 3 and 18 deleted, $W' = 0.979$.*

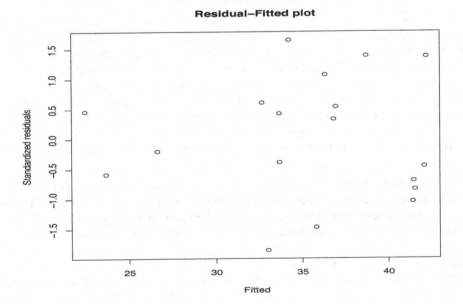

Figure 10.6: *Standardized residuals versus predicted values, cases 3 and 18 deleted.*

are based on defining a criterion for a best model and then finding the models that are best by this criterion. Section 10.3 considers three methods of making sequential selections of variables. Obviously, it is better to consider all reduced models whenever feasible rather than making sequential selections. Sequential methods are flawed but they are cheap and easy.

10.2.1 R^2 statistic

The fundamental statistic in comparing all possible reduced models is the R^2 statistic. This is appropriate but we should recall some of the weaknesses of R^2. The numerical size of R^2 is more related

Table 10.8: *Best subset regression: R^2 statistic.*

Vars.	R^2	$\sqrt{MSE}$	x_1	x_2	x_3	x_4	x_5
1	86.0	2.2392			X		
1	56.8	3.9299		X			
2	88.7	2.0641			X	X	
2	86.2	2.2866			X		X
3	90.1	1.9974	X		X	X	
3	88.9	2.1137			X	X	X
4	90.2	2.0514	X		X	X	X
4	90.1	2.0603	X	X	X	X	
5	90.6	2.0743	X	X	X	X	X

to predictive ability than to model adequacy. The perfect model can have small predictive ability and thus a small R^2, while demonstrably inadequate models can still have substantial predictive ability and thus a high R^2. Fortunately, we are typically more interested in prediction than in finding the perfect model, especially since our models are typically empirical approximations for which no perfect model exists. In addition, when considering transformations of the dependent variable, the R^2 values for different models are not comparable (unless predictions are back transformed to the original scale and correlated with the original data to obtain R^2).

In the present context, the most serious drawback of R^2 is that it typically goes up when more predictor variables are added to a model. (It cannot go down.) Thus it is not really appropriate to compare the R^2 values of two models with different numbers of predictors. However, we can use R^2 to compare models with *the same* number of predictor variables. In fact, for models with the same number of predictors, we can use R^2 to order them from best to worse; the largest R^2 value then corresponds to the best model. R^2 is the fundamental model comparison statistic for best subset methods in that, *for comparing models with the same number of predictors*, the other methods considered give the same relative orderings for models as R^2. The essence of the other methods is to develop a criterion for comparing models that have *different* numbers of predictors, i.e., the methods incorporate penalties for adding more regression parameters.

Table 10.8 contains the two best models for the *Coleman Report* data based on the R^2 statistic for each number of predictor variables. The best single variable is x_3; the second best is x_2. This information could be obtained from the correlations between y and the predictor variables given in Section 9.1. Note the drastic difference between the R^2 for using x_3 and that for x_2. The best pair of variables for predicting y is x_3 and x_4, while the second best pair is x_3 and x_5. The best three-variable model contains x_1, x_3, and x_4. Note that the largest R^2 values go up very little when a forth or fifth variable is added. Moreover, all the models in Table 10.8 that contain three or more variables include x_3 and x_4. We could conduct F tests to compare models with different numbers of predictor variables, as long as the smaller models are contained in the larger ones.

Any models that we think are good candidates should be examined for influential and outlying observations, consistency with assumptions, and subject matter implications. Any model that makes particularly good sense to a subject matter specialist warrants special consideration. Models that make particularly poor sense to subject matter specialists may be dumb luck but they may also be the springboard for new insights into the process generating the data. We also need to concern ourselves with the role of observations that are influential or outlying in the original (full) model. We will examine this in more detail later. Finally, recall that when making predictions based on reduced models, the point at which we are making the prediction generally needs to be consistent with the original data on all variables, not just the variables in the reduced model. When we drop a variable, we do not conclude that the variable is not important, we conclude that it is not important *for this set of data*. For different data, a dropped variable may become important. We cannot presume to make predictions from a reduced model for new cases that are substantially different from the original data.

10.2.2 Adjusted R^2 statistic

The adjusted R^2 statistic is simply an adjustment of R^2 that allows comparisons to be made between models with different numbers of predictor variables. Let p be the number of predictor variables in a regression equation (including the intercept), then the adjusted R^2 is defined to be

$$\text{Adj } R^2 \equiv 1 - \frac{n-1}{n-p}\left(1 - R^2\right).$$

For the *Coleman Report* example with all predictor variables, this becomes

$$0.873 = 1 - \frac{20-1}{20-6}\left(1 - 0.9063\right),$$

or, as it is commonly written, 87.3%.

It is not too difficult to see that

$$\text{Adj } R^2 = 1 - \frac{MSE}{s_y^2}$$

where s_y^2 is the sample variance of the y_is, i.e., $s_y^2 = SSTot/(n-1)$. This is a much simpler statement than the defining relationship. For the *Coleman Report* example with all predictor variables, this is

$$0.873 = 1 - \frac{4.30}{(642.92)/19}.$$

Note that *when comparing two models, the model with the smaller MSE has the larger adjusted R^2*.

R^2 is always between 0 and 1, but while the adjusted R^2 cannot get above 1, it can get below 0. It is possible to find models that have $MSE > s_y^2$. In these cases, the adjusted R^2 is actually less than 0.

Models with large adjusted R^2s are precisely models with small mean squared errors. At first glance, this seems like a reasonable way to choose models, but upon closer inspection the idea seems flawed. The problem is that when comparing some model with a reduced model, the adjusted R^2 is greater for the larger model whenever the mean squared error of the larger model is less than the numerator mean square for testing the adequacy of the smaller model. In other words, the adjusted R^2 is greater for the larger model whenever the F statistic for comparing the models is greater than 1. Typically, we want the F statistic to be substantially larger than 1 before concluding that the extra variables in the larger model are important.

To see that the adjusted R^2 is larger for the larger model whenever $F > 1$, consider the simplest example, that of comparing the full model to the model that contains just an intercept. For the *Coleman Report* data, the mean squared error for the intercept model is

$$SSTot/19 = 642.92/19 = (SSReg + SSE)/19$$

$$= (5MSReg + 14MSE)/19 = \frac{5}{19}116.54 + \frac{14}{19}4.30.$$

Thus $SSTot/19$ is a weighted average of *MSReg* and *MSE*. The *MSReg* is greater than the *MSE* $(F > 1)$, so the weighted average of the terms must be greater than the smaller term, *MSE*. The weighted average is $SSTot/19$, which is the mean squared error for the intercept model, while *MSE* is the mean squared error for the full model. Thus $F > 1$ implies that the mean squared error for the smaller model is greater than the mean squared error for the larger model and the model with the smaller mean squared error has the higher adjusted R^2.

Table 10.9: *Best subset regression: Adjusted R^2 statistic.*

Vars.	Adj. R^2	$\sqrt{MSE}$	x_1	x_2	x_3	x_4	x_5
3	88.2	1.9974	X		X	X	
4	87.6	2.0514	X		X	X	X
4	87.5	2.0603	X	X	X	X	
2	87.4	2.0641			X	X	
5	87.3	2.0743	X	X	X	X	X
3	86.8	2.1137			X	X	X

Included variables

In general, the mean squared error for the smaller model is a weighted average of the mean square for the variables being added and the mean squared error of the larger model. If the mean square for the variables being added is greater than the mean squared error of the larger model, i.e., if $F > 1$, the mean squared error for the smaller model must be greater than that for the larger model. If we add variables to a model whenever the F statistic is greater than 1, we will include a lot of unnecessary variables.

Table 10.9 contains the six best-fitting models as judged by the adjusted R^2 criterion. As advertised, the ordering of the models from best to worst is consistent whether one maximizes the adjusted R^2 or minimizes the MSE (or equivalently, $\sqrt{MSE}$). The best model based on the adjusted R^2 is the model with variables x_1, x_3, and x_4, but a number of the best models are given. Presenting a number of the best models reinforces the idea that selection of one or more final models should be based on many more considerations than just the value of one model selection statistic. Moreover, the *best* model as determined by the adjusted R^2 often contains too many variables.

Note also that the two models in Table 10.9 with three variables are precisely the two three-variable models with the highest R^2 values from Table 10.8. The same is true about the two four-variable models that made this list. As indicated earlier, when the number of variables is fixed, ordering models by their R^2s is equivalent to ordering models by their adjusted R^2s. The comments about model checking and prediction made in the previous subsection continue to apply.

10.2.3 Mallows's C_p statistic

Mallows's C_p statistic estimates a measure of the difference between the fitted regression surface from a reduced model and the actual regression surface. The idea is to compare the points

$$z_i = \beta_0 + \beta_1 x_{i1} + \beta_2 x_{i2} + \beta_3 x_{i3} + \ldots + \beta_{p-1} x_{i,p-1}$$

on the actual regression surface of the full model (Full) to the corresponding predictions $\hat{y}_{iR}$ from some reduced model (Red.) with, say, r predictor variables (including the constant). The comparisons are made at the locations of the original data. The model comparison is based on the sum of standardized squared differences,

$$\kappa \equiv \sum_{i=1}^{n} (\hat{y}_{iR} - z_i)^2 / \sigma^2.$$

The term σ^2 serves only to provide some standardization. Small values of κ indicate good reduced models. Note that κ is not directly useful because it is unknown. It depends on the z_i values and they depend on the unknown full model regression parameters. However, if we think of the $\hat{y}_{iR}$s as functions of the random variables y_i, the comparison value κ is a function of the y_is and thus is a random variable with an expected value. Mallows's C_p statistic is an estimate of the expected value of κ. In particular, Mallows's C_p statistic is

$$C_p = \frac{SSE(Red.)}{MSE(Full)} - (n - 2r).$$

Table 10.10: *Best subset regression: C_p statistic.*

			Included variables				
Vars	C_p	$\sqrt{MSE}$	x_1	x_2	x_3	x_4	x_5
2	2.8	2.0641			X	X	
3	2.8	1.9974	X		X	X	
3	4.6	2.1137			X	X	X
4	4.7	2.0514	X		X	X	X
3	4.8	2.1272		X	X	X	
4	4.8	2.0603	X	X	X	X	

For a derivation of this statistic see Christensen (2011, Section 14.1). The smaller the C_p value, the better the model (up to the variability of the estimation). If the C_p statistic is computed for the full model, the result is always p, the number of predictor variables including the intercept. For general linear models r is the number of functionally distinct mean parameters in the reduced model.

In multiple regression, estimated regression surfaces are identical to prediction surfaces, so models with Mallows's C_p statistics that are substantially less than p can be viewed as reduced models that are estimated to be better at prediction than the full model. Of course this comparison between predictions from the full and reduced models is restricted to the actual combinations of predictor variables in the observed data.

For the *Coleman Report* data, Table 10.10 contains the best six models based on the C_p statistic. The best model is the one with variables x_3 and x_4, but the model including x_1, x_3, and x_4 has essentially the same value of C_p. There is a substantial increase in C_p for any of the other four models. Clearly, we would focus attention on the two best models to see if they are adequate in terms of outliers, influential observations, agreement with assumptions, and subject matter implications. As always, predictions can only be made with safety from the reduced models when the new cases are to be obtained in a similar fashion to the original data. In particular, new cases must have similar values to those in the original data for all of the predictor variables, not just those in the reduced model. Note that the ranking of the best models is different here than for the adjusted R^2. The full model is not included here, while it was in the adjusted R^2 table. Conversely, the model with x_2, x_3, and x_4 is included here but was not included in the adjusted R^2 table. Note also that among models with three variables, the C_p rankings agree with the R^2 rankings and the same holds for four-variable models.

It is my impression that Mallows's C_p statistic is the most popular method for selecting a best subset of the predictor variables. It is certainly my favorite. Mallows's C_p statistic is closely related to Akaike's information criterion (AIC), which is a general criterion for model selection. AIC and the relationship between C_p and AIC are examined in Christensen (1997, Section 4.8).

10.2.4 A combined subset selection table

Table 10.11 lists the three best models based on R^2 for each number of predictor variables. In addition, the adjusted R^2 and C_p values for each model are listed in the table. It is easy to identify the best models based on any of the model selection criteria. The output is extensive enough to include a few notably bad models. Rather than asking for the best 3, one might ask for the best 4, or 5, or 6 models for each number of predictor variables but it is difficult to imagine a need for any more extensive summary of the models when beginning a search for good reduced models.

Note that the model with x_1, x_3, and x_4 is the best model as judged by adjusted R^2 and is nearly the best model as judged by the C_p statistic. (The model with x_3 and x_4 has a slightly smaller C_p value.) The model with x_2, x_3, x_4 has essentially the same C_p statistic as the model with x_1, x_2, x_3, x_4 but the latter model has a larger adjusted R^2.

Table 10.11: *Best subset regression.*

Vars.	R^2	Adj. R^2	C_p	$\sqrt{MSE}$	x_1	x_2	x_3	x_4	x_5
1	86.0	85.2	5.0	2.2392			X		
1	56.8	54.4	48.6	3.9299		X			
1	53.7	51.2	53.1	4.0654					X
2	88.7	87.4	2.8	2.0641			X	X	
2	86.2	84.5	6.7	2.2866			X		X
2	86.0	84.4	6.9	2.2993		X	X		
3	90.1	88.2	2.8	1.9974	X		X	X	
3	88.9	86.8	4.6	2.1137			X	X	X
3	88.7	86.6	4.8	2.1272		X	X	X	
4	90.2	87.6	4.7	2.0514	X		X	X	X
4	90.1	87.5	4.8	2.0603	X	X	X	X	
4	89.2	86.3	6.1	2.1499		X	X	X	X
5	90.6	87.3	6.0	2.0743	X	X	X	X	X

10.3 Stepwise model selection

Best subset selection methods evaluate all the possible subsets of variables from a full model and identify the best reduced regression models based on some criterion. Evaluating all possible models is the most reasonable way to proceed in variable selection but the computational demands of evaluating every model can be staggering. Every additional variable in a model doubles the number of reduced models that can be constructed. In our example with five variables, there are $2^5 = 32$ reduced models to be considered; in an example with 8 variables there are $2^8 = 256$ reduced models to be fitted. Years ago, when computation was slow and expensive, fitting large numbers of models was not practical, and even now, when one has a very large number of predictor variables, fitting all models can easily overwhelm a computer. (Actually, improved computer algorithms allow us to avoid fitting all models, but even with the improved algorithms, computational limits can be exceeded.)

An alternative to fitting all models is to evaluate the variables one at a time and look at a sequence of models. Stepwise variable selection methods do this. The best of these methods begin with a full model and sequentially identify variables that can be eliminated. In some procedures, variables that have been eliminated may be put back into the model if they meet certain criteria. The virtue of starting with the full model is that if you start with an adequate model and only do reasonable things, you should end up with an adequate model. A less satisfactory procedure is to begin with no variables and see which ones can be added into the model. This begins with an inadequate model and there is no guarantee that an adequate model will ever be achieved. We consider three methods: backwards elimination in which variables are deleted from the full model, forward selection in which variables are added to a model (typically the model that includes only the intercept), and stepwise methods in which variables can be both added and deleted. Because these methods only consider the deletion or addition of one variable at a time, they may never find the best models as determined by best subset selection methods.

10.3.1 Backwards elimination

Backwards elimination begins with the full model and sequentially eliminates from the model the least important variable. The importance of a variable is judged by the size of the t (or equivalent F) statistic for dropping the variable from the model, i.e., the t statistic for testing whether the corresponding regression coefficient is 0. After the variable with the smallest absolute t statistic is dropped, the model is refitted and the t statistics recalculated. Again, the variable with the smallest absolute t statistic is dropped. The process ends when all of the absolute values of the t statistics are greater than some predetermined level. The predetermined level can be a fixed number for all steps

Table 10.12: *Backwards elimination of y on 5 predictors with N = 20.*

Step		Const.	x_1	x_2	x_3	x_4	x_5	R^2	$\sqrt{MSE}$
1	$\hat{\beta}$	19.95	−1.8	0.044	0.556	1.11	−1.8	90.63	2.07
	t_{obs}		−1.45	0.82	5.98	2.56	−0.89		
2	$\hat{\beta}$	15.47	−1.7		0.582	1.03	−0.5	90.18	2.05
	t_{obs}		−1.41		6.75	2.46	−0.41		
3	$\hat{\beta}$	12.12	−1.7		0.553	1.04		90.07	2.00
	t_{obs}		−1.47		11.27	2.56			
4	$\hat{\beta}$	14.58			0.542	0.75		88.73	2.06
					10.82	2.05			

or it can change depending on the step. When allowing it to change depending on the step, we could set up the process so that it stops when all of the P values are below a fixed level.

Table 10.12 illustrates backwards elimination for the *Coleman Report* data. In this example, the predetermined level for stopping the procedure is 2. If all $|t|$ statistics are greater than 2, elimination of variables halts. Step 1 includes all 5 predictor variables. The table gives estimated regression coefficients, t statistics, the R^2 value, and the square root of the MSE. In step 1, the smallest absolute t statistic is 0.82, so variable x_2 is eliminated from the model. The statistics in step 2 are similar to those in step 1 but now the model includes only variables x_1, x_3, x_4, and x_5. In step 2, the smallest absolute t statistic is $|-0.41|$, so variable x_5 is eliminated from the model. Step 3 is based on the model with x_1, x_3, and x_4. The smallest absolute t statistic is the $|-1.47|$ for variable x_1, so x_1 is dropped. Step 4 uses the model with only x_3 and x_4. At this step, the t statistics are both greater than 2, so the process halts. Note that the intercept is not considered for elimination.

The final model given in Table 10.12 happens to be the best model as determined by the C_p statistic and the model at stage 3 is the second-best model as determined by the C_p statistic. This is a fortuitous event; there is no reason that this should happen other than these data being particularly clear about the most important variables.

10.3.2 Forward selection

Forward selection begins with an initial model and adds variables to the model one at a time. Most often, the initial model contains only the intercept, but many computer programs have options for including other variables in the initial model. Another reasonable starting point is to include all variables with large t statistics when fitting the full model containing all predictors. Logically, variables that are important in the full model should never lose their importance in reduced models.

To determine which variable to add at any step in the process, a candidate variable is added to the current model and the t statistic is computed for the candidate variable. This is done for each candidate variable and the candidate variable with the largest $|t|$ statistic is added to the model. The procedure stops when none of the absolute t statistics is greater than a predetermined level. The predetermined level can be a fixed number for all steps or it can change with the step. When allowing it to change depending on the step, we could set the process so that it stops when none of the P values for the candidate variables is below a fixed level.

Table 10.13 gives an abbreviated summary of the procedure for the *Coleman Report* data using 2 as the predetermined $|t|$ level for stopping the process and starting with the intercept-only model. At the first step, the five models $y_i = \gamma_{0j} + \gamma_j x_{ij} + \varepsilon_i$, $j = 1, \ldots, 5$ are fitted to the data. The variable x_j with the largest absolute t statistic for testing $\gamma_j = 0$ is added to the model. Table 10.13 indicates that this was variable x_3. At step 2, the four models $y_i = \beta_{0j} + \beta_{3j}x_{i3} + \beta_j x_{ij} + \varepsilon_i$, $j = 1, 2, 4, 5$ are fitted to the data and the variable x_j with the largest absolute t statistic for testing $\beta_j = 0$ is added to the model. In the example, the largest absolute t statistic belongs to x_4. At this point, the table stops,

Table 10.13: *Forward selection of y on 5 predictors with N = 20.*

Step		Const.	x_1	x_2	x_3	x_4	x_5	R^2	$\sqrt{MSE}$
1	$\hat{\beta}$	33.32			0.560			85.96	2.24
	t_{obs}				10.50				
2	$\hat{\beta}$	14.58			0.542	0.75		88.73	2.06
	t_{obs}				10.82	2.05			

indicating that when the three models $y_i = \eta_{0j} + \eta_{3j}x_{i3} + \eta_{4j}x_{i4} + \eta_j x_{ij} + \varepsilon_i$, $j = 1,2,5$ were fitted to the model, none of the absolute t statistics for testing $\eta_j = 0$ were greater than 2.

The final model selected is the model with predictor variables x_3 and x_4. This is the same model obtained from backwards elimination and the model that has the smallest C_p statistic. Again, this is a fortuitous circumstance. There is no assurance that such agreement between methods will occur.

Rather than using t statistics, the decisions could be made using the equivalent F statistics. The stopping value of 2 for t statistics corresponds to a stopping value of 4 for F statistics. In addition, this same procedure can be based on sample correlations and partial correlations. The decision in step 1 is equivalent to adding the variable that has the largest absolute sample correlation with y. The decision in step 2 is equivalent to adding the variable that has the largest absolute sample partial correlation with y after adjusting for x_3. Step 3 is not shown in the table, but the computations for step 3 must be made in order to know that the procedure stops after step 2. The decision in step 3 is equivalent to adding the variable that has the largest absolute sample partial correlation with y after adjusting for x_3 and x_4, provided this value is large enough.

The author has a hard time imagining any situation where forward selection from the intercept-only model is a reasonable thing to do, except possibly as a screening device when there are more predictor variables than there are observations. In such a case, the full model cannot be fitted meaningfully, so best subset methods and backwards elimination do not work.

10.3.3   Stepwise methods

Stepwise methods alternate between forward selection and backwards elimination. Suppose you have just arrived at a model by dropping a variable. A stepwise method will then check to see if any variable can be added to the model. If you have just arrived at a model by adding a variable, a stepwise method then checks to see if any variable can be dropped. The value of the absolute t statistic required for dropping a variable is allowed to be different from the value required for adding a variable. Stepwise methods often start with an initial model that contains only an intercept, but many computer programs allow starting the process with the full model. In the *Coleman Report* example, the stepwise method beginning with the intercept model gives the same results as forward selection and the stepwise method beginning with the full model gives the same results as backwards elimination. (The absolute t statistics for both entering and removing were set at 2.) Other initial models can also be used. Christensen (2011, Section 14.2) discusses some alternative rules for conducting stepwise regression.

10.4   Model selection and case deletion

In this section we examine how the results of the previous two sections change when influential cases are deleted. Before beginning, we make a crucial point. *Both variable selection and the elimination of outliers cause the resulting model to appear better than it probably should. Both tend to give MSEs that are unrealistically small. It follows that confidence and prediction intervals are unrealistically narrow and test statistics are unrealistically large.* Outliers tend to be cases with large residuals; any policy of eliminating the largest residuals obviously makes the SSE, which is the sum of the squared residuals, and the MSE smaller. Some large residuals occur by chance even when

Table 10.14: *Best subset regression: Case 18 deleted.*

		Adj.			Included variables				
Vars	R^2	R^2	C_p	$\sqrt{MSE}$	x_1	x_2	x_3	x_4	x_5
1	89.6	89.0	21.9	1.9653			X		
1	56.0	53.4	140.8	4.0397		X			
1	53.4	50.6	150.2	4.1595					X
2	92.3	91.3	14.3	1.7414			X	X	
2	91.2	90.1	18.2	1.8635			X		X
2	89.8	88.6	23.0	2.0020		X	X		
3	93.7	92.4	11.4	1.6293			X	X	X
3	93.5	92.2	12.1	1.6573	X		X	X	
3	92.3	90.8	16.1	1.7942		X	X	X	
4	95.2	93.8	8.1	1.4766		X	X	X	X
4	94.7	93.2	9.8	1.5464	X		X	X	X
4	93.5	91.6	14.1	1.7143	X	X	X	X	
5	96.3	94.9	6.0	1.3343	X	X	X	X	X

the model is correct. Systematically eliminating these large residuals makes the estimate of the variance too small. Variable selection methods tend to identify as good reduced models those with small MSEs. The most extreme case is that of using the adjusted R^2 criterion, which identifies as the best model the one with the smallest MSE. Confidence and prediction intervals based on models that are arrived at after variable selection or outlier deletion should be viewed as the smallest reasonable intervals available, with the understanding that more appropriate intervals would probably be wider. Tests performed after variable selection or outlier deletion should be viewed as giving the greatest reasonable evidence against the null hypothesis, with the understanding that more appropriate tests would probably display a lower level of significance.

Recall that in Section 10.1, case 18 was identified as an influential point in the *Coleman Report* data and then case 3 was identified as highly influential. Table 10.14 gives the results of a best subset selection when case 18 has been eliminated. The full model is the best model as measured by either the C_p statistic or the adjusted R^2 value. This is a far cry from the full data analysis in which the models with x_3, x_4 and with x_1, x_3, x_4 had the smallest C_p statistics. These two models are only the seventh and fifth best models in Table 10.14. The two closest competitors to the full model in Table 10.14 involve dropping one of variables x_1 and x_2. The fourth and fifth best models involve dropping x_2 and one of variables x_1 and x_5. In this case, the adjusted R^2 ordering of the five best models agrees with the C_p ordering.

Table 10.15 gives the best subset summary when cases 3 and 18 have both been eliminated. Once again, the best model as judged by either C_p or adjusted R^2 is the full model. The second best model drops x_1 and the third best model drops x_2. However, the subsequent ordering changes substantially.

Now consider backwards elimination and forward selection with influential observations deleted. In both cases, we continue to use the $|t|$ value 2 as the cutoff to stop addition and removal of variables.

Table 10.16 gives the results of a backwards elimination when case 18 is deleted and when cases 3 and 18 are deleted. In both situations, all five of the variables remain in the model. The regression coefficients are similar in the two models with the largest difference being in the coefficients for x_5. Recall that when all of the cases were included, the backwards elimination model included only variables x_3 and x_4, so we see a substantial difference due to the deletion of one or two cases.

The results of forward selection are given in Table 10.17. With case 18 deleted, the process stops with a model that includes x_3 and x_4. With case 3 also deleted, the model includes x_1, x_3, and x_4. While these happen to agree quite well with the results from the complete data, they agree poorly with the results from best subset selection and from backwards elimination, both of which indicate that all variables are important. Forward selection gets hung up after a few variables and cannot deal

Table 10.15: *Best subset regression: Cases 3 and 18 deleted.*

Vars	R^2	Adj. R^2	C_p	$\sqrt{MSE}$	x_1	x_2	x_3	x_4	x_5
1	92.2	91.7	66.5	1.7548			X		
1	57.9	55.3	418.8	4.0688		X			
1	55.8	53.0	440.4	4.1693					X
2	95.3	94.7	36.1	1.4004			X	X	
2	93.2	92.2	58.3	1.6939			X		X
2	92.3	91.2	67.6	1.8023		X	X		
3	96.6	95.8	25.2	1.2412	X		X	X	
3	96.1	95.2	30.3	1.3269			X	X	X
3	95.3	94.3	38.0	1.4490		X	X	X	
4	97.5	96.8	17.3	1.0911		X	X	X	X
4	97.2	96.3	20.8	1.1636	X		X	X	X
4	96.6	95.6	27.0	1.2830	X	X	X	X	
5	98.8	98.3	6.0	0.78236	X	X	X	X	X

Table 10.16: *Backwards elimination.*

		Case 18 deleted							
Step		Const.	x_1	x_2	x_3	x_4	x_5	R^2	$\sqrt{MSE}$
1	$\hat{\beta}$	34.29	−1.62	0.085	0.674	1.11	−4.6	96.33	1.33
	t_{obs}		−2.04	2.41	10.34	3.98	−3.18		

		Cases 18 and 3 deleted							
Step		Const.	x_1	x_2	x_3	x_4	x_5	R^2	$\sqrt{MSE}$
1	$\hat{\beta}$	29.76	−1.70	0.085	0.666	1.18	−4.07	98.83	0.782
	t_{obs}		−3.64	4.09	17.42	7.21	−4.79		

with the fact that adding several variables (rather than one at a time) improves the fit of the model substantially.

10.5 Lasso regression

An alternative to least squares estimation that has become quite popular is *lasso regression*, which was proposed by Tibshirani (1996). "Lasso" stands for *least absolute shrinkage and selection op-*

Table 10.17: *Forward selection.*

		Case 18 deleted							
Step		Const.	x_1	x_2	x_3	x_4	x_5	R^2	$\sqrt{MSE}$
1	$\hat{\beta}$	32.92			0.604			89.59	1.97
	t_{obs}				12.10				
2	$\hat{\beta}$	14.54			0.585	0.74		92.31	1.74
	t_{obs}				13.01	2.38			

		Cases 18 and 3 deleted							
Step		Const.	x_1	x_2	x_3	x_4	x_5	R^2	$\sqrt{MSE}$
1	$\hat{\beta}$	33.05			0.627			92.17	1.75
	t_{obs}				13.72				
2	$\hat{\beta}$	13.23			0.608	0.79		95.32	1.40
	t_{obs}				16.48	3.18			
3	$\hat{\beta}$	10.86	−1.66		0.619	1.07		96.57	1.24
	t_{obs}		−2.26		18.72	4.23			

Table 10.18: *Lasso and least squares estimates:* The Coleman Report *data.*

Predictor	1	0.6	Lasso λ 0.56348	0.5	0	Reduced Model Least Squares	
Constant	19.95	18.79306	20.39486	26.51564	35.0825	12.1195	14.58327
x_1	-1.793	-0.33591	0.00000	0.00000	0.0000	-1.7358	0.00000
x_2	0.04360	0.00000	0.00000	0.00000	0.0000	0.00000	0.00000
x_3	0.55576	0.51872	0.51045	0.47768	0.0000	0.5532	0.54156
x_4	1.1102	0.62140	0.52194	0.28189	0.0000	1.0358	0.74989
x_5	-1.811	0.00000	0.00000	0.00000	0.0000	0.00000	0.00000

erator. The interesting thing about lasso, and the reason for its inclusion in this chapter, is that it automatically performs variable selection while it is estimating the regression parameters.

As discussed in Subsection 9.1.2, the least squares estimates $\hat{\beta}_j$ satisfy

$$\sum_{i=1}^{n} \left(y_i - \hat{\beta}_0 - \hat{\beta}_1 x_{i1} - \hat{\beta}_2 x_{i2} - \cdots - \hat{\beta}_{p-1} x_{i,p-1} \right)^2 =$$

$$\min_{\beta_0,\dots,\beta_{p-1}} \sum_{i=1}^{n} (y_i - \beta_0 - \beta_1 x_{i1} - \beta_2 x_{i2} - \cdots - \beta_{p-1} x_{i,p-1})^2.$$

There are various ways that one can present the lasso criterion for estimation. One of them is to minimize the least squares criterion

$$\sum_{i=1}^{n} (y_i - \beta_0 - \beta_1 x_{i1} - \beta_2 x_{i2} - \cdots - \beta_{p-1} x_{i,p-1})^2$$

subject to an upper bound on the sum of the absolute values of the regression coefficients. We define the upper bound in terms of the least squares estimates so that the lasso estimates must satisfy

$$\sum_{j=1}^{p-1} |\beta_j| \le \lambda \sum_{j=1}^{p-1} |\hat{\beta}_j| \tag{10.5.1}$$

for some λ with $0 \le \lambda \le 1$. The lasso estimates depend on the choice of λ. The least squares estimates obviously satisfy the inequality when $\lambda = 1$, so $\lambda = 1$ gives least squares estimates. When $\lambda = 0$, all the regression coefficients in the inequality must be zero, but notice that the intercept is not subject to the upper bound in (10.5.1). Thus, $\lambda = 0$ gives the least squares estimates for the intercept-only model, i.e., it zeros out all the regression coefficients except the intercept, which it estimates with $\bar{y}_{..}$.

EXAMPLE 10.5.1.   We examine the effect of lasso regression on *The Coleman Report* data. Table 10.18 contains results for five values of λ and least squares estimates for two reduced models. For $\lambda = 1$, the estimates are identical to the least squares estimates for the full model.

R's *lasso2* package has a default value of $\lambda = 0.5$, which zeros out the coefficients for x_1, x_2, and x_5. The reduced model that only includes x_3 and x_4 is the model that we liked in Section 9.3. The lasso estimates of β_3 and β_4 are noticeably smaller than the least squares estimates from the reduced model given in the last column of Table 10.18. I also found the largest value of λ that zeros out the coefficients for x_1, x_2, and x_5. That value is $\lambda = 0.56348$. With this larger value of λ, the lasso estimates are closer to the reduced model least squares estimates but still noticeably different.

For $\lambda \ge 0.56349$, lasso produces a nonzero coefficient for x_1. From Section 9.3, if we were going to add another variable to the model containing only x_3 and x_4, the best choice is to add x_1. Table 10.18 includes results for $\lambda = 0.6$ and least squares on the three-variable model. $\lambda = 0.6$ still has the coefficients for x_2 and x_5 zeroed out. Again, the nonzero lasso estimates for β_1, β_3, and β_4 are all closer to zero than the least squares estimates from the model with just x_1, x_3, and x_4.   □

Lasso seems to do a good job of identifying the important variables and it does it pretty automatically. That can be both a blessing and a curse. It is far less obvious how well lasso is estimating the regression coefficients. The least squares estimates seem more stable across reduced models than do the lasso estimates. And there is the whole issue of choosing λ.

Notice that the inequality (10.5.1) uses the same weight λ on all of the regression coefficients. That is not an obviously reasonable thing to do when the predictor variables are measured in different units, so lasso is often applied to standardized predictor variables, i.e., variables that have their sample mean subtracted and are then divided by their standard deviation. (This is the default in R's lasso2 package.) The regression estimates can then be transformed back to their original scales to be comparable to the least squares estimates. Section 11.6 illustrates this standardization procedure for another regression technique, principal components regression. Lasso applied to the unstandardized *Coleman Report* data gives very different, and less appealing, results.

10.6   Exercises

EXERCISE 10.6.1.   Reconsider the advertising data of Exercise 9.12.1.

(a) Are there any high-leverage points? Why or why not?

(b) Test whether each case is an outlier using an overall significance level no greater than $\alpha = 0.05$. Completely state the appropriate reference distribution.

(c) Discuss the importance of Cook's distances in regard to these data.

(d) Using only analysis of variance tables, compute R^2, the adjusted R^2, and the C_p statistic for $y_i = \beta_0 + \beta_1 x_{i1} + \beta_2 x_{i2} + \varepsilon_i$. Show your work.

(e) In the three-variable model, which if any variable would be deleted by a backwards elimination method? Why?

EXERCISE 10.6.2.   Consider the information given in Table 10.19 on diagnostic statistics for the wood data of Exercise 9.12.2.

(a) Are there any outliers in the predictor variables? Why are these considered outliers?

(b) Are there any outliers in the dependent variable? If so, why are these considered outliers?

(c) What are the most influential observations in terms of the predictive ability of the model?

EXERCISE 10.6.3.   Consider the information in Table 10.20 on best subset regression for the wood data of Exercise 9.12.2.

(a) In order, what are the three best models as measured by the C_p criterion?

(b) What is the mean squared error for the model with variables x_1, x_3, and x_4?

(c) In order, what are the three best models as measured by the adjusted R^2 criterion? (Yes, it is possible to distinguish between the best four!)

(d) What do you think are the best models and what would you do next?

EXERCISE 10.6.4.   Consider the information in Table 10.21 on stepwise regression for the wood data of Exercise 9.12.2.

(a) What is being given in the rows labeled x_1, x_2, x_3, and x_4? What is being given in the rows labeled t?

(b) Is this table for forward selection, backwards elimination, stepwise regression, or some other procedure?

(c) Describe the results of the procedure.

Table 10.19: *Diagnostics for wood data.*

Obs.	Leverage	r	t	C	Obs.	Leverage	r	t	C
1	0.085	−0.25	−0.25	0.001	29	0.069	0.27	0.26	0.001
2	0.055	1.34	1.35	0.021	30	0.029	0.89	0.89	0.005
3	0.021	0.57	0.57	0.001	31	0.204	0.30	0.30	0.005
4	0.031	0.35	0.35	0.001	32	0.057	0.38	0.37	0.002
5	0.032	2.19	2.28	0.032	33	0.057	0.05	0.05	0.000
6	0.131	0.20	0.19	0.001	34	0.085	−2.43	−2.56	0.109
7	0.027	1.75	1.79	0.017	35	0.186	−2.17	−2.26	0.215
8	0.026	1.23	1.24	0.008	36	0.184	1.01	1.01	0.046
9	0.191	0.52	0.52	0.013	37	0.114	0.85	0.85	0.019
10	0.082	0.47	0.46	0.004	38	0.022	0.19	0.19	0.000
11	0.098	−3.39	−3.82	0.250	39	0.022	−0.45	−0.45	0.001
12	0.066	0.32	0.32	0.001	40	0.053	−1.15	−1.15	0.015
13	0.070	−0.09	−0.09	0.000	41	0.053	0.78	0.78	0.007
14	0.059	0.08	0.08	0.000	42	0.136	−0.77	−0.76	0.018
15	0.058	−0.91	−0.91	0.010	43	0.072	−0.78	−0.77	0.009
16	0.085	−0.09	−0.09	0.000	44	0.072	−0.27	−0.26	0.001
17	0.113	1.28	1.29	0.042	45	0.072	−0.40	−0.40	0.002
18	0.077	−1.05	−1.05	0.018	46	0.063	−0.62	−0.62	0.005
19	0.167	0.38	0.38	0.006	47	0.025	0.46	0.46	0.001
20	0.042	0.24	0.23	0.000	48	0.021	0.18	0.18	0.000
21	0.314	−0.19	−0.19	0.003	49	0.050	−0.44	−0.44	0.002
22	0.099	0.56	0.55	0.007	50	0.161	−0.66	−0.66	0.017
23	0.093	0.47	0.46	0.004	51	0.042	−0.44	−0.43	0.002
24	0.039	−0.60	−0.60	0.003	52	0.123	−0.26	−0.26	0.002
25	0.098	−1.07	−1.07	0.025	53	0.460	1.81	1.86	0.558
26	0.033	0.14	0.13	0.000	54	0.055	0.50	0.50	0.003
27	0.042	1.19	1.19	0.012	55	0.093	−1.03	−1.03	0.022
28	0.185	−1.41	−1.42	0.090					

Table 10.20: *Best subset regression of wood data.*

Vars	R^2	Adj. R^2	C_p	$\sqrt{MSE}$	x_1	x_2	x_3	x_4
1	97.9	97.9	12.9	18.881	X			
1	63.5	62.8	1064.9	78.889				X
1	32.7	31.5	2003.3	107.04			X	
2	98.3	98.2	3.5	17.278	X		X	
2	97.9	97.8	14.3	18.969	X	X		
2	97.9	97.8	14.9	19.061	X			X
3	98.3	98.2	5.3	17.419	X	X	X	
3	98.3	98.2	5.4	17.430	X		X	X
3	98.0	97.9	13.7	18.763	X	X		X
4	98.4	98.2	5.0	17.193	X	X	X	X

Table 10.21: *Stepwise regression on wood data.*

STEP	1	2	3
Constant	23.45	41.87	43.85
x_1	0.932	1.057	1.063
t	10.84	38.15	44.52
x_2	0.73	0.09	
t	1.56	0.40	
x_3	−0.50	−0.50	−0.51
t	−3.28	−3.27	−3.36
x_4	3.5		
t	1.53		
$\sqrt{MSE}$	17.2	17.4	17.3
R^2	98.36	98.29	98.28

EXERCISE 10.6.5.   Reanalyze the Prater data of Atkinson (1985) and Hader and Grandage (1958) from Exercise 9.12.3. Examine residuals and influential observations. Explore the use of the various model selection methods.

EXERCISE 10.6.6.   Reanalyze the Chapman data of Exercise 9.12.4. Examine residuals and influential observations. Explore the use of the various model selection methods.

EXERCISE 10.6.7.   Reanalyze the pollution data of Exercise 9.12.5. Examine residuals and influential observations. Explore the use of various model selection methods.

EXERCISE 10.6.8.   Repeat Exercise 9.12.6 on the body fat data with special emphasis on diagnostics and model selection.

Chapter 11

Multiple Regression: Matrix Formulation

In this chapter we use matrices to write regression models. Properties of matrices are reviewed in Appendix A. The economy of notation achieved through using matrices allows us to arrive at some interesting new insights and to derive several of the important properties of regression analysis.

11.1 Random vectors

In this section we discuss vectors and matrices that are made up of random variables rather than just numbers. For simplicity, we focus our discussion on vectors that contain 3 rows, but the results are completely general.

Let y_1, y_2, and y_3 be random variables. From these, we can construct a 3×1 random vector, say

$$Y = \begin{bmatrix} y_1 \\ y_2 \\ y_3 \end{bmatrix}.$$

The expected value of the random vector is just the vector of expected values of the random variables. For the random variables write $E(y_i) = \mu_i$, then

$$E(Y) \equiv \begin{bmatrix} E(y_1) \\ E(y_2) \\ E(y_3) \end{bmatrix} = \begin{bmatrix} \mu_1 \\ \mu_2 \\ \mu_3 \end{bmatrix} \equiv \mu.$$

In other words, expectation of a random vector is performed elementwise. In fact, the expected value of any random matrix (a matrix consisting of random variables) is the matrix made up of the expected values of the elements in the random matrix. Thus if $w_{ij}, i = 1, 2, 3, j = 1, 2$ is a collection of random variables and we write

$$W = \begin{bmatrix} w_{11} & w_{12} \\ w_{21} & w_{22} \\ w_{31} & w_{33} \end{bmatrix},$$

then

$$E(W) \equiv \begin{bmatrix} E(w_{11}) & E(w_{12}) \\ E(w_{21}) & E(w_{22}) \\ E(w_{31}) & E(w_{33}) \end{bmatrix}.$$

We also need a concept for random vectors that is analogous to the variance of a random variable. This is the *covariance matrix*, sometimes called the *dispersion matrix*, the *variance matrix*, or the *variance-covariance matrix*. The covariance matrix is simply a matrix consisting of all the variances and covariances associated with the vector Y. Write

$$\text{Var}(y_i) = E(y_i - \mu_i)^2 \equiv \sigma_{ii}$$

and

$$\text{Cov}(y_i, y_j) = E[(y_i - \mu_i)(y_j - \mu_j)] \equiv \sigma_{ij}.$$

Two subscripts are used on σ_{ii} to indicate that it is the variance of y_i *rather than* writing $\mathrm{Var}(y_i) = \sigma_i^2$. The covariance matrix of our 3×1 vector Y is

$$\mathrm{Cov}(Y) = \begin{bmatrix} \sigma_{11} & \sigma_{12} & \sigma_{13} \\ \sigma_{21} & \sigma_{22} & \sigma_{23} \\ \sigma_{31} & \sigma_{32} & \sigma_{33} \end{bmatrix}.$$

When Y is 3×1, the covariance matrix is 3×3. If Y were 20×1, $\mathrm{Cov}(Y)$ would be 20×20. The covariance matrix is always symmetric because $\sigma_{ij} = \sigma_{ji}$ for any i, j. The variances of the individual random variables lie on the diagonal that runs from the top left to the bottom right. The covariances lie off the diagonal.

In general, if Y is an $r \times 1$ random vector and $\mathrm{E}(Y) = \mu$, then $\mathrm{Cov}(Y) = \mathrm{E}[(Y - \mu)(Y - \mu)']$. In other words, $\mathrm{Cov}(Y)$ is the expected value of the random matrix $(Y - \mu)(Y - \mu)'$.

11.2 Matrix formulation of regression models

11.2.1 Simple linear regression in matrix form

The usual model for simple linear regression is

$$y_i = \beta_0 + \beta_1 x_i + \varepsilon_i \quad i = 1, \dots, n, \tag{11.2.1}$$

$\mathrm{E}(\varepsilon_i) = 0$, $\mathrm{Var}(\varepsilon_i) = \sigma^2$, and $\mathrm{Cov}(\varepsilon_i, \varepsilon_j) = 0$ for $i \neq j$. In matrix terms this can be written as

$$\begin{bmatrix} y_1 \\ y_2 \\ \vdots \\ y_n \end{bmatrix} = \begin{bmatrix} 1 & x_1 \\ 1 & x_2 \\ \vdots & \vdots \\ 1 & x_n \end{bmatrix} \begin{bmatrix} \beta_0 \\ \beta_1 \end{bmatrix} + \begin{bmatrix} \varepsilon_1 \\ \varepsilon_2 \\ \vdots \\ \varepsilon_n \end{bmatrix}.$$

$$Y_{n \times 1} \quad = \quad X_{n \times 2} \quad \beta_{2 \times 1} \quad + \quad e_{n \times 1}$$

Multiplying and adding the matrices on the right-hand side gives

$$\begin{bmatrix} y_1 \\ y_2 \\ \vdots \\ y_n \end{bmatrix} = \begin{bmatrix} \beta_0 + \beta_1 x_1 + \varepsilon_1 \\ \beta_0 + \beta_1 x_2 + \varepsilon_2 \\ \vdots \\ \beta_0 + \beta_1 x_n + \varepsilon_n \end{bmatrix}.$$

These two vectors are equal if and only if the corresponding elements are equal, which occurs if and only if Model (11.2.1) holds. The conditions on the ε_is translate into matrix terms as

$$\mathrm{E}(e) = 0$$

where 0 is the $n \times 1$ matrix containing all zeros and

$$\mathrm{Cov}(e) = \sigma^2 I$$

where I is the $n \times n$ identity matrix. By definition, the covariance matrix $\mathrm{Cov}(e)$ has the variances of the ε_is down the diagonal. The variance of each individual ε_i is σ^2, so all the diagonal elements of $\mathrm{Cov}(e)$ are σ^2, just as in $\sigma^2 I$. The covariance matrix $\mathrm{Cov}(e)$ has the covariances of distinct ε_is as its off-diagonal elements. The covariances of distinct ε_is are all 0, so all the off-diagonal elements of $\mathrm{Cov}(e)$ are zero, just as in $\sigma^2 I$.

Table 11.1: *Weights for various heights.*

Ht.	Wt.	Ht.	Wt.
65	120	63	110
65	140	63	135
65	130	63	120
65	135	72	170
66	150	72	185
66	135	72	160

EXAMPLE 11.2.1.   Height and weight data are given in Table 11.1 for 12 individuals. In matrix terms, the SLR model for regressing weights (y) on heights (x) is

$$
\begin{bmatrix} y_1 \\ y_2 \\ y_3 \\ y_4 \\ y_5 \\ y_6 \\ y_7 \\ y_8 \\ y_9 \\ y_{10} \\ y_{11} \\ y_{12} \end{bmatrix}
=
\begin{bmatrix} 1 & 65 \\ 1 & 65 \\ 1 & 65 \\ 1 & 65 \\ 1 & 66 \\ 1 & 66 \\ 1 & 63 \\ 1 & 63 \\ 1 & 63 \\ 1 & 72 \\ 1 & 72 \\ 1 & 72 \end{bmatrix}
\begin{bmatrix} \beta_0 \\ \beta_1 \end{bmatrix}
+
\begin{bmatrix} \varepsilon_1 \\ \varepsilon_2 \\ \varepsilon_3 \\ \varepsilon_4 \\ \varepsilon_5 \\ \varepsilon_6 \\ \varepsilon_7 \\ \varepsilon_8 \\ \varepsilon_9 \\ \varepsilon_{10} \\ \varepsilon_{11} \\ \varepsilon_{12} \end{bmatrix}.
$$

The observed data for this example are

$$
\begin{bmatrix} y_1 \\ y_2 \\ y_3 \\ y_4 \\ y_5 \\ y_6 \\ y_7 \\ y_8 \\ y_9 \\ y_{10} \\ y_{11} \\ y_{12} \end{bmatrix}
=
\begin{bmatrix} 120 \\ 140 \\ 130 \\ 135 \\ 150 \\ 135 \\ 110 \\ 135 \\ 120 \\ 170 \\ 185 \\ 160 \end{bmatrix}.
$$

We could equally well rearrange the order of the observations to write

$$
\begin{bmatrix} y_7 \\ y_8 \\ y_9 \\ y_1 \\ y_2 \\ y_3 \\ y_4 \\ y_4 \\ y_6 \\ y_{10} \\ y_{11} \\ y_{12} \end{bmatrix}
=
\begin{bmatrix} 1 & 63 \\ 1 & 63 \\ 1 & 63 \\ 1 & 65 \\ 1 & 65 \\ 1 & 65 \\ 1 & 65 \\ 1 & 66 \\ 1 & 66 \\ 1 & 72 \\ 1 & 72 \\ 1 & 72 \end{bmatrix}
\begin{bmatrix} \beta_0 \\ \beta_1 \end{bmatrix}
+
\begin{bmatrix} \varepsilon_7 \\ \varepsilon_8 \\ \varepsilon_9 \\ \varepsilon_1 \\ \varepsilon_2 \\ \varepsilon_3 \\ \varepsilon_4 \\ \varepsilon_5 \\ \varepsilon_6 \\ \varepsilon_{10} \\ \varepsilon_{11} \\ \varepsilon_{12} \end{bmatrix}
$$

in which the x_i values are ordered from smallest to largest.   □

11.2.2   The general linear model

The general linear model is a generalization of the matrix form for the simple linear regression model. The general linear model is

$$Y = X\beta + e, \quad E(e) = 0, \quad Cov(e) = \sigma^2 I.$$

Y is an $n \times 1$ vector of observable random variables. X is an $n \times p$ matrix of known constants. β is a $p \times 1$ vector of unknown (regression) parameters. e is an $n \times 1$ vector of unobservable random errors. It will be assumed that $n \geq p$. Regression is any general linear model where the rank of X is p. In a general linear model, the number of functionally distinct mean parameters is the rank of X, cf. Section 3.1.

EXAMPLE 11.2.2.   *Multiple regression*
In non-matrix form, the multiple regression model is

$$y_i = \beta_0 + \beta_1 x_{i1} + \beta_2 x_{i2} + \cdots + \beta_{p-1} x_{i,p-1} + \varepsilon_i, \quad i = 1, \ldots, n, \qquad (11.2.2)$$

where

$$E(\varepsilon_i) = 0, \quad Var(\varepsilon_i) = \sigma^2, \quad Cov(\varepsilon_i, \varepsilon_j) = 0, \ i \neq j.$$

In matrix terms this can be written as

$$
\begin{bmatrix} y_1 \\ y_2 \\ \vdots \\ y_n \end{bmatrix}
=
\begin{bmatrix}
1 & x_{11} & x_{12} & \cdots & x_{1,p-1} \\
1 & x_{21} & x_{22} & \cdots & x_{2,p-1} \\
\vdots & \vdots & \vdots & \ddots & \vdots \\
1 & x_{n1} & x_{n2} & \cdots & x_{n,p-1}
\end{bmatrix}
\begin{bmatrix} \beta_0 \\ \beta_1 \\ \beta_2 \\ \vdots \\ \beta_{p-1} \end{bmatrix}
+
\begin{bmatrix} \varepsilon_1 \\ \varepsilon_2 \\ \vdots \\ \varepsilon_n \end{bmatrix}.
$$

$$Y_{n\times 1} \qquad = \qquad X_{n\times p} \qquad\qquad \beta_{p\times 1} \qquad + \qquad e_{n\times 1}$$

Multiplying and adding the right-hand side gives

$$
\begin{bmatrix} y_1 \\ y_2 \\ \vdots \\ y_n \end{bmatrix}
=
\begin{bmatrix}
\beta_0 + \beta_1 x_{11} + \beta_2 x_{12} + \cdots + \beta_{p-1} x_{1,p-1} + \varepsilon_1 \\
\beta_0 + \beta_1 x_{21} + \beta_2 x_{22} + \cdots + \beta_{p-1} x_{2,p-1} + \varepsilon_2 \\
\vdots \\
\beta_0 + \beta_1 x_{n1} + \beta_2 x_{n2} + \cdots + \beta_{p-1} x_{n,p-1} + \varepsilon_n
\end{bmatrix},
$$

which holds if and only if (11.2.2) holds. The conditions on the ε_is translate into

$$E(e) = 0,$$

where 0 is the $n \times 1$ matrix consisting of all zeros, and

$$Cov(e) = \sigma^2 I,$$

where I is the $n \times n$ identity matrix.   □

EXAMPLE 11.2.3.   In Example 11.2.1 we illustrated the matrix form of a SLR using the data on heights and weights. We now illustrate some of the models from Chapter 8 applied to these data.
    The cubic model

$$y_i = \beta_0 + \beta_1 x_i + \beta_2 x_i^2 + \beta_3 x_i^3 + \varepsilon_i \qquad (11.2.3)$$

is

$$
\begin{bmatrix} y_1 \\ y_2 \\ y_3 \\ y_4 \\ y_5 \\ y_6 \\ y_7 \\ y_8 \\ y_9 \\ y_{10} \\ y_{11} \\ y_{12} \end{bmatrix}
=
\begin{bmatrix}
1 & 65 & 65^2 & 65^3 \\
1 & 65 & 65^2 & 65^3 \\
1 & 65 & 65^2 & 65^3 \\
1 & 65 & 65^2 & 65^3 \\
1 & 66 & 66^2 & 66^3 \\
1 & 66 & 66^2 & 66^3 \\
1 & 63 & 63^2 & 63^3 \\
1 & 63 & 63^2 & 63^3 \\
1 & 63 & 63^2 & 63^3 \\
1 & 72 & 72^2 & 72^3 \\
1 & 72 & 72^2 & 72^3 \\
1 & 72 & 72^2 & 72^3
\end{bmatrix}
\begin{bmatrix} \beta_0 \\ \beta_1 \\ \beta_2 \\ \beta_3 \\ \beta_4 \end{bmatrix}
+
\begin{bmatrix} \varepsilon_1 \\ \varepsilon_2 \\ \varepsilon_3 \\ \varepsilon_4 \\ \varepsilon_5 \\ \varepsilon_6 \\ \varepsilon_7 \\ \varepsilon_8 \\ \varepsilon_9 \\ \varepsilon_{10} \\ \varepsilon_{11} \\ \varepsilon_{12} \end{bmatrix}.
$$

Some of the numbers in X are getting quite large, i.e., $65^3 = 274{,}625$. The model has better numerical properties if we compute $\bar{x}. = 69.41666$ and replace Model (11.2.3) with the equivalent model

$$
y_i = \gamma_0 + \gamma_1(x_i - \bar{x}.) + \gamma_2(x_i - \bar{x}.)^2 + \beta_3(x_i - \bar{x}.)^3 + \varepsilon_i
$$

and its matrix form

$$
\begin{bmatrix} y_1 \\ y_2 \\ y_3 \\ y_4 \\ y_5 \\ y_6 \\ y_7 \\ y_8 \\ y_9 \\ y_{10} \\ y_{11} \\ y_{12} \end{bmatrix}
=
\begin{bmatrix}
1 & (65 - \bar{x}.) & (65 - \bar{x}.)^2 & (65 - \bar{x}.)^3 \\
1 & (65 - \bar{x}.) & (65 - \bar{x}.)^2 & (65 - \bar{x}.)^3 \\
1 & (65 - \bar{x}.) & (65 - \bar{x}.)^2 & (65 - \bar{x}.)^3 \\
1 & (65 - \bar{x}.) & (65 - \bar{x}.)^2 & (65 - \bar{x}.)^3 \\
1 & (66 - \bar{x}.) & (66 - \bar{x}.)^2 & (66 - \bar{x}.)^3 \\
1 & (66 - \bar{x}.) & (66 - \bar{x}.)^2 & (66 - \bar{x}.)^3 \\
1 & (63 - \bar{x}.) & (63 - \bar{x}.)^2 & (63 - \bar{x}.)^3 \\
1 & (63 - \bar{x}.) & (63 - \bar{x}.)^2 & (63 - \bar{x}.)^3 \\
1 & (63 - \bar{x}.) & (63 - \bar{x}.)^2 & (63 - \bar{x}.)^3 \\
1 & (72 - \bar{x}.) & (72 - \bar{x}.)^2 & (72 - \bar{x}.)^3 \\
1 & (72 - \bar{x}.) & (72 - \bar{x}.)^2 & (72 - \bar{x}.)^3 \\
1 & (72 - \bar{x}.) & (72 - \bar{x}.)^2 & (72 - \bar{x}.)^3
\end{bmatrix}
\begin{bmatrix} \gamma_0 \\ \gamma_1 \\ \gamma_2 \\ \beta_3 \end{bmatrix}
+
\begin{bmatrix} \varepsilon_1 \\ \varepsilon_2 \\ \varepsilon_3 \\ \varepsilon_4 \\ \varepsilon_5 \\ \varepsilon_6 \\ \varepsilon_7 \\ \varepsilon_8 \\ \varepsilon_9 \\ \varepsilon_{10} \\ \varepsilon_{11} \\ \varepsilon_{12} \end{bmatrix}.
$$

This third-degree polynomial is the largest polynomial that we can fit to these data. Two points determine a line, three points determine a quadratic, and with only four district x values in the data, we cannot fit a model greater than a cubic.

Define $\tilde{x} = (x - 63)/9$ so that

$$
(x_1, \ldots, x_{12}) = (65, 65, 65, 65, 66, 66, 63, 63, 63, 72, 72, 72)
$$

transforms to

$$
(\tilde{x}_1, \ldots, \tilde{x}_{12}) = (2/9, 2/9, 2/9, 2/9, 1/3, 1/3, 0, 0, 0, 1, 1, 1).
$$

The basis function model based on cosines

$$
y_i = \beta_0 + \beta_1 x_i + \beta_2 \cos(\pi \tilde{x}_i) + \beta_3 \cos(\pi 2 \tilde{x}_i) + \varepsilon_i
$$

becomes

$$
\begin{bmatrix} y_1 \\ y_2 \\ y_3 \\ y_4 \\ y_5 \\ y_6 \\ y_7 \\ y_8 \\ y_9 \\ y_{10} \\ y_{11} \\ y_{12} \end{bmatrix}
=
\begin{bmatrix}
1 & 65 & \cos(2\pi/9) & \cos(4\pi/9) \\
1 & 65 & \cos(2\pi/9) & \cos(4\pi/9) \\
1 & 65 & \cos(2\pi/9) & \cos(4\pi/9) \\
1 & 65 & \cos(2\pi/9) & \cos(4\pi/9) \\
1 & 66 & \cos(\pi/3) & \cos(2\pi/3) \\
1 & 66 & \cos(\pi/3) & \cos(2\pi/3) \\
1 & 63 & \cos(0) & \cos(0) \\
1 & 63 & \cos(0) & \cos(0) \\
1 & 63 & \cos(0) & \cos(0) \\
1 & 72 & \cos(\pi) & \cos(2\pi) \\
1 & 72 & \cos(\pi) & \cos(2\pi) \\
1 & 72 & \cos(\pi) & \cos(2\pi)
\end{bmatrix}
\begin{bmatrix} \beta_0 \\ \beta_1 \\ \beta_2 \\ \beta_3 \end{bmatrix}
+
\begin{bmatrix} \varepsilon_1 \\ \varepsilon_2 \\ \varepsilon_3 \\ \varepsilon_4 \\ \varepsilon_5 \\ \varepsilon_6 \\ \varepsilon_7 \\ \varepsilon_8 \\ \varepsilon_9 \\ \varepsilon_{10} \\ \varepsilon_{11} \\ \varepsilon_{12} \end{bmatrix} .
$$

The "Haar wavelet" model

$$
y_i = \beta_0 + \beta_1 x_i + \beta_2 I_{[0,.50)}(\tilde{x}_i) + \beta_3 I_{[.5,1]}(\tilde{x}_i) + \varepsilon_i
$$

becomes

$$
\begin{bmatrix} y_1 \\ y_2 \\ y_3 \\ y_4 \\ y_5 \\ y_6 \\ y_7 \\ y_8 \\ y_9 \\ y_{10} \\ y_{11} \\ y_{12} \end{bmatrix}
=
\begin{bmatrix}
1 & 65 & 1 & 0 \\
1 & 65 & 1 & 0 \\
1 & 65 & 1 & 0 \\
1 & 65 & 1 & 0 \\
1 & 66 & 1 & 0 \\
1 & 66 & 1 & 0 \\
1 & 63 & 1 & 0 \\
1 & 63 & 1 & 0 \\
1 & 63 & 1 & 0 \\
1 & 72 & 0 & 1 \\
1 & 72 & 0 & 1 \\
1 & 72 & 0 & 1
\end{bmatrix}
\begin{bmatrix} \beta_0 \\ \beta_1 \\ \beta_2 \\ \beta_3 \end{bmatrix}
+
\begin{bmatrix} \varepsilon_1 \\ \varepsilon_2 \\ \varepsilon_3 \\ \varepsilon_4 \\ \varepsilon_5 \\ \varepsilon_6 \\ \varepsilon_7 \\ \varepsilon_8 \\ \varepsilon_9 \\ \varepsilon_{10} \\ \varepsilon_{11} \\ \varepsilon_{12} \end{bmatrix} .
$$

Notice that the last two columns of the X matrix add up to a column of 1s, like the first column. This causes the rank of the 12×4 model matrix X to be only 3, so the model is not a regression model. Dropping either of the last two columns (or the first column) does not change the model in any meaningful way but makes the model a regression.

If we partition the SLR model into points below 65.5 and above 65.5, the matrix model becomes

$$
\begin{bmatrix} y_1 \\ y_2 \\ y_3 \\ y_4 \\ y_5 \\ y_6 \\ y_7 \\ y_8 \\ y_9 \\ y_{10} \\ y_{11} \\ y_{12} \end{bmatrix}
=
\begin{bmatrix}
1 & 65 & 0 & 0 \\
1 & 65 & 0 & 0 \\
1 & 65 & 0 & 0 \\
1 & 65 & 0 & 0 \\
0 & 0 & 1 & 66 \\
0 & 0 & 1 & 66 \\
1 & 63 & 0 & 0 \\
1 & 63 & 0 & 0 \\
1 & 63 & 0 & 0 \\
0 & 0 & 1 & 72 \\
0 & 0 & 1 & 72 \\
0 & 0 & 1 & 72
\end{bmatrix}
\begin{bmatrix} \beta_1 \\ \beta_2 \\ \beta_3 \\ \beta_4 \end{bmatrix}
+
\begin{bmatrix} \varepsilon_1 \\ \varepsilon_2 \\ \varepsilon_3 \\ \varepsilon_4 \\ \varepsilon_5 \\ \varepsilon_6 \\ \varepsilon_7 \\ \varepsilon_8 \\ \varepsilon_9 \\ \varepsilon_{10} \\ \varepsilon_{11} \\ \varepsilon_{12} \end{bmatrix} .
$$

Alternatively, we could rewrite the model as

$$
\begin{bmatrix} y_7 \\ y_8 \\ y_9 \\ y_1 \\ y_2 \\ y_3 \\ y_4 \\ y_4 \\ y_6 \\ y_{10} \\ y_{11} \\ y_{12} \end{bmatrix}
=
\begin{bmatrix}
1 & 63 & 0 & 0 \\
1 & 63 & 0 & 0 \\
1 & 63 & 0 & 0 \\
1 & 65 & 0 & 0 \\
1 & 65 & 0 & 0 \\
1 & 65 & 0 & 0 \\
1 & 65 & 0 & 0 \\
0 & 0 & 1 & 66 \\
0 & 0 & 1 & 66 \\
0 & 0 & 1 & 72 \\
0 & 0 & 1 & 72 \\
0 & 0 & 1 & 72
\end{bmatrix}
\begin{bmatrix} \beta_1 \\ \beta_2 \\ \beta_3 \\ \beta_4 \end{bmatrix}
+
\begin{bmatrix} \varepsilon_7 \\ \varepsilon_8 \\ \varepsilon_9 \\ \varepsilon_1 \\ \varepsilon_2 \\ \varepsilon_3 \\ \varepsilon_4 \\ \varepsilon_5 \\ \varepsilon_6 \\ \varepsilon_{10} \\ \varepsilon_{11} \\ \varepsilon_{12} \end{bmatrix}.
$$

This makes it a bit clearer that we are fitting a SLR to the points with small x values and a separate SLR to cases with large x values. The pattern of 0s in the X matrix ensure that the small x values only involve the intercept and slope parameters β_1 and β_2 for the line on the first partition set and that the large x values only involve the intercept and slope parameters β_3 and β_4 for the line on the second partition set.

Fitting this model can also be accomplished by fitting the model

$$
\begin{bmatrix} y_7 \\ y_8 \\ y_9 \\ y_1 \\ y_2 \\ y_3 \\ y_4 \\ y_4 \\ y_6 \\ y_{10} \\ y_{11} \\ y_{12} \end{bmatrix}
=
\begin{bmatrix}
1 & 63 & 0 & 0 \\
1 & 63 & 0 & 0 \\
1 & 63 & 0 & 0 \\
1 & 65 & 0 & 0 \\
1 & 65 & 0 & 0 \\
1 & 65 & 0 & 0 \\
1 & 65 & 0 & 0 \\
1 & 66 & 1 & 66 \\
1 & 66 & 1 & 66 \\
1 & 72 & 1 & 72 \\
1 & 72 & 1 & 72 \\
1 & 72 & 1 & 72
\end{bmatrix}
\begin{bmatrix} \beta_0 \\ \beta_1 \\ \gamma_0 \\ \gamma_1 \end{bmatrix}
+
\begin{bmatrix} \varepsilon_7 \\ \varepsilon_8 \\ \varepsilon_9 \\ \varepsilon_1 \\ \varepsilon_2 \\ \varepsilon_3 \\ \varepsilon_4 \\ \varepsilon_5 \\ \varepsilon_6 \\ \varepsilon_{10} \\ \varepsilon_{11} \\ \varepsilon_{12} \end{bmatrix}.
$$

Here we have changed the first two columns to make them agree with the SLR of Example 11.2.1. However, notice that if we subtract the third column from the first column we get the first column of the previous version. Similarly, if we subtract the fourth column from the second column we get the second column of the previous version. This model has intercept and slope parameters β_0 and β_1 for the first partition and intercept and slope parameters $(\beta_0 + \gamma_0)$ and $(\beta_1 + \gamma_1)$ for the second partition.

Because of the particular structure of these data with 12 observations but only four distinct values of x, except for the Haar wavelet model, all of these models are equivalent to one another and all of them are equivalent to a model with the matrix formulation

$$
\begin{bmatrix} y_1 \\ y_2 \\ y_3 \\ y_4 \\ y_5 \\ y_6 \\ y_7 \\ y_8 \\ y_9 \\ y_{10} \\ y_{11} \\ y_{12} \end{bmatrix}
=
\begin{bmatrix}
1 & 0 & 0 & 0 \\
1 & 0 & 0 & 0 \\
1 & 0 & 0 & 0 \\
1 & 0 & 0 & 0 \\
0 & 1 & 0 & 0 \\
0 & 1 & 0 & 0 \\
0 & 0 & 1 & 0 \\
0 & 0 & 1 & 0 \\
0 & 0 & 1 & 0 \\
0 & 0 & 0 & 1 \\
0 & 0 & 0 & 1 \\
0 & 0 & 0 & 1
\end{bmatrix}
\begin{bmatrix} \beta_0 \\ \beta_1 \\ \beta_2 \\ \beta_3 \end{bmatrix}
+
\begin{bmatrix} \varepsilon_1 \\ \varepsilon_2 \\ \varepsilon_3 \\ \varepsilon_4 \\ \varepsilon_5 \\ \varepsilon_6 \\ \varepsilon_7 \\ \varepsilon_8 \\ \varepsilon_9 \\ \varepsilon_{10} \\ \varepsilon_{11} \\ \varepsilon_{12} \end{bmatrix} .
$$

The models are equivalent in that they all give the same fitted values, residuals, and degrees of freedom for error. We will see in the next chapter that this last matrix model has the form of a one-way analysis of variance model. $\qquad\square$

Other models to be discussed later such as analysis of variance and analysis of covariance models can also be written as general linear models. However, they are frequently not regression models in that they frequently have the rank of X less than the number of columns p.

11.3   Least squares estimation of regression parameters

The regression estimates given by standard computer programs are least squares estimates. For simple linear regression, the least squares estimates are the values of β_0 and β_1 that minimize

$$
\sum_{i=1}^{n} (y_i - \beta_0 - \beta_1 x_i)^2 . \tag{11.3.1}
$$

For multiple regression, the least squares estimates of the β_js minimize

$$
\sum_{i=1}^{n} (y_i - \beta_0 - \beta_1 x_{i1} - \beta_2 x_{i2} - \cdots - \beta_{p-1} x_{i,p-1})^2 .
$$

In matrix terms these can both be written as minimizing

$$
(Y - X\beta)'(Y - X\beta). \tag{11.3.2}
$$

The form in (11.3.2) is just the sum of the squares of the elements in the vector $(Y - X\beta)$. See also Exercise 11.7.1.

We now give the general form for the least squares estimate of β in regression problems.

Proposition 11.3.1.   If $r(X) = p$, then $\hat{\beta} = (X'X)^{-1}X'Y$ is the least squares estimate of β.

PROOF: *The proof is optional material.*
    Note that $(X'X)^{-1}$ exists only because in a regression problem the rank of X is p. The proof stems from rewriting the function to be minimized.

$$
\begin{aligned}
(Y - X\beta)'(Y - X\beta) &= \left(Y - X\hat{\beta} + X\hat{\beta} - X\beta\right)' \left(Y - X\hat{\beta} + X\hat{\beta} - X\beta\right) \tag{11.3.3} \\
&= \left(Y - X\hat{\beta}\right)' \left(Y - X\hat{\beta}\right) + \left(Y - X\hat{\beta}\right)' \left(X\hat{\beta} - X\beta\right) \\
&\quad + \left(X\hat{\beta} - X\beta\right)' \left(Y - X\hat{\beta}\right) + \left(X\hat{\beta} - X\beta\right)' \left(X\hat{\beta} - X\beta\right).
\end{aligned}
$$

Consider one of the two cross-product terms from the last expression, say $\left(X\hat{\beta} - X\beta\right)'\left(Y - X\hat{\beta}\right)$. Using the definition of $\hat{\beta}$ given in the proposition,

$$
\begin{aligned}
\left(X\hat{\beta} - X\beta\right)'\left(Y - X\hat{\beta}\right) &= \left[X\left(\hat{\beta} - \beta\right)\right]'\left(Y - X\hat{\beta}\right) \\
&= \left(\hat{\beta} - \beta\right)'X'\left(Y - X(X'X)^{-1}X'Y\right) \\
&= \left(\hat{\beta} - \beta\right)'X'\left(I - X(X'X)^{-1}X'\right)Y
\end{aligned}
$$

but

$$
X'\left(I - X(X'X)^{-1}X'\right) = X' - (X'X)(X'X)^{-1}X' = X' - X' = 0.
$$

Thus

$$
\left(X\hat{\beta} - X\beta\right)'\left(Y - X\hat{\beta}\right) = 0
$$

and similarly

$$
\left(Y - X\hat{\beta}\right)'\left(X\hat{\beta} - X\beta\right) = 0.
$$

Eliminating the two cross-product terms in (11.3.3) gives

$$
(Y - X\beta)'(Y - X\beta) = \left(Y - X\hat{\beta}\right)'\left(Y - X\hat{\beta}\right) + \left(X\hat{\beta} - X\beta\right)'\left(X\hat{\beta} - X\beta\right).
$$

This form is easily minimized. The first of the terms on the right-hand side does not depend on β, so the β that minimizes $(Y - X\beta)'(Y - X\beta)$ is the β that minimizes the second term $\left(X\hat{\beta} - X\beta\right)'\left(X\hat{\beta} - X\beta\right)$. The second term is non-negative because it is the sum of squares of the elements in the vector $X\hat{\beta} - X\beta$ and it is minimized by making it zero. This is accomplished by choosing $\beta = \hat{\beta}$. $\square$

EXAMPLE 11.3.2. *Simple linear regression*
We now show that Proposition 11.3.1 gives the usual estimates for simple linear regression. Readers should refamiliarize themselves with the results in Section 6.10. They should also be warned that the algebra in the first half of the example is a bit more sophisticated than that used elsewhere in this book.

Assume the model

$$
y_i = \beta_0 + \beta_1 x_i + \varepsilon_i \quad i = 1, \dots, n.
$$

and write

$$
X = \begin{bmatrix} 1 & x_1 \\ 1 & x_2 \\ \vdots & \vdots \\ 1 & x_n \end{bmatrix}
$$

so

$$
X'X = \begin{bmatrix} n & \sum_{i=1}^{n} x_i \\ \sum_{i=1}^{n} x_i & \sum_{i=1}^{n} x_i^2 \end{bmatrix}.
$$

Inverting this matrix gives

$$
(X'X)^{-1} = \frac{1}{n\sum_{i=1}^{n} x_i^2 - \left(\sum_{i=1}^{n} x_i\right)^2} \begin{bmatrix} \sum_{i=1}^{n} x_i^2 & -\sum_{i=1}^{n} x_i \\ -\sum_{i=1}^{n} x_i & n \end{bmatrix}.
$$

The denominator in this term can be simplified by observing that

$$
n\sum_{i=1}^{n} x_i^2 - \left(\sum_{i=1}^{n} x_i\right)^2 = n\left(\sum_{i=1}^{n} x_i^2 - n\bar{x}_{\cdot}^2\right) = n\sum_{i=1}^{n} (x_i - \bar{x}_{\cdot})^2.
$$

Note also that

$$X'Y = \begin{bmatrix} \sum_{i=1}^{n} y_i \\ \sum_{i=1}^{n} x_i y_i \end{bmatrix}.$$

Finally, we get

$$
\begin{aligned}
\hat{\beta} &= (X'X)^{-1} X'Y \\
&= \frac{1}{n \sum_{i=1}^{n} (x_i - \bar{x}.)^2} \begin{bmatrix} \sum_{i=1}^{n} x_i^2 \sum_{i=1}^{n} y_i - \sum_{i=1}^{n} x_i \sum_{i=1}^{n} x_i y_i \\ -\sum_{i=1}^{n} x_i \sum_{i=1}^{n} y_i + n \sum_{i=1}^{n} x_i y_i \end{bmatrix} \\
&= \frac{1}{\sum_{i=1}^{n} (x_i - \bar{x}.)^2} \begin{bmatrix} \bar{y} \sum_{i=1}^{n} x_i^2 - \bar{x}. \sum_{i=1}^{n} x_i y_i \\ (\sum_{i=1}^{n} x_i y_i) - n \bar{x}. \bar{y} \end{bmatrix} \\
&= \frac{1}{\sum_{i=1}^{n} (x_i - \bar{x}.)^2} \begin{bmatrix} \bar{y} \sum_{i=1}^{n} x_i^2 - n \bar{x}.^2 \bar{y} - \{ \bar{x}. (\sum_{i=1}^{n} x_i y_i) - (n \bar{x}.^2 \bar{y}) \} \\ \hat{\beta}_1 \sum_{i=1}^{n} (x_i - \bar{x}.)^2 \end{bmatrix} \\
&= \frac{1}{\sum_{i=1}^{n} (x_i - \bar{x}.)^2} \begin{bmatrix} \bar{y} (\sum_{i=1}^{n} x_i^2 - n \bar{x}.^2) - \bar{x}. (\sum_{i=1}^{n} x_i y_i - n \bar{x}. \bar{y}) \\ \hat{\beta}_1 \sum_{i=1}^{n} (x_i - \bar{x}.)^2 \end{bmatrix} \\
&= \begin{bmatrix} \bar{y} - \hat{\beta}_1 \bar{x}. \\ \hat{\beta}_1 \end{bmatrix} = \begin{bmatrix} \hat{\beta}_0 \\ \hat{\beta}_1 \end{bmatrix}.
\end{aligned}
$$

As usual, the alternative regression model

$$y_i = \beta_{*0} + \beta_1 (x_i - \bar{x}.) + \varepsilon_i \quad i = 1, \ldots, n$$

is easier to work with. Write the model in matrix form as

$$Y = Z \beta_* + e$$

where

$$Z = \begin{bmatrix} 1 & (x_1 - \bar{x}.) \\ 1 & (x_2 - \bar{x}.) \\ \vdots & \vdots \\ 1 & (x_n - \bar{x}.) \end{bmatrix}$$

and

$$\beta_* = \begin{bmatrix} \beta_{*0} \\ \beta_1 \end{bmatrix}.$$

We need to compute $\hat{\beta}_* = (Z'Z)^{-1} Z'Y$. Observe that

$$Z'Z = \begin{bmatrix} n & 0 \\ 0 & \sum_{i=1}^{n} (x_i - \bar{x}.)^2 \end{bmatrix},$$

$$(Z'Z)^{-1} = \begin{bmatrix} \frac{1}{n} & 0 \\ 0 & 1 \big/ \sum_{i=1}^{n} (x_i - \bar{x}.)^2 \end{bmatrix},$$

$$Z'Y = \begin{bmatrix} \sum_{i=1}^{n} y_i \\ \sum_{i=1}^{n} (x_i - \bar{x}.) y_i \end{bmatrix},$$

and

$$\hat{\beta}_* = (Z'Z)^{-1} Z'Y = \begin{bmatrix} \bar{y} \\ \sum_{i=1}^{n} (x_i - \bar{x}.) y_i \big/ \sum_{i=1}^{n} (x_i - \bar{x}.)^2 \end{bmatrix} = \begin{bmatrix} \hat{\beta}_{*0} \\ \hat{\beta}_1 \end{bmatrix}.$$

These are the usual estimates.   □

Recall that least squares estimates have a number of other properties. If the errors are independent with mean zero, constant variance, and are normally distributed, the least squares estimates are maximum likelihood estimates, cf. Subsection 23.2.2, and minimum variance unbiased estimates. If the errors are merely uncorrelated with mean zero and constant variance, the least squares estimates are best (minimum variance) linear unbiased estimates.

In multiple regression, simple algebraic expressions for the parameter estimates are not possible. The only nice equations for the estimates are the matrix equations.

We now find expected values and covariance matrices for the data Y and the least squares estimate $\hat{\beta}$. Two simple rules about expectations and covariance matrices can take one a long way in the theory of regression. These are matrix analogues of Proposition 1.2.11. In fact, to prove these matrix results, one really only needs Proposition 1.2.11, cf. Exercise 11.7.3.

Proposition 11.3.3. Let A be a fixed $r \times n$ matrix, let c be a fixed $r \times 1$ vector, and let Y be an $n \times 1$ random vector, then

1. $E(AY + c) = AE(Y) + c$

2. $\text{Cov}(AY + c) = A\text{Cov}(Y)A'$.

Applying these results allows us to find the expected value and covariance matrix for Y in a linear model. The linear model has $Y = X\beta + e$ where $X\beta$ is a fixed vector (even though β is unknown), $E(e) = 0$, and $\text{Cov}(e) = \sigma^2 I$. Applying the proposition gives

$$E(Y) = E(X\beta + e) = X\beta + E(e) = X\beta + 0 = X\beta$$

and

$$\text{Cov}(Y) = \text{Cov}(e) = \sigma^2 I.$$

We can also find the expected value and covariance matrix of the least squares estimate $\hat{\beta}$. In particular, we show that $\hat{\beta}$ is an *unbiased* estimate of β by showing

$$E\left(\hat{\beta}\right) = E\left((X'X)^{-1}X'Y\right) = (X'X)^{-1}X'E(Y) = (X'X)^{-1}X'X\beta = \beta.$$

To find variances and standard errors we need $\text{Cov}\left(\hat{\beta}\right)$. To obtain this matrix, we use the rules in Proposition A.7.1. In particular, recall that the inverse of a symmetric matrix is symmetric and that $X'X$ is symmetric.

$$
\begin{aligned}
\text{Cov}\left(\hat{\beta}\right) &= \text{Cov}\left[(X'X)^{-1}X'Y\right] \\
&= \left[(X'X)^{-1}X'\right]\text{Cov}(Y)\left[(X'X)^{-1}X'\right]' \\
&= \left[(X'X)^{-1}X'\right]\text{Cov}(Y)X\left[(X'X)^{-1}\right]' \\
&= (X'X)^{-1}X'\text{Cov}(Y)X(X'X)^{-1} \\
&= \sigma^2(X'X)^{-1}X'X(X'X)^{-1} \\
&= \sigma^2(X'X)^{-1}.
\end{aligned}
$$

EXAMPLE 11.3.2 CONTINUED. For simple linear regression the covariance matrix becomes

$$\text{Cov}\left(\hat{\beta}\right) = \sigma^2(X'X)^{-1}$$

$$= \sigma^2 \frac{1}{n\sum_{i=1}^{n}(x_i-\bar{x}_.)^2} \begin{bmatrix} \sum_{i=1}^{n}x_i^2 & -\sum_{i=1}^{n}x_i \\ -\sum_{i=1}^{n}x_i & n \end{bmatrix}$$

$$= \sigma^2 \frac{1}{n\sum_{i=1}^{n}(x_i-\bar{x}_.)^2} \begin{bmatrix} \sum_{i=1}^{n}x_i^2 - n\bar{x}_.^2 + n\bar{x}_.^2 & -n\bar{x}_. \\ -n\bar{x}_. & n \end{bmatrix}$$

$$= \sigma^2 \frac{1}{n\sum_{i=1}^{n}(x_i-\bar{x}_.)^2} \begin{bmatrix} \sum_{i=1}^{n}(x_i-\bar{x}_.)^2 + n\bar{x}_.^2 & -n\bar{x}_. \\ -n\bar{x}_. & n \end{bmatrix}$$

$$= \sigma^2 \begin{bmatrix} \frac{1}{n} + \frac{\bar{x}_.^2}{\sum_{i=1}^{n}(x_i-\bar{x}_.)^2} & \frac{-\bar{x}_.}{\sum_{i=1}^{n}(x_i-\bar{x}_.)^2} \\ \frac{-\bar{x}_.}{\sum_{i=1}^{n}(x_i-\bar{x}_.)^2} & \frac{1}{\sum_{i=1}^{n}(x_i-\bar{x}_.)^2} \end{bmatrix},$$

which agrees with results given earlier for simple linear regression.

11.4   Inferential procedures

We begin by examining the analysis of variance table for the regression model (11.2.2). We then discuss tests, confidence intervals, and prediction intervals.

There are two frequently used forms of the ANOVA table:

Source	df	SS	MS
β_0	1	$n\bar{y}^2 \equiv C$	$n\bar{y}^2$
Regression	$p-1$	$\hat{\beta}'X'X\hat{\beta}-C$	$SSReg/(p-1)$
Error	$n-p$	$Y'Y-C-SSReg$	$SSE/(n-p)$
Total	n	$Y'Y$	

and the more often used form

Source	df	SS	MS
Regression	$p-1$	$\hat{\beta}'X'X\hat{\beta}-C$	$SSReg/(p-1)$
Error	$n-p$	$Y'Y-C-SSReg$	$SSE/(n-p)$
Total	$n-1$	$Y'Y-C$	

Note that $Y'Y = \sum_{i=1}^{n}y_i^2$, $C = n\bar{y}^2 = \left(\sum_{i=1}^{n}y_i\right)^2/n$, and $\hat{\beta}'X'X\hat{\beta} = \hat{\beta}'X'Y$. The difference between the two tables is that the first includes a line for the intercept or grand mean while in the second the total has been corrected for the grand mean.

The coefficient of determination can be computed as

$$R^2 = \frac{SSReg}{Y'Y-C}.$$

This is the ratio of the variability explained by the predictor variables to the total variability of the data. Note that $(Y'Y-C)/(n-1) = s_y^2$, the sample variance of the ys without adjusting for any structure except the existence of a possibly nonzero mean.

EXAMPLE 11.4.1.   *Simple linear regression*
    For simple linear regression, we know that

$$SSReg = \hat{\beta}_1^2 \sum_{i=1}^{n}(x_i-\bar{x}_.)^2 = \hat{\beta}_1 \sum_{i=1}^{n}(x_i-\bar{x}_.)^2 \hat{\beta}_1.$$

We will examine the alternative model

$$y_i = \beta_{*0} + \beta_1(x_i - \bar{x}_.) + \varepsilon_i.$$

Note that $C = n\hat{\beta}_{*0}^2$, so the general form for *SSReg* reduces to the simple linear regression form because

$$
\begin{aligned}
SSReg &= \hat{\beta}_*'Z'Z\hat{\beta}_* - C \\
&= \begin{bmatrix} \hat{\beta}_{*0} \\ \hat{\beta}_1 \end{bmatrix}' \begin{bmatrix} n & 0 \\ 0 & \sum_{i=1}^n (x_i - \bar{x}.)^2 \end{bmatrix} \begin{bmatrix} \hat{\beta}_{*0} \\ \hat{\beta}_1 \end{bmatrix} - C \\
&= \hat{\beta}_1^2 \sum_{i=1}^n (x_i - \bar{x}.)^2 .
\end{aligned}
$$

The same result can be obtained from $\hat{\beta}'X'X\hat{\beta} - C$ but the algebra is more tedious. □

To obtain tests and confidence regions we need to make additional distributional assumptions. In particular, we assume that the y_is have independent normal distributions. Equivalently, we take

$$
\varepsilon_1, \ldots, \varepsilon_n \text{ indep. } N(0, \sigma^2).
$$

To test the hypothesis

$$
H_0 : \beta_1 = \beta_2 = \cdots = \beta_{p-1} = 0,
$$

use the analysis of variance table test statistic

$$
F = \frac{MSReg}{MSE}.
$$

Under H_0,

$$
F \sim F(p-1, n-p).
$$

We can also perform a variety of t tests for individual regression parameters β_k. The procedures fit into the general techniques of Chapter 3 based on identifying 1) the parameter, 2) the estimate, 3) the standard error of the estimate, and 4) the distribution of $(Est - Par)/SE(Est)$. The parameter of interest is β_k. Having previously established that

$$
E \begin{bmatrix} \hat{\beta}_0 \\ \hat{\beta}_1 \\ \vdots \\ \hat{\beta}_{p-1} \end{bmatrix} = \begin{bmatrix} \beta_0 \\ \beta_1 \\ \vdots \\ \beta_{p-1} \end{bmatrix},
$$

it follows that for any $k = 0, \ldots, p-1$,

$$
E\left(\hat{\beta}_k\right) = \beta_k.
$$

This shows that $\hat{\beta}_k$ is an unbiased estimate of β_k. Before obtaining the standard error of $\hat{\beta}_k$, it is necessary to identify its variance. The covariance matrix of $\hat{\beta}$ is $\sigma^2 (X'X)^{-1}$, so the variance of $\hat{\beta}_k$ is the $(k+1)$st diagonal element of $\sigma^2 (X'X)^{-1}$. The $(k+1)$st diagonal element is appropriate because the first diagonal element is the variance of $\hat{\beta}_0$ not $\hat{\beta}_1$. If we let a_k be the $(k+1)$st diagonal element of $(X'X)^{-1}$ and estimate σ^2 with *MSE*, we get a standard error for $\hat{\beta}_k$ of

$$
SE\left(\hat{\beta}_k\right) = \sqrt{MSE}\sqrt{a_k}.
$$

Under normal errors, the appropriate reference distribution is

$$
\frac{\hat{\beta}_k - \beta_k}{SE(\hat{\beta}_k)} \sim t(n-p).
$$

Standard techniques now provide tests and confidence intervals. For example, a 95% confidence interval for β_k has endpoints

$$\hat{\beta}_k \pm t(.975, n-p)\,\mathrm{SE}(\hat{\beta}_k)$$

where $t(.975, n-p)$ is the 97.5th percentile of a t distribution with $n-p$ degrees of freedom.

A $(1-\alpha)100\%$ simultaneous confidence region for $\beta_0, \beta_1, \ldots, \beta_{p-1}$ consists of all the β vectors that satisfy

$$\frac{\left(\hat{\beta}-\beta\right)' X'X \left(\hat{\beta}-\beta\right)/p}{MSE} \leq F(1-\alpha, p, n-p).$$

This region also determines joint $(1-\alpha)100\%$ confidence intervals for the individual β_ks with limits

$$\hat{\beta}_k \pm \sqrt{pF(1-\alpha, p, n-p)}\,\mathrm{SE}(\hat{\beta}_k).$$

These intervals are an application of Scheffé's method of multiple comparisons, cf. Section 13.3.

We can also use the Bonferroni method to obtain joint $(1-\alpha)100\%$ confidence intervals with limits

$$\hat{\beta}_k \pm t\left(1-\frac{\alpha}{2p}, n-p\right)\mathrm{SE}(\hat{\beta}_k).$$

Finally, we consider estimation of the point on the surface that corresponds to a given set of predictor variables and the prediction of a new observation with a given set of predictor variables. Let the predictor variables be $x_1, x_2, \ldots, x_{p-1}$. Combine these into the row vector

$$x' = (1, x_1, x_2, \ldots, x_{p-1}).$$

The point on the surface that we are trying to estimate is the parameter $x'\beta = \beta_0 + \sum_{j=1}^{p-1}\beta_j x_j$. The least squares estimate is $x'\hat{\beta}$, which can be thought of as a 1×1 matrix. The variance of the estimate is

$$\mathrm{Var}\left(x'\hat{\beta}\right) = \mathrm{Cov}\left(x'\hat{\beta}\right) = x'\mathrm{Cov}\left(\hat{\beta}\right)x = \sigma^2 x' (X'X)^{-1}x,$$

so the standard error is

$$\mathrm{SE}\left(x'\hat{\beta}\right) = \sqrt{MSE}\sqrt{x'(X'X)^{-1}x} \equiv \mathrm{SE}(Surface).$$

This is the standard error of the estimated regression surface. The appropriate reference distribution is

$$\frac{x'\hat{\beta} - x'\beta}{\mathrm{SE}\left(x'\hat{\beta}\right)} \sim t(n-p)$$

and a $(1-\alpha)100\%$ confidence interval has endpoints

$$x'\hat{\beta} \pm t\left(1-\frac{\alpha}{2}, n-p\right)\mathrm{SE}(x'\hat{\beta}).$$

When predicting a new observation, the point prediction is just the estimate of the point on the surface but the standard error must incorporate the additional variability associated with a new observation. The original observations were assumed to be independent with variance σ^2. It is reasonable to assume that a new observation is independent of the previous observations and has the same variance. Thus, in the prediction we have to account for the variance of the new observation, which is σ^2, plus the variance of the estimate $x'\hat{\beta}$, which is $\sigma^2 x' (X'X)^{-1}x$. This leads to a variance for the prediction of $\sigma^2 + \sigma^2 x' (X'X)^{-1}x$ and a standard error of

$$\sqrt{MSE + MSE\,x'(X'X)^{-1}x} = \sqrt{MSE\left[1 + x'(X'X)^{-1}x\right]} \equiv \mathrm{SE}(Prediction).$$

Note that

$$SE(Prediction) = \sqrt{MSE + [SE(Surface)]^2}.$$

The $(1-\alpha)100\%$ prediction interval has endpoints

$$x'\hat{\beta} \pm t\left(1-\frac{\alpha}{2}, n-p\right)\sqrt{MSE\left[1+x'(X'X)^{-1}x\right]}.$$

Results of this section constitute the theory behind most of the applications in Sections 9.1 and 9.2.

11.5 Residuals, standardized residuals, and leverage

Let $x_i' = (1, x_{i1}, \ldots, x_{i,p-1})$ be the ith row of X, then the ith fitted value is

$$\hat{y}_i = \hat{\beta}_0 + \hat{\beta}_1 x_{i1} + \cdots + \hat{\beta}_{p-1} x_{i,p-1} = x_i'\hat{\beta}$$

and the corresponding residual is

$$\hat{\varepsilon}_i = y_i - \hat{y}_i = y_i - x_i'\hat{\beta}.$$

The vector of predicted (fitted) values is

$$\hat{Y} = \begin{bmatrix} \hat{y}_1 \\ \vdots \\ \hat{y}_n \end{bmatrix} = \begin{bmatrix} x_1'\hat{\beta} \\ \vdots \\ x_n'\hat{\beta} \end{bmatrix} = X\hat{\beta}.$$

The vector of residuals is

$$\begin{aligned}
\hat{e} &= Y - \hat{Y} \\
&= Y - X\hat{\beta} \\
&= Y - X(X'X)^{-1}X'Y \\
&= \left(I - X(X'X)^{-1}X'\right)Y \\
&= (I - M)Y
\end{aligned}$$

where

$$M \equiv X(X'X)^{-1}X'.$$

M is called the perpendicular projection operator (matrix) onto $C(X)$, the column space of X. M is the key item in the analysis of the general linear model, cf. Christensen (2011). Note that M is symmetric, i.e., $M = M'$, and idempotent, i.e., $MM = M$, so it is a perpendicular projection operator as discussed in Appendix A. Using these facts, observe that

$$\begin{aligned}
SSE &= \sum_{i=1}^{n} \hat{\varepsilon}_i^2 \\
&= \hat{e}'\hat{e} \\
&= [(I - M)Y]'[(I - M)Y] \\
&= Y'(I - M' - M + M'M)Y \\
&= Y'(I - M)Y.
\end{aligned}$$

Another common way of writing SSE is

$$SSE = \left[Y - X\hat{\beta}\right]'\left[Y - X\hat{\beta}\right].$$

Having identified M, we can define the standardized residuals. First we find the covariance matrix of the residual vector $\hat{e}$:

$$
\begin{aligned}
\text{Cov}(\hat{e}) &= \text{Cov}([I-M]Y) \\
&= [I-M]\text{Cov}(Y)[I-M]' \\
&= [I-M]\sigma^2 I[I-M]' \\
&= \sigma^2(I-M-M'+MM') \\
&= \sigma^2(I-M).
\end{aligned}
$$

The last equality follows from $M = M'$ and $MM = M$. Typically, the covariance matrix is not diagonal, so the residuals are not uncorrelated.

The variance of a particular residual $\hat{\varepsilon}_i$ is σ^2 times the ith diagonal element of $(I-M)$. The ith diagonal element of $(I-M)$ is the ith diagonal element of I, 1, minus the ith diagonal element of M, say, m_{ii}. Thus

$$
\text{Var}(\hat{\varepsilon}_i) = \sigma^2(1-m_{ii})
$$

and the standard error of $\hat{\varepsilon}_i$ is

$$
\text{SE}(\hat{\varepsilon}_i) = \sqrt{MSE(1-m_{ii})}.
$$

The ith standardized residual is defined as

$$
r_i \equiv \frac{\hat{\varepsilon}_i}{\sqrt{MSE(1-m_{ii})}}.
$$

The leverage of the ith case is defined to be m_{ii}, the ith diagonal element of M. Some people like to think of M as the 'hat' matrix because it transforms Y into $\hat{Y}$, i.e., $\hat{Y} = X\hat{\beta} = MY$. More common than the name 'hat matrix' is the consequent use of the notation h_i for the ith leverage. This notation was used in Chapter 7 but the reader should realize that $h_i \equiv m_{ii}$. In any case, the leverage can be interpreted as a measure of how unusual x_i' is relative to the other rows of the X matrix, cf. Christensen (2011, Section 13.1).

Christensen (2011, Chapter 13) discusses the computation of standardized deleted residuals and Cook's distance.

11.6 Principal components regression

In Section 9.7 we dealt with the issue of collinearity. Four points were emphasized as the effects of collinearity.

1. The estimate of any parameter, say $\hat{\beta}_2$, depends on *all* the variables that are included in the model.

2. The sum of squares for any variable, say x_2, depends on *all* the other variables that are included in the model. For example, none of $SSR(x_2)$, $SSR(x_2|x_1)$, and $SSR(x_2|x_3,x_4)$ would typically be equal.

3. In a model such as $y_i = \beta_0 + \beta_1 x_{i1} + \beta_2 x_{i2} + \beta_3 x_{i3} + \varepsilon_i$, small t statistics for both $H_0 : \beta_1 = 0$ and $H_0 : \beta_2 = 0$ are not sufficient to conclude that an appropriate model is $y_i = \beta_0 + \beta_3 x_{i3} + \varepsilon_i$. To arrive at a reduced model, one must compare the reduced model to the full model.

4. A moderate amount of collinearity has little effect on predictions and therefore little effect on SSE, R^2, and the explanatory power of the model. Collinearity increases the variance of the $\hat{\beta}_j$s, making the estimates of the parameters less reliable. Depending on circumstances, sometimes a large amount of collinearity can have an effect on predictions. Just by chance one may get a better fit to the data than can be justified scientifically.

At its worst, collinearity involves near redundancies among the predictor variables. An exact redundancy among the predictor variables occurs when we can find a $p \times 1$ vector $d \neq 0$ so that

Table 11.2: *Eigen analysis of the correlation matrix.*

Eigenvalue	2.8368	1.3951	0.4966	0.2025	0.0689
Proportion	0.567	0.279	0.099	0.041	0.014
Cumulative	0.567	0.846	0.946	0.986	1.000

$Xd = 0$. When this happens the rank of X is not p, so we cannot find $(X'X)^{-1}$ and we cannot find the estimates of β in Proposition 11.3.1. *Near* redundancies occur when we can find a vector d that is not too small, say with $d'd = 1$, having $Xd \doteq 0$. Principal components (PC) regression is a method designed to identify near redundancies among the predictor variables. Having identified near redundancies, they can be eliminated if we so choose. In Section 10.7 we mentioned that having small collinearity requires more than having small correlations among all the predictor variables, it requires all partial correlations among the predictor variables to be small as well. For this reason, eliminating near redundancies cannot always be accomplished by simply dropping well-chosen predictor variables from the model.

The basic idea of principal components is to find new variables that are linear combinations of the x_js and that are *best able to (linearly) predict the entire set of x_js*; see Christensen (2001, Chapter 3). Thus the first principal component variable is the one linear combination of the x_js that is best able to predict all of the x_js. The second principal component variable is the linear combination of the x_js that is best able to predict all the x_js among those linear combinations having a sample correlation of 0 with the first principal component variable. The third principal component variable is the best predictor that has sample correlations of 0 with the first two principal component variables. The remaining principal components are defined similarly. With $p-1$ predictor variables, there are $p-1$ principal component variables. The full collection of principal component variables always predicts the full collection of x_js perfectly. The last few principal component variables are least able to predict the original x_j variables, so they are the least useful. They are also the aspects of the predictor variables that are most redundant; see Christensen (2011, Section 15.1). The best (linear) predictors used in defining principal components can be based on either the covariances between the x_js or the correlations between the x_js. Unless the x_js are measured on the same scale (with similarly sized measurements), it is generally best to use principal components defined using the correlations.

For *The Coleman Report* data, a matrix of sample correlations between the x_js was given in Example 9.7.1. Principal components are derived from the eigenvalues and eigenvectors of this matrix, cf. Section A.8. (Alternatively, one could use eigenvalues and eigenvectors of the matrix of sample covariances.) An eigenvector corresponding to the largest eigenvalue determines the first principal component variable.

The eigenvalues are given in Table 11.2 along with proportions and cumulative proportions. The proportions in Table 11.2 are simply the eigenvalues divided by the sum of the eigenvalues. The cumulative proportions are the sum of the first group of eigenvalues divided by the sum of all the eigenvalues. In this example, the sum of the eigenvalues is

$$5 = 2.8368 + 1.3951 + 0.4966 + 0.2025 + 0.0689.$$

The sum of the eigenvalues must equal the sum of the diagonal elements of the original matrix. The sum of the diagonal elements of a correlation matrix is the number of variables in the matrix. The third eigenvalue in Table 11.2 is .4966. The proportion is $.4966/5 = .099$. The cumulative proportion is $(2.8368 + 1.3951 + 0.4966)/5 = 0.946$. With an eigenvalue proportion of 9.9%, the third principal component variable accounts for 9.9% of the variance associated with predicting the x_js. Taken together, the first three principal components account for 94.6% of the variance associated with predicting the x_js because the third cumulative eigenvalue proportion is 0.946.

For the school data, the principal component (PC) variables are determined by the coefficients in Table 11.3. The first principal component variable is

Table 11.3: *Principal component variable coefficients.*

Variable	PC1	PC2	PC3	PC4	PC5
x_1	−0.229	−0.651	0.723	0.018	−0.024
x_2	−0.555	0.216	0.051	−0.334	0.729
x_3	−0.545	0.099	−0.106	0.823	−0.060
x_4	−0.170	−0.701	−0.680	−0.110	0.075
x_5	−0.559	0.169	−0.037	−0.445	−0.678

Table 11.4: *Table of Coefficients: Principal component regression.*

Predictor	$\hat{\gamma}$	SE($\hat{\gamma}$)	t	P
Constant	35.0825	0.4638	75.64	0.000
PC1	−2.9419	0.2825	−10.41	0.000
PC2	0.0827	0.4029	0.21	0.840
PC3	−2.0457	0.6753	−3.03	0.009
PC4	4.380	1.057	4.14	0.001
PC5	1.433	1.812	0.79	0.442

$$PC1_i = -0.229(x_{i1} - \bar{x}_{.1})/s_1 - 0.555(x_{i2} - \bar{x}_{.2})/s_2$$
$$- 0.545(x_{i3} - \bar{x}_{.3})/s_3 - 0.170(x_{i4} - \bar{x}_{.5})/s_4 - 0.559(x_{i5} - \bar{x}_{.5})/s_5 \quad (11.6.1)$$

for $i = 1, \ldots, 20$ where s_1 is the sample standard deviation of the x_{i1}s, etc. The columns of coefficients given in Table 11.3 are actually eigenvectors for the correlation matrix of the x_js. The PC1 coefficients are an eigenvector corresponding to the largest eigenvalue, the PC2 coefficients are an eigenvector corresponding to the second largest eigenvalue, etc.

We can now perform a regression on the new principal component variables. The table of coefficients is given in Table 11.4. The analysis of variance is given in Table 11.5. The value of R^2 is 0.906. The analysis of variance table and R^2 are identical to those for the original predictor variables given in Section 9.1. The plot of standardized residuals versus predicted values from the principal component regression is given in Figure 11.1. This is identical to the plot given in Figure 10.2 for the original variables. All of the predicted values and all of the standardized residuals are identical.

Since Table 11.5 and Figure 11.1 are unchanged, any usefulness associated with principal component regression must come from Table 11.4. The principal component variables display no collinearity. Thus, contrary to the warnings given earlier about the effects of collinearity, we can make final conclusions about the importance of variables directly from Table 11.4. We do not have to worry about fitting one model after another or about which variables are included in which models. From examining Table 11.4, it is clear that the important variables are PC1, PC3, and PC4. We can construct a reduced model with these three; the estimated regression surface is simply

$$\hat{y} = 35.0825 - 2.9419(PC1) - 2.0457(PC3) + 4.380(PC4), \quad (11.6.2)$$

where we merely used the estimated regression coefficients from Table 11.4. Refitting the reduced model is unnecessary because there is no collinearity.

Table 11.5: *Analysis of Variance: Principal component regression.*

Source	df	SS	MS	F	P
Regression	5	582.69	116.54	27.08	0.000
Error	14	60.24	4.30		
Total	19	642.92			

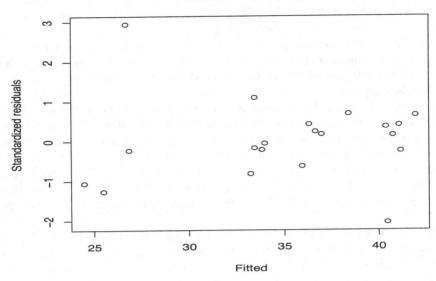

Residual–Fitted plot

Figure 11.1: *Standardized residuals versus predicted values for principal component regression.*

To get predictions for a new set of x_js, just compute the corresponding PC1, PC3, and PC4 variables using formulae similar to those in Equation (11.6.1) and make the predictions using the fitted model in Equation (11.6.2). When using equations like (11.6.1) to obtain new values of the principal component variables, continue to use the $\bar{x}_{.j}$s and s_js computed from only the original observations.

As an alternative to this prediction procedure, we could use the definitions of the principal component variables, e.g., Equation (11.6.1), and substitute for PC1, PC3, and PC4 in Equation (11.6.2) to obtain estimated coefficients on the original x_j variables.

$$\hat{y} = 35.0825 + [-2.9419, -2.0457, 4.380] \begin{bmatrix} \text{PC1} \\ \text{PC3} \\ \text{PC4} \end{bmatrix}$$

$$= 35.0825 + [-2.9419, -2.0457, 4.380] \times$$

$$\begin{bmatrix} -0.229 & -0.555 & -0.545 & -0.170 & -0.559 \\ 0.723 & 0.051 & -0.106 & -0.680 & -0.037 \\ 0.018 & -0.334 & 0.823 & -0.110 & -0.445 \end{bmatrix} \begin{bmatrix} (x_1 - \bar{x}_{.1})/s_1 \\ (x_2 - \bar{x}_{.2})/s_2 \\ (x_3 - \bar{x}_{.3})/s_3 \\ (x_4 - \bar{x}_{.4})/s_4 \\ (x_5 - \bar{x}_{.5})/s_5 \end{bmatrix}$$

$$= 35.0825 + [-0.72651, 0.06550, 5.42492, 1.40940, -0.22889] \times$$

$$\begin{bmatrix} (x_1 - 2.731)/0.454 \\ (x_2 - 40.91)/25.90 \\ (x_3 - 3.14)/9.63 \\ (x_4 - 25.069)/1.314 \\ (x_5 - 6.255)/0.654 \end{bmatrix}.$$

Obviously this can be simplified into a form $\hat{y} = \tilde{\beta}_0 + \tilde{\beta}_1 x_1 + \tilde{\beta}_2 x_2 + \tilde{\beta}_3 x_3 + \tilde{\beta}_4 x_4 + \tilde{\beta}_5 x_5$, which in turn simplifies the process of making predictions and provides new estimated regression coefficients for the x_js that correspond to the fitted principal component model. In this case they become $\hat{y} = 12.866 - 1.598x_1 + 0.002588x_2 + 0.5639x_3 + 1.0724x_4 - 0.3484x_5$. These PC regression estimates of the original β_js can be compared to the least squares estimates. Many computer programs for performing PC regression report these estimates of the β_js and their corresponding standard errors. *A similar method is used to obtain lasso estimates when the lasso procedure is performed on standardized predictor variables, cf. Section 10.5.*

It was mentioned earlier that collinearity tends to increase the variance of regression coefficients. The fact that the later principal component variables are more nearly redundant is reflected in Table 11.4 by the fact that the standard errors for their estimated regression coefficients increase (excluding the intercept).

One rationale for using PC regression is that you just don't believe in using nearly redundant variables. The exact nature of such variables can be changed radically by small errors in the x_js. For this reason, one might choose to ignore PC5 because of its small eigenvalue proportion, regardless of any importance it may display in Table 11.4. If the t statistic for PC5 appeared to be significant, it could be written off as a chance occurrence or, perhaps more to the point, as something that is unlikely to be reproducible. If you don't believe redundant variables, i.e., if you don't believe that they are themselves reproducible, any predictive ability due to such variables will not be reproducible either.

When considering PC5, the case is pretty clear. PC5 accounts for only about 1.5% of the variability involved in predicting the x_js. It is a very poorly defined aspect of the predictor variables x_j and, anyway, it is not a significant predictor of y. The case is less clear when considering PC4. This variable has a significant effect for explaining y, but it accounts for only 4% of the variability in predicting the x_js, so PC4 is reasonably redundant within the x_js. If this variable is measuring some reproducible aspect of the original x_j data, it should be included in the regression. If it is not reproducible, it should not be included. From examining the PC4 coefficients in Table 11.3, we see that PC4 is roughly the average of the percent white-collar fathers x_2 and the mothers' education x_5 contrasted with the socio-economic variable x_3. (Actually, this comparison is between the variables after they have been adjusted for their means and standard deviation as in Equation (11.6.1).) If PC4 strikes the investigator as a meaningful, reproducible variable, it should be included in the regression.

In our discussion, we have used PC regression both to eliminate questionable aspects of the predictor variables and as a method for selecting a reduced model. We dropped PC5 primarily because it was poorly defined. We dropped PC2 solely because it was not a significant predictor. Some people might argue against this second use of PC regression and choose to take a model based on PC1, PC2, PC3, and possibly PC4.

On occasion, PC regression is based on the sample covariance matrix of the x_js rather than the sample correlation matrix. Again, eigenvalues and eigenvectors are used, but in using relationships like Equation (11.6.1), the s_js are deleted. The eigenvalues and eigenvectors for the covariance matrix typically differ from those for the correlation matrix. The relationship between estimated principal component regression coefficients and original least squares regression coefficient estimates is somewhat simpler when using the covariance matrix.

It should be noted that PC regression is just as sensitive to violations of the assumptions as regular multiple regression. Outliers and high-leverage points can be very influential in determining the results of the procedure. Tests and confidence intervals rely on the independence, homoscedasticity, and normality assumptions. Recall that in the full principal components regression model, the residuals and predicted values are identical to those from the regression on the original predictor variables. Moreover, highly influential points in the original predictor variables typically have a large influence on the coefficients in the principal component variables.

11.7 Exercises

EXERCISE 11.7.1.   Show that the form (11.3.2) simplifies to the form (11.3.1) for simple linear regression.

EXERCISE 11.7.2.   Show that $\text{Cov}(Y) = E[(Y - \mu)(Y - \mu)']$.

EXERCISE 11.7.3. Use Proposition 1.2.11 to show that $E(AY + c) = A E(Y) + c$ and $Cov(AY + c) = ACov(Y)A'$.

EXERCISE 11.7.4. Using eigenvalues, discuss the level of collinearity in:

(a) the Younger data from Exercise 9.12.1,
(b) the Prater data from Exercise 9.12.3,
(c) the Chapman data of Exercise 9.12.4,
(d) the pollution data from Exercise 9.12.5,
(e) the body fat data of Exercise 9.12.6.

EXERCISE 11.7.5. Do a principal components regression on the Younger data from Exercise 9.12.1.

EXERCISE 11.7.6. Do a principal components regression on the Prater data from Exercise 9.12.3.

EXERCISE 11.7.7. Do a principal components regression on the Chapman data of Exercise 9.12.4.

EXERCISE 11.7.8. Do a principal components regression on the pollution data of Exercise 9.12.5.

EXERCISE 11.7.9. Do a principal components regression on the body fat data of Exercise 9.12.6.

Chapter 12

One-Way ANOVA

Analysis of variance (ANOVA) involves comparing random samples from several populations (groups). Often the samples arise from observing experimental units with different treatments applied to them and we refer to the populations as treatment groups. The sample sizes for the groups are possibly different, say, N_i and we assume that the samples are all independent. Moreover, we assume that each population has the same variance and is normally distributed. Assuming different means for each group we have a model

$$y_{ij} = \mu_i + \varepsilon_{ij}, \quad \varepsilon_{ij}\text{s independent } N(0, \sigma^2)$$

or, equivalently,

$$y_{ij}\text{s independent } N(\mu_i, \sigma^2),$$

where with a groups, $i = 1, \ldots, a$, and with N_i observations in the ith group, $j = 1, \ldots, N_i$. There is one mean parameter μ_i for each group and it is estimated by the sample mean of the group, say, $\bar{y}_{i\cdot}$. Relating this model to the general models of Section 3.9, we have replaced the single subscript h that identifies all observations with a double subscript ij in which i identifies a group and j identifies an observation within the group. The group identifier i is our (categorical) predictor variable. The fitted values are $\hat{y}_h \equiv \hat{y}_{ij} = \bar{y}_{i\cdot}$, i.e., the point prediction we make for any observation is just the sample mean from the observation's group. The residuals are $\hat{\varepsilon}_h \equiv \hat{\varepsilon}_{ij} = y_{ij} - \bar{y}_{i\cdot}$. The total sample size is $n = N_1 + \cdots + N_a$. The model involves estimating a mean values, one for each group, so $dfE = n - a$. The SSE is

$$SSE = \sum_{h=1}^{n} \hat{\varepsilon}_h^2 = \sum_{1=1}^{a}\sum_{j=1}^{N_i} \hat{\varepsilon}_{ij}^2,$$

and the MSE is SSE/dfE.

12.1 Example

EXAMPLE 12.1.1. Table 12.1 gives data from Koopmans (1987, p. 409) on the ages at which suicides were committed in Albuquerque during 1978. Ages are listed by ethnic group. The data are plotted in Figure 12.1. The assumption is that the observations in each group are a random sample from some population. While it is not clear what these populations would be, we proceed to examine the data. Note that there are fewer Native Americans in the study than either Hispanics or non-Hispanic Caucasians (Anglos); moreover the ages for Native Americans seem to be both lower and less variable than for the other groups. The ages for Hispanics seem to be a bit lower than for non-Hispanic Caucasians. Summary statistics follow for the three groups.

Sample statistics: Suicide ages

Group	N_i	$\bar{y}_{i\cdot}$	s_i^2	s_i
Caucasians	44	41.66	282.9	16.82
Hispanics	34	35.06	268.3	16.38
Native Am.	15	25.07	74.4	8.51

Table 12.1: *Suicide ages.*

Non-Hispanic Caucasians				Hispanics			Native Americans	
21	31	28	52	50	27	45	26	23
55	31	24	27	31	22	57	17	25
42	32	53	76	29	20	22	24	23
25	43	66	44	21	51	48	22	22
48	57	90	35	27	60	48	16	
22	42	27	32	34	15	14	21	
42	34	48	26	76	19	52	36	
53	39	47	51	35	24	29	18	
21	24	49	19	55	24	21	48	
21	79	53	27	24	18	28	20	
31	46	62	58	68	43	17	35	
				38				

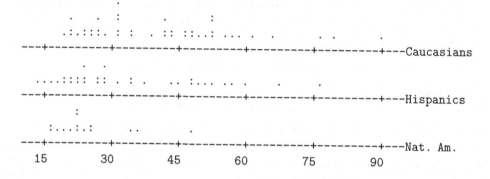

Figure 12.1: *Dot plots of suicide age data.*

The sample standard deviation for the Native Americans is about half the size of the others.

To evaluate the combined normality of the data, we did a normal plot of the standardized residuals. One normal plot for all of the y_{ij}s would not be appropriate because they have different means, μ_i. The residuals adjust for the different means. Of course with the reasonably large samples available here for each group, it would be permissible to do three separate normal plots, but in other situations with small samples for each group, individual normal plots would not contain enough observations to be of much value. The normal plot for the standardized residuals is given as Figure 12.2. The plot is based on $n = 44 + 34 + 15 = 93$ observations. This is quite a large number, so if the data are normal the plot should be quite straight. In fact, the plot seems reasonably curved.

In order to improve the quality of the assumptions of equal variances and normality, we consider transformations of the data. In particular, consider taking the log of each observation. Figure 12.3 contains the plot of the transformed data. The variability in the groups seems more nearly the same. This is confirmed by the following sample statistics.

Sample statistics: Log of suicide ages

Group	N_i	$\bar{y}_{i\cdot}$	s_i^2	s_i
Caucasians	44	3.6521	0.1590	0.3987
Hispanics	34	3.4538	0.2127	0.4612
Native Am.	15	3.1770	0.0879	0.2965

The largest sample standard deviation is only about 1.5 times the smallest. The normal plot of standardized residuals for the transformed data is given in Figure 12.4; it seems considerably straighter than the normal plot for the untransformed data.

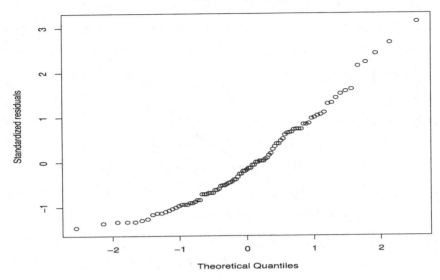

Figure 12.2: *Normal plot of suicide residuals, W' = 0.945.*

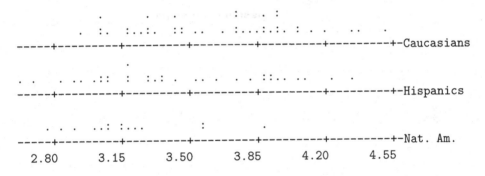

Figure 12.3: *Dotplots of log suicide age data.*

All in all, the logs of the original data seem to satisfy the assumptions reasonably well and considerably better than the untransformed data. The square roots of the data were also examined as a possible transformation. While the square roots seem to be an improvement over the original scale, they do not seem to satisfy the assumptions nearly as well as the log transformed data.

A basic assumption in analysis of variance is that the variance is the same for all groups. Although we can find the *MSE* as the sum of the squared residuals divided by the degrees of freedom for error, equivalently, as we did for two independent samples with the same variance, we can also compute it as a pooled estimate of the variance. This is a weighted average of the variance estimates from the individual groups with weights that are the individual degrees of freedom. For the logs of the suicide age data, the mean squared error is

$$MSE = \frac{(44-1)(0.1590) + (34-1)(0.2127) + (15-1)(0.0879)}{(44-1) + (34-1) + (15-1)} = 0.168.$$

The degrees of freedom for this estimate are the sum of the degrees of freedom for the individual variance estimates, s_i^2, so the degrees of freedom for error are

$$dfE = (44-1) + (34-1) + (15-1) = (44 + 34 + 15) - 3 = 90.$$

Normal Q–Q Plot

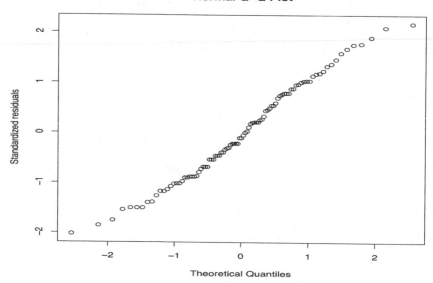

Figure 12.4: *Normal plot of suicide residuals, log data, $W' = 0.986$.*

Residual–Fitted plot

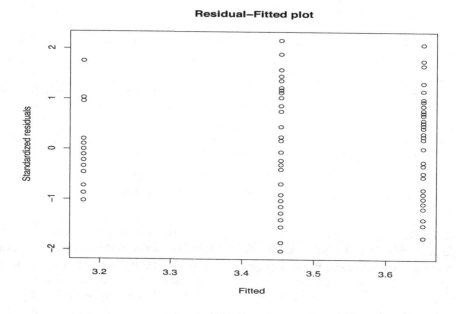

Figure 12.5: *Suicide residuals versus fitted values, log data.*

This is also the total number of observations, $n = 93$, minus the number of mean parameters we have to estimate, $a = 3$. The data have an approximate normal distribution, so we can use $t(90)$ as the reference distribution for statistical inferences on a single parameter. The sum of squares error is $SSE \equiv dfE \times MSE$.

For completeness, we also include the residual-fitted value plot as Figure 12.5.

We can now perform statistical inferences for a variety of parameters using our standard procedure involving a *Par*, an *Est*, a SE(*Est*), and a $t(dfE)$ distribution for $[Est - Par]/\text{SE}(Est)$. In this example, perhaps the most useful things to look at are whether there is evidence of any age differences in the three groups. Let $\mu_C \equiv \mu_1$, $\mu_H \equiv \mu_2$, and $\mu_N \equiv \mu_3$ denote the population means for the log ages of the non-Hispanic Caucasian (Anglo), Hispanic, and Native American groups, respectively. First, we briefly consider inferences for one of the group means. Our most lengthy discussion is for differences between group means. We then discuss more complicated linear functions of the group means. Finally, we discuss testing $\mu_C = \mu_H = \mu_N$.

12.1.1 Inferences on a single group mean

In constructing confidence intervals, prediction intervals, or tests for an individual mean μ_i, we use the methods of Chapter 2 except that the variance is estimated with MSE so that the reference distribution is the $t(dfE)$. In particular we might choose $Par = \mu_H \equiv \mu_2$, $Est = \bar{y}_2. =$ 3.4538, $\mathrm{SE}(\bar{y}_2.) = \sqrt{MSE/34} = \sqrt{0.168/34}$, and a $t(90)$ distribution because $dfE = 90$. The value $t(0.995, 90) = 2.631$ is needed for $\alpha = 0.01$ tests and 99% confidence intervals. This t table value appears repeatedly in our discussion.

The endpoints of a 99% confidence interval for μ_H, the mean of the log suicide age for this Hispanic population, are

$$3.4538 \pm 2.631 \sqrt{\frac{0.168}{34}}$$

for an interval of $(3.269, 3.639)$. Transforming the interval back to the original scale gives $(e^{3.269}, e^{3.639})$ or $(26.3, 38.1)$, i.e., we are 99% confident that the median age of suicides for this Hispanic population is between 26.3 years old and 38.1 years old. By assumption, μ_H is the mean of the Hispanic log-suicide ages but, under normality, it is also the median. (The median has half the observations above it and half below.) The interval $(e^{3.269}, e^{3.639}) = (26.3, 38.1)$ is a 99% confidence interval for e^{μ_H}, which is the median of the Hispanic suicide ages, even though e^{μ_H} is not the mean of the Hispanic suicide ages. (99% confident means that the values in the interval would not be rejected by an $\alpha = 0.01$ test.)

A 99% prediction interval for the age of a future log-suicide from this Hispanic population has endpoints

$$3.4538 \pm 2.631 \sqrt{0.168 + \frac{0.168}{34}}$$

for an interval of $(2.360, 4.548)$. Transforming the interval back to the original scale gives $(10.6, 94.4)$, i.e., we are 99% confident that a future suicide from this Hispanic population would be between 10.6 years old and 94.4 years old. This interval happens to include all of the observed suicide ages for Hispanics in Table 12.1; that seems reasonable, if not terribly informative.

12.1.2 Inference on pairs of means

The primary parameters of interest for these data are probably the differences between the group population means. These parameters, with their estimates and the variances of the estimates, are given below.

Par	Est	Var(Est)
$\mu_C - \mu_H$	$3.6521 - 3.4538$	$\sigma^2 \left(\frac{1}{44} + \frac{1}{34}\right)$
$\mu_C - \mu_N$	$3.6521 - 3.1770$	$\sigma^2 \left(\frac{1}{44} + \frac{1}{15}\right)$
$\mu_H - \mu_N$	$3.4538 - 3.1770$	$\sigma^2 \left(\frac{1}{34} + \frac{1}{15}\right)$

The estimates and variances are obtained exactly as in Section 4.2. The standard errors of the estimates are obtained by substituting MSE for σ^2 in the variance formula and taking the square root. Below are given the estimates, standard errors, the t_{obs} values for testing $H_0 : Par = 0$, the P values, and the 99% confidence intervals for Par. Computing the confidence intervals requires the value $t(0.995, 90) = 2.632$.

Table of Coefficients

Par	Est	SE(Est)	t_{obs}	P	99% CI
$\mu_C - \mu_H$	0.1983	0.0936	2.12	0.037	$(-0.04796, 0.44456)$
$\mu_C - \mu_N$	0.4751	0.1225	3.88	0.000	$(0.15280, 0.79740)$
$\mu_H - \mu_N$	0.2768	0.1270	2.18	0.032	$(-0.05734, 0.61094)$

While the estimated difference between Hispanics and Native Americans is half again as large as the difference between non-Hispanic Caucasians and Hispanics, the t_{obs} values, and thus the significance levels of the differences, are almost identical. This occurs because the standard errors are substantially different. The standard error for the estimate of $\mu_C - \mu_H$ involves only the reasonably large samples for non-Hispanic Caucasians and Hispanics; the standard error for the estimate of $\mu_H - \mu_N$ involves the comparatively small sample of Native Americans, which is why this standard error is larger. On the other hand, the standard errors for the estimates of $\mu_C - \mu_N$ and $\mu_H - \mu_N$ are very similar. The difference in the standard error between having a sample of 34 or 44 is minor by comparison to the effect on the standard error of having a sample size of only 15.

The hypothesis $H_0 : \mu_C - \mu_N = 0$, or equivalently $H_0 : \mu_C = \mu_N$, is the only one rejected at the 0.01 level. Summarizing the results of the tests at the 0.01 level, we have no strong evidence of a difference between the ages at which non-Hispanic Caucasians and Hispanics commit suicide, we have no strong evidence of a difference between the ages at which Hispanics and Native Americans commit suicide, but we do have strong evidence that there is a difference in the ages at which non-Hispanic Caucasians and Native Americans commit suicide. Of course, all of these statements about null hypotheses presume that the underlying model is correct.

Establishing a difference between non-Hispanic Caucasians and Native Americans does little to explain why that difference exists. The reason that Native Americans committed suicide at younger ages could be some complicated function of socio-economic factors or it could be simply that there were many more young Native Americans than old ones in Albuquerque at the time. The test only indicates that the two groups were different; it says nothing about why the groups were different.

The confidence interval for the difference between non-Hispanic Caucasians and Native Americans was constructed on the log scale. Back transforming the interval gives $(e^{0.1528}, e^{0.7974})$ or $(1.2, 2.2)$. We are 99% confident that the median age of suicides is between 1.2 and 2.2 times higher for non-Hispanic Caucasians than for Native Americans. Note that examining differences in log ages transforms to the original scale as a multiplicative factor between groups. The parameters μ_C and μ_N are both means and medians for the logs of the suicide ages. When we transform the interval $(0.1528, 0.7974)$ for $\mu_C - \mu_N$ into the interval $(e^{0.1528}, e^{0.7974})$, we obtain a confidence interval for $e^{\mu_C - \mu_N}$ or equivalently for e^{μ_C}/e^{μ_N}. The values e^{μ_C} and e^{μ_N} are median values for the age distributions of the non-Hispanic Caucasians and Native Americans although they are not the expected values (population means) of the distributions. Obviously, $e^{\mu_C} = (e^{\mu_C}/e^{\mu_N})e^{\mu_N}$, so e^{μ_C}/e^{μ_N} is the number of times greater the median suicide age is for non-Hispanic Caucasians. That is the basis for the interpretation of the interval $(e^{0.1528}, e^{0.7974})$.

With these data, the tests for differences in means do not depend crucially on the log transformation but interpretations of the confidence intervals do. For the untransformed data, the mean squared error is $MSE_u = 245$ and the observed value of the test statistic for comparing non-Hispanic Caucasians and Native Americans is

$$t_u = 3.54 = \frac{41.66 - 25.07}{\sqrt{245 \left(\frac{1}{44} + \frac{1}{15}\right)}},$$

which is not far from the transformed value 3.88. However, the untransformed 99% confidence interval is $(4.3, 28.9)$, indicating a 4-to-29-year-higher age for the mean non-Hispanic Caucasian suicide, rather than the transformed interval $(1.2, 2.2)$, indicating that typical non-Hispanic Caucasian suicide ages are 1.2 to 2.2 times greater than those for Native Americans.

12.1.3 Inference on linear functions of means

The data do not strongly suggest that the means for Hispanics and Native Americans are different, so we *might* wish to compare the mean of the non-Hispanic Caucasians with the average of these groups. Typically, *averaging means will only be of interest if we feel comfortable treating those means as the same.* The parameter of interest is $Par = \mu_C - (\mu_H + \mu_N)/2$ or

$$Par = \mu_C - \frac{1}{2}\mu_H - \frac{1}{2}\mu_N$$

with

$$Est = \bar{y}_C - \frac{1}{2}\bar{y}_H - \frac{1}{2}\bar{y}_N = 3.6521 - \frac{1}{2}3.4538 - \frac{1}{2}3.1770 = 0.3367.$$

It is not really appropriate to use our standard methods to test this *contrast* between the means because the contrast was suggested by the data. Nonetheless, we will illustrate the standard methods. From the independence of the data in the three groups and Proposition 1.2.11, the variance of the estimate is

$$
\begin{aligned}
\text{Var}\left(\bar{y}_C - \frac{1}{2}\bar{y}_H - \frac{1}{2}\bar{y}_N\right) &= \text{Var}(\bar{y}_C) + \left(\frac{-1}{2}\right)^2 \text{Var}(\bar{y}_H) + \left(\frac{-1}{2}\right)^2 \text{Var}(\bar{y}_N) \\
&= \frac{\sigma^2}{44} + \left(\frac{-1}{2}\right)^2 \frac{\sigma^2}{34} + \left(\frac{-1}{2}\right)^2 \frac{\sigma^2}{15} \\
&= \sigma^2 \left[\frac{1}{44} + \left(\frac{-1}{2}\right)^2 \frac{1}{34} + \left(\frac{-1}{2}\right)^2 \frac{1}{15}\right].
\end{aligned}
$$

Substituting the *MSE* for σ^2 and taking the square root, the standard error is

$$0.0886 = \sqrt{0.168 \left[\frac{1}{44} + \left(\frac{-1}{2}\right)^2 \frac{1}{34} + \left(\frac{-1}{2}\right)^2 \frac{1}{15}\right]}.$$

Note that the standard error happens to be smaller than any of those we have considered when comparing pairs of means. To test the null hypothesis that the mean for non-Hispanic Caucasians equals the average of the other groups, i.e., $H_0 : \mu_C - \frac{1}{2}\mu_H - \frac{1}{2}\mu_N = 0$, the test statistic is

$$t_{obs} = \frac{0.3367 - 0}{0.0886} = 3.80,$$

so the null hypothesis is easily rejected. This is an appropriate test statistic for evaluating H_0, but when letting the data suggest the parameter, the $t(90)$ distribution is no longer appropriate for quantifying the level of significance. Similarly, we could construct the 99% confidence interval with endpoints

$$0.3367 \pm 2.631(0.0886)$$

but again, the confidence coefficient 99% is not really appropriate for a parameter suggested by the data.

While the parameter $\mu_C - \frac{1}{2}\mu_H - \frac{1}{2}\mu_N$ was suggested by the data, the theory of inference in Chapter 3 assumes that the parameter of interest does not depend on the data. In particular, the reference distributions we have used are invalid when the parameters depend on the data. Moreover, performing numerous inferential procedures complicates the analysis. Our standard tests are set up to check on one particular hypothesis. In the course of analyzing these data we have performed several tests. Thus we have had multiple opportunities to commit errors. In fact, the reason we have been discussing 0.01 level tests rather than 0.05 level tests is to help limit the number of errors made when all of the null hypotheses are true. In Chapter 13, we discuss methods of dealing with the problems that arise from making *multiple comparisons* among the means.

Table 12.2: *Analysis of Variance: Logs of suicide age data.*

Source	df	SS	MS	F	P
Groups	2	2.655	1.328	7.92	0.001
Error	90	15.088	0.168		
Total	92	17.743			

12.1.4 Testing $\mu_1 = \mu_2 = \mu_3$

To test $H_0 : \mu_1 = \mu_2 = \mu_3$ we test the one-way ANOVA model against the reduced model that fits only the grand mean (intercept), $y_{ij} = \mu + \varepsilon_{ij}$. The results are summarized in Table 12.2. Subject to round-off error, the information for the Error line is as given previously for the one-way ANOVA model, i.e., $dfE = 90$, $MSE = 0.168$, and $SSE = 90(0.168) = 15.088$. *The information in the Total line is the dfE and SSE for the grand-mean model.* For the grand-mean model, $dfE = n - 1 = 92$, $MSE = s_y^2 = 0.193$, i.e., the sample variance of all $n = 93$ observation, and the SSE is found by multiplying the two, $SSE = 92(0.193) = 17.743$. The dfE and SSE for Groups are found by subtracting the entries in the Error line from the Total line, so the df and SS are precisely what we need to compute the numerator of the F statistic, $dfGrps = 92 - 90 = 2$, $SSGrps = 17.743 - 15.008 = 2.655$. The reported F statistic

$$7.92 = \frac{1.328}{0.168} = \frac{MSGrps}{MSE} = \frac{[SSTot - SSE]/[dfTot - dfE]}{MSE}$$

is the statistic for testing our reduced (null) model.

The extremely small P value for the analysis of variance F test, as reported in Table 12.2, establishes clear differences among the mean log suicide ages. More detailed comparisons are needed to identify which particular groups are different. We established earlier that at the 0.01 level, only non-Hispanic Caucasians and Native Americans display a pairwise difference.

12.2 Theory

In analysis of variance, we assume that we have independent observations on, say, a different normal populations with the same variance. In particular, we assume the following data structure.

Sample	Data		Distribution
1	$y_{11}, y_{12}, \dots, y_{1N_1}$	iid	$N(\mu_1, \sigma^2)$
2	$y_{21}, y_{22}, \dots, y_{2N_2}$	iid	$N(\mu_2, \sigma^2)$
$\vdots$	$\vdots$	$\vdots$	$\vdots$
a	$y_{a1}, y_{a2}, \dots, y_{aN_a}$	iid	$N(\mu_a, \sigma^2)$

Here each sample is independent of the other samples. These assumptions are written more succinctly as the one-way analysis of variance model

$$y_{ij} = \mu_i + \varepsilon_{ij}, \quad \varepsilon_{ij}\text{s independent } N(0, \sigma^2) \tag{12.2.1}$$

$i = 1, \dots, a, \ j = 1, \dots, N_i$. The ε_{ij}s are unobservable random errors. Alternatively, Model (12.2.1) is often written as

$$y_{ij} = \mu + \alpha_i + \varepsilon_{ij}, \quad \varepsilon_{ij}\text{s independent } N(0, \sigma^2) \tag{12.2.2}$$

where $\mu_i \equiv \mu + \alpha_i$. The parameter μ is viewed as a grand mean, while α_i is an effect for the ith group. The μ and α_i parameters are not well defined. In Model (12.2.2) they only occur as the sum $\mu + \alpha_i$, so for any choice of μ and α_i the choices, say, $\mu + 5$ and $\alpha_i - 5$, are equally valid. The 5 can be replaced by any number we choose. The parameters μ and α_i are not completely specified by the model. There would seem to be little point in messing around with Model (12.2.2) except that it has useful relationships with other models that will be considered later.

Alternatively, using the notation of Chapter 3, we could write the model

$$y_h = m(x_h) + \varepsilon_h, \qquad h = 1, \ldots, n, \qquad (12.2.3)$$

where $n \equiv N_1 + \cdots + N_a$. In this case the predictor variable x_h takes on one of a distinct values to identify the group for each observation. Suppose x_h takes on the values $1, 2, \ldots, a$, then we identify

$$\mu_1 \equiv m(1), \quad \ldots, \quad \mu_a \equiv m(a).$$

The model involves a distinct mean parameters, so $dfE = n - a$. Switching from the h subscripts to the ij subscripts gives Model (12.2.1) with $x_h = i$.

To analyze the data, we compute summary statistics from each sample. These are the sample means and sample variances. For the ith group of observations, the sample mean is

$$\bar{y}_{i\cdot} \equiv \frac{1}{N_i} \sum_{j=1}^{N_i} y_{ij}$$

and the sample variance is

$$s_i^2 \equiv \frac{1}{N_i - 1} \sum_{j=1}^{N_i} (y_{ij} - \bar{y}_{i\cdot})^2.$$

With independent normal errors having the same variance, all *of the summary statistics are independent of one another*. Except for checking the validity of our assumptions, these summary statistics are more than sufficient for the entire analysis. Typically, we present the summary statistics in tabular form.

Sample statistics

Group	Size	Mean	Variance
1	N_1	$\bar{y}_{1\cdot}$	s_1^2
2	N_2	$\bar{y}_{2\cdot}$	s_2^2
$\vdots$	$\vdots$	$\vdots$	$\vdots$
a	N_a	$\bar{y}_{a\cdot}$	s_a^2

The sample means, the $\bar{y}_{i\cdot}$s, are estimates of the corresponding μ_is and the s_i^2s all estimate the common population variance σ^2. With unequal sample sizes an efficient pooled estimate of σ^2 must be a weighted average of the s_i^2s. The weights are the degrees of freedom associated with the various estimates. The pooled estimate of σ^2 is the *mean squared error (MSE)*,

$$
MSE \equiv s_p^2 \equiv \frac{(N_1 - 1)s_1^2 + (N_2 - 1)s_2^2 + \cdots + (N_a - 1)s_a^2}{\sum_{i=1}^{a}(N_i - 1)}
$$

$$
= \frac{1}{(n - a)} \sum_{i=1}^{a} \sum_{j=1}^{N_i} (y_{ij} - \bar{y}_{i\cdot})^2.
$$

The degrees of freedom for the *MSE* are the *degrees of freedom for error*,

$$dfE \equiv n - a = \sum_{i=1}^{a}(N_i - 1).$$

This is the sum of the degrees of freedom for the individual variance estimates. Note that the *MSE* depends only on the sample variances, so, with independent normal errors having the same variance, *MSE is independent of the $\bar{y}_{i\cdot}$s*.

A simple average of the sample variances s_i^2 is not reasonable. If we had $N_1 = 1,000,000$ observations in the first sample and only $N_2 = 5$ observations in the second sample, obviously the variance estimate from the first sample is much better than that from the second and we want to give it more weight.

In Model (12.2.3) the fitted values for group i are

$$\hat{y}_h \equiv \hat{m}(i) = \hat{\mu}_i = \bar{y}_{i\cdot}.$$

and the residuals are

$$\hat{\varepsilon}_h = y_h - \hat{y}_h = y_{ij} - \bar{y}_{i\cdot} = \hat{\varepsilon}_{ij}.$$

As usual,

$$MSE = \frac{\sum_{h=1}^{n} \hat{\varepsilon}_h^2}{n-a} = \frac{\sum_{i=1}^{a} \sum_{j=1}^{N_i} \hat{\varepsilon}_{ij}^2}{n-a} = \frac{1}{(n-a)} \sum_{i=1}^{a} \sum_{j=1}^{N_i} (y_{ij} - \bar{y}_{i\cdot})^2.$$

We need to check the validity of our assumptions. The errors in models (12.2.1) and (12.2.2) are assumed to be independent normals with mean 0 and variance σ^2, so we would like to use them to evaluate the distributional assumptions, e.g., equal variances and normality. Unfortunately, the errors are unobservable; we only see the y_{ij}s and we do not know the μ_is, so we cannot compute the ε_{ij}s. However, since $\varepsilon_{ij} = y_{ij} - \mu_i$ and we can estimate μ_i, we can estimate the errors with the residuals, $\hat{\varepsilon}_{ij} = y_{ij} - \bar{y}_{i\cdot}$. The residuals can be plotted against *fitted values* $\bar{y}_{i\cdot}$ to check whether the variance depends in some way on the means μ_i. They can also be plotted against rankits (normal scores) to check the normality assumption. More often we use the standardized residuals,

$$r_{ij} = \frac{\hat{\varepsilon}_{ij}}{\sqrt{MSE \left(1 - \frac{1}{N_i}\right)}},$$

see Sections 7.2 and 11.5.

If we are satisfied with the assumptions, we proceed to examine the parameters of interest. The basic parameters of interest in analysis of variance are the μ_is, which have natural estimates, the $\bar{y}_{i\cdot}$s. We also have an estimate of σ^2, so we are in a position to draw a variety of statistical inferences. The main problem in obtaining tests and confidence intervals is in finding appropriate standard errors. To do this we need to observe that each of the a samples are independent. The $\bar{y}_{i\cdot}$s are computed from different samples, so they are independent of each other. Moreover, $\bar{y}_{i\cdot}$ is the sample mean of N_i normal observations, so

$$\bar{y}_{i\cdot} \sim N\left(\mu_i, \frac{\sigma^2}{N_i}\right).$$

For inferences about a single mean, say, μ_2, use the general procedures with $Par = \mu_2$ and $Est = \bar{y}_{2\cdot}$. The variance of $\bar{y}_{2\cdot}$ is σ^2/N_2, so $SE(\bar{y}_{2\cdot}) = \sqrt{MSE/N_2}$. The reference distribution is $[\bar{y}_{2\cdot} - \mu_2]/SE(\bar{y}_{2\cdot}) \sim t(dfE)$. Note that the degrees of freedom for the t distribution are precisely the degrees of freedom for the MSE. The general procedures also provide prediction intervals using the MSE and $t(dfE)$ distribution.

For inferences about the difference between two means, say, $\mu_2 - \mu_1$, use the general procedures with $Par = \mu_2 - \mu_1$ and $Est = \bar{y}_{2\cdot} - \bar{y}_{1\cdot}$. The two means are independent, so the variance of $\bar{y}_{2\cdot} - \bar{y}_{1\cdot}$ is the variance of $\bar{y}_{2\cdot}$ plus the variance of $\bar{y}_{1\cdot}$, i.e., $\sigma^2/N_2 + \sigma^2/N_1$. The standard error of $\bar{y}_{2\cdot} - \bar{y}_{1\cdot}$ is

$$SE(\bar{y}_{2\cdot} - \bar{y}_{1\cdot}) = \sqrt{\frac{MSE}{N_2} + \frac{MSE}{N_1}} = \sqrt{MSE\left[\frac{1}{N_1} + \frac{1}{N_2}\right]}.$$

The reference distribution is

$$\frac{(\bar{y}_{2\cdot} - \bar{y}_{1\cdot}) - (\mu_2 - \mu_1)}{\sqrt{MSE\left[\frac{1}{N_1} + \frac{1}{N_2}\right]}} \sim t(dfE).$$

We might wish to compare one mean, μ_1, with the average of two other means, $(\mu_2 + \mu_3)/2$. In this case, the parameter can be taken as $Par = \mu_1 - (\mu_2 + \mu_3)/2 = \mu_1 - \frac{1}{2}\mu_2 - \frac{1}{2}\mu_3$. The estimate is $Est = \bar{y}_{1.} - \frac{1}{2}\bar{y}_{2.} - \frac{1}{2}\bar{y}_{3.}$. By the independence of the sample means, the variance of the estimate is

$$
\begin{aligned}
\mathrm{Var}\left(\bar{y}_{1.} - \frac{1}{2}\bar{y}_{2.} - \frac{1}{2}\bar{y}_{3.}\right) &= \mathrm{Var}(\bar{y}_{1.}) + \mathrm{Var}\left(\frac{-1}{2}\bar{y}_{2.}\right) + \mathrm{Var}\left(\frac{-1}{2}\bar{y}_{3.}\right) \\
&= \frac{\sigma^2}{N_1} + \left(\frac{-1}{2}\right)^2 \frac{\sigma^2}{N_2} + \left(\frac{-1}{2}\right)^2 \frac{\sigma^2}{N_3} \\
&= \sigma^2 \left[\frac{1}{N_1} + \frac{1}{4}\frac{1}{N_2} + \frac{1}{4}\frac{1}{N_3}\right].
\end{aligned}
$$

The standard error is

$$
\mathrm{SE}\left(\bar{y}_{1.} - \frac{1}{2}\bar{y}_{2.} - \frac{1}{2}\bar{y}_{3.}\right) = \sqrt{MSE\left[\frac{1}{N_1} + \frac{1}{4N_2} + \frac{1}{4N_3}\right]}.
$$

The reference distribution is

$$
\frac{\left(\bar{y}_{1.} - \frac{1}{2}\bar{y}_{2.} - \frac{1}{2}\bar{y}_{3.}\right) - \left(\mu_1 - \frac{1}{2}\mu_2 - \frac{1}{2}\mu_3\right)}{\sqrt{MSE\left[\frac{1}{N_1} + \frac{1}{4N_2} + \frac{1}{4N_3}\right]}} \sim t(dfE).
$$

In general, we are concerned with parameters that are *linear combinations* of the μ_is. For *known* coefficients $\lambda_1, \ldots, \lambda_a$, interesting parameters are defined by

$$
Par = \lambda_1 \mu_1 + \cdots + \lambda_a \mu_a = \sum_{i=1}^{a} \lambda_i \mu_i.
$$

For example, μ_2 has $\lambda_2 = 1$ and all other λ_is equal to 0. The difference $\mu_2 - \mu_1$ has $\lambda_1 = -1, \lambda_2 = 1$, and all other λ_is equal to 0. The parameter $\mu_1 - \frac{1}{2}\mu_2 - \frac{1}{2}\mu_3$ has $\lambda_1 = 1, \lambda_2 = -1/2, \lambda_3 = -1/2$, and all other λ_is equal to 0.

The natural estimate of $Par = \sum_{i=1}^{a} \lambda_i \mu_i$ substitutes the sample means for the population means, i.e., the natural estimate is

$$
Est = \lambda_1 \bar{y}_{1.} + \cdots + \lambda_a \bar{y}_{a.} = \sum_{i=1}^{a} \lambda_i \bar{y}_{i.}.
$$

In fact, Proposition 1.2.11 gives

$$
\mathrm{E}\left(\sum_{i=1}^{a} \lambda_i \bar{y}_{i.}\right) = \sum_{i=1}^{a} \lambda_i \mathrm{E}(\bar{y}_{i.}) = \sum_{i=1}^{a} \lambda_i \mu_i,
$$

so by definition this is an *unbiased* estimate of the parameter.

Using the independence of the sample means and Proposition 1.2.11,

$$
\begin{aligned}
\mathrm{Var}\left(\sum_{i=1}^{a} \lambda_i \bar{y}_{i.}\right) &= \sum_{i=1}^{a} \lambda_i^2 \mathrm{Var}(\bar{y}_{i.}) \\
&= \sum_{i=1}^{a} \lambda_i^2 \frac{\sigma^2}{N_i} \\
&= \sigma^2 \sum_{i=1}^{a} \frac{\lambda_i^2}{N_i}.
\end{aligned}
$$

The standard error is

$$\text{SE}\left(\sum_{i=1}^{a}\lambda_i\bar{y}_{i\cdot}\right) = \sqrt{MSE\left[\frac{\lambda_1^2}{N_1} + \cdots + \frac{\lambda_a^2}{N_a}\right]} = \sqrt{MSE\sum_{i=1}^{a}\frac{\lambda_i^2}{N_i}}$$

and the reference distribution is

$$\frac{\left(\sum_{i=1}^{a}\lambda_i\bar{y}_{i\cdot}\right) - \left(\sum_{i=1}^{a}\lambda_i\mu_i\right)}{\sqrt{MSE\sum_{i=1}^{a}\lambda_i^2/N_i}} \sim t(dfE),$$

see Exercise 12.8.14. If the independence and equal variance assumptions hold, then the central limit theorem and law of large numbers can be used to justify a $N(0,1)$ reference distribution even when the data are not normal as long as all the N_is are large, although I would continue to use the t distribution since the normal is clearly too optimistic.

In analysis of variance, we are most interested in *contrasts* (comparisons) among the μ_is. These are characterized by having $\sum_{i=1}^{a}\lambda_i = 0$. The difference $\mu_2 - \mu_1$ is a contrast as is the parameter $\mu_1 - \frac{1}{2}\mu_2 - \frac{1}{2}\mu_3$. If we use Model (12.2.2) rather than Model (12.2.1) we get

$$\sum_{i=1}^{a}\lambda_i\mu_i = \sum_{i=1}^{a}\lambda_i(\mu + \alpha_i) = \mu\sum_{i=1}^{a}\lambda_i + \sum_{i=1}^{a}\lambda_i\alpha_i = \sum_{i=1}^{a}\lambda_i\alpha_i,$$

thus contrasts in Model (12.2.2) involve only the group effects. This is of some importance later when dealing with more complicated models.

Having identified a parameter, an estimate, a standard error, and an appropriate reference distribution, inferences follow the usual pattern. A 95% confidence interval for $\sum_{i=1}^{a}\lambda_i\mu_i$ has endpoints

$$\sum_{i=1}^{a}\lambda_i\bar{y}_{i\cdot} \pm t(0.975, dfE)\sqrt{MSE\sum_{i=1}^{a}\lambda_i^2/N_i}.$$

An $\alpha = .05$ test of $H_0 : \sum_{i=1}^{a}\lambda_i\mu_i = 0$ rejects H_0 if

$$\frac{\left|\sum_{i=1}^{a}\lambda_i\bar{y}_{i\cdot} - 0\right|}{\sqrt{MSE\sum_{i=1}^{a}\lambda_i^2/N_i}} > t(0.975, dfE). \tag{12.2.4}$$

An equivalent procedure to the test in (12.2.4) is often useful. If we square both sides of (12.2.4), the test rejects if

$$\left(\frac{\left|\sum_{i=1}^{a}\lambda_i\bar{y}_{i\cdot} - 0\right|}{\sqrt{MSE\sum_{i=1}^{a}\lambda_i^2/N_i}}\right)^2 > [t(0.975, dfE)]^2.$$

The square of the test statistic leads to another common statistic, the sum of squares for the parameter. Rewrite the test statistic as

$$\left(\frac{\left|\sum_{i=1}^{a}\lambda_i\bar{y}_{i\cdot} - 0\right|}{\sqrt{MSE\sum_{i=1}^{a}\lambda_i^2/N_i}}\right)^2 = \frac{\left(\sum_{i=1}^{a}\lambda_i\bar{y}_{i\cdot} - 0\right)^2}{MSE\sum_{i=1}^{a}\lambda_i^2/N_i}$$

$$= \frac{\left(\sum_{i=1}^{a}\lambda_i\bar{y}_{i\cdot}\right)^2 / \sum_{i=1}^{a}\lambda_i^2/N_i}{MSE}$$

and define the *sum of squares for the parameter* as

$$SS\left(\sum_{i=1}^{a}\lambda_i\mu_i\right) \equiv \frac{\left(\sum_{i=1}^{a}\lambda_i\bar{y}_{i\cdot}\right)^2}{\sum_{i=1}^{a}\lambda_i^2/N_i}. \tag{12.2.5}$$

The $\alpha = .05$ t test of $H_0 : \sum_{i=1}^{a} \lambda_i \mu_i = 0$ is equivalent to rejecting H_0 if

$$\frac{SS\left(\sum_{i=1}^{a} \lambda_i \mu_i\right)}{MSE} > [t(0.975, dfE)]^2.$$

It is a mathematical fact that for any α between 0 and 1 and any dfE,

$$\left[t\left(1 - \frac{\alpha}{2}, dfE\right)\right]^2 = F(1 - \alpha, 1, dfE).$$

Thus the test based on the sum of squares is an F test with 1 degree of freedom in the numerator. *Any parameter of this type has 1 degree of freedom associated with it.*

In Section 12.1 we transformed the suicide age data so that they better satisfy the assumptions of equal variances and normal distributions. In fact, analysis of variance tests and confidence intervals are frequently useful even when these assumptions are violated. Scheffé (1959, p. 345) concludes that (a) nonnormality is not a serious problem for inferences about means but it is a serious problem for inferences about variances, (b) unequal variances are not a serious problem for inferences about means from samples of the same size but are a serious problem for inferences about means from samples of unequal sizes, and (c) lack of independence can be a serious problem. Of course, any such rules depend on just how bad the nonnormality is, how unequal the variances are, and how bad the lack of independence is. My own interpretation of these rules is that if you check the assumptions and they do not look too bad, you can probably proceed with a fair amount of assurance.

12.2.1 *Analysis of variance tables*

To test the (null) hypothesis

$$H_0 : \mu_1 = \mu_2 = \cdots = \mu_a,$$

we test Model (12.2.1) against the reduced model

$$y_{ij} = \mu + \varepsilon_{ij}, \quad \varepsilon_{ij}\text{s independent } N(0, \sigma^2) \tag{12.2.6}$$

in which each group has the same mean. Recall that the variance estimate for this model is the sample variance, i.e., $MSE(Red.) = s_y^2$, with $dfE(Red.) = n - 1$.

The computations are typically summarized in an analysis of variance table. The commonly used form for the analysis of variance table is given below.

Analysis of Variance

Source	df	SS	MS	F
Groups	$a - 1$	$\sum_{i=1}^{a} N_i (\bar{y}_{i.} - \bar{y}_{..})^2$	$SSGrps/(a-1)$	$\frac{MSGrps}{MSE}$
Error	$n - a$	$\sum_{i=1}^{a} \sum_{j=1}^{N_i} (y_{ij} - \bar{y}_{i.})^2$	$SSE/(n-a)$	
Total	$n - 1$	$\sum_{i=1}^{a} \sum_{j=1}^{N_i} (y_{ij} - \bar{y}_{..})^2$		

The entries in the Error line are just dfE, SSE, and MSE for Model (12.2.1). The entries for the Total line are dfE and SSE for Model (12.2.6). These are often referred to as $dfTot$ and $SSTot$ and sometimes as $dfTot - C$ and $SSTot - C$. The Groups line is obtained by subtracting the Error df and SS from the Total df and SS, respectively, so that $MSGroups \equiv SSGrps/dfGrps$ gives precisely the numerator of the F statistic for testing our hypothesis. It is some work to show that the algebraic formula given for $SSGrps$ is correct.

The total line is corrected for the grand mean. An obvious meaning for the phrase "sum of squares total" would be the sum of the squares of all the observations, $\sum_{ij} y_{ij}^2$. The reported sum of squares total is $SSTot = \sum_{i=1}^{a} \sum_{j=1}^{N_i} y_{ij}^2 - C$, which is the sum of the squares of all the observations minus the correction factor for fitting the grand mean, $C \equiv n\bar{y}_{..}^2$. Similarly, an obvious meaning for the phrase "degrees of freedom total" would be n, the number of observations: one degree of freedom for each observation. The reported $dfTot$ is $n-1$, which is corrected for fitting the grand mean μ in Model (12.2.6).

EXAMPLE 12.2.1.   We now illustrate direct computation of $SSGrps$, the only part of the analysis of variance table computations that we have not illustrated for the logs of the suicide data. The sample statistics are repeated below.

Sample statistics: Log of suicide ages

Group	N_i	$\bar{y}_{i\cdot}$	s_i^2
Caucasians	44	3.6521	0.1590
Hispanics	34	3.4538	0.2127
Native Am.	15	3.1770	0.0879

The sum of squares groups is

$$SSGrps = 2.655 = 44(3.6521-3.5030)^2 + 34(3.4538-3.5030)^2 + 15(3.1770-3.5030)^2$$

where

$$3.5030 = \bar{y}_{\cdot\cdot} = \frac{44(3.6521)+34(3.4538)+15(3.1770)}{44+34+15}.$$

The ANOVA table was presented as Table 12.2.   □

If the data happen to be *balanced* in the sense that $N_1 = \cdots = N_a \equiv N$, a convenient way to compute the mean square for groups is as

$$MSGrps = Ns_{\bar{y}}^2,$$

where $s_{\bar{y}}^2$ is the sample variance of the group means, i.e.,

$$s_{\bar{y}}^2 \equiv \frac{1}{a-1}\sum_{i=1}^{a}(\bar{y}_{i\cdot}-\bar{y}_{\cdot\cdot})^2.$$

This idea can be used as the basis for analyzing virtually any balanced multifactor ANOVA. Recall from Section 3.9 that a multifactor ANOVA is simply a model that involves more than one categorical predictor variable. Christensen (1996) examined this idea in detail.

12.3   Regression analysis of ANOVA data

We now discuss how to use multiple regression to analyze ANOVA data. Table 12.3 presents the suicide age data with the categorical predictor variable "Group" taking on the values 1, 2, 3. The predictor Group identifies which observations belong to each of the three groups. To analyze the data as a regression, we need to replace the three-*category (factor)* predictor Group with a series of three *indicator variables*, x_1, x_2, and x_3; see Table 12.3. Each of these x variables consist of 0s and 1s, with the 1s indicating membership in one of the three groups. Thus, for any observation that is in group 1 (Anglo), $x_1 = 1$ and for any observation that is in group 2 (Hisp.) or group 3 (N.A.), $x_1 = 0$. Similarly, x_2 is a 0-1 indicator variable that is 1 for Hispanics and 0 for any other group. Finally, x_3 is the indicator variable for Native Americans. Many computer programs will generate indicator variables like x_1, x_2, x_3 corresponding to a categorical variable like Group.

We fit the multiple regression model without an intercept

$$y_h = \mu_1 x_{h1} + \mu_2 x_{h2} + \mu_3 x_{h3} + \varepsilon_h, \qquad h = 1,\ldots,n. \qquad (12.3.1)$$

Table 12.3: *Suicide age data file.*

Age	Group	$x_1 = $ Anglo	$x_2 = $ Hisp.	$x_3 = $ N.A.
			Indicator Variables	
21	1	1	0	0
55	1	1	0	0
42	1	1	0	0
⋮	⋮	⋮	⋮	⋮
19	1	1	0	0
27	1	1	0	0
58	1	1	0	0
50	2	0	1	0
31	2	0	1	0
29	2	0	1	0
⋮	⋮	⋮	⋮	⋮
21	2	0	1	0
28	2	0	1	0
17	2	0	1	0
26	3	0	0	1
17	3	0	0	1
24	3	0	0	1
⋮	⋮	⋮	⋮	⋮
23	3	0	0	1
25	3	0	0	1
23	3	0	0	1
22	3	0	0	1

It does not matter that we are using Greek μs for the regression coefficients rather than βs. Model (12.3.1) is precisely the same model as

$$y_{ij} = \mu_i + \varepsilon_{ij}, \qquad i = 1,2,3, \; j = 1,2,\dots,N_i,$$

i.e., Model (12.2.1). They give the same fitted values, residuals, and dfE.

Model (12.3.1) is fitted without an intercept (constant). Fitting the regression model to the log suicide age data gives a Table of Coefficients and an ANOVA table. The tables are adjusted for the fact that the model was fitted without an intercept. Obviously, the Table of Coefficients cannot contain a constant term, since we did not fit one.

Table of Coefficients: Model (12.3.1)

Predictor	$\hat{\mu}_i$	SE($\hat{\mu}_i$)	t	P
Anglo	3.65213	0.06173	59.17	0.000
Hisp.	3.45377	0.07022	49.19	0.000
N.A.	3.1770	0.1057	30.05	0.000

The estimated regression coefficients are just the group sample means as displayed in Section 12.1. The reported standard errors are the standard errors appropriate for performing confidence intervals and tests on a single population mean as discussed in Subsection 12.1.1, i.e., $\hat{\mu}_i = \bar{y}_{i\cdot}$ and SE($\hat{\mu}_i$) = $\sqrt{MSE/N_i}$. The table also provides test statistics and P values for $H_0 : \mu_i = 0$ but these are not typically of much interest. The 95% confidence interval for, say, the Hispanic mean μ_2 has endpoints

$$3.45377 \pm 2.631(0.0.07022)$$

for an interval of (3.269, 3.639), just as in Subsection 12.1.1. Prediction intervals are easily obtained from most software by providing the corresponding 0-1 input for x_1, x_2, and x_3, e.g., to predict a Native American log suicide age, $(x_1, x_2, x_3) = (0, 0, 1)$.

In the ANOVA table

Analysis of Variance: Model (12.3.1)

Source	df	SS	MS	F	P
Regression	3	1143.84	381.28	2274.33	0.000
Error	90	15.09	0.17		
Total	93	1158.93			

The Error line is the same as that given in Section 12.1, up to round-off error. Without fitting an intercept (grand mean) in the model, most programs report the Total line in the ANOVA table without correcting for the grand mean. Here the Total line has $n = 93$ degrees of freedom, rather than the usual $n - 1$. Also, the Sum of Squares Total is the sum of the squares of all 93 observations, rather than the usual corrected number $(n - 1)s_y^2$. Finally, the F test reported in the ANOVA table is for testing the regression model against the relatively uninteresting model $y_h = 0 + \varepsilon_h$. It provides a simultaneous test of $0 = \mu_C = \mu_H = \mu_N$ rather than the usual test of $\mu_C = \mu_H = \mu_N$.

12.3.1 Testing a pair of means

In Subsection 12.1.2, we tested all three of the possible pairs of means. By reintroducing an intercept into the multiple regression model, we can get immediate results for testing any two of them. Rewrite the multiple regression model as

$$y_h = \mu + \alpha_1 x_{h1} + \alpha_2 x_{h2} + \alpha_3 x_{h3} + \varepsilon_h, \tag{12.3.2}$$

similar to Model (12.2.2). Remember, the Greek letters we choose to use as regression coefficients make no difference to the substance of the model. Model (12.3.2) is no longer a regression model because the parameters are redundant. The data have three groups, so we need no more than three model parameters to explain them. Model (12.3.2) contains four parameters. To make it into a regression model, we need to drop one of the predictor variables. In most important ways, which predictor variable we drop makes no difference. The fitted values, the residuals, the dfE, SSE, and MSE all remain the same. However, the meaning of the parameters changes depending on which variable we drop.

At the beginning of this section, we dropped the constant term from Model (12.3.2) to get Model (12.3.1) and discussed the parameter estimates. Now we leave in the intercept but drop one of the other variables. Let's drop x_3, the indicator variable for Native Americans. This makes the Native Americans into a baseline group with the other two groups getting compared to it. As mentioned earlier, we will obtain two of the three comparisons from Subsection 12.1.2, specifically the comparisons between Anglo and N.A., $\mu_C - \mu_N$, and Hisp. and N.A., $\mu_H - \mu_N$. Fitting the regression model with an intercept but without x_3, i.e.,

$$y_h = \beta_0 + \beta_1 x_{h1} + \beta_2 x_{h2} + \varepsilon_h, \tag{12.3.3}$$

gives the Table of Coefficients and ANOVA table.

Table of Coefficients: Model (12.3.3)

Predictor	$\hat{\beta}_k$	SE($\hat{\beta}_k$)	t	P
Constant	3.1770	0.1057	30.05	0.000
Anglo	0.4752	0.1224	3.88	0.000
Hisp.	0.2768	0.1269	2.18	0.032

Analysis of Variance: Model (12.3.3)

Source	df	SS	MS	F	P
Regression	2	2.6553	1.3276	7.92	0.001
Error	90	15.0881	0.1676		
Total	92	17.7434			

The estimate for the Constant is just the mean for the Native Americans and the rest of the Constant line provides results for evaluating μ_N. The results for the Anglo and Hisp. lines agree with the results from Subsection 12.1.2 for evaluating $\mu_C - \mu_N$ and $\mu_H - \mu_N$, respectively. Up to round-off error, the ANOVA table is the same as presented in Table 12.2.

Fitting Model (12.3.3) gives us results for inference on μ_N, $\mu_C - \mu_N$, and $\mu_H - \mu_N$. To make inferences for $\mu_C - \mu_H$, the estimate is easily obtained as $0.4752 - 0.2768$ but the standard error and other results are not easily obtained from fitting Model (12.3.3).

We can make inferences on $\mu_C - \mu_H$ by fitting another model. If we drop x_2, the indicator for Hispanics, and fit

$$y_h = \gamma_0 + \gamma_1 x_{h1} + \gamma_3 x_{h3} + \varepsilon_h,$$

Hispanic becomes the baseline group, so the constant term γ_0 corresponds to μ_H, the Anglo term γ_1 corresponds to $\mu_C - \mu_H$, and the N.A. term γ_3 corresponds to $\mu_N - \mu_H$. Similarly, if we drop the Anglo predictor and fit

$$y_h = \delta_0 + \delta_2 x_{h2} + \delta_3 x_{h3} + \varepsilon_h,$$

the constant term δ_0 corresponds to μ_C, the Hisp. term δ_2 corresponds to $\mu_H - \mu_C$, and the N.A. term δ_3 corresponds to $\mu_N - \mu_C$.

Dropping a predictor variable from Model (12.3.2) is equivalent to imposing a *side condition* on the parameters μ, α_1, α_2, α_3. In particular, dropping the intercept corresponds to assuming $\mu = 0$, dropping x_1 amounts to assuming $\alpha_1 = 0$, dropping x_2 amounts to assuming $\alpha_2 = 0$, and dropping x_3 amounts to assuming $\alpha_3 = 0$. In Subsection 12.3.3 we will look at a regression model that amounts to assuming that $\alpha_1 + \alpha_2 + \alpha_3 = 0$.

12.3.2 Model testing

It will not always be possible or convenient to manipulate the model as we did here so that the Table of Coefficients gives us interpretable results. Alternatively, we can use model testing to provide a test of, say, $\mu_C - \mu_N = 0$. Begin with our original no-intercept model (12.3.1), i.e.,

$$y_h = \mu_1 x_{h1} + \mu_2 x_{h2} + \mu_3 x_{h3} + \varepsilon_h.$$

To test $\mu_1 - \mu_3 \equiv \mu_C - \mu_N = 0$, rewrite the hypothesis as $\mu_1 = \mu_3$ and substitute this relation into Model (12.3.1) to get a reduced model

$$y_h = \mu_1 x_{h1} + \mu_2 x_{h2} + \mu_1 x_{h3} + \varepsilon_h$$

or

$$y_h = \mu_1 (x_{h1} + x_{h3}) + \mu_2 x_{h2} + \varepsilon_h.$$

The Greek letters change their meaning in this process, so we could just as well write the model as

$$y_h = \gamma_1 (x_{h1} + x_{h3}) + \gamma_2 x_{h2} + \varepsilon_h. \tag{12.3.4}$$

This reduced model still only involves indicator variables: x_2 is the indicator variable for group 2 (Hisp.) but $x_1 + x_3$ is now an indicator variable that is 1 if an individual is either Anglo or N.A. and 0 otherwise. We have reduced our three-group model with Anglos, Hispanics, and Native Americans to a two-group model that lumps Anglos and Native Americans together but distinguishes Hispanics. The question is whether this reduced model fits adequately relative to our full model that distinguishes all three groups. Fitting the model gives a Table of Coefficients and an ANOVA table.

Table of Coefficients: Model (12.3.4)

Predictor	$\hat{\gamma}_k$	SE($\hat{\gamma}_k$)	t	P
$x_1 + x_3$	3.53132	0.05728	61.65	0.000
Hisp.	3.45377	0.07545	45.78	0.000

The Table of Coefficients is not very interesting. It gives the same mean for Hisp. as Model (12.3.1) but provides a standard error based on a *MSE* from Model (12.3.4) that does not distinguish between Anglos and N.A.s. The other estimate in the table is the average of all the Anglos and N.A.s. Similarly, the ANOVA table is not terribly interesting except for its Error line.

Analysis of Variance: Model (12.3.4)

Source	df	SS	MS	F	P
Regression	2	1141.31	570.66	2948.27	0.000
Error	91	17.61	0.19		
Total	93	1158.93			

From this Error line and the Error for Model (12.3.1), the model testing statistic for the hypothesis $\mu_1 - \mu_3 \equiv \mu_C - \mu_N = 0$ is

$$F_{obs} = \frac{[17.61 - 15.09]/[91 - 90]}{15.09/90} = 15.03 \doteq (3.88)^2.$$

The last (almost) equality between 15.03 and $(3.88)^2$ demonstrates that this F statistic is the square of the t statistic reported in Subsection 12.1.2 for testing $\mu_1 - \mu_3 \equiv \mu_C - \mu_N = 0$. Rejecting an $F(1,90)$ test for large values of F_{obs} is equivalent to rejecting a $t(90)$ test for t_{obs} values far from zero. The lack of equality between 15.03 and $(3.88)^2$ is entirely due to round-off error. To reduce round-off error, in computing F_{obs} we used $MSE(Full) = 15.09/90$ as the full model mean squared error, rather than the reported value from Model (12.3.1) of $MSE(Full) = 0.17$. To further reduce round-off error, we could use even more accurate numbers reported earlier for Model (12.3.3),

$$F_{obs} = \frac{[17.61 - 15.0881]/[91 - 90]}{15.0881/90} = 15.04 \doteq (3.88)^2.$$

Two final points. First, $17.61 - 15.0881 = SS(\mu_1 - \mu_3)$, the sum of squares for the contrast as defined in (12.2.5). Second, to test $\mu_1 - \mu_3 = 0$, rather than manipulating the indicator variables, the next section discusses how to get the same results by manipulating the group subscript.

Similar to testing $\mu_1 - \mu_3 = 0$, we could test $\mu_1 - \frac{1}{2}\mu_2 - \frac{1}{2}\mu_3 = 0$. Rewrite the hypothesis as $\mu_1 = \frac{1}{2}\mu_2 + \frac{1}{2}\mu_3$ and obtain the reduced model by substituting this relationship into Model (12.3.1) to get

$$y_h = \left(\frac{1}{2}\mu_2 + \frac{1}{2}\mu_3\right) x_{h1} + \mu_2 x_{h2} + \mu_3 x_{h3} + \varepsilon_h$$

or

$$y_h = \mu_2 \left(\frac{1}{2}x_{h1} + x_{h2}\right) + \mu_3 \left(\frac{1}{2}x_{h1} + x_{h3}\right) + \varepsilon_h.$$

This is just a no-intercept regression model with two predictor variables $\tilde{x}_1 = \left(\frac{1}{2}x_1 + x_2\right)$ and $\tilde{x}_2 = \left(\frac{1}{2}x_1 + x_3\right)$, say,

$$y_i = \gamma_1 \tilde{x}_{i1} + \gamma_2 \tilde{x}_{i2} + \varepsilon_i.$$

Fitting the reduced model gives the usual tables. Our interest is in the ANOVA table Error line.

Table of Coefficients

Predictor	$\hat{\gamma}_k$	$SE(\hat{\gamma}_k)$	t	P
$\tilde{x}_1$	3.55971	0.06907	51.54	0.000
$\tilde{x}_2$	3.41710	0.09086	37.61	0.000

Analysis of Variance

Source	df	SS	MS	F	P
Regression	2	1141.41	570.71	2965.31	0.000
Error	91	17.51	0.19		
Total	93	1158.93			

From this Error line and the Error for Model (12.3.1), the model testing statistic for the hypothesis $\mu_1 - \frac{1}{2}\mu_2 - \frac{1}{2}\mu_3 = 0$ is

$$F_{obs} = \frac{[17.51 - 15.09]/[91 - 90]}{15.09/90} = 14.43 \doteq (3.80)^2.$$

Again, 3.80 is the t_{obs} that was calculated in Subsection 12.1.3 for testing this hypothesis.

Finally, suppose we wanted to test $\mu_1 - \mu_3 \equiv \mu_C - \mu_N = 1.5$. Using results from the Table of Coefficients for fitting Model (12.3.3), the t statistic is

$$t_{obs} = \frac{0.4752 - 1.5}{0.1224} = -8.3725.$$

The corresponding model-based test reduces the full model (12.3.1) by incorporating $\mu_1 = \mu_3 + 1.5$ to give the reduced model

$$y_h = (\mu_3 + 1.5)x_{h1} + \mu_2 x_{h2} + \mu_3 x_{h3} + \varepsilon_h$$

or

$$y_h = 1.5x_{h1} + \mu_2 x_{h2} + \mu_3(x_{h1} + x_{h3}) + \varepsilon_h.$$

The term $1.5x_{i1}$ is completely known (not multiplied by an unknown parameter) and is called an *offset*. To analyze the model, we take the offset to the left-hand side of the model, rewriting it as

$$y_h - 1.5x_{h1} = \mu_2 x_{h2} + \mu_3(x_{h1} + x_{h3}) + \varepsilon_h. \tag{12.3.5}$$

This regression model has a different dependent variable than Model (12.3.1), but because the offset is a known multiple of a variable that is in Model (12.3.1), the offset model can be compared to Model (12.3.1) in the usual way, cf. Christensen (2011, Subsection 3.2.1). The predictor variables in the reduced model (12.3.5) are exactly the same as the predictor variables used in Model (12.3.4) to test $\mu_1 - \mu_3 \equiv \mu_C - \mu_N = 0$.

Fitting Model (12.3.5) gives the usual tables. Our interest is in the Error line.

Table of Coefficients: Model (12.3.5)

Predictor	$\hat{\gamma}_k$	SE($\hat{\gamma}_k$)	t	P
Hisp.	3.45377	0.09313	37.08	0.000
$x_1 + x_3$	2.41268	0.07070	34.13	0.000

Analysis of Variance: Model (12.3.5)

Source	df	SS	MS	F	P
Regression	2	749.01	374.51	1269.87	0.000
Error	91	26.84	0.29		
Total	93	775.85			

From this Error line and the Error for Model (12.3.1), the F statistic for the hypothesis $\mu_1 - \mu_3 \equiv \mu_C - \mu_N = 1.5$ is

$$F_{obs} = \frac{[26.84 - 15.09]/[91 - 90]}{15.09/90} = 70.08 \doteq (-8.3725)^2.$$

Again, the lack of equality is entirely due to round-off error and rejecting the $F(1,90)$ test for large values of F_{obs} is equivalent to rejecting the $t(90)$ test for t_{obs} values far from zero.

Instead of writing $\mu_1 = \mu_3 + 1.5$ and substituting for μ_1 in Model (12.3.1), we could just as well have written $\mu_1 - 1.5 = \mu_3$ and substituted for μ_3 in Model (12.3.1). It is not obvious, but this leads to exactly the same test. Try it!

12.3.3 Another choice

Another variation on fitting regression models involves subtracting out the last predictor. In some programs, notably Minitab, the overparameterized model

$$y_h = \mu + \alpha_1 x_{h1} + \alpha_2 x_{h2} + \alpha_3 x_{h3} + \varepsilon_h,$$

is fitted as the equivalent regression model

$$y_h = \gamma_0 + \gamma_1(x_{h1} - x_{h3}) + \gamma_2(x_{h2} - x_{h3}) + \varepsilon_h. \tag{12.3.6}$$

Up to round-off error, the ANOVA table is the same as Table 12.2. Interpreting the Table of Coefficients is a bit more complicated.

Table of Coefficients: Model (12.3.6)

Predictor	$\hat{\gamma}_k$	SE($\hat{\gamma}_k$)	t	P
Constant (γ_0)	3.42762	0.04704	72.86	0.000
Group				
C (γ_1)	0.22450	0.05902	3.80	0.000
H (γ_2)	0.02615	0.06210	0.42	0.675

The relationship between this regression model (12.3.6) and the easiest model

$$y_h = \mu_1 x_{h1} + \mu_2 x_{h2} + \mu_3 x_{h3} + \varepsilon_h$$

is

$$\gamma_0 + \gamma_1 = \mu_1, \quad \gamma_0 + \gamma_2 = \mu_2, \quad \gamma_0 - (\gamma_1 + \gamma_2) = \mu_3,$$

so

$$\hat{\gamma}_0 + \hat{\gamma}_1 = 3.42762 + 0.22450 = 3.6521 = \bar{y}_{1.} = \hat{\mu}_1,$$

$$\hat{\gamma}_0 + \hat{\gamma}_2 = 3.42762 + 0.02615 = 3.4538 = \bar{y}_{2.} = \hat{\mu}_2,$$

and

$$\hat{\gamma}_0 - (\hat{\gamma}_1 + \hat{\gamma}_2) = 3.42762 - (0.22450 + 0.02615) = 3.1770 = \bar{y}_{3.} = \hat{\mu}_3.$$

Alas, this table of coefficients is not really very useful. We can see that $\gamma_0 = (\mu_1 + \mu_2 + \mu_3)/3$. The table of coefficients provides the information needed to perform inferences on this *not very interesting* parameter. Moreover, it allows inference on

$$\gamma_1 = \mu_1 - \gamma_0 = \mu_1 - \frac{\mu_1 + \mu_2 + \mu_3}{3} = \frac{2}{3}\mu_1 + \frac{-1}{3}\mu_2 + \frac{-1}{3}\mu_3,$$

another, not tremendously interesting, parameter. The interpretation of γ_2 is similar to γ_1.

The relationship between Model (12.3.6) and the overparameterized model (12.3.2) is

$$\gamma_0 + \gamma_1 = \mu + \alpha_1, \quad \gamma_0 + \gamma_2 = \mu + \alpha_2, \quad \gamma_0 - (\gamma_1 + \gamma_2) = \mu + \alpha_3,$$

which leads to

$$\gamma_0 = \mu, \quad \gamma_1 = \alpha_1, \quad \gamma_2 = \alpha_2, \quad -(\gamma_1 + \gamma_2) = \alpha_3,$$

provided that the side conditions $\alpha_1 + \alpha_2 + \alpha_3 = 0$ hold. Under this side condition, the relationship between the γs and the parameters of Model (12.3.2) is very simple. That is probably the motivation for considering Model (12.3.6). But the most meaningful parameters are clearly the μ_is, and there is no simple relationship between the γs and them.

Table 12.4: *Subscripts for ANOVA on* log(*y*)*: Mandel data.*

					Columns			
1	2	3	4	5	6	7	8	9
133	1	1	1	1	1	1	1	1
129	1	1	1	1	1	1	1	1
123	1	1	1	1	1	1	1	1
156	1	1	1	1	1	1	1	1
129	2	1	2	1	1	2	2	1
125	2	1	2	1	1	2	2	1
136	2	1	2	1	1	2	2	1
127	2	1	2	1	1	2	2	1
121	3	3	3	3	1	3	3	1
125	3	3	3	3	1	3	3	1
109	3	3	3	3	1	3	3	1
128	3	3	3	3	1	3	3	1
57	4	4	3	3	1	4	4	1
58	4	4	3	3	1	4	4	1
59	4	4	3	3	1	4	4	1
67	4	4	3	3	1	4	4	1
122	5	5	5	5	5	5	5	2
98	5	5	5	5	5	5	5	2
107	5	5	5	5	5	5	5	2
110	5	5	5	5	5	5	5	2
109	6	6	6	6	6	6	5	2
120	6	6	6	6	6	6	5	2
112	6	6	6	6	6	6	5	2
107	6	6	6	6	6	6	5	2
80	7	7	7	7	7	6	5	2
72	7	7	7	7	7	6	5	2
76	7	7	7	7	7	6	5	2
64	7	7	7	7	7	6	5	2
y	*i*	1=2	1=2; 3=4	1=2; 3=4	1=2= 3=4	6=7	5=6=7	1=2=3=4; 5=6=7

12.4 Modeling contrasts

In one-way ANOVA we have simple methods available for examining contrasts. These were discussed in Sections 1 and 2. However, in more complicated models, like the unbalanced multifactor ANOVAs discussed in Chapters 14 and 16 and the models for count data discussed later, these simple methods do not typically apply. In fact, we will see that in such models, examining a series of contrasts can be daunting. We now introduce modeling methods, based on relatively simple manipulations of the group subscript, that allow us to test a variety of interesting contrasts in some very general models. In fact, what this section does is present ways to manipulate the indicator variables of the previous section without ever actually defining the indicator variables.

EXAMPLE 12.4.1. Mandel (1972) and Christensen (1996, Chapter 6) presented data on the stress at 600% elongation for natural rubber with a 40-minute cure at 140°C. Stress was measured in 7 laboratories and each lab measured it four times. The dependent variable was measured in kilograms per centimeter squared (kg/cm^2). Following Christensen (1996) the analysis is conducted on the logs of the stress values. The data are presented in the first column of Table 12.4 with column 2 indicating the seven laboratories. The other seven columns will be discussed later.

The model for the one-way ANOVA is

$$y_{ij} = \mu_i + e_{ij} \tag{12.4.1}$$
$$= \mu + \alpha_i + e_{ij} \quad i = 1,2,3,4,5,6,7, \ j = 1,2,3,4.$$

The ANOVA table is given as Table 12.5. Clearly, there are some differences among the laboratories.

Table 12.5: *Analysis of Variance: Model (12.4.1), C2, seven groups.*

Source	df	SS	MS	F	P
Groups	6	2.26921	0.37820	62.72	0.000
Error	21	0.12663	0.00603		
Total	27	2.39584			

If these seven laboratories did not have any natural structure to them, about the only thing of interest would be to compare all of the pairs of labs to see which ones are different. This involves looking at $7(6)/2 = 21$ pairs of means, a process discussed more in Chapter 13.

As in Christensen (1996), suppose that the first two laboratories are in San Francisco, the second two are in Seattle, the fifth is in New York, and the sixth and seventh are in Boston. This structure suggests that there are six interesting questions to ask. On the West Coast, is there any difference between the San Francisco labs and is there any difference between the Seattle labs? If there are no such differences, it makes sense to discuss the San Francisco and Seattle population means, in which case, is there any difference between the San Francisco labs and the Seattle labs? On the East Coast, is there any difference between the Boston labs and if not, do they differ from the New York lab? Finally, if the West Coast labs have the same mean, and the East Coast labs have the same mean, is there a difference between labs on the West Coast and labs on the East Coast?

12.4.1 A hierarchical approach

We present what seems like a reasonable approach to looking at the six comparisons discussed earlier. For this, we use columns 3 through 9 in Table 12.4. Every column from 2 to 9 in Table 12.4 can be used as the index to define a one-way ANOVA for these data. Each column incorporates different assumptions (null hypotheses) about the group means.

Columns 3 through 6 focus attention on the West Coast labs. Column 3 has the same index for both the San Francisco labs so it defines a one-way ANOVA that incorporates $\mu_1 = \mu_2$, i.e, that the San Francisco labs have the same mean. Column 4 has the same index for both Seattle labs and gives a model that incorporates $\mu_3 = \mu_4$, i.e, that the Seattle labs have the same mean. Using column 5 gives a model that simultaneously incorporates equality of the San Francisco labs and the Seattle labs, i.e., $\mu_1 = \mu_2$ and $\mu_3 = \mu_4$. Using column 6 goes a step further to give a model that has equality among all of the West Coast labs, i.e., $\mu_1 = \mu_2 = \mu_3 = \mu_4$.

With column 7 attention switches to the East Coast labs. It gives a model that incorporates $\mu_6 = \mu_7$, i.e, that the Boston labs have the same mean. Using column 8 goes a step further to give a model that incorporates equality among all of the East Coast labs, i.e., $\mu_5 = \mu_6 = \mu_7$. Finally, column 9 is a model in which all West Coast labs have the same mean and all East Coast labs have the same mean, i.e., $\mu_1 = \mu_2 = \mu_3 = \mu_4$ and $\mu_5 = \mu_6 = \mu_7$.

Many of these models are not comparable, but we can view them as a structured hierarchy of models as indicated below. All models assume the validity of Model (12.4.1). Any assumption of pairwise equality is evaluated relative to the original model (12.4.1), so these three noncomparable models are in the second row, just beneath Model (12.4.1). We then build more structured reduced models from these initial pairwise equalities.

$$(12.4.1)$$

$$\mu_1 = \mu_2 \quad \Big| \quad \mu_3 = \mu_4 \quad \Big| \quad \mu_6 = \mu_7$$

$$\mu_1 = \mu_2; \mu_3 = \mu_4$$
$$\mu_1 = \mu_2 = \mu_3 = \mu_4 \qquad\qquad \mu_5 = \mu_6 = \mu_7$$

$$\mu_1 = \mu_2 = \mu_3 = \mu_4; \mu_5 = \mu_6 = \mu_7$$

$$\mu_1 = \cdots = \mu_7$$

Models separated by vertical bars are not comparable, but other than that, models in any row can be tested against as a full model against a reduced model in a lower row or as a reduced model against a full model in a higher row. The last model in the hierarchy is just the grand-mean (intercept-only) model.

The hierarchy has six rows. In the second row down, primary interest is in comparing the models to Model (12.4.1) in the top row.

The third row involves a semicolon! Comparing the third row to the first row merely involves doing a simultaneous test of two hypotheses that we have already looked at. By involving the second row, we can look at these hypotheses in different orders. But the real interest in a row with a semi-colon is in comparing it to the model below it. The real interest in looking at the model with both $\mu_1 = \mu_2$ and $\mu_3 = \mu_4$ is to see if it fits better than the model in row four with $\mu_1 = \mu_2 = \mu_3 = \mu_4$.

Similarly, the other model that involves a semicolon is in row five, i.e., $\mu_1 = \mu_2 = \mu_3 = \mu_4$; $\mu_5 = \mu_6 = \mu_7$, and the real interest is in whether it fits better than the model below it, the row six grand-mean model. This is not to say that there are not worthwhile comparisons to be made between the model in row five and models in higher rows.

As a shorthand, it is convenient to refer to the models in the hierarchy by their column numbers from Table 12.4. This makes the hierarchy

$$
\begin{array}{ccccc}
 & & C2 & & \\
C3 & | \quad C4 & | & C7 & \\
 & C5 & | & & \\
 & C6 & | & C8 & \\
 & & C9 & & \\
 & & GM & & \\
\end{array}
$$

While this hierarchy of models was designed in response to the structure of our specific treat-ments, the hierarchical approach is pretty general. Suppose our groups were five diets: Control, Beef A, Beef B, Pork, and Beans. With five diets, we might be interested in four comparisons suggested by the structure of the diets. First, we might compare the two beef diets. Second, compare the beef diets with the pork diet. (If the beef diets are the same, are they different from pork?) Third, compare the meat diets with the Bean diet. (If the meat diets are the same, are they different from beans?) Fourth, is the control different from the rest? These four comparisons suggest a hierarchy of models, cf. Exercise 12.7.15. Other nonhierarchical options would be to compare the control to each of the other diets or to compare each diet with every other diet.

Now suppose our five diets are: Control, Beef, Pork, Lima Beans, and Soy Beans. Again, we could compare the control to each of the other diets or compare all pairs of diets, but the structure of the treatments suggests the four comparisons 1) beef with pork, 2) lima beans with soy beans, 3) meat with beans, and 4) control with the others, which suggests a hierarchy of models, cf. Exer-cise 12.7.16.

12.4.2   Evaluating the hierarchy

With seven groups, there are six degrees of freedom for groups. The structure of the groups has suggested a hierarchy of models, which in turn suggests six F tests, each with one degree of freedom in the numerator.

To start off our analysis of Mandel's data, suppose we wanted to evaluate whether there is any demonstrable difference between labs 1 and 2 (the two in San Francisco). From a modeling point of view, this is very easy. We currently have Model (12.4.1) that distinguishes between all 7 labs. To test whether there are differences between labs 1 and 2, all we have to do is compare Model (12.4.1) to a model in which there are no differences between labs 1 and 2. In other words, our reduced model makes no distinction between labs 1 and 2. To perform such an ANOVA, rather than using the indices in column 2 of Table 12.4, the reduced model is an ANOVA using the indices in column 3. The fact that labs 1 and 2 are being equated is indicated at the bottom of column 3 with the notation $1 = 2$. The ANOVA table for this reduced model is

Table 12.6: *West Coast.*

Analysis of Variance: $H_0 : \mu_1 = \mu_2, C3$				
Source	df	SS	MS	F	P
Groups, 1=2	5	2.26572	0.45314	76.61	0.000
Error	22	0.13012	0.00591		
Total	27	2.39584			

Analysis of Variance: $H_0 : \mu_3 = \mu_4, C4$				
Source	df	SS	MS	F	P
Groups, 3=4	5	1.30194	0.26039	5.24	0.003
Error	22	1.09390	0.04972		
Total	27	2.39584			

Analysis of Variance: $H_0 : \mu_1 = \mu_2; \mu_3 = \mu_4, C5$				
Source	df	SS	MS	F	P
Groups, 1=2,3=4	4	1.29844	0.32461	6.80	0.001
Error	23	1.09740	0.04771		
Total	27	2.39584			

Analysis of Variance: $H_0 : \mu_1 = \mu_2 = \mu_3 = \mu_4, C6$				
Source	df	SS	MS	F	P
Groups, 1=2=3=4	3	0.53108	0.17703	2.28	0.105
Error	24	1.86476	0.07770		
Total	27	2.39584			

Analysis of Variance: $H_0 : \mu_1 = \mu_2, C3$

Source	df	SS	MS	F	P
Groups, 1=2	5	2.26572	0.45314	76.61	0.000
Error	22	0.13012	0.00591		
Total	27	2.39584			

Comparing the reduced model C3 to the full model C2 [equivalently, Model (12.4.1)] whose ANOVA table is given as Table 12.5, we get the F statistic

$$\frac{[SSE(C3) - SSE(C2)]}{[dfE(C3) - dfE(C2)]} \Big/ MSE(C2) = \frac{[0.13012 - 0.12663]/[22 - 21]}{0.00603} = 0.58.$$

There is no evidence of differences between the San Francisco labs. Note that the numerator sum of squares is $0.13012 - 0.12663 = 0.00349 = SS(\mu_1 - \mu_2)$ as defined in (12.2.5).

When fitting an intermediate ANOVA model from our hierarchy, like C3, our primary interest is in using the Error line of the fitted model to construct the F statistic that is our primary interest. But the ANOVA table for model C3 also provides a test of the intermediate model against the grand-mean model. In the case of fitting model C3: $\mu_1 = \mu_2$, the F statistic of 76.61 reported in the ANOVA table with 5 numerator degrees of freedom suggests that, even when the first two labs are treated as the same, differences exist somewhere among this pair and the other five labs for which we have made no assumptions. We probably would not want to perform this 5 df test if we got a significant result in the test of model C3 versus model C2 because the 5 df test would then be based on an assumption that is demonstrably false. (Ok, "demonstrably false" is a little strong.) Also, since we are looking at a variety of models, all of which are special cases of model C2, our best practice uses $MSE(C2)$ in the denominator of all tests, including this 5 df test. In particular, it would be better to replace $F = 0.45314/0.00591 = 76.61$ from the C3 ANOVA table with $F = 0.45314/0.00603 = 75.15$ having 5 and 21 degrees of freedom.

Table 12.6 contains ANOVA tables for the models that impose restrictions on the West Coast labs. The reduced models have all incorporated some additional conditions on the μ_is relative to model C2 and these are given at the top of each ANOVA table. The first model is the one we have just examined, the model in which no distinction is made between labs 1 and 2.

The second ANOVA table in Table 12.6 is based on the model in which no distinction is made between labs 3 and 4, the two in Seattle. (This is model C4.) A formal test for equality takes the form

$$\frac{[SSE(C4) - SSE(C2)]}{[dfE(C4) - dfE(C2)]} \bigg/ MSE(C2) = \frac{[1.09390 - 0.12663]/[22 - 21]}{0.00603} = 155.26$$

There is great evidence for differences between the two Seattle labs. At this point, it does not make much sense to look further at any models that incorporate $\mu_3 = \mu_4$, but we plunge forward just to demonstrate the complete process.

The third ANOVA table in Table 12.6 is based on the model in which no distinction is made between the two labs within San Francisco and also no distinction is made between the two labs within Seattle. (This model uses index column 5.) The difference in SSE between this ANOVA model and the SSE for model C2 is $1.09740 - 0.12663 = 0.97077$. The degrees of freedom are $23 - 21 = 2$. A formal test of $H_0: \mu_1 = \mu_2; \mu_3 = \mu_4$ takes the form

$$\frac{[SSE(C5) - SSE(C2)]/[dfE(C5) - dfE(C2)]}{MSE(C2)}$$
$$= \frac{[1.09740 - 0.12663]/[23 - 21]}{0.00603} = \frac{0.97077/2}{0.00603} = 80.50.$$

This is compared to an $F(2, 21)$ distribution and provides a simultaneous test for no differences in San Francisco as well as no differences in Seattle. Note that we could also test full models C3 and C4 against the reduced model C5. We leave it to the reader to see that when using $MSE(C2)$ in the denominator these comparisons agree with the tests for $H_0: \mu_3 = \mu_4$ and $H_0: \mu_1 = \mu_2$ given earlier.

To test for no differences between San Francisco and Seattle, compare the full model C5 that has no differences within either city but distinguishes labs in the two cities to the reduced model C6 that makes no distinctions between any labs on the West Coast. The ANOVA table for the model with no distinctions between any of the labs on the West Coast is the last one in Table 12.6. The sum of squares for the test is $1.86476 - 1.09740 = .76736$. This is obtained from the last two ANOVA tables in Table 12.6. A formal test takes the form

$$\frac{[SSE(C6) - SSE(C5)]/[dfE(C6) - dfE(C5)]}{MSE(C2)}$$
$$= \frac{[1.86476 - 1.09740]/[24 - 23]}{0.00603} = \frac{0.76736}{0.00603} = 127.26.$$

In the denominator of the test we have incorporated our best practice of using the mean square error from model C2, which is a more general model than either the reduced or full models being compared, cf. Subsection 3.1.1.

We can make similar comparisons for the East Coast laboratories. Table 12.7 gives ANOVA tables. The first table is for a model that incorporates no distinctions between the two labs in Boston, i.e., uses column 7 as subscripts. To test for no differences, compare that model to model C2.

$$\frac{[SSE(C7) - SSE(C2)]/[dfE(C7) - dfE(C2)]}{MSE(C2)}$$
$$= \frac{[0.49723 - 0.12663]/[22 - 21]}{0.00603} = \frac{0.3706}{0.00603} = 61.46.$$

There is clear evidence for a difference between the labs in Boston, so it makes little sense to consider any models that incorporate $\mu_6 = \mu_7$, but, as we did for the West Coast labs, we carry on to illustrate the process.

The second ANOVA table in Table 12.7 is for a model that incorporates no distinctions between any of the labs on the East Coast (column 8). To test for no differences between any of the three,

Table 12.7: *East Coast.*

Analysis of Variance: $H_0 : \mu_6 = \mu_7, C7$					
Source	df	SS	MS	F	P
Groups, 6=7	5	1.89861	0.37972	16.80	0.000
Error	22	0.49723	0.02260		
Total	27	2.39584			

Analysis of Variance: $H_0 : \mu_5 = \mu_6 = \mu_7, C8$					
Source	df	SS	MS	F	P
Groups, 5=6=7	4	1.80410	0.45102	17.53	0.000
Error	23	0.59174	0.02573		
Total	27	2.39584			

Table 12.8: *West Coast versus East Coast.*

Analysis of Variance: $H_0 : \mu_1 = \mu_2 = \mu_3 = \mu_4; \mu_5 = \mu_6 = \mu_7, C9$					
Source	df	SS	MS	F	P
Groups, 1=2=3=4, 5=6=7	1	0.06597	0.06597	0.74	0.399
Error	26	2.32987	0.08961		
Total	27	2.39584			

compare the model to model C2.

$$\frac{[SSE(C8) - SSE(C2)]/[dfE(C8) - dfE(C2)]}{MSE(C2)}$$

$$= \frac{[0.59174 - 0.12663]/[23 - 21]}{0.00603} = \frac{0.46511/2}{0.00603} = 38.57.$$

In addition, one can test the full model that has no differences between labs in Boston but has distinctions with New York against the model that has no distinctions between any of the three labs. The test uses both ANOVAs in Table 12.7 and is

$$\frac{[SSE(C8) - SSE(C7)]/[dfE(C8) - dfE(C7)]}{MSE(C2)}$$

$$= \frac{[0.59174 - 0.49723]/[23 - 22]}{0.00603} = \frac{0.09451}{0.00603} = 15.67.$$

Once again we are using a denominator from model C2.

Table 12.8 contains an ANOVA table based on a model that includes only two treatments, one for the West Coast and one for the east. The indices are in column 9 of Table 12.4. Table 12.8 also illustrates the condition on the μ_is that is incorporated into model C2 to get the ANOVA table for this reduced model, i.e., the ANOVA table F test with value $F_{obs} = 0.74$ examines whether there is any difference between the West and East Coast labs, implicitly treating all labs on each coast the same. However, this test is biased by the fact that there really are differences among the East Coast labs and among the West Coast labs. A better test would use $MSE(C2)$ in the denominator, hence $F = 0.06597/0.00603 = 10.94$ with 1 and 21 degrees of freedom, which provides a hugely different result, but then neither test is easily interpretable since both are based on an assumption that is pretty clearly false, i.e., that the means on each coast are all the same.

The methods illustrated in this section are useful regardless of whether the ANOVA is balanced or unbalanced. Moreover, the methods can be easily extended to two-factor ANOVAs, higher-order ANOVAs, and count data.

Altogether, we have looked primarily at six F tests to go along with our six degrees of freedom for groups. To test $H_0 : \mu_1 = \mu_2$ we compared model C3 to model C2. To test $H_0 : \mu_3 = \mu_4$ we compared models C4 and C2. To test $H_0 : \mu_6 = \mu_7$ we compared models C7 and C2. To test $H_0 :$

$\mu_1 = \mu_2 = \mu_3 = \mu_4$ we assumed $\mu_1 = \mu_2$ and $\mu_3 = \mu_4$ and compared model C6 to model C5. Normally, to test $H_0 : \mu_5 = \mu_6 = \mu_7$, we would assume $\mu_6 = \mu_7$ and test model C8 against C7, but I deviated from the traditional path and tested model C8 against model C2, a test that has two degrees of freedom in the numerator, while all these others only have one. Finally, to test $H_0 : \mu_1 = \mu_2 = \mu_3 = \mu_4$; $\mu_5 = \mu_6 = \mu_7$ we assumed $\mu_1 = \mu_2 = \mu_3 = \mu_4$ and $\mu_5 = \mu_6 = \mu_7$ and compared model C9 to the grand-mean model. The only other test we did was the somewhat redundant test of model C5 versus C2, which was a simultaneous test of $H_0 : \mu_1 = \mu_2$, $\mu_3 = \mu_4$ and also had two degrees of freedom in the numerator.

12.4.3 Regression analysis

The key to performing analysis of variance as regression is creating indicator variables. For Mandel's data, we need seven indicator variables, one for each lab: $x_1, x_2, x_3, x_4, x_5, x_6, x_7$. There is a very simple relationship between the hierarchy of models we have considered in this section and these indicator variables. Each ANOVA model defined by a column of Table 12.4, i.e.,

$$(12.4.1)$$

$$
\begin{array}{c|c|c}
\mu_1 = \mu_2 & \mu_3 = \mu_4 & \mu_6 = \mu_7 \\
\mu_1 = \mu_2; \mu_3 = \mu_4 & & \mu_5 = \mu_6 = \mu_7 \\
\mu_1 = \mu_2 = \mu_3 = \mu_4 & & \\
& \mu_1 = \mu_2 = \mu_3 = \mu_4; \mu_5 = \mu_6 = \mu_7 & \\
& \mu_1 = \cdots = \mu_7 &
\end{array}
$$

has a corresponding model defined by adding together the indicators that correspond to any means that have been set equal.

$$
\begin{array}{c|c|c}
& x_1, \ldots, x_7 & \\
x_1 + x_2, x_3, \ldots, x_7 & x_1, x_2, x_3 + x_4, x_5, x_6, x_7 & x_1, \ldots, x_5, x_6 + x_7 \\
x_1 + x_2, x_3 + x_4, x_5, x_6, x_7 & & \\
x_1 + \cdots + x_4, x_5, x_6, x_7 & & x_1, \ldots, x_4, x_5 + x_6 + x_7 \\
& x_1 + \cdots + x_4, x_5 + x_6 + x_7 & \\
& x_1 + \cdots + x_7 &
\end{array}
$$

In each case, we fit a regression through the origin (no intercept) using all of the variables indicated.

12.4.4 Relation to orthogonal contrasts

The maxim in unbalanced analyses is that if you change anything, you change everything. The beauty of balanced analyses is that the maxim does not apply. A great many things remain invariant to changes in balanced analyses.

Mandel's data are balanced, with equal numbers of observations on each group and normally distributed data, so there is a beautiful analysis that can be made using orthogonal contrasts, cf. Christensen (1996, Chapter 7). It was long my intention to demonstrate how the hierarchical approach displayed here relates to the balanced analysis. Everything that one would look at when exploiting the balance of Mandel's data also appears in our hierarchical analysis. But after writing the next subsection, I have come to realize just how exceptional the balanced analysis is. It is so exceptional, that I no longer think it is worth the effort to relate it to the unbalanced analysis that is our focus. Christensen (1996) treats balanced analyses in great detail, so if you want to learn about them, you could look there.

12.4.5 Theory: Difficulties in general unbalanced analyses

We have presented a reasonable model-based method for exploring group effects that would traditionally be explored by examining a series of contrasts in an unbalanced one-way ANOVA. To keep the discussion as simple as possible, I have been hiding how complicated these issues can really be. Fortunately, we will see in Chapters 14 and 15 and later chapters that when examining real data we can often find good-fitting models without subjecting ourselves to the pain suggested by the remainder of this subsection.

When looking at unbalanced analyses, if you change anything, you change everything. In Subsection 12.1.3 we looked at inference for the parameter $\mu_1 - (\mu_2 + \mu_3)/2$, which has an estimate of $\bar{y}_1. - (\bar{y}_2. + \bar{y}_3.)/2$. This uses the standard method for looking at contrasts, which is pretty much the only one ever taught. This contrast is of primary interest when $\mu_2 = \mu_3$ because otherwise you are comparing μ_1 to an average that does not represent any particular group. What almost never gets pointed out is that if you actually incorporate $\mu_2 = \mu_3$ for unbalanced data it changes the estimate of the contrast. For the unbalanced suicide data, if we look at the same parameter $\mu_1 - (\mu_2 + \mu_3)/2$ *after incorporating* $\mu_2 = \mu_3$, the estimate changes to $\bar{y}_1. - (34\bar{y}_2. + 15\bar{y}_3.)/(34 + 15)$. If you change anything, you change everything.

If Mandel's seven laboratories did not have any natural structure to them, we would compare all of the pairs of labs to see which ones are different. But in complicated unbalanced models this activity can be surprisingly difficult because the results can depend on the order in which you do the comparisons. The number of pairwise comparisons of 7 labs is 21. The number of orders in which you can choose to look at these 21 comparisons is huge, 5×10^{19}, and for unbalanced data the results can depend on the order in which you choose to look at the pairwise comparisons. For example, when we compare μ_1 with μ_2 we typically do not modify the estimates of these parameters based on what we have previously decided about whether the two means equal μ_3 or μ_4 because it is much simpler if we do not worry about such things. In more complicated models, we have to pick some method that we think is reasonable and not worry about the fact that we cannot examine every possible method for evaluating all the pairs. In a one-way ANOVA, the accepted process of testing 21 pairs of means is to look at each one of them as if they were the only thing being tested, which is what we will do in the next chapter. The same device works pretty well in more general models.

Fortunately, we have a structure to the 7 groups in Mandel's data that allows us to focus on just 6 comparisons rather than the 21 pairs of means. Nonetheless, for complicated models in which results depend on the order in which we evaluate things, there could be 720 different orderings to consider. For example, there are two obvious but different tests for $\mu_1 = \mu_2$. It can be tested by comparing Model (12.4.1) to model C3 that only incorporates $\mu_1 = \mu_2$ but it can also be tested by comparing model C4 that assumes $\mu_3 = \mu_4$ to model C5 with both $\mu_1 = \mu_2$ and $\mu_3 = \mu_4$. Other tests could also be constructed for $\mu_1 = \mu_2$ that depend on other relationships among our six contrasts of interest. For example, we could test a reduced model with $\mu_1 = \mu_2$ and $\mu_5 = \mu_6 = \mu_7$ against a model with just $\mu_5 = \mu_6 = \mu_7$. In a one-way ANOVA, if we always use MSE(C2) in the denominator, the tests will remain the same, but if we use different denominators or in more complicated models than one-way ANOVA, the tests can differ; see Subsection 14.2.1

In complicated unbalanced models, these orderings typically lead to different tests, cf. Section 14.2. The 720 orderings are far too many for us to evaluate them all. We need to pick some reasonable method and not worry about the fact that we cannot examine every ordering. The hierarchical approach displayed earlier provides one such method. Moreover, the exact results for examining a hierarchy of complicated unbalanced models depend on the exact way in which we have modeled other aspects of the problem that are not directly related to the hierarchy.

12.5 Polynomial regression and one-way ANOVA

We now exploit the relationships between polynomial regression and analysis of variance. In some analysis of variance problems, the treatment groups are determined by quantitative levels of a factor. For example, one might take observations on the depth of hole made by a drill press in a given amount of time with 20, 30, or 40 pounds of downward thrust applied. The groups are determined by the quantitative levels, 20, 30, and 40. In such a situation we could fit a one-way analysis of variance with three groups, or we could fit a simple linear regression model. Simple linear regression is appropriate because all the data come as pairs. The pairs are (x_i, y_{ij}), where x_i is the numerical

Table 12.9: *Axial stiffness index data.*

Plate	ASI	Plate	ASI	Plate	ASI	Plate	ASI	Plate	ASI
4	309.2	6	402.1	8	392.4	10	346.7	12	407.4
4	409.5	6	347.2	8	366.2	10	452.9	12	441.8
4	311.0	6	361.0	8	351.0	10	461.4	12	419.9
4	326.5	6	404.5	8	357.1	10	433.1	12	410.7
4	316.8	6	331.0	8	409.9	10	410.6	12	473.4
4	349.8	6	348.9	8	367.3	10	384.2	12	441.2
4	309.7	6	381.7	8	382.0	10	362.6	12	465.8

Table 12.10: *ASI summary statistics.*

Plate	N	$\bar{y}_{i\cdot}$	s_i^2	s_i
4	7	333.2143	1338.6981	36.59
6	7	368.0571	816.3629	28.57
8	7	375.1286	433.7990	20.83
10	7	407.3571	1981.1229	44.51
12	7	437.1714	675.8557	26.00

level of thrust and y_{ij} is the depth of the hole on the jth trial with x_i pounds of downward thrust. Not only can we fit a simple linear regression, but we can fit polynomials to the data. In this example, we could fit no polynomial above second-degree (quadratic), because three points determine a parabola and we only have three distinct x values. If we ran the experiment with $20, 25, 30, 35,$ and 40 pounds of thrust, we could fit at most a fourth-degree (quartic) polynomial because five points determine a fourth-degree polynomial and we would only have five x values.

In general, some number a of distinct x values allows fitting of an $a - 1$ degree polynomial. Moreover, fitting the $a - 1$ degree polynomial is equivalent to fitting the one-way ANOVA with groups defined by the a different x values. However, as discussed in Section 8.2, fitting high-degree polynomials is often a very questionable procedure. The problem is not with how the model fits the observed data but with the suggestions that a high-degree polynomial makes about the behavior of the process for x values other than those observed. In the example with $20, 25, 30, 35,$ and 40 pounds of thrust, the quartic polynomial will fit as well as the one-way ANOVA model but the quartic polynomial may have to do some very weird things in the areas between the observed x values. Of course, the ANOVA model gives no indications of behavior for x values other than those that were observed. When performing regression, we usually like to have some smooth-fitting model giving predictions that, in some sense, interpolate between the observed data points. High-degree polynomials often fail to achieve this goal.

Much of the discussion that follows, other than observing the equivalence of fitting a one-way ANOVA and an $a - 1$ degree polynomial, is simply a discussion of fitting a polynomial. It is very similar to the discussion in Section 8.1 but with fewer possible values for the predictor variable x. However, we will find the concept of replacing a categorical variable with a polynomial to be a very useful one in higher-order ANOVA and in modeling count data.

EXAMPLE 12.5.1. Beineke and Suddarth (1979) and Devore (1991, p. 380) consider data on roof supports involving trusses that use light-gauge metal connector plates. Their dependent variable is an axial stiffness index (ASI) measured in kips per inch. The predictor variable is the length of the light-gauge metal connector plates. The data are given in Table 12.9.

Viewed as regression data, we might think of fitting a simple linear regression model

$$y_h = \beta_0 + \beta_1 x_h + \varepsilon_h,$$

$h = 1, \ldots, 35$. Note that while h varies from 1 to 35, there are only five distinct values of x_h that

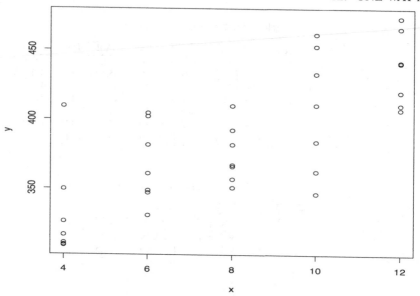

Figure 12.6: *ASI data versus plate length.*

occur in the data. The data could also be considered as an analysis of variance with plate lengths being different groups and with seven observations on each treatment. Table 12.10 gives the usual summary statistics for a one-way ANOVA. As an analysis of variance, we usually use two subscripts to identify an observation: one to identify the group and one to identify the observation within the group. The ANOVA model is often written

$$y_{ij} = \mu_i + \varepsilon_{ij}, \tag{12.5.1}$$

where $i = 1,2,3,4,5$ and $j = 1,\dots,7$. We can also rewrite the regression model using the two subscripts i and j in place of h,

$$y_{ij} = \beta_0 + \beta_1 x_i + \varepsilon_{ij},$$

where $i = 1,2,3,4,5$ and $j = 1,\dots,7$. Note that all of these models account for exactly 35 observations.

Figure 12.5 contains a scatter plot of the data. With multiple observations at each x value, the regression is really only fitted to the mean of the y values at each x value. The means of the ys are plotted against the x values in Figure 12.6. The overall trend of the data is easier to evaluate in this plot than in the full scatter plot. We see an overall increasing trend that is very nearly linear except for a slight anomaly with 6-inch plates. We need to establish if these visual effects are real or just random variation, i.e., we would also like to establish whether the simple regression model is appropriate for the trend that exists.

A more general model for trend is a polynomial. With only five distinct x values, we can fit at most a quartic (fourth-degree) polynomial, say,

$$y_{ij} = \beta_0 + \beta_1 x_i + \beta_2 x_i^2 + \beta_3 x_i^3 + \beta_4 x_i^4 + \varepsilon_{ij}. \tag{12.5.2}$$

We would prefer a simpler model, something smaller than a quartic, i.e., a cubic, quadratic, or a linear polynomial.

Table 12.11 contains ANOVA tables for fitting the linear, quadratic, cubic, and quartic polynomial regressions and for fitting the one-way ANOVA model. From our earlier discussion, the F test in the simple linear regression ANOVA table strongly suggests that there is an overall trend in the data. From Figure 12.6 we see that this trend must be increasing, i.e., as lengths go up, by and large

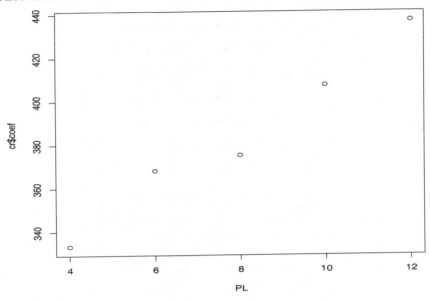

Figure 12.7: *ASI means versus plate length.*

the ASI readings go up. ANOVA tables for higher-degree polynomial models have been discussed in Chapter 8 but for now the key point to recognize is that *the ANOVA table for the quartic polynomial is identical to the ANOVA table for the one-way analysis of variance.* This occurs because models (12.5.1) and (12.5.2) are equivalent, i.e. they have the same fitted values, residuals, SSE, dfE, and MSE. Note however that the ANOVA model provides predictions (fitted values) only for the five plate lengths in the data whereas the polynomial models provide predictions for any plate length, although predictions that have dubious value when fitting high-order polynomials at lengths other than those in the data.

The first question of interest is whether a quartic polynomial is needed or whether a cubic model would be adequate. This is easily evaluated from the Table of Coefficients for the quartic fit that follows. For computational reasons, the results reported are for a polynomial involving powers of $x - \bar{x}$. rather than powers of x, cf. Section 8.1. This has *no* effect on our subsequent discussion.

Table of Coefficients

Predictor	$\hat{\beta}_k$	$SE(\hat{\beta}_k)$	t	P
Constant	375.13	12.24	30.64	0.000
$(x - \bar{x}.)$	8.768	5.816	1.51	0.142
$(x - \bar{x}.)^2$	3.983	4.795	0.83	0.413
$(x - \bar{x}.)^3$	0.2641	0.4033	0.65	0.517
$(x - \bar{x}.)^4$	-0.2096	0.2667	-0.79	0.438

There is little evidence ($P = 0.438$) that $\beta_4 \neq 0$, so a cubic polynomial seems to be an adequate explanation of the data.

This Table of Coefficients is inappropriate for evaluating β_3 in the cubic model (even the cubic model based on $x - \bar{x}$.). To evaluate β_3, we need to fit the cubic model. If we then decide that a parabola is an adequate model, evaluating β_2 in the parabola requires one to fit the quadratic model. *In general, regression estimates are only valid for the model fitted. A new model requires new estimates and standard errors.*

If we fit the sequence of polynomial models: intercept-only, linear, quadratic, cubic, quartic, we could look at testing whether the coefficient of the highest-order term is zero in each model's Table of Coefficients or we could compare the models by comparing SSEs. The latter is often more con-

Table 12.11: *Analysis of variance tables for ASI data.*

Analysis of Variance: simple linear regression					
Source	df	SS	MS	F	P
Regression	1	42780	42780	43.19	0.000
Error	33	32687	991		
Total	34	75468			

Analysis of Variance: quadratic polynomial					
Source	df	SS	MS	F	P
Regression	2	42894	21447	21.07	0.000
Error	32	32573	1018		
Total	34	75468			

Analysis of Variance: cubic polynomial					
Source	df	SS	MS	F	P
Regression	3	43345	14448	13.94	0.000
Error	31	32123	1036		
Total	34	75468			

Analysis of Variance: quartic polynomial					
Source	df	SS	MS	F	P
Regression	4	43993	10998	10.48	0.000
Error	30	31475	1049		
Total	34	75468			

Analysis of Variance: one-way ANOVA					
Source	df	SS	MS	F	P
Trts(plates)	4	43993	10998	10.48	0.000
Error	30	31475	1049		
Total	34	75468			

venient. The degrees of freedom and sums of squares for error for the four polynomial regression models and the model with only an intercept β_0 (grand mean) follow. The differences in sums of squares error for adjacent models are also given as sequential sums of squares (Seq. SS), cf. Section 9.4; the differences in degrees of freedom error are just 1.

Model comparisons

Model	dfE	SSE	(Difference) Seq. SS
Intercept	34	75468	—-
Linear	33	32687	42780
Quadratic	32	32573	114
Cubic	31	32123	450
Quartic	30	31475	648

Note that the dfE and SSE for the intercept model are those from the corrected Total lines in the ANOVAs of Table 12.11. The dfEs and SSEs for the other models also come from Table 12.11. One virtue of using this method is that many regression programs will report the Seq. SS when fitting Model (12.5.2), so we need not fit all four polynomial models, cf. Subsection 8.1.1.

To test the quartic model against the cubic model we take

$$F_{obs} = \frac{648/1}{31475/30} = 0.62 = (-0.79)^2.$$

This is just the square of the t statistic for testing $\beta_4 = 0$ in the quartic model. The reference distribution for the F statistic is $F(1,30)$ and the P value is 0.438, as it was for the t test.

If we decide that we do not need the quartic term, we can test whether we need the cubic term. We can test the quadratic model against the cubic model with

$$F_{obs} = \frac{450/1}{32123/31} = 0.434.$$

The reference distribution is $F(1,31)$. This test is equivalent to the t test of $\beta_3 = 0$ *in the cubic model. The t test of $\beta_3 = 0$ in the quartic model is inappropriate.*

Our best practice is an alternative to this F test. The denominator of this test is $32123/31$, the mean squared error from the cubic model. If we accepted the cubic model only after testing the quartic model, the result of the quartic test is open to question and thus using the MSE from the cubic model to estimate of σ^2 is open to question. It might be better just to use the estimate of σ^2 from the quartic model, which is the mean squared error from the one-way ANOVA. If we do this, the test statistic for the cubic term becomes

$$F_{obs} = \frac{450/1}{31475/30} = 0.429.$$

The reference distribution for the alternative test is $F(1,30)$. In this example the two F tests give essentially the same answers. This should, by definition, almost always be the case. If, for example, one test were significant at 0.05 and the other were not, they are both likely to have P values near 0.05 and the fact that one is a bit larger than 0.05 and the other is a bit smaller than 0.05 should not be a cause for concern. The only time these tests would be very different is if we performed them when there was considerable evidence that $\beta_4 \neq 0$, something that would be silly to do.

As originally discussed in Subsection 3.1.1, when making a series of tests related to Model (12.5.2), we recommend using its mean squared error, say $MSE(2)$, as the denominator of all the F statistics. We consider this preferable to actually fitting all four polynomial models and looking at the Table of Coefficients t statistics for the highest-order term, because the tables of coefficients from the four models will not all use $MSE(2)$.

If we decide that neither the quartic nor the cubic terms are important, we can test whether we need the quadratic term. Testing the quadratic model against the simple linear model gives

$$F_{obs} = \frac{114/1}{32573/32} = 0.112,$$

which is compared to an $F(1,32)$ distribution. This test is equivalent to the t test of $\beta_2 = 0$ in the quadratic model. Again, we prefer the quadratic term test statistic

$$F_{obs} = \frac{114/1}{31475/30} = 0.109$$

with a reference distribution of $F(1,30)$.

If we decide that we need none of the higher-order terms, we can test whether we need the linear term. Testing the intercept model against the simple linear model gives

$$F_{obs} = \frac{42780/1}{32687/33} = 43.190.$$

This is just the test for zero slope *in the simple linear regression model*. Again, the preferred test for the linear term has

$$F_{obs} = \frac{42780/1}{31475/30} = 40.775.$$

Table 12.12 contains an expanded analysis of variance table for the one-way ANOVA that incorporates the information from this sequence of comparisons.

Table 12.12: *Analysis of Variance: ASI data.*

Source	df	(Seq.) SS	MS	F	P
Treatments	4	43993	10998	10.48	0.000
(linear)	(1)	(42780)	(42780)	(40.78)	
(quadratic)	(1)	(114)	(114)	(0.11)	
(cubic)	(1)	(450)	(450)	(0.43)	
(quartic)	(1)	(648)	(648)	(0.62)	
Error	30	31475	1049		
Total	34	75468			

Residual–Plate Length plot

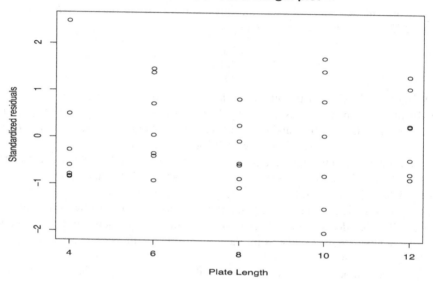

Figure 12.8: *ASI SLR standardized residuals versus plate length.*

From Table 12.12, the P value of 0.000 indicates strong evidence that the five groups are different, i.e., there is strong evidence for the quartic polynomial helping to explain the data. The results from the sequential terms are so clear that we did not bother to report P values for them. There is a huge effect for the linear term. The other three F statistics are all much less than 1, so there is no evidence of the need for a quartic, cubic, or quadratic polynomial. As far as we can tell, a line fits the data just fine. For completeness, some residual plots are presented as Figures 12.7 through 12.12. Note that the normal plot for the simple linear regression in Figure 12.8 is less than admirable, while the normal plot for the one-way ANOVA in Figure 12.12 is only slightly better. It appears that one should not put great faith in the normality assumption. □

12.5.1 Fisher's lack-of-fit test

We now give a more extensive discussion of Fisher's lack-of-fit test that was introduced in Section 8.6.

Comparing one of the reduced polynomial models against the one-way ANOVA model is often referred to as a test of *lack of fit*. This is especially true when the reduced model is the simple linear regression model. In these tests, the degrees of freedom, sums of squares, and mean squares used in the numerator of the tests are all described as being for *lack of fit*. The denominator of the test is based on the error from the one-way ANOVA. The mean square, sum of squares, and degrees of

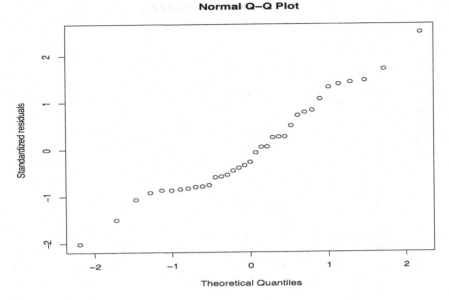

Figure 12.9: *ASI SLR standardized residuals normal plot, $W' = 0.961$.*

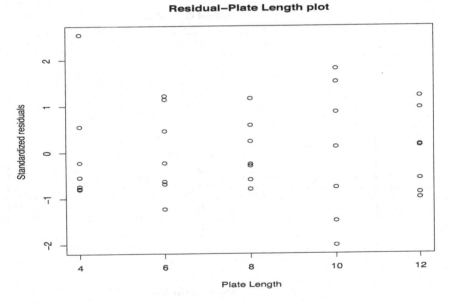

Figure 12.10: *ASI ANOVA standardized residuals versus plate length.*

freedom for error in the one-way ANOVA are often referred to as the mean square, sum of squares, and degrees of freedom for *pure error*. This lack-of-fit test can be performed *whenever* the data contain multiple observations at *any* x values. Often the appropriate unbalanced one-way ANOVA includes groups with only one observation on them. These groups do not provide an estimate of σ^2, so they simply play no role in obtaining the mean square for pure error. In Section 8.6 the Hooker data had only two x values with multiple observations and both groups only had two observations in them. Thus, the $n = 31$ cases are divided into $a = 29$ groups but only four cases were involved in finding the pure error.

For testing lack of fit in the simple linear regression model with the ASI data, the numerator

Residual–Fitted plot

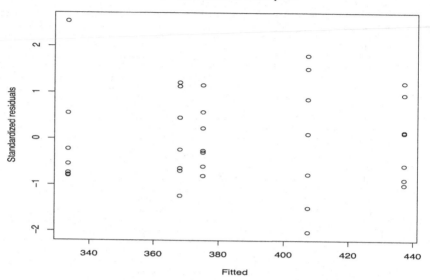

Figure 12.11: *ASI ANOVA standardized residuals versus predicted values.*

Normal Q–Q Plot

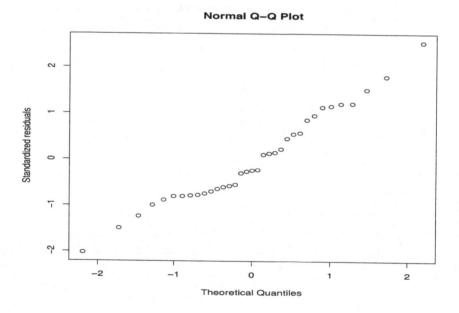

Figure 12.12: *ASI ANOVA standardized residuals normal plot, $W' = 0.965$.*

sum of squares can be obtained by differencing the sums of squares for error in the simple linear regression model and the one-way ANOVA model. Here the sum of squares for lack of fit is $32687 - 31475 = 1212$ and the degrees of freedom for lack of fit are $33 - 30 = 3$. The mean square for lack of fit is $1212/3 = 404$. The pure error comes from the one-way ANOVA table. The lack-of-fit test F statistic for the simple linear regression model is

$$F_{obs} = \frac{404}{1049} = 0.39$$

which is less than 1, so there is no evidence of a lack of fit in the simple linear regression model. If an $\alpha = 0.05$ test were desired, the test statistic would be compared to $F(0.95, 3, 30)$.

A similar lack-of-fit test is available for any of the reduced polynomial models relative to the one-way ANOVA model.

12.5.2 More on R^2

Consider a balanced one-way ANOVA,

$$y_{ij} = \mu_i + \varepsilon_{ij}, \quad E(\varepsilon_{ij}) = 0, \quad Var(\varepsilon_{ij}) = \sigma^2,$$

$i = 1, \ldots, a$, $j = 1, \ldots, N$, wherein all observations are independent and group i is associated with a scalar x_i. Now consider the simple linear regression model

$$y_{ij} = \beta_0 + \beta_1 x_i + \varepsilon_{ij} \tag{12.5.3}$$

and the model obtained by averaging within groups

$$\bar{y}_{i\cdot} = \beta_0 + \beta_1 x_i + \bar{\varepsilon}_{i\cdot}. \tag{12.5.4}$$

The two regression models give the same least squares estimates and predictions.

Exercise 12.7.17 is to show that R^2 is always at least as large for Model (12.5.4) as for Model (12.5.3) by showing that for Model (12.5.3),

$$R^2 = \frac{[N \sum_{i=1}^{a}(x_i - \bar{x}_\cdot)(\bar{y}_{i\cdot} - \bar{y}_{\cdot\cdot})]^2}{[N \sum_{i=1}^{a}(x_i - \bar{x}_\cdot)^2][N \sum_{i=1}^{a}(\bar{y}_{i\cdot} - \bar{y}_{\cdot\cdot})^2 + \sum_{ij}(y_{ij} - \bar{y}_{i\cdot})^2]} \tag{12.5.5}$$

and for Model (12.5.4),

$$R^2 = \frac{[\sum_{i=1}^{a}(x_i - \bar{x}_\cdot)(\bar{y}_{i\cdot} - \bar{y}_{\cdot\cdot})]^2}{[\sum_{i=1}^{a}(x_i - \bar{x}_\cdot)^2][\sum_{i=1}^{a}(\bar{y}_{i\cdot} - \bar{y}_{\cdot\cdot})^2]}.$$

The two regressions are equivalent in terms of finding a good model but it is easier to predict averages than individual observations because averages have less variability. R^2 for a model depends both on how good the model is relative to a perfect prediction model and how much variability there is in y when using a perfect model. Remember, perfect models can have low R^2 when there is much variability and demonstrably wrong models can have high R^2 when the variability of a perfect model is small, but the wrong model captures the more important features of a perfect model in relating x to y.

Discussing R^2 in the context of one-way ANOVA is useful because the one-way ANOVA provides a perfect model for predicting y based on x, whereas the simple linear regression model may or may not be a perfect model. For a given value of $SSTot = \sum_{ij}(y_{ij} - \bar{y}_{\cdot\cdot})^2$, the size of R^2 for the one-way ANOVA depends only on the within-group variability, that is, the variability of y in the perfect model. The size of R^2 for the simple linear regression depends on both the variability of y in the perfect model as well as how well the simple linear regression model approximates the perfect model. The term $\sum_{ij}(y_{ij} - \bar{y}_{i\cdot})^2$ in the denominator of (12.5.5) is the sum of squares error from the one-way ANOVA, so it is large precisely when the variability in the perfect model is large. (The one-way ANOVA's MSE is an estimate of the variance from a perfect prediction model.) Averaging observations within a group makes the variability of a perfect model smaller, i.e., the variance is smaller in

$$\bar{y}_{i\cdot} = \mu_i + \bar{\varepsilon}_{i\cdot}, \quad E(\bar{\varepsilon}_{i\cdot}) = 0, \quad Var(\bar{\varepsilon}_{i\cdot}) = \sigma^2/N,$$

so the R^2 of Model (12.5.4) is larger because, while the component of R^2 due to approximating the perfect model remains the same, the component due to variability in the perfect model is reduced.

12.6 Weighted least squares

In general, weighted regression is a method for dealing with observations that have nonconstant variances and nonzero correlations. In this section, we deal with the simplest form of weighted regression in which we assume zero correlations between observations. Weighted regression has some interesting connections to fitting polynomials to one-way ANOVA data that we will examine here, and it has connections to analyzing the count data considered later.

Our standard regression model from Chapter 11 has

$$Y = X\beta + e, \quad E(e) = 0, \quad \text{Cov}(e) = \sigma^2 I.$$

We now consider a model for data that do not all have the same variance. In this model, we assume that the *relative* sizes of the variances are known but that the variances themselves are unknown. In this simplest form of weighted regression, we have a covariance structure that changes from $\text{Cov}(e) = \sigma^2 I$ to $\text{Cov}(e) = \sigma^2 D(w)^{-1}$. Here $D(w)$ is a diagonal matrix with *known* weights $w = (w_1, \ldots, w_n)'$ along the diagonal. The covariance matrix involves $D(w)^{-1}$, which is just a diagonal matrix having diagonal entries that are $1/w_1, \ldots, 1/w_n$. The variance of an observation y_i is σ^2/w_i. If w_i is large relative to the other weights, the relative variance of y_i is small, so it contains more information than other observations and we should place more weight on it. Conversely, if w_i is relatively small, the variance of y_i is large, so it contains little information and we should place little weight on it. For all cases, w_i is a measure of how much relative weight should be placed on case i. Note that the weights are relative, so we could multiply or divide them all by a constant and obtain essentially the same analysis. Obviously, in standard regression the weights are all taken to be 1.

In matrix form, our new model is

$$Y = X\beta + e, \quad E(e) = 0, \quad \text{Cov}(e) = \sigma^2 D(w)^{-1}. \tag{12.6.1}$$

In this model all the observations are uncorrelated because the covariance matrix is diagonal. We do not know the variance of any observation because σ^2 is unknown. However, we do know the relative sizes of the variances because we know the weights w_i. It should be noted that when Model (12.6.1) is used to make predictions, it is necessary to specify weights for any future observations.

Before giving a general discussion of weighted regression models, we examine some examples of their application. A natural application of weighted regression is to data for a one-way analysis of variance with groups that are quantitative levels of some factor. With a quantitative factor, we can perform either a one-way ANOVA or a regression on the data. However, if for some reason the full data are not available, we can still obtain an appropriate simple linear regression by performing a weighted regression analysis on the treatment means. The next examples explore the relationships between regression on the full data and weighted regression on the treatment means.

In the weighted regression, the weights turn out to be the treatment group sample sizes from the ANOVA. In a standard unbalanced ANOVA $y_{ij} = \mu_i + \varepsilon_{ij}$, $i = 1, \ldots, a$, $j = 1, \ldots, N_i$, the sample means have $\text{Var}(\bar{y}_{i\cdot}) = \sigma^2/N_i$. Thus, if we perform a regression on the means, the observations have different variances. In particular, from our earlier discussion of variances and weights, it is appropriate to take the sample sizes as the weights, i.e., $w_i = N_i$.

EXAMPLE 12.6.1.   In Section 12.5 we considered the axial stiffness data of Table 12.9. A simple linear regression on the full data gives the following:

Table of Coefficients: SLR

Predictor	$\hat{\beta}_k$	SE($\hat{\beta}_k$)	t	P
Constant	285.30	15.96	17.88	0.000
x (plates)	12.361	1.881	6.57	0.000

The analysis of variance table for the simple linear regression is given below. The usual error line would have 33 degrees of freedom but, as per Subsection 12.5.1, we have broken this into two components, one for lack of fit and one for pure error.

Analysis of Variance: SLR

Source	df	SS	MS
Regression	1	42780	42780
Lack of fit	3	1212	404
Pure error	30	31475	1049
Total	34	75468	

In Section 12.5 we presented group summary statistics, $\bar{y}_{i\cdot}$ and s_i^2, for the four plate lengths. The mean squared pure error is just the pooled estimate of the variance and the sample sizes and sample means are given below.

Plate	4	6	8	10	12
N	7	7	7	7	7
$\bar{y}_{i\cdot}$	333.2143	368.0571	375.1286	407.3571	437.1714

As mentioned in Section 12.5, one can get the same estimated line by just fitting a simple linear regression to the means. For an unbalanced ANOVA, getting the correct regression line from the means requires a weighted regression. In this balanced case, if we use a weighted regression we get not only the same fitted line but also some interesting relationships in the ANOVA tables. Below are given the Table of Coefficients and the ANOVA table for the weighted regression on the means. The weights are the sample sizes for each mean.

Table of Coefficients: Weighted SLR

Predictor	$\hat{\beta}_k$	SE($\hat{\beta}_k$)	t	P
Constant	285.30	10.19	27.99	0.000
x (plates)	12.361	1.201	10.29	0.002

Analysis of Variance: Weighted SLR

Source	df	SS	MS	F	P
Regression	1	42780	42780	105.88	0.002
Error	3	1212	404		
Total	4	43993			

The estimated regression coefficients are identical to those given in Section 12.5. The standard errors and thus the other entries in the table of coefficients differ. In the ANOVA tables, the regression lines agree while the error line from the weighted regression is identical to the lack-of-fit line in the ANOVA table for the full data. In the weighted regression, all standard errors use the lack of fit as an estimate of the variance. In the regression on the full data, the standard errors use a variance estimate obtained from pooling the lack of fit and the pure error. The ultimate point is that by using weighted regression on the summarized data, we can still get most relevant summary statistics for simple linear regression. Of course, this assumes that the simple linear regression model is correct, and unfortunately the weighted regression does not allow us to test for lack of fit.

If we had taken all the weights to be one, i.e., if we had performed a standard regression on the means, the parameter estimate table would be the same but the ANOVA table would not display the identities discussed above. The sums of squares would all have been off by a factor of 7.   □

Unbalanced weights

We now examine an unbalanced one-way ANOVA and again compare a simple linear regression including identification of pure error and lack of fit to a weighted regression on sample means.

EXAMPLE 12.6.2.   Consider the data of Exercise 6.11.1 and Table 6.8. These involve ages of truck tractors and the costs of maintaining the tractors. A simple linear regression on the full data yields the tables given below.

Table of Coefficients: SLR

Predictor	$\hat{\beta}_k$	$SE(\hat{\beta}_k)$	t	P
Constant	323.6	146.9	2.20	0.044
Age	131.72	35.61	3.70	0.002

Analysis of Variance: SLR

Source	df	SS	MS
Regression	1	1099635	1099635
Lack of fit	5	520655	104131
Pure error	10	684752	68475
Total	16	2305042	

The weighted regression analysis is based on the sample means and sample sizes given below. The means serve as the y variable, the ages are the x variable, and the sample sizes are the weights.

Age	0.5	1.0	4.0	4.5	5.0	5.5	6.0
N_i	2	3	3	3	3	1	2
$\bar{y}_{i\cdot}$	172.5	664.3	633.0	900.3	1202.0	987.0	1068.5

The Table of Coefficients and ANOVA table for the weighted regression are

Table of Coefficients: Weighted SLR

Predictor	$\hat{\beta}_k$	$SE(\hat{\beta}_k)$	t	P
Constant	323.6	167.3	1.93	0.111
Age	131.72	40.53	3.25	0.023

and

Analysis of Variance: Weighted SLR

Source	df	SS	MS	F	P
Regression	1	1099635	1099635	10.56	0.023
Error	5	520655	104131		
Total	6	1620290			

Note that, as in the previous example, the regression estimates agree with those from the full data, that the regression sum of squares from the ANOVA table agrees with the full data, and that the lack of fit line from the full data ANOVA agrees with the error line from the weighted regression. *For an unbalanced ANOVA, you cannot obtain a correct simple linear regression analysis from the group means without using weighted regression.*   □

12.6.1 Theory

The analysis of the weighted regression model (12.6.1) is based on changing it into a standard regression model. The trick is to create a new diagonal matrix that has entries $\sqrt{w_i}$. In a minor abuse of notation, we write this matrix as $D(\sqrt{w})$. We now multiply Model (12.6.1) by this matrix to obtain

$$D(\sqrt{w})Y = D(\sqrt{w})X\beta + D(\sqrt{w})e. \qquad (12.6.2)$$

It is not difficult to see that

$$E\big(D(\sqrt{w})e\big) = D(\sqrt{w})E(e) = D(\sqrt{w})0 = 0$$

and

$$Cov\big(D(\sqrt{w})e\big) = D(\sqrt{w})Cov(e)D(\sqrt{w})' = D(\sqrt{w})\left[\sigma^2 D(w)^{-1}\right]D(\sqrt{w}) = \sigma^2 I.$$

Thus Equation (12.6.2) defines a standard regression model. For example, by Proposition 11.3.1, the least squares regression estimates from Model (12.6.2) are

$$\begin{aligned}
\hat{\beta} &= \left([D(\sqrt{w})X]'[D(\sqrt{w})X]\right)^{-1}[D(\sqrt{w})X]'[D(\sqrt{w})Y] \\
&= (X'D(w)X)^{-1}X'D(w)Y.
\end{aligned}$$

Table 12.13: *Rubber stress at five laboratories.*

Lab.	Sample size	Sample mean	Sample variance
1	4	57.00	32.00
2	4	67.50	46.33
3	4	40.25	14.25
4	4	56.50	5.66
5	4	52.50	6.33

The estimate of β given above is referred to as a weighted least squares estimate because rather than minimizing $[Y - X\beta]'[Y - X\beta]$, the estimates are obtained by minimizing

$$[D(\sqrt{w})Y - D(\sqrt{w})X\beta]'[D(\sqrt{w})Y - D(\sqrt{w})X\beta] = [Y - X\beta]'D(w)[Y - X\beta].$$

Thus the original minimization problem has been changed into a similar minimization problem that incorporates the weights. The sum of squares for error from Model (12.6.2) is

$$SSE = \left[D(\sqrt{w})Y - D(\sqrt{w})X\hat{\beta}\right]'\left[D(\sqrt{w})Y - D(\sqrt{w})X\hat{\beta}\right] = \left[Y - X\hat{\beta}\right]'D(w)\left[Y - X\hat{\beta}\right].$$

The dfE are unchanged from a standard model and MSE is simply SSE divided by dfE. Standard errors are found in much the same manner as usual except now

$$\text{Cov}\left(\hat{\beta}\right) = \sigma^2(X'D(w)X)^{-1}.$$

Because the $D(w)$ matrix is diagonal, it is very simple to modify a computer program for standard regression to allow the analysis of models like (12.6.1). Of course, to make a prediction, a weight must now be specified for the new observation. Essentially the same idea of rewriting Model (12.6.1) as the standard regression model (12.6.2) works even when $D(w)$ is not a diagonal matrix, cf. Christensen (2011, Sections 2.7 and 2.8).

12.7 Exercises

EXERCISE 12.7.1.   In addition to the data in Table 12.4, Mandel (1972) reported stress test data from five additional laboratories. Summary statistics are given in Table 12.13. Based on just these five additional labs, compute the analysis of variance table and test for differences in means between all pairs of labs. Use $\alpha = .01$. Is there any reason to worry about the assumptions of the analysis of variance model?

EXERCISE 12.7.2.   Snedecor and Cochran (1967, Section 6.18) presented data obtained in 1942 from South Dakota on the relationship between the size of farms (in acres) and the number of acres planted in corn. Summary statistics are presented in Table 12.14. Note that the sample standard deviations rather than the sample variances are given. In addition, the pooled standard deviation is 0.4526.

(a) Give the one-way analysis of variance model with all of its assumptions. Can any problems with the assumptions be identified?

(b) Give the analysis of variance table for these data. Test whether there are any differences in corn acreages due to the different sized farms. Use $\alpha = 0.01$.

(c) Test for differences between all pairs of farm sizes using $\alpha = 0.01$ tests.

(d) Find the sum of squares for the contrast defined by the following coefficients:

Table 12.14: *Acreage in corn for different sized farms.*

Farm acres	Sample size	Sample mean	Sample std. dev.
80	5	2.9957	0.4333
160	5	3.6282	0.4056
240	5	4.1149	0.4169
320	5	4.0904	0.4688
400	5	4.4030	0.5277

Table 12.15: *Weights (in pounds) for various heights (in inches).*

Height	Sample size	Sample mean	Sample variance
63	3	121.66	158.333
65	4	131.25	72.913
66	2	142.50	112.500
72	3	171.66	158.333

Farm	80	160	240	320	400
Coeff.	−2	−1	0	1	2

What percentage is this of the treatment sum of squares?

(e) Give 95% confidence and prediction intervals for the number of acres in corn for each farm size.

EXERCISE 12.7.3.   Table 12.15 gives summary statistics on heights and weights of people. Give the analysis of variance table and test for differences among the four groups. Give a 99% confidence interval for the mean weight of people in the 72-inch height group.

EXERCISE 12.7.4.   In addition to the data discussed earlier, Mandel (1972) reported data from one laboratory on four different types of rubber. Four observations were taken on each type of rubber. The means are given below.

Material	A	B	C	D
Mean	26.4425	26.0225	23.5325	29.9600

The sample variance of the 16 observations is 14.730793. Compute the analysis of variance table, the overall F test, and test for differences between each pair of rubber types. Use $\alpha = .05$.

EXERCISE 12.7.5.   In Exercise 12.7.4 on the stress of four types of rubber, the observations on material B were 22.96, 22.93, 22.49, and 35.71. Redo the analysis, eliminating the outlier. The sample variance of the 15 remaining observations is 9.3052838.

EXERCISE 12.7.6.   Bethea et al. (1985) reported data on an experiment to determine the effectiveness of four adhesive systems for bonding insulation to a chamber. The data are a measure of the peel-strength of the adhesives and are presented in Table 12.16. A disturbing aspect of these data is that the values for adhesive system 3 are reported with an extra digit.

(a) Compute the sample means and variances for each group. Give the one-way analysis of variance model with all of its assumptions. Are there problems with the assumptions? If so, does an analysis on the square roots or logs of the data reduce these problems?

(b) Give the analysis of variance table for these (possibly transformed) data. Test whether there are any differences in adhesive systems. Use $\alpha = 0.01$.

Table 12.16: *Peel-strength of various adhesive systems.*

Adhesive system	Observations					
1	60	63	57	53	56	57
2	57	52	55	59	56	54
3	19.8	19.5	19.7	21.6	21.1	19.3
4	52	53	44	48	48	53

Table 12.17: *Weight gains of rats.*

Thyroxin	Thiouracil		Control	
132	68	68	107	115
84	63	52	90	117
133	80	80	91	133
118	63	61	91	115
87	89	69	112	95
88				
119				

(c) Test for differences between all pairs of adhesive systems using $\alpha = 0.01$ tests.

(d) Find the sums of squares i) for comparing system 1 with system 4 and ii) for comparing system 2 with system 3.

(e) Assuming that systems 1 and 4 have the same means and that systems 2 and 3 have the same means, perform a 0.01 level F test for whether the peel-strength of systems 1 and 4 differs from the peel-strength of systems 2 and 3.

(f) Give a 99% confidence interval for the mean of every adhesive system.

(g) Give a 99% prediction interval for every adhesive system.

(h) Give a 95% confidence interval for the difference between systems 1 and 2.

EXERCISE 12.7.7.   Table 12.17 contains weight gains of rats from Box (1950). The rats were given either Thyroxin or Thiouracil or were in a control group. Do a complete analysis of variance on the data. Give the model, check assumptions, make residual plots, give the ANOVA table, and examine appropriate relationships among the means.

EXERCISE 12.7.8.   Aitchison and Dunsmore (1975) presented data on Cushing's syndrome. Cushing's syndrome is a condition in which the adrenal cortex overproduces cortisol. Patients are divided into one of three groups based on the cause of the syndrome: a—adenoma, b— bilateral hyperplasia, and c—carcinoma. The data are amounts of tetrahydrocortisone in the urine of the patients. The data are given in Table 12.18. Give a complete analysis.

EXERCISE 12.7.9.   Draper and Smith (1966, p. 41) considered data on the relationship between

Table 12.18: *Tetrahydrocortisone values for patients with Cushing's syndrome.*

a	b		c
3.1	8.3	15.4	10.2
3.0	3.8	7.7	9.2
1.9	3.9	6.5	9.6
3.8	7.8	5.7	53.8
4.1	9.1	13.6	15.8
1.9			

Table 12.19: *Age and costs of maintenance for truck tractors.*

Age	Costs		
0.5	163	182	
1.0	978	466	549
4.0	495	723	681
4.5	619	1049	1033
5.0	890	1522	1194
5.5	987		
6.0	764	1373	

the age of truck tractors (in years) and the cost (in dollars) of maintaining them over a six-month period. The data are given in Table 12.19.

Note that there is only one observation at 5.5 years of age. This group does not yield an estimate of the variance and can be ignored for the purpose of computing the mean squared error. In the weighted average of variance estimates, the variance of this group is undefined but the variance gets 0 weight, so there is no problem.

Give the analysis of variance table for these data. Does cost differ with age? Is there a significant difference between the cost at 0.5 years as opposed to 1.0 year? Determine whether there are any differences between costs at 4, 4.5, 5, 5.5, and 6 years. Are there differences between the first two ages and the last five? How well do polynomials fit the data?

EXERCISE 12.7.10. Lehmann (1975), citing Heyl (1930) and Brownlee (1960), considered data on determining the gravitational constant of three elements: gold, platinum, and glass. The data Lehmann gives are the third and fourth decimal places in five determinations of the gravitational constant. Analyze the following data.

Gold	Platinum	Glass
83	61	78
81	61	71
76	67	75
79	67	72
76	64	74

EXERCISE 12.7.11. Shewhart (1939, p. 69) also presented the gravitational constant data of Heyl (1930) that was considered in the previous problem, but Shewhart reports six observations for gold instead of five. Shewhart's data are given below. Analyze these data and compare your results to those of the previous exercise.

Gold	Platinum	Glass
83	61	78
81	61	71
76	67	75
79	67	72
78	64	74
72		

EXERCISE 12.7.12. Recall that if $Z \sim N(0,1)$ and $W \sim \chi^2(r)$ with Z and W independent, then by Definition 2.1.3, $Z/\sqrt{W/r}$ has a $t(r)$ distribution. Also recall that in a one-way ANOVA with independent normal errors, a contrast has

$$\sum_{i=1}^{a} \lambda_i \bar{y}_{i\cdot} \sim N\left(\sum_{i=1}^{a} \lambda_i \mu_i, \sigma^2 \sum_{i=1}^{a} \frac{\lambda_i^2}{N_i} \right),$$

$$\frac{SSE}{\sigma^2} \sim \chi^2(dfE),$$

and MSE independent of all the $\bar{y}_{i\cdot}$s. Show that

$$\frac{\sum_{i=1}^{a} \lambda_i \bar{y}_{i\cdot} - \sum_{i=1}^{a} \lambda_i \mu_i}{\sqrt{MSE \sum_{i=1}^{a} \lambda_i^2/N_i}} \sim t(dfE).$$

EXERCISE 12.7.13.   Suppose a one-way ANOVA involves four diet treatments: Control, Beef A, Beef B, Pork, and Beans. As in Subsection 12.4.1, construct a reasonable hierarchy of models to examine that involves five rows and no semicolons.

EXERCISE 12.7.14.   Suppose a one-way ANOVA involves four diet treatments: Control, Beef, Pork, Lima Beans, and Soy Beans. As in Subsection 12.4.1, construct a reasonable hierarchy of models that involves four rows, one of which involves a semicolon.

EXERCISE 12.7.15.   Conover (1971, p. 326) presented data on the amount of iron found in the livers of white rats. Fifty rats were randomly divided into five groups of ten and each group was given a different diet. We analyze the logs of the original data. The total sample variance of the 50 observations is 0.521767 and the means for each diet are given below.

Diet	A	B	C	D	E
Mean	1.6517	0.87413	0.89390	0.40557	0.025882

Compute the analysis of variance table and test whether there are differences due to diet.

If diets A and B emphasize beef and pork, respectively, diet C emphasizes poultry, and diets D and E are based on dried beans and oats, the following contrasts may be of interest.

	Diet				
Contrast	A	B	C	D	E
Beef vs. pork	1	−1	0	0	0
Mammals vs. poultry	1	1	−2	0	0
Beans vs. oats	0	0	0	1	−1
Animal vs. vegetable	2	2	2	−3	−3

Compute sums of squares for each contrast. Construct a hierarchy of models based on the diet labels and figure out how to test them using weighted least squares and the mean squared error for pure error that you found to construct the ANOVA table. What conclusions can you draw about the data?

EXERCISE 12.7.16.   Prove formulas (12.5.3) and (12.5.4).

Chapter 13

Multiple Comparison Methods

As illustrated in Chapter 12, the most useful information from a one-way ANOVA is obtained through examining contrasts. That can be done either by estimating contrasts and performing tests and confidence intervals or by incorporating contrasts directly into reduced models. The first technique is convenient for one-way ANOVA and also for balanced multifactor ANOVA but it is difficult to apply to unbalanced multifactor ANOVA or to models for count data. In the latter cases, modeling contrasts is easier. In either case, the trick is in picking interesting contrasts to consider. Interesting contrasts are determined by the structure of the groups or are suggested by the data.

The structure of the groups often suggests contrasts that are of interest. We introduced this idea in Section 12.4. For example, if one of the groups is a standard group or a control, it is of interest to compare all of the other groups to the standard. With a groups, this leads to $a - 1$ contrasts. Later we will consider factorial group structures. These include situations such as four fertilizer groups, say,

$$n_0 p_0 \quad n_0 p_1 \quad n_1 p_0 \quad n_1 p_1$$

where $n_0 p_0$ is no fertilizer, $n_0 p_1$ consists of no nitrogen fertilizer but application of a phosphorous fertilizer, $n_1 p_0$ consists of a nitrogen fertilizer but no phosphorous fertilizer, and $n_1 p_1$ indicates both types of fertilizer. Again the group structure suggests contrasts to examine. One interesting contrast compares the two groups having nitrogen fertilizer against the two without nitrogen fertilizer, another compares the two groups having phosphorous fertilizer against the two without phosphorous fertilizer, and a third contrast compares the effect of nitrogen fertilizer when phosphorous is not applied with the effect of nitrogen fertilizer when phosphorous is applied. Again, we have a groups and $a - 1$ contrasts. Even when there is an apparent lack of structure in the groups, the very lack of structure suggests a set of contrasts. If there is no apparent structure, the obvious thing to do is compare all of the groups with all of the other groups. With three groups, there are three distinct pairs of groups to compare. With four groups, there are six distinct pairs of groups to compare. With five groups, there are ten pairs. With seven groups, there are 21 pairs. With 13 groups, there are 78 pairs.

One problem is that, with a moderate number of groups, there are many contrasts to examine. When we do tests or confidence intervals, there is a built-in chance for error. The more statistical inferences we perform, the more likely we are to commit an error. The purpose of the multiple comparison methods examined in this chapter is to control the probability of making a specific type of error. When testing many contrasts, we have many null hypotheses. This chapter considers *multiple comparison methods that control (i.e., limit) the probability of making an error in any of the tests, when* all *of the null hypotheses are correct*. Limiting this probability is referred to as weak control of the *experimentwise error rate*. It is referred to as weak control because the control only applies under the very stringent assumption that all null hypotheses are correct. Some authors consider a different approach and define strong control of the experimentwise error rate as control of the probability of falsely rejecting any null hypothesis. Thus strong control limits the probability of false rejections even when some of the null hypotheses are false. Not everybody distinguishes between weak and strong control, so the definition of experimentwise error rate depends on whose work you are reading. One argument against weak control of the experimentwise error rate is that in designed experiments, you choose groups that you expect to have different effects. In such cases, it makes little sense to concentrate on controlling the error under the assumption that all groups have the same effect. On the other hand, strong control is more difficult to establish.

Our discussion of multiple comparisons focuses on testing whether contrasts are equal to 0. In all but one of the methods considered in this chapter, the experimentwise error rate is (weakly) controlled by first doing a test of the hypothesis $\mu_1 = \mu_2 = \cdots = \mu_a$. If this test is not rejected, we do not claim that any individual contrast is different from 0. In particular, if $\mu_1 = \mu_2 = \cdots = \mu_a$, any contrast among the means must equal 0, so all of the null hypotheses are correct. Since the error rate for the test of $\mu_1 = \mu_2 = \cdots = \mu_a$ is controlled, the weak experimentwise error rate for the contrasts is also controlled.

Many multiple testing procedures can be adjusted to provide multiple confidence intervals that have a guaranteed simultaneous coverage. Several such methods will be presented in this chapter.

Besides the group structure suggesting contrasts, the other source of interesting contrasts is having the data suggest them. If the data suggest a contrast, then the 'parameter' in our standard theory for statistical inferences is a function of the data and not a parameter in the usual sense of the word. When the data suggest the parameter, the standard theory for inferences does not apply. To handle such situations we can often include the contrasts suggested by the data in a broader class of contrasts and develop a procedure that applies to *all* contrasts in the class. In such cases we can ignore the fact that the data suggested particular contrasts of interest because these are still contrasts in the class and the method applies for all contrasts in the class. Of the methods considered in the current chapter, only Scheffé's method (discussed in Section 13.3) is generally considered appropriate for this kind of data dredging.

A number of books have been published on multiple comparison methods, e.g., Hsu (1996), Hochberg and Tamhane (1987). A classic discussion is Miller (1981), who also focuses on weak control of the experimentwise error rate, cf. Miller's Section 1.2.

We present multiple comparison methods in the context of the one-way ANOVA model (12.2.1) but the methods extend to many other situations. We will use Mandel's (1972) data from Section 12.4 to illustrate the methods.

13.1 "Fisher's" least significant difference method

The easiest way to adjust for multiple comparisons is to use the least significant difference method. To put it as simply as possible, with this method you first look at the analysis of variance F test for whether there are differences between the groups. If this test provides no evidence of differences, you quit and go home. If the test is significant at, say, the $\alpha = 0.05$ level, you just ignore the multiple comparison problem and do all other tests in the usual way at the 0.05 level. *This method is generally considered inappropriate for use with contrasts suggested by the data.* While the theoretical basis for excluding contrasts suggested by the data is not clear (at least relative to weak control of the experimentwise error rate), experience indicates that the method rejects far too many individual null hypotheses if this exclusion is not applied. In addition, many people would not apply the method unless the number of comparisons to be made was quite small.

EXAMPLE 13.1.1.   For Mandel's laboratory data, Subsection 12.4.2 discussed six F tests to go along with our six degrees of freedom for groups. To test $H_0 : \mu_1 = \mu_2$ we compared model C3 to model C2. To test $H_0 : \mu_3 = \mu_4$ we compared models C4 and C2. To test $H_0 : \mu_6 = \mu_7$ we compared models C7 and C2. To test $H_0 : \mu_1 = \mu_2 = \mu_3 = \mu_4$, we assumed $\mu_1 = \mu_2$ and $\mu_3 = \mu_4$ and compared model C6 to model C5. Normally, to test $H_0 : \mu_5 = \mu_6 = \mu_7$, we would assume $\mu_6 = \mu_7$ and test model C8 against C7. Finally, to test $H_0 : \mu_1 = \mu_2 = \mu_3 = \mu_4 = \mu_5 = \mu_6 = \mu_7$ we assumed $\mu_1 = \mu_2 = \mu_3 = \mu_4$ and $\mu_5 = \mu_6 = \mu_7$ and compared model C9 to the grand-mean model.

Under the least significant difference method with $\alpha = 0.05$, first check that the P value in Table 12.5 is no greater than 0.05, and, if it is, perform the six tests in the usual way at the 0.05 level. In Subsection 12.4.2 we did not test model C8 against C7, we tested model C8 against C2, and we also performed a test of model C4 against C2. These changes in what were tested cause no change in procedure. However, if the P value in Table 12.5 is greater than 0.05, you simply do not perform any of the other tests. □

The name "least significant difference" comes from comparing pairs of means in a balanced ANOVA. With N observations in each group, there is a number, the least significant difference (LSD), such that the difference between two means must be greater than the LSD for the corresponding groups to be considered significantly different. Generally, we have a significant difference between μ_i and μ_j if

$$\frac{|\bar{y}_{i\cdot} - \bar{y}_{j\cdot}|}{\sqrt{MSE\left[\frac{1}{N} + \frac{1}{N}\right]}} > t\left(1 - \frac{\alpha}{2}, dfE\right).$$

Multiplying both sides by the standard error leads to rejection if

$$|\bar{y}_{i\cdot} - \bar{y}_{j\cdot}| > t\left(1 - \frac{\alpha}{2}, dfE\right)\sqrt{MSE\left[\frac{1}{N} + \frac{1}{N}\right]}.$$

The number on the right is defined as the least significant difference,

$$LSD \equiv t\left(1 - \frac{\alpha}{2}, dfE\right)\sqrt{MSE\frac{2}{N}}.$$

Note that the LSD depends on the choice of α but does not depend on which means are being examined. If the absolute difference between two sample means is greater than the LSD, the population means are declared significantly different. Recall, however, that these comparisons are never attempted unless the analysis of variance F test is rejected at the α level. The reason that a single number exists for comparing all pairs of means is that in a balanced ANOVA the standard error is the same for any comparison between a pair of means.

EXAMPLE 13.1.1 CONTINUED.   For Mandel's laboratory data, the analysis of variance F test is highly significant, so we can proceed to make individual comparisons among pairs of means. With $\alpha = 0.05$,

$$LSD = t(0.975, 39)\sqrt{0.00421\left[\frac{1}{4} + \frac{1}{4}\right]} = 2.023(0.0459) = 0.093.$$

Means that are greater than 0.093 apart are significantly different. Means that are less than 0.093 apart are not significantly different. We display the results visually. Order the sample means from smallest to largest and indicate groups of means that are not significantly different by underlining the group. Such a display follows for comparing laboratories 1 through 7.

Lab.	4	7	5	6	3	2	1
Mean	4.0964	4.2871	4.6906	4.7175	4.7919	4.8612	4.9031

Laboratories 4 and 7 are distinct from all other laboratories. All the other consecutive pairs of labs are insignificantly different. Thus labs 5 and 6 cannot be distinguished. Similarly, labs 6 and 3 cannot be distinguished, 3 and 2 cannot be distinguished, and labs 2 and 1 cannot be distinguished. However, lab 5 is significantly different from labs 3, 2, and 1. Lab 6 is significantly different from labs 2 and 1. Also, lab 3 is different from lab 1.
    An alternative display is often used by computer programs.

Lab.	Mean		
4	4.0964	A	
7	4.2871	B	
5	4.6906	C	
6	4.7175	C	D
3	4.7919	E	D
2	4.8612	E	F
1	4.9031		F

Displays such as these may not be possible when dealing with unbalanced data. What makes them possible is that, with balanced data, the standard error is the same for comparing every pair of means. To illustrate their impossibility, we modify the log suicide sample means while leaving their standard errors alone. Suppose that the means are

Sample statistics: log of suicide ages
modified data

Group	N_i	$\bar{y}_{i\cdot}$
Caucasians	44	3.6521
Hispanics	34	3.3521
Native Am.	15	3.3321

(The fact that all three sample means have the same last two digits are a clue that the data are made up.) Now if we test whether all pairs of differences are zero, at $\alpha = 0.01$ the critical value is 2.632.

Table of Coefficients

Par	Est	SE(Est)	t_{obs}
$\mu_C - \mu_H$	0.3000	0.0936	3.21
$\mu_C - \mu_N$	0.3200	0.1225	2.61
$\mu_H - \mu_N$	0.0200	0.1270	0.16

The Anglo mean is farther from the Native American mean than it is from the Hispanic mean, but the Anglos and Hispanics are significantly different whereas the Anglos and the Native Americans are not.   □

Apparently some people have taken to calling this method the Fisher significant difference (FSD) method. One suspects that this is a reaction to another meaning commonly associated with the letters LSD. I, for one, would *never* suggest that only people who are hallucinating would believe all differences declared by LSD are real.

The least significant difference method has traditionally been ascribed to R. A. Fisher and is often called "Fisher's least significant difference method." However, from my own reading of Fisher, I am unconvinced that he either suggested the method or would have approved of it.

13.2 Bonferroni adjustments

The Bonferroni method is the one method we consider that *does not* stem from a test of $\mu_1 = \mu_2 = \cdots = \mu_a$. Rather, it controls the experimentwise error rate by employing a simple adjustment to the significance level of each individual test. If you have planned to do s tests, you just perform each test at the α/s level rather than at the α level. This method is *absolutely not appropriate for contrasts that are suggested by the data*.

The justification for Bonferroni's method relies on a very simple result from probability: for two events, the probability that one or the other event occurs is no more than the sum of the probabilities for the individual events. Thus with two tests, say A and B, the probability that we reject A or reject B is less than or equal to the probability of rejecting A plus the probability of rejecting B. In particular, if we fix the probability of rejecting A at $\alpha/2$ and the probability of rejecting B at $\alpha/2$, then the probability of rejecting A or B is no more than $\alpha/2 + \alpha/2 = \alpha$. More generally, if we have s tests and control the probability of type I error for each test at α/s, then the probability of rejecting any of the tests when all s null hypotheses are true is no more than $\alpha/s + \cdots + \alpha/s = \alpha$. This is precisely what we did in Subsection 7.2.2 to deal with testing multiple standardized deleted (t) residuals.

EXAMPLE 13.2.1.   For Mandel's laboratory data, using the structure exploited in Section 12.4, we had six F tests to go along with our six degrees of freedom for groups. To test $H_0 : \mu_1 = \mu_2$ we compared model C3 to model C2. To test $H_0 : \mu_3 = \mu_4$ we compared models C4 and C2. To test $H_0 : \mu_6 = \mu_7$ we compared models C7 and C2. To test $H_0 : \mu_1 = \mu_2 = \mu_3 = \mu_4$, we assumed $\mu_1 = \mu_2$ and $\mu_3 = \mu_4$ and compared model C6 to model C5. Normally, to test $H_0 : \mu_5 = \mu_6 = \mu_7$, we would assume $\mu_6 = \mu_7$ and test model C8 against C7. Finally, to test $H_0 : \mu_1 = \mu_2 = \mu_3 = \mu_4 = \mu_5 = \mu_6 = \mu_7$ we assumed $\mu_1 = \mu_2 = \mu_3 = \mu_4$ and $\mu_5 = \mu_6 = \mu_7$ and compared model C9 to the grand-mean model. Under the Bonferroni method with $\alpha = 0.05$ and six tests to perform, you simply perform each one at the $\alpha/6 = 0.05/6 = 0.0083$ level. Personally, with six tests, I would instead pick $\alpha = 0.06$ so that $\alpha/6 = 0.06/6 = 0.01$. Rather than these six tests, in Subsection 12.4.2 we actually performed seven tests, so for an $\alpha = 0.05$ Bonferroni procedure we need to perform each one at the $\alpha/7 = 0.05/7 = 0.0071$ level. Again, I would personally just raise the Bonferroni level to 0.07 and do all the tests at the 0.01 level. If I had nine tests, I would **not** raise the Bonferroni level all the way to 0.09, but I might lower it to 0.045 so that I could do the individual tests at the 0.005 level. $\Box$

To compare pairs of means in a balanced ANOVA, as with the least significant difference method, there is a single number to which we can compare the differences in means. For a fixed α, this number is called the *Bonferroni significant difference* and takes on the value

$$BSD \equiv t\left(1 - \frac{\alpha}{2s}, dfE\right)\sqrt{MSE\left[\frac{1}{N} + \frac{1}{N}\right]}.$$

Recall, for comparison, that with the least significant difference method, the necessary tabled value is $t(1 - \alpha/2, dfE)$, which is always smaller than the tabled value used in the BSD. Thus the BSD is always larger than the LSD and the BSD tends to display fewer differences among the means than the LSD.

Bonferroni adjustments can also be used to obtain confidence intervals that have a simultaneous confidence of $(1 - \alpha)100\%$ for covering all of the contrasts. The endpoints of these intervals are

$$\sum_{i=1}^{a} \lambda_i \bar{y}_{i\cdot} \pm t\left(1 - \frac{\alpha}{2s}, dfE\right) SE\left(\sum_{i=1}^{a} \lambda_i \bar{y}_{i\cdot}\right).$$

Recall that for an unbalanced ANOVA,

$$SE\left(\sum_{i=1}^{a} \lambda_i \bar{y}_{i\cdot}\right) = \sqrt{MSE \sum_{i=1}^{a} \frac{\lambda_i^2}{N_i}}.$$

Only the t distribution value distinguishes this interval from a standard confidence interval for $\sum_{i=1}^{a} \lambda_i \mu_i$. In the special case of comparing pairs of means in a balanced ANOVA, the Bonferroni confidence interval for, say, $\mu_i - \mu_j$ reduces to

$$(\bar{y}_{i\cdot} - \bar{y}_{j\cdot}) \pm BSD.$$

For these intervals, we are $(1 - \alpha)100\%$ confident that the collection of all such intervals simultaneously contain all of the corresponding differences between pairs of population means.

EXAMPLE 13.2.1.   In comparing Mandel's 7 laboratories, we have $\binom{7}{2} = 21$ pairs of laboratories to contrast. The Bonferroni significant difference for $\alpha = 0.05$ is

$$BSD = t\left(1 - \frac{0.025}{21}, 39\right)\sqrt{0.00421\left[\frac{1}{4} + \frac{1}{4}\right]}$$

$$= t(0.99881, 39)0.04588 = 3.2499(0.04588) = 0.149.$$

Means that are greater than 0.149 apart are significantly different. Means that are less than 0.149 apart are not significantly different. Once again, we display the results visually. We order the sample means from smallest to largest and indicate groups of means that are not significantly different by underlining the group.

Lab.	4	7	5	6	3	2	1
Mean	4.0964	4.2871	4.6906	4.7175	4.7919	4.8612	4.9031

Laboratories 4 and 7 are distinct from all other laboratories. Labs 5, 6, and 3 cannot be distinguished. Similarly, labs 6, 3, and 2 cannot be distinguished; however, lab 5 is significantly different from lab 2 and also lab 1. Labs 3, 2, and 1 cannot be distinguished, but lab 1 is significantly different from lab 6. Remember there is no assurance that such a display can be constructed for unbalanced data.

The Bonferroni simultaneous 95% confidence interval for, say, $\mu_2 - \mu_5$ has endpoints

$$(4.8612 - 4.6906) \pm 0.149,$$

which gives the interval (0.021, 0.320). Transforming back to the original scale from the logarithmic scale, we are 95% confident that the median for lab 2 is being between $e^{0.021} = 1.02$ and $e^{0.320} = 1.38$ times greater than the median for lab 5. Similar conclusions are drawn for the other twenty comparisons between pairs of means.   □

13.3 Scheffé's method

Scheffé's method is very general. Suppose we have some hierarchy of models that includes a biggest model (Big.), some full model (Full), a reduced model (Red.), and a smallest model (Sml.). In most hierarchies of models, there are many choices for Full and Red. but Big. and Sml. are fixed. Scheffé's method can be used to perform tests on a fixed set of choices for Full and Red., or on all possible choices for Full and Red., or on a few choices determined by the data.

In Chapter 3, we introduced model testing for a full and reduced model using the F statistic

$$F = \frac{[SSE(Red.) - SSE(Full)]/[dfE(Red.) - dfE(Full)]}{MSE(Full)}$$

with reference distribution $F(dfE(Red.) - dfE(Full), dfE(Full))$. As we got into hierarchies of models, we preferred the statistic

$$F = \frac{[SSE(Red.) - SSE(Full)]/[dfE(Red.) - dfE(Full)]}{MSE(Big.)}$$

with reference distribution $F(dfE(Red.) - dfE(Full), dfE(Big.))$. Scheffé's method requires a further modification of the test statistic.

If the smallest model is true, then all of the other models are also true. The experimentwise error rate is the probability of rejecting any reduced model Red. (relative to a full model Full) when model Sml. is true. Scheffé's method allows us to compare any and all full and reduced models, those we even pick by looking at the data, and controls the experimentwise error rate at α by rejecting the reduced model only when

$$F = \frac{[SSE(Red.) - SSE(Full)]/[dfE(Sml.) - dfE(Big.)]}{MSE(Big.)}$$
$$> F(1 - \alpha, dfE(Sml.) - dfE(Big.), dfE(Big.)).$$

To justify this procedure, note that the test of the smallest model versus the biggest model rejects when

$$F = \frac{[SSE(Sml.) - SSE(Big.)]/[dfE(Sml.) - dfE(Big.)]}{MSE(Big.)}$$
$$> F(1 - \alpha, dfE(Sml.) - dfE(Big.), dfE(Big.))$$

and when the smallest model is true, this has only an α chance of occurring. Because

$$SSE(Sml.) \geq SSE(Red.) \geq SSE(Full) \geq SSE(Big.),$$

we have

$$[SSE(Sml.) - SSE(Big.)] \geq [SSE(Red.) - SSE(Full)]$$

and

$$\frac{[SSE(Sml.) - SSE(Big.)]/[dfE(Sml.) - dfE(Big.)]}{MSE(Big.)}$$
$$\geq \frac{[SSE(Red.) - SSE(Full)]/[dfE(Sml.) - dfE(Big.)]}{MSE(Big.)}.$$

It follows that you cannot reject Red. relative to Full unless you have already rejected Big. relative to Sml., and rejecting Big. relative to Sml. occurs only with probability α when Sml. is true. In other words, there is no more than an α chance of rejecting any of the reduced models when they are true.

Scheffé's method is valid for examining any and all contrasts simultaneously. *This method is primarily used with contrasts that were suggested by the data.* Scheffé's method should not be used for comparing pairs of means in a balanced ANOVA because the *HSD* method presented in the next section has properties comparable to Scheffé's but is better for comparing pairs of means.

In one-way ANOVA, the analysis of variance F test is rejected when

$$\frac{SSGrps/(a-1)}{MSE} > F(1 - \alpha, a - 1, dfE). \tag{13.3.1}$$

It turns out that for any contrast $\sum_i \lambda_i \mu_i$,

$$SS\left(\sum_i \lambda_i \mu_i\right) \leq SSGrps. \tag{13.3.2}$$

It follows immediately that

$$\frac{SS\left(\sum_i \lambda_i \mu_i\right)/(a-1)}{MSE} \leq \frac{SSGrps/(a-1)}{MSE}.$$

Scheffé's method is to replace $SSGrps$ in (13.3.1) with $SS\left(\sum_i \lambda_i \mu_i\right)$ and to reject $H_0 : \sum_i \lambda_i \mu_i = 0$ if

$$\frac{SS\left(\sum_i \lambda_i \mu_i\right)/(a-1)}{MSE} > F(1 - \alpha, a - 1, dfE).$$

From (13.3.1) and (13.3.2), Scheffé's test cannot possibly be rejected unless the ANOVA test is rejected. This controls the experimentwise error rate for multiple tests. Moreover, there always exists a contrast that contains all of the $SSGrps$, i.e., there is always a contrast that achieves equality in relation (13.3.2), so if the ANOVA test is rejected, there is always some contrast that can be rejected using Scheffé's method. This contrast may not be interesting but it exists.

Scheffé's method can be adapted to provide simultaneous $(1 - \alpha)100\%$ confidence intervals for contrasts. These have the endpoints

$$\sum_{i=1}^{a} \lambda_i \bar{y}_{i\cdot} \pm \sqrt{(a-1)F(1-\alpha, a-1, dfE)} \; \text{SE}\left(\sum_{i=1}^{a} \lambda_i \bar{y}_{i\cdot}\right).$$

13.4 Studentized range methods

Studentized range methods are generally used *only for comparing pairs of means in balanced analysis of variance problems*. (This includes the balanced multifactor ANOVAs to be discussed later.) They are not based on the analysis of variance F test but on an alternative test of $\mu_1 = \mu_2 = \cdots = \mu_a$. The method is not really appropriate for unbalanced data, but it is so common that we discuss it anyway.

The *range* of a random sample is the difference between the largest observation and the smallest observation. For a known variance σ^2, the *range* of a random sample from a normal population has a distribution that can be worked out. This distribution depends on σ^2 and the number of observations in the sample. It is only reasonable that the distribution depend on the number of observations because the difference between the largest and smallest observations ought to be larger in a sample of 75 observations than in a sample of 3 observations. Just by chance, we would expect the extreme observations to become more extreme in larger samples.

Knowing the distribution of the range is not very useful because the distribution depends on σ^2, which we do not know. To eliminate this problem, divide the range by an independent estimate of the standard deviation, say, $\hat{\sigma}$ having $r\hat{\sigma}^2/\sigma^2 \sim \chi^2(r)$. The distribution of such a *studentized range* no longer depends on σ^2 but rather it depends on the degrees of freedom for the variance estimate. For a sample of n observations and a variance estimate with r degrees of freedom, the distribution of the studentized range is written as

$$Q(n, r).$$

Tables are given in Appendix B.5. The α percentile is denoted $Q(\alpha, n, r)$.

If $\mu_1 = \mu_2 = \cdots = \mu_a$ in a balanced ANOVA, the $\bar{y}_{i\cdot}$s form a random sample of size a from a $N(\mu_1, \sigma^2/N)$ population. Looking at the range of this sample and dividing by the natural independent chi-squared estimate of the standard deviation leads to the statistic

$$Q = \frac{\max \bar{y}_{i\cdot} - \min \bar{y}_{i\cdot}}{\sqrt{MSE/N}}.$$

If the observed value of this studentized range statistic Q is consistent with its coming from a $Q(a, dfE)$ distribution, then the data are consistent with the null hypothesis of equal means μ_i. If the μ_is are not all equal, the studentized range Q tends to be larger than if the means were all equal; the difference between the largest and smallest observations will involve not only random variation but also the differences in the μ_is. Thus, for an $\alpha = 0.05$ level test, if the observed value of Q is larger than $Q(0.95, a, dfE)$, we reject the claim that the means are all equal.

In applying these methods to a higher-order ANOVA, the key ideas are to compare a set of sample means using the MSE appropriate to the model and taking N as the number of observations that go into each mean.

The studentized range multiple comparison methods discussed in this section begin with this studentized range test.

13.4.1 Tukey's honest significant difference

John Tukey's honest significant difference method is to reject the equality of a pair of means, say, μ_i and μ_j at the $\alpha = 0.05$ level, if

$$\frac{|\bar{y}_{i\cdot} - \bar{y}_{j\cdot}|}{\sqrt{MSE/N}} > Q(0.95, a, dfE).$$

Obviously, this test cannot be rejected for any pair of means unless the test based on the maximum and minimum sample means is also rejected. For an equivalent way of performing the test, reject equality of μ_i and μ_j if

$$|\bar{y}_{i\cdot} - \bar{y}_{j\cdot}| > Q(0.95, a, dfE)\sqrt{MSE/N}.$$

With a fixed α, the honest significant difference is

$$HSD \equiv Q(1 - \alpha, a, dfE)\sqrt{MSE/N}.$$

For any pair of sample means with an absolute difference greater than the HSD, we conclude that the corresponding population means are significantly different. The HSD is the number that an observed difference must be greater than in order for the population means to have an 'honestly' significant difference. The use of the word 'honest' is a reflection of the view that the LSD method allows 'too many' rejections.

Tukey's method can be extended to provide simultaneous $(1 - \alpha)100\%$ confidence intervals for all differences between pairs of means. The interval for the difference $\mu_i - \mu_j$ has end points

$$\bar{y}_{i\cdot} - \bar{y}_{j\cdot} \pm HSD$$

where HSD depends on α. For $\alpha = 0.05$, we are 95% confident that the collection of all such intervals simultaneously contains all of the corresponding differences between pairs of population means.

EXAMPLE 13.4.1.   For comparing the 7 laboratories in Mandel's data with $\alpha = 0.05$, the honest significant difference is approximately

$$HSD = Q(0.95, 7, 40)\sqrt{MSE/4} = 4.39\sqrt{0.00421/4} = 0.142.$$

Here we have used $Q(0.95, 7, 40)$ rather than the correct value $Q(0.95, 7, 39)$ because the correct value was not available in the table used. Group means that are more than 0.142 apart are significantly different. Means that are less than 0.142 apart are not significantly different. Note that the HSD value is similar in size to the corresponding BSD value of 0.149; this frequently occurs. Once again, we display the results visually.

Lab.	4	7	5	6	3	2	1
Mean	4.0964	4.2871	4.6906	4.7175	4.7919	4.8612	4.9031

These results are nearly the same as for the BSD except that labs 6 and 2 are significantly different by the HSD criterion. Many Statistics packages will either perform Tukey's procedure or allow you to find $Q(1 - \alpha, a, dfE)$.

The HSD simultaneous 95% confidence interval for, say, $\mu_2 - \mu_5$ has endpoints

$$(4.8612 - 4.6906) \pm 0.142,$$

which gives the interval $(0.029, 0.313)$. Transforming back to the original scale from the logarithmic scale, we are 95% confident that the median for lab 2 is between $e^{0.029} = 1.03$ and $e^{0.313} = 1.37$ times greater than the median for lab 5. Again, there are 20 more intervals to examine.   □

The *Newman–Keuls multiple range method* involves repeated use of the honest significant difference method with some minor adjustments; see Christensen (1996) for an example.

Table 13.1: *Rubber stress at five laboratories.*

Lab.	Sample size	Sample mean	Sample variance
1	4	57.00	32.00
2	4	67.50	46.33
3	4	40.25	14.25
4	4	56.50	5.66
5	4	52.50	6.33

13.5 Summary of multiple comparison procedures

The least significant difference, the Bonferroni, and the Scheffé methods can be used for arbitrary sets of preplanned hypotheses. They are listed in order from least conservative (most likely to reject an individual null hypothesis) to most conservative (least likely to reject). Scheffé's method can also be used for examining contrasts suggested by the data. Bonferroni's method has the advantage that it can easily be applied to almost any multiple testing problem.

To compare all of the groups in a balanced analysis of variance, we can use the least significant difference, the Bonferroni, and the Tukey methods. Again, these are (roughly) listed in the order from least conservative to most conservative. In some cases, for example when comparing Bonferroni and Tukey, an exact statement of which is more conservative is not possible.

To decide on a method, you need to decide on how conservative you want to be. If it is very important not to claim differences when there are none, you should be very conservative. If it is most important to identify differences that *may* exist, then you should choose less conservative methods.

Many methods other than those discussed have been proposed for balanced ANOVA models. Some of those are discussed in Christensen (1996, Chapter 6).

Note that methods for balanced ANOVA models are much better developed than for unbalanced models, so with our emphasis on unbalanced models, our discussion is relatively short. Also, multiple comparison methods seem to be closely tied to Neyman–Pearson theory, something I sought to avoid. Fisher used similar adjustments, but apparently for different philosophical reasons, cf. Fisher (1935, Section 24).

13.6 Exercises

EXERCISE 13.6.1.   Exercise 12.7.1 involved measurements from different laboratories on the stress at 600% elongation for a certain type of rubber. The summary statistics are repeated in Table 13.1. Ignoring any reservations you may have about the appropriateness of the analysis of variance model for these data, compare all pairs of laboratories using $\alpha = 0.10$ for the LSD, Bonferroni, Tukey, and Newman–Keuls methods. Give joint 95% confidence intervals using Tukey's method for all differences between pairs of labs.

EXERCISE 13.6.2.   Use Scheffé's method with $\alpha = 0.01$ to test whether the contrast in Exercise 12.7.2d is zero.

EXERCISE 13.6.3.   Use Bonferroni's method with an α near 0.01 to give simultaneous confidence intervals for the mean weight in each height group for Exercise 12.7.3.

EXERCISE 13.6.4.   Exercise 12.7.4 contained data on stress measurements for four different types of rubber. Four observations were taken on each type of rubber; the means are repeated below,

Material	A	B	C	D
Mean	26.4425	26.0225	23.5325	29.9600

and the sample variance of the 16 observations is 14.730793. Test for differences between all pairs of materials using $\alpha = 0.05$ for the LSD, Bonferroni, and Tukey methods. Give 95% confidence intervals for the differences between all pairs of materials using the BSD method.

EXERCISE 13.6.5.   In Exercise 12.7.5 on the stress of four types of rubber, an outlier was noted in material B. Redo the multiple comparisons of the previous problem eliminating the outlier and using only the methods that are still applicable.

EXERCISE 13.6.6.   In Exercise 12.7.6 on the peel-strength of different adhesive systems, parts (b) and (c) amount to doing LSD multiple comparisons for all pairs of systems. Compare the LSD results with the results obtained using Tukey's methods with $\alpha = .01$.

EXERCISE 13.6.7.   For the weight gain data of Exercise 12.7.7, use the LSD, Bonferroni, and Scheffé methods to test whether the following contrasts are zero: 1) the contrast that compares the two drugs and 2) the contrast that compares the control with the average of the two drugs. Pick an α level but clearly state the level chosen.

EXERCISE 13.6.8.   For the Cushing's syndrome data of Exercise 12.7.8, use all appropriate methods to compare all pairwise differences among the three groups. Pick an α level but clearly state the level chosen.

EXERCISE 13.6.9.   Use Scheffé's method with $\alpha = 0.05$ and the data of Exercise 12.7.9 to test the significance of the contrast

Age	0.5	1.0	4.0	4.5	5.0	5.5	6.0
Coeff.	−5	−5	2	2	2	2	2

EXERCISE 13.6.10.   Restate the least significant difference method in terms of testing Biggest, Full, Reduced, and Smallest models.

Chapter 14

Two-Way ANOVA

This chapter involves many model comparisons, so, for simplicity within a given section, say 14.2, equation numbers such as (14.2.1) that redundantly specify the section number are referred to in the text without the section number, hence simply as (1). When referring to an equation number outside the current section, the full equation number is given.

14.1   Unbalanced two-way analysis of variance

Bailey (1953), Scheffé (1959), and Christensen (2011) examined data on infant female rats that were given to foster mothers for nursing. The variable of interest was the weight of the rat at 28 days. Weights were measured in grams. Rats are classified into four genotypes: A, F, I, and J. Specifically, rats from litters of each genotype were given to a foster mother of each genotype. The data are presented in Table 14.1.

Table 14.1: *Infant rats weight gain with foster mothers.*

Genotype of Litter	Genotype of Foster Mother			
	A	F	I	J
A	61.5	55.0	52.5	42.0
	68.2	42.0	61.8	54.0
	64.0	60.2	49.5	61.0
	65.0		52.7	48.2
	59.7			39.6
F	60.3	50.8	56.5	51.3
	51.7	64.7	59.0	40.5
	49.3	61.7	47.2	
	48.0	64.0	53.0	
		62.0		
I	37.0	56.3	39.7	50.0
	36.3	69.8	46.0	43.8
	68.0	67.0	61.3	54.5
			55.3	
			55.7	
J	59.0	59.5	45.2	44.8
	57.4	52.8	57.0	51.5
	54.0	56.0	61.4	53.0
	47.0			42.0
				54.0

14.1.1   Initial analysis

One way to view these data is as a one-way ANOVA with $4 \times 4 = 16$ groups. Specifically,

$$y_{hk} = \mu_h + \varepsilon_{hk},$$

with $h = 1, \ldots, 16, k = 1, \ldots, N_h$. It is convenient to replace the subscript h with the pair of subscripts (i, j) and write

$$y_{ijk} = \mu_{ij} + \varepsilon_{ijk}, \tag{14.1.1}$$

$$\varepsilon_{ijk}\text{s independent } N(0, \sigma^2),$$

where $i = 1, \ldots, 4$ indicates the litter genotype and $j = 1, \ldots, 4$ indicates the foster mother genotype so that, together, i and j identify the 16 groups. The index $k = 1, \ldots, N_{ij}$ indicates the various observations in each group.

Equivalently, we can write an overparameterized version of Model (1) called the *interaction model*,

$$y_{ijk} = \mu + \alpha_i + \eta_j + \gamma_{ij} + \varepsilon_{ijk}. \tag{14.1.2}$$

The idea is that μ is an overall effect (grand mean) to which we add α_i, an effect for the ith litter genotype, plus η_j, an effect for the j foster mother genotype, plus an effect γ_{ij} for each combination of a litter genotype and foster mother genotype. Comparing the interaction model (2) with the one-way ANOVA model (1), we see that the γ_{ij}s in (2) play the same role as the μ_{ij}s in (1), making all of the μ, α_i and η_j parameters completely redundant. There are 16 groups so we only need 16 parameters to explain the group means and there are 16 γ_{ij}s. In particular, all of the μ, α_i and η_j parameters could be 0 and the interaction model would explain the data exactly as well as Model (1). In fact, we could set these parameters to be any numbers at all and still have a free γ_{ij} parameter to explain each group mean. It is equally true that any data features that the μ, α_i and η_j parameters could explain could already be explained by the γ_{ij}s.

So why bother with the interaction model? Simply because dropping the γ_{ij}s out of the model gives us a much simpler, more interpretable *no-interaction model*

$$y_{ijk} = \mu + \alpha_i + \eta_j + \varepsilon_{ijk}, \quad \varepsilon_{ij}\text{s independent } N(0, \sigma^2) \tag{14.1.3}$$

in which we have structured the effects of the litter and foster mother genotypes so that each adds some fixed amount to our observations. Model (3) is actually a special case of the general additive-effects model (9.9.2), which did not specify whether predictors were categorical or measurement variables. In Model (3), the population mean difference between litter genotypes A and F must be the same, regardless of the foster mother genotype, i.e.,

$$(\mu + \alpha_1 + \eta_j) - (\mu + \alpha_2 + \eta_j) = \alpha_1 - \alpha_2.$$

Similarly, the difference between foster mother genotypes F and J must be the same regardless of the litter genotype, i.e.,

$$(\mu + \alpha_i + \eta_2) - (\mu + \alpha_i + \eta_4) = \eta_2 - \eta_4.$$

Model (3) has *additive effects* for the two factors: litter genotype and foster mother genotype. The effect for either factor is consistent across the other factor. This property is also referred to as the absence of *interaction* or as the absence of *effect modification*. Model (3) requires that the effect of any foster mother genotype be the same for every litter genotype, and also that the effect of any litter genotype be the same for every foster mother genotype. Without this property, one could not meaningfully speak about the effect of a litter genotype, because it would change from foster mother genotype to foster mother genotype. Similarly, foster mother genotype effects would depend on the litter genotypes.

Model (2) imposes no such restrictions on the factor effects. Model (2) would happily allow the

Table 14.2: *Statistics from fitting models to the data of Table 14.1.*

Model	Model	SSE	df	C_p
(14.1.2): $G+L+M+LM$	[LM]	2440.82	45	16.0
(14.1.3): $G+L+M$	[L][M]	3264.89	54	13.2
(14.1.4): $G+L$	[L]	4039.97	57	21.5
(14.1.5): $G+M$	[M]	3328.52	57	8.4
(14.1.6): G	[G]	4100.13	60	16.6

foster mother genotype that has the highest weight gains for litter type A to be also the foster mother genotype that corresponds to the smallest weight gains for Litter J, a dramatic interaction. Model (2) does not require that the effect of a foster mother genotype be consistent for every litter type or that the effect of a litter genotype be consistent for every foster mother genotype. If the effect of a litter genotype can change depending on the foster mother genotype, the model is said to display *effect modification* or *interaction*.

The γ_{ij}s in Model (2) are somewhat erroneously called *interaction effects*. Although they can explain much more than interaction, eliminating the γ_{ij}s in Model (2) eliminates any interaction. (Whereas eliminating the equivalent μ_{ij} effects in Model (1) eliminates far more than just interaction; it leads to a model in which every group has mean 0.)

The test for whether interaction exists is simply the test of the full, interaction, model (2) against the reduced, no-interaction, model (3). Remember that Model (2) is equivalent to the one-way ANOVA model (1), so models (1) and (2) have the same fitted values $\hat{y}_{ijk}$ and residuals $\hat{\varepsilon}_{ijk}$ and $dfE(1) = dfE(2)$. The analysis for models like (1) was given in Chapter 12. While it may not be obvious that Model (3) is a reduced model relative to Model (1), Model (3) is obviously a reduced model relative to the interaction model (2). Computationally, the fitting of Model (3) is much more complicated than fitting a one-way ANOVA.

If Model (3) does not fit the data, there is often little one can do except go back to analyzing Model (1) using the one-way ANOVA techniques of Chapters 12 and 13. In later chapters, depending on the nature of the factors, we will explore ways to model interaction by looking at models that are intermediate between (2) and (3).

Table 14.2 contains results for fitting models (2) and (3) along with results for fitting other models to be discussed anon. In our example, a test of whether Model (3) is an adequate substitute for Model (2) rejects Model (3) if

$$F = \frac{[SSE(3) - SSE(2)]/[dfE(3) - dfE(2)]}{SSE(2)/dfE(2)}$$

is too large. The F statistic is compared to an $F(dfE(3) - dfE(2), dfE(2))$ distribution. Specifically, we get

$$F_{obs} = \frac{[3264.89 - 2440.82]/[54 - 45]}{2440.82/45} = \frac{91.56}{54.24} = 1.69,$$

with a one-sided P value of 0.129, i.e., 1.69 is the 0.871 percentile of an $F(9,45)$ distribution denoted $1.69 = F(.871, 9, 45)$.

If Model (3) fits the data adequately, we can explore further to see if even simpler models adequately explain the data. Using Model (3) as a working model, we might be interested in whether there are really any effects due to Litters, or any effects due to Mothers. Remember that in the interaction model (2), it makes little sense even to talk about a Litter effect or a Mother effect without specifying a particular level for the other factor, so this discussion requires that Model (3) be reasonable.

The effect of Mothers can be measured in two ways. First, by comparing the no-interaction model (3) with a model that eliminates the effect for Mothers

$$y_{ijk} = \mu + \alpha_i + \varepsilon_{ijk}. \qquad (14.1.4)$$

This model comparison *assumes* that there is an effect for Litters because the α_is are included in both models. Using Table 14.2, the corresponding F statistic is

$$F_{obs} = \frac{[4039.97 - 3264.89]/[57 - 54]}{3264.89/54} = \frac{258.36}{60.46} = 4.27,$$

with a one-sided P value of 0.009, i.e., $4.27 = F(.991, 3, 54)$. *There is substantial evidence for differences in Mothers* **after accounting** *for any differences due to Litters.* We constructed this F statistic in the usual way for comparing the reduced model (4) to the full model (3) but when examining a number of models that are all smaller than a largest model, in this case Model (2), our preferred practice is to use the *MSE* from the largest model in the denominator of all the F statistics, thus we compute

$$F_{obs} = \frac{[4039.97 - 3264.89]/[57 - 54]}{2440.82/45} = \frac{258.36}{54.24} = 4.76$$

and compare the result to an $F(3, 45)$ distribution.

An alternative test for Mother effects assumes that there are no Litter effects and bases our evaluation of Mother effects on comparing the model with Mother effects but no Litter effects,

$$y_{ijk} = \mu + \eta_j + \varepsilon_{ijk} \tag{14.1.5}$$

to the model that contains no group effects at all,

$$y_{ijk} = \mu + \varepsilon_{ijk}. \tag{14.1.6}$$

In this case, using Table 14.2 gives the appropriate F as

$$F_{obs} = \frac{[4100.13 - 3328.52]/[57 - 54]}{2440.82/45} = \frac{257.20}{54.24} = 4.74,$$

so *there is substantial evidence for differences in Mothers* **when ignoring** *any differences due to Litters.* The two F statistics for Mothers, 4.74 and 4.76, are very similar in this example, but the difference is real; it is not round-off error. Special cases exist where the two F statistics will be identical, cf. Christensen (2011, Chapter 7).

Similarly, the effect of Litters can be measured by comparing the no-interaction model (3) with Model (5) that eliminates the effect for Litters. Here Mothers are included in both the full and reduced models, because the η_js are included in both models. Additionally, we could assume that there are no Mother effects and base our evaluation of Litter effects on comparing Model (4) with Model (6). Using Table 14.2, both of the corresponding F statistics turn out very small, below 0.4, so there is no evidence of a Mother effect whether accounting for or ignoring effects due to Litters.

In summary, both of the tests for Mothers show Mother effects and neither test for Litters shows Litter effects, so the one-way ANOVA model (5), the model with Mother effects but no Litter effects, seems to be the best-fitting model. Of course the analysis is not finished by identifying Model (5). Having identified that the Mother effects are the interesting ones, we should explore how the four foster mother groups behave. Which genotype gives the largest weight gains? Which gives the smallest? Which genotypes are significantly different? If you accept Model (5) as a working model, all of these issues can be addressed as in any other one-way ANOVA. However, it would be good practice to use $MSE(2)$ when constructing any standard errors, in which case the $t(dfE(2))$ distribution must be used. Moreover, we have done nothing yet to check our assumptions. *We should have checked assumptions on Model (2) before doing any tests.* Diagnostics will be considered in Subsection 14.1.4.

All of the models considered have their SSE, dfE, and C_p statistic (cf. Subsection 10.2.3) reported in Table 14.2. Tests of various models constitute the traditional form of analysis. These tests are further summarized in the next subsection. But all of this testing seems like a lot of work to identify a model that the C_p statistic immediately identifies as the best model. Table 14.2 also incorporates some shorthand notations for the models. First, we replace the Greek letters with Roman letters that remind us of the effects being fitted, i.e., G for the grand mean, L for Litter effects, M for Mother effects, and LM for interaction effects. Model (2) is thus rewritten as

$$y_{ijk} = G + L_i + M_j + (LM)_{ij} + \varepsilon_{ijk}.$$

A second form of specifying models eliminates any group of parameters that is completely redundant and assumes that distinct terms in square brackets are added together. Thus, Model (2) is [LM] because it requires only the $(LM)_{ij}$ terms and Model (3) is written [L][M] because in Model (3) the G (μ) term is redundant and the L (α) and M (η) terms are added together. Model (3) is the most difficult to fit of the models in Table 14.2. Model (6) is a one-sample model, and models (1)=(2), (4), and (5) are all one-way ANOVA models. When dealing with Model (3), you have to be able to coax a computer program into giving you all the results that you want and need. With the other models, you could easily get what you need from a hand calculator.

14.1.2 Hierarchy of models

All together we fitted a hierarchy of models that we can display from the largest model to the smallest as

$$[LM]$$
$$[L][M]$$
$$[L] \qquad\qquad [M]$$
$$[G]$$

or, in terms of numbered models,

$$(1) \quad = \quad (2)$$
$$(3)$$
$$(4) \qquad\qquad (5)$$
$$(6).$$

Models (4) and (5) are not directly comparable, but both are reductions of (3) and both contain (6) as a special case. Any model in a row of this hierarchy can be tested as a full model relative to any (reduced) model in a lower row or tested as a reduced model relative to any (full) model in a higher row. However, we typically modify our testing procedure so that in the denominator of the F statistic we always use $MSE(2)$, the MSE from the model at the top of the hierarchy, i.e., the MSE from the largest model being considered. In other words,

$$F = \frac{[SSE(Full) - SSE(Red.)] / [dfE(Full) - dfE(Red.)]}{SSE(2)/dfE(2)}$$

and is compared to an $F(dfE(Full) - dfE(Red.), dfE(2))$ distribution.

With this hierarchy of models, there are only two sequences of models that go from the smallest model to the largest model. We can fit the sequence (6), (5), (3), (2) or fit the sequence (6), (4), (3), (2). Table 14.3 provides results from doing model comparisons in both of the two sequential fitting schemes. The first ANOVA table results from fitting the sequence of models (6), (5), (3), (2). The second ANOVA results from fitting (6), (4), (3), (2). Together, they provide all of the tests that we performed in Subsection 14.1.1. The first ANOVA table looks at Mothers (ignoring Litters), Litters (after Mothers), and interaction, while the second looks at Litters (ignoring Mothers), Mothers (after

Table 14.3: *Analyses of variance for rat weight gains.*

Source	df	Seq SS	MS	F	P
Mothers	3	771.61	257.20	4.74	0.006
Litters	3	63.63	21.21	0.39	0.761
Mothers∗Litters	9	824.07	91.56	1.69	0.120
Error	45	2440.82	54.24		
Total	60	4100.13			

Source	df	Seq SS	MS	F	P
Litters	3	60.16	20.05	0.37	0.776
Mothers	3	775.08	258.36	4.76	0.006
Litters∗Mothers	9	824.07	91.56	1.69	0.120
Error	45	2440.82	54.24		
Total	60	4100.13			

Litters), and interaction. In the first ANOVA table, Mothers are fitted to the data before Litters. In the second table, Litters are fitted before Mothers.

Although models are fitted from smallest to largest and, in ANOVA tables, results are reported from smallest model to largest, *a sequence of models is evaluated from largest model to smallest.* Thus, we begin the analysis of Table 14.3 at the bottom, looking for interaction. The rows for Mother∗Litter interaction are identical in both tables. The sum of squares and degrees of freedom for Mother∗Litter interaction in the table is obtained by differencing the error sums of squares and degrees of freedom for models (3) and (2). *If the interaction is significant, there is little point in looking at the rest of the ANOVA table.* One can either analyze the data as a one-way ANOVA or, as will be discussed in later chapters, try to model the interaction by developing models intermediate between models (2) and (3), cf. Subsection 15.3.2.

Our interaction F statistic is quite small, so there is little evidence of interaction and we proceed with an analysis of Model (3). In particular, we now examine the main effects. Table 14.3 shows clear effects for both Mothers ignoring Litters ($F = 4.74$) and Mothers after fitting Litters ($F = 4.76$) with little evidence for Litters fitted after Mothers ($F = 0.39$) or Litters ignoring Mothers ($F = 0.37$).

The difference in the error sums of squares for models (4) [L] and (3) [L][M] is the sum of squares reported for Mothers in the second of the two ANOVA tables in Table 14.3. The difference in the error sums of squares for models (6) [G] and (5) [M] is the sum of squares reported for Mothers in the first of the two ANOVA tables in Table 14.3. The difference in the error sums of squares for models (5) [M] and (3) [L][M] is the sum of squares reported for Litters in the first of the two ANOVA tables in Table 14.3. The difference in the error sums of squares for models (6) [G] and (4) [L] is the sum of squares reported for Litters in the second of the ANOVA tables in Table 14.3.

Balanced two-way ANOVA is the special case where $N_{ij} = N$ for all i and j. For balanced ANOVA the two ANOVA tables (cf. Table 14.3) would be identical.

14.1.3 Computing issues

Many computer programs for fitting general linear models readily provide the ANOVA tables in Table 14.3. Recall that the interaction model (2) was written

$$y_{ijk} = \mu + \alpha_i + \eta_j + \gamma_{ij} + \varepsilon_{ijk}$$

where μ is an overall effect (grand mean), the α_is are effects for litter genotype, the η_js are effects for foster mother genotype, and the γ_{ij}s are effects for each combination of a litter genotype and foster mother genotype. *Just like regression programs, general linear models programs typically fit a sequence of models where the sequence is determined by the order in which the terms are specified.*

Thus, specifying Model (2) causes the sequence (6), (4), (3), (2) to be fitted and the second ANOVA table in Table 14.3 to be produced. Specifying the equivalent but reordered model

$$y_{ijk} = \mu + \eta_j + \alpha_i + \gamma_{ij} + \varepsilon_{ijk}$$

causes the sequence (6), (5), (3), (2) to be fitted and the first ANOVA table in Table 14.3 to be produced.

When obtaining an analysis of Model (2), many computer programs give ANOVA tables with either the *sequential sums of squares* or *"adjusted" sums of squares*. Adjusted sums of squares are for adding a term to the model last. Thus, in Model (2) the adjusted sums of squares for Litters is the sum of squares for dropping Litters out of the model

$$y_{ijk} = \mu + \eta_j + \gamma_{ij} + \alpha_i + \varepsilon_{ijk}.$$

This is idiotic! As we have mentioned, the γ_{ij} terms can explain anything the α_i or η_j terms can explain, so the model without Litter main effects

$$y_{ijk} = \mu + \eta_j + \gamma_{ij} + \varepsilon_{ijk}$$

is equivalent to Model (2).

What do these adjusted sums of squares really mean? Unfortunately, you have to enter the bowels of the computer program to find out. Most computer programs build in *side conditions* that allow them to give some form of parameter estimates. Only Model (1) really allows all the parameters to be estimated. In any of the other models, parameters cannot be estimated without imposing some arbitrary side conditions. In the interaction model (2) the adjusted sums of squares for main effects depend on these side conditions, so programs that use different side conditions (*and programs DO use different side conditions*) give different adjusted sums of squares for main effects after interaction. These values are worthless! Unfortunately, many programs, by default, produce mean squares, F statistics, and P values using these adjusted sums of squares. The interaction sum of squares and F test are not affected by this issue.

To be fair, if you are dealing with Model (3) instead of Model (2), the adjusted sums of squares are perfectly reasonable. In Model (3),

$$y_{ijk} = \mu + \alpha_i + \eta_j + \varepsilon_{ijk},$$

the adjusted sum of squares for Litters just compares Model (3) to Model (5) and the adjusted sum of squares for Mothers compares Model (3) to Model (4). Adjusted sums of squares are only worthless when you fit main effects after having already fit an interaction that involves the main effect.

14.1.4 Discussion of model fitting

If there is no interaction but an effect for Mothers after accounting for Litters and an effect for Litters after accounting for Mothers, both Mothers and Litters would have to appear in the final model, i.e.,

$$y_{ijk} = \mu + \alpha_i + \eta_j + \varepsilon_{ijk},$$

because neither effect could be dropped.

If there were an effect for Mothers after accounting for Litters, but no effect for Litters after accounting for Mothers, we could drop the effect of Litters from the model. Then *if the effect for Mothers was still apparent* when Litters were ignored, a final model

$$y_{ijk} = \mu + \alpha_i + \varepsilon_{ijk}$$

that includes Mother effects but not Litter effects would be appropriate. Similar reasoning with the roles of Mothers and Litters reversed would lead one to the model

$$y_{ijk} = \mu + \eta_j + \varepsilon_{ijk}.$$

Unfortunately, except in special cases, it is possible to get contradictory results. If there were an effect for Mothers after accounting for Litters but no effect for Litters after accounting for Mothers we could drop the effect of Litters from the model and consider the model

$$y_{ijk} = \mu + \alpha_i + \varepsilon_{ijk}.$$

However, it is possible that in this model *there may be no apparent effect for Mothers* (when Litters are ignored), so dropping Mothers is suggested and we get the model

$$y_{ijk} = \mu + \varepsilon_{ijk}.$$

This model contradicts our first conclusion that there is an effect for Mothers, albeit one that only shows up after adjusting for Litters. These issues are discussed more extensively in Christensen (2011, Section 7.5).

14.1.5 Diagnostics

It is necessary to consider the validity of our assumptions. Table 14.4 contains many of the standard diagnostic statistics used in regression analysis. They are computed from the interaction model (2). Model (2) is equivalent to a one-way ANOVA model, so the leverage associated with y_{ijk} in Table 14.3 is just $1/N_{ij}$.

Figures 14.1 and 14.2 contain diagnostic plots. Figure 14.1 contains a normal plot of the standardized residuals, a plot of the standardized residuals versus the fitted values, and boxplots of the residuals versus Litters and Mothers, respectively. Figure 14.2 plots the leverages, the t residuals, and Cook's distances against case numbers. The plots identify one potential outlier. From Table 14.4 this is easily identified as the observed value of 68.0 for Litter I and Foster Mother A. This case has by far the largest standardized residual r, standardized deleted residual t, and Cook's distance C. We can test whether this case is consistent with the other data. The t residual of 4.02 has an unadjusted P value of 0.000225. If we use a Bonferroni adjustment for having made $n = 61$ tests, the P value is $61 \times 0.000225 \doteq 0.014$. There is substantial evidence that this case does not belong with the other data.

14.1.6 Outlier deleted analysis

We now consider the results of an analysis with the outlier deleted. Fitting the interaction model (2) we get

$$dfE(2) = 44, \qquad SSE(2) = 1785.60, \qquad MSE(2) = 40.58$$

and fitting the additive model (3) gives

$$dfE(3) = 53, \qquad SSE(3) = 3049,$$

so

$$F_{obs} = \frac{1263.48/9}{40.58} = \frac{140.39}{40.58} = 3.46,$$

with a one-sided P value of .003. The interaction in significant, so we could reasonably go back to treating the data as a one-way ANOVA with 16 groups. Typically, we would print out the 16 group means and try to figure out what is going on. But in this case, most of the story is determined by the plot of the standardized residuals versus the fitted values for the deleted data, Figure 14.3.

Case 12 was dropped from the Litter I, Mother A group that contained three observations. After dropping case 12, that group has two observations and, as can been seen from Figure 14.3, that group has a far lower sample mean and has far less variability than any other group. In this example, deleting the one observation that does not seem consistent with the other data makes the entire group inconsistent with the rest of the data.

Table 14.4: *Diagnostics for rat weight gains: Model (14.1.2).*

Case	Litter	Mother	y	$\hat{y}$	Leverage	r	t	C
1	A	A	61.5	63.680	0.20	−0.33	−0.33	0.002
2	A	A	68.2	63.680	0.20	0.69	0.68	0.007
3	A	A	64.0	63.680	0.20	0.05	0.04	0.000
4	A	A	65.0	63.680	0.20	0.20	0.20	0.001
5	A	A	59.7	63.680	0.20	−0.60	−0.60	0.006
6	F	A	60.3	52.325	0.25	1.25	1.26	0.033
7	F	A	51.7	52.325	0.25	−0.10	−0.10	0.000
8	F	A	49.3	52.325	0.25	−0.47	−0.47	0.005
9	F	A	48.0	52.325	0.25	−0.68	−0.67	0.010
10	I	A	37.0	47.100	0.33	−1.68	−1.72	0.088
11	I	A	36.3	47.100	0.33	−1.89	−1.84	0.101
12	I	A	68.0	47.100	0.33	3.48	4.02	0.377
13	J	A	59.0	54.350	0.25	0.73	0.73	0.011
14	J	A	57.4	54.350	0.25	0.48	0.47	0.005
15	J	A	54.0	54.350	0.25	−0.05	−0.05	0.000
16	J	A	47.0	54.350	0.25	−1.15	−1.16	0.028
17	A	F	55.0	52.400	0.33	0.43	0.43	0.006
18	A	F	42.0	52.400	0.33	−1.73	−1.77	0.093
19	A	F	60.2	52.400	0.33	1.30	1.31	0.053
20	F	F	50.8	60.640	0.20	−1.49	−1.52	0.035
21	F	F	64.7	60.640	0.20	0.62	0.61	0.006
22	F	F	61.7	60.640	0.20	0.16	0.16	0.000
23	F	F	64.0	60.640	0.20	0.51	0.51	0.004
24	F	F	62.0	60.640	0.20	0.21	0.20	0.001
25	I	F	56.3	64.367	0.33	−1.34	−1.35	0.056
26	I	F	69.8	64.367	0.33	0.90	0.90	0.026
27	I	F	67.0	64.367	0.33	0.44	0.43	0.006
28	J	F	59.5	56.100	0.33	0.57	0.56	0.010
29	J	F	52.8	56.100	0.33	−0.55	−0.54	0.009
30	J	F	56.0	56.100	0.33	−0.02	−0.02	0.000
31	A	I	52.5	54.125	0.25	−0.25	−0.25	0.001
32	A	I	61.8	54.125	0.25	1.20	1.21	0.030
33	A	I	49.5	54.125	0.25	−0.73	−0.72	0.011
34	A	I	52.7	54.125	0.25	−0.22	−0.22	0.001
35	F	I	56.5	53.925	0.25	0.40	0.49	0.003
36	F	I	59.0	53.925	0.25	0.80	0.79	0.013
37	F	I	47.2	53.925	0.25	−1.05	−1.06	0.023
38	F	I	53.0	53.925	0.25	−0.15	−0.14	0.000
39	I	I	39.7	51.600	0.20	−1.81	−1.85	0.051
40	I	I	46.0	51.600	0.20	−0.85	−0.85	0.011
41	I	I	61.3	51.600	0.20	1.47	1.49	0.034
42	I	I	55.3	51.600	0.20	0.56	0.56	0.005
43	I	I	55.7	51.600	0.20	0.62	0.62	0.006
44	J	I	45.2	54.533	0.33	−1.55	−1.58	0.075
45	J	I	57.0	54.533	0.33	0.41	0.41	0.005
46	J	I	61.4	54.533	0.33	1.14	1.15	0.041
47	A	J	42.0	48.960	0.20	−1.06	−1.06	0.017
48	A	J	54.0	48.960	0.20	0.77	0.76	0.009
49	A	J	61.0	48.960	0.20	1.83	1.88	0.052
50	A	J	48.2	48.960	0.20	−0.12	−0.11	0.000
51	A	J	39.6	48.960	0.20	−1.42	−1.44	0.032
52	F	J	51.3	45.900	0.50	1.04	1.04	0.067
53	F	J	40.5	45.900	0.50	−1.04	−1.04	0.067
54	I	J	50.0	49.433	0.33	0.09	0.09	0.000
55	I	J	43.8	49.433	0.33	−0.94	−0.94	0.027
56	I	J	54.5	49.433	0.33	0.84	0.84	0.022
57	J	J	44.8	49.060	0.20	−0.65	−0.64	0.007
58	J	J	51.5	49.060	0.20	0.37	0.37	0.002
59	J	J	53.0	49.060	0.20	0.60	0.59	0.006
60	J	J	42.0	49.060	0.20	−1.07	−1.07	0.018
61	J	J	54.0	49.060	0.20	0.75	0.75	0.009

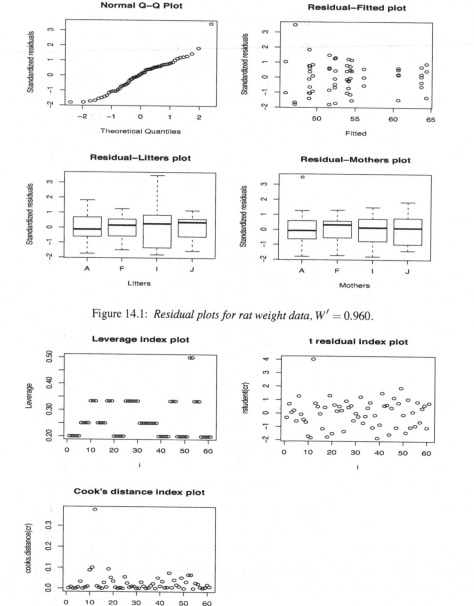

Figure 14.1: *Residual plots for rat weight data, $W' = 0.960$.*

Figure 14.2: *Diagnostic index plots for rat weight data.*

The small mean value for the Litter I, Mother A group after deleting case 12 is causing the interaction. If we delete the entire group, the interaction test becomes

$$F_{obs} = \frac{578.74/8}{1785.36/43} = 1.74,$$ (14.1.7)

which gives a P value of 0.117. Note that by dropping the Litter I, Mother A group we go from our original 61 observations to 58 observations, but we also go from 16 groups to 15 groups, so $dfE(2) = 58 - 15 = 43$. On the other hand, the number of free parameters in Model (3) is unchanged, so $dfE(3) = 58 - 7 = 51$, which leaves us 8 degrees of freedom in the numerator of the test.

Residual–Fitted plot

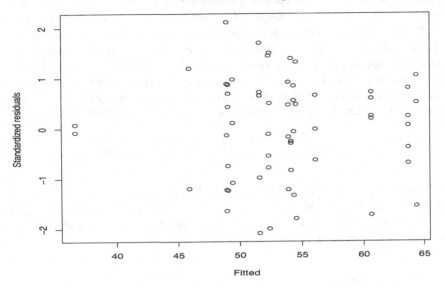

Figure 14.3: *Standardized residuals versus predicted values: Case 12 deleted.*

The Litter I, Mother A group is just weird. It contains three cases, the two smallest along with the third largest case out of 61 total cases. It is weird if we leave case 12 in the data and it is weird if we take case 12 out of the data. With all the data, the best-fitting model is (5). Deleting the Litter I, Mother A group, the best-fitting model again turns out to be (5).

For the full data and Model (5), LSD at the 5% level can be summarized as

Mother	Mean		
F	58.700	A	
A	55.400	A	
I	53.362	A	B
J	48.680		B

The residual plots and diagnostics look reasonably good for this model. The plots and diagnostics are different from those given earlier for Model (2).

For Model (5) with the Litter I, Mother A group removed, LSD at the 5% level can be summarized as

Mother	Mean		
F	58.700	A	
A	57.315	A	B
I	53.362	C	B
J	48.680	C	

(These are unbalanced one-way ANOVAs, so there is no guarantee that such displays can be constructed.) Again, the residual plots and diagnostics look reasonably good but are different from those for the full data models (2) and (5).

The main difference between these two analyses is that one has Mothers F and I significantly different and the other does not. Given that the change to the analysis consisted of deleting observations from Mother A leaving groups F and I alone, that is somewhat strange. The confidence intervals for $\mu_I - \mu_F$ are $(-10.938, 0.263)$ for the full data and $(-10.248, -0.427)$ for the deleted data, so one is just barely insignificant for testing $H_0 : \mu_I - \mu_F = 0$ and the other is just barely significant. The discrepancy comes from using different MSEs. Both are based on Model (5) but they are based on different data. It would be preferable to base the LSDs on MSEs from Model (2), but the results would still be different for the different data. For all of the weirdness of the Litter I, Mother A group, in the end, the results are remarkably consistent whether we delete the group or not.

Finally, we have enough information to test whether the three observations in the Litter I, Mother A group are collectively a group of outliers. We do that by testing a full model that is Model (2) defined for the deleted data against a reduced model that is Model (2) for the full data. This may seem like a backwards definition of a full and reduced model, but the deleted data version of Model (2) can be obtained from the full data Model (2) by adding a separate parameter for each of the three points we want to delete. Using information from Equation (7) and either of Table 14.2 or 14.3,

$$F_{obs} = \frac{[2440.82 - 1785.36]/[45 - 43]}{1785.36/43} = 7.89,$$

which is highly significant: statistical evidence that Litter I, Mother A is a weird group. The numerator degrees of freedom is 2. Model (2) for the full data already has one parameter for the Litter I, Mother A group, so we need add only two more free parameters to have a separate parameter for every observation in the group.

14.2   Modeling contrasts

The interesting part of any analysis is figuring out how the groups really differ. To do that, you need to look at contrasts. We examined contrasts for one-way ANOVA models in Chapter 12, and all the models we have looked at in this chapter, except the additive-effects model, have been essentially one-way ANOVA models. In particular, our final conclusions about Mothers in the previous section came from the one-way ANOVA that ignored Litters.

But what if we could not ignore Litters? What if we needed to see how Mothers differed in the additive-effects model rather than a one-way ANOVA? As mentioned earlier, when dealing with the additive-effects model you cannot just compute what you want on a hand calculator. You have to be able to coax whatever information you need out of a computer program. These issues are addressed in this section and the next. You can generally get everything you need by fitting equivalent models in a regression program as discussed in the next section. Here we focus on extracting information from an ANOVA program, i.e., we focus on manipulating the subscripts that are fed into an ANOVA program.

When the treatments have no structure to exploit, one way to start is by looking for evidence of differences between all pairs of means.

Parameter	Est	SE(Est)	t	Bonferroni P
$\eta_F - \eta_A$	3.516	2.862	1.229	1.0000
$\eta_I - \eta_A$	−1.832	2.767	−0.662	1.0000
$\eta_J - \eta_A$	−6.755	2.810	−2.404	0.1182
$\eta_I - \eta_F$	−5.35	2.863	−1.868	0.4029
$\eta_J - \eta_F$	−10.27	2.945	−3.488	0.0059
$\eta_J - \eta_I$	−4.923	2.835	−1.736	0.5293

If there are b levels of the second factor, as there are 4 levels of Mother, there are $b(b-1)/2 = 4(3)/2 = 6$ pairwise comparisons to make. Of these, we will see in the next section that we can get $b - 1 = 4 - 1 = 3$ of them by fitting a regression model. Some programs, like Minitab, will provide all of these pairwise comparisons.

It is tempting to just summarize these results and be done with them. For an LSD procedure with $\alpha = 0.05$ (actually specified in Minitab as a Bonferonni procedure with $\alpha = 0.3$), these results can be summarized by

Mother	Mean		
F	58.8	A	
A	55.2	A	
I	53.4	A	B
J	48.5		B

It is by no means clear what the "Mean" values are. (They are explained in the next section.) But what is important, and is reported correctly, are the relative differences among the "Mean" values. From the display, we see no differences among Mothers F, A, and I and no difference between Mothers I and J. We do see differences between F and J and between A and J.

Unfortunately, as discussed in Chapter 13, it is possible that no such display can be generated because it is possible to have, say, a significant difference between F and A but no significant difference between F and I. This is possible, for example, if $SE(\hat{\eta}_F - \hat{\eta}_A)$ is much smaller than $SE(\hat{\eta}_F - \hat{\eta}_I)$.

Based on the pairwise testing results, one could perform a backwards elimination. The pair of means with the least evidence for a difference from 0 is $\eta_I - \eta_A$ with $t_{obs} = -0.662$. We could incorporate the assumption $\eta_I = \eta_A$ into the model and look for differences between the remaining three groups: Mothers F, Mothers J, and the combined group Mothers A or I and continue the process of finding groups that could be combined. If we followed this procedure, at the next step we would combine Mothers A, F and I and then finally conclude that J was different from the other three. Another plausible model might be to combine J with A and I and leave F separate. These additive models with $\eta_A = \eta_I$, $\eta_A = \eta_I = \eta_F$, and $\eta_A = \eta_I = \eta_J$ have respective C_p values of 11.7, 13.2, and 16.0. Only $\eta_A = \eta_I$ is a minor improvement over the full two-factor additive-effects model, which has $C_p = 13.2$ as reported in Table 14.2.

The other methods of Section 12.4 extend easily to two-factor models but the results depend on the specific model in which we incorporate the hypotheses.

14.2.1 Nonequivalence of tests

The general rule for unbalanced data is that if you change anything about a model you change everything about the model. We illustrate this by showing that the tests for $\eta_F = \eta_A$ change between the one-way model (14.1.5), the additive two-way model (14.1.3), and Model (14.1.3) with the additional assumption that $\eta_J = \eta_I$, even when we use the same denominator for all three F tests, namely the MSE from the interaction model (14.1.2).

The pairwise comparison estimates are determined as though the parameter is the last thing being added to the model. If we assumed that $\eta_J - \eta_I = 0$, it could effect the estimate of the completely unrelated parameter $\eta_F - \eta_A$, something that does not happen in one-way ANOVA. In fact, we will show that for the rat data the test for $\eta_F = \eta_A$ in the additive-effects model is different depending on whether you assume $\eta_J = \eta_I$. First we illustrate that the test depends on whether or not we keep Litters in the model.

Assuming that there is no interaction, we might want to test that Mothers A and F have the same effect, i.e., $\eta_A = \eta_F$ or $\eta_1 = \eta_2$. We can incorporate this hypothesis into either the additive model (14.1.3) or the Mothers-only model (14.1.5). As is good practice, our tests will all use the MSE from Model (14.1.2).

To incorporate $\eta_A = \eta_F$, when using a data file like the first four columns of Table 14.4, we merely change the Mother column so that it contains the same symbol for Mothers A and F. I just changed all the Fs to As. Now we refit models (14.1.3) and (14.1.5) using this new "subscript" for the Mothers.

Refitting the one-way model (14.1.5) incorporating $\eta_A = \eta_F$ leads to the ANOVA table

Analysis of Variance

Source	df	SS	MS	F	P
Mother A=F	2	690.29	345.15	5.87	0.005
Error	58	3409.83	58.79		
Total	60	4100.13			

Using results from Table 14.2, to test $\eta_A = \eta_F$ in Model (14.1.5) the statistic is

$$F_{obs} = \frac{[3409.83 - 3328.52]/[58 - 57]}{2440.82/45} = \frac{81.31}{54.24} = 1.50,$$

so the data are consistent with $\eta_A = \eta_F$ in Model (14.1.5). If we used the MSE from Model (14.1.5) rather than Model (14.1.2), this would be equivalent to performing the LSD test as we did in Subsection 14.1.5. The ANOVA table F test for Mother A=F ($F_{obs} = 5.87$) suggests that even when treating Mothers A and F as the same group, there remain noticeable differences in the three remaining groups: A=F, I, and J.

To test $\eta_A = \eta_F$ in the additive-effects model (14.1.3) we must refit the model incorporating $\eta_A = \eta_F$. As in Table 14.3, refitting could lead to either the ANOVA table

Analysis of Variance

Source	df	SS	MS	F	P
Mother A=F	2	690.29	345.15	5.66	0.006
Litter	3	53.69	17.90	0.29	0.830
Error	55	3356.15	61.02		
Total	60	4100.13			

or

Analysis of Variance

Source	df	SS	MS	F	P
Litter	3	60.16	20.05	0.33	0.805
Mother A=F	2	683.82	341.91	5.60	0.006
Error	55	3356.15	61.02		
Total	60	4100.13			

All we really care about is the Error term, and that is the same in both tables. Using results from Table 14.2, to test $\eta_A = \eta_F$ in Model (14.1.3) the statistic is

$$F_{obs} = \frac{[3356.15 - 3264.89]/[55 - 54]}{2440.82/45} = \frac{91.26}{54.24} = 1.68,$$

so the data are again consistent with $\eta_A = \eta_F$, but now the result is for Model (14.1.3). The ANOVA table F statistic for Mother A=F after fitting Litters ($F_{obs} = 5.60$) suggests that even when treating Mothers A and F as the same group, there remain noticeable differences in the three remaining groups: A=F, I, and J.

The key point is that, as expected, the two F statistics for testing $\eta_A = \eta_F$ in models (14.1.5) and (14.1.3) are noticeably different (even using the same denominator). In the former, it is 1.50 and in the latter it is 1.68. Note however that if we modify the denominator of the test for Model (14.1.3) by using its own MSE, we get

$$F_{obs} = \frac{[3356.15 - 3264.89]/[55 - 54]}{3264.89/54} = 1.509 = (1.2286)^2,$$

which agrees with the (paired comparisons) t test given earlier for $\eta_F = \eta_A$ in Model (14.1.3).

Unlike the one-way model, in the two-way additive model even the test for $\eta_A = \eta_F$ depends on our assumptions about the other Mother effects. To demonstrate, we show that the test for $\eta_A = \eta_F$ changes when we assume $\eta_I = \eta_J$. Building $\eta_I = \eta_J$ into the additive model (14.1.3) yields an ANOVA table

Analysis of Variance

Source	df	Seq. SS	MS	F	P
Litter	3	60.16	20.05	0.32	0.811
Mother I=J	2	592.81	296.41	4.73	0.013
Error	55	3447.16	62.68		
Total	60	4100.13			

Now, if we also incorporate our hypothesis $\eta_A = \eta_F$ we get an ANOVA table

Analysis of Variance

Source	df	Seq. SS	MS	F	P
Litter	3	60.16	20.05	0.32	0.813
Mother A=F;I=J	1	505.27	505.27	8.00	0.006
Error	56	3534.70	63.12		
Total	60	4100.13			

Comparing the error terms and using our usual denominator gives a different F statistic for testing $\eta_A = \eta_F$ assuming $\eta_I = \eta_J$ in the additive model,

$$F_{obs} = \frac{[3534.70 - 3447.16]/[56 - 55]}{2440.82/45} = \frac{87.54}{54.24} = 1.61,$$

rather than the 1.68 we got from the additive model without assuming that $\eta_I = \eta_J$.

In this example, the test statistics are noticeably, but not substantially, different. With other data, the differences can be much more substantial.

In a balanced ANOVA, the numerators for these three tests would all be identical and the only differences in the tests would be due to alternative choices of a MSE for the denominator.

14.3 Regression modeling

The additive-effects model

$$y_{ijk} = \mu + \alpha_i + \eta_j + \varepsilon_{ijk}$$

is the only new model that we have considered in this chapter. All of the other models reduce to fitting a one-way ANOVA. If we create four 0-1 indicator variables, say, x_1, x_2, x_3, x_4 for the four Litter categories and another four indicator variables, say, x_5, x_6, x_7, x_8 for the four Mother categories, we can rewrite the additive model as

$$y_{ijk} = \mu + \alpha_1 x_{ij1} + \alpha_2 x_{ij2} + \alpha_3 x_{ij3} + \alpha_4 x_{ij4} + \eta_1 x_{ij5} + \eta_2 x_{ij6} + \eta_3 x_{ij7} + \eta_4 x_{ij8} + \varepsilon_{ijk}.$$

The model is overparameterized; largely because for any ij,

$$x_{ij1} + x_{ij2} + x_{ij3} + x_{ij4} = 1 = x_{ij5} + x_{ij6} + x_{ij7} + x_{ij8}.$$

Also, associated with the grand mean μ is a predictor variable that always takes the value 1, say, $x_0 \equiv 1$. To make a regression model out of the additive-effects model we need to drop one variable from two of the three sets of variables $\{x_0\}$, $\{x_1, x_2, x_3, x_4\}$, $\{x_5, x_6, x_7, x_8\}$. We illustrate the procedures by dropping two of the three variables, x_0, x_2 (the indicator for Litter F), and x_8 (the indicator for Mother J).

If we drop x_2 and x_8 the model becomes

$$y_{ijk} = \delta + \gamma_1 x_{ij1} + \gamma_3 x_{ij3} + \gamma_4 x_{ij4} + \beta_1 x_{ij5} + \beta_2 x_{ij6} + \beta_3 x_{ij7} + \varepsilon_{ijk}. \tag{14.3.1}$$

In this model, the Litter F, Mother J group becomes a baseline group and

$$\delta = \mu + \alpha_2 + \eta_4, \quad \gamma_1 = \alpha_1 - \alpha_2, \quad \gamma_3 = \alpha_3 - \alpha_2, \quad \gamma_4 = \alpha_4 - \alpha_2,$$

$$\beta_1 = \eta_1 - \eta_4, \quad \beta_2 = \eta_2 - \eta_4, \quad \beta_3 = \eta_3 - \eta_4.$$

After fitting Model (1), the Table of Coefficients gives immediate results for testing whether differences exist between Mother J and each of Mothers A, F, and I. It also gives immediate results for testing no difference between Litter F and each of Litters A, I, and J.

If we drop x_0, x_8 the model becomes

$$y_{ijk} = \gamma_1 x_{ij1} + \gamma_2 x_{ij2} + \gamma_3 x_{ij3} + \gamma_4 x_{ij4} + \beta_1 x_{ij5} + \beta_2 x_{ij6} + \beta_3 x_{ij7} + \varepsilon_{ijk}$$

but now

$$\gamma_1 = \mu + \alpha_1 + \eta_4, \quad \gamma_2 = \mu + \alpha_2 + \eta_4, \quad \gamma_3 = \mu + \alpha_3 + \eta_4, \quad \gamma_4 = \mu + \alpha_4 + \eta_4,$$

$$\beta_1 = \eta_1 - \eta_4, \quad \beta_2 = \eta_2 - \eta_4, \quad \beta_3 = \eta_3 - \eta_4,$$

so the Table of Coefficients still gives immediate results for testing whether differences exist between Mother J and Mothers A, F, and I.

If we drop x_0, x_2 the model becomes

$$y_{ijk} = \gamma_1 x_{ij1} + \gamma_3 x_{ij3} + \gamma_4 x_{ij4} + \beta_1 x_{ij5} + \beta_2 x_{ij6} + \beta_3 x_{ij7} + \beta_4 x_{ij8} + \varepsilon_{ijk}.$$

Now

$$\beta_1 = \mu + \eta_1 + \alpha_2, \quad \beta_2 = \mu + \eta_2 + \alpha_2, \quad \beta_3 = \mu + \eta_3 + \alpha_2, \quad \beta_4 = \mu + \eta_4 + \alpha_2,$$

$$\gamma_1 = \alpha_1 - \alpha_2, \quad \gamma_3 = \alpha_3 - \alpha_2, \quad \gamma_4 = \alpha_4 - \alpha_2.$$

The Table of Coefficients still gives immediate results for testing whether differences exist between Litter F and Litters A, I, and J.

To illustrate these claims, we fit Model (1) to the rat data to obtain the following Table of Coefficients.

Table of Coefficients

Predictor	Est	SE(Est)	t	P
Constant (δ)	48.129	2.867	16.79	0.000
x_1:L-A (γ_1)	2.025	2.795	0.72	0.472
x_3:L-I (γ_3)	−0.628	2.912	−0.22	0.830
x_4:L-J (γ_4)	0.004	2.886	0.00	0.999
x_5:M-A (β_1)	6.755	2.810	2.40	0.020
x_6:M-F (β_2)	10.271	2.945	3.49	0.001
x_7:M-I (β_3)	4.923	2.835	1.74	0.088

If, for example, you ask Minitab's GLM procedure to test all pairs of Mother effects using a Bonferroni adjustment, you get the table reported earlier,

Parameter	Est	SE(Est)	t	Bonferroni P
$\eta_F - \eta_A$	3.516	2.862	1.229	1.0000
$\eta_I - \eta_A$	−1.832	2.767	−0.662	1.0000
$\eta_J - \eta_A$	−6.755	2.810	−2.404	0.1182
$\eta_I - \eta_F$	−5.35	2.863	−1.868	0.4029
$\eta_J - \eta_F$	−10.27	2.945	−3.488	0.0059
$\eta_J - \eta_I$	−4.923	2.835	−1.736	0.5293

Note that the estimate, say, $\hat{\beta}_2 = \hat{\eta}_2 - \hat{\eta}_4 = \hat{\eta}_F - \hat{\eta}_J = 10.271$, is the negative of the estimate of $\eta_J - \eta_F$, that they have the same standard error, that the t statistics are the negatives of each other, and that the Bonferroni P values are 6 times larger than the Table of Coefficients P values. Similar results hold for $\beta_1 = \eta_1 - \eta_4 = \eta_A - \eta_J$ and $\beta_3 = \eta_3 - \eta_4 = \eta_I - \eta_J$.

A display of results given earlier was

Mother	Mean		
F	58.8	A	
A	55.2	A	
I	53.4	A	B
J	48.5		B

The problems with the display are that the column of "Mean" values has little meaning and that no meaningful display may be possible because standard errors depend on the difference being estimated. As for the Mean values, the relative differences among the Mother effects are portrayed correctly, but the actual numbers are arbitrary. The relative sizes of Mother effects must be the same for any Litter, but there is nothing one could call an overall Mother effect. You could add any constant to each of these four Mean values and they would be just as meaningful.

To obtain these "Mean" values as given, fit the model

$$y_{ijk} = \delta + \gamma_1(x_{ij1} - x_{ij4}) + \gamma_2(x_{ij2} - x_{ij4}) + \gamma_3(x_{ij3} - x_{ij4})$$
$$+ \beta_1(x_{ij7} - x_{ij8}) + \beta_2(x_{ij6} - x_{ij8}) + \beta_3(x_{ij7} - x_{ij8}) + \varepsilon_{ijk} \quad (2)$$

to get the following Table of Coefficients.

Table of Coefficients

Predictor	Est	SE(Est)	t	P
Constant (δ)	53.9664	0.9995	53.99	0.000
Litter				
A (γ_1)	1.675	1.675	1.00	0.322
F (γ_2)	−0.350	1.763	−0.20	0.843
I (γ_3)	−0.979	1.789	−0.55	0.587
Mother				
A (β_1)	1.268	1.702	0.75	0.459
F (β_2)	4.784	1.795	2.66	0.010
I (β_3)	−0.564	1.712	−0.33	0.743

The "Mean" values in the display are obtained from the Table of Coefficients, wherein the estimated effect for Mother F is $58.8 = \hat{\delta} + \hat{\beta}_2 = 53.9664 + 4.784$, for Mother A is $55.2 = \hat{\delta} + \hat{\beta}_1 = 53.9664 + 1.268$, for Mother I is $53.4 = \hat{\delta} + \hat{\beta}_3 = 53.9664 - 0.564$, and for Mother J is $48.5 = \hat{\delta} - (\hat{\beta}_1 + \hat{\beta}_2 + \hat{\beta}_3) = 53.9664 - (1.268 + 4.784 - 0.564)$.

Dropping the two indicator variables is equivalent to imposing *side conditions* on the parameters. Dropping x_2 and x_8 amounts to assuming $\alpha_2 = 0 = \eta_4$. Dropping the intercept x_0 and x_8 amounts to assuming that $\mu = 0 = \eta_4$. Dropping x_0 and x_2 amounts to assuming that $\mu = 0 = \alpha_2$. The regression model (2) is equivalent to assuming that $\alpha_1 + \alpha_2 + \alpha_3 + \alpha_4 = 0 = \eta_1 + \eta_2 + \eta_3 + \eta_4$.

14.4 Homologous factors

An interesting aspect of having two factors is dealing with factors that have comparable levels. For example, the two factors could be mothers and fathers and the factor levels could be a categorization of their educational level, perhaps: not a high school graduate, high school graduate, some college, college graduate, post graduate work. In addition to the issues raised already, we might be interested in whether fathers' education has the same effect as mothers' education. Alternatively, the two factors might be a nitrogen-based fertilizer and a phosphorus-based fertilizer and the levels might be multiples of a standard dose. In that case we might be interested in whether nitrogen and phosphorus have the same effect. Factors with comparable levels are called *homologous* factors. Example 14.1.1 involves genotypes of mothers and genotypes of litters where the genotypes are identical for the mothers and the litters, so it provides an example of homologous factors.

14.4.1 Symmetric additive effects

We have talked about father's and mother's educational levels having the same effect. To do this we must have reasonable definitions of the effects for a father's educational level and a mother's educational level. As discussed in Section 1, factor level effects are used in the additive two-way model

$$y_{ijk} = \mu + \alpha_i + \eta_j + e_{ijk}. \quad (14.4.1)$$

Here the α_is represent, say, father's education effects or litter genotype effects and the η_js represent mother's education effects or foster mother genotype effects. Most often with homologous factors we assume that the number of levels is the same for each factor. For the education example, fathers and mothers both have 5 levels. For the rat genotypes, both factors have 4 levels. We call this number t. (Occasionally, we can extend these ideas to unequal numbers of levels.)

If fathers' and mothers' educations have the same effect, or if litters' and foster mothers' genotypes have the same effect, then

$$\alpha_1 = \eta_1, \ldots, \alpha_t = \eta_t.$$

Incorporating this hypothesis into the additive-effects model (1) gives the *symmetric-additive-effects model*

$$y_{ijk} = \mu + \alpha_i + \alpha_j + e_{ijk}. \tag{14.4.2}$$

Alas, not many ANOVA computer programs know how to fit such a model, so we will have to do it ourselves in a regression program. *The remainder of the discussion in this subsection is for the rat weight data.*

We begin by recasting the additive-effects model (1) as a regression model just as we did in Section 14.3 but now relabeling the indicator variables. The factor variable Litters has 4 levels, so we can replace it with 4 indicator variables, say, L_A, L_F, L_I, L_J. We can also replace the 4 level factor variable Mothers with 4 indicator variables, M_A, M_F, M_I, M_J. Now the no-interaction model (1) can be written

$$y_h = \mu + \alpha_1 L_{hA} + \alpha_2 L_{hF} + \alpha_3 L_{hI} + \alpha_4 L_{hJ} + \eta_1 M_{hA} + \eta_2 M_{hF} + \eta_3 M_{hI} + \eta_4 M_{hJ} + \varepsilon_h, \tag{14.4.3}$$

$h = 1, \ldots, 61$. This model is overparameterized. If we just run the model, most good regression programs are smart enough to throw out redundant parameters (predictor variables). Performing this operation ourselves, we fit the model

$$y_h = \mu + \alpha_1 L_{hA} + \alpha_2 L_{hF} + \alpha_3 L_{hI} + \eta_1 M_{hA} + \eta_2 M_{hF} + \eta_3 M_{hI} + \varepsilon_h \tag{14.4.4}$$

that eliminates L_J and M_J. Remember, Model (4) is equivalent to (1) and (3). Fitting Model (4), we have little interest in the Table of Coefficients but the ANOVA table follows.

Analysis of Variance: Model (14.4.4)

Source	df	SS	MS	F	P
Regression	6	835.24	139.21	2.30	0.047
Error	54	3264.89	60.46		
Total	60	4100.13			

As advertised, the Error line agrees with the results given for the no-interaction model (14.1.2) in Section 1.

To fit the symmetric-additive-effects model (2), we incorporate the assumption $\alpha_1 = \eta_1, \ldots, \alpha_4 = \eta_4$ into Model (3) getting

$$y_h = \mu + \alpha_1 L_{hA} + \alpha_2 L_{hF} + \alpha_3 L_{hI} + \alpha_4 L_{hJ} + \alpha_1 M_{hA} + \alpha_2 M_{hF} + \alpha_3 M_{hI} + \alpha_4 M_{hJ} + \varepsilon_h$$

or

$$y_h = \mu + \alpha_1 (L_{hA} + M_{hA}) + \alpha_2 (L_{hF} + M_{hF}) + \alpha_3 (L_{hI} + M_{hI}) + \alpha_4 (L_{hJ} + M_{hJ}) + \varepsilon_h.$$

Fitting this model requires us to construct new regression variables, say,

$$
\begin{aligned}
A &= L_A + M_A \\
F &= L_F + M_F \\
I &= L_I + M_I \\
J &= L_J + M_J.
\end{aligned}
$$

The symmetric-additive-effects model (2) is then written as

$$y_h = \mu + \alpha_1 A_h + \alpha_2 F_h + \alpha_3 I_h + \alpha_4 J_h + \varepsilon_h,$$

or, emphasizing that the parameters mean different things,

$$y_h = \gamma_0 + \gamma_1 A_h + \gamma_2 F_h + \gamma_3 I_h + \gamma_4 J_h + \varepsilon_h.$$

This model is also overparameterized, so we actually fit

$$y_h = \gamma_0 + \gamma_1 A_h + \gamma_2 F_h + \gamma_3 I_h + \varepsilon_h, \qquad (14.4.5)$$

giving

Table of Coefficients: Model (14.4.5)

Predictor	$\hat{\gamma}_k$	SE($\hat{\mu}_k$)	t	P
Constant	48.338	2.595	18.63	0.000
A	4.159	1.970	2.11	0.039
F	5.049	1.912	2.64	0.011
I	1.998	1.927	1.04	0.304

and

Analysis of Variance: Model (14.4.5)

Source	df	SS	MS	F	P
Regression	3	513.64	171.21	2.72	0.053
Error	57	3586.49	62.92		
Total	60	4100.13			

We need the ANOVA table Error line to test whether the symmetric-additive-effects model (2) fits the data adequately relative to the additive-effects model (1). The test statistic is

$$F_{obs} = \frac{[3586.49 - 3264.89]/[57 - 54]}{60.46} = 1.773$$

with $P = 0.164$, so the model seems to fit. As discussed in Section 14.1, it would be reasonable to use the interaction model MSE in the denominator of the F statistic, which makes

$$F_{obs} = \frac{[3586.49 - 3264.89]/[57 - 54]}{54.24} = 1.976,$$

but the P value remains a relatively high 0.131.

Presuming that the symmetric-additive-effects model (4) fits, we can interpret the Table of Coefficients. We dropped variable J in the model, so the constant term $\hat{\gamma}_0 = 48.338$ estimates the effect of having genotype J for both Litters and Mothers, i.e., $\gamma_0 = \mu + 2\alpha_4$. The estimated regression coefficient for A, $\hat{\gamma}_1 = 4.159$, is the estimated effect for the difference between the genotype A effect and the genotype J effect. The P value of 0.039 indicates weak evidence for a difference between genotypes A and J. Similarly, there is pretty strong evidence for a difference between genotypes F and J ($P = 0.011$) but little evidence for a difference between genotypes I and J ($P = 0.304$). From the table of coefficients, the estimated effect for having, say, litter genotype A and mother genotype I is $48.338 + 4.159 + 1.998 = 54.495$, which is that same as for litter genotype I and mother genotype A.

14.4.2   Skew symmetric additive effects

Thinking of parents' education and genotypes, it is possible that fathers' and mothers' education could have exact opposite effects or that litters' and mothers' genotypes could have exact opposite effects, i.e.,

$$\alpha_1 = -\eta_1, \ldots, \alpha_4 = -\eta_t.$$

Incorporating this hypothesis into the additive-effects model (1) gives the *skew-symmetric-additive-effects model*

$$y_{ijk} = \mu + \alpha_i - \alpha_j + e_{ijk}. \tag{14.4.6}$$

Sometimes this is called the *alternating-additive-effects model*.

In Model (6), μ is a well-defined parameter and it is the common mean value for the four groups that have the same genotype for litters and mothers. Nonetheless, the skew symmetric additive model is overparameterized but only in that the αs are redundant, i.e., $(L_{hA} - M_{hA}) + (L_{hF} - M_{hF}) + (L_{hI} - M_{hI}) + (L_{hJ} - M_{hJ}) = 0$ for all h.

To fit the model, we write it in regression form

$$y_h = \mu + \alpha_1(L_{hA} - M_{hA}) + \alpha_2(L_{hF} - M_{hF}) + \alpha_3(L_{hI} - M_{hI}) + \alpha_4(L_{hJ} - M_{hJ}) + \varepsilon_h \tag{14.4.7}$$

and drop the last predictor $L_J - M_J$, i.e., fit

$$y_h = \gamma_0 + \gamma_1(L_{hA} - M_{hA}) + \gamma_2(L_{hF} - M_{hF}) + \gamma_3(L_{hI} - M_{hI}) + \varepsilon_h.$$

This yields standard output:

Table of Coefficients: Model (14.4.7)

Predictor	$\hat{\gamma}_k$	SE($\hat{\mu}_k$)	t	P
Constant	53.999	1.048	51.54	0.000
$(L_A - M_A)$	-2.518	2.098	-1.20	0.235
$(L_F - M_F)$	-4.917	2.338	-2.10	0.040
$(L_I - M_I)$	-2.858	2.273	-1.26	0.214

Analysis of Variance: Model (14.4.7)

Source	df	SS	MS	F	P
Regression	3	297.73	99.24	1.49	0.228
Error	57	3802.39	66.71		
Total	60	4100.13			

If the model fitted the data, we could interpret the table of coefficients. Relative to Model (7), the parameter estimates are $\hat{\gamma}_0 = \hat{\mu}$, $\hat{\gamma}_1 = \hat{\alpha}_1 - \hat{\alpha}_4 \equiv \hat{\alpha}_A - \hat{\alpha}_J$, $\hat{\gamma}_2 = \hat{\alpha}_3 - \hat{\alpha}_4 \equiv \hat{\alpha}_F - \hat{\alpha}_J$, $\hat{\gamma}_3 = \hat{\alpha}_3 - \hat{\alpha}_4 \equiv \hat{\alpha}_I - \hat{\alpha}_J$; see Exercise 14.7. But the skew symmetric additive model does not fit very well because, relative to the additive model (1),

$$F_{obs} = \frac{[3802.39 - 3264.89]/[57 - 54]}{60.46} = 2.93,$$

which gives $P = 0.042$.

It is of some interest to note that the model that includes both symmetric additive effects and skew symmetric additive effects,

$$y_h = \mu + \alpha_1 A_h + \alpha_2 F_h + \alpha_3 I_h + \alpha_4 J_h + \tilde{\alpha}_1(L_{hA} - M_{hA})$$
$$+ \tilde{\alpha}_2(L_{hF} - M_{hF}) + \tilde{\alpha}_3(L_{hI} - M_{hI}) + \tilde{\alpha}_4(L_{hJ} - M_{hJ}) + \varepsilon_h$$

is actually equivalent to the no-interaction model (1). Thus, our test for whether the symmetric additive model fits can also be thought of as a test for whether skew symmetric additive effects exist after fitting symmetric additive effects and our test for whether the skew symmetric additive model fits can also be thought of as a test for whether symmetric additive effects exist after fitting skew symmetric additive effects. Neither the symmetric additive model (2) nor the skew symmetric additive model (6) is comparable to either of the single-effects-only models (14.1.4) and (14.1.5).

Table 14.5: *Symmetric and symmetric additive education effects.*

Educ. Level of Fathers	Symmetric Additive Effects					Father Effect
	Education Level of Mother					
	<HS	HS Grad	<Coll	Coll Grad	Post	
<HS	10	12	13	15	15	0
HS Grad	12	14	15	17	17	2
<Coll	13	15	16	18	18	3
Coll Grad	15	17	18	20	20	5
Post	15	17	18	20	20	5
Mother Effect	0	2	3	5	5	

Educ. Level of Fathers	Symmetric Nonadditive Effects					
	Education Level of Mother					
	<HS	HS Grad	<Coll	Coll Grad	Post	
<HS	8	10	13	15	15	
HS Grad	10	12	15	17	17	
<Coll	13	15	19	21	21	
Coll Grad	15	17	21	23	23	
Post	15	17	21	23	23	

14.4.3 Symmetry

The assumption of symmetry is that the two factors are interchangeable. Think again about our fathers' and mothers' education. Under symmetry, there is no difference between having a college graduate father and postgraduate mother as opposed to having a postgraduate father and college graduate mother. Symmetric additive models display this symmetry but impose the structure that there is some consistent effect for, say, being a college graduate and for being a high school graduate. But symmetry can exist even when no overall effects for educational levels exist. For overall effects to exist, the effects must be additive.

Table 14.5 gives examples of symmetric additive and symmetric nonadditive effects. The symmetric additive effects have a "grand mean" of 10, an effect of 0 for being less than a HS Grad, an effect of 2 for a HS Grad, an effect of 3 for some college, and an effect of 5 for both college grad and postgrad. The nonadditive effects were obtained by modifying the symmetric additive effects. In the nonadditive effects, any pair where both parents have any college is up 3 units and any pair where both parents are without any college is down 2 units.

In Subsection 14.4.1 we looked carefully at the symmetric-additive-effects model, which is a special case of the additive-effects (no-interaction) model. Now we impose symmetry on the interaction model.

Rather than the interaction model (14.1.2) we focus on the equivalent one-way ANOVA model (14.1.1), i.e.,

$$y_{ijk} = \mu_{ij} + \varepsilon_{ijk}. \tag{14.4.8}$$

For rat genotypes, $i = 1,\ldots,4$ and $j = 1,\ldots,4$ are used together to indicate the 16 groups. Alternatively, we can replace the pair of subscripts ij with a single subscript $r = 1,\ldots,16$,

$$y_{rk} = \mu_r + \varepsilon_{rk}. \tag{14.4.9}$$

The top half of Table 14.6 shows how the subscripts r identify the 16 groups. The error for this model should agree with the error for Model (14.1.2), which was given in Section 1. You can see from the ANOVA table that it does.

Analysis of Variance: Model (14.4.9)

Source	df	SS	MS	F	P
Rat groups	15	1659.3	110.6	2.04	0.033
Error	45	2440.8	54.2		
Total	60	4100.1			

Table 14.6: *Rat indices*.

One-Way ANOVA: subscripts r				
Genotype of Litter	Genotype of Foster Mother			
	A	F	I	J
A	1	5	9	13
F	2	6	10	14
I	3	7	11	15
J	4	8	12	16
Symmetric group effects: subscripts s				
Genotype of Litter	Genotype of Foster Mother			
	A	F	I	J
A	1	2	3	4
F	2	6	7	8
I	3	7	11	12
J	4	8	12	16

To impose symmetry, in Model (8) we require that for all i and j,

$$\mu_{ij} = \mu_{ji}.$$

This places no restrictions on the four groups with $i = j$. Translating the symmetry restriction into Model (9) with the identifications of Table 14.6, symmetry becomes

$$\mu_2 = \mu_5, \mu_3 = \mu_9, \mu_4 = \mu_{13}, \mu_7 = \mu_{10}, \mu_8 = \mu_{14}, \mu_{12} = \mu_{15}.$$

Imposing these restrictions on the one-way ANOVA model (9) amounts to constructing a new one-way ANOVA model with only 10 groups. Symmetry forces the 6 pairs of groups for which $i \neq j$ and $(i, j) = (j, i)$ to act like 6 single groups and the 4 groups with $i = j$ are unaffected by symmetry. The bottom half of Table 14.6 provides subscripts s for the one-way ANOVA model that incorporates symmetry

$$y_{sm} = \mu_s + \varepsilon_{sm}. \tag{14.4.10}$$

Note that in the nonadditive symmetry model (10), the second subscript for identifying observations within a group also has to change. There are still 61 observations, so if we use fewer groups, we must have more observations in some groups.

Fitting the nonadditive symmetry model gives

Analysis of Variance: Model (14.4.10)

Source	df	SS	MS	F	P
Symmetric groups	9	1159.4	128.8	2.23	0.034
Error	51	2940.8	57.7		
Total	60	4100.1			

Testing this against the interaction model (6) provides

$$F_{obs} = \frac{[2940.8 - 2440.8]/[51 - 45]}{54.24} = 1.54$$

with $P = 0.188$. The nonadditive symmetry model is consistent with the data.

We can also test the symmetry model (10) versus the reduced symmetric-additive-effects model (1) giving

$$F_{obs} = \frac{[3264.89 - 2940.8]/[54 - 51]}{54.24} = 1.99$$

with $P = 0.129$, so there is no particular reason to choose the nonadditive symmetry model over the additive symmetry model.

Finally, it would also be possible to define a *skew symmetry model*. With respect to Model (8), skew symmetry imposes

$$\mu_{ij} = -\mu_{ji},$$

which implies that $\mu_{ii} = 0$. To fit this model we would need to construct indicator variables for all 16 groups, say $I_1, \ldots, I_{16}$. From these we can construct both the symmetric model and the skew symmetric model. Using the notation of the top half of Table 14.6, the symmetry model would have 10 indicator variables

$$I_1, I_6, I_{11}, I_{16}, I_2 + I_5, I_3 + I_9, I_4 + I_{13}, I_7 + I_{10}, I_8 + I_{14}, I_{12} + I_{15}.$$

The skew symmetric model would use the 6 predictor variables

$$I_2 - I_5, I_3 - I_9, I_4 - I_{13}, I_7 - I_{10}, I_8 - I_{14}, I_{12} - I_{15}.$$

If we used all 16 of the predictor variables in the symmetry and skew symmetry model, we would have a model equivalent to the interaction model (8), so the test for the adequacy of the symmetry model is also a test for whether skew symmetry adds anything after fitting symmetry.

It is hard to imagine good applications of skew symmetry. Perhaps less so if we added an intercept to the skew symmetry variables, since that would make it a generalization of the skew symmetric additive model. In fact, to analyze homologous factors without replications, i.e., when Model (8) has 0 degrees of freedom for error, McCullagh (2000) suggests using the *MSE* from the symmetry model as a reasonable estimate of error, i.e., take as error the mean square for skew symmetry after fitting symmetry.

14.4.4 Hierarchy of models

The most interesting hierarchy of models involving symmetric effects is

Interaction

Nonadditive Symmetry   Additive Effects

Symmetric Additive
Intercept-Only

or, in terms of numbered models,

$(14.1.2) = (14.4.9)$

(14.4.10)   (14.1.3)

(14.4.2)

(14.1.6).

If the additive two-factor model does not fit, we might try the symmetry model. The symmetry model is a reduced model relative to the interaction model but is not comparable to the additive two-factor model. As in Subsection 14.1.2, there are two sequences of models that go from the smallest model to the largest and we could use an ANOVA table similar to Table 14.3.

Rather than doing all this testing, we could look at C_p statistics as given in Table 14.7. The model with only Mother effects is still looking good.

14.5 Exercises

EXERCISE 14.5.1.   Cochran and Cox (1957) presented data from Pauline Paul on the effect of cold storage on roast beef tenderness. Treatments are labeled A through F and consist of 0, 1, 2, 4, 9, and 18 days of storage, respectively. The data are tenderness scores and are presented in Table 14.8.

Table 14.7: C_p statistics for fitting models to the data of Table 14.1.

Model	Model	SSE	df	C_p
(14.1.2)	[LM]	2440.82.	45	16.0
(14.1.3)	[L][M]	3264.89	54	13.2
(14.1.4)	[L]	4039.97	57	21.5
(14.1.5)	[M]	3328.52	57	8.4
(14.1.6)	[G]	4100.13	60	16.6
(14.4.2)	Symmetric Additive	3586.49	57	13.1
(14.4.6)	Skew Symmetric Additive	3802.39	57	17.1
(14.4.10)	Symmetric Nonadditive	2940.8	51	13.2

Table 14.8: Beef tenderness scores.

Block	Trt, Score		Block	Trt, Score	
1	A, 7	B, 17	9	A, 17	C, 27
2	C, 26	D, 25	10	B, 23	E, 27
3	E, 33	F, 29	11	D, 29	F, 30
4	A, 25	E, 40	12	A, 11	F, 27
5	B, 25	D, 34	13	B, 24	C, 21
6	C, 34	F, 32	14	D, 26	E, 32
7	A, 10	D, 25	15	B, 26	F, 37
8	C, 24	E, 26			

Analyze the data using a additive two-way ANOVA model involving blocks and treatments. Focus your analysis on identifying differences between treatments.

EXERCISE 14.5.2. Inman et al. (1992) report data on the percentages of Manganese (Mn) in various samples as determined by a spectrometer. Ten samples were used and the percentage of Mn in each sample was determined by each of 4 operators. The data are given in Table 14.9. The operators actually made two readings; the data presented are the averages of the two readings for each sample–operator combination.

Using an additive two-way ANOVA model, analyze the data. Include in your analysis an evaluation of whether any operators are significantly different. Identify a potential outlier, delete that outlier, reanalyze the data, and compare the results of the two analyses.

EXERCISE 14.5.3. Nelson (1993) presents data on the average access times for various disk drives. The disk drives are five brands of half-height fixed drives. The performance of disk drives depends on the computer in which they are installed. The computers could only hold four disk

Table 14.9: Percentage of manganese concentrations.

Sample	Operator			
	1	2	3	4
1	0.615	0.620	0.600	0.600
2	0.635	0.635	0.660	0.630
3	0.590	0.605	0.600	0.590
4	0.745	0.740	0.735	0.745
5	0.695	0.695	0.680	0.695
6	0.640	0.635	0.635	0.630
7	0.655	0.665	0.650	0.650
8	0.640	0.645	0.620	0.610
9	0.670	0.675	0.670	0.665
10	0.655	0.660	0.645	0.650

Table 14.10: *Access times (ms) for disk drives.*

Computer	1	2	Brand 3	4	5
A	35	42	31	30	—
B	41	45	—	32	40
C	—	40	42	33	39
D	32	—	33	35	36
E	40	38	35	—	37

Table 14.11: *Tensile strength of uniform twill.*

Fabric strips	m_1	Machines m_2	m_3	m_4
s_1	18	7	5	9
s_2	9	11	12	3
s_3	7	11	11	1
s_4	6	4	10	8
s_5	10	8	6	10
s_6	7	12	3	15
s_7	13	5	15	16
s_8	1	11	8	12

drives. The data are given in Table 14.10. Analyze the data using an additive two-factor model. Focus your analysis on identifying differences among brands.

EXERCISE 14.5.4. Garner (1956) presented data on the tensile strength of fabrics. Here we consider a subset of the data. The complete data and a more extensive discussion of the experimental procedure are given in Exercise 18.7.2. The experiment involved testing fabric strengths on four different machines. Eight homogeneous strips of cloth were divided into four samples. Each sample was tested on one of four machines. The data are given in Table 14.11.

(a) Analyze the data using an additive two-way model focusing on machine differences. Give an appropriate analysis of variance table. Examine appropriate contrasts using Bonferroni's method with $\alpha = 0.05$.

(b) Check the assumptions of the model and adjust the analysis appropriately.

EXERCISE 14.5.5. Repeat the analyses of this chapter after eliminating Litter I and Mother I from the data in Table 14.1.

EXERCISE 14.5.6. Repeat the analyses of this chapter after eliminating Litter I and Mother J from the data in Table 14.1.

EXERCISE 14.5.7. Repeat the analyses of this chapter after eliminating Litter I from the data in Table 14.1. This requires extending the ideas on homologous factors to situations with unequal numbers of factor levels.

EXERCISE 14.5.8. Explain how dropping the last term out of Model (14.4.7) gives results that are different from dropping the indicator for the last factor level out of a one-way ANOVA model with four groups. Focus on the intercept.

Chapter 15

ACOVA and Interactions

Analysis of covariance (ACOVA) incorporates one or more regression variables into an analysis of variance. As such, we can think of it as analogous to the two-way ANOVA of Chapter 14 except that instead of having two different factor variables as predictors, we have one factor variable and one continuous variable. The regression variables are referred to as covariates (relative to the dependent variable), hence the name analysis of covariance. Covariates are also known as *supplementary* or *concomitant observations*. Cox (1958, Chapter 4) gives a particularly nice discussion of the experimental design ideas behind analysis of covariance and illustrates various useful plotting techniques; also see Figure 15.4 below. In 1957 and 1982, *Biometrics* devoted entire issues to the analysis of covariance. We begin our discussion with an example that involves a two-group one-way analysis of variance and one covariate. Section 15.2 looks at an example where the covariate can also be viewed as a factor variable. Section 15.3 uses ACOVA to look at lack-of-fit testing.

15.1 One covariate example

Fisher (1947) gives data on the body weights (in kilograms) and heart weights (in grams) for domestic cats of both sexes that were given digitalis. A subset of the data is presented in Table 15.1. Our primary interest is to determine whether females' heart weights differ from males' heart weights when both have received digitalis.

As a first step, we might fit a one-way ANOVA model with sex groups,

$$
\begin{aligned}
y_{ij} &= \mu_i + \varepsilon_{ij} \\
&= \mu + \alpha_i + \varepsilon_{ij},
\end{aligned}
\tag{15.1.1}
$$

where the y_{ij}s are the heart weights, $i = 1, 2$, and $j = 1, \ldots, 24$. This model yields the analysis of

Table 15.1: *Body weights (kg) and heart weights (g) of domestic cats.*

Females				Males			
Body	Heart	Body	Heart	Body	Heart	Body	Heart
2.3	9.6	2.0	7.4	2.8	10.0	2.9	9.4
3.0	10.6	2.3	7.3	3.1	12.1	2.4	9.3
2.9	9.9	2.2	7.1	3.0	13.8	2.2	7.2
2.4	8.7	2.3	9.0	2.7	12.0	2.9	11.3
2.3	10.1	2.1	7.6	2.8	12.0	2.5	8.8
2.0	7.0	2.0	9.5	2.1	10.1	3.1	9.9
2.2	11.0	2.9	10.1	3.3	11.5	3.0	13.3
2.1	8.2	2.7	10.2	3.4	12.2	2.5	12.7
2.3	9.0	2.6	10.1	2.8	13.5	3.4	14.4
2.1	7.3	2.3	9.5	2.7	10.4	3.0	10.0
2.1	8.5	2.6	8.7	3.2	11.6	2.6	10.5
2.2	9.7	2.1	7.2	3.0	10.6	2.5	8.6

Table 15.2: *One-way analysis of variance on heart weights: Model (15.1.1).*

Source	df	SS	MS	F	P
Sex	1	56.117	56.117	23.44	0.0000
Error	46	110.11	2.3936		
Total	47	166.223			

Table 15.3: *Analysis of variance for heart weights: Model (15.1.2).*

Source	df	Adj. SS	MS	F	P
Body weights	1	37.828	37.828	23.55	0.000
Sex	1	4.499	4.499	2.80	0.101
Error	45	72.279	1.606		
Total	47	166.223			

variance given in Table 15.2. Note the overwhelming effect due to sexes. We now develop a model for both sex and weight that is analogous to the additive model (14.1.3).

15.1.1   Additive regression effects

Fisher provided both heart weights and body weights, so we can ask a more complex question, "Is there a sex difference in the heart weights over and above the fact that male cats are naturally larger than female cats?" To examine this we add a regression term to Model (15.1.1) and fit the traditional *analysis of covariance model,*

$$y_{ij} = \mu_i + \gamma z_{ij} + \varepsilon_{ij} \qquad (15.1.2)$$
$$= \mu + \alpha_i + \gamma z_{ij} + \varepsilon_{ij}.$$

Here the zs are the body weights and γ is a slope parameter associated with body weights. For this example the mean model is

$$m(\text{sex}, z) = \begin{cases} \mu_1 + \gamma z, & \text{if sex} = \text{female} \\ \mu_2 + \gamma z & \text{if sex} = \text{male}. \end{cases}$$

Model (15.1.2) is a special case of the general additive-effects model (9.9.2). It is an extension of the simple linear regression between the ys and the zs in which we allow a different intercept μ_i for each sex but the same slope. In many ways, it is analogous to the two-way additive-effects model (14.1.3). In Model (15.1.2) the effect of sex on heart weight is always the same for any fixed body weight, i.e.,

$$(\mu_1 + \gamma z) - (\mu_2 + \gamma z) = \mu_1 - \mu_2.$$

Thus we can talk about $\mu_1 - \mu_2$ being the sex effect regardless of body weight. The means for females and males are parallel lines with common slope γ and $|\mu_1 - \mu_2|$ the distance between the lines.

An analysis of variance table for Model (15.1.2) is given as Table 15.3. The interpretation of this table is different from the ANOVA tables examined earlier. For example, the sums of squares for body weights, sex, and error *do not* add up to the sum of squares total. The sums of squares in Table 15.3 are referred to as *adjusted sums of squares (Adj. SS)* because the body weight sum of squares is adjusted for sexes and the sex sum of squares is adjusted for body weights.

The error line in Table 15.3 is simply the error from fitting Model (15.1.2). The body weights line comes from comparing Model (15.1.2) with the reduced model (15.1.1). Note that the only difference between models (15.1.1) and (15.1.2) is that (15.1.1) does not involve the regression on body weights, so by testing the two models we are testing whether there is a significant effect due to the regression on body weights. The standard way of comparing a full and a reduced model is by comparing their error terms. Model (15.1.2) has one more parameter, γ, than Model (15.1.1), so there is one more degree of freedom for error in Model (15.1.1) than in Model (15.1.2), hence one degree of freedom for body weights. The adjusted sum of squares for body weights is the difference between the sum of squares error in Model (15.1.1) and the sum of squares error in Model (15.1.2). Given the sum of squares and the mean square, the F statistic for body weights is constructed in the usual way and is reported in Table 15.3. We see a major effect due to the regression on body weights.

The Sex line in Table 15.3 provides a test of whether there are differences in sexes *after adjusting for the regression on body weights*. This comes from comparing Model (15.1.2) to a similar model in which sex differences have been eliminated. In Model (15.1.2), the sex differences are incorporated as μ_1 and μ_2 in the first version and as α_1 and α_2 in the second version. To eliminate sex differences in Model (15.1.2), we simply eliminate the distinctions between the μs (the αs). Such a model can be written as

$$y_{ij} = \mu + \gamma z_{ij} + \varepsilon_{ij}. \tag{15.1.3}$$

The analysis of covariance model without treatment effects is just a simple linear regression of heart weight on body weight. We have reduced the two sex parameters to one overall parameter, so the difference in degrees of freedom between Model (15.1.3) and Model (15.1.2) is 1. The difference in the sums of squares error between Model (15.1.3) and Model (15.1.2) is the adjusted sum of squares for sex reported in Table 15.3. We see that the evidence for a sex effect over and above the effect due to the regression on body weights is not great.

While ANOVA table Error terms are always the same for equivalent models, the table of coefficients depends on the particular parameterization of a model. I prefer the ACOVA model parameterization

$$y_{ij} = \mu_i + \gamma z_{ij} + \varepsilon_{ij}.$$

Some computer programs insist on using the equivalent model

$$y_{ij} = \mu + \alpha_i + \gamma z_{ij} + \varepsilon_{ij}, \tag{15.1.4}$$

which is overparameterized. To get estimates of the parameters in Model (15.1.4), one must impose *side conditions* on them. My choice would be to make $\mu = 0$ and get a model equivalent to the first one. Other common choices of side conditions are: (a) $\alpha_1 = 0$, (b) $\alpha_2 = 0$, and (c) $\alpha_1 + \alpha_2 = 0$. Some programs are flexible enough to let you specify the side conditions yourself. Minitab, for example, uses the side conditions (c) and reports

Covariate	$\hat{\gamma}$	SE($\hat{\gamma}$)	t	P
Constant	2.755	1.498	1.84	0.072
Sex				
1	−0.3884	0.2320	−1.67	0.101
Body Wt	2.7948	0.5759	4.85	0.000

Relative to Model (15.1.4) the parameter estimates are $\hat{\mu} = 2.755$, $\hat{\alpha}_1 = -0.3884$, $\hat{\alpha}_2 = 0.3884$, $\hat{\gamma} = 2.7948$, so the estimated regression line for females is

$$E(y) = [2.755 + (-0.3884)] + 2.7948z = 2.3666 + 2.7948z$$

and for males

$$E(y) = [2.755 - (-0.3884)] + 2.7948z = 3.1434 + 2.7948z,$$

e.g., the predicted values for females are

$$\hat{y}_{1j} = [2.755 + (-0.3884)] + 2.7948z_{1j} = 2.3666 + 2.7948z_{1j}$$

and for males are

$$\hat{y}_{2j} = [2.755 - (-0.3884)] + 2.7948z_{2j} = 3.1434 + 2.7948z_{2j}.$$

Note that the t statistic for sex 1 is the square root of the F statistic for sex in Table 15.3. The P values are identical. Similarly, the tests for body weights are equivalent. Again, we find clear evidence for the effect of body weights after fitting sexes.

A 95% confidence interval for γ has end points

$$2.7948 \pm 2.014(0.5759),$$

which yields the interval $(1.6, 4.0)$. We are 95% confident that, for data comparable to the data in this study, an increase in body weight of one kilogram corresponds to a mean increase in heart weight of between 1.6g and 4.0g. (An increase in body weight *corresponds* to an increase in heart weight. Philosophically, we have no reason to believe that increasing body weights by one kg will *cause* an increase in heart weight.)

In Model (15.1.2), comparing treatments by comparing the treatment means $\bar{y}_{i\cdot}$ is inappropriate because of the complicating effect of the covariate. Adjusted means are often used to compare treatments. The formula and the actual values for the adjusted means are given below along with the raw means for body weights and heart rates.

$$\text{Adjusted means} \equiv \bar{y}_{i\cdot} - \hat{\gamma}(\bar{z}_{i\cdot} - \bar{z}_{\cdot\cdot})$$

Sex	N	Body	Heart	Adj. Heart
Female	24	2.333	8.887	9.580
Male	24	2.829	11.050	10.357
Combined	48	2.581	9.969	

Note that the difference in adjusted means is

$$9.580 - 10.357 = \hat{\alpha}_1 - \hat{\alpha}_2 = 2(-0.3884).$$

We have seen previously that there is little evidence of a differential effect on heart weights due to sexes after adjusting for body weights. Nonetheless, from the adjusted means what evidence exists suggests that, even after adjusting for body weights, a typical heart weight for males, 10.357, is larger than a typical heart weight for females, 9.580.

Figures 15.1 through 15.3 contain residual plots for Model (15.1.2). The plot of residuals versus predicted values looks exceptionally good. The plot of residuals versus sexes shows slightly less variability for females than for males. The difference is probably not enough to worry about. The normal plot of the residuals is alright with W' above the appropriate percentile.

The models that we have fitted form a hierarchy similar to that discussed in Chapter 14. The ACOVA model is larger than both the one-way and simple linear regression models, which are not comparable, but both are larger than the intercept-only model.

ACOVA

One-Way ANOVA   Simple Linear Regression

Intercept-Only

In terms of numbered models the hierarchy is

(15.1.2)

(15.1.1)   (15.1.3)

(14.1.6).

Such a hierarchy leads to two sequential ANOVA tables that are displayed in Table 15.4. All of the results in Table 15.3 appear in Table 15.4.

15.1.2   Interaction models

With these data, there is little reason to assume that when regressing heart weight on body weight the linear relationships are the same for females and males. Model (15.1.2) allows different intercepts

Residual–Fitted plot

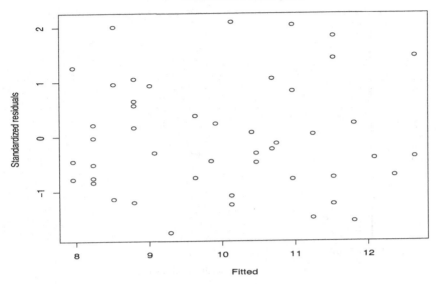

Figure 15.1: *Residuals versus predicted values, cat data.*

Residual–Socio plot

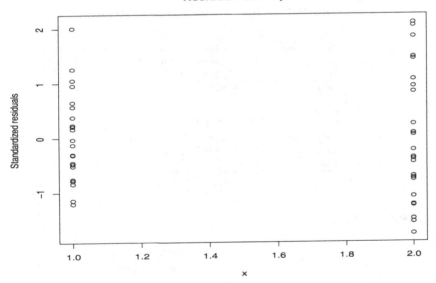

Figure 15.2: *Residuals versus sex, cat data.*

for these regressions but uses the same slope γ. We should test the assumption of a common slope by fitting the more general model that allows different slopes for females and males, i.e.,

$$
\begin{aligned}
y_{ij} &= \mu_i + \gamma_i z_{ij} + \varepsilon_{ij} \\
&= \mu + \alpha_i + \gamma_i z_{ij} + \varepsilon_{ij}.
\end{aligned}
\tag{15.1.5}
$$

In Model (15.1.5) the γs depend on i and thus the slopes are allowed to differ between the sexes. While Model (15.1.5) may look complicated, it consists of nothing more than fitting a simple linear regression to each group: one to the female data and a separate simple linear regression to the male

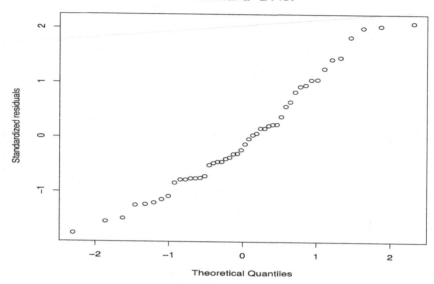

Figure 15.3: *Normal plot for cat data, $W' = 0.970$.*

Table 15.4: *Analyses of variance for rat weight gains.*

Source	df	Seq SS	MS	F	P
Body weights	1	89.445	89.445	55.69	0.000
Sex	1	4.499	4.499	2.80	0.101
Error	45	72.279	1.606		
Total	47	166.223			

Source	df	Seq SS	MS	F	P
Sex	1	56.117	56.117	39.94	0.000
Body Weights	1	37.828	37.828	23.55	0.000
Error	45	72.279	1.606		
Total	47	166.223			

data. The means model is

$$m(\text{sex}, z) = \begin{cases} \mu_1 + \gamma_1 z, & \text{if sex} = \text{female} \\ \mu_2 + \gamma_2 z & \text{if sex} = \text{male}. \end{cases}$$

Figure 15.4 contains some examples of how Model (15.1.2) and Model (15.1.5) might look when plotted. In Model (15.1.2) the lines are always parallel. In Model (15.1.5) they can have several appearances.

The sum of squares error for Model (15.1.5) can be found directly but it also comes from adding the error sums of squares for the separate female and male simple linear regressions. It is easily seen that for females the simple linear regression has an error sum of squares of 22.459 on 22 degrees of freedom and the males have an error sum of squares of 49.614 also on 22 degrees of freedom. Thus Model (15.1.5) has an error sum of squares of $22.459 + 49.614 = 72.073$ on $22 + 22 = 44$ degrees of freedom. The mean squared error for Model (15.1.5) is

$$MSE(5) = \frac{72.073}{44} = 1.638$$

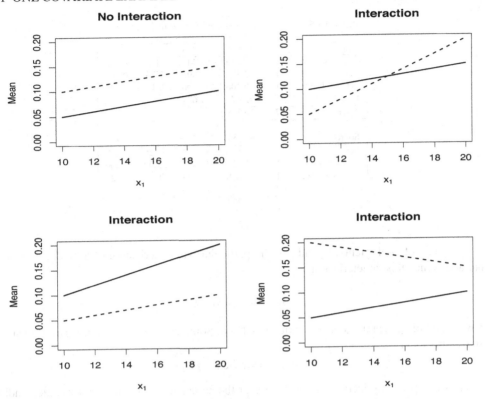

Figure 15.4 *Patterns of interaction (effect modification) between a continuous predictor x_1 and a binary predictor x_2.*

and, using results from Table 15.3, the test of Model (15.1.5) against the reduced model (15.1.2) has

$$F = \frac{[72.279 - 72.073]/[45 - 44]}{1.638} = \frac{0.206}{1.638} = 0.126.$$

The F statistic is very small; there is no evidence that we need to fit different slopes for the two sexes. Fitting Model (15.1.5) gives us no reason to question our analysis of Model (15.1.2). The interaction model is easily incorporated into our previous hierarchy of models.

Interaction
ACOVA

One-Way ANOVA   Simple Linear Regression

Intercept-Only

or, in terms of numbered models,

(15.1.5)
(15.1.2)

(15.1.1)   (15.1.3)

(14.1.6).

The hierarchy leads to the two ANOVA tables given in Table 15.5. We could also report C_p statistics for all five models relative to the interaction model (15.1.5).

Table 15.5: *Analyses of variance for rat weight gains.*

Source	df	Seq SS	MS	F	P
Body weights	1	89.445	89.445	54.61	0.000
Sex	1	4.499	4.499	2.75	0.105
Sex*Body Wt	1	0.206	0.206	0.13	0.725
Error	44	72.073	1.638		
Total	47	166.223			

Source	df	Seq SS	MS	F	P
Sex	1	56.117	56.117	34.26	0.000
Body Weights	1	37.828	37.828	23.09	0.000
Sex*Body Wt	1	0.206	0.206	0.13	0.725
Error	44	72.073	1.638		
Total	47	166.223			

The table of coefficients depends on the particular parameterization of a model. I prefer the interaction model parameterization

$$y_{ij} = \mu_i + \gamma_i z_{ij} + \varepsilon_{ij},$$

in which all of the parameters are uniquely defined. Some computer programs insist on using the equivalent model

$$y_{ij} = \mu + \alpha_i + \beta z_{ij} + \gamma_i z_{ij} + \varepsilon_{ij} \tag{15.1.6}$$

which is overparameterized. To get estimates of the parameters, one must impose side conditions on them. My choice would be to make $\mu = 0 = \beta$ and get a model equivalent to the first one. Other common choices of side conditions are: (a) $\alpha_1 = 0 = \gamma_1$, (b) $\alpha_2 = 0 = \gamma_2$, and (c) $\alpha_1 + \alpha_2 = 0 = \gamma_1 + \gamma_2$. Some programs are flexible enough to let you specify the model yourself. Minitab, for example, uses the side conditions (c) and reports

Covariate	$\hat{\gamma}$	SE($\hat{\gamma}$)	t	P
Constant	2.789	1.516	1.84	0.072
Sex				
1	0.142	1.516	0.09	0.926
Body Wt	2.7613	0.5892	4.69	0.000
Body Wt*Sex				
1	−0.2089	0.5892	−0.35	0.725

Relative to Model (15.1.6) the parameter estimates are $\hat{\mu} = 2.789$, $\hat{\alpha}_1 = 0.142$, $\hat{\alpha}_2 = -0.142$, $\hat{\beta} = 2.7613$, $\hat{\gamma}_1 = -0.2089$, $\hat{\gamma}_2 = 0.2089$, so the estimated regression line for females is

$$E(y) = (2.789 + 0.142) + [2.7613 + (-0.2089)]z = 2.961 + 2.5524z$$

and for males

$$E(y) = (2.789 - 0.142) + [2.7613 - (-0.2089)]z = 2.647 + 2.7613z,$$

i.e., the fitted values for females are

$$\hat{y}_{1j} = (2.789 + 0.142) + [2.7613 + (-0.2089)]z_{1j} = 2.961 + 2.5524z_{1j}$$

and for males

$$\hat{y}_{2j} = (2.789 - 0.142) + [2.7613 - (-0.2089)]z_{2j} = 2.647 + 2.7613z_{2j}.$$

15.1.3 Multiple covariates

In our cat example, we had one covariate, but it would be very easy to extend Model (15.1.2) to include more covariates. For example, with three covariates, x_1, x_2, x_3, the ACOVA model becomes

$$y_{ij} = \mu_i + \gamma_1 x_{ij1} + \gamma_2 x_{ij2} + \gamma_3 x_{ij3} + \varepsilon_{ij}.$$

We could even apply this idea to the cat example by considering a polynomial model. Incorporating into Model (15.1.2) a cubic polynomial for one predictor z gives

$$y_{ij} = \mu_i + \gamma_1 z_{ij} + \gamma_2 z_{ij}^2 + \gamma_3 z_{ij}^3 + \varepsilon_{ij}.$$

The key point is that ACOVA models are additive-effects models because none of the γ parameters depend on sex (i). If we have three covariates x_1, x_2, x_3, an ACOVA model has

$$y_{ij} = \mu_i + h(x_{ij1}, x_{ij2}, x_{ij3}) + \varepsilon_{ij},$$

for some function $h(\cdot)$. In this case $\mu_1 - \mu_2$ is the differential effect for the two groups regardless of the covariate values.

One possible interaction model allows completely different regressions functions for each group,

$$y_{ij} = \mu_i + \gamma_{i1} x_{ij1} + \gamma_{i2} x_{ij2} + \gamma_{i3} x_{ij3} + \varepsilon_{ij}.$$

Here we allow the slope parameters to depend on i. For the cat example we might consider separate cubic polynomials for each sex, i.e.,

$$y_{ij} = \mu_i + \gamma_{i1} z_{ij} + \gamma_{i2} z_{ij}^2 + \gamma_{i3} z_{ij}^3 + \varepsilon_{ij}.$$

15.2 Regression modeling

Consider again the ACOVA model (15.1.2) based on the factor variable sex (i) and the measurement variable body weight (z). To make life more interesting, let's consider a third sex category, say, herm (for hermaphrodite). If we create 0-1 indicator variables for each of our three categories, say, x_1, x_2, x_3, we can rewrite both the one-way ANOVA model (15.1.1) and Model (15.1.2) as linear models. (The SLR model (15.1.3) is already in linear model form.) The first form for the means of Model (15.1.1) becomes a no-intercept multiple regression model

$$
\begin{aligned}
m(x_1, x_2, x_3) &= \mu_1 x_1 + \mu_2 x_2 + \mu_3 x_3 \qquad &(15.2.1)\\
&= \begin{cases} \mu_1, & \text{female} \\ \mu_2, & \text{male} \\ \mu_3, & \text{herm} \end{cases}
\end{aligned}
$$

and the second form for the means is the overparameterized model

$$
\begin{aligned}
m(x_1, x_2, x_3) &= \mu + \alpha_1 x_1 + \alpha_2 x_2 + \alpha_3 x_3 \qquad &(15.2.2)\\
&= \begin{cases} (\mu + \alpha_1), & \text{female} \\ (\mu + \alpha_2), & \text{male} \\ (\mu + \alpha_3), & \text{herm.} \end{cases}
\end{aligned}
$$

The first form for the means of Model (15.1.2) is the parallel lines regression model

$$
\begin{aligned}
m(x_1, x_2, x_3, z) &= \mu_1 x_1 + \mu_2 x_2 + \mu_3 x_3 + \gamma z \qquad &(15.2.3)\\
&= \begin{cases} \mu_1 + \gamma z, & \text{female} \\ \mu_2 + \gamma z, & \text{male} \\ \mu_3 + \gamma z, & \text{herm} \end{cases}
\end{aligned}
$$

and the second form is the overparameterized parallel lines model

$$m(x_1,x_2,x_3,z) = \mu + \alpha_1 x_1 + \alpha_2 x_2 + \alpha_3 x_3 + \gamma z \qquad (15.2.4)$$

$$= \begin{cases} (\mu+\alpha_1)+\gamma z, & \text{female} \\ (\mu+\alpha_2)+\gamma z, & \text{male} \\ (\mu+\alpha_3)+\gamma z, & \text{herm.} \end{cases}$$

Similarly, we could have "parallel" polynomials. For quadratics that would be

$$m(x_1,x_2,x_3,z) = \mu_1 x_1 + \mu_2 x_2 + \mu_3 x_3 + \gamma_1 z + \gamma_2 z^2$$

$$= \begin{cases} \mu_1+\gamma_1 z+\gamma_2 z^2, & \text{female} \\ \mu_2+\gamma_1 z+\gamma_2 z^2, & \text{male} \\ \mu_3+\gamma_1 z+\gamma_2 z^2, & \text{herm} \end{cases}$$

wherein only the intercepts are different.

The interaction model (15.1.5) gives separate lines for each group and can be written as

$$m(x_1,x_2,x_3,z) = \mu_1 x_1 + \mu_2 x_2 + \mu_3 x_3 + \gamma_1 z x_1 + \gamma_2 z x_2 + \gamma_3 z x_3$$

$$= \begin{cases} \mu_1+\gamma_1 z, & \text{female} \\ \mu_2+\gamma_2 z, & \text{male} \\ \mu_3+\gamma_3 z, & \text{herm} \end{cases}$$

and the second form is the overparameterized model

$$m(x_1,x_2,x_3,z) = \mu + \alpha_1 x_1 + \alpha_2 x_2 + \alpha_3 x_3 + \beta z + \gamma_1 z x_1 + \gamma_2 z x_2 + \gamma_3 z x_3$$

$$= \begin{cases} (\mu+\alpha_1)+(\beta+\gamma_1)z, & \text{female} \\ (\mu+\alpha_2)+(\beta+\gamma_2)z, & \text{male} \\ (\mu+\alpha_3)+(\beta+\gamma_3)z, & \text{herm.} \end{cases}$$

Every sex category has a completely separate line with different slopes and intercepts. Interaction parabolas would be completely separate parabolas for each group

$$m(x_1,x_2,x_3,z) = \mu_1 x_1 + \mu_2 x_2 + \mu_3 x_3 + \gamma_{11} z x_1 + \gamma_{21} z^2 x_1$$
$$+ \gamma_{12} z x_2 + \gamma_{22} z^2 x_2 + \gamma_{13} z x_3 + \gamma_{23} z^2 x_3$$

$$= \begin{cases} \mu_1+\gamma_{11} z+\gamma_{21} z^2, & \text{female} \\ \mu_2+\gamma_{12} z+\gamma_{22} z^2, & \text{male} \\ \mu_3+\gamma_{13} z+\gamma_{23} z^2, & \text{herm.} \end{cases}$$

15.2.1   Using overparameterized models

As discussed in Chapter 12, Model (15.2.2) can be made into a regression model by dropping any one of the predictor variables, say x_1,

$$m(x_1,x_2,x_3) = \mu + \alpha_2 x_2 + \alpha_3 x_3 \qquad (15.2.5)$$

$$= \begin{cases} \mu, & \text{female} \\ (\mu+\alpha_2), & \text{male} \\ (\mu+\alpha_3), & \text{herm.} \end{cases}$$

Using an intercept and indicators x_2 and x_3 for male and herm makes female the baseline category. Similarly, if we fit the ACOVA model (15.2.4) but drop out x_1 we get parallel lines

$$m(x_1,x_2,x_3,z) = \mu + \alpha_2 x_2 + \alpha_3 x_3 + \gamma z \qquad (15.2.6)$$

$$= \begin{cases} \mu+\gamma z, & \text{female} \\ (\mu+\alpha_2)+\gamma z, & \text{male} \\ (\mu+\alpha_3)+\gamma z, & \text{herm.} \end{cases}$$

If, in the one-way ANOVA, we thought that males and females had the same mean, we could drop both x_1 and x_2 from Model (15.2.2) to get

$$m(x_1,x_2,x_3) = \mu + \alpha_3 x_3 = \begin{cases} \mu, & \text{female or male} \\ \mu + \alpha_3, & \text{herm.} \end{cases}$$

If we thought that males and herms had the same mean, since neither male nor herm is the baseline, we could replace x_2 and x_3 with a new variable $\tilde{x} = x_2 + x_3$ that indicates membership in either group to get

$$m(x_1,x_2,x_3) = \mu + \alpha\tilde{x} = \begin{cases} \mu, & \text{female} \\ \mu + \alpha, & \text{male or herm.} \end{cases}$$

We could equally well fit the model

$$m(x_1,x_2,x_3) = \mu_1 x_1 + \mu_3 \tilde{x} = \begin{cases} \mu_1, & \text{female} \\ \mu_3, & \text{male or herm.} \end{cases}$$

In these cases, the analysis of covariance (15.2.4) behaves similarly. For example, without both x_1 and x_2 Model (15.2.4) becomes

$$\begin{aligned} m(x_1,x_2,x_3,z) &= \mu + \alpha_3 x_3 + \gamma z \qquad\qquad\qquad\qquad (15.2.7) \\ &= \begin{cases} \mu + \gamma z, & \text{female or male} \\ (\mu + \alpha_3) + \gamma z, & \text{herm} \end{cases} \end{aligned}$$

and involves only two parallel lines, one that applies to both females and males, and another one for herms.

Dropping both x_1 and x_2 from Model (15.2.2) gives very different results than dropping the intercept and x_2 from Model (15.2.2). That statement may seem obvious, but if you think about the fact that dropping x_1 alone does not actually affect how the model fits the data, it might be tempting to think that further dropping x_2 could have the same effect after dropping x_1 as dropping x_2 has in Model (15.2.1). We have already examined dropping both x_1 and x_2 from Model (15.2.2), now consider dropping both the intercept and x_2 from Model (15.2.2), i.e., dropping x_2 from Model (15.2.1). The model becomes

$$m(x) = \mu_1 x_1 + \mu_3 x_3 = \begin{cases} \mu_1, & \text{female} \\ 0, & \text{male} \\ \mu_3, & \text{herm.} \end{cases}$$

This occurs because all of the predictor variables in the model take the value 0 for male. If we incorporate the covariate age into this model we get

$$m(x) = \mu_1 x_1 + \mu_3 x_3 + \gamma z = \begin{cases} \mu_1 + \gamma z, & \text{female} \\ 0 + \gamma z, & \text{male} \\ \mu_3 + \gamma z, & \text{herm,} \end{cases}$$

which are three parallel lines but male has an intercept of 0.

15.3 ACOVA and two-way ANOVA

The material in Section 15.1 is sufficiently complex to warrant another example. This time we use a covariate that also defines a grouping variable and explore the relationships between fitting an ACOVA and fitting a two-way ANOVA.

EXAMPLE 15.3.1.   *Hopper Data.*
The data in Table 15.6 were provided by Schneider and Pruett (1994). They were interested in

Table 15.6: *Multiple weighings of a hopper car.*

Day	First	Second	Third	Day	First	Second	Third
1	5952	5944	6004	11	5986	5920	5944
2	5930	5873	5895	12	6036	6084	6054
3	6105	6113	6101	13	6035	6136	6128
4	5943	5878	5931	14	6070	6016	6111
5	6031	6009	6000	15	6015	5990	5950
6	6064	6030	6070	16	6049	5988	6000
7	6093	6129	6154	17	6139	6153	6151
8	5963	5978	5966	18	6077	6012	6005
9	5982	6005	5970	19	5932	5899	5944
10	6052	6046	6029	20	6115	6087	6078

Table 15.7: *Summary statistics for hopper data.*

DAY	N	MEAN	STDEV	DAY	N	MEAN	STDEV
1	3	5966.7	32.6	11	3	5950.0	33.4
2	3	5899.3	28.7	12	3	6058.0	24.2
3	3	6106.3	6.1	13	3	6099.7	56.1
4	3	5917.3	34.6	14	3	6065.7	47.6
5	3	6013.3	15.9	15	3	5985.0	32.8
6	3	6054.7	21.6	16	3	6012.3	32.3
7	3	6125.3	30.7	17	3	6147.7	7.6
8	3	5969.0	7.9	18	3	6031.3	39.7
9	3	5985.7	17.8	19	3	5925.0	23.3
10	3	6042.3	11.9	20	3	6093.3	19.3

whether the measurement system for the weight of railroad hopper cars was under control. A standard hopper car weighing about 266,000 pounds was used to obtain the first 3 weighings of the day on each of 20 days. The process was to move the car onto the scales, weigh the car, move the car off, move the car on, weigh the car, move it off, move it on, and weigh it a third time. The tabled values are the weight of the car minus 260,000.

As we did with the cat data, the first thing we might do is treat the three repeat observations as replications and do a one-way ANOVA on the days,

$$y_{ij} = \mu_i + \varepsilon_{ij}, \qquad i = 1,\ldots,20, \ j = 1,2,3.$$

Summary statistics are given in Table 15.7 and the ANOVA table follows.

Analysis of Variance				
Source	df	SS	MS	F	P
Day	19	296380	15599	18.25	0.000
Error	40	34196	855		
Total	59	330576			

Obviously, there are differences in days.

15.3.1   Additive effects

The three repeat observations on the hopper could be subject to trends. Treat the three observations as measurements of time with values 1, 2, 3. This now serves as a covariate z. With three distinct covariate values, we could fit a parabola.

$$y_{ij} = \mu_i + \gamma_1 z_{ij} + \gamma_2 z_{ij}^2 + \varepsilon_{ij}, \qquad i = 1,\ldots,20, \ j = 1,2,3.$$

The software I used actually fits

$$y_{ij} = \mu + \alpha_i + \gamma_1 z_{ij} + \gamma_2 z_{ij}^2 + \varepsilon_{ij}, \qquad i = 1,\ldots,20,\; j = 1,2,3$$

with the additional constraint that $\alpha_1 + \cdots + \alpha_{20} = 0$, so that $\hat{\alpha}_{20} = -(\hat{\alpha}_1 + \cdots + \hat{\alpha}_{19})$. The output then only presents $\hat{\alpha}_1, \ldots, \hat{\alpha}_{19}$

Table of Coefficients

Predictor	Est	SE(Est)	t	P
Constant	6066.10	28.35	213.98	0.000
z	−49.50	32.19	−1.54	0.132
z^2	11.850	7.965	1.49	0.145
Day				
1	−55.73	16.37	−3.41	0.002
2	−123.07	16.37	5.13	0.000
4	−105.07	16.37	−6.42	0.000
5	−9.07	16.37	−0.55	0.583
6	32.27	16.37	1.97	0.056
7	102.93	16.37	6.29	0.000
8	−53.40	16.37	−3.26	0.002
9	−36.73	16.37	−2.24	0.031
10	19.93	16.37	1.22	0.231
11	−72.40	16.37	−4.42	0.000
12	35.60	16.37	2.18	0.036
13	77.27	16.37	4.72	0.000
14	43.27	16.37	2.64	0.012
15	−37.40	16.37	−2.29	0.028
16	−10.07	16.37	−0.62	0.542
17	125.27	16.37	7.65	0.000
18	8.93	16.37	0.55	0.588
19	−97.40	16.37	−5.95	0.000

The table of coefficients is ugly, especially because there are so many days, but the main point is that the z^2 term in not significant ($P = 0.145$).

The corresponding ANOVA table is a little strange. The only really important thing is that it gives the Error line. There is also some interest in the fact that the F statistic reported for z^2 is the square of the t statistic, having identical P values.

Analysis of Variance

Source	df	SS	MS	F	P
z	1	176	176	2.36	0.132
Day	19	296380	15599	18.44	0.000
z^2	1	1872	1872	2.21	0.145
Error	38	32147	846		
Total	59	330576			

Similar to Section 12.5, instead of fitting a maximal polynomial (we only have three times so can fit at most a quadratic in time), we could alternatively treat z as a factor variable and do a two-way ANOVA as in Chapter 14, i.e., fit

$$y_{ij} = \mu + \alpha_i + \eta_j + \varepsilon_{ij}, \qquad i = 1,\ldots,20,\; j = 1,2,3.$$

The quadratic ACOVA model is equivalent to this two-way ANOVA model, so the two-way ANOVA model should have an equivalent ANOVA table.

Analysis of Variance

Source	df	SS	MS	F	P
Day	19	296380	15599	18.44	0.000
Time	2	2049	1024	1.21	0.309
Error	38	32147	846		
Total	59	330576			

This has the same Error line as the quadratic ACOVA model.

With a nonsignificant z^2 term in the quadratic model, it makes sense to check whether we need the linear term in z. The model is

$$y_{ij} = \mu_i + \gamma_1 z_{ij} + \varepsilon_{ij}, \qquad i = 1,\ldots,20, \ j = 1,2,3$$

or

$$y_{ij} = \mu + \alpha_i + \gamma_1 z_{ij} + \varepsilon_{ij}, \qquad i = 1,\ldots,20, \ j = 1,2,3$$

subject to the constraint that $\alpha_1 + \cdots + \alpha_{20} = 0$. The table of coefficients is

Table of Coefficients

Predictor	Est	SE(Est)	t	P
Constant	6026.60	10.09	597.40	0.000
Time	−2.100	4.670	−0.45	0.655
Day				
1	−55.73	16.62	−3.35	0.002
2	−123.07	16.62	−7.40	0.000
3	83.93	16.62	5.05	0.000
4	−105.07	16.62	−6.32	0.000
5	−9.07	16.62	−0.55	0.588
6	32.27	16.62	1.94	0.059
7	102.93	16.62	6.19	0.000
8	−53.40	16.62	−3.21	0.003
9	−36.73	16.62	−2.21	0.033
10	19.93	16.62	1.20	0.238
11	−72.40	16.62	−4.36	0.000
12	35.60	16.62	2.14	0.038
13	77.27	16.62	4.65	0.000
14	43.27	16.62	2.60	0.013
15	−37.40	16.62	−2.25	0.030
16	−10.07	16.62	−0.61	0.548
17	125.27	16.62	7.54	0.000
18	8.93	16.62	0.54	0.594
19	−97.40	16.62	−5.86	0.000

and we find no evidence that we need the linear term ($P = 0.655$). For completeness, an ANOVA table is

Analysis of Variance

Source	df	SS	MS	F	P
z	1	176	176	0.20	0.655
Day	19	296380	15599	17.88	0.000
Error	39	34020	872		
Total	59	330576			

It might be tempting to worry about interaction in this model. Resist the temptation! First, there are not enough observations for us to fit a full interaction model and still estimate σ^2. If we fit separate quadratics for each day, we would have 60 mean parameters and 60 observations, so zero

Table 15.8: *Hooker data.*

Case	Temperature	Pressure	Near Rep.	Case	Temperature	Pressure	Near Rep.
1	180.6	15.376	1	17	191.1	19.490	9
2	181.0	15.919	1	18	191.4	19.758	9
3	181.9	16.106	2	19	193.4	20.480	10
4	181.9	15.928	2	20	193.6	20.212	10
5	182.4	16.235	2	21	195.6	21.605	11
6	183.2	16.385	3	22	196.3	21.654	12
7	184.1	16.959	4	23	196.4	21.928	12
8	184.1	16.817	4	24	197.0	21.892	13
9	184.6	16.881	4	25	199.5	23.030	14
10	185.6	17.062	5	26	200.1	23.369	15
11	185.7	17.267	5	27	200.6	23.726	15
12	186.0	17.221	5	28	202.5	24.697	16
13	188.5	18.507	6	29	208.4	27.972	17
14	188.8	18.356	6	30	210.2	28.559	18
15	189.5	18.869	7	31	210.8	29.211	19
16	190.6	19.386	8				

degrees of freedom for error. Exactly the same thing would happen if we fit a standard interaction model from Chapter 14. But more importantly, it just makes sense to think of interaction as error for these data. What does it mean for there to be a time trend in these data? Surely we have no interest in time trends that go up one day and down another day without any rhyme or reason. For a time trend to be meaningful, it needs to be something that we can spot on a consistent basis. It has to be something that is strong enough that we can see it over and above the natural day-to-day variation of the weighing process. Well, the natural day-to-day variation of the weighing process is precisely the Day-by-Time interaction, so the interaction is precisely what we want to be using as our error term. In the model

$$y_{ij} = \mu + \alpha_i + \gamma_1 z_{ij} + \gamma_2 z_{ij}^2 + \varepsilon_{ij},$$

changes that are inconsistent across days and times, terms that depend on both i and j, are what we want to use as error. (An exception to this claim is if, say, we noticed that time trends go up one day, down the next, then up again, etc. That is a form of interaction that we could be interested in, but its existence requires additional structure for the Days because it involves modeling effects for alternate days.)

15.4 Near replicate lack-of-fit tests

In Section 8.6 and Subsection 12.5.1 we discussed Fisher's lack-of-fit test. Fisher's test is based on there being duplicate cases among the predictor variables. Often, there are few or none of these. Near replicate lack-of-fit tests were designed to ameliorate that problem by clustering together cases that are nearly replicates of one another.

With the Hooker data of Table 7.1, Fisher's lack-of-fit test suffers from few degrees of freedom for pure error. Table 15.8 contains a list of near replicates. These were obtained by grouping together cases that were within 0.5 degrees F. We then construct an F test by fitting 3 models. First, reindex the observations y_i, $i = 1,\ldots,31$ into y_{jk} with $j = 1,\ldots,19$ identifying the near replicate groups and $k = 1,\ldots,N_i$ identifying observations within the near replicate group. Thus the simple linear regression model

$$y_i = \beta_0 + \beta_1 x_i + \varepsilon_i$$

can be rewritten as

$$y_{jk} = \beta_0 + \beta_1 x_{jk} + \varepsilon_{jk}.$$

The first of the three models in question is the simple linear regression performed on the near replicate cluster means $\bar{x}_{j\cdot}$.

$$y_{jk} = \beta_0 + \beta_1 \bar{x}_{j\cdot} + \varepsilon_{jk}. \tag{15.4.1}$$

This is sometimes called the *artificial means model* because it is a regression on the near replicate cluster means $\bar{x}_{j\cdot}$ but the clusters are artificially constructed. The second model is a one-way analysis of variance model with groups defined by the near replicate clusters,

$$y_{jk} = \mu_j + \varepsilon_{jk}. \tag{15.4.2}$$

As a regression model, define the predictor variables δ_{hj} for $h = 1, \ldots, 19$, that are equal to 1 if $h = j$ and 0 otherwise. Then the model can be rewritten as a multiple regression model through the origin

$$y_{jk} = \mu_1 \delta_{1j} + \mu_2 \delta_{2j} + \cdots + \mu_{19}\delta_{19,j} + \varepsilon_{jk}.$$

The last model is called an analysis of covariance model because it incorporates the original predictor (covariate) x_{jk} into the analysis of variance model (15.4.2). The model is

$$y_{jk} = \mu_j + \beta_1 x_{jk} + \varepsilon_{jk}, \tag{15.4.3}$$

which can alternatively be written as a regression

$$y_{jk} = \mu_1 \delta_{1j} + \mu_2 \delta_{2j} + \cdots + \mu_{19}\delta_{19,j} + \beta_1 x_{jk} + \varepsilon_{jk}.$$

Fitting these three models gives

Analysis of Variance: Artificial means model (15.4.1).

Source	df	SS	MS	F	P
Regression	1	444.05	444.05	3389.06	0.000
Error	29	3.80	0.13		
Total	30	447.85			

Analysis of Variance: Near replicate groups (15.4.2).

Source	df	SS	MS	F	P
Near Reps	18	447.437	24.858	722.79	0.000
Error	12	0.413	0.034		
Total	30	447.850			

Analysis of Covariance: (15.4.3).

Source	df	SS	MS	F	P
x	1	444.167	0.118	4.43	0.059
Near Reps	18	3.388	0.188	7.04	0.001
Error	11	0.294	0.027		
Total	30	447.850			

The lack-of-fit test uses the difference in the sums of squares error for the first two models in the numerator of the test and the mean squared error for the analysis of covariance model in the denominator of the test. The lack-of-fit test statistic is

$$F = \frac{(3.80 - 0.413)/(29 - 12)}{0.027} = 7.4.$$

This can be compared to an $F(17,11)$ distribution, that yields a P value of 0.001. This procedure is known as *Shillington's test*, cf. Christensen (2011).

Table 15.9: *Compressive strength of hoop pine trees (y) with moisture contents (z).*

					Temperature					
	−20° C		0° C		20° C		40° C		60° C	
Tree	z	y	z	y	z	y	z	y	z	y
1	42.1	13.14	41.1	12.46	43.1	9.43	41.4	7.63	39.1	6.34
2	41.0	15.90	39.4	14.11	40.3	11.30	38.6	9.56	36.7	7.27
3	41.1	13.39	40.2	12.32	40.6	9.65	41.7	7.90	39.7	6.41
4	41.0	15.51	39.8	13.68	40.4	10.33	39.8	8.27	39.3	7.06
5	41.0	15.53	41.2	13.16	39.7	10.29	39.0	8.67	39.0	6.68
6	42.0	15.26	40.0	13.64	40.3	10.35	40.9	8.67	41.2	6.62
7	40.4	15.06	39.0	13.25	34.9	10.56	40.1	8.10	41.4	6.15
8	39.3	15.21	38.8	13.54	37.5	10.46	40.6	8.30	41.8	6.09
9	39.2	16.90	38.5	15.23	38.5	11.94	39.4	9.34	41.7	6.26
10	37.7	15.45	35.7	14.06	36.7	10.74	38.9	7.75	38.2	6.29

15.5 Exercises

EXERCISE 15.5.1.   Table 15.9 contains data from Sulzberger (1953) and Williams (1959) on y, the maximum compressive strength parallel to the grain of wood from ten hoop pine trees. The data also include the temperature of the evaluation and a covariate z, the moisture content of the wood. Analyze the data. Examine polynomials in the temperatures.

EXERCISE 15.5.2.   Smith, Gnanadesikan, and Hughes (1962) gave data on urine characteristics of young men. The men were divided into four categories based on obesity. The data contain a covariate z that measures specific gravity. The dependent variable is y_1; it measures pigment creatinine. These variables are included in Table 15.10. Perform an analysis of covariance on y_1. How do the conclusions about obesity effects change between the ACOVA and the results of the ANOVA that ignores the covariate?

EXERCISE 15.5.3.   Smith, Gnanadesikan, and Hughes (1962) also give data on the variable y_2 that measures chloride in the urine of young men. These data are also reported in Table 15.10. As in the previous problem, the men were divided into four categories based on obesity. Perform an analysis of covariance on y_2 again using the specific gravity as the covariate z. Compare the results of the ACOVA to the results of the ANOVA that ignores the covariate.

EXERCISE 15.5.4.   The data of Exercise 14.5.1 and Table 14.8 involved two factors, one of which had unequally spaced quantitative levels. Find the smallest polynomial that gives an adequate fit in place of treating "days in storage" as a factor variable.

EXERCISE 15.5.5.   Apply Shillington's test to the data of Exercise 9.12.3 and Table 9.3. The challenge is to come up with some method of identifying near replicates. (A clustering algorithm is a good idea but beyond the scope of this book.)

EXERCISE 15.5.6.   Referring back to Subsection 7.3.3, test the need for a power transformation in each of the following problems from the previous chapter. Use all three constructed variables on each data set and compare results.

(a) Exercise 14.5.1.

(b) Exercise 14.5.2.

(c) Exercise 14.5.3.

(d) Exercise 14.5.4.

Table 15.10: *Excretory characteristics.*

Group I			Group II		
z	y_1	y_2	z	y_1	y_2
24	17.6	5.15	31	18.1	9.00
32	13.4	5.75	23	19.7	5.30
17	20.3	4.35	32	16.9	9.85
30	22.3	7.55	20	23.7	3.60
30	20.5	8.50	18	19.2	4.05
27	18.5	10.25	23	18.0	4.40
25	12.1	5.95	31	14.8	7.15
30	12.0	6.30	28	15.6	7.25
28	10.1	5.45	21	16.2	5.30
24	14.7	3.75	20	14.1	3.10
26	14.8	5.10	15	17.5	2.40
27	14.4	4.05	26	14.1	4.25
			24	19.1	5.80
			16	22.5	1.55

Group III			Group IV		
z	y_1	y_2	z	y_1	y_2
18	17.0	4.55	32	12.5	2.90
10	12.5	2.65	25	8.7	3.00
33	21.5	6.50	28	9.4	3.40
25	22.2	4.85	27	15.0	5.40
35	13.0	8.75	23	12.9	4.45
33	13.0	5.20	25	12.1	4.30
31	10.9	4.75	26	13.2	5.00
34	12.0	5.85	34	11.5	3.40
16	22.8	2.85			
31	16.5	6.55			
28	18.4	6.60			

(e) Exercise 14.5.5.

(f) Exercise 14.5.6.

EXERCISE 15.5.7.   Write models (15.1.1), (15.1.2), and (15.1.3) in matrix form. For each model use a regression program on the heart weight data of Table 15.1 to find 95% and 99% prediction intervals for a male and a female each with body weight of 3.0. Hint: Use models without intercepts whenever possible.

EXERCISE 15.5.8.   Consider the analysis of covariance for a one-way ANOVA with one covariate. Find the form for a 99% prediction interval for an observation, say, from the first treatment group with a given covariate value z.

EXERCISE 15.5.9.   Assume that in Model (15.1.2), $\mathrm{Cov}(\bar{y}_{i\cdot}, \hat{\gamma}) = 0$. Show that

$$\mathrm{Var}\left(\sum_{i=1}^{a} \lambda_i \left(\bar{y}_{i\cdot} - \bar{z}_{i\cdot} \hat{\gamma} \right) \right) = \sigma^2 \left[\frac{\sum_{i=1}^{a} \lambda_i^2}{b} + \frac{\left(\sum_{i=1}^{a} \lambda_i \bar{z}_{i\cdot} \right)^2}{SSE_{zz}} \right]$$

where SSE_{zz} is the sum of squares error from doing a one-way ANOVA on z.

Chapter 16

Multifactor Structures

In this chapter we introduce analysis of variance models that involve more than two factors and examine interactions between two factors.

16.1 Unbalanced three-factor analysis of variance

Most of the material of this section was originally published as Example 7.6.1 in Christensen (1987). It is reprinted with the kind permission of Springer-Verlag.

Table 16.1 is derived from Scheffé (1959) and gives the moisture content (in grams) for samples of a food product made with three kinds of salt (A), three amounts of salt (B), and two additives (C). The amounts of salt, as measured in moles, are equally spaced. The two numbers listed for some treatment combinations are replications. We wish to analyze these data.

We will consider these data as a three-factor ANOVA. From the structure of the replications the ANOVA has unequal numbers. The general model for a three-factor ANOVA with replications is

$$y_{ijkm} = G + A_i + B_j + C_k + [AB]_{ij} + [AC]_{ik} + [BC]_{jk} + [ABC]_{ijk} + e_{ijkm}.$$

Our first priority is to find out which interactions are important.

Table 16.2 contains the sum of squares for error and the degrees of freedom for error for all the ANOVA models that include all of the main effects. Each model is identified in the table by the highest-order terms in the model. For example, $[AB][AC]$ indicates the model

$$y_{ijkm} = G + A_i + B_j + C_k + [AB]_{ij} + [AC]_{ik} + e_{ijkm}$$

with only the $[AB]$ and $[AC]$ interactions. In $[AB][AC]$, the grand mean and all of the main effects are redundant; it does not matter whether these terms are included in the model. Similarly, $[AB][C]$ indicates the model

$$y_{ijkm} = G + A_i + B_j + C_k + [AB]_{ij} + e_{ijkm}$$

with the $[AB]$ interaction and the C main effect. In $[AB][C]$, the grand mean and the A and B main effects are redundant. Readers familiar with methods for fitting log-linear models (cf. Christensen, 1997 or Fienberg, 1980) will notice a correspondence between Table 16.2 and similar displays used

Table 16.1: *Moisture content of a food product.*

A (salt)		1			2			3		
B (amount salt)		1	2	3	1	2	3	1	2	3
	1	8	17	22	7	26	34	10	24	39
			13	20	10	24		9		36
C (additive)										
	2	5	11	16	3	17	32	5	16	33
		4	10	15	5	19	29	4		34

Table 16.2: *Statistics for fitting models to the data of Table 16.1.*

Model	SSE	dfE	F*	C_p
[ABC]	32.50	14		18.0
[AB][AC][BC]	39.40	18	.743	13.0
[AB][AC]	45.18	20	.910	11.5
[AB][BC]	40.46	20	.572	9.4
[AC][BC]	333.2	22	16.19	131.5
[AB][C]	45.75	22	.713	7.7
[AC][B]	346.8	24	13.54	133.4
[BC][A]	339.8	24	13.24	130.4
[A][B][C]	351.1	26	11.44	131.2

The F statistics are for testing each model against the model with a three-factor interaction, i.e., [ABC]. The denominator of each F statistic is $MSE([ABC]) = 32.50/14 = 2.3214$.

in fitting three-dimensional contingency tables. The analogies between selecting log-linear models and selecting models for unbalanced ANOVA are pervasive.

All of the models have been compared to the full model using F statistics in Table 16.2. It takes neither a genius nor an F table to see that the only models that fit the data are the models that include the [AB] interaction. The C_p statistics tell the same story.

In addition to testing models against the three-factor interaction model, there are a number of other comparisons that can be made among models that include [AB]. These are [AB][AC][BC] versus [AB][AC], [AB][AC][BC] versus [AB][BC], [AB][AC][BC] versus [AB][C], [AB][AC] versus [AB][C], and [AB][BC] versus [AB][C]. None of the comparisons show any lack of fit. The last two comparisons are illustrated below.

$$[AB][AC] \text{ versus } [AB][C]$$

$$R(AC|AB,C) = 45.75 - 45.18 = 0.57$$

$$F_{obs} = (0.57/2)/2.3214 = 0.123$$

$$[AB][BC] \text{ versus } [AB][C]$$

$$R(BC|AB,C) = 45.75 - 40.46 = 5.29$$

$$F_{obs} = (5.29/2)/2.3214 = 1.139.$$

Here we use the $R(\cdot|\cdot)$ notation introduced in Subsection 9.3.1 that is similar to the $SSR(\cdot|\cdot)$ notation. The denominator in each test is $MSE([ABC])$, i.e., the variance estimate from the biggest model under consideration.

The smallest model that seems to fit the data adequately is [AB][C]. This is indicated by the C_p statistic but also the F statistics for comparing [AB][C] to the larger models are all extremely small. Writing out the model [AB][C], it is

$$y_{ijkm} = G + A_i + B_j + C_k + [AB]_{ij} + e_{ijkm}.$$

We need to examine the [AB] interaction. Since the levels of B are quantitative, a model that is equivalent to [AB][C] is a model that includes the main effects for C, but, instead of fitting an interaction in A and B, fits a separate regression equation in the levels of B for each level of A. Let x_j, $j = 1, 2, 3$ denote the levels of B. There are three levels of B, so the most general polynomial we can fit is a second-degree polynomial in x_j. Since the amounts of salt were equally spaced, it does not matter

Table 16.3: *Additional statistics for data of Table 16.1.*

Model	SSE	dfE
$[A_0][A_1][A_2][C]$	45.75	22
$[A_0][A_1][C]$	59.98	25
$[A_0][A_1]$	262.0	26
$[A_0][C]$	3130.	28

much what we use for the x_js. The computations were performed using $x_1 = 1, x_2 = 2, x_3 = 3$. In particular, the model $[AB][C]$ was reparameterized as

$$y_{ijkm} = A_{i0} + A_{i1}x_j + A_{i2}x_j^2 + C_k + e_{ijkm}. \tag{16.1.1}$$

The nature of this model is that for a fixed additive, the three curves for the three salts can take any shapes at all. However, if you change to the other additive all three of the curves will shift, either up or down, exactly the same amount due to the change in additive. The shapes of the curves do not change.

With a notation similar to that used in Table 16.2, the *SSE* and the *dfE* are reported in Table 16.3 for Model (16.1.1) and three reduced models. Note that the *SSE* and *dfE* reported in Table 16.3 for $[A_0][A_1][A_2][C]$ are identical to the values reported in Table 16.2 for $[AB][C]$. This, of course, must be true if the models are merely reparameterizations of one another. First we want to establish whether the quadratic effects are necessary in the regressions. To do this we drop the A_{i2} terms from Model (16.1.1) and test

$$[A_0][A_1][A_2][C] \text{ versus } [A_0][A_1][C]$$

$$R(A_2|A_1, A_0, C) = 59.98 - 45.75 = 14.23$$

$$F_{obs} = (14.23/3)/2.3214 = 2.04.$$

Since $F(.95, 3, 14) = 3.34$, there is no evidence of any nonlinear effects.

At this point it might be of interest to test whether there are any linear effects. This is done by testing $[A_0][A_1][C]$ against $[A_0][C]$. The statistics needed for this test are in Table 16.3. Instead of actually doing the test, recall that no models in Table 16.2 fit the data unless they included the $[AB]$ interaction. If we eliminated the linear effects we would have a model that involved none of the $[AB]$ interaction. (The model $[A_0][C]$ is identical to the ANOVA model $[A][C]$.) We already know that such models do not fit.

Finally, we have never explored the possibility that there is no main effect for C. This can be done by testing

$$[A_0][A_1][C] \text{ versus } [A_0][A_1]$$

$$R(C|A_1, A_0) = 262.0 - 59.98 = 202$$

$$F_{obs} = (202/1)/2.3214 = 87.$$

Obviously, there is a substantial main effect for C, the type of food additive.

Our conclusion is that the model $[A_0][A_1][C]$ is the smallest model that has been considered that adequately fits the data. This model indicates that there is an effect for the type of additive and a linear relationship between amount of salt and moisture content. The slope and intercept of the line may depend on the type of salt. (The intercept of the line also depends on the type of additive.) Table 16.4 contains parameter estimates and standard errors for the model. All estimates in the example use the side condition $C_1 = 0$.

Note that, in lieu of the F test given earlier, the test for the main effect C could be performed from Table 16.4 by looking at $t = -5.067/.5522 = -9.176$. Moreover, we should have $t^2 = F$. The t statistic squared is 84, while the F statistic reported earlier is 87. The difference is due to the fact

Table 16.4: $y_{ijkm} = A_{i0} + A_{i1}x_j + C_k + e_{ijkm}$.

Table of Coefficients		
Parameter	Estimate	SE
A_{10}	3.35	1.375
A_{11}	5.85	.5909
A_{20}	-3.789	1.237
A_{21}	13.24	.5909
A_{30}	-4.967	1.231
A_{31}	14.25	.5476
C_1	0.	none
C_2	-5.067	.5522

Table 16.5: $y_{ijkm} = A_{i0} + A_{i1}x_j + C_k + e_{ijkm}$, $A_{21} = A_{31}$, $A_{20} = A_{30}$.

Table of Coefficients		
Parameter	Estimate	SE
A_{10}	3.395	1.398
A_{11}	5.845	.6008
A_{20}	-4.466	.9030
A_{21}	13.81	.4078
C_1	0.	none
C_2	-5.130	.5602

that the SE reported in Table 16.4 uses the *MSE* for the model being fitted, while in performing the *F* test we used $MSE([ABC])$.

Are we done yet? No. The parameter estimates suggest some additional questions. Are the slopes for salts 2 and 3 the same, i.e., is $A_{21} = A_{31}$? In fact, are the entire lines for salts 2 and 3 the same, i.e., are $A_{21} = A_{31}$, $A_{20} = A_{30}$? We can fit models that incorporate these assumptions.

Model	SSE	dfE
$[A_0][A_1][C]$	59.98	25
$[A_0][A_1][C]$, $A_{21} = A_{31}$	63.73	26
$[A_0][A_1][C]$, $A_{21} = A_{31}$, $A_{20} = A_{30}$	66.97	27

It is a small matter to check that there is no lack of fit displayed by any of these models. The smallest model that fits the data is now $[A_0][A_1][C]$, $A_{21} = A_{31}$, $A_{20} = A_{30}$. Thus there seems to be no difference between salts 2 and 3, but salt 1 has a different regression than the other two salts. (We did not actually test whether salt 1 is different, but if salt 1 had the same slope as the other two then there would be no $[AB]$ interaction and we know that interaction exists.) There is also an effect for the food additives. The parameter estimates and standard errors for the final model are given in Table 16.5.

Figure 16.1 shows the fitted values as functions of the amount of salt for each combination of a salt (with salts 2 and 3 treated as the same) and the additive. The fact that the slope for salt 1 is different from the slope for salts 2 and 3 constitutes an AB interaction. The vertical distances between the two lines for each salt are the same due to the simple main effect for C (additive). The two lines are shockingly close at $B = x_1 = 1$, which makes one wonder if perhaps $B = 1$ is an indication of no salt being used.

If the level $B = 1$ really consists of not adding salt, then, when $B = 1$, the means should be identical for the three salts. The additives can still affect the moisture contents and positive salt amounts can affect the moisture contents. To incorporate these ideas, we subtract one from the salt amounts and eliminate the intercepts from the lines in the amount of salt. That makes the effects for the additive the de facto intercepts, and they are no longer overparameterized,

$$y_{ijkm} = C_k + A_{i1}(x_j - 1) + e_{ijkm}, \qquad A_{21} = A_{31}.$$

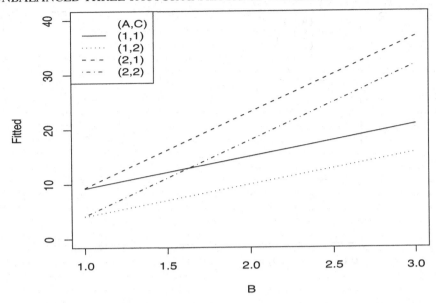

Figure 16.1: *Fitted values for moisture content data treating salts 2 and 3 as the same.*

Table 16.6: $y_{ijkm} = C_k + A_{il}(x_j - 1) + e_{ijkm}$, $A_{21} = A_{31}$.

Table of Coefficients			
Parameter	Estimate	SE	t_{obs}
C_1	9.3162	0.5182	17.978
C_2	4.1815	0.4995	8.371
A_{11}	5.8007	0.4311	13.456
A_{21}	13.8282	0.3660	37.786

This model has $dfE = 28$ and $SSE = 67.0$ so it fits the data almost as well as the previous model but with one less parameter. The estimated coefficients are given in Table 16.6 and the results are plotted in Figure 16.2. The figure is almost identical to Figure 16.1. Note that the vertical distances between the two lines with "the same" salt in Figure 16.2 are $5.1347 = 9.3162 - 4.1815$, almost identical to the 5.130 in Figure 16.1.

Are we done yet? Probably not. We have not even considered the validity of the assumptions. Are the errors normally distributed? Are the variances the same for every treatment combination? Technically, we need to ask whether $C_1 = C_2$ in this new model. A quick look at the estimates and standard errors answers the question in the negative.

Exercise 16.4.7 examines the process of fitting the more unusual models found in this section.

16.1.1 Computing

Because it is the easiest program I know, most of the analyses in this book were done in Minitab. We now present and contrast R and SAS code for fitting $[AB][C]$ and discuss the fitting of other models from this section. Table 16.7 illustrates the variables needed for a full analysis. The online data file contains only the y values and indices for the three groups. Creating X and X2 is generally easy. Creating the variable A2 that does not distinguish between salts 2 and 3 can be trickier. If we had a huge number of observations, we would want to write a program to modify A into A2. With the data we have, in Minitab it is easy to make a copy of A and modify it appropriately in the spreadsheet.

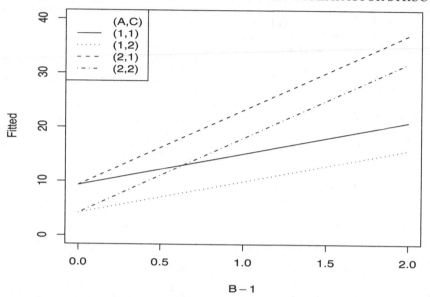

Figure 16.2: *Fitted values for moisture content data treating salts 2 and 3 as the same and $B = 1$ as 0 salt.*

Table 16.7: *Moisture data, indices, and predictors.*

y	A i	B j	C k	X x	X2 x^2	A2	y	A i	B j	C k	X x	X2 x^2	A2
8	1	1	1	1	1	1	11	1	2	2	2	4	1
17	1	2	1	2	4	1	16	1	3	2	3	9	1
22	1	3	1	3	9	1	3	2	1	2	1	1	2
7	2	1	1	1	1	2	17	2	2	2	2	4	2
26	2	2	1	2	4	2	32	2	3	2	3	9	2
34	2	3	1	3	9	2	5	3	1	2	1	1	2
10	3	1	1	1	1	2	16	3	2	2	2	4	2
24	3	2	1	2	4	2	33	3	3	2	3	9	2
39	3	3	1	3	9	2	4	1	1	2	1	1	1
13	1	2	1	2	4	1	10	1	2	2	2	4	1
20	1	3	1	3	9	1	15	1	3	2	3	9	1
10	2	1	1	1	1	2	5	2	1	2	1	1	2
24	2	2	1	2	4	2	19	2	2	2	2	4	2
9	3	1	1	1	1	2	29	2	3	2	3	9	2
36	3	3	1	3	9	2	4	3	1	2	1	1	2
5	1	1	2	1	1	1	34	3	3	2	3	9	2

Similarly, it is easy to create A2 in R using A2=A followed by A2[(A2 == 3)] <- 2. For SAS, I would probably modify the data file so that I could read A2 with the rest of the data.

An R script for fitting $[AB][C]$ follows. R needs to locate the data file, which in this case is located at E:\Books\ANREG2\DATA2\tab16-1.dat.

```
scheffe <- read.table("E:\\Books\\ANREG2\\DATA2\\tab16-1.dat",
        sep="",col.names=c("y","a","b","c"))
attach(scheffe)
scheffe
summary(scheffe)

#Summary tables
A=factor(a)
```

```
B=factor(b)
C=factor(c)
X=b
X2=X*X
sabc <- lm(y ~ A:B + C)
coef=summary(sabc)
coef
anova(sabc)
```

SAS code for fitting $[AB][C]$ follows. The code assumes that the data file is the same directory (folder) as the SAS file.

```
options ps=60 ls=72 nodate;
data anova;
    infile 'tab16-1.dat';
    input y A B C;
    X = B;
    X2=X*X;
proc glm data=anova;
    class A B C ;
    model y = A*B C ;
    means C / lsd alpha=.01 ;
    output out=new r=ehat p=yhat cookd=c h=hi rstudent=tresid student=sr;
proc plot;
    plot ehat*yhat sr*R/ vpos=16 hpos=32;
proc rank data=new normal=blom;
    var sr;
    ranks nscores;
proc plot;
    plot sr*nscores/vpos=16 hpos=32;
run;
```

To fit the other models, one needs to modify the part of the code that specifies the model. In R this involves changes to "sabc <- lm(y ~ A:B + C)" and in SAS it involves changes to "model y = A*B C;". Alternative model specifications follow.

Model	Minitab	R	SAS
$[ABC]$	$A\|B\|C$	A:B:C	A*B*C
$[AB][BC]$	$A\|B$ $B\|C$	A:B+B:C	A*B B*C
$[AB][C]$	$A\|B$ C	A:B+C	A*B C
$[A_0][A_1][A_2][C]$	$A\|X$ $A\|X2$ C	A+A:X+A:X2+C	A A*X A*X2 C
$[A_0][A_1][C], A_{21} = A_{31}$	A $A2\|X$ C	A+A2:X+C-1	A A2*X C
$[A_0][A_1][C], A_{21} = A_{31}, A_{20} = A_{30}$	$A2$ $A2\|X$ C	A2+A2:X+C-1	A2 A2*X C

16.1.2 Regression fitting

We start by creating 0-1 indicator variables for the factor variables A, B, and C. Call these, A_1, A_2, A_3, B_1, B_2, B_3, C_1, C_2, respectively. The values used to identify groups in factor variable B are measured quantities, so create a measurement variable $x \equiv B$ and another x^2. We can construct all of the models from these 10 predictor variables by multiplying them together judiciously. Of course there are many equivalent ways of specifying these models; we present only one. None of the models contain an intercept.

Table 16.8: *Abrasion resistance data.*

		Surface treatment					
		Yes			No		
Proportions		25%	50%	75%	25%	50%	75%
	A	194	233	265	155	198	235
	A	208	241	269	173	177	229
Fill							
	B	239	224	243	137	129	155
	B	187	243	226	160	98	132

Model	Variables
$[ABC]$	$A_1B_1C_1, A_1B_1C_2, A_1B_2C_1, A_1B_2C_2, A_1B_3C_1, \ldots, A_3B_3C_1, A_3B_3C_2$
$[AB][AC][BC]$	$A_1B_1, A_1B_2, \ldots, A_3B_3, A_1C_2, A_2C_2, A_3C_2, B_2C_2, B_3C_2$
$[AB][BC]$	$A_1B_1, A_1B_2, \ldots, A_3B_3, B_1C_2, B_2C_2, B_3C_2$
$[AB][C]$	$A_1B_1, A_1B_2, \ldots, A_3B_3, C_2$
$[A][B][C]$	$A_1, A_2, A_3, B_2, B_3, C_2$
$[A_0][A_1][A_2][C]$	$A_1, A_2, A_3, A_1x, A_2x, A_3x, A_1x^2, A_2x^2, A_3x^2, C_2$
$[A_0][A_1][C]$	$A_1, A_2, A_3, A_1x, A_2x, A_3x, C_2$
$[A_0][A_1]$	$A_1, A_2, A_3, A_1x, A_2x, A_3x$
$[A_0][C]$	A_1, A_2, A_3, C_2

Constructing the models in which salts 2 and 3 are treated alike requires some additional algebra.

Model	Variables
$[A_0][A_1][C], A_{21} = A_{31}$	$A_1, A_2, A_3, A_1x, (A_2 + A_3)x, C_2$
$[A_0][A_1][C], A_{21} = A_{31}, A_{20} = A_{30}$	$A_1, (A_2 + A_3), A_1x, (A_2 + A_3)x, C_2$

16.2 Balanced three-factors

In this section we consider another three-way ANOVA. This time the data are balanced, but we will not let that affect our analysis very much.

EXAMPLE 16.2.1. Box (1950) considers data on the abrasion resistance of a fabric. The data are weight loss of a fabric that occurs during the first 1000 revolutions of a machine designed to test abrasion resistance. A piece of fabric is weighed, put on the machine for 1000 revolutions, and weighed again. The measurement is the change in weight. Fabrics of several different types are compared. They differ by whether a surface treatment was applied, the type of filler used, and the proportion of filler used. Two pieces of fabric of each type are examined, giving two replications in the analysis of variance. The data, given in Table 16.8, are balanced because they have the same number of observations for each group.

The three factors are referred to as "*surf*," "*fill*," and "*prop*," respectively. The factors have 2, 2, and 3 levels, so there are $2 \times 2 \times 3 = 12$ groups. This can also be viewed as just a one-way ANOVA with 12 groups. Using the three subscripts ijk to indicate a treatment by indicating the levels of $surf, fill$, and $prop$, respectively, the one-way ANOVA model is

$$y_{ijkm} = \mu_{ijk} + \varepsilon_{ijkm} \qquad (16.2.1)$$

$i = 1, 2, j = 1, 2, k = 1, 2, 3, m = 1, 2$. Equivalently, we can break the treatment effects into main effects for each factor, interactions between each pair of factors, and an interaction between all three factors, i.e.,

$$y_{ijkm} = G + S_i + F_j + P_k + (SF)_{ij} + (SP)_{ik} + (FP)_{jk} + (SFP)_{ijk} + \varepsilon_{ijkm}. \qquad (16.2.2)$$

Here the **S**, **F**, and **P** effects indicate main effects for *surf*, *fill*, and *prop*, respectively. (We hope no confusion occurs between the factor **F** and the use of F statistics or between the factor **P** and the use of P values!) The (**SF**)s are effects that allow for the two-factor interaction between *surf* and *fill*;

Residual–Fitted plot

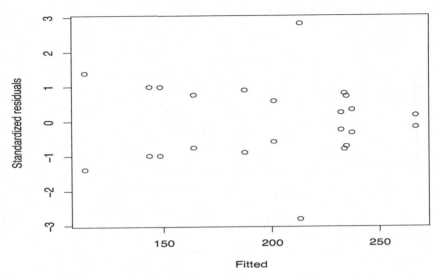

Figure 16.3: *Plot of residuals versus predicted values, 1000-rotation Box data.*

(**SP**) and (**FP**) are defined similarly. The (**SFP**)s are effects that allow for three-factor interaction. A three-factor interaction can be thought of as a two-factor interaction that changes depending on the level of the third factor. *The main effects, two-factor interactions, and three-factor interaction simply provide a structure that allows us to proceed in a systematic fashion.*

We begin by considering the one-way analysis of variance.

Analysis of Variance

Source	df	SS	MS	F
Treatments	11	48183	4380	16.30
Error	12	3225	269	
Total	23	51408		

The F statistic is very large. If the standard one-way ANOVA assumptions are reasonably valid, there is clear evidence that not all of the 12 treatments have the same effect.

Now consider the standard residual checks for a one-way ANOVA. Figure 16.3 contains the residuals plotted against the predicted values. The symmetry of the plot about a horizontal line at 0 is due to the model fitting, which forces the two residuals in each group to add to 0. Except for one pair of observations, the variability seems to decrease as the predicted values increase. The residual pattern is not one that clearly suggests heteroscedastic variances. We simply note the pattern and would bring it to the attention of the experimenter to see if it suggests something to her. In the absence of additional information, we proceed with the analysis. Figure 16.4 contains a normal plot of the residuals. It does not look too bad. Note that with 24 residuals and only 12 dfE, we may want to use dfE as the sample size should we choose to perform a W' test.

Table 16.9 results from fitting a variety of models to the data. It is constructed just like Table 16.2. From the C_p statistics and the tests of each model against the three-factor interaction model, the obvious candidate models are [**SF**][**SP**][**FP**] and its reduced model [**SF**][**FP**]. Using $MSE([\mathbf{SFP}])$ in the denominator, testing them gives

$$F_{obs} = \frac{[4889.7 - 3703.6]/[16-14]}{3225.0/12} = \frac{593.05}{268.75} = 2.21,$$

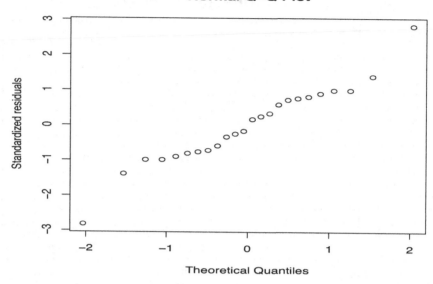

Figure 16.4: *Normal plot of residuals,* $W' = 0.97$, *1000-rotation Box data.*

Table 16.9: *Statistics for fitting models to the 1000-rotation abrasion resistance data of Table 16.8.*

Model	SSE	dfE	F*	C_p
[SFP]	3225.0	12	—	12.0
[SF][SP][FP]	3703.6	14	0.89	9.8
[SF][SP]	7232.7	16	3.73	18.9
[SF][FP]	4889.7	16	1.55	10.2
[SP][FP]	7656.3	15	5.50	22.5
[SF][P]	8418.7	18	3.22	19.3
[SP][F]	11185.3	17	5.92	31.6
[FP][S]	8842.3	17	4.18	22.9
[S][F][P]	12371.4	19	4.86	32.0

which has a P value of 0.153. This is a test for whether we need the [SP] interaction in a model that already includes [SF][FP]. We will tentatively go with the smaller model,

$$y_{ijkm} = [\mathbf{SF}]_{ij} + [\mathbf{FP}]_{jk} + e_{ijkm}$$

or its more overparameterized version,

$$y_{ijkm} = G + \mathbf{S}_i + \mathbf{F}_j + \mathbf{P}_k + [\mathbf{SF}]_{ij} + [\mathbf{FP}]_{jk} + e_{ijkm}.$$

The test for adding the [SP] interaction to this model was the one test we really needed to perform, but there are several tests available for [SP] interaction. In addition to the test we performed, one could test [SF][SP] versus [SF][P] as well as [SP][FP] versus [S][FP]. Normally, these would be three distinct tests but with balanced data like the 1000-rotation data, the tests are all identical. Because of this and similar simplifications due to balanced data, we can present a unique ANOVA table, in lieu of Table 16.9, that provides a comprehensive summary of all ANOVA model tests. This is given as Table 16.10. Note that the F statistic and P value for testing *surf * prop* in Table 16.10 agree with our values from the previous paragraph. For a two-factor model, we presented ANOVA tables like Table 14.3 that depended on fitting both of the two reasonable sequences of models. In an unbalanced three-factor ANOVA, there are too many possible model sequences to present them all,

Table 16.10: *Analysis of Variance: Abrasion resistance.*

Source	df	SS	MS	F	P
surf	1	26268.2	26268.2	97.74	0.000
fill	1	6800.7	6800.7	25.30	0.000
prop	2	5967.6	2983.8	11.10	0.002
surf * *fill*	1	3952.7	3952.7	14.71	0.002
surf * *prop*	2	1186.1	593.0	2.21	0.153
fill * *prop*	2	3529.1	1764.5	6.57	0.012
surf * *fill* * *prop*	2	478.6	239.3	0.89	0.436
Error	12	3225.0	268.8		
Total	23	51407.8			

Table 16.11: *Abrasion resistance under* $[SF][F_1][F_2]$.

		Analysis of Variance			
Source	df	SS	MS	F	P
[SF]	4	977126	244281	799.3393	0.000
[F_1]	2	9114	4557	14.9117	0.000
[F_2]	2	383	191	0.6259	0.547
Error	16	4890	306		

so we use tables like 16.2 and 16.9, except in the balanced case where everything can be summarized as in Table 16.10.

In the previous section, our best model for the moisture data had only one two-factor term. For the abrasion data our working model has two two-factor terms: [SF] and [FP]. Both two-factor terms involve F, so if we fix a level of *fill*, we will have an additive model in *surf* and *prop*. In other words, for each level of *fill* there will be some effect for *surf* that is added to some effect for the proportions. The interaction comes about because the *surf* effect can change depending on the *fill*, and the *prop* effects can also change depending on the *fill*. Moreover, *prop* is a quantitative factor with three levels, so an equivalent model will be to fit separately, for each level of fill, the surface effects as well as a parabola in proportions. Let p_k, $k = 1, 2, 3$ denote the levels of *prop*. Since the proportions were equally spaced, it does not matter much what we use for the p_ks. We take $p_1 = 1$, $p_2 = 2$, $p_3 = 3$, although another obvious set of values would be $25, 50, 75$. The model, equivalent to [SF][FP], is

$$y_{ijkm} = SF_{ij} + F_{j1}p_k + F_{j2}p_k^2 + e_{ijkm}.$$

Denote this model $[SF][F_1][F_2]$. An ANOVA table is given as Table 16.11. Note that the Error line agrees, up to round-off error, with the Error information on [SF][FP] in Table 16.9.

Table 16.12: $[SF][F_1][F_2]$.

	Table of Coefficients			
Parameter	Estimate	SE	t	P
SF_{11}	180.50	38.43	4.696	0.000
SF_{21}	140.00	38.43	3.643	0.002
SF_{12}	256.67	38.43	6.678	0.000
SF_{22}	164.83	38.43	4.289	0.001
F_{11}	18.50	43.27	0.428	0.675
F_{21}	−41.38	43.27	−0.956	0.353
F_{12}	3.75	10.71	0.350	0.731
F_{22}	11.38	10.71	1.063	0.304

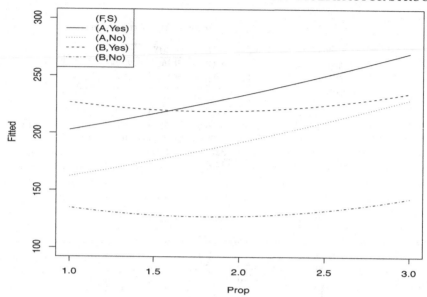

Figure 16.5: $[\mathbf{SF}][\mathbf{F}_1][\mathbf{F}_2]$, *1000-rotation Box data.*

The table of coefficients for $[\mathbf{SF}][\mathbf{F}_1][\mathbf{F}_2]$ is given as Table 16.12. It provides our fitted model

$$\hat{m}(i,j,p) = \begin{cases} 180.50 + 18.50p + 3.75p^2 & \text{Surf = Yes, Fill = A} \\ 140.00 + 18.50p + 3.75p^2 & \text{Surf = No, Fill = A} \\ 256.67 - 41.38p + 11.38p^2 & \text{Surf = Yes, Fill = B} \\ 164.83 - 41.38p + 11.38p^2 & \text{Surf = No, Fill = B,} \end{cases}$$

which is graphed in Figure 16.5. The two parabolas for Fill = A are parallel and remarkably straight. The two parabolas for Fill = B are also parallel and not heavily curved. That the curves are parallel for a fixed Fill is indicative of there being no $[\mathbf{SP}]$ or $[\mathbf{SFP}]$ interactions in the model. The fact that the shapes of the Fill = A parabolas are different from the shapes of the Fill = B parabolas is indicative of the $[\mathbf{FP}]$ interaction. The fact that the distance between the two parallel Fill = A parabolas is different from the distance between the two parallel Fill = B parabolas is indicative of the $[\mathbf{SF}]$ interaction.

Both quadratic terms have large P values in Table 16.12. We might consider fitting a reduced model that eliminates the curvatures, i.e., fits straight lines. The reduced model is

$$y_{ijkm} = \mathbf{SF}_{ij} + \mathbf{F}_{j1}p_k + e_{ijkm}$$

denoted $[\mathbf{SF}][\mathbf{F}_1]$. Table 16.13 gives the ANOVA table which, when compared to Table 16.11, allows us to test simultaneously whether we need the two quadratic terms. With

$$F_{obs} = \frac{[5272 - 4890]/[18 - 16]}{3225.0/12} = \frac{191}{268.75} = 0.71,$$

we have no evidence of curvature.

The table of coefficients in Table 16.14 provides us with our fitted model for $[\mathbf{SF}][\mathbf{F}_1]$,

$$\hat{m}(i,j,p) = \begin{cases} 168.000 + 33.50p & \text{Surf = Yes, Fill = A} \\ 127.500 + 33.50p & \text{Surf = No, Fill = A} \\ 218.750 + 4.125p & \text{Surf = Yes, Fill = B} \\ 126.917 + 4.125p & \text{Surf = No, Fill = B.} \end{cases}$$

Table 16.13: *Abrasion resistance under* $[\mathbf{SF}][\mathbf{F}_1]$.

		Analysis of Variance			
Source	df	SS	MS	F	P
S:F	4	977126	244281	834.008	0.000
F:p	2	9114	4557	15.558	0.000
Error	18	5272	293		

Table 16.14: *Abrasion resistance under* $[\mathbf{SF}][\mathbf{F}_1]$.

	Table of Coefficients			
Parameter	Estimate	SE	t	P
$\mathbf{SF}_{11}$	168.000	13.974	12.023	0.000
$\mathbf{SF}_{21}$	127.500	13.974	9.124	0.000
$\mathbf{SF}_{12}$	218.750	13.974	15.654	0.000
$\mathbf{SF}_{22}$	126.917	13.974	9.082	0.000
$\mathbf{F}_{11}$	33.500	6.051	5.536	0.000
$\mathbf{F}_{21}$	4.125	6.051	0.682	0.504

This is graphed in Figure 16.6. The difference in the slopes for Fills A and B indicate the $[\mathbf{FP}]$ interaction. The fact that the distance between the two parallel lines for Fill A is different from the distance between the two parallel lines for Fill B indicates the presence of $[\mathbf{SF}]$ interaction. The nature of this model is that for a fixed Fill the proportion curves will be parallel but when you change fills both the shape of the curves and the distance between the curves can change.

The slope for Fill B looks to be nearly 0. The P value in Table 16.14 is 0.504. We could incorporate $\mathbf{F}_{21} = 0$ into a model $y_{ijkm} = m(i, j, p_k) + \varepsilon_{ijkm}$ so that

$$m(i,j,p) = \begin{cases} [\mathbf{SF}]_{11} + \mathbf{F}_{11}p & \text{Surf = Yes, Fill = A} \\ [\mathbf{SF}]_{21} + \mathbf{F}_{11}p & \text{Surf = No, Fill = A} \\ [\mathbf{SF}]_{12} & \text{Surf = Yes, Fill = B} \\ [\mathbf{SF}]_{22} & \text{Surf = No, Fill = B.} \end{cases}$$

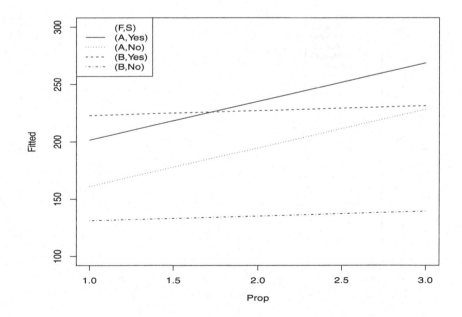

Figure 16.6: $[\mathbf{SF}][\mathbf{F}_1]$, *Box data*.

Table 16.15: *Abrasion resistance under* $[\mathbf{SF}][\mathbf{F}_{11}]$.

Analysis of Variance					
Source	df	SS	MS	F	P
$\mathbf{F}_{11}$	1	524966	524966	1844.26	0.000
$\mathbf{S{:}F}$	4	461138	115284	405.01	0.000
Error	19	5408	285		

Table 16.16: *Abrasion resistance coefficients under* $[\mathbf{SF}][\mathbf{F}_{11}]$.

Table of Coefficients				
Parameter	Estimate	SE	t	P
$\mathbf{F}_{11}$	33.500	5.965	5.616	0.000
$\mathbf{SF}_{11}$	168.000	13.776	12.196	0.000
$\mathbf{SF}_{21}$	127.500	13.776	9.256	0.000
$\mathbf{SF}_{12}$	227.000	6.888	32.957	0.000
$\mathbf{SF}_{22}$	135.167	6.888	19.624	0.000

Denote this $[\mathbf{SF}][\mathbf{F}_{11}]$. The ANOVA table and the Table of Coefficients are given as Tables 16.15 and 16.16. The fitted model is

$$\hat{m}(i,j,p) = \begin{cases} 168.000 + 33.500p & \text{Surf = Yes, Fill = A} \\ 127.500 + 33.500p & \text{Surf = No, Fill = A} \\ 227.000 & \text{Surf = Yes, Fill = B} \\ 135.167 & \text{Surf = No, Fill = B,} \end{cases}$$

which is graphed as Figure 16.7.

Finally, we could take the modeling another step (too far) by noticing that in Table 16.15 the estimated effects of $[\mathbf{SF}]_{21}$ and $[\mathbf{SF}]_{22}$ are close. Incorporating their equality into a model $[\mathbf{SF}(3)][\mathbf{F}_{11}]$

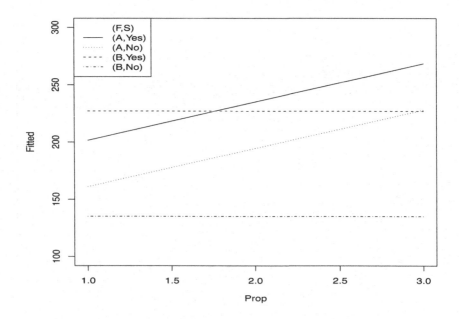

Figure 16.7: $[\mathbf{SF}][\mathbf{F}_{11}]$, *Box data.*

Table 16.17: *Abrasion resistance under* $[\mathbf{SF}(3)][\mathbf{F}_{11}]$.

		Analysis of Variance			
Source	df	SS	MS	F	P
SF	3	966564	322188	1176.112	0.000
$\mathbf{F}_{11}$	1	19469	19469	71.069	0.000
Error	20	5479	274		

Table 16.18: *Abrasion resistance coefficients under* $[\mathbf{SF}(3)][\mathbf{F}_{11}]$.

	Table of Coefficients			
Parameter	Estimate	SE	t	P
$\mathbf{SF}_{11}$	172.600	10.022	17.22	0.000
$\mathbf{SF}_{12}$	227.000	6.757	33.59	0.000
$\mathbf{SF}_{21}$	133.633	6.044	22.11	0.000
$\mathbf{F}_{11}$	31.200	3.701	8.43	0.000

with

$$
m(i,j,p) = \begin{cases} [\mathbf{SF}]_{11} + \mathbf{F}_{11}p & \text{Surf = Yes, Fill = A} \\ [\mathbf{SF}]_{21} + \mathbf{F}_{11}p & \text{Surf = No, Fill = A} \\ [\mathbf{SF}]_{12} & \text{Surf = Yes, Fill = B} \\ [\mathbf{SF}]_{21} & \text{Surf = No, Fill = B} \end{cases}
$$

fits well but is rather dubious. Extrapolating to 0% fill, the estimated weight losses would be the same for no surface treatment and both fills. But as the proportion increases, the weight loss remains flat for Fill B but increases for Fill A. With a surface treatment, the extrapolated weight losses at 0% fill are different, but for Fill B it remains flat while for Fill A it increases. The ANOVA table and Table of Coefficients are given as Tables 16.17 and 16.18.

16.3 Higher-order structures

Unbalanced data with four or more factors are difficult because there are too many ANOVA type models even to make tables like Table 14.2, Table 16.2, or Table 16.9 (much less list all sequences of models like Table 14.3). Various methods developed for log-linear models can be exploited in the analysis, cf. Christensen (1997, Chapter 6). Balanced data can provide an ANOVA table like Table 16.10 to identify important effects for a reduced model. A four-factor model will be examined in Chapter 19 in conjunction with a split-plot analysis.

16.4 Exercises

EXERCISE 16.4.1. Baten (1956) presented data on lengths of steel bars. An excessive number of bars had recently failed to meet specifications and the experiment was conducted to identify the causes of this problem. The bars were made with one of two heat treatments (W, L) and cut on one of four screw machines (A, B, C, D) at one of three times of day (8 am, 11 am, 3 pm). The three times were used to investigate the possibility of worker fatigue during the course of the day. The bars were intended to be between 4.380 and 4.390 inches long. The data presented in Table 16.19 are thousandths of an inch in excess of 4.380. Treating the data as a $2 \times 3 \times 4$ ANOVA, give an analysis of the data.

EXERCISE 16.4.2. Bethea et al. (1985) reported data on an experiment to determine the effectiveness of four adhesive systems for bonding insulation to a chamber. The adhesives were applied both with and without a primer. Tests of peel-strength were conducted on two different thicknesses

Table 16.19: *Steel bar lengths*.

	Heat treatment W				Heat treatment L			
Machine	A	B	C	D	A	B	C	D
Time 1	6	7	1	6	4	6	−1	4
	9	9	2	6	6	5	0	5
	1	5	0	7	0	3	0	5
	3	5	4	3	1	4	1	4
Time 2	6	8	3	7	3	6	2	9
	3	7	2	9	1	4	0	4
	1	4	1	11	1	1	−1	6
	−1	8	0	6	−2	3	1	3
Time 3	5	10	−1	10	6	8	0	4
	4	11	2	5	0	7	−2	3
	9	6	6	4	3	10	4	7
	6	4	1	8	7	0	−4	0

Table 16.20: *Peel-strength of various adhesive systems*.

	Adhesive				Adhesive			
	1	2	3	4	1	2	3	4
With Primer	60	57	19.8	52	73	52	32.0	77
	63	52	19.5	53	79	56	33.0	78
	57	55	19.7	44	76	57	32.0	70
	53	59	21.6	48	69	58	34.0	74
	56	56	21.1	48	78	52	31.0	74
	57	54	19.3	53	74	53	27.3	81
Without Primer	59	51	29.4	49	78	52	37.8	77
	48	44	32.2	59	72	42	36.7	76
	51	42	37.1	55	72	51	35.4	79
	49	54	31.5	54	75	47	40.2	78
	45	47	31.3	49	71	57	40.7	79
	48	56	33.0	58	72	45	42.6	79
	Thickness A				Thickness B			

of rubber. Using two thicknesses of rubber was not part of the original experimental design. *The existence of this factor was only discovered by inquiring about a curious pattern of numbers in the laboratory report.* The data are presented in Table 16.20. Another disturbing aspect of these data is that the values for adhesive system 3 are reported with an extra digit. Presumably, a large number of rubber pieces were available and the treatments were randomly assigned to these pieces, but, given the other disturbing elements in these data, I wouldn't bet the house on it. A subset of these data was examined earlier in Exercise 12.7.6.

(a) Give an appropriate model. List all the assumptions made in the model.

(b) Check the assumptions of the model and adjust the analysis appropriately.

(c) Analyze the data. Give an appropriate analysis of variance table. Examine appropriate contrasts.

EXERCISE 16.4.3. The data of Table 16.21 were presented in Finney (1964) and Bliss (1947). The observations are serum calcium values of dogs after they have been injected with a dose of parathyroid extract. The doses are the treatments and they have factorial structure. One factor involves using either the standard preparation (S) or a test preparation (T). The other factor is the amount of a dose; it is either low (L) or high (H). Low doses are 0.125 cc and high doses are 0.205 cc. Each dog is subjected to three injections at about 10 day intervals. Serum calcium is measured on the day after an injection. Analyze the data using a three-factor model with dogs, preparations, and amounts but do not include any interactions involving dogs. Should day effects be incorporated? Can this be done conveniently? If so, do so.

Table 16.21: *Serum calcium for dogs.*

Dog	Day I	Day II	Day III
1	TL, 14.7	TH, 15.4	SH, 14.8
2	TL, 15.1	TH, 15.0	SH, 15.8
3	TH, 14.4	SH, 13.8	TL, 14.4
4	TH, 16.2	TL, 14.0	SH, 13.0
5	TH, 15.8	SH, 16.0	TL, 15.0
6	TH, 15.8	TL, 14.3	SL, 14.8
7	TH, 17.0	TL, 16.5	SL, 15.0
8	TL, 13.6	SL, 15.3	TH, 17.2
9	TL, 14.0	TH, 13.8	SL, 14.0
10	TL, 13.0	SL, 13.4	TH, 13.8
11	SL, 13.8	SH, 17.0	TH, 16.0
12	SL, 12.0	SH, 13.8	TH, 14.0
13	SH, 14.6	TH, 15.4	SL, 14.0
14	SH, 13.0	SL, 14.0	TH, 14.0
15	SH, 15.2	TH, 16.2	SL, 15.0
16	SH, 15.0	SL, 14.5	TL, 14.0
17	SH, 15.0	SL, 14.0	TL, 14.6
18	SL, 15.8	TL, 15.0	SH, 15.2
19	SL, 13.2	SH, 16.0	TL, 14.9
20	SL, 14.2	TL, 14.1	SH, 15.0

EXERCISE 16.4.4. Using the notation of Section 16.1, write the models $[A_0][A_1][C]$, $[A_0][A_1][C] \ A_{21} = A_{31}$, and $[A_0][A_1][C] \ A_{21} = A_{31}, A_{20} = A_{30}$ in matrix form. (Hint: To obtain $[A_0][A_1][C] \ A_{21} = A_{31}$ from $[A_0][A_1][C]$, replace the two columns of X corresponding to A_{21} and A_{31} with one column consisting of their sum.) Use a regression program to fit these three models. (Hint: Eliminate the intercept, and to impose the side condition $C_1 = 0$, drop the column corresponding to C_1.)

Chapter 17

Basic Experimental Designs

In this chapter we examine basic experimental designs: completely randomized designs (CRDs), randomized complete block (RCB) designs, Latin square (LS) designs, balanced incomplete block (BIB) designs, and more. The focus of this chapter is on ideas of experimental design and how they determine the analysis of data. We have already examined in the text and in the exercises data from many of these experimental designs.

17.1 Experiments and causation

The basic object of running an experiment is to determine causation. Determining causation is difficult. We regularly collect data and find relationships between "dependent" variables and predictor variables. But this does not imply causation. One can predict air pressure extremely well from the boiling point of water, but does the boiling point of water *cause* the air pressure? Isn't it the other way around? We found that females scored lower in a particular Statistics class than males, but does being female *cause* that result? Doesn't it seem plausible that something that is correlated with sexes might cause the result? Interest in Statistics? Time devoted to studying Statistics? Understanding the instructor's teaching style? Being Native American in Albuquerque in 1978 was highly associated with lower suicide ages. But to claim that being Native American *caused* lower suicide ages would be incredibly simplistic. Causation is fundamentally tied to the idea that if you change one thing (the cause), you will change something else (the result). If that is true, can sex or race ever cause anything, since we cannot really change them?

In constructing an experiment we *randomly assign treatments to experimental units*. For example, we can randomly assign (many kinds of) drugs to people. We can randomly assign which employee will operate a particular machine or use a particular process. Unfortunately, there are many things we cannot perform experiments on. We cannot randomly assign sexes or races to people. As a practical matter, we cannot assign the use of illegal drugs to people.

The key point in determining causation is randomization. If we have a collection of experimental units and randomly assign the treatments to them, then (on average) there can be no systematic differences between the treatment groups other than the treatments. Therefore, any differences we see among the treatment groups must be caused by the treatments.

Alas, there are still problems. The randomization argument works on average. Experimental units, whether they be people, rats, or plots of ground, are subject to variability. One can get unlucky with a particular assignment of treatments to experimental units. If by chance one treatment happens to get far more of the "bad" experimental units it will look like a bad treatment. For example, if we want to know whether providing milk to elementary school students improves their performances, we cannot let the teachers decide who gets the milk. The teachers may give it to the poorest students in which case providing milk could easily look like it harms student performances. Similar things can happen by chance when randomly assigning treatments. To infer causation, the experiment should be repeated often enough that chance becomes a completely implausible explanation for the results.

Moreover, if we measure a huge number of items on each experimental unit, there is a good chance that one of the treatment groups will randomly have an inordinate number of good units for some variable, and hence show an effect that is really due to chance. In other words, if we measure enough variables, just by chance, some of them will display a relationship to the treatment groups, regardless of how the treatment groups were chosen.

A particularly disturbing problem is that the experimental treatments are often not what we think they are. *An experimental treatment is everything we do differently to a group of experimental units.* If we give a drug to a bunch of rats and then stick them into an asbestos filled attic, the fact that those rats have unusually high cancer rates does not mean that the drug caused it. The treatment caused it, but just because we call the treatment by the name of the drug does not make the drug the treatment.

Alternatively, suppose we want to test whether artificial sweeteners made with a new chemical cause cancer. We get some rats, randomly divide them into a treatment group and a control. We inject the treatment rats with a solution of the sweetener combined with another (supposedly benign) chemical. We leave the control rats alone. For simplicity we keep the treatment rats in one cage and the control rats in another cage. Eventually, we find an increased risk of cancer among the treatment rats as compared to the control rats. We can reasonably conclude that the treatments caused the increased cancer rate. Unfortunately, we do not really know whether the sweetener or the supposedly benign chemical or the combination of the two caused the cancer. In fact, we do not really know that it was the chemicals that caused the cancer. Perhaps the process of injecting the rats caused the cancer or perhaps something about the environment in the treatment rats' cage caused the cancer. A treatment consists of *all the ways* in which a group is treated differently from other groups. It is crucially important to *treat all experimental units as similarly as possible so that (as nearly as possible) the only differences between the units are the agents that were meant to be investigated.*

Random assignment of treatments is fundamental to conducting an experiment but it does not mean haphazard assignment of treatments to experimental units. Haphazard assignment is subject to the (unconscious) biases of the person making the assignments. Random assignment uses a reliable table of random numbers or a reliable computer program to generate random numbers. It then uses these numbers to assign treatments. For example, suppose we have four experimental units labeled u_1, u_2, u_3, and u_4 and four treatments labeled A, B, C, and D. Given a program or table that provides random numbers between 0 and 1 (i.e., random samples from a Uniform(0,1) distribution), we associate numbers between 0 and .25 with treatment A, numbers between .25 and .50 with treatment B, numbers between .50 and .75 with treatment C, and numbers between .75 and 1 with treatment D. The first random number selected determines the treatment for u_1. If the first number is .6321, treatment C is assigned to u_1 because $.50 < .6321 < .75$. If the second random number is .4279, u_2 gets treatment B because $.25 < .4279 < .50$. If the third random number is .2714, u_3 would get treatment B, but we have already assigned treatment B to u_2, so we throw out the third number. If the fourth number is .9153, u_3 is assigned treatment D. Only one unit and one treatment are left, so u_4 gets treatment A. Any reasonable rule (decided ahead of time) can be used to make the assignment if a random number hits a boundary, e.g., if a random number comes up, say, .2500.

By definition, treatments must be amenable to change. As discussed earlier, things like sex and race are not capable of change, but in addition many viable treatments cannot be randomly assigned for social reasons. If we want to know if smoking causes cancer in humans, running an experiment is difficult. In our society we cannot force some people to smoke a specific amount for a long period of time and force others not to smoke at all. Nonetheless, we are very interested in whether smoking causes cancer. What are we to do?

When experiments cannot be run, the other common method for inferring causation is the "What else could it be?" approach. For smoking, the idea is that we measure *everything else* that could possibly be causing cancer and *appropriately adjust* for those measurements. If, after fitting all of those variables, smoking still has a significant effect on predicting cancer, then smoking must be causing the cancer. The catch is that this is extremely difficult to do. How do we even identify, much less measure, everything else that could be causing cancer? And even if we do measure everything, how do we know that we have adjusted for those variables appropriately? The key to this argument is independent replication of the studies! If there are many such *observational studies* with many different ideas of what other variables could be causing the effect (cancer) and many ways of adjusting for those variables, and if the studies consistently show that smoking remains an important predictor, at some point it would seem foolish to ignore the possibility that smoking causes cancer.

I have long contended that one cannot infer causation from data analysis. Certainly data analysis speaks to the relative validity of competing causal models but that is a far cry from actually determining causation. I believe that causation must be determined by some external argument. I find randomization to be the most compelling external argument. In "What else can it be?" the external argument is that all other variables of importance have been measured and appropriately considered.

My contention that data analysis cannot lead to causation may be wrong. I have not devoted my life to studying causal models. And I know that people study causation by the consideration of counterfactuals. But for now, I stand by my contention.

Although predictive ability does not imply causation, for many (perhaps most) purposes, predictive ability is more important. Do we really care why the lights go on when we flip a switch? Or do we care that our prediction comes true? We probably only care about causation when the lights stop working. How many people really understand the workings of an automobile? How many can successfully predict how automobiles will behave?

17.2 Technical design considerations

As a technical matter, the first object in designing an experiment is to construct one that allows for a valid estimate of σ^2, the variance of the observations. Without a valid estimate of error, we cannot know whether the treatment groups are exhibiting any real differences. Obtaining a valid *estimate of error* requires appropriate replication of the experiment. Having one observation on each treatment is not sufficient. All of the basic designs considered in this chapter allow for a valid estimate of the variance. (In my experience, failure to replicate is the most common sin committed on the television show *Mythbusters*.)

The simplest experimental design is the *completely randomized design (CRD)*. With four drug treatments and observations on eight animals, a valid estimate of the error can be obtained by randomly assigning each of the drugs to two animals. *If the treatments are assigned completely at random to the experimental units* (animals), *the design is a completely randomized design*. The fact that there are more animals than treatments provides our replication.

It is not crucial that the design be balanced, i.e., it is not crucial that we have the same number of replications on each treatment. But it is useful to have more than one observation on each unit to help check our assumption of equal variances.

A second important consideration is to construct a design that yields a small variance. A smaller variance leads to sharper statistical inferences, i.e., narrower confidence intervals and more powerful tests. The basic idea is to examine the treatments on homogeneous experimental material. The people of Bergen, Norway are probably more homogenous than the people of New York City. It will be easier to find treatment effects when looking at people from Bergen. Of course the downside is that we end up with results that apply to the people of Bergen. The results may or may not apply to the people of New York City.

A fundamental tool for reducing variability is *blocking*. The people of New York City may be more variable than the people of Bergen but we might be able to divide New Yorkers into subgroups that are just as homogeneous as the people of Bergen. With our drugs and animals illustration, a smaller variance for treatment comparisons is generally obtained when the eight animals consist of two litters of four siblings and each treatment is applied to one randomly selected animal from each litter. With each treatment applied in every litter, all comparisons among treatments can be performed *within* each litter. Having at least two litters is necessary to get a valid estimate of the variance of the comparisons. *Randomized complete block designs (RCBs) : 1) identify blocks of homogeneous experimental material (units) and 2) randomly assign each treatment to an experi-*

mental unit within each block. The blocks are complete in the sense that each block contains all of the treatments.

The key point in blocking on litters is that, if we randomly assigned treatments to experimental units without consideration of the litters, our measurements on the treatments would be subject to all of the litter-to-litter variability. By blocking on litters, we can eliminate the litter-to-litter variability so that our comparisons of treatments are subject only to the variability within litters (which, presumably, is smaller). Blocking has completely changed the nature of the variability in our observations.

The focus of block designs is in isolating groups of experimental units that are homogeneous: litters, identical twins, plots of ground that are close to one another. If we have three treatments and four animals to a litter, we can simply not use one animal. If we have five treatments and four animals to a litter, a randomized complete block experiment becomes impossible.

A *balanced incomplete block (BIB)* design is one in which every pair of treatments occur together in a block the same number of times. For example, if our experimental material consists of identical twins and we have the drugs A, B, and C, we might give the first set of twins drugs A and B, the second set B and C, and the third set C and A. Here every pair of treatments occurs together in one of the three blocks.

BIBs do not provide balanced data in our usual sense of the word "balanced" but they do have a relatively simple analysis. RCBs are balanced in the usual sense. Unfortunately, losing any observations from either design destroys the balance that they display. Our focus is in analyzing unbalanced data, so we use techniques for analyzing block designs that do not depend on any form of balance.

The important ideas here are replication and blocking. RCBs and BIBs make very efficient designs but keeping their balance is not crucial. In olden days, before good computing, the simplicity of their analyses was important. But simplicity of analysis was never more than a side effect of the good experimental designs.

Latin squares use two forms of blocking at once. For example, if we suspect that birth order within the litter might also have an important effect on our results, we continue to take observations on each treatment within every litter, but we also want to have each treatment observed in every birth order. This is accomplished by having four litters with treatments arranged in a Latin square design. Here we are simultaneously blocking on litter and birth order.

Another method for reducing variability is incorporating covariates into the analysis. This topic is discussed in Section 17.8.

Ideas of blocking can also be useful in observational studies. While one cannot really create blocks in observational studies, one can adjust for important groupings.

EXAMPLE 17.2.1.   If we wish to run an experiment on whether cocaine users are more paranoid than other people, we may decide that it is important to block on socioeconomic status. This is appropriate if the underlying level of paranoia in the population differs by socioeconomic status. Conducting an experiment in this setting is difficult. Given groups of people of various socioeconomic statuses, it is a rare researcher who has the luxury of deciding which subjects will ingest cocaine and which will not.   □

The seminal work on experimental design was written by Fisher (1935). It is still well worth reading. My favorite source on the ideas of experimentation is Cox (1958). The books by Cochran and Cox (1957) and Kempthorne (1952) are classics. Cochran and Cox is more applied. Kempthorne is more theoretical. Kempthorne has been supplanted by Hinkelmann and Kempthorne (2008, 2005). There is a huge literature in both journal articles and books on the general subject of designing experiments. The article by Coleman and Montgomery (1993) is interesting in that it tries to formalize many aspects of planning experiments that are often poorly specified. Two other useful books are Cox and Reid (2000) and Casella (2008).

17.3 Completely randomized designs

In a completely randomized design, a group of experimental units are available and the experimenter randomly assigns treatments to the experimental units. The data consist of a group of observations on each treatment. Typically, these groups of observations are subjected to a one-way analysis of variance.

EXAMPLE 17.3.1.   In Example 12.4.1, we considered data from Mandel (1972) on the elasticity measurements of natural rubber made by 7 laboratories. While Mandel did not discuss how the data were obtained, it could well have been the result of a completely randomized design. For a CRD, we would need 28 pieces of the type of rubber involved. These should be randomly divided into 7 groups (using a table of random numbers or random numbers generated by a reliable computer program). The first group of samples is then sent to the first lab, the second group to the second lab, etc. For a CRD, it is important that a sample is not sent to a lab because the sample somehow seems appropriate for that particular lab.

Personally, I would also be inclined to send the four samples to a given lab at different times. If the four samples are sent at the same time, they might be analyzed by the same person, on the same machines, at the same time. Samples sent at different times might be treated differently. If samples are treated differently at different times, this additional source of variation should be included in any predictive conclusions we wish to make about the labs.

When samples sent at different times are treated differently, sending a batch of four samples at the same time constitutes *subsampling*. There are two sources of variation to deal with: variation from time to time and variation within a given time. The values from four samples at a given time collectively help reduce the effect on treatment comparisons due to variability at a given time, but samples analyzed at different times are still *required* if we are to obtain a valid estimate of the error. In fact, with subsampling, a perfectly valid analysis can be based on the means of the four subsamples. In our example, such an analysis gives only one 'observation' at each time, so the need for sending samples at more than one time is obvious. If the four samples were sent at the same time, there would be no replication, hence no estimate of error. Subsection 19.4.1 and Christensen (2011, Section 9.4) discuss subsampling in more detail.   □

EXAMPLE 17.3.2.   In Chapter 12, we considered suicide age data. A designed experiment would require that we take a group of people who we know will commit suicide and randomly assign one of the ethnic groups to the people. Obviously a difficult task.   □

17.4 Randomized complete block designs

In a randomized complete block design the experimenter obtains (constructs) blocks of homogeneous material that contain as many experimental units as there are treatments. The experimenter then randomly assigns a different treatment to each of the units in the block. The random assignments are performed independently for each block. The advantage of this procedure is that treatment comparisons are subject only to the variability within the blocks. Block-to-block variation is eliminated in the analysis. In a completely randomized design applied to the same experimental material, the treatment comparisons would be subject to both the within-block and the between-block variability.

The key to a good blocking design is in obtaining blocks that have little within-block variability. Often this requires that the blocks be relatively small. A difficulty with RCB designs is that the blocks must be large enough to allow all the treatments to be applied within each block. This can be a serious problem if there is a substantial number of treatments or if maintaining homogeneity within blocks requires the blocks to be very small. If the treatments cannot all be fitted into each block, we need some sort of *incomplete block* design.

Table 17.1: *Spectrometer data.*

Treatment	Block 1	Block 2	Block 3
New-clean	0.9331	0.8664	0.8711
New-soiled	0.9214	0.8729	0.8627
Used-clean	0.8472	0.7948	0.7810
Used-soiled	0.8417	0.8035	

Table 17.2: *Analysis of Variance: Spectrometer data.*

Source	df	SS	MS	F	P
Block	2	0.0063366	0.0031683	62.91	0.000
Treatments	3	0.0166713	0.0055571	110.34	0.000
Error	5	0.0002518	0.0000504		
Total	10	0.0232598			

The typical analysis of a randomized complete block design is a two-way ANOVA without replication or interaction. Except for the experimental design considerations, the analysis is like that of the Hopper Data from Example 15.3.1. A similar analysis is illustrated below. As with the Hopper data, block-by- treatment interaction is properly considered to be error. *If the treatment effects are not large enough to be detected above any interaction, then they are not large enough to be interesting.*

EXAMPLE 17.4.1.   Inman, Ledolter, Lenth, and Niemi (1992) studied the performance of an optical emission spectrometer. Table 17.1 gives some of their data on the percentage of manganese (Mn) in a sample. The data were collected using a sharp counterelectrode tip with the sample to be analyzed partially covered by a boron nitride disk. Data were collected under three temperature conditions. Upon fixing a temperature, the sample percentage of Mn was measured using 1) a new boron nitride disk with light passing through a clean window (new-clean), 2) a new boron nitride disk with light passing through a soiled window (new-soiled), 3) a used boron nitride disk with light passing through a clean window (used-clean), and 4) a used boron nitride disk with light passing through a soiled window (used-soiled). The four conditions, new-clean, new-soiled, used-clean, and used-soiled are the treatments. The temperature was then changed and data were again collected for each of the four treatments. A block is always made up of experimental units that are homogeneous. The temperature conditions were held constant while observations were taken on the four treatments so the temperature levels identify blocks. Presumably, the treatments were considered in random order. Christensen (1996) analyzed these data including the data point for Block 3 and used-soiled. We have dropped that point to illustrate an analysis for unbalanced data.

The two-factor additive-effects model for these data is

$$y_{ij} = \mu + \beta_i + \eta_j + \varepsilon_{ij},$$

$i = 1, 2, 3$, $j = 1, 2, 3, 4$; however, the $i = 3$, $j = 4$ observation is missing. Here β_i denotes a block effect and η_j a treatment effect. As usual, we assume the errors are independent and $N(0, \sigma^2)$.

Unlike the analysis for two factors in Chapter 14, *in blocking experiments we always examine the treatments after the blocks.* We constructed the blocks, so we know they should have effects. The only relevant ANOVA table is given as Table 17.2.

For now, we just perform all pairwise comparisons of the treatments.

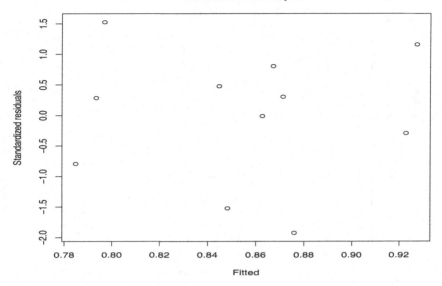

Figure 17.1: *Plot of residuals versus predicted values, spectrometer data.*

| | | | | Bonferroni |
Parameter	*Est*	SE(*Est*)	*t*	*P*
$\eta_2 - \eta_1$	−0.00453	0.005794	−0.78	1.0000
$\eta_3 - \eta_1$	−0.08253	0.005794	−14.24	0.0002
$\eta_4 - \eta_1$	−0.07906	0.006691	−11.82	0.0005
$\eta_3 - \eta_2$	−0.07800	0.005794	−13.46	0.0002
$\eta_4 - \eta_2$	−0.07452	0.006691	−11.14	0.0006
$\eta_4 - \eta_3$	0.003478	0.006691	0.5198	1.000

The one missing observation is from treatment 4 so the standard errors that involve treatment 4 are larger. Although we have different standard errors, the results can be summarized as follows.

New-clean	New-soiled	Used-soiled	Used-clean
$\hat{\eta}_1$	$\hat{\eta}_2$	$\hat{\eta}_4$	$\hat{\eta}_3$
0	−0.00453	−0.07906	−0.08253

The new disk treatments are significantly different from the used disk treatments but the new disk treatments are not significantly different from each other nor are the used disk treatments significantly different from each other. The structure of the treatments suggests an approach to analyzing the data that will be exploited in the next chapter. Here we used a side condition of $\eta_1 = 0$ because it made the estimates readily agree with the table of pairwise comparisons.

Table 17.2 contains an *F* test for blocks. In a true blocking experiment, there is not much interest in testing whether block means are different. After all, one *chooses* the blocks so that they have different means. Nonetheless, the *F* statistic *MSBlks/MSE* is of some interest because it indicates how effective the blocking was, i.e., it indicates how much the variability was reduced by blocking. For this example, *MSBlks* is 63 times larger than *MSE*, indicating that blocking was definitely worthwhile. In our model for block designs, there is no reason not to test for blocks, but some models used for block designs do not allow a test for blocks.

Residual plots for the data are given in Figures 17.1 through 17.4. Figure 17.1 is a plot of the residuals versus the predicted values. Figure 17.2 plots the residuals versus indicators of the treatments. While the plot looks something like a bow tie, I am not overly concerned. Figure 17.3 contains a plot of the residuals versus indicators of blocks. The residuals look pretty good. From

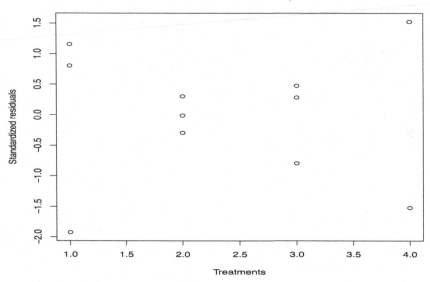

Figure 17.2: *Plot of residuals versus treatment groups, spectrometer data.*

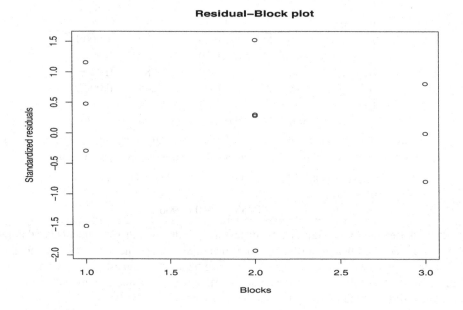

Figure 17.3: *Plot of residuals versus blocks, spectrometer data.*

Figure 17.4, the residuals look reasonably normal. In the normal plot there are 11 residuals but the analysis has only 5 degrees of freedom for error. If we want to do a W' test for normality, we might use a sample size of 11 and compare the value $W' = 0.966$ to $W'(\alpha, 11)$, but it may be appropriate to use the dfE as the sample size for the test and use $W'(\alpha, 5)$.

The leverages (not shown) are all reasonable. The largest t residual is -3.39 for Block 2, Treatment 1, which gives a Bonferonni adjusted P value of 0.088.   □

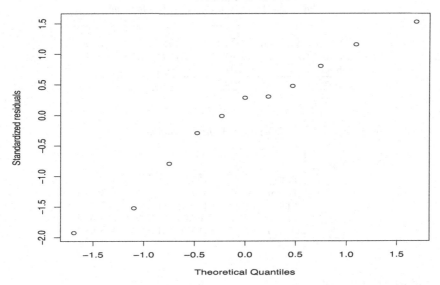

Figure 17.4: *Normal plot of residuals, spectrometer data, $W' = 0.966$.*

17.4.1 Paired comparisons

An interesting special case of complete block data is paired comparison data as discussed in Section 4.1. In paired comparison data, there are two treatments to contrast and each pair constitutes a complete block.

EXAMPLE 17.4.2. *Shewhart's hardness data.*
In Section 4.1, we examined Shewhart's data from Table 4.1 on hardness of two items that were welded together. In this case, it is impossible to group arbitrary formless pairs of parts and then randomly assign a part to be either part 1 or part 2, so the data do not actually come from an RCB experiment. Nonetheless, the two-way ANOVA model remains reasonable with pairs playing the role of blocks.

The data were given in Section 4.1 along with the means for each of the two parts. The two-way ANOVA analysis also requires the mean for each pair of parts. The analysis of variance table for the blocking analysis is given in Table 17.3. In comparing the blocking analysis to the paired comparison analysis given earlier, allowance for round-off errors must be made. The *MSE* is exactly half the value of $s_d^2 = 17.77165$ given in Section 4.1. The Table of Coefficients (from Minitab) gives a test for no Part effects of

$$t_{obs} = \frac{6.3315}{0.4057} = 15.61\,.$$

This is exactly the same t statistic as obtained in Section 4.1. The reference distribution is $t(26)$, again exactly the same. The analysis of variance F statistic is just the square of the t_{obs} and gives equivalent results for two-sided tests. Confidence intervals for the difference in means are also exactly the same in the blocking analysis and the paired comparison analysis. The one real difference between this analysis and the analysis of Section 4.1 is that this analysis provides an indication of whether the effort used to account for pairing was worthwhile. In this case, with a P value of 0.006, it was worthwhile to account for pairing. □

Table 17.3: *Analysis of Variance: Hardness data.*

Source	df	SS	MS	F	P
Pairs(Blocks)	26	634.94	24.42	2.75	0.006
Parts(Trts)	1	2164.73	2164.73	243.62	0.000
Error	26	231.03	8.89		
Total	53	3030.71			

Table 17.4: *Mangold root data.*

	Columns				
Rows	1	2	3	4	5
1	D(376)	E(371)	C(355)	B(356)	A(335)
2	B(316)	D(338)	E(336)	A(356)	C(332)
3	C(326)	A(326)	B(335)	D(343)	E(330)
4	E(317)	B(343)	A(330)	C(327)	D(336)
5	A(321)	C(332)	D(317)	E(318)	B(306)

17.5 Latin square designs

Latin square designs involve two simultaneous but distinct definitions of blocks. The treatments are arranged so that every treatment is observed in every block for both kinds of blocks.

EXAMPLE 17.5.1.   Mercer and Hall (1911) and Fisher (1925, Section 49) consider data on the weights of mangold roots. They used a Latin square design with 5 rows, columns, and treatments. The rectangular field on which the experiment was run was divided into five rows and five columns. This created 25 plots, arranged in a square, on which to apply the treatments A, B, C, D, and E. Each row of the square was viewed as a block, so every treatment was applied in every row. The unique feature of Latin square designs is that there is a second set of blocks. Every column was also considered a block, so every treatment was also applied in every column. The data are given in Table 17.4, arranged by rows and columns with the treatment given in the appropriate place and the observed root weight given in parentheses.

Table 17.5 contains the analysis of variance table including the analysis of variance F test for the null hypothesis that the effects are the same for every treatment. The F statistic $MSTrts/MSE$ is very small, 0.56, so there is no evidence that the treatments behave differently. Blocking on columns was not very effective as evidenced by the F statistic of 1.20, but blocking on rows was very effective, $F = 7.25$.

Many experimenters are less than thrilled when told that there is no evidence for their treatments having any differential effects. Inspection of the table of coefficients (not given) leads to an obvious conclusion that most of the treatment differences are due to the fact that treatment D has a much larger effect than the others, so we look at this a bit more.

We created a new factor variable called "Contrast" that has the same code for all of treatments A, B, C, E but a different code for D. Fitting a model with Columns and Rows but Contrast in lieu

Table 17.5: *Analysis of Variance: Mangold root data.*

Source	df	SS	MS	F	P
Columns	4	701.8	175.5	1.20	.360
Rows	4	4240.2	1060.1	7.25	.003
Trts	4	330.2	82.6	0.56	.696
Error	12	1754.3	146.2		
Total	24	7026.6			

Table 17.6: *Analysis of Variance: Mangold root data.*

Source	df	SS	MS	F	P
Columns	4	701.8	175.5	1.47	0.260
Rows	4	4240.2	1060.1	8.89	0.001
Contrast	1	295.8	295.8	2.48	0.136
Error	15	1788.7	119.2		
Total	24	7026.6			

of Treatments gives the ANOVA table in Table 17.6. The ANOVA table F statistic for Contrast is $295.8/119.2 = 2.48$ with a P value of 0.136. It provides a test of whether treatment D is different from the other treatments, when the other treatments are taken to have identical effects. Using our best practice, we would actually compute the F statistic with the MSE from Table 17.5 in the denominator giving $F_{obs} = 295.8/146.2 = 2.02$, which looks even less significant. This contrast was chosen by looking at the data so as to appear as significant as possible and yet it still has a large P value. Testing the two models against each other by using Tables 17.5 and 17.6 provides a test of whether there are any differences among treatments A, B, C, and E. The F statistic of 0.08 is so small that it would be suspiciously small if it had not been chosen, by looking at the data, to be small.

The standard residual plots were given in Christensen (1996). They look quite good.

If these data were unbalanced, i.e., if we lost some observations, it would be important to look at an ANOVA table that fits Treatments after both Columns and Rows. Fitted in the current order, the F test for Rows indicates that blocking on rows after blocking on Columns was worthwhile but the F test for Columns indicates that blocking on Columns alone would have been a waste of time. In an unbalanced experiment, if we cared enough, we might fit Columns after Rows to see whether blocking on Columns was a complete waste of time. Because the data are balanced, the two tests for Columns are the same and we can safely say from Table 17.5 that blocking on Columns was a waste of time. □

17.5.1 Latin square models

The model for an $r \times r$ Latin square design is a three-way analysis of variance,

$$y_{ijk} = \mu + \kappa_i + \rho_j + \tau_k + \varepsilon_{ijk}, \quad \varepsilon_{ijk}\text{s independent } N(0, \sigma^2). \quad (17.5.1)$$

The parameter μ is viewed as a grand mean, κ_i is an effect for the ith column, ρ_j is an effect for the jth row, and τ_k is an effect for the kth treatment. The subscripting for this model is peculiar. All of the subscripts run from 1 to r but not freely. If we specify a row and a column, the design tells you the treatment. Thus, if we know j and i, the design tells you k. If we specify a row and a treatment, the design tells you the column, so j and k dictate i. In fact, if we know any two of the subscripts, the design tells you the third.

17.5.2 Discussion of Latin squares

The idea of simultaneously having two distinct sets of complete blocks is quite useful. For example, suppose we wish to compare the performance of four machines in producing something. Productivity is notorious for depending on the day of the week, with Mondays and Fridays often having low productivity; thus we may wish to block on days. The productivity of the machine is also likely to depend on who is operating the machine, so we may wish to block on operators. Thus we may decide to run the experiment on Monday through Thursday with four machine operators and using each operator on a different machine each day. One possible design is

	Operator			
Day	1	2	3	4
Mon	A	B	C	D
Tue	B	C	D	A
Wed	C	D	A	B
Thu	D	A	B	C

where the numbers 1 through 4 are randomly assigned to the four people who will operate the machines and the letters A through D are randomly assigned to the machines to be examined. Moreover, the days of the week should actually be randomly assigned to the rows of the Latin square. In general, the rows, columns, and treatments should all be randomized in a Latin square.

Another distinct Latin square design for this situation is

	Operator			
Day	1	2	3	4
Mon	A	B	C	D
Tue	B	A	D	C
Wed	C	D	B	A
Thu	D	C	A	B

This square cannot be obtained from the first one by any interchange of rows, columns, and treatments. Typically, one would randomly choose a possible Latin square design from a list of such squares (see, for example, Cochran and Cox, 1957) in addition to randomly assigning the numbers, letters, and rows to the operators, machines, and days.

The use of Latin square designs can be extended in numerous ways. One modification is the incorporation of a third kind of block; such designs are called *Graeco-Latin squares*. The use of Graeco-Latin squares is explored in the exercises for this chapter. A problem with Latin squares is that small squares give poor variance estimates because they provide few degrees of freedom for error. For example, a 3×3 Latin square gives only 2 degrees of freedom for error. In such cases, the Latin square experiment is often performed several times, giving additional replications that provide improved variance estimation. Section 18.6 presents an example in which several Latin squares are used.

17.6   Balanced incomplete block designs

Balanced incomplete block (BIB) designs are not balanced in the same way that balanced ANOVAs are balanced. Balanced incomplete block designs are balanced in the sense that *every pair of treatments occurs together in the same block some fixed number of times*, say, λ. In a BIB the analysis of blocks is conducted ignoring treatments and the analysis of treatments is conducted after adjusting for blocks. This is the only order of fitting models that we need to consider. Blocks are designed to have effects and these effects are of no intrinsic interest, so there is no reason to worry about fitting treatments first and then examining blocks after adjusting for treatments. Blocks are nothing more than an adjustment factor.

The analysis being discussed here is known as the *intrablock* analysis of a BIB; it is appropriate when the block effects are viewed as fixed effects. If the block effects are viewed as random effects with mean 0, there is an alternative analysis that is known as the recovery of *interblock* information. Cochran and Cox (1957) and Christensen (2011, Section 12.11) discuss this analysis; we will not.

EXAMPLE 17.6.1.   A simple balanced incomplete block design is given below for four treatments A, B, C, D in four blocks of three units each.

Block	Treatments		
1	A	B	C
2	B	C	D
3	C	D	A
4	D	A	B

Note that every pair of treatments occurs together in the same block exactly $\lambda = 2$ times. Thus, for example, the pair A, B occurs in blocks 1 and 4. There are $b = 4$ blocks each containing $k = 3$ experimental units. There are $t = 4$ treatments and each treatment is observed $r = 3$ times. □

There are two relationships that must be satisfied by the numbers of blocks, b, units per block, k, treatments, t, replications per treatment, r, and λ. Recall that λ is the number of times two treatments occur together in a block. First, the total number of observations is the number of blocks times the number of units per block, but the total number of observations is also the number of treatments times the number of replications per treatment, thus

$$bk = rt.$$

The other key relationship in balanced incomplete block designs involves the number of comparisons that can be made between a given treatment and the other treatments *within the same block*. Again, there are two ways to count this. The number of comparisons is the number of other treatments, $t - 1$, multiplied by the number of times each other treatment is in the same block as the given treatment, λ. Alternatively, the number of comparisons within blocks is the number of other treatments within each block, $k - 1$, times the number of blocks in which the given treatment occurs, r. Thus we have

$$(t - 1)\lambda = r(k - 1).$$

In Example 17.6.1, these relationships reduce to

$$(4)3 = 3(4)$$

and

$$(4 - 1)2 = 3(3 - 1).$$

The nice thing about balanced incomplete block designs is that the theory behind them works out so simply that the computations can all be done on a hand calculator. I know, I did it once; see Christensen (2011, Section 9.4). But once was enough for this lifetime! We will rely on a computer program to provide the computations. We illustrate the techniques with an example.

EXAMPLE 17.6.2. John (1961) reported data on the number of dishes washed prior to losing the suds in the wash basin. Dishes were soiled in a standard way and washed one at a time. Three operators and three basins were available for the experiment, so at any one time only three treatments could be applied. Operators worked at the same speed, so no effect for operators was necessary nor should there be any effect due to basins. Nine detergent treatments were evaluated in a balanced incomplete block design. The treatments and numbers of dishes washed are given in Table 17.7. There were $b = 12$ blocks with $k = 3$ units in each block. Each of the $t = 9$ treatments was replicated $r = 4$ times. Each pair of treatments occurred together $\lambda = 1$ time. The three treatments assigned to a block were randomly assigned to basins as were the operators. The blocks were run in random order.

The analysis of variance is given in Table 17.8. The F test for treatment effects is clearly significant. We now need to examine contrasts in the treatments.

The treatments were constructed with a structure that leads to interesting effects. Treatments A, B, C, and D all consisted of detergent I using, respectively, 3, 2, 1, and 0 doses of an additive. Similarly, treatments E, F, G, and H used detergent II with 3, 2, 1, and 0 doses of the additive.

Table 17.7: *Balanced incomplete block design investigating detergents; data are numbers of dishes washed.*

Block	Treatment, Observation		
1	A, 19	B, 17	C, 11
2	D, 6	E, 26	F, 23
3	G, 21	H, 19	J, 28
4	A, 20	D, 7	G, 20
5	B, 17	E, 26	H, 19
6	C, 15	F, 23	J, 31
7	A, 20	E, 26	J, 31
8	B, 16	F, 23	G, 21
9	C, 13	D, 7	H, 20
10	A, 20	F, 24	H, 19
11	B, 17	D, 6	J, 29
12	C, 14	E, 24	G, 21

Table 17.8: *Analysis of Variance: BIB.*

Source	df	Seq SS	MS	F	P
Blocks	11	412.750	37.523	45.54	0.000
Trts	8	1086.815	135.852	164.85	0.000
Error	16	13.185	0.824		
Total	35	1512.750			

Treatment J was a control. We return to this example for a more detailed analysis of the treatments in the next chapter.

As always, we need to evaluate our assumptions. The normal plot looks less than thrilling but is not too bad. The fifth percentile of W' for 36 observations is .940, whereas the observed value is .953. Alternatively, the residuals have only 16 degrees of freedom and $W'(.95, 16) = .886$. The data are counts, so a square root or log transformation might be appropriate, but we continue with the current analysis. A plot of standardized residuals versus predicted values looks good.

Table 17.9 contains diagnostic statistics for the example. Note that the leverages are all identical for the BIB design. Some of the standardized deleted residuals (ts) are near 2 but none are so large as to indicate an outlier. The Cook's distances bring to one's attention exactly the same points as the standardized residuals and the ts.   □

The data in Exercises 14.5.1, 14.5.3, and 16.4.3 were all balanced incomplete block designs. Note that in those exercises we specifically indicated that block-by-treatment interactions should not be entertained.

17.6.1 Special cases

Balanced lattice designs are BIBs with $t = k^2$, $r = k + 1$, and $b = k(k + 1)$. Table 17.10 gives an example for $k = 3$. These designs can be viewed as $k + 1$ squares in which each treatment occurs once. Each row of a square is a block, each block contains k units, there are k rows in a square, so all of the $t = k^2$ treatments can appear in each square. To achieve a BIB, $k + 1$ squares are required, so there are $r = k + 1$ replications of each treatment. With $k + 1$ squares and k blocks (rows) per square, there are $b = k(k + 1)$ blocks. The analysis follows the standard form for a BIB. In fact, the design in Example 17.6.2 is a balanced lattice with $k = 3$.

Youden squares are a generalization of BIBs that allows a second form of blocking and a very similar analysis. These designs are discussed in the next section.

Table 17.9: *Diagnostics for the detergent data.*

Block	Trt.	y	$\hat{y}$	Leverage	r	t	C
1	A	19	18.7	0.56	0.49	0.48	0.01
1	B	17	16.1	0.56	1.41	1.46	0.12
1	C	11	12.1	0.56	−1.90	−2.09	0.22
2	D	6	6.6	0.56	−0.98	−0.98	0.06
2	E	26	25.4	0.56	1.04	1.04	0.07
2	F	23	23.0	0.56	−0.06	−0.06	0.00
3	G	21	20.5	0.56	0.86	0.85	0.05
3	H	19	18.6	0.56	0.67	0.66	0.03
3	J	28	28.9	0.56	−1.53	−1.60	0.15
4	A	20	19.6	0.56	0.61	0.60	0.02
4	D	7	6.4	0.56	0.98	0.98	0.06
4	G	20	21.0	0.56	−1.59	−1.68	0.16
5	B	17	17.3	0.56	−0.49	−0.48	0.01
5	E	26	25.4	0.56	0.98	0.98	0.06
5	F	19	19.3	0.56	−0.49	−0.48	0.01
6	C	15	14.3	0.56	1.16	1.18	0.08
6	F	23	24.1	0.56	−1.77	−1.92	0.20
6	J	31	30.6	0.56	0.61	0.60	0.02
7	A	20	20.6	0.56	−0.92	−0.91	0.05
7	E	26	26.1	0.56	−0.18	−0.18	0.00
7	J	31	30.3	0.56	1.10	1.11	0.08
8	B	16	16.8	0.56	−1.29	−1.31	0.10
8	F	23	22.6	0.56	0.73	0.72	0.03
8	G	21	20.7	0.56	0.55	0.54	0.02
9	C	13	13.6	0.56	−0.92	−0.91	0.05
9	D	7	6.9	0.56	0.18	0.18	0.00
9	H	20	19.6	0.56	0.73	0.72	0.03
10	A	20	20.1	0.56	−0.18	−0.18	0.00
10	F	24	23.3	0.56	1.10	1.11	0.08
10	H	19	19.6	0.56	−0.92	−0.91	0.05
11	B	17	16.8	0.56	0.37	0.36	0.01
11	D	6	6.1	0.56	−0.18	−0.18	0.00
11	J	29	29.1	0.56	−0.18	−0.18	0.00
12	C	14	13.0	0.56	1.65	1.76	0.17
12	E	24	25.1	0.56	−1.84	−2.00	0.21
12	G	21	20.9	0.56	0.18	0.18	0.00

Table 17.10: *Balanced lattice design for 9 treatments.*

Block				Block			
1	A	B	C	7	A	H	F
2	D	E	F	8	D	B	I
3	G	H	I	9	G	E	C
4	A	D	G	10	A	E	I
5	B	E	H	11	G	B	F
6	C	F	I	12	D	H	C

Table 17.11: *Mangold root data.*

Row	Columns			
	1	2	3	4
1	D(376)	E(371)	C(355)	B(356)
2	B(316)	D(338)	E(336)	A(356)
3	C(326)	A(326)	B(335)	D(343)
4	E(317)	B(343)	A(330)	C(327)
5	A(321)	C(332)	D(317)	E(318)

Table 17.12: *Analysis of Variance.*

Source	df	Seq SS	MS	F	P
Rows	4	4247.2	1061.8	6.87	
Column	3	367.0	122.3	0.79	
Trts	4	224.1	56.0	0.36	0.829
Error	8	1236.7	154.6		
Total	19	6075.0			

17.7 Youden squares

Consider the data on mangold roots in Table 17.11. There are five rows, four columns, and five treatments. If we ignore the columns, the rows and the treatments form a balanced incomplete block design, in which every pair of treatments occurs together three times. The key feature of Youden squares is that additionally the treatments are also set up in such a way that every treatment occurs once in each column. Since every row also occurs once in each column, the analysis for columns can be conducted independently of the analysis for rows and treatments. Columns are balanced relative to both treatments and rows.

Table 17.12 contains the analysis of variance for these data. Rows need to be fitted before Treatments. As long as balance is maintained, it does not matter where Columns are fitted. If the data become unbalanced, Treatments need to be fitted last. From the ANOVA table, there is no evidence for a difference between treatments.

Evaluation of assumptions is carried out as in all unbalanced ANOVAs. Diagnostic statistics are given in Table 17.13. The diagnostic statistics look reasonably good.

A normal plot looks very reasonable. A predicted value plot may indicate increasing variability as predicted values increase. One could attempt to find a transformation that would improve the plot but there is so little evidence of any difference between treatments that it hardly seems worth the bother.

The reader may note that the data in this section consist of the first four columns of the Latin square examined in Example 17.5.1. Dropping one column (or row) from a Latin square is a simple way to produce a Youden square. As Youden square designs do not give a square array of numbers (our example had 4 columns and 5 rows), one presumes that the name Youden *square* derives from this relationship to Latin squares. Table 17.14 presents an alternative method of presenting the data in Table 17.11 that is often used.   □

17.7.1 Balanced lattice squares

The key idea in *balanced lattice square designs* is that if we look at every row as a block, the treatments form a balanced incomplete block design and simultaneously if every column is viewed as a block, the treatments again form a balanced incomplete block design. In other words, each pair of treatments occurs together in the same row *or* column the same number of times. Of course every row appears with every column and vice versa. Balanced lattice square designs are similar to balanced lattices in that the number of treatments is $t = k^2$ and that the treatments are arranged in

Table 17.13: *Diagnostics.*

Row	Col	Trt	y	$\hat{y}$	Leverage	r	t	C
1	1	D	376	364.5	0.6	1.46	1.59	0.27
2	1	B	316	326.8	0.6	−1.37	−1.47	0.24
3	1	C	326	323.9	0.6	0.27	0.25	0.01
4	1	E	317	322.0	0.6	−0.64	−0.61	0.05
5	1	A	321	318.8	0.6	0.28	0.26	0.01
1	2	E	371	367.7	0.6	0.42	0.40	0.02
2	2	D	338	345.9	0.6	−1.01	−1.01	0.13
3	2	A	326	340.3	0.6	−1.81	−2.21	0.41
4	2	B	343	332.1	0.6	1.38	1.48	0.24
5	2	C	332	324.0	0.6	1.02	1.02	0.13
1	3	C	355	360.8	0.6	−0.74	−0.71	0.07
2	3	E	336	330.9	0.6	0.65	0.63	0.05
3	3	B	335	326.1	0.6	1.14	1.16	0.16
4	3	A	330	331.5	0.6	−0.19	−0.18	0.00
5	3	D	317	323.7	0.6	−0.86	−0.84	0.09
1	4	B	356	365.0	0.6	−1.14	−1.17	0.16
2	4	A	356	342.4	0.6	1.73	2.04	0.37
3	4	D	343	339.8	0.6	0.41	0.38	0.02
4	4	C	327	331.3	0.6	−0.55	−0.53	0.04
5	4	E	318	321.5	0.6	−0.44	−0.42	0.02

Table 17.14: *Mangold root data: Column (observation).*

	Treatments				
Row	A	B	C	D	E
1		4(356)	3(355)	1(376)	2(371)
2	4(356)	1(316)		2(338)	3(336)
3	2(326)	3(335)	1(326)	4(343)	
4	3(330)	2(343)	4(327)		1(317)
5	1(321)		2(332)	3(317)	4(318)

$k \times k$ squares. Table 17.15 gives an example for $k = 3$. If k is odd, one can typically get by with $(k+1)/2$ squares. If k is even, $k+1$ squares are generally needed.

17.8 Analysis of covariance in designed experiments

In Section 17.2 we discussed blocking as a method of variance reduction. Blocks were then incorporated as a factor variable into an additive-effects model with blocks and treatments, cf. Chapter 14. An alternative method of variance reduction is to incorporate a properly defined covariate into an additive ACOVA model with treatments and the covariate, cf. Chapter 15. This section focuses on choosing proper covariates.

In designing an experiment to investigate a group of treatments, concomitant observations can be used to reduce the error of treatment comparisons. One way to use the concomitant observations is to define blocks based on them. For example, income, IQ, and heights can all be used to collect

Table 17.15: *Balanced lattice square design for 9 treatments.*

	Column					Column		
Row	1	2	3	Row	4	5	6
1	A	B	C	4	A	F	H
2	D	E	F	5	I	B	D
3	G	H	I	6	E	G	C

people into similar groups for a block design. In fact, any construction of blocks must be based on information not otherwise incorporated into the ANOVA model, so any experiment with blocking uses concomitant information. In analysis of covariance we use the concomitant observations more directly, as regression variables in the statistical model.

Obviously, for a covariate to help our analysis it must be related to the dependent variable. Unfortunately, improper use of concomitant observations can invalidate, or at least alter, comparisons among the treatments. In the example of Section 15.1, the original ANOVA demonstrated an effect on heart weights associated with sex but after adjusting for body weights, there was little evidence for a sex difference. The very nature of what we were comparing changed when we adjusted for body weights. Originally, we investigated whether heart weights were different for females and males. The analysis of covariance examined whether there were differences between female heart weights and male heart weights *beyond what could be accounted for by the regression on body weights*. These are very different interpretations. In a designed experiment, we want to investigate the effects of the treatments and not the treatments adjusted for some covariates. To this end, in a designed experiment we require that the covariates be logically independent of the treatments. In particular, we require that

the concomitant observations be made before assigning the treatments to the experimental units,

the concomitant observations be made after assigning treatments to experimental units but before the effect of the treatments has developed, or

the concomitant observations be such that they are unaffected by treatment differences.

For example, suppose the treatments are five diets for cows and we wish to investigate milk production. Milk production is related to the size of the cow, so we might pick height of the cow as a covariate. For immature cows over a long period of time, diet may well affect both height and milk production. Thus to use height as a covariate we should measure heights before treatments begin or we could measure heights, say, two days after treatments begin. Two days on any reasonable diet should not affect a cow's height. Alternatively, if we use only mature cows their heights should be unaffected by diet and thus the heights of mature cows could be measured at any time during the experiment. Typically, *one should be very careful when claiming that a covariate measured near the end of an experiment is unaffected by treatments*.

The requirements listed above on the nature of covariates in a designed experiment are imposed so that the treatment effects do not depend on the presence or absence of covariates in the analysis. The treatment effects are logically identical regardless of whether covariates are actually measured or incorporated into the analysis. Recall that in the observational study of Section 15.1, the nature of the group (sex) effects changed depending on whether covariates were incorporated in the model. (Intuitively, the covariate body weight depends on the sex "treatment.") The role of the covariates in the analysis of a designed experiment is solely to reduce the error. In particular, using good covariates should reduce both the variance of the observations σ^2 and its estimate, the *MSE*. On the other hand, one pays a price for using covariates. Variances of treatment comparisons are σ^2 times a constant. With covariates in the model, the constant is larger than when they are not present. However, with well-chosen covariates the appropriate value of σ^2 should be sufficiently smaller that the reduction in *MSE* overwhelms the increase in the multiplier. Nonetheless, in designing an experiment we need to play these aspects off against one another. We need covariates whose reduction in *MSE* more than makes up for the increase in the constant.

The requirements imposed on the nature of the covariates in a designed experiment have little affect on the analysis illustrated in Section 15.1. The analysis focuses on a model such as (15.1.2). In Section 15.1, we also considered Model (15.1.3) that has different slope parameters for the different treatments (sexes). The requirements on the covariates in a designed experiment imply that the relationship between the dependent variable y and the covariate z *cannot* depend on the treatments. Thus with covariates chosen for a designed experiment *it is inappropriate to have slope parameters that depend on the treatment*. There is one slope that is valid for the entire analysis and the treatment effects do not depend on the presence or absence of the covariates. If a model such as (15.1.3) fits better than (15.1.2) when the covariate has been chosen appropriately, it suggests that the effects of treatments may differ from experimental unit to experimental unit. In such cases a treatment cannot really be said to have *an* effect, it has a variety of effects depending on which units it is applied to. A suitable transformation of the dependent variable may alleviate the problem.

17.9   Discussion of experimental design

Data are frequently collected with the intention of evaluating a change in the current system of doing things. If we really want to know the effect of a change in the system, we have to execute the change. It is not enough to look at conditions in the past that were similar to the proposed change because, along with the past similarities, there were dissimilarities. For example, suppose we think that instituting a good sex education program in schools will decrease teenage pregnancies. To evaluate this, it is not enough to compare schools that currently have such programs with schools that do not, because along with the differences in sex education programs there are other differences in the schools that affect teen pregnancy rates. Such differences may include parents' average socio-economic status and education. While adjustments can be made for any such differences that can be identified, there is no assurance that all important differences can be found. Moreover, initiating the proposed program involves making a change and the very act of change can affect the results. For example, current programs may exist and be effective because of the enthusiasm of the school staff that initiated them. Such enthusiasm is not likely to be duplicated when the new program is mandated from above.

To establish the effect of instituting a sex education program in a population of schools, we really need to (randomly) choose schools and actually institute the program. The schools at which the program is instituted should be chosen randomly, so no (unconscious) bias creeps in due to the selection of schools. For example, the people conducting the investigation are likely to favor or oppose the project. They could (perhaps unconsciously) choose the schools in such a way that makes the evaluation likely to reflect their prior attitudes. Unconscious bias occurs frequently and should *always* be assumed. Other schools without the program should be monitored to establish a base of comparison. These other schools should be treated as similarly as possible to the schools with the new program. For example, if the district school administration or the news media pay a lot of attention to the schools with the new program but ignore the other schools, we will be unable to distinguish the effect of the program from the effect of the attention. In addition, blocking similar schools together can improve the precision of the experimental results.

One of the great difficulties in learning about human populations is that obtaining the best data often requires morally unacceptable behavior. We object to having our lives randomly changed for the benefit of experimental science and typically the more important the issue under study, the more we object to such changes. Thus we find that in studying humans, the best data available are often historical. In our example we might have to accept that the best data available will be an historical record of schools with and without sex education programs. We must then try to identify and adjust for *all* differences in the schools that could potentially affect our conclusions. It is the extreme difficulty of doing this that leads to the relative unreliability of many studies in the social sciences. On the other hand, it would be foolish to give up the study of interesting and important phenomena just because they are difficult to study.

Analytic and enumerative studies

In one-sample, two-sample, and one-way ANOVA problems, we assume that we have random samples from various populations. In more sophisticated models we continue to assume that at least the errors are a random sample from a $N(0, \sigma^2)$ population. The statistical inferences we draw are valid for the populations that were sampled. Often it is not clear what the sampled populations are. What

Table 17.16: *Dead adult flies*.

Medium	Units of active ingredient			
	0	4	8	16
A	423	445	414	247
B	326	113	127	147
C	246	122	206	138
D	141	227	78	148
E	208	132	172	356
F	303	31	45	29
G	256	177	103	63

are the populations from which the Albuquerque suicide ages were sampled? Presumably, our data were all of the suicides reported in 1978 for these ethnic groups.

When we analyze data, we assume that the measurements are subject to errors and that the errors are consistent with our models. However, the populations from which these samples are taken may be nothing more than mental constructs. In such cases, it requires extrastatistical reasoning to justify applying the statistical conclusions to whatever issues we really wish to address. Moreover, the desire to predict the future underlies virtually all studies and, unfortunately, one can never be sure that data collected now will apply to the conditions of the future. So what can we do? Only our best. We can try to make our data as relevant as possible to our anticipation of future conditions. We can try to collect data for which the assumptions will be reasonably true. We can try to validate our assumptions. Studies in which it is not clear that the data are random samples from the population of immediate interest are often called *analytic studies*.

About the only time one can be really sure that statistical conclusions apply directly to the population of interest is when one has control of the population of interest. If we have a list of all the elements in the population, we can choose a random sample from the population. Of course, choosing a random sample is still very different from obtaining a random sample of observations. Without control or total cooperation, we may not be able to take measurements on the sample. (Even when we can find people that we want for a sample, many will not submit to a measurement process.) Studies in which one can arrange to have the assumptions met are often called *enumerative studies*. See Hahn and Meeker (1993) and Deming (1986) for additional discussion of these issues.

17.10   Exercises

EXERCISE 17.10.1.   Snedecor (1945b) presented data on a spray for killing adult flies as they emerged from a breeding medium. The data were numbers of adults found in cages that were set over the medium containers. The treatments were different levels of the spray's active ingredient, namely 0, 4, 8, and 16 units. (Actually, it is not clear whether a spray with 0 units was actually applied or whether no spray was applied. The former might be preferable.) Seven different sources for the breeding mediums were used and each spray was applied on each distinct breeding medium. The data are presented in Table 17.16.

(a) Identify the design for this experiment and give an appropriate model. List all the assumptions made in the model.

(b) Analyze the data. Give an appropriate analysis of variance table. Ccompare the treatment with no active ingredient to the average of the three treatments that contain the active ingredient. Ignoring the treatment with no active ingredient, the other three treatments are quantitative levels of the active ingredient. On the log scale, these levels are equally spaced.

(c) Check the assumptions of the model and adjust the analysis appropriately.

Table 17.17: *Cornell's scaled vinyl thickness values.*

Formulation	Production setting							
	1	2	3	4	5	6	7	8
1	8	7	12	10	7	8	12	11
2	6	5	9	8	7	6	10	9
3	10	11	13	12	9	10	14	12
4	4	5	6	3	5	4	6	5
5	11	10	15	11	9	7	13	9

Table 17.18: *Phosphorous fertilizer data.*

Fertilizer	Laboratory				
	1	2	3	4	5
F	20.20	19.92	20.91	20.65	19.94
G	30.20	30.09	29.10	29.85	30.29
H	31.40	30.42	30.18	31.34	31.11
I	45.88	45.48	45.51	44.82	44.63
J	46.75	47.14	48.00	46.37	46.63

EXERCISE 17.10.2. Cornell (1988) considered data on scaled thickness values for five formulations of vinyl designed for use in automobile seat covers. Eight groups of material were prepared. The production process was then set up and the five formulations run with the first group. The production process was then reset and another group of five was run. In all, the production process was set eight times and a group of five formulations was run with each setting. The data are displayed in Table 17.17.

(a) From the information given, identify the design for this experiment and give an appropriate model. List all the assumptions made in the model.

(b) Analyze the data. Give an appropriate analysis of variance table. Examine appropriate contrasts using the Bonferroni method with an α of about .05.

(c) Check the assumptions of the model and adjust the analysis appropriately.

EXERCISE 17.10.3. In data related to that of the previous problem, Cornell (1988) has scaled thickness values for vinyl under four different process conditions. The process conditions were A, high rate of extrusion, low drying temperature; B, low rate of extrusion, high drying temperature; C, low rate of extrusion, low drying temperature; D, high rate of extrusion, high drying temperature. An initial set of data with these conditions was collected and later a second set was obtained. The data are given below.

	Treatments			
	A	B	C	D
Rep 1	7.8	11.0	7.4	11.0
Rep 2	7.6	8.8	7.0	9.2

Identify the design, give the model, check the assumptions, give the analysis of variance table and interpret the F test for treatments. The structure of the treatments suggests looking at average rates, average temperatures, and interaction between rates and temperatures.

EXERCISE 17.10.4. Johnson (1978) and Mandel and Lashof (1987) present data on measurements of P_2O_5 (phosphorous pentoxide) in fertilizers. Table 17.18 presents data for five fertilizers, each analyzed in five labs. Our interest is in differences among the labs. Analyze the data.

Table 17.19: *Cowpea hay yields.*

Treatment	Block 1	Block 2	Block 3	Trt. means
I4	45	43	46	44.666
I8	50	45	48	47.666
II4	61	60	63	61.333
II8	58	56	60	58.000
Block means	53.50	51.00	54.25	52.916

Table 17.20: *Hydrostatic pressure tests: Operator, yield.*

A	B	C	D
40.0	43.5	39.0	44.0
B	A	D	C
40.0	42.0	40.5	38.0
C	D	A	B
42.0	40.5	38.0	40.0
D	C	B	A
40.0	36.5	39.0	38.5

EXERCISE 17.10.5.   Table 17.19 presents data on yields of cowpea hay. Four treatments are of interest, variety I of hay was planted 4 inches apart (I4), variety I of hay was planted 8 inches apart (I8), variety II of hay was planted 4 inches apart (II4), and variety II of hay was planted 8 inches apart (II8). Three blocks of land were each divided into four plots and one of the four treatments was randomly applied to each plot. These data are actually a subset of a larger data set given by Snedecor and Cochran (1980, p. 309) that involves three varieties and three spacings in four blocks. Analyze the data. Check your assumptions. Examine appropriate contrasts.

EXERCISE 17.10.6.   In the study of the optical emission spectrometer discussed in Example 17.4.1 and Table 17.1, the target value for readings was 0.89. Subtract 0.89 from each observation and repeat the analysis. What new questions are of interest? Which aspects of the analysis have changed and which have not?

EXERCISE 17.10.7.   An experiment was conducted to examine differences among operators of Suter hydrostatic testing machines. These machines are used to test the water repellency of squares of fabric. One large square of fabric was available but its water repellency was thought to vary along the length (warp) and width (fill) of the fabric. To adjust for this, the square was divided into four equal parts along the length of the fabric and four equal parts along the width of the fabric, yielding 16 smaller pieces. These pieces were used in a Latin square design to investigate differences among four operators: A, B, C, D. The data are given in Table 17.20. Construct an analysis of variance table. What, if any, differences can be established among the operators? Compare the results of using the Tukey and Bonferroni methods for comparing the operators.

EXERCISE 17.10.8.   Table 17.21 contains data similar to that in the previous exercise except that in this Latin square differences among four machines: 1, 2, 3, 4, were investigated rather than differences among operators. Machines 1 and 2 were operated with a hand lever, while machines 3 and 4 were operated with a foot lever. Construct an analysis of variance table. What, if any, differences can be established among the machines?

EXERCISE 17.10.9.   Table 17.21 is incomplete. The data were actually obtained from a Graeco-Latin square that incorporates four different operators as well as the four different machines. The

Table 17.21: *Hydrostatic pressure tests: Machine, yield.*

2	4	3	1
39.0	39.0	41.0	41.0
1	3	4	2
36.5	42.5	40.5	38.5
4	2	1	3
40.0	39.0	41.5	41.5
3	1	2	4
41.5	39.5	39.0	44.0

Table 17.22: *Hydrostatic pressure tests: Operator, machine.*

B,2	A,4	D,3	C,1
A,1	B,3	C,4	D,2
D,4	C,2	B,1	A,3
C,3	D,1	A,2	B,4
Operators are A, B, C, D.			
Machines are 1, 2, 3, 4.			

correct design is given in Table 17.22. Note that this is a Latin square for machines when we ignore the operators and a Latin square for operators when we ignore the machines. Moreover, every operator works once with every machine. Give the new analysis of variance table. How do the results on machines change? What evidence is there for differences among operators? Was the analysis for machines given earlier incorrect or merely inefficient?

EXERCISE 17.10.10.   Table 17.23 presents data given from Nelson (1993) on disk drives from a Graeco-Latin square design (see Exercise 17.10.9). The experiment was planned to investigate the effect of four different substrates on the drives. The dependent variable is the amplitude of a signal read from the disk where the signal written onto the disk had a fixed amplitude. Blocks were constructed from machines, operators, and day of production. (In Table 17.23, days are indicated by lower case Latin letters.) The substrata consist of A, aluminum; B, nickel-plated aluminum; and two types of glass, C and D. Analyze the data. In particular, check for differences between aluminum and glass, between the two types of glass, and between the two types of aluminum. Check your assumptions.

EXERCISE 17.10.11.   George Snedecor (1945a) asked for the appropriate variance estimate in the following problem. One of six treatments was applied to the 10 hens contained in each of 12 cages. Each treatment was randomly assigned to two cages. The data were the number of eggs laid by each hen.

(a) What should you tell Snedecor? Were the treatments applied to the hens or to the cages? How will the analysis differ depending on the answer to this question?

Table 17.23: *Amplitudes of disk drives.*

Operator	Machine			
	1	2	3	4
I	Aa 8	Cd 7	Db 3	Bc 4
II	Cc 11	Ab 5	Bd 9	Da 5
III	Dd 2	Ba 2	Ac 7	Cb 9
IV	Bb 8	Dc 4	Ca 9	Ad 3

(b) The mean of the 12 sample variances computed from the 10 hens in each cage was 297.8. The average of the 6 sample variances computed from the two cage means for each treatment was 57.59. The sample variance of the 6 treatment means was 53.725. How should you construct an F test? Remember that the numbers reported above are not necessarily mean squares.

EXERCISE 17.10.12. The data in Exercises 14.5.1, 14.5.3, and 16.4.3 were all balanced incomplete block designs. Determine the values of t, r, b, k, and λ for each experiment.

## Chapter 18

Factorial Treatments

Factorial treatment structures are simply an efficient way of defining the treatments used in an experiment. They can be used with any of the standard experimental designs discussed in Chapter 17. Factorial treatment structures have two great advantages, they give information that is not readily available from other methods and they use experimental material very efficiently. Section 18.1 introduces factorial treatment structures with an examination of treatments that involve two factors. Section 18.2 illustrates the analysis of factorial structures on the data of Example 17.4.1. Section 18.3 addresses some modeling issues involved with factorial structures. Section 18.4 looks at modeling interaction in the context of a designed experiment. Section 18.5 looks at a treatment structure that is slightly more complicated than factorial structure. Section 18.6 examines extensions of the Latin square designs that were discussed in Section 17.5.

18.1   Factorial treatment structures

The effect of alcohol and sleeping pills taken together is much greater than one would suspect based on examining the effects of alcohol and sleeping pills separately. If we did one experiment with 20 subjects to establish the effect of a 'normal' dose of alcohol and a second experiment with 20 subjects to establish the effect of a 'normal' dose of sleeping pills, the temptation would be to conclude (incorrectly) that the effect of taking a normal dose of both alcohol and sleeping pills would be just the sum of the individual effects. Unfortunately, the two separate experiments provide no basis for either accepting or rejecting such a conclusion.

We can redesign the investigation to be both more efficient and more informative by using a *factorial treatment structure*. The alcohol experiment would involve 10 people getting no alcohol (a_0) and 10 people getting a normal dose of alcohol (a_1). Similarly, the sleeping pill experiment would have 10 people given no sleeping pills (s_0) and 10 people getting a normal dose of sleeping pills (s_1). The two *factors* in this investigation are alcohol (A) and sleeping pills (S). Each factor is at two *levels*, no drug (a_0 and s_0, respectively) and a normal dose (a_1 and s_1, respectively). A factorial treatment structure uses treatments that are all combinations of the different levels of the factors. Thus a factorial experiment to investigate alcohol and sleeping pills might have 5 people given no alcohol and no sleeping pills ($a_0 s_0$), 5 people given no alcohol but a normal dose of sleeping pills ($a_0 s_1$), 5 people given alcohol but no sleeping pills ($a_1 s_0$), and 5 people given both alcohol and sleeping pills ($a_1 s_1$).

Assigning the treatments in this way has two major advantages. First, it is more informative in that it provides direct evidence about the effect of taking alcohol and sleeping pills together. If the joint effect is different from the sum of the effect of alcohol plus the effect of sleeping pills, the factors are said to *interact*. If the factors interact, there does not exist a single effect for alcohol; the effect of alcohol depends on whether the person has taken sleeping pills or not. Similarly, there is no one effect for sleeping pills; the effect depends on whether a person has taken alcohol or not. Note that if the factors interact, the separate experiments described earlier have very limited value.

The second advantage of using factorial treatments is that *if the factors do not interact*, the factorial experiment is more efficient than performing the two separate experiments. The two separate

Table 18.1: *Spectrometer data.*

Treatment	Block 1	Block 2	Block 3
New-clean	0.9331	0.8664	0.8711
New-soiled	0.9214	0.8729	0.8627
Used-clean	0.8472	0.7948	0.7810
Used-soiled	0.8417	0.8035	

experiments involve the use of 40 people, the factorial experiment involves the use of only 20 people, yet the factorial experiment contains just as much information about both alcohol effects and sleeping pill effects as the two separate experiments. The effect of alcohol can be studied by contrasting the 5 $a_0 s_0$ people with the 5 $a_1 s_0$ people and also by comparing the 5 $a_0 s_1$ people with the 5 $a_1 s_1$ people. Thus we have a total of 10 no-alcohol people to compare with 10 alcohol people, just as we had in the separate experiment for alcohol. Recall that with no interaction, the effect of factor A is the same regardless of the dose of factor S, so we have 10 valid comparisons of the effect of alcohol. A similar analysis shows that we have 10 no-sleeping-pill people to compare with 10 people using sleeping pills, the same as in the separate experiment for sleeping pills. Thus, *when there is no interaction*, the 20 people in the factorial experiment are as informative about the effects of alcohol and sleeping pills as the 40 people in the two separate experiments. Moreover, the factorial experiment provides information about possible interactions between the factors that is unavailable from the separate experiments.

The factorial treatment concept involves only the definition of the treatments. *Factorial treatment structure can be used in any design*, e.g., completely randomized designs, randomized block designs, and in Latin square designs. All of these designs allow for arbitrary treatments, so the treatments can be chosen to have factorial structure.

Experiments involving factorial treatment structures are often referred to as *factorial experiments* or *factorial designs*. A useful notation for factorial experiments identifies the number of factors and the number of levels of each factor. For example, the alcohol–sleeping pill experiment has 4 treatments because there are 2 levels of alcohol times 2 levels of sleeping pills. This is described as a 2×2 factorial experiment. If we had 3 levels of alcohol and 4 doses (levels) of sleeping pills we would have a 3×4 experiment involving 12 treatments.

18.2 Analysis

A CRD is analyzed as a one-way ANOVA with the treatments defining the groups. However, if the CRD has treatments defined by two factors, it can also be analyzed as a two-way ANOVA with interaction. Similarly, if the CRD has treatments defined by three factors, it can be analyzed as a three-way ANOVA as illustrated in Chapter 16. Similarly, an RCB design uses a two-way model with no interaction between treatments and blocks. For treatments based on two factors, an equivalent model for an RCB is a three-way model but the only interaction is between the two treatment factors. We now illustrate a two-factor treatment structure in a randomized block design.

EXAMPLE 18.2.1.   *A 2×2 factorial in 3 randomized blocks*
Consider again the spectroscopy data of Example 17.4.1. The treatments were all combinations of two disks (new, used) and two windows (clean, soiled), so the treatments have a 2×2 factorial structure. The data are repeated in Table 18.1. The analysis of variance table for the four treatments is

Analysis of Variance

Source	df	Seq SS	MS	F	P
Blocks	2	0.0063366	0.0031683	62.91	0.000
Treatments	3	0.0166713	0.0055571	110.34	0.000
Error	5	0.0002518	0.0000504		
Total	10	0.0232598			

In Chapter 14, when analyzing the rat weights, we replaced the 15 degrees of freedom for the $4 \times 4 = 16$ Litter–Mother treatments with 3 degrees of freedom for Litters, 3 degrees of freedom for Mothers, and 9 degrees of freedom for Litter–Mother interaction. Employing the current factorial structure, we can similarly replace the three degrees of freedom for the four treatments with one degree of freedom for Disks, one degree of freedom for Windows, and one degree of freedom for Disk–Window interaction. Since the data are unbalanced, we should look at Disks fitted before Windows

Analysis of Variance

Source	df	Seq SS	MS	F	P
Blocks	2	0.0063366	0.0031683	62.91	0.000
Disk	1	0.0166269	0.0166269	330.14	0.000
Window	1	0.0000032	0.0000032	0.06	0.812
Disk*Window	1	0.0000413	0.0000413	0.82	0.407
Error	5	0.0002518	0.0000504		
Total	10	0.0232598			

as well as Windows fitted before Disks.

Analysis of Variance

Source	df	Seq SS	MS	F	P
Blocks	2	0.0063366	0.0031683	62.91	0.000
Window	1	0.0002059	0.0002059	4.09	0.099
Disk	1	0.0164241	0.0164241	326.12	0.000
Window*Disk	1	0.0000413	0.0000413	0.82	0.407
Error	5	0.0002518	0.0000504		
Total	10	0.0232598			

The procedure is analogous to Chapter 14 except that we fit blocks prior to considering the other effects.

The F statistic for Disk–Window interaction is not significant. This indicates a lack of evidence that Disks behave differently with clean windows than with soiled windows. In examining the four Disk–Window effects in Example 17.4.1 we considered the paired comparisons

New-clean	New-soiled	Used-soiled	Used-clean
$\hat{\eta}_1$	$\hat{\eta}_2$	$\hat{\eta}_4$	$\hat{\eta}_3$
0	−0.00453	−0.07906	−0.08253

The estimated difference between clean and soiled windows differs in sign between new and used disks. Clean windows give higher readings for new disks but soiled windows give higher readings for used disks. Although this is some indication of interaction, the ANOVA table F test for interaction makes it clear that the effect is not significant.

The F statistics for Windows show little evidence in either ANOVA table that the window types affect yield. If interaction existed, this would be merely an artifact. Windows would have to be important because the interaction would imply that disks behave differently with the different types of windows. However, we possess no evidence of interaction.

The F statistics for Disks indicate that disk types have different effects. From the table of coefficients for either the no Disk–Window interaction model or the model that completely ignores Windows, the positive coefficient for "New" (or negative coefficient for "Used") indicates that new disks give greater yields than used disks. In fact, that is even clear from the pairwise comparisons given earlier for the interaction model.

If the effects contain multiple degrees of freedom, it would be wise to investigate the components of the ANOVA table (Interaction, Disks, and Windows) further as we illustrated in Chapter 16. However, for these data each source has only one degree of freedom, thus the analysis of variance table provides F statistics for all the interesting effects and the analysis given is complete.

The factorial treatment structure also suggests two residual plots that were not examined earlier. These are plots of the residuals versus Disks and the residuals versus Windows. The plots give no particular cause for alarm.

$\square$

18.3 Modeling factorials

The general model for a block design is

$$y_{ij} = \mu + \beta_i + \tau_j + \varepsilon_{ij}, \quad \varepsilon_{ij}\text{s independent } N(0,\sigma^2),$$

$i = 1,\ldots,b, j = 1,\ldots,a$, where i denotes the blocks and j denotes treatments. In incomplete block designs, not all combinations of i and j appear in the data. For a factorial treatment structure involving two factors, one factor with levels $g = 1,\ldots,G$ and the other with levels $h = 1,\ldots,H$, we must have $a = GH$ and we can replace the single subscript j for treatments with the pair of subscripts gh. For example, with $G = 3$ and $H = 2$ we might use the following correspondence.

j	1	2	3	4	5	6
(g,h)	$(1,1)$	$(1,2)$	$(2,1)$	$(2,2)$	$(3,1)$	$(3,2)$

We can now rewrite the block model as an equivalent model,

$$y_{igh} = \mu + \beta_i + \tau_{gh} + \varepsilon_{ghj}, \quad \varepsilon_{igh}\text{s independent } N(0,\sigma^2), \tag{18.3.1}$$

$i = 1,\ldots,b, g = 1,\ldots,G, h = 1,\ldots,H$, where τ_{gh} is the effect due to the treatment combination having level g of the first factor and level h of the second. Changing the subscripts really does nothing to the model; the subscripting is merely a convenience.

We can also rewrite the model to display factorial effects similar to those used earlier. This is done by expanding the treatment effects into effects corresponding to the ANOVA table lines. Write

$$y_{igh} = \mu + \beta_i + \gamma_g + \xi_h + (\gamma\xi)_{gh} + \varepsilon_{igh}, \tag{18.3.2}$$

where the γ_gs are main effects for the first factor, the ξ_hs are main effects for the second factor, and the $(\gamma\xi)_{gh}$s are effects that allow interaction between the factors.

Changing from Model (18.3.1) to Model (18.3.2) is accomplished by making the substitution

$$\tau_{gh} \equiv \gamma_g + \xi_h + (\gamma\xi)_{gh}.$$

There is *less* going on here than meets the eye. The only difference between the parameters τ_{gh} and $(\gamma\xi)_{gh}$ is the choice of Greek letters and the presence of parentheses. They accomplish exactly the same things for the two models. The parameters γ_g and ξ_h are completely redundant. Anything one could explain with these parameters could be explained equally well with the $(\gamma\xi)_{gh}$s. As they stand, models (18.3.1) and (18.3.2) are equivalent. The point of using Model (18.3.2) is that it lends itself nicely to an interesting reduced model. If we drop the τ_{gh}s from Model (18.3.1), we drop all of the treatment effects, so testing Model (18.3.1) against this reduced model is a test of whether there are any treatment effects. If we drop the $(\gamma\xi)_{gh}$s from Model (18.3.2), we get

$$y_{ghj} = \mu + \beta_i + \gamma_g + \xi_h + \varepsilon_{igh}. \tag{18.3.3}$$

This still has the γ_gs and the ξ_hs in the model. Thus, dropping the $(\gamma\xi)_{gh}$s does not eliminate all of the treatment effects, it only eliminates effects that cannot be explained as the sum of an effect for the first factor plus an effect for the second factor. In other words, it only eliminates the interaction effects. The reduced model (18.3.3) is the model without interaction and consists of *additive* factor effects. The test for interaction is the test of Model (18.3.3) against the larger model (18.3.2). By definition, interaction is any effect that can be explained by Model (18.3.2) but not by Model (18.3.3).

As discussed, testing for interaction is a test of whether the $(\gamma\xi)_{gh}$s can be dropped from Model (18.3.2). If there is no interaction, a test for main effects, say the γ_gs after fitting the ξ_hs, examines whether the γ_gs can be dropped from Model (18.3.3), i.e., whether the factor has any effect or whether $\gamma_1 = \gamma_2 = \cdots = \gamma_G$ in a model with the ξ_hs. *If the interaction terms* $(\gamma\xi)_{gh}$ *are present*, there is no test of main effects. Dropping the γ_gs from Model (18.3.2) leaves a model equivalent to Model (18.3.2). Any test that a computer program might report for $\gamma_1 = \gamma_2 = \cdots = \gamma_G$ in Model (18.3.2) will depend crucially on arbitrary side conditions that the program has imposed to obtain estimates of parameters that cannot otherwise be estimated. Different programs that use different side conditions will give different results. *Never trust "adjusted" F tests for main effects in models with interaction.*

If interactions are important, they must be dealt with. Either we give up on Model (18.3.2), go back to Model (18.3.1), and simply examine the various treatments as best we can, or we examine the nature of the interaction directly. Note that we did *not* say that whenever interactions are significant they must be dealt with. Whether an interaction is important or not depends on the particular application. For example, if interactions are statistically significant but are an order of magnitude smaller than the main effects, one *might* be able to draw useful conclusions while ignoring the interactions.

The procedure for incorporating factorial treatment structures is largely independent of the experimental design. The basic Latin square model is

$$y_{ijk} = \mu + \kappa_i + \rho_j + \tau_k + \varepsilon_{ijk}, \quad \varepsilon_{ijk}\text{s independent } N(0,\sigma^2),$$

where the subscripts i, j, and k indicate columns, rows, and treatments, respectively. With two factors, we can again replace the treatment subscript k with the pair (g,h) and write

$$y_{ijgh} = \mu + + \kappa_i + \rho_j + \tau_{gh} + \varepsilon_{ijgh}, \quad \varepsilon_{ijgh}\text{s independent } N(0,\sigma^2).$$

Again, we can expand the treatment effects τ_{gh} to correspond to the factorial treatment structure as

$$y_{ijgh} = \mu + \kappa_i + \rho_j + \gamma_g + \xi_h + (\gamma\xi)_{gh} + \varepsilon_{ghjk}.$$

18.4 Interaction in a Latin square

We have examined the process of modeling interactions earlier, but for completeness we reexamine the process in a designed experiment with factorial treatment structure.

EXAMPLE 18.4.2. *A 2×3 factorial in a 6×6 Latin square.*
Fisher (1935, Sections 36, 64) presented data on the pounds of potatoes harvested from a piece of ground that was divided into a square consisting of 36 plots. Six treatments were randomly assigned to the plots in such a way that each treatment occurred once in every row and once in every column of the square. The treatments involved two factors, a nitrogen-based fertilizer (N) and a phosphorous-based fertilizer (P). The nitrogen fertilizer had two levels, none (n_0) and a standard dose (n_1). The phosphorous fertilizer had three levels, none (p_0), a standard dose (p_1), and double the standard dose (p_2). We identify the six treatments for this 2×3 experiment as follows:

A	B	C	D	E	F
n_0p_0	n_0p_1	n_0p_2	n_1p_0	n_1p_1	n_1p_2

Table 18.2: *Potato data: Treatment(Yield).*

Row	Column					
	1	2	3	4	5	6
1	E(633)	B(527)	F(652)	A(390)	C(504)	D(416)
2	B(489)	C(475)	D(415)	E(488)	F(571)	A(282)
3	A(384)	E(481)	C(483)	B(422)	D(334)	F(646)
4	F(620)	D(448)	E(505)	C(439)	A(323)	B(384)
5	D(452)	A(432)	B(411)	F(617)	E(594)	C(466)
6	C(500)	F(505)	A(259)	D(366)	B(326)	E(420)

Table 18.3: *Analysis of Variance: Potato data.*

Source	df	SS	MS	F	P
Rows	5	54199	10840	7.10	.001
Columns	5	24467	4893	3.20	.028
Treatments	5	248180	49636	32.51	.000
Error	20	30541	1527		
Total	35	357387			

The data are presented in Table 18.2. The basic ANOVA table is presented as Table 18.3. The ANOVA F test indicates substantial differences between the treatments. Blocking on rows of the square was quite effective with an F ratio of 7.10. Blocking on columns was considerably less effective with an F of only 3.20, but it was still worthwhile. For unbalanced data, the rows and columns can be fitted in either order (with appropriate interpretations of test results) but the treatments should be fitted last.

We begin by fitting the Latin square model for six treatments $k = 1, \ldots, 6$,

$$y_{ijk} = \mu + \rho_i + \kappa_j + \tau_k + \varepsilon_{ijk}.$$

Switching to factorial subscripts $k \to gh$ gives the model

$$y_{ijgh} = \mu + \rho_i + \kappa_j + \tau_{gh} + \varepsilon_{ijgh},$$

$g = 0, 1$ and $h = 0, 1, 2$. Adding main effect parameters for nitrogen and phosphorous leads to

$$y_{ijgh} = \mu + \rho_i + \kappa_j + N_g + P_h + (NP)_{gh} + \varepsilon_{ijgh} \qquad (18.4.1)$$

and fitting a sequence of models that successively adds each term from left to right in Model (18.4.1), we get the following ANOVA table.

Analysis of Variance: Model (18.4.1)

Source	df	SS	MS	F	P
Rows	5	54199	10840	7.10	0.001
Columns	5	24467	4893	3.20	0.028
N	1	77191	77191	50.55	0.000
P	2	164872	82436	53.98	0.000
N*P	2	6117	3059	2.00	0.161
Error	20	30541	1527		
Total	35	357387			

The Error line is the same as in Table 18.3, as should be the case for equivalent models. There does not seem to be much evidence for interaction with a P value of 0.161. (But we may soon change our minds about that.)

The three levels of phosphorous are quantitative, so we can fit separate quadratic models in lieu of fitting interaction. Letting $p_h = 0, 1, 2$ be the known quantitative levels, an equivalent model is

$$y_{ijgh} = \mu + \rho_i + \kappa_j + N_g + \delta_{g1}p_h + \delta_{g2}p_h^2 + \varepsilon_{ijgh}.$$

An alternative parameterization of this model is

$$y_{ijgh} = \mu + \rho_i + \kappa_j + N_g + \gamma_1 p_h + \delta_{g1}p_h + \gamma_2 p_h^2 + \delta_{g2}p_h^2 + \varepsilon_{ijgh}, \tag{18.4.2}$$

which provides the ANOVA table

Analysis of Variance: Model (18.4.2)

Source	df	SS	MS	F	P
Rows	5	54199	10840	7.10	0.001
Columns	5	24467	4893	3.20	0.028
N	1	77191	77191	50.55	0.000
p	1	162855	162855	106.65	0.000
$N*p$	1	6112	6112	4.00	0.059
p^2	1	2016	2016	1.32	0.264
$N*p^2$	1	5	5	0.00	0.955
Error	20	30541	1527		
Total	35	357387			

Again, there is no change in the Error line as the models are equivalent. The ANOVA table comes from fitting the terms in (18.4.2) successively, thus determining a sequence of models.

The P value for the last line of the ANOVA table, $N*p^2$, is 0.955 and suggests that we do not need different quadratic terms for the two levels of nitrogen. The terms are fitted sequentially, so the large P value for p^2, 0.264, suggests that there is no quadratic effect at all in phosphorous, i.e., neither quadratic term is significant. Finally, the P value for $N*p$ is a relatively small 0.059, which suggests that perhaps there is some interaction between phosphorous and nitrogen in that the linear coefficients (slopes) are different for n_0 and n_1. Although the parameterization of Model (18.4.2) may seem awkward, sequential fitting of terms leads to a very useful ANOVA table.

Incorporating these suggestions, we fit separate straight lines in phosphorous for each level of nitrogen,

$$y_{ijgh} = \mu + \rho_i + \kappa_j + N_g + \gamma_1 p_h + \delta_{g1}p_h + \varepsilon_{ijgh}. \tag{18.4.3}$$

This gives the following ANOVA table.

Analysis of Variance: Model (18.4.3)

Source	df	SS	MS	F	P
Rows	5	54199	10840	7.32	0.000
Columns	5	24467	4893	3.31	0.022
N	1	77191	77191	52.15	0.000
p	1	162855	162855	110.03	0.000
$N*p$	1	6112	6112	4.13	0.054
Error	22	32562	1480		
Total	35	357387			

With sequential fitting, the ANOVA table provides no new information; however, a table of coefficients is presented as Table 18.4. What we care about are the differential effects of the treatments. Using Table 18.4 in which $0 = \hat{N}_0 + \hat{N}_1 = \hat{\delta}_{01} + \hat{\delta}_{11}$, we can summarize the fitted model as two lines in the amount of phosphorous, one line for no nitrogen and another line for a single dose of nitrogen,

$$\hat{m}(i,j,g,p) = \begin{cases} \hat{\mu}_{ij} - 30.35 + (82.375 - 15.958)p & \text{for } n_0 \\ \hat{\mu}_{ij} + 30.35 + (82.375 + 15.958)p & \text{for } n_1. \end{cases}$$

Table 18.4: *Table of Coefficients: Model (18.4.3).*

Predictor	Est	SE	t	P
Constant	380.37	10.14	37.52	0.000
Rows				
1	57.58	14.34	4.02	0.001
2	−9.42	14.34	−0.66	0.518
3	−4.42	14.34	−0.31	0.761
4	−9.58	14.34	−0.67	0.511
5	32.58	14.34	2.27	0.033
Columns				
1	50.25	14.34	3.50	0.002
2	15.25	14.34	1.06	0.299
3	−8.58	14.34	−0.60	0.556
4	−9.08	14.34	−0.63	0.533
5	−20.75	14.34	−1.45	0.162
N				
0	−30.35	10.14	−2.99	0.007
p	82.375	7.853	10.49	0.000
p*N				
0	−15.958	7.853	−2.03	0.054

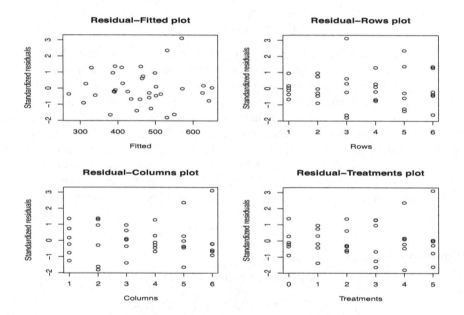

Figure 18.1: *Residual plots for potato data.*

Of course the predicted lines also depend on the row i and the column j. For no phosphorous, no nitrogen is estimated to yield $60.70 = 2(30.35)$ pounds less than a dose of nitrogen. For a single dose of phosphorous, no nitrogen is estimated to yield $2(30.35 + 15.958) \doteq 93$ pounds less than a dose of nitrogen. For a double dose of phosphorous, no nitrogen is estimated to yield $2(30.35 + 15.958 \times 2) \doteq 125$ pounds less than a dose of nitrogen. Estimated yields go up as you add more phosphorous, but estimated yields go up faster (at a rate of about $32 \doteq 2(15.958)$ pounds per dose) if you are also applying nitrogen.

Figures 18.1 and 18.2 contain residual plots from the full interaction model. They show some interesting features but nothing so outstanding that I, personally, find them disturbing.

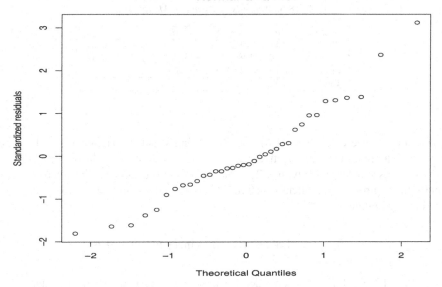

Figure 18.2: *Normal plot for potato data,* $W' = 0.967$.

Table 18.5: *Balanced incomplete block design investigating detergents; data are numbers of dishes washed.*

Block	Treatment, Observation		
1	A, 19	B, 17	C, 11
2	D, 6	E, 26	F, 23
3	G, 21	H, 19	J, 28
4	A, 20	D, 7	G, 20
5	B, 17	E, 26	H, 19
6	C, 15	F, 23	J, 31
7	A, 20	E, 26	J, 31
8	B, 16	F, 23	G, 21
9	C, 13	D, 7	H, 20
10	A, 20	F, 24	H, 19
11	B, 17	D, 6	J, 29
12	C, 14	E, 24	G, 21

18.5   A balanced incomplete block design

In Example 17.6.2 we considered a balanced incomplete block design with an unusual structure to the treatments. We now return to that example.

EXAMPLE 18.5.1.   John (1961) reported data on the number of dishes washed prior to losing the suds in the wash basin. Dishes were soiled in a standard way and washed one at a time. Three operators and three basins were available for the experiment, so at any one time only three treatments could be applied. Operators worked at the same speed, so no effect for operators was necessary nor should there be any effect due to basins. Nine detergent treatments were evaluated in a balanced incomplete block design. The treatments and numbers of dishes washed are repeated in Table 18.5. There were $b = 12$ blocks with $k = 3$ units in each block. Each of the $t = 9$ treatments was replicated $r = 4$ times. Each pair of treatments occurred together $\lambda = 1$ time. The three treatments assigned to a block were randomly assigned to basins as were the operators. The blocks were run in random order. The analysis of variance is given in Table 18.6. The F test for treatment effects is clearly significant. We now examine the treatments more carefully.

Table 18.6: *Analysis of Variance: Model (18.5.1)*.

Source	df	Seq SS	MS	F	P
Blocks	11	412.750	37.523	45.54	0.000
Treatments	8	1086.815	135.852	164.85	0.000
Error	16	13.185	0.824		
Total	35	1512.750			

The treatments were constructed with an interesting structure. Treatments A, B, C, and D all consisted of detergent I using, respectively, 3, 2, 1, and 0 doses of an additive. Similarly, treatments E, F, G, and H used detergent II with 3, 2, 1, and 0 doses of the additive. Treatment J was a control. Except for the control, the treatment structure is factorial in detergents and amounts of additive. The general blocking model is

$$y_{ij} = \mu + \beta_i + \tau_j + \varepsilon_{ij}, \quad \varepsilon_{ij}\text{s independent } N(0, \sigma^2). \tag{18.5.1}$$

Here $i = 1, \ldots, 12$, $j = 1, \ldots, 9$, where i denotes the blocks and j denotes treatments.

As a first step, I created three new factor variables to replace the treatment factor. Control (k) takes the value 1 if a treatment is not the control and a value 2 if it is the control. Detergent (g) takes the value 1 if a treatment involves detergent I, the value 2 if it involves detergent II, and the value 3 if it is the control. Amount (h) takes on the values 0 through 3 for the doses of additive with each treatment other than the control and is 0 for the control. The correspondence between treatments and subscripts is as follows.

Treatment	j	kgh
A	1	113
B	2	112
C	3	111
D	4	110
E	5	123
F	6	122
G	7	121
H	8	120
J	9	230

The three subscripts kgh uniquely identify the treatments so we can refit the original blocking model (18.5.1) as

$$y_{ikgh} = \mu + \beta_i + \tau_{kgh} + \varepsilon_{ikgh}.$$

Going a step further, we can identify Control effects, Detergent effects, Amount effects, and a Detergent–Amount interaction,

$$y_{ikgh} = \mu + \beta_i + C_k + D_g + A_h + (DA)_{gh} + \varepsilon_{ikgh}. \tag{18.5.2}$$

This is equivalent to Model (18.5.1), but is somewhat awkward to fit for many general linear model programs because both the variable Control and the variable Detergent uniquely identify the control treatment. For example, Minitab only gives the following output in the ANOVA table.

Analysis of Variance: Model (18.5.2)

Source	df	Seq. SS	MS	F	P
Blocks	11	412.750			
Control	1	345.042			
Detergent	1	381.338			
Amount	3	311.384			
Det*Amt	3	49.051			
Error	16	13.185			
Total	35	1512.750			

As usual, this is obtained by sequentially fitting the terms in Model (18.5.2) from left to right. Minitab is also somewhat dubious about the degrees of freedom. But never fear, all is well. The Error line agrees with that in Table 18.6, which is good evidence for my claim of equivalence. On the other hand, R just fills out the ANOVA table.

We now exploit the quantitative nature of the amounts by recasting our Amount factor variable as a quantitative variable a, which we also square and cube. Fitting a separate cubic polynomial for both detergents other than the control gives

$$y_{ikgh} = \mu + \beta_i + C_k + \delta_{g0} + \delta_{g1}a_h + \delta_{g2}a_h^2 + \delta_{g3}a_h^3 + \varepsilon_{ikgh},$$ (18.5.3)

or the equivalent but even more overparameterized version,

$$y_{ikgh} = \mu + \beta_i + C_k + \delta_{g0} + \gamma_1 a_h + \delta_{g1}a_h + \gamma_2 a_h^2 + \delta_{g2}a_h^2 + \gamma_3 a_h^3 + \delta_{g3}a_h^3 + \varepsilon_{ikgh}.$$ (18.5.4)

Models (18.5.3) and (18.5.4) should give us a model equivalent to Model (18.5.2). But these models still involve both the Control (k) and Detergent (g) effects, so they remain computer unfriendly. Sequentially fitting the terms in Model (18.5.4), gives

Analysis of Variance: Model (18.5.4)

Source	df	Seq. SS	MS	F	P
Blocks	11	412.750			
Control	1	345.042			
Detergent	1	381.338			
a	1	306.134			
Det*a	1	41.223			
a^2	1	5.042			
Det*a^2	1	7.782			
a^3	1	0.208			
Det*a^3	1	0.045			
Error	16	13.185			
Total	35	1512.750			

Again, as should be the case for equivalent models, Model (18.5.4) gives the same Error term as models (18.5.1) and (18.5.2). We could easily fill out the blank columns of the ANOVA table with a hand calculator because all the difficult computations have been made. With $MSE = 0.824$ from Table 18.6, and most of the terms having one degree of freedom, it is easy to glace at the ANOVA table for Model (18.5.4) and see what effects are important. The Det*a^3 term checks whether the cubic coefficients are different for detergents I and II, and a^3 checks whether the overall cubic coefficient is different from 0. Both are small, so there is no evidence for any cubic effects. On the other hand, Det*a^2 is almost 10 times the size of MSE, so we need different parabolas for detergents I and II.

Rather than continuing to work with a general linear model program, I refit Model (18.5.3) as a regression. I used 11 indictor variables for blocks 2 through 12, $B_2, \ldots, B_{12}$, I created three indicator variables for detergents, d_1, d_2, d_3, where d_3 is the indicator of the control, and I used the quantitative amount variable a to create variables to fit separate polynomials for detergents I and II by defining the products d_1a, d_1a^2, d_1a^3 and d_2a, d_2a^2, d_2a^3. Fitting the regression model

$$y_r = \beta_0 + \sum_{j=2}^{12} \beta_{0j} B_{rj} + \delta_{30} d_{r3} + \delta_{10} d_{r1}$$

$$+ \delta_{11} d_{r1} a_r + \delta_{21} d_{r2} a_r + \delta_{12} d_{r1} a_r^2 + \delta_{22} d_{r2} a_r^2 + \delta_{13} d_{r1} a_r^3 + \delta_{23} d_{r2} a_r^3 + \varepsilon_r, \quad (18.5.5)$$

gives

Analysis of Variance: Model (18.5.5)

Source	df	SS	MS	F	P
Regression	19	1499.56	78.924	95.774	0.000000
Error	16	13.19	0.824		
Total	35	1512.75			

from which, with the sequential sums of squares, we can construct

Analysis of Variance: Model (18.5.5)

Source	df	Seq. SS	MS	F	P
Blocks	11	412.75	37.523	45.533	0.000000
d_3	1	345.04	345.042	418.702	0.000000
d_1	1	381.34	381.338	462.747	0.000000
$d_1 * a$	1	286.02	286.017	347.076	0.000000
$d_2 * a$	1	61.34	61.341	74.436	0.000000
$d_1 * a^2$	1	12.68	12.676	15.382	0.001216
$d_2 * a^2$	1	0.15	0.148	0.180	0.677212
$d_1 * a^3$	1	0.22	0.224	0.272	0.609196
$d_2 * a^3$	1	0.03	0.030	0.036	0.851993
Error	16	13.19	0.824		
Total	35	1512.75			

Both of the cubic terms are small, so we need neither. This agrees with the Model (18.5.4) ANOVA table where we had sequential sums of squares for a^3 of 0.208 and for Det*a^3 of 0.045, which add to 0.253. The cubic terms here have sums of squares 0.22 and 0.03, which sum to the same thing (except for round-off error). In the Model (18.5.4) ANOVA table we could only identify that there was quadratic interaction. From the term $d_2 * a^2$ we see that detergent II needs no quadratic term, while from $d_1 * a^2$ we see that detergent II does need such a term. This implies a difference in the quadratic coefficients, hence the significant Det*a^2 quadratic interaction in the ANOVA table of Model (18.5.4). All of the other effects look important. The table of coefficients for Model (18.5.5) [not given] actually has a large P value for $d_1 * a^2$ but that provides a test for dropping $d_1 * a^2$ out of a model that still contains the cubic terms, so the result is irrelevant.

Incorporating these conclusions, the table of coefficients for fitting

$$y_r = \beta_0 + \sum_{j=2}^{12} \beta_{0j} B_{rj} + \delta_{30} d_{r3} + \delta_{10} d_{r1} + \delta_{11}(d_{r1} a_r) + \delta_{21}(d_{r2} a_r) + \delta_{12}(d_{r1} a_r^2) + \varepsilon_r \quad (18.5.6)$$

is given as Table 18.7. Alternatively, we present the fitted model as

$$\hat{m}(i,k,g,a) = \begin{cases} 18.0796 + 10.4222 = 28.5018 & \text{for Block 1, Control} \\ 18.0796 + \hat{\beta}_{0i} + 10.4222 & \text{for Block } i \neq 1, \text{Control} \\ 18.0796 - 12.5167 + 7.4500a - 1.0278a^2 & \text{for Block 1, Det. I} \\ 18.0796 + \hat{\beta}_{0i} - 12.5167 + 7.4500a - 1.0278a^2 & \text{for Block } i \neq 1, \text{Det. I} \\ 18.0796 + 2.0222a & \text{for Block 1, Det. II} \\ 18.0796 + \hat{\beta}_{0i} + 2.0222a & \text{for Block } i \neq 1, \text{Det. II.} \end{cases}$$

Table 18.7: *Table of Coefficients: Model (18.5.6).*

Predictor	$\hat{\mu}_k$	SE($\hat{\mu}_k$)	t	P
Constant	18.0796	0.695241	26.0049	0.000
blk				
2	1.0556	0.790605	1.3351	0.198
3	0.4389	0.793114	0.5534	0.586
4	0.8907	0.750211	1.1873	0.250
5	1.1407	0.731441	1.5596	0.135
6	2.1296	0.747558	2.8488	0.010
7	1.8963	0.758112	2.5013	0.022
8	0.4741	0.731441	0.6481	0.525
9	1.4574	0.757238	1.9246	0.069
10	1.3778	0.749327	1.8387	0.082
11	0.5278	0.741332	0.7119	0.485
12	0.9222	0.745784	1.2366	0.231
d_3	10.4222	0.636575	16.3723	0.000
d_1	−12.5167	0.627143	−19.9582	0.000
$d_1 * a$	7.4500	0.764202	9.7487	0.000
$d_2 * a$	2.0222	0.218343	9.2617	0.000
$d_1 * a^2$	−1.0278	0.244115	−4.2102	0.000

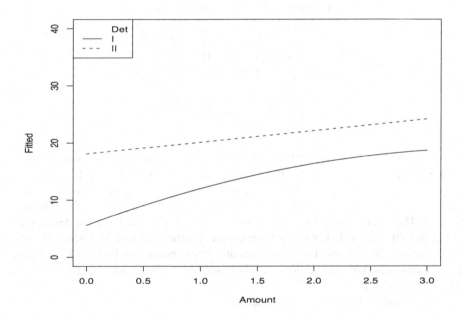

Figure 18.3 *Predicted dishes washed for two detergents as a function of the amounts of an additive in Block 1.*

Figure 18.3 displays the results for Block 1 and detergents I and II.

From inspection of Figure 18.3, suds last longer when there is more additive (up to a triple dose). Detergent II works uniformly better than detergent I. The effect of a dose of the additive is greater at low levels for detergent I than at high levels but the effect of a dose is steady for detergent II. The control is easily better than any of the new treatments with $\hat{m}(1,2,3,a) = 28.5018$.

18.6 Extensions of Latin squares

Section 17.5 discussed Latin square designs and mentioned that an effective experimental design often requires the use of several small Latin squares. We now present an example of such a design. The example does not actually involve factorial treatment structures but it uses many related ideas.

Table 18.8: *Milk production data.*

Period	Cow 1	2	3	Cow 4	5	6
1	A 768	B 662	C 731	A 669	B 459	C 624
2	B 600	C 515	A 680	C 550	A 409	B 462
3	C 411	A 506	B 525	B 416	C 222	A 426

Period	Cow 7	8	9	Cow 10	11	12
1	A 1091	B 1234	C 1300	A 1105	B 891	C 859
2	B 798	C 902	A 1297	C 712	A 830	B 617
3	C 534	A 869	B 962	B 453	C 629	A 597

Period	Cow 12	14	15	Cow 16	17	18
1	A 941	B 794	C 779	A 933	B 724	C 749
2	B 718	C 603	A 718	C 658	A 649	B 594
3	C 548	A 613	B 515	B 576	C 496	A 612

Table 18.9: *Analysis of Variance: Model (18.6.1).*

Source	df	SS
Latin squares	5	1392534
Cows(Squares)	12	318241
Periods(Squares)	12	872013
Trts	2	121147
Error	22	52770
Total	53	2756704

EXAMPLE 18.6.1.   Patterson (1950) and John (1971) considered the milk production of cows that were given three different diets. The three feed regimens were A, good hay; B, poor hay; and C, straw. Eighteen cows were used and milk production was measured during three time periods for each cow. Each cow received a different diet during each time period. The data are given in Table 18.8. The cows were divided into six groups of 3. A 3×3 Latin square design was used for each group of three cows along with the three periods and the three feed treatments. Having eighteen cows, we get 6 Latin squares. The six squares are clearly marked in Table 18.8 by double vertical and horizontal lines. We will not do a complete analysis of these data, rather we point out salient features of the analysis.

The basic model for multiple Latin squares is

$$y_{hijk} = \mu + S_h + C_{hi} + P_{hj} + \tau_k + \varepsilon_{hijk}, \qquad (18.6.1)$$

where S indicates the 6 Square effects, C and P indicate 3 Cow and 3 Period effects within a Latin square, but the effects change between Latin squares (2 degrees of freedom per square times 6 squares), and τ indicates 3 treatment effects that do not change between Latin squares. The analysis of variance table is presented in Table 18.9. In general, all the ANOVA tables should be obtained by fitting a sequence of hierarchical models where the terms are added to the sequence in the same order that we have placed them in the model. These data are balanced, which makes the order of fitting less important.

So far we have acted as though the model presumes that the columns are different in every Latin square, as are the rows. This is true for the columns, no cow is ever used in more than one square. It is less clear whether, say, period 1 is the same in the first Latin square as it is in the second and other squares. We will return to this issue later. It is clear, however, that the treatments are the same in every Latin square.

Table 18.10: *Analysis of Variance: Model (18.6.2)*.

Source	df	SS
Squares	5	1392534
Cows(Squares)	12	318241
Periods	2	814222
Period*Square	10	57790
Trts	2	121147
Error	22	52770
Total	53	2756704

From Table 18.9, mean squares and F statistics are easily obtained. If this was a classic application of multiple Latin squares, the only F test of real interest would be that for treatments, since the other lines of Table 18.9 denote various forms of blocking. The F statistic for treatments is about 25, so, with 22 degrees of freedom for Error, the test is highly significant. One should then compare the three treatments using contrasts and check the validity of the assumptions using residual plots.

The basic model (18.6.1) and analysis of variance Table 18.9 can be modified in many ways. We now present some of those ways.

As a standard practice, John (1971, Section 6.5) includes a square-by-treatment interaction to examine whether the treatments behave the same in the various Latin squares,

$$y_{hijk} = \mu + S_h + C_{hi} + P_{hj} + \tau_k + (S\tau)_{hk} + \varepsilon_{hijk}.$$

In our example with 6 squares and 3 treatments such a term would typically have $(6-1) \times (3-1) = 10$ degrees of freedom.

We mentioned earlier that periods might be considered the same from square to square. If so, we should fit

$$y_{hijk} = \mu + S_h + C_{hi} + P_j + (SP)_{hj} + \tau_k + \varepsilon_{hijk}. \tag{18.6.2}$$

We will want to test this against the no-interaction model to examine whether the periods behave the same from square to square. The analysis of variance table incorporating this change is presented as Table 18.10. Our current data are balanced but for unbalanced data one could debate whether the appropriate test for square-by-period interaction should be conducted before or after fitting treatments. I would always fit treatments after everything that involves blocks.

If the Latin squares were constructed using the complete randomization discussed in Section 17.5, one could argue that the period-by-squares interaction must really be error and that the 10 degrees of freedom and corresponding sum of squares should be pooled with the current error. Such an analysis is equivalent to simply thinking of the design as one large rectangle with three terms to consider: the 3 periods (rows), the 18 cows (columns), and the 3 treatments. For this design,

$$y_{hijk} = \mu + C_{hi} + P_j + \tau_k + \varepsilon_{hijk}. \tag{18.6.3}$$

Such an analysis is illustrated in Table 18.11. The sum of squares for Cows in Table 18.11 equals the sum of squares for Cows within Squares plus the sum of squares for Squares from the earlier ANOVA tables. The 17 degrees of freedom for Cows are also the 12 degrees of freedom for cows within squares plus the 5 degrees of freedom for Squares.

In this example, choosing between the analyses of Tables 18.10 and 18.11 is easy because of additional structure in the design that we have not yet considered. This particular design was chosen because consuming a particular diet during one period might have an effect that carries over into the next time period. In the three Latin squares on the left of Table 18.8, treatment A is always followed by treatment B, treatment B is always followed by treatment C, and treatment C is always followed by treatment A. In the three Latin squares on the right of Table 18.8, treatment A is always followed by treatment C, treatment B is followed by treatment A, and treatment C is followed by treatment B.

Table 18.11: *Analysis of Variance: Model(18.6.3).*

Source	df	SS
Cows	17	1710775
Periods	2	814222
Trts	2	121147
Error	32	110560
Total	53	2756704

This is referred to as a *cross-over* or *change-over* design. Since there are systematic changes in the squares, it is reasonable to investigate whether the period effects differ from square to square and so we should use Table 18.10. In particular, we would like to isolate 2 degrees of freedom from the period-by-square interaction to look at whether the period effects differ between the three squares on the left as compared to the three squares on the right. To do this, we replace the Squares subscript $h = 1, \ldots, 6$ with two subscripts: $f = 1, 2$ and $g = 1, 2, 3$ where f identifies right and left squares. We then fit the model

$$y_{fgijk} = \mu + C_{fgi} + P_j + (SdP)_{fj} + \tau_k + \varepsilon_{fgijk}$$

where $(SdP)_{fj}$ is a side-by-period interaction. When the data are balanced, we don't need to worry about whether to fit this interaction before or after treatments. These issues are addressed in Exercise 18.7.6. $\qquad\qquad\square$

18.7 Exercises

EXERCISE 18.7.1.   The process condition treatments in Exercise 17.11.3 on vinyl thickness had factorial treatment structure. Give the factorial analysis of variance table for the data. The data are repeated below.

Rate	High	Low	Low	High
Temp	Low	High	Low	High
Rep. 1	7.8	11.0	7.4	11.0
Rep. 2	7.6	8.8	7.0	9.2

EXERCISE 18.7.2.   Garner (1956) presented data on the tensile strength of fabrics as measured with Scott testing machines. The experimental procedure involved selecting eight 4×100-inch strips from available stocks of uniform twill, type I. Each strip was divided into sixteen 4×6 inch samples (with some left over). Each of three operators selected four samples at random and, assigning each sample to one of four machines, tested the samples. The four extra samples from each strip were held in reserve in case difficulties arose in the examination of any of the original samples. It was considered that each 4×100 inch strip constituted a relatively homogeneous batch of material. Effects due to operators include differences in the details of preparation of samples for testing and mannerisms of testing. Machine differences include differences in component parts, calibration, and speed. The data are presented in Table 18.12. Entries in Table 18.12 are values of the strengths in excess of 180 pounds.

(a) Identify the design for this experiment and give an appropriate model. List all the assumptions made in the model.

(b) Analyze the data. Give an appropriate analysis of variance table.

(c) Check the assumptions of the model and adjust the analysis appropriately.

EXERCISE 18.7.3.   Consider the milk production data in Table 18.8 and the corresponding analysis of variance in Table 18.9. Relative to the periods, the squares on the left of Table 18.8 always

Table 18.12: *Tensile strength of uniform twill.*

	o_1				o_2				o_3			
	m_1	m_2	m_3	m_4	m_1	m_2	m_3	m_4	m_1	m_2	m_3	m_4
s_1	18	7	5	9	12	16	15	9	18	13	10	22
s_2	9	11	12	3	16	4	21	19	25	13	19	12
s_3	7	11	11	1	7	14	12	6	17	20	19	20
s_4	6	4	10	8	15	10	16	12	10	16	12	18
s_5	10	8	6	10	17	12	12	22	18	16	21	22
s_6	7	12	3	15	18	22	14	19	18	23	22	14
s_7	13	5	15	16	14	18	18	9	16	16	10	15
s_8	1	11	8	12	7	13	11	13	15	14	14	11
o = operator, m = machine, s = strip												

Table 18.13: *Hydrostatic pressure tests: Operator, yield.*

Square I				Square II			
C	D	A	B	D	C	B	A
41.0	38.5	39.0	43.0	43.0	40.5	43.5	39.5
D	C	B	A	C	D	A	B
41.0	38.5	41.5	41.0	41.0	39.0	39.5	41.5
A	B	C	D	B	A	D	C
39.5	42.0	41.5	42.0	42.0	41.0	40.5	37.5
B	A	D	C	A	B	C	D
41.5	41.0	40.5	41.5	40.5	42.5	44.0	41.0
Operators are A, B, C, D.							

have treatment A followed by B, B followed by C, and C followed by A. The squares on the right always have treatment A followed by C, B followed by A, and C followed by B. Test whether there is an interaction between periods and left–right square differences.

EXERCISE 18.7.4.   As in Exercise 17.11.7, we consider differences in hydrostatic pressure tests due to operators. Table 18.13 contains two Latin squares. Analyzing these together, give an appropriate analysis of variance table and report on any differences that can be established among the operators.

EXERCISE 18.7.5.   Exercises 17.11.7, 17.11.8, 17.11.9, and the previous exercise used subsets of data reported in Garner (1956). The experiment was designed to examine differences among operators and machines when using Suter hydrostatic pressure-testing machines. No interaction between machines and operators was expected.

A one-foot square of cloth was placed in a machine. Water pressure was applied using a lever until the operator observed three droplets of water penetrating the cloth. The pressure was then relieved using the same lever. The observation was the amount of water pressure consumed and it was measured as the number of inches that water rose up a cylindrical tube with radial area of 1 square inch. Operator differences are due largely to differences in their ability to spot the droplets and their reaction times in relieving the pressure. Machines 1 and 2 were operated with a hand lever. Machines 3 and 4 were operated with at foot lever.

A 52×200-inch strip of water-repellant cotton Oxford was available for the experiment. From this, four 48×48-inch squares were cut successively along the warp (length) of the fabric. It was decided to adjust for heterogeneity in the application of the water repellant along the warp and fill (width) of the fabric, so each 48×48 square was divided into four equal parts along the warp and four equal parts along the fill, yielding 16 smaller squares. The design involves four replications of a Graeco-Latin square. In each 48×48 square, every operator worked once with every row and

Table 18.14: *Hydrostatic pressure tests: Operator, machine, yield.*

Square 1				Square 2			
A,1	B,3	C,4	D,2	B,2	A,4	D,3	C,1
40.0	43.5	39.0	44.0	39.0	39.0	41.0	41.0
B,2	A,4	D,3	C,1	A,1	B,3	C,4	D,2
40.0	42.0	40.5	38.0	36.5	42.5	40.5	38.5
C,3	D,1	A,2	B,4	D,4	C,2	B,1	A,3
42.0	40.5	38.0	40.0	40.0	39.0	41.5	41.5
D,4	C,2	B,1	A,3	C,3	D,1	A,2	B,4
40.0	36.5	39.0	38.5	41.5	39.5	39.0	44.0

Square 3				Square 4			
C,3	D,1	A,2	B,4	D,4	C,2	B,1	A,3
41.0	38.5	39.0	43.0	43.0	40.5	43.5	39.5
D,4	C,2	B,1	A,3	C,3	D,1	A,2	B,4
41.0	38.5	41.5	41.0	41.0	39.0	39.5	41.5
A,1	B,3	C,4	D,2	B,2	A,4	D,3	C,1
39.5	42.0	41.5	42.0	42.0	41.0	40.5	37.5
B,2	A,4	D,3	C,1	A,1	B,3	C,4	D,2
41.5	41.0	40.5	41.5	40.5	42.5	44.0	41.0

Operators are A, B, C, D. Machines are 1, 2, 3, 4.

column of the larger square and once with every machine. Similarly, every row and column of the 48×48 square was used only once on each machine. The data are given in Table 18.14.

Analyze the data. Give an appropriate analysis of variance table. Give a model and check your assumptions. Use the Bonferonni method to determine differences among operators and to determine differences among machines.

The cuts along the warp of the fabric were apparently the rows. Should the rows be considered the same from square to square? How would doing this affect the analysis?

Look at the means for each square. Is there any evidence of a trend in the water repellency as we move along the warp of the fabric? How should this be tested?

EXERCISE 18.7.6.   Consider the milk production data in Table 18.8 and the corresponding analysis of variance in Table 18.10. Relative to the periods, the squares on the left of Table 11.8 always have treatment A followed by B, B followed by C, and C followed by A. The squares on the right always have treatment A followed by C, B followed by A, and C followed by B. Test whether there is an average difference between the squares on the left and those on the right. Test whether there is an interaction between periods and left–right square differences.

Chapter 19

Dependent Data

In this chapter we examine methods for performing analysis of variance on data that are not completely independent. The two methods considered are appropriate for similar data but they are based on different assumptions. All the data involved have independent groups of observations but the observations within groups are not independent. In terms of analyzing unbalanced data, both of these procedures easily handle unbalanced groups of observations but the statistical theory breaks down when the observations within the groups become unbalanced. The first method was developed for analyzing the results of *split-plot designs*. The corresponding models involve constant variance for all observations and the lack of independence consists of a constant correlation between observations within each group. The second method is *multivariate analysis of variance*. Multivariate ANOVA allows an arbitrary variance and correlation structure among the observations within groups but assumes that the same structure applies for every group. These are two extremes in terms of modeling dependence among observations in a group and many useful models can be fitted that have other interesting variance-correlation structures, cf. Christensen et al. (2010, Section 10.3). However, the two variance-correlation structures considered here are the most amenable to further statistical analysis.

Section 19.1 introduces unbalanced split-plot models and illustrates the highlights of the analysis. Section 19.2 gives a detailed analysis for a complicated balanced split-plot model using methods that are applicable to unbalanced groups. Subsection 19.2.1 even discusses methods that apply for unbalanced observations within the groups but such imbalance requires us to abandon comparisons between groups. Section 19.3 introduces multivariate analysis of variance. Section 19.4 considers some special cases of the model examined in Sections 19.1 and 19.2; these are *subsampling models* and one-way analysis of variance models in which group effects are random.

19.1 The analysis of split-plot designs

Split-plot designs involve simultaneous use of different sized experimental units. The corresponding models involve more than one error term. According to Casella (2008, p.171), "Split-plot experiments are the workhorse of statistical design. There is a saying that if the only tool you own is a hammer, then everything in the world looks like a nail. It might be fair to say that, from now on, almost every design that you see will be some sort of split plot."

Suppose we produce an agricultural commodity and are interested in the effects of two factors: an insecticide and a fertilizer. The fertilizer is applied using a tractor and the insecticide is applied via crop dusting. The method of applying the chemicals is part of the treatment. Crop dusting involves using an airplane to spread the material. Obviously, you need a fairly large piece of land for crop dusting, so the number of replications on the crop-dusted treatments will be relatively few. On the other hand, different fertilizers can be applied with a tractor to reasonably small pieces of land, so we can obtain more replications. If our primary interest is in the main effects of the crop-dusted insecticides, we are stuck. Accurate results require a substantial number of large fields to obtain replications on the crop-dusting treatments. However, if our primary interest is in the fertilizers or the interaction between fertilizers and insecticides, we can design a good experiment using only a few large fields.

To construct a split-plot design, start with the large fields and design an experiment that is appropriate for examining just the insecticides. Depending on the information available about the fields, this can be a CRD, a RCB design, a Latin square design, or pretty much any design you think is appropriate. Suppose there are three levels of insecticide to be investigated. If we have three fields in the Gallatin Valley of Montana, three fields near Willmar, Minnesota, and three fields along the Rio Grande River in New Mexico, it is appropriate to set up a block in each state so that we see each insecticide in each location. Alternatively, if we have one field near Bozeman, MT, one near Cedar City, UT, one near Twin Peaks, WA, one near Winters, CA, one near Fields, OR, and one near Grand Marais, MN, a CRD seems more appropriate. We need a valid design for this experiment on insecticides, but often it will not have enough replications to yield a very precise analysis. Each of the large fields used for insecticides is called a *whole plot*. The insecticides are randomly applied to the whole plots, so they are referred to as the *whole-plot treatments*. Any complete blocks used in the whole-plot design are typically called "Replications" or just "Reps."

Regardless of the design for the insecticides, the key to a split-plot design is using each whole plot (large field) as a block for examining the *subplot treatments* (fertilizers). If we have four fertilizer treatments, we divide each whole plot into four subplots. The fertilizers are randomly assigned to the subplots. The analysis for the *subplot treatments* is just a modification of the RCB analysis with each whole plot treated as a block.

We have a much more accurate experiment for fertilizers than for insecticides. If, as alluded to earlier, the insecticide (whole-plot) experiment was set up with 3 blocks (MT, MN, NM) each containing 3 whole plots, we have just 3 replications on each insecticide, but each of the 9 whole plots is a block for the fertilizers, so we have 9 replications of the fertilizers. Moreover, fertilizers are compared within whole plots, so they are not subject to the whole-plot-to-whole-plot variation.

Perhaps the most important aspect of the design is the interaction. It is easy to set up a mediocre design for insecticides and a good experiment for fertilizers; the difficulty is in getting to look at them together and the primary point in looking at them together is to investigate interaction. The most important single fact in the analysis is that the interaction between insecticides and fertilizers is subject to exactly the same variability as fertilizer comparisons. Thus we have eliminated a major source of variation, the whole-plot-to-whole-plot variability. Interaction effects are only subject to the subplot variability, i.e., the variability *within* whole plots.

The basic idea behind split-plot designs is very general. The key idea is that an observational unit (whole plot, large field) is broken up to allow several distinct measurements on the unit. These are often called *repeated measures*. In an example in the next section, the weight loss due to abrasion of *one* piece of fabric is measured after 1000, 2000, and 3000 revolutions of a machine designed to cause abrasion. Another possibility is giving drugs to people and measuring their heart rates after 10, 20, and 30 minutes. When repeated measurements are made on the same observational unit, these measurements are more likely to be similar than measurements taken on different observational units. Thus the measurements on the same unit are correlated. This correlation needs to be modeled in the analysis. Note, however, that with the weight loss and heart rate examples, the "treatments" (rotations, minutes) cannot be randomly assigned to the units. In such cases the variance-correlation structure of a split-plot model may be less appropriate than that of the multivariate ANOVA model or various other models. In terms of balance, the methods of analysis presented hold if we lose all data on an observational unit (piece of fabric, person) but break down if we lose some but not all of the information on a unit.

We now consider an example of a simple split-plot design. Section 19.2 presents a second example that considers the detailed analysis of a study with four factors.

EXAMPLE 19.1.1.   Garner (1956) and Christensen (1996, Section 12.1) present data on the amount of moisture absorbed by water-repellant cotton Oxford material. Two 24-yard strips of cloth were obtained. Each strip is a replication and was divided into four 6-yard strips. The 6-yard strips

Table 19.1: *Garner's dynamic absorption data.*

| Laundry | Rep 1 | | | | | Rep 2 | | | | |
| | Test | | | | | Test | | | | |
	A	B	C	D	$\bar{y}_{1j\cdot}$	A	B	C	D	$\bar{y}_{2j\cdot}$
1	7.20	11.70	15.12	8.10	10.5300	9.06	11.79	14.38	8.12	10.8375
2	2.40	7.76	6.13	2.64	4.7325	2.14	7.76	6.89	3.17	4.9900
3	2.19	4.92	5.34	2.47	3.7300					
4	1.22	2.62	5.50	2.74	3.0200	2.43	3.90	5.27	2.31	3.4775

were randomly assigned to one of four laundries. After laundering and drying, the 6-yard strips were further divided into four 1.5-yard strips and randomly assigned to one of four laboratories for determination of dynamic water absorption. The data presented in Table 19.1 are actually the means of two determinations of dynamic absorption made for each 1.5-yard strip. The label "Test" is used to identify different laboratories (out of fear that the words laundry and laboratory might get confused). To illustrate the analysis of unbalanced data we have removed the data for Laundry 3 from the second replication.

First consider how the experimental design deals with laundries. There are two blocks (Reps) of material available, the 24-yard strips. These are subdivided into four sections and randomly assigned to laundries. Thus we have a randomized complete block (RCB) design for laundries with two blocks and four treatments from which we are missing the information on the third treatment in the second block. The 6-yard strips are the whole-plot experimental units, laundries are whole-plot treatments, and the 24-yard strips are whole-plot blocks.

After the 6-yard strips have been laundered, they are further subdivided into 1.5-yard strips and these are randomly assigned to laboratories for testing. In other words, each experimental unit in the whole-plot design for laundries is split into subunits for further treatment. The whole-plot experimental units (6-yard strips) serve as blocks for the subplot treatments. The 1.5-yard strips are subplot experimental units and the tests are subplot treatments.

The peculiar structure of the design leads us to analyze the data almost as two separate experiments. There is a whole-plot analysis focusing on laundries and a subplot analysis focusing on tests. The subplot analysis also allows us to investigate interaction.

Consider the effects of the laundries. The analysis for laundries is called the *whole-plot analysis*. We have a block design for laundries but a block analysis requires just one number for each laundry observation (whole plot). The one number used for each whole plot is the mean absorption averaged over the four subplots (tests) contained in the whole plot. These 7 mean values are reported in Table 19.1. With two reps, four treatments, and a missing whole plot we get

Whole-plot ANOVA for laundries using subplot means.

Source	df	Seq. SS	MS	F	P
Reps	1	1.489	1.488	274.83	0.004
Laundry (after Reps)	3	65.379	21.793	4023.34	0.000
Error 1	2	0.011	0.005		
Total	6	66.879			

As usual, we fit (whole-plot) treatments after reps (whole-plot blocks). With one minor exception, this provides the whole-plot analysis section of a split-plot model ANOVA table. The degrees of freedom are Reps, 1; Laundry, 3; and the *whole-plot error*, Error 1, with 2 *df*. The minor exception is that when we present the combined split-plot model ANOVA in Table 19.2, the sums of squares and mean squares presented here are all multiplied by the number of subplot treatments, four. This multiplication has no effect on significance tests, e.g., in an *F* test the numerator mean square and the denominator mean square are both multiplied by the same number, so the multiplications cancel. Multiplying these mean squares and sums of squares by the number of subplot treatments maintains consistency with the subplot model computations.

Now consider the analysis of the subplot treatments, i.e., the absorption tests. The *subplot analysis* is largely produced by treating each whole plot as a block. Note that we observe every subplot treatment within each whole plot, so the blocks are complete. There will be, however, one notable exception to treating the subplot analysis as an RCB analysis, i.e., the identification of interaction effects.

RCB ANOVA for tests: Whole plots as subplot blocks.

Source	df	Seq. SS	MS	F	P
Whole plots	6	267.515	44.586	31.79	0.000
Test	3	105.290	35.097	25.02	0.000
Error	18	25.246	1.403		
Total	27	398.050			

In a blocking analysis with whole plots taken as subplot blocks there are 7 whole plots, so there are 6 degrees of freedom for subplot blocks. In addition there are 3 degrees of freedom for tests, so the degrees of freedom for error are $28 - 6 - 3 - 1 = (6)(3) = 18$.

The subplot analysis differs from the standard blocking analysis in the handling of the 18 degrees of freedom for error. A standard blocking analysis takes the block-by-treatment interaction as error. This is appropriate because the extent to which treatment effects vary from block to block is an appropriate measure of error for treatment effects. However, in a split-plot design the subplot blocks are not obtained haphazardly, they have consistencies due to the whole-plot treatments. We can identify structure within the subplot-block-by-subplot-treatment interaction. Some of the block-by-treatment interaction can be ascribed to whole-plot-treatment-by-subplot-treatment interaction. In this experiment, the laundry-by-test interaction has $3 \times 3 = 9$ degrees of freedom. This is extracted from the 18 degrees of freedom for error in the subplot RCB analysis to give a *subplot error* term (Error 2) with only $18 - 9 = 9$ degrees of freedom. Finally, it is of interest to note that the 6 degrees of freedom for subplot blocks correspond to the 6 degrees of freedom in the whole-plot analysis: 1 for Reps, 3 for Laundries, and 2 for Whole Plot Error. In addition, up to round-off error, the sum of squares for subplot blocks is also the total of the sums of squares for Reps, Laundries (after fitting Reps), and Whole-plot Error (Error 1) reported earlier after multiplying those sums of squares by the number of subplot treatments.

Table 19.2 combines the whole-plot analysis and the subplot analysis into a common analysis of variance table. *Error 1 indicates the whole-plot error term and its mean square is used for inferences about laundries and Reps* (if you think it is appropriate to draw inferences about Reps). *Error 2 indicates the subplot error term and its mean square is used for inferences about tests and laundry-by-test interaction.* A subplot blocks line does not appear in the table; the whole-plot analysis replaces it. Note that for the given whole-plot design, Error 1 is computationally equivalent to a Rep $*$ Laundry interaction. Again, the sums of squares and mean squares for Reps, Laundry, and Error 1 in Table 19.2 are, up to round-off error, equal to 4 times the values given earlier in the analysis based on the 7 Rep–Laundry means.

As for comparing the "whole plots as subplot blocks" ANOVA table given earlier to Table 19.2, in the first row the whole plots degrees of freedom and sums of squares are the sums of the Reps, Laundry, and Error 1 degrees of freedom and sum of squares in Table 19.2. The Test lines in the second row are identical. The "whole plots as subplot blocks" Error term is broken into the Laundry$*$Test interaction and Error 2 degrees of freedom and sum of squares of Table 19.2. The Total lines are identical.

From Table 19.2 the Laundry $*$ Test interaction is clearly significant, so the analysis would typically focus there. On the other hand, while the interaction is statistically important, its F statistic is an order of magnitude smaller than the F statistic for tests, so the person responsible for the experiment might decide that interaction is not of practical importance. The analysis might then ignore the interaction and focus on the main effects for tests and laundries. Since I am not responsible for the experiment (only for its inclusion in this book), I will not presume to declare a highly significant interaction unimportant. Modeling the interaction will be considered in the next subsection.   □

Table 19.2: *Analysis of Variance: Dynamic absorption data.*

Source	df	Seq. SS	MS	F	P
Reps	1	5.955	5.955	275.60	0.004
Laundry	3	261.517	87.172	4036.04	0.000
Error 1	2	0.043	0.021		
Test	3	105.290	35.097	92.876	0.000
Laundry * Test	9	21.845	2.427	6.423	0.005
Error 2	9	3.401	0.378		
Total	27	398.050			

We now examine the assumptions behind this analysis. The basic split-plot model for a whole-plot design with Reps is

$$y_{ijk} = \mu + r_i + w_j + \eta_{ij} + s_k + (ws)_{jk} + \varepsilon_{ijk} \tag{19.1.1}$$

where $i = 1, \ldots, a$ indicates the replication, $j = 1, \ldots, b$ indicates the whole-plot treatment, $k = 1, \ldots, c$ indicates the subplot treatment, μ is a grand mean, r_i, w_j, and s_k indicate Rep, whole-plot treatment, and subplot treatment effects, and $(ws)_{jk}$ indicates whole-plot treatment–subplot treatment effects that allow whole-plot treatment–subplot treatment interaction in the model. Note that in our example we do not have data for all combinations of i and j but for every pair ij that we observe, we have all c of the observations indexed by k. The model has two sets of random error terms: the η_{ij}s, that are errors specific to a given whole plot, and the ε_{ijk}s, that are errors specific to a given subplot. All of the errors are assumed to be independent with the η_{ij}s distributed $N(0, \sigma_w^2)$ and the ε_{ijk}s distributed $N(0, \sigma_s^2)$.

As advertised earlier, $\text{Var}(y_{ijk}) = \sigma_w^2 + \sigma_s^2$ is the same for all observations and for $k \neq k'$ within a group, $\text{Corr}(y_{ijk}, y_{ijk'}) = \sigma_w^2/(\sigma_w^2 + \sigma_s^2)$ is constant. Observations in different groups have both η_{ij} and ε_{ijk} different, so are independent.

The trick to split-plot models is that (for theoretical reasons related to seeing every subplot treatment exactly once within every whole plot) for any model that includes whole-plot treatments, subplot treatments and their interaction, *the analysis can proceed by treating the whole-plot error terms, the η_{ij}s, as though they were standard fixed effects.* We then fit the model sequentially to give us an overall ANOVA table. That is how Table 19.2 was constructed except we relabeled the Reps*Laundry interaction as Error 1 and the usual Error as Error 2. Inferences largely follow the usual patterns. There are two exceptions. First, in any modeling of whole-plot treatments, or other whole-plot effects like replications, we need to replace our usual MSE, SSE, and dfE with the corresponding values from fitting the η_{ij}s after all the whole-plot effects. This gives Error 1. Second, in modeling subplot treatments or interactions between subplot effects and whole-plot effects, we can create models for relationships between subplot treatments (tests) for fixed whole-plot treatments (laundries) and we can contrast these for different whole-plot treatments, but the mathematics does not allow us to look at relationships between whole-plot treatments for a fixed subplot treatment. This is discussed in more detail later.

As always with models for two grouping factors, the first thing to do is check for interaction by testing the reduced model

$$y_{ijk} = \mu + r_i + w_j + \eta_{ij} + s_k + \varepsilon_{ijk}. \tag{19.1.2}$$

This test is executed as if the η_{ij} terms were fixed effects, rather than random, in both models (19.1.1) and (19.1.2), so the test is just our usual test of two linear models, i.e., is based on Error 2.

If interaction exists, we need to explore the relationships between all 16 treatments. Because of its mathematical tractability, a useful approach is to look at the relationships between subplot treatments for each fixed whole-plot treatment. Interactions involve seeing whether such relationships change from whole-plot treatment to whole-plot treatment. For example, if the subplot treatments have quantitative levels, we can fit a polynomial for each whole-plot treatment. Any comparisons among polynomial coefficients other than the intercepts are subject to the subplot Error 2. Comparisons among the intercepts are subject to the whole-plot Error 1. These issues are treated in detail in Subsection 19.1.1.

If we decide that interaction is not important, we can look at relationships between the subplot treatments, which will be performed in the usual way treating the η_{ij}s as fixed. Indeed, this amounts to analyzing the RCB with whole plots as subplot blocks but using Error 2 rather than the full subplot-block-by-subplot-treatment interaction as the error term. Moreover, we can examine the whole-plot treatments (and other whole-plot effects) in the usual way but using results for the whole-plot Error 1 rather than the usual Error 2. Alternatively, the entire analysis for the whole-plot treatments could be obtained from a reanalysis of the $\bar{y}_{ij}$.s. In other words, to examine the relationships among whole-plot treatments (something that is of little interest if we believe that interaction exists), we simply replace the usual MSE with the mean square for fitting the η_{ij} whole-plot error terms after fitting any fixed effects that occur in the whole-plot analysis. In this example it is the mean square for fitting the η_{ij}s after fitting the Reps and the whole-plot treatments.

The model assumes normality and equal variances for each set of error terms. These assumptions should be checked using residual plots. We can get Error 2 residuals as in the usual way by treating the η_{ij}s as fixed. We can get Error 1 residuals from doing the whole-plot analysis on the $\bar{y}_{ij}$.s.

The $MSE(1)$ turns out to be an estimate of

$$E[MSE(1)] = \sigma_s^2 + c\,\sigma_w^2$$

and $MSE(2)$ is an estimate of

$$E[MSE(2)] = \sigma_s^2.$$

If there is no whole-plot-to-whole-plot variability over and above the variability due to the subplots within the whole plots, i.e., if $\sigma_w^2 = 0$, then the two error terms are estimating the same thing and their ratio has an F distribution. In other words, we can test $H_0 : \sigma_w^2 = 0$ by rejecting H_0 when

$$MSE(1)/MSE(2) > F(1 - \alpha, dfE(1), dfE(2)).$$

In the laundries example we get $0.011/0.378 = 0.029$ on 2 and 9 degrees of freedom and a P value of 0.97, which may be suspiciously large. This is rather like testing for blocks in a randomized complete block design. Both tests merely tell you if you wasted your time. An insignificant test for blocks indicates that blocking was a waste of time. Similarly, an insignificant test for whole-plot variability indicates that forming a split-plot design was a waste of time. In each case, it is too late to do anything about it. The analysis should follow the design that was actually used. However, the information may be of value in designing future studies.

EXAMPLE 19.1.1 CONTINUED.

Figure 19.1 contains a series of Error 1 residual plots. These were obtained from averaging Model (19.1.1) over the subplot treatments and fitting

$$\begin{aligned}
\bar{y}_{ij\cdot} &= \mu + r_i + w_j + \eta_{ij} + \bar{s}_\cdot + \overline{(ws)}_{j\cdot} + \bar{\varepsilon}_{ij\cdot} \\
&= \mu + r_i + w_j + \eta_{ij} + \bar{\varepsilon}_{ij\cdot},
\end{aligned}$$

where we drop $\bar{s}_\cdot$ and $\overline{(ws)}_{j\cdot}$ because they are indistinguishable from the μ and w_j effects. The top left panel of Figure 19.1 contains a normal plot for the Error 1 residuals; it looks reasonably straight. The top right panel has the Error 1 residuals versus predicted values. Note the wider spread for predicted values near 3. The bottom left panel plots the Error 1 residuals against Reps and shows nothing startling. The bottom right panel is a plot against Laundries in which we see that the spread for Laundry 4 is much wider than the spread for the other laundries. This seems to be worth discussing with the experimenter. Of course there are only 6 residuals (with only 2 degrees of freedom), so it is difficult to draw any firm conclusions.

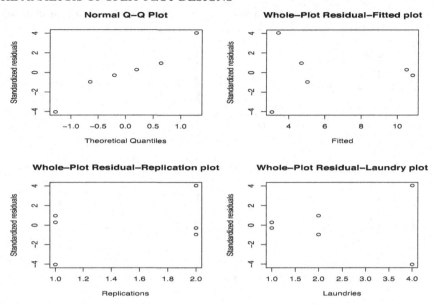

Figure 19.1 *Normal plot of whole-plot residuals. Whole-plot residuals versus predicted values, replications, and laundries. Absorption data.*

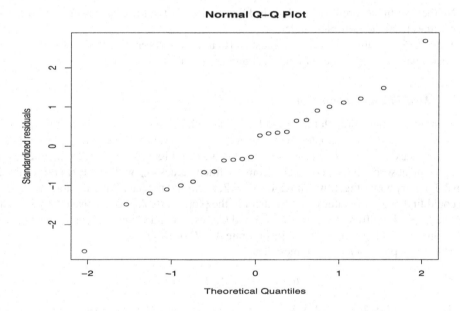

Figure 19.2: *Normal plot of subplot residuals, absorption data.*

There are seven whole plots, so why are there only six whole-plot residuals in the graphs? Rep 2, Laundry 3 is missing, so the only observation on Laundry 3 is that from Rep 1, Laundry 3. It follows that Rep 1, Laundry 3 has a leverage of one and the fitted value always equals the data point. There is little value to a residual that the model forces to be zero.

Figures 19.2 and 19.3 contain a series of Error 2 residual plots obtained from Model (19.1.1) fitted with η fixed. Figure 19.2 contains the normal plot; it looks alright. The top left panel of Figure 19.3 plots Error 2 residuals versus predicted values (treating the ηs as fixed). The other panels are plots against Reps, Laundries, and Tests. There is nothing startling. □

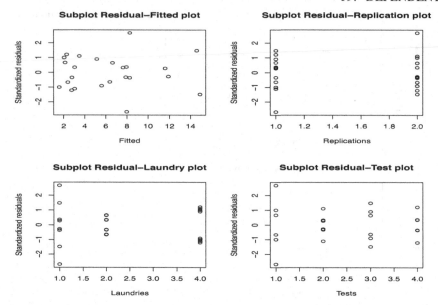

Figure 19.3: *Subplot residuals versus predicted values, replications, laundries, and tests. Absorption data.*

Neither the fitted (predicted) values in Figure 19.1 nor the fitted values in Figure 19.3 are the fitted values from the split-plot model.

Table 19.3 contains the usual diagnostics (treating the ηs as fixed). All the observations on Laundry 3 have leverage 1 because Rep 2, Laundry 3 is missing.

19.1.1 Modeling with interaction

To construct an interaction plot for Model (19.1.1) with unbalanced data, we plot the pairs $(k, \hat{y}_{ijk})$ where $k = 1, \ldots, c$ is the subplot treatment index. We do this for every value of the whole-plot treatment index $j = 1, \ldots, b$ but for some fixed value of i. Figure 19.4 gives the interaction plot for the dynamic absorption data computed from the fitted values $\hat{y}_{ijk}$ with $i = 1$. (In our example with Rep 2, Laundry 3 missing, if we chose to fix $i = 2$, we would not have "fitted" values when $j = 3$ but we could find predicted values and use them in the plot.) Tests A and D behave very similarly and they behave quite differently from Tests B and C. Tests B and C also behave somewhat similarly. This suggests looking at models that incorporate $A = D$ or $B = C$.

The basic split-plot model for these data,

$$y_{ijk} = \mu + r_i + w_j + \eta_{ij} + s_k + (ws)_{jk} + \varepsilon_{ijk},$$

$i = 1, 2$, $j = 1, 2, 3, 4$, $k = 1, 2, 3, 4$, $(i, j) \neq (2, 3)$ can conveniently be switched to a model that recasts the pairs of numbers jk (that identify all treatment combinations) as integers $h = 1, \ldots, 16$. The equivalent model is then

$$y_{ih} = \mu + r_i + w_j + \eta_{ij} + (ws)_h + \varepsilon_{ih}, \qquad j = \mathrm{mod}_4(h-1) + 1. \tag{19.1.3}$$

(The $\mathrm{mod}_a(b)$ function for integers a and b is the remainder when b is divided by a, thus $\mathrm{mod}_4(7) = 3$.) The data are presented again in Table 19.4 in a form suitable for modeling. Rep, Laundry, and Test correspond to i, j, k while "inter" corresponds to h.

We fit two hierarchies of four models for each Laundry. First, we successively fit models that incorporate $A = D$; $A = D$ and $B = C$; $A = D = B = C$ for one laundry at a time. The other hierarchy successively incorporates $B = C$; $A = D$ and $B = C$; $A = D = B = C$. In these two hierarchies, only

Table 19.3: *Diagnostics for dynamic absorption data.*

Rep	Laundry	Test	y	$\hat{y}$	leverage	r	t	C
1	1	1	7.20	7.97625	0.625	−2.062	−2.677	0.373
1	2	1	2.40	2.14125	0.625	0.687	0.666	0.041
1	3	1	2.19	2.19000	1.000			
1	4	1	1.22	1.59625	0.625	−1.000	−0.999	0.088
1	1	2	11.70	11.59125	0.625	0.289	0.274	0.007
1	2	2	7.76	7.63125	0.625	0.342	0.325	0.010
1	3	2	4.92	4.92000	1.000			
1	4	2	2.62	3.03125	0.625	−1.093	−1.106	0.105
1	1	3	15.12	14.59625	0.625	1.391	1.481	0.170
1	2	3	6.13	6.38125	0.625	−0.667	−0.645	0.039
1	3	3	5.34	5.34000	1.000			
1	4	3	5.50	5.15625	0.625	0.913	0.904	0.073
1	1	4	8.10	7.95625	0.625	0.382	0.363	0.013
1	2	4	2.64	2.77625	0.625	−0.362	−0.344	0.011
1	3	4	2.47	2.47000	1.000			
1	4	4	2.74	2.29625	0.625	1.179	1.209	0.122
2	1	1	9.06	8.28375	0.625	2.062	2.677	0.373
2	2	1	2.14	2.39875	0.625	−0.687	−0.666	0.041
2	4	1	2.43	2.05375	0.625	1.000	0.999	0.088
2	1	2	11.79	11.89875	0.625	−0.289	−0.274	0.007
2	2	2	7.76	7.88875	0.625	−0.342	−0.325	0.010
2	4	2	3.90	3.48875	0.625	1.093	1.106	0.105
2	1	3	14.38	14.90375	0.625	−1.391	−1.481	0.170
2	2	3	6.89	6.63875	0.625	0.667	0.645	0.039
2	4	3	5.27	5.61375	0.625	−0.913	−0.904	0.073
2	1	4	8.12	8.26375	0.625	−0.382	−0.363	0.013
2	2	4	3.17	3.03375	0.625	0.362	0.344	0.011
2	4	4	2.31	2.75375	0.625	−1.179	−1.209	0.122

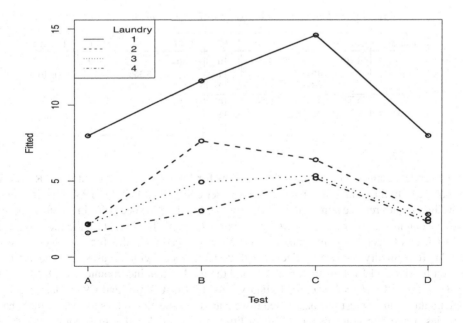

Figure 19.4: *Interaction plot for dynamic absorption data. Plot of $\hat{y}_{1jk}$ versus k for each j.*

Table 19.4: *Model fitting for dynamic absorption data.*

C1	C2	C3	C4	C5	C6	C7	C8	C9	C10
i	*j*	*k*		*h*	L1	L1	L1	L1	L:2-4
Rep	Laundry	Test	*y*	inter	A=D	B=C	both	ABCD	T:B-C
1	1	1	7.20	1	1	1	1	1	1
1	1	2	11.70	2	2	2	2	1	2
1	1	3	15.12	3	3	2	2	1	3
1	1	4	8.10	4	1	4	1	1	4
2	1	1	9.06	1	1	1	1	1	1
2	1	2	11.79	2	2	2	2	1	2
2	1	3	14.38	3	3	2	2	1	3
2	1	4	8.12	4	1	4	1	1	4
1	2	1	2.40	5	5	5	5	5	5
1	2	2	7.76	6	6	6	6	6	6
1	2	3	6.13	7	7	7	7	7	7
1	2	4	2.64	8	8	8	8	8	8
2	2	1	2.14	5	5	5	5	5	5
2	2	2	7.76	6	6	6	6	6	6
2	2	3	6.89	7	7	7	7	7	7
2	2	4	3.17	8	8	8	8	8	8
1	3	1	2.19	9	9	9	9	9	9
1	3	2	4.92	10	10	10	10	10	10
1	3	3	5.34	11	11	11	11	11	11
1	3	4	2.47	12	12	12	12	12	12
1	4	1	1.22	13	13	13	13	13	13
1	4	2	2.62	14	14	14	14	14	6
1	4	3	5.50	15	15	15	15	15	7
1	4	4	2.74	16	16	16	16	16	16
2	4	1	2.43	13	13	13	13	13	13
2	4	2	3.90	14	14	14	14	14	6
2	4	3	5.27	15	15	15	15	15	7
2	4	4	2.31	16	16	16	16	16	16

Table 19.5: *SSE and dfE for dynamic absorption hierarchical models.*

Laundry	A = D		B = C	dfE	A = D		B = C	Laundry
	3.401		12.431	10	3.804		4.963	
Laundry 1		12.431		11		5.366		Laundry 2
		65.014		12		46.726		
	3.440		3.489	10	3.891		7.916	
Laundry 3		3.528		11		8.406		Laundry 4
		11.368		12		17.630		

the first model changes. Fitting these four models for Laundry 1 involves fitting a Reps-Laundry effect along with fitting an effect for one of the other columns from Table 19.4. The first models in the two hierarchies replace *h* in Model (19.1.3) with C6 and C7, respectively, in which to incorporate $A = D$ in Laundry 1 we have replaced the index for Laundry 1 — Test D by the index for Laundry 1 — Test A and to get $B = C$ in Laundry 1 we have replaced the index for Laundry 1 — Test C by the index for Laundry 1 — Test B. The second model in each hierarchy uses C8. The last uses C9. We refer to Model (19.1.4) as the C5 model and the models that incorporate $A = D$; $B = C$; $A = D$ and $B = C$; and $A = D = B = C$ for Laundry 1 as the C6, C7, C8, and C9 models, respectively. Similar data columns must be constructed if we want to fit the hierarchies to each of the other three Laundries. Table 19.5 contains the results of fitting the 16 models that incorporate $A = D$; $B = C$; $A = D$ and $B = C$; $A = D = B = C$ for each laundry (leaving the other laundries unmodeled).

The sums of squares for various model tests are obtained by comparing the error sums of squares with each other and with that from the interaction model (19.1.1) given in Table 19.2. Consider modeling the Laundry 1 behavior. Note that, to three decimal places, the SSE is the same regardless of whether we incorporate $A = D$ for Laundry 1. The sum of squares for testing $A = D$ in Laundry 1 can be obtained by comparing the full model (19.1.1), i.e., model C5, to the C6 model. With the subplots balanced in each laundry, the same sum of squares for testing $A = D$ is obtained after fitting $B = C$ by comparing the C7 model to the C8 model. Similarly, $B = C$ can be tested by comparing models C5 and C7 or models C6 and C8. We can also test $A = D = B = C$ after fitting $A = D$ and $B = C$ by comparing the full model C8 to the reduced model C9. Again, all of these models also include a Reps-Laundry effect. These comparisons give the Laundry 1 sums of squares that follow.

| | Sums of squares for testing reduced models | | |
	$A = D$	$B = C$	$A = D = B = C$
Laundry 1	0.000	9.030	52.583
Laundry 2	0.403	1.563	41.360
Laundry 3	0.039	0.088	7.840
Laundry 4	0.490	4.516	9.224

From these sums of squares there is little evidence against $A = D$ for any laundry. For every laundry there is considerable evidence against $B = C$ except for Laundry 3, where we have half as much data. There is very considerable evidence against all four tests being equal (given that $A = D$ and $B = C$, the latter of which is unlikely to be true). Except for the Laundry 3 results, these sums of squares agree, up to round-off error, with the balanced analysis presented in Christensen (1996, Section 12.2). These sums of squares can be compared with $MSE(2) = 0.378$ from Table 19.2 to obtain tests with a null $F(1,9)$ distribution. Unfortunately, similar techniques for comparing the Laundries for each fixed Test will not lead to a test statistic that can be compared to a known F distribution. [Actually, such a test is possible by refitting the whole-plot model but eliminating the data from all Tests except the one of interest. This leads to different error terms for each fixed Test. This possibility is again mentioned in the next section.]

Finally, we can formally examine interactions. Since there is no evidence against $A = D$ for any laundry, we will not find evidence that the differential effect of A versus D changes from laundry to laundry. Let's examine the interaction effect of whether the difference between tests B and C is the same for Laundry 2 as it is for Laundry 4. We can test this by fitting a model with main effects for Laundries 2 and 4 and main effects for Tests B and C, but that allows every other combination of a laundry and test to have its own effect. Comparing this reduced model to the full model gives the desired test for interaction. However, since the model we are fitting already (implicitly) contains main effects for Laundries, the first step of fitting main effects for Laundries 2 and 4 is redundant. Fitting the model using column C10 after Rep-Laundry effects accomplishes our goal of having main effects only for Tests B and C in Laundries 2 and 4 but possible interaction for any other factor combination and gives $dfE = 10$ and $SSE = 9.096$. Comparing this to the full model (19.1.1) with $dfE = 9$ and $SSE = 3.401$, the test mean square is $9.096 - 3.401 = 5.695$ on 1 degree of freedom, which is much larger than $MSE(2) = 0.378$, so there is substantial evidence that the B-C effect changes from Laundry 2 to Laundry 4. Note that C10 is similar to the full interaction model C5 except that for Laundry 4, instead of having distinct indices for Tests B and C, C10 uses the same indices for Tests B and C as were used for them with Laundry 2.

One last note. Because the whole-plot design is allowed to be unbalanced, rather than just having one whole-plot ANOVA table as in Table 19.2, we might need to consider fitting different sequences of models for the whole plots, similar to Chapters 14 and 16. However, because the subplots are balanced within whole plots, typically there will be only one form for the subplot entries in ANOVA tables like Table 19.2.

Table 19.6: *Abrasion resistance data.*

Surf. treat.	Fill	Proportions 25% 1000	25% 2000	25% 3000	50% 1000	50% 2000	50% 3000	75% 1000	75% 2000	75% 3000
Yes	A	194	192	141	233	217	171	265	252	207
	A	208	188	165	241	222	201	269	283	191
	B	239	127	90	224	123	79	243	117	100
	B	187	105	85	243	123	110	226	125	75
No	A	155	169	151	198	187	176	235	225	166
	A	173	152	141	177	196	167	229	270	183
	B	137	82	77	129	94	78	155	76	91
	B	160	82	83	98	89	48	132	105	67

19.2 A four-factor example

We now consider a split-plot analysis involving four factors, detailed examination of three-factor interactions, and a whole-plot design that is a CRD. Christensen (1996, Section 12.2) uses contrasts to analyze these balanced data. We use model fitting ideas that apply to unbalanced data.

The illustration in this section is a split-plot analysis because it is based on split-plot models; however, the data are clearly not collected from a split-plot experimental design because the "subplot treatments" cannot be randomly assigned to subplot units. More properly, this is called a *repeated measures* design because it is an experiment in which multiple measurements were taken on the experimental units. Because the multiple measurements are similar (all involve weight loss on a piece of fabric), a split-plot model provides a viable, if not necessarily laudable, analysis. Our purpose is to illustrate such an analysis. We begin this section by introducing the data and the balanced analysis. In Subsection 19.2.1 we discuss an analysis of subplot effects that remains valid even when the subplots are unbalanced. In Subsection 19.2.2 we discuss the whole-plot analysis, that allows unbalanced whole plots but requires balanced subplots. Subsection 19.2.3 looks at the useful device of fixing the level for one factor and examining relationships among the other factors. Subsection 19.2.4 draws some final conclusions. These data are reanalyzed in Section 19.3 with the more appropriate multivariate ANOVA model.

EXAMPLE 19.2.1. In Section 16.2 we considered data from Box (1950) on fabric abrasion. The data consisted of three factors: Surface treatment (yes, no), Fill (A, B), and Proportion of fill (25%, 50%, 75%). These are referred to as **S**, **F**, and **P**, respectively. (Again, we hope no confusion occurs between the factor **F** and the use of F statistics or between the factor **P** and the use of P values!) In Section 16.2 we restricted our attention to the weight loss that occurred during the first 1000 revolutions of a machine designed for evaluating abrasion resistance, but data are also available on each piece of cloth for weight loss between 1000 and 2000 rotations and weight loss occurring between 2000 and 3000 rotations. The full data are given in Table 19.6. In analyzing the full data, many aspects are just simple extensions of the analysis given earlier in Section 16.2. There are now four factors, **S**, **F**, **P**, and one for rotations, say, **R**. With four factors, there are many more effects to deal with. There is one more main effect, **R**, three more two-factor interactions, $\mathbf{S} * \mathbf{R}$, $\mathbf{F} * \mathbf{R}$, and $\mathbf{P} * \mathbf{R}$, three more three-factor interactions, $\mathbf{S} * \mathbf{F} * \mathbf{R}$, $\mathbf{S} * \mathbf{P} * \mathbf{R}$, and $\mathbf{F} * \mathbf{P} * \mathbf{R}$, and a four-factor interaction, $\mathbf{S} * \mathbf{F} * \mathbf{P} * \mathbf{R}$.

In addition to having more factors than we have considered before, what makes these data worthy of our further attention is the fact that not all of the observations are independent. Observations on different pieces of fabric may be independent, but the three observations on the same piece of fabric, one after 1000, one after 2000, and one after 3000 revolutions, should behave similarly as compared to observations on different pieces of fabric. In other words, the three observations on one piece of fabric should display positive correlations. The analysis considered in this section assumes that the correlation is the same between any two of the three observations on a piece of fabric. To achieve this, we consider a model that includes two error terms,

$$
\begin{aligned}
y_{hijkm} = \ & \mu + s_h + f_i + p_j && (19.2.1)\\
& + (sf)_{hi} + (sp)_{hj} + (fp)_{ij} + (sfp)_{hij}\\
& + \eta_{hijk}\\
& + r_m + (sr)_{hm} + (fr)_{im} + (pr)_{jm}\\
& + (sfr)_{him} + (spr)_{hjm} + (fpr)_{ijm} + (sfpr)_{hijm}\\
& + \varepsilon_{hijkm}.
\end{aligned}
$$

$h = 1, 2, i = 1, 2, j = 1, 2, 3, k = 1, 2, m = 1, 2, 3$. The error terms are the η_{hijk}s and the ε_{hijkm}s. These are all assumed to be independent of each other with

$$
\eta_{hijk} \sim N(0, \sigma_w^2) \quad \text{and} \quad \varepsilon_{hijkm} \sim N(0, \sigma_s^2).
$$

The η_{hijk}s are error terms due to the use of a particular piece of fabric and the ε_{hijkm}s are error terms due to taking the observations after $1000, 2000$, and 3000 rotations. While we have two error terms, and thus two variances, the variances are assumed to be constant for each error term, so that all observations have the same variance, $\sigma_w^2 + \sigma_s^2$. Observations on the same piece of fabric are identically correlated because they all involve the same fabric error term η_{hijk}. Note that Model (19.2.1) could also be written in a form more similar to the previous section as

$$
y_{hijkm} = \mu + (sfp)_{hij} + \eta_{hijk} + r_m + (sfpr)_{hijm} + \varepsilon_{hijkm},
$$

where the terms $(sfp)_{hij}$ are whole-plot treatment effects.

Split plot terminology devolves from analyses on plots of ground. In this application, a whole plot is a piece of fabric. The subplots correspond to the three observations on each piece of fabric. The **S**, **F**, and **P** treatments are all applied to an entire piece of fabric, so they are referred to as whole-plot treatment factors. The three levels of rotation are "applied" within a piece of fabric and are called subplot treatments.

Our data are weight losses due to the first, second, and third 1000 rotations. The split-plot model seems at least plausible for the differences. Another possible model, one that we will not address, uses an 'autoregressive' correlation structure, cf. Christensen et al. (2010, Section 10.3.1). In Section 19.3 we will briefly consider a more general (multivariate) model that can be applied and includes both the split-plot model and the autoregressive structure as special cases. Of course when the split-plot model is appropriate, the split-plot analysis is more powerful than the general multivariate analysis.

We will concern ourselves with checking the assumptions of equal variances and normality later. We now consider the analysis of variance given in Table 19.7. Because the whole-plot model is balanced, we do not have to worry about alternative orders for fitting the terms. The rotation effects should be fitted after the whole-plot terms as should the interactions that involve rotations. Again, due to balance we do not need to consider alternative orders for fitting the interaction effects that include rotations. Just as there are two error terms in Model (19.2.1), there are two error terms in the analysis of variance table. Both error terms are used to construct tests and it is crucial to understand which error term is used for which tests. The mean square from Error 1 is the whole-plot error term and is used for any inferences that exclusively involve whole-plot treatments and their interactions. Thus, in Table 19.7, the $MSE(1)$ from the Error 1 line is used for all inferences relating exclusively to the whole-plot treatment factors **S**, **F**, and **P**. This includes examination of interactions. The Error 2 line that yields $MSE(2)$ is used for all inferences involving the subplot treatments. This includes all effects involving **R**: the main effects and all interactions.

Because the data are balanced, we have a unique ANOVA in Table 19.7 on which we could base the entire analysis, cf. Christensen (1996, Section 12.2). As always, the analysis proceeds from

Table 19.7: *Analysis of Variance.*

Source	df	SS	MS	F	P
S	1	24494.2	24494.2	78.58	0.000
F	1	107802.7	107802.7	345.86	0.000
P	2	13570.4	6785.2	21.77	0.000
S $*$ F	1	1682.0	1682.0	5.40	0.039
S $*$ P	2	795.0	397.5	1.28	0.315
F $*$ P	2	9884.7	4942.3	15.86	0.000
S $*$ F $*$ P	2	299.3	149.6	0.48	0.630
Error 1	12	3740.3	311.7		
R	2	60958.5	30479.3	160.68	0.000
S $*$ R	2	8248.0	4124.0	21.74	0.000
F $*$ R	2	18287.7	9143.8	48.20	0.000
P $*$ R	4	1762.8	440.7	2.32	0.086
S $*$ F $*$ R	2	2328.1	1164.0	6.14	0.007
S $*$ P $*$ R	4	686.0	171.5	0.90	0.477
F $*$ P $*$ R	4	1415.6	353.9	1.87	0.149
S $*$ F $*$ P $*$ R	4	465.9	116.5	0.61	0.657
Error 2	24	4552.7	189.7		
Total	71	260973.9			

examining the highest-order interactions down to the two-factor interactions and main effects. From Table 19.7 we see that the four-factor interaction has a test statistic of 0.61 and a very large P value, 0.657. We will see that the same results arise from methods for unbalanced data. Even if all of the four-factor interaction sum of squares was ascribed to one degree of freedom, an unadjusted F test would not be significant. There is no evidence for a four-factor interaction.

The next step is to consider three-factor interactions. There is one three-factor interaction in the whole plots and three of them in the subplots. We need different methods to evaluate these. Nonetheless, in this balanced case we easily see that the only clearly important three-factor interaction is $S * F * R$ whereas the $F * P * R$ interaction, with 4 degrees of freedom, has a P value that is small enough that we might want to investigate whether some interesting, interpretable interaction effect is being hidden by the overall test. In the absence of $F * P * R$ interaction, we would want to explore all of the corresponding two-factor effects, in particular the $F * P$ and $P * R$ interactions, which Table 19.7 tells us are clearly significant and marginally significant, respectively. The other two-factor effect subsumed by $F * P * R$ is $F * R$, but that is also subsumed by the significant $S * F * R$ effects, so $F * R$ does not warrant separate consideration. However, our goal is to illustrate techniques that can be used for unbalanced observations, and we examine these interactions using such methods (as opposed to using contrasts, which is what one would traditionally do for a balanced analysis).

19.2.1 Unbalanced subplot analysis

The strength of split-plot designs/models is their ability to analyze the subplot effects and interactions between subplot effects and whole-plot effects. Typically, there is less information available on the whole-plot effects (including their interactions among themselves). If the subplots are unbalanced, it is not possible to perform a clean analysis of the whole-plot effects (regardless of whether the whole plots are balanced). The methods illustrated earlier in Section 19.1 and later in Subsection 19.2.2 require balanced subplots, i.e., no missing subplots (but missing an entire whole plot is OK). The discussion in this subsection provides methods that can be used to analyze just the subplot effects and interactions between subplot effects and whole-plot effects for unbalanced subplots and also gives the correct analysis for balanced subplots, cf. Christensen (2011, Chapter 11). As alluded to earlier, for balanced subplots, simpler analyses that exploit the balance can be used.

To analyze the subplot effects and interactions between subplot effects and whole-plot effects,

Table 19.8: *Subplot high-order interaction models for data of Table 19.6.*

Model	SSE	dfE	F*	C_p
[SFPW][SFPR]	4552.7	24		48.0
[SFPW][SFR][SPR][FPR]	5018.6	28	0.61	42.5
[SFPW][SFR][SPR]	6434.2	32	1.24	41.9
[SFPW][SFR][FPR]	5704.6	32	0.76	38.1
[SFPW][SPR][FPR]	7346.7	30	2.45	50.7
[SFPW][SFR][PR]	7120.2	36	1.13	37.5
[SFPW][FPR][SR]	8032.6	34	1.83	46.3
[SFPW][SPR][FR]	8762.3	34	2.22	50.2
[SFPW][SR][PR][FR]	9448.3	38	1.84	45.8

*The F statistics are for testing each model against the model
with whole-plot effects and a four-factor interaction, i.e., **[SFPW][SFPR]**.
The denominator of each F statistic is
$MSE(2) \equiv MSE(\textbf{[SFPW][SFPR]}) = 4552.7/24 = 189.696$.

we examine models that include a separate fixed effect for each whole plot, i.e., we treat the η terms as fixed effects. We label these effects as **SFPW**. Rotations are the only effect not included in this term, so any interesting additional effects must include rotations. The model with all the subplot effects and subplot–whole-plot interactions is

$$y_{hijkm} = (sfpw)_{hijk} + r_m + (sr)_{hm} + (fr)_{im} + (pr)_{jm}$$
$$+ (sfr)_{him} + (spr)_{hjm} + (fpr)_{ijm} + (sfpr)_{hijm} + \varepsilon_{hijkm}. \quad (19.2.2)$$

Table 19.8 gives results for fitting this model and various reduced models using shorthand notation to denote models. Note the similarity between the models considered in Table 19.8 and the models considered in Table 16.2. The models in Table 19.8 all include **[SFPW]** and the other terms all include an **R** but otherwise the nine models are similar. For balanced subplots, the information in Table 19.8 can be obtained from Table 19.7, but Table 19.8 is also appropriate for unbalanced subplots. Table 19.8 also includes the C_p statistics for these models. The C_p statistics can be treated in the usual way but would not be appropriate for comparing models that do not have a separate fixed effect for each whole plot. The best-fitting models are **[SFPW][SFR][FPR]** and **[SFPW][SFR][PR]**, both of which include the **S** ∗ **F** ∗ **R** interaction between surface treatments, fills, and rotations.

The two best C_p models are hierarchical and the test of them, that is, of **[SFPW][SFR][FPR]** versus **[SFPW][SFR][PR]**, provides a test of **F** ∗ **P** ∗ **R** interaction with statistic

$$F_{obs} = \frac{[7120.2 - 5704.6]/[36 - 32]}{189.696} = 1.87,$$

which, when compared to an $F(4, 24)$ distribution, gives a one-sided P value of 0.149 as reported in Table 19.7 for these balanced data. There is no strong evidence for an **F** ∗ **P** ∗ **R** interaction, but that is not proof that it does not exist.

In our analysis from Section 16.2 of the 1000-rotation data, we found an **S** ∗ **F** interaction but similar analyses for the 2000 and 3000 rotation data show no **S** ∗ **F** interaction. (All three ANOVA tables are given in Section 19.3.) Tables 19.7 and 19.8 confirm the importance of fitting an **S** ∗ **F** ∗ **R** interaction. Assuming no four-factor interaction, the three different tests for **S** ∗ **F** ∗ **R** available from Table 19.8 give

$$F_{obs} = \frac{[7346.7 - 5018.6]/[30 - 28]}{189.696} = \frac{[8762.3 - 6434.2]/[34 - 32]}{189.696}$$
$$= \frac{[9448.3 - 7120.2]/[38 - 35]}{189.696} = \frac{[2328.1]/2}{189.696} = 6.14,$$

Table 19.9: *Subplot interaction models for examining* $\mathbf{S} * \mathbf{F} * \mathbf{R}$ *interaction.*

Model	SSE	dfE	SS Diff.	C_p
[SFPW][SFR][SPR][FPR]	5018.6	28		42.5
[SFPW][SFR$_{2=3}$][SPR][FPR]	5042.7	29	24.1	40.6
[SFPW][SPR][FPR]	7346.7	30	2304.0	50.7
[SFPW][SFR][FPR]	5704.6	32		38.1
[SFPW][SFR$_{2=3}$][FPR][SR]	5728.6	33	24.0	36.2
[SFPW][FPR][SR]	8032.6	34	2304.0	46.3
[SFPW][SFR][PR]	7120.2	36		37.5
[SFPW][SFR$_{2=3}$][SR][FR][PR]	7144.3	37	24.1	35.7
[SFPW][SR][PR][FR]	9448.3	38	2304.0	45.8

which all agree because of subplot balance. Thus the $\mathbf{S} * \mathbf{F}$ interaction depends on the number of rotations. It might be of interest if we could find a natural interpretation for this interaction. We now proceed to examine what is driving the $\mathbf{S} * \mathbf{F} * \mathbf{R}$ interaction and the more dubious $\mathbf{F} * \mathbf{P} * \mathbf{R}$ interaction.

We begin by looking at the $\mathbf{S} * \mathbf{F} * \mathbf{R}$ interaction. Normally, with rotations at quantitative levels, we would use linear and quadratic models in rotations to examine interaction. However, we previously analyzed the data from each number of rotations separately and discovered no $\mathbf{S} * \mathbf{F}$ interaction at 2000 and 3000 rotations, so we will use a model that does not distinguish between these levels of rotations, i.e., in a new categorical variable that we will call $\mathbf{R}_{2=3}$, 2000 and 3000 rotations have the same index. These models continue to include terms $(sr)_{hm} + (fr)_{im} + (pr)_{jm}$ or their equivalents but replace $(sfr)_{him}$ with $(sfr)_{hi\tilde{m}}$, which uses a new index variable $\tilde{m}$ that does not distinguish between rotations 2000 and 3000. These models *do not* incorporate the idea that there is no $\mathbf{S} * \mathbf{F}$ interaction at 2000 or 3000 rotations, but they do incorporate the idea that the $\mathbf{S} * \mathbf{F}$ interaction is the same at 2000 and 3000 rotations yet is possibly different from the $\mathbf{S} * \mathbf{F}$ interaction at 1000 rotations. We can investigate these terms in any model that includes the $\mathbf{S} * \mathbf{F} * \mathbf{R}$ interaction but not the four-factor interaction. The most reasonable choices for evaluating $\mathbf{S} * \mathbf{F} * \mathbf{R}$ with unbalanced subplot data are in the model with all of the three-factor interactions or in the two good models identified by the C_p statistic.

Table 19.9 gives the model fitting information. In particular, Table 19.9 gives three sets of three models, one that includes the $\mathbf{S} * \mathbf{F} * \mathbf{R}$ interaction, one that posits no change in the $\mathbf{S} * \mathbf{F}$ interaction for rotations 2000 and 3000, and one that eliminates the $\mathbf{S} * \mathbf{F} * \mathbf{R}$ interaction. It also gives the differences in sums of squares for the three models. The models that posit no change in the $\mathbf{S} * \mathbf{F}$ interaction for rotations 2000 and 3000 fit the data well with a difference in sums of squares of 24.1. Because of subplot balance, these numbers do not depend on which of the three particular sets of model comparisons are being made. (The value of 24.0 rather than 24.1 is round-off error.) However, models that posit no difference between the $\mathbf{S} * \mathbf{F}$ interaction at 1000 rotations and the common $\mathbf{S} * \mathbf{F}$ interaction at 2000 and 3000 rotations have a substantial difference in sums of squares of 2304.0, which leads to a significant F test. Using Scheffé's multiple comparison method is appropriate because the data suggested the model. (Previous analysis showed no $\mathbf{S} * \mathbf{F}$ interaction for 2000 or 3000 rotations but some for 1000.) The test statistic that compares the $\mathbf{S} * \mathbf{F}$ interaction at 1000 rotations with the others is

$$F_{obs} = \frac{2304/2}{189.7} = 6.07,$$

which is significant at the 0.01 level because $F(0.99, 2, 24) = 5.61$. Again, for unbalanced data the three model comparisons could differ. This model-based analysis that is applicable to unbalanced subplot data reproduces the results in Christensen (1996, Section 12.2) that examine orthogonal contrasts in the $\mathbf{S} * \mathbf{F} * \mathbf{R}$ interaction for balanced data.

Table 19.10: *Subplot interaction models for examining* $\mathbf{F} * \mathbf{P} * \mathbf{R}$ *interaction.*

Model	SSE	dfE	SS Diff.	C_p
[SFPW][SFR][FPR]	5704.6	32		38.1
[SFPW][SFR][PR][FP^{1}R^1][FP^{1}R^2][FP^{2}R^1][FP^{2}R^2]	5704.6	32		38.1
[SFPW][SFR][PR][FP^{1}R^1][FP^{1}R^2][FP^{2}R^1]	6364.6	33	660.0	39.6
[SFPW][SFR][PR][FP^{1}R^1][FP^{1}R^2]	6404.7	34	40.1	37.8
[SFPW][SFR][PR][FP^{1}R^1]	6899.7	35	495.0	38.4
[SFPW][SFR][PR]	7120.2	36	220.5	37.5
[SFPW][SFR][PR][FP^{1}R^1][FP^{1}R^2][FP^{2}R^1][FP^{2}R^2]	5704.6	32		38.1
[SFPW][SFR][PR][FP^{1}R^1][FP^{1}R^2][FP^{2}R^1]	6364.6	33	660.0	39.6
[SFPW][SFR][PR][FP^{1}R^1][FP^{2}R^1]	6859.7	34	495.1	40.2
[SFPW][SFR][PR][FP^{1}R^1]	6899.7	35	40.0	38.4
[SFPW][SFR][PR]	7120.2	36	220.5	37.5

Recall that in our earlier analysis from Section 16.2 based on just the 1000-rotation data, we also found an $\mathbf{F} * \mathbf{P}$ interaction. An $\mathbf{F} * \mathbf{P} * \mathbf{R}$ interaction indicates that the $\mathbf{F} * \mathbf{P}$ interaction changes with the number of rotations. If we conclude that no $\mathbf{F} * \mathbf{P} * \mathbf{R}$ interaction exists, we need to consider the corresponding two-factor interactions involving $\mathbf{P}$. We need to focus on $\mathbf{P}$ because it is the only factor that is not included in the significant $\mathbf{S} * \mathbf{F} * \mathbf{R}$ interaction. The possible two-factor interactions are $\mathbf{F} * \mathbf{P}$ and $\mathbf{P} * \mathbf{R}$.

It is not clear that an $\mathbf{F} * \mathbf{P} * \mathbf{R}$ interaction exists but, to be safe, we will examine some reasonable reduced interaction models. If some interpretable interaction effect has a large sum of squares, it suggests that an important interaction may be hidden within the 4-degree-of-freedom interaction test. To examine $\mathbf{F} * \mathbf{P} * \mathbf{R}$ interaction, we consider polynomial models in both proportions and rotations,

$$y_{hijkm} = (sfpw)_{hijk} + (sfr)_{him} + (pr)_{jm} + \beta_{11i}jm + \beta_{12i}jm^2 + \beta_{21i}j^2m + \beta_{22i}j^2m^2 + \varepsilon_{hijkm}. \quad (19.2.3)$$

Relative to this quadratic-by-quadratic interaction model, the first reduced model fitted drops the term $\beta_{22i}j^2m^2$. From the model without $\beta_{22i}j^2m^2$, we can drop either $\beta_{21i}j^2m$ or $\beta_{12i}jm^2$ determining two hierarchies of models. The last reduced model in the hierarchy will include only the linear-by-linear interaction term $\beta_{11i}jm$. Dropping this term leads to a model without $\mathbf{F} * \mathbf{P} * \mathbf{R}$ interaction.

The results of fitting the two hierarchies are given in Table 19.10. Because these subplot data are balanced, the differential effects for the intermediate regression terms are identical (up to round-off error) in the two hierarchies, i.e., 660 for fitting a quadratic-by-quadratic term after fitting the others, 495 for fitting a proportion-linear-by-rotation-quadratic term, regardless of whether a proportion-quadratic-by-rotation-linear term has already been fitted, 40 for fitting a proportion-quadratic-by-rotation-linear term, regardless of whether a proportion-linear-by-rotation-quadratic term has already been fitted, and 220.5 for fitting a proportion-linear-by-rotation-linear term. These results provide a model-based reproduction of results obtained using orthogonal interaction contrasts for balanced data in Christensen (1996, Section 12.2). Note also that these hierarchies involve dropping pairs of regression coefficients, e.g., β_{22i}, $i = 1, 2$, but dropping these pairs only reduces the error degrees of freedom by 1. This is a result of having $(sfpw)_{hijk}$ in every model.

These models were not chosen by looking at the data, so less stringent multiple comparison methods than Scheffé's can be used on them. On the other hand, the models are not particularly informative. None of these models suggests a particularly strong source of interaction. F tests are constructed by dividing each of the four sums of squares by $MSE(2)$. None of the F ratios is significant when compared to $F(0.95, 1, 24) = 4.26$. This analysis seems consistent with the hypothesis of no $\mathbf{F} * \mathbf{P} * \mathbf{R}$ interaction.

If we accept the working assumption of no $\mathbf{F} * \mathbf{P} * \mathbf{R}$ interaction, we need to examine the two-

Table 19.11: *Subplot interaction models for examining* $\mathbf{P} * \mathbf{R}$ *interaction.*

Model	SSE	dfE	SS Diff.	C_p
[SFPW][SFR][PR]	7120.2	36	—	37.5
[SFPW][SFR][P^1R^1][P^1R^2][P^2R^1][P^2R^2]	7120.2	36	—	37.5
[SFPW][SFR][P^1R^1][P^1R^2][P^2R^1]	7121.9	37	1.7	35.5
[SFPW][SFR][P^1R^1][P^1R^2]	7415.9	38	294.0	35.1
[SFPW][SFR][P^1R^1]	8141.9	39	726.0	36.9
[SFPW][SFR]	8883.0	40	741.1	38.8
[SFPW][SFR][P^1R^1][P^1R^2][P^2R^1][P^2R^2]	7120.2	36	—	37.5
[SFPW][SFR][P^1R^1][P^1R^2][P^2R^1]	7121.9	37	1.7	35.5
[SFPW][SFR][P^1R^1][P^2R^1]	7847.9	38	726.0	37.5
[SFPW][SFR][P^1R^1]	8141.9	39	294.0	36.9
[SFPW][SFR]	8883.0	40	741.1	38.8

factor interactions that can be constructed from the three factors. These are $\mathbf{F} * \mathbf{P}$, $\mathbf{F} * \mathbf{R}$, and $\mathbf{P} * \mathbf{R}$. The $\mathbf{F} * \mathbf{R}$ effects are, however, not worth further consideration because they are subsumed within the $\mathbf{S} * \mathbf{F} * \mathbf{R}$ effects that have already been established as important. Another way of looking at this is that in Model (19.2.1), the $(fr)_{im}$ effects are unnecessary in a model that already has $(sfr)_{him}$ effects. Thus we focus our attention on $\mathbf{F} * \mathbf{P}$ and $\mathbf{P} * \mathbf{R}$. The $\mathbf{F} * \mathbf{P}$ interaction is a whole-plot effect, so it will be considered in the next subsection.

We now examine the $\mathbf{P} * \mathbf{R}$ interaction. Information for testing whether [$\mathbf{PR}$] can be dropped from [$\mathbf{SFPW}$][$\mathbf{SFR}$][$\mathbf{PR}$] is given at the top and bottom of Table 19.11. The F statistic becomes,

$$F_{obs} = \frac{[8883.0 - 7120.2]/[40 - 36]}{189.7} = 2.32,$$

which agrees with Table 19.7.

The 4 degrees of freedom for $\mathbf{P} * \mathbf{R}$ in the interaction test have the potential of hiding one or two important, *interpretable* interaction effects. We explore this possibility by investigating $\mathbf{P} * \mathbf{R}$ interaction models based on the linear and quadratic effects in both $\mathbf{P}$ and $\mathbf{R}$.

We used Model (19.2.3) to examine $\mathbf{F} * \mathbf{P} * \mathbf{R}$ interaction; if there is no $\mathbf{F} * \mathbf{P} * \mathbf{R}$ interaction, a similar model can be used to examine the $\mathbf{P} * \mathbf{R}$ interaction,

$$y_{hijkm} = (sfpw)_{hijk} + (sfr)_{him} + \beta_{11}jm + \beta_{12}jm^2 + \beta_{21}j^2m + \beta_{22}j^2m^2 + \varepsilon_{hijkm}. \qquad (19.2.4)$$

Results for fitting reduced models are given in Table 19.11. There are two hierarchies but due to subplot balance they give the same results. We find that the sequential sum of squares for dropping β_{12} is 726.0 and for dropping β_{11} is 741.1. Comparing them to $MSE(2)$, these sums of squares are not small but neither are they clearly significant. The interaction plot in Figure 19.5 of $\hat{y}_{11j1m}$ values from Model (19.2.4) seems to confirm that there is no obvious interaction being overlooked by the four degrees of freedom test. We remain unconvinced that there is any substantial $\mathbf{P} * \mathbf{R}$ interaction. These are exact analogues to results in Christensen (1996, Section 12.2.).

19.2.2   Whole-plot analysis

There is no simple whole-plot analysis unless the subplots are balanced. Losing any subplot observations (short of losing an entire whole plot) causes mathematical difficulties that preclude a simple whole-plot analysis. With balanced subplots, we can perform the whole-plot analysis on the subplot means or, as demonstrated here, we can accomplish the same thing by simply ignoring the subplot effects. *We cannot overemphasize that the methods in this subsection are inappropriate for unbalanced subplots.* The analysis *is* appropriate for unbalanced whole plots.

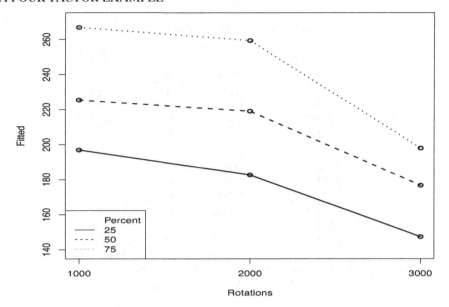

Figure 19.5: *Proportion−Rotation interaction plot for abrasion data. Plot of* $\hat{y}_{11j1m}$ *versus* m *for each* j.

We begin by dropping all the subplot effects out of Model (19.2.2) and fitting

$$y_{hijkm} = (sfpw)_{hijk} + \varepsilon_{hijkm},$$

cf. Table 19.12. To obtain the whole-plot Error, we compare this to a model with all whole-plot effects but no subplot effects.

$$\begin{aligned}
y_{hijkm} &= \mu + s_h + f_i + p_j \\
&\quad + (sf)_{hi} + (sp)_{hj} + (fp)_{ij} + (sfp)_{hij} + \varepsilon_{hijkm}.
\end{aligned} \tag{19.2.5}$$

From Table 19.12, $MSE(1) = [102446 - 98705]/[60 - 48] = 3741/12 = 311.7$, which agrees with Table 19.7.

As we create reduced models relative to Model (19.2.5) in the whole plots we can get test degrees of freedom and sums of squares by differencing the errors of various reduced models in the usual way. Table 19.12 includes the usual 9 models for 3 factors and modified C_P statistics. From Table 19.12 we can reproduce the whole-plot tests of Table 19.7. For example, $SS(\mathbf{S} * \mathbf{F} * \mathbf{P}) = 102745 - 102446 = 299$ with $df(\mathbf{S} * \mathbf{F} * \mathbf{P}) = 62 - 60 = 2$. Moreover, one sum of squares for $\mathbf{S} * \mathbf{P}$ is

$$R\left([\mathbf{SP}]\Big|[\mathbf{FP}][\mathbf{SF}]\right) \equiv SSE([\mathbf{SP}][\mathbf{FP}][\mathbf{SF}]) - SSE([\mathbf{FP}][\mathbf{SF}]) = 103540 - 102745 = 795$$

with

$$dfE\left([\mathbf{SP}][\mathbf{FP}][\mathbf{SF}]\right) - dfE\left([\mathbf{FP}][\mathbf{SF}]\right) = 64 - 62 = 2.$$

The best among the usual 9 models appears to be $[\mathbf{SF}][\mathbf{FP}]$, which is equivalent to a quadratic model in proportions

$$y_{hijkm} = (sf)_{hi} + p_j + \beta_{1i}j + \beta_{2i}j^2 + \varepsilon_{hijkm},$$

a model that we denote $[\mathbf{SF}][\mathbf{P}][\mathbf{FP}^1][\mathbf{FP}^2]$. Dropping the quadratic terms in proportions gives $[\mathbf{SF}][\mathbf{P}][\mathbf{FP}^1]$ and dropping the linear term reduces us to $[\mathbf{SF}][\mathbf{P}]$. The sum of squares for the quadratic $\mathbf{F} * \mathbf{P}$ interaction term is

$$\begin{aligned}
R\left([\mathbf{FP}^2]\Big|[\mathbf{SF}][\mathbf{P}][\mathbf{FP}^1]\right) &\equiv SSE([\mathbf{SF}][\mathbf{P}][\mathbf{FP}^1]) - SSE([\mathbf{SF}][\mathbf{P}][\mathbf{FP}^1][\mathbf{FP}^2]) \\
&= 103563 - 103540 = 23
\end{aligned}$$

Table 19.12: *Whole-plot interaction models.*

Model	SSE	dfE	C_{p*}
[SFPW]	98705	48	—
[SFP]	102446	60	12.0
[SF][SP][FP]	102745	62	8.9
[SF][SP]	112630	64	36.7
[SF][FP]	103540	64	7.5
[SP][FP]	104427	63	12.4
[P][SF]	113425	66	35.2
[F][SP]	114312	65	40.1
[S][FP]	105222	65	10.9
[S][F][P]	115107	67	38.6
[SF][P][FP^1][FP^2]	103540	64	7.5
[SF][P][FP^1]	103563	65	5.6
[SF][P]	113425	66	35.2

$C_{p*} = [SSE - 98705]/[3741/12] - 2(dfE - 48)) + 24$ where
$n = 72$ observations, $N = 3$ observations per whole plot,
$24 = n/N$ and $48 = n - n/N$.

and the sum of squares for the linear $\mathbf{F} * \mathbf{P}$ interaction term is

$$R\left([\mathbf{FP}^1]\big|[\mathbf{SF}][\mathbf{P}]\right) \equiv SSE([\mathbf{SF}][\mathbf{P}]) - SSE([\mathbf{SF}][\mathbf{P}][\mathbf{FP}^1])$$
$$= 113425 - 103563 = 9862,$$

each on 1 degree of freedom. The $\mathbf{F} * \mathbf{P}$ interaction is a whole-plot effect, so the appropriate error is $MSE(1) = 311.7$ and the F ratios are 0.075 and 31.64, respectively. There is no evidence that the curvatures in proportions are different for Fills A and B. However, there is evidence that the slopes are different for Fills A and B.

In fact, we can take this further. The data are consistent with there being not only no change in curvature but no curvature at all and, although the slopes are different, there is no evidence of a nonzero slope for Fill B.

To fit separate quadratic models in j for each fill, we need to manipulate the indices. A 0-1 indicator variable for Fill B is $f_B \equiv i - 1$ and an indicator variable for Fill A is $f_A \equiv 2 - i$. Define a linear term in proportions for Fill A only as $p_A \equiv j * f_A$ and the quadratic term is $p_A^2 \equiv j^2 * f_A$. Similarly, the linear and quadratic terms for Fill B are $p_B \equiv j * f_B$ and $p_B^2 \equiv j^2 * f_B$. The following model is equivalent to [$\mathbf{SF}$][$\mathbf{FP}$],

$$y_{hijkm} = (sf)_{hi} + \beta_{A1}p_A + \beta_{A2}p_A^2 + \beta_{B1}p_B + \beta_{B2}p_B^2 + \varepsilon_{hijkm}.$$

As expected, it has $dfE = 64$ and $SSE = 103540$, just like [$\mathbf{SF}$][$\mathbf{FP}$].

Dropping the quadratic term for Fill B gives

$$y_{hijkm} = (sf)_{hi} + \beta_{A1}p_A + \beta_{A2}p_A^2 + \beta_{B1}p_B + \varepsilon_{hijkm},$$

with $dfE = 65$ and $SSE = 103652$ for a difference in sums of squares of $103652 - 103540 = 112$ for the Fill B quadratic term. Further dropping the linear term for Fill B gives

$$y_{hijkm} = (sf)_{hi} + \beta_{A1}p_A + \beta_{A2}p_A^2 + \varepsilon_{hijkm},$$

with $dfE = 66$ and $SSE = 103793$ for a difference in sums of squares of $103793 - 103652 = 141$ due to the Fill B linear term. As far as we can tell, the weight loss does not change as a function of proportion of filler when using Fill B.

A similar analysis for Fill A shows that weight loss increases with proportion and there is again no evidence of curvature. In particular, for Fill A, the quadratic term has a sum of squares of 14. For Fill A, the linear term has sum of squares 23202.

The other significant whole-plot effect is the $S * F$ interaction but those effects are subsumed by incorporating the $S * F * R$ subplot effects.

19.2.3 Fixing effect levels

In a split-plot model, we can examine subplot effects for fixed levels of the whole-plot effects but, within the model, we cannot examine whole-plot effects for fixed levels of a subplot effect. For example, when investigating the $S * F * R$ interaction, within a split-plot model, we can fix the Fill as A ($i = 1$) and examine the corresponding $S * R$ interaction, but we cannot, without going outside the split-plot model, fix the Rotation at 1000 ($m = 1$) and examine the $S * F$ interaction.

Continuing these examples, if we fix $i = 1$ in Model (19.2.2), the model becomes

$$y_{h1jkm} = (sfpw)_{h1jk} + r_m + (sr)_{hm} + (fr)_{1m} + (pr)_{jm}$$
$$+ (sfr)_{h1m} + (spr)_{hjm} + (fpr)_{1jm} + (sfpr)_{h1jm} + \varepsilon_{h1jkm},$$

or equivalently,

$$y_{h1jkm} = (spw)_{hjk} + r_m + (sr)_{hm} + (pr)_{jm} + (sr)_{hm} + (spr)_{hjm} + \varepsilon_{hjkm},$$

which is a perfectly reasonable model to fit and one that allows exploration of $S * R$ interaction for Fill A. Any inferences we choose to make can continue to be based on $MSE(2)$ as our estimate of variance. Similarly, there is no problem with fixing the level of a whole-plot effect in the whole-plot analysis, similar to what we did in the previous subsection.

On the other hand, if we fix $m = 1$ in Model (19.2.2), the subplot model becomes

$$y_{hijk1} = (sfpw)_{hijk} + r_1 + (sr)_{h1} + (fr)_{i1} + (pr)_{j1}$$
$$+ (sfr)_{hi1} + (spr)_{hj1} + (fpr)_{ij1} + (sfpr)_{hij1} + \varepsilon_{hijk1},$$

or equivalently,

$$y_{hijk1} = (sfpw)_{hijk} + \varepsilon_{hijk1},$$

which is not a model that allows us to examine $S * F$ interaction.

Fortunately, we can examine $S * F$ interaction for a fixed level of m; we just cannot do it in the split-plot model context. If we go back to the original split-plot model (19.2.1) and fix $m = 1$ we get

$$
\begin{aligned}
y_{hijk1} &= \mu + s_h + f_i + p_j \\
&\quad + (sf)_{hi} + (sp)_{hj} + (fp)_{ij} + (sfp)_{hij} \\
&\quad + \eta_{hij1} \\
&\quad + r_1 + (sr)_{h1} + (fr)_{i1} + (pr)_{j1} \\
&\quad + (sfr)_{hi1} + (spr)_{hj1} + (fpr)_{ij1} + (sfpr)_{hij1} \\
&\quad + \varepsilon_{hijk1}
\end{aligned}
$$

or equivalently

$$y_{hijk1} = \mu + s_h + f_i + p_j + (sf)_{hi} + (sp)_{hj} + (fp)_{ij} + (sfp)_{hij} + \eta_{hij1} + \varepsilon_{hijk1},$$

which is really just the model that we analyzed in Section 16.2 where the two independent error terms η_{hijk} and ε_{hijk1} are added and treated as a single error term. Our test in Section 16.2 of the $S * F$ interaction using just the 1000-rotation data is perfectly appropriate and similar tests using just the 2000-rotation data and just the 3000-rotation data would also be appropriate. ANOVA tables for the separate analyses of the 1000-, 2000-, and 3000-rotation data are given in Section 19.3 as Tables 19.14, 19.15, and 19.16. Note, though, that the separate analyses are not independent, because the observations at 2000 rotations are not independent of the observations at 3000 rotations, etc.

On occasion, when examining models for a fixed subplot treatment, rather than using the MSEs from the separate analyses, the degrees of freedom and sums of squares for Error 1 and Error 2 are pooled and these are used instead. This is precisely the error estimate obtained by pooling the error estimates from the three separate ANOVAs. Such a pooled estimate should be better than the estimates from the separate analyses but it is difficult to quantify the effect of pooling. The three separate ANOVAs are not independent, so pooling the variance estimates does not have the nice properties of the pooled estimate of the variance used in, say, one-way ANOVA. As alluded to above, we cannot get exact F tests based on the pooled variance estimate. If the three ANOVA's were independent, the pooled error would have $12 + 12 + 12 = 36$ degrees of freedom, but we do not have independence, so we do not even know an appropriate number of degrees of freedom to use with the pooled estimate, much less the appropriate distribution.

19.2.4   Final models and estimates

If the subplots are unbalanced, the final models and estimates are whatever comes out of the analysis of Subsection 19.2.1. If we have balanced subplots, we can also incorporate a whole-plot analysis in our final results.

We have found two important interaction effects, $\mathbf{S} * \mathbf{F} * \mathbf{R}$ from the subplot analysis and $\mathbf{F} * \mathbf{P}$ from the whole-plot analysis. These two interactions are the highest-order terms that are significant and they include all four of the factors. The only factor contained in both interactions is $\mathbf{F}$, so the simplest overall explanation of the data can be arrived at by giving separate explanations for the two fills. To do this, we need to re-evaluate the $\mathbf{S} * \mathbf{F} * \mathbf{R}$ interaction in terms of how the $\mathbf{S} * \mathbf{R}$ interaction changes from Fill A to Fill B; previously, we focused on how the $\mathbf{S} * \mathbf{F}$ interaction changed with rotations. One benefit of this change in emphasis is that, as discussed earlier, we can use $MSE(2)$ for valid tests of the $\mathbf{S} * \mathbf{R}$ interactions effects for a fixed level of $\mathbf{F}$ because we are fixing a whole-plot factor, not a subplot factor.

Separate the data into Fill A and Fill B and fit models by rewriting

$$y_{hijkm} = (fp)_{ij} + \eta_{hijk} + (sfr)_{him} + \varepsilon_{hijkm}$$

as

$$y_{h1jkm} = p_j + \eta_{h1jk} + (sr)_{hm} + \varepsilon_{h1jkm} \qquad (19.2.6)$$

and

$$y_{h2jkm} = p_j + \eta_{h2jk} + (sr)_{hm} + \varepsilon_{h2jkm}. \qquad (19.2.7)$$

Models (19.2.6) and (19.2.7) can be written in split plot form as

$$y_{h1jkm} = s_h + p_j + \eta_{h1jk} + r_m + (sr)_{hm} + \varepsilon_{h1jkm}$$

and

$$y_{h2jkm} = s_h + p_j + \eta_{h2jk} + r_m + (sr)_{hm} + \varepsilon_{h2jkm}.$$

In the $\mathbf{S} * \mathbf{F} * \mathbf{R}$ interaction, rotations 2000 and 3000 are similar, so, as alternates to models (19.2.6) and (19.2.7), we could fit models that do not distinguish between them using the index $\tilde{m}$,

$$y_{h1jkm} = s_h + p_j + \eta_{h1jk} + r_m + (sr)_{h\tilde{m}} + \varepsilon_{h1jkm} \qquad (19.2.8)$$

and

$$y_{h2jkm} = s_h + p_j + \eta_{h2jk} + r_m + (sr)_{h\tilde{m}} + \varepsilon_{h2jkm}. \qquad (19.2.9)$$

And we can also fit the no-interaction models

$$y_{h1jkm} = s_h + p_j + \eta_{h1jk} + r_m + \varepsilon_{h1jkm} \qquad (19.2.10)$$

and

$$y_{h2jkm} = s_h + p_j + \eta_{h2jk} + r_m + \varepsilon_{h2jkm}. \qquad (19.2.11)$$

The sum of squares for comparing models (19.2.6) and (19.2.8) is

$$SS(S*(2000\,vs\,3000)R;\,fill\,A) = 165.69$$

and for comparing models (19.2.8) and (19.2.10) is

$$SS(S*(1000\,vs\,others)R;\,fill\,A) = 754.01.$$

The sum of squares for comparing models (19.2.7) and (19.2.9) is

$$SS(S*(2000\,vs\,3000)R;\,fill\,B) = 391.72$$

and for comparing models (19.2.8) and (19.2.10) is

$$SS(S*(1000\,vs\,others)R;\,fill\,B) = 9225.35.$$

All of these are compared to $MSE(2) = 189.7$. There is no evidence of interactions involving 2000 and 3000 rotations with surface treatments, regardless of fill type. With Fill A, there is marginal evidence of an interaction in which the effect of **S** is different at 1000 rotations than at 2000 and 3000 rotations. With Fill B, there is clear evidence of an interaction where the effect of **S** is different at 1000 rotations than at 2000 and 3000 rotations.

We earlier established that there is no quadratic effect in proportions for fill A, so Model (19.2.6) can be replaced by

$$y_{h1jkm} = \gamma j + \eta_{h1jk} + (sr)_{hm} + \varepsilon_{h1jkm}.$$

We earlier showed that there is no linear or quadratic effects in proportions for fill B so Model (19.2.7) can be replaced by

$$y_{h2jkm} = \eta_{h2jk} + (sr)_{hm} + \varepsilon_{h2jkm}.$$

Incorporating the earlier subplot models gives us the split-plot models

$$y_{h1jkm} = s_h + \gamma j + \eta_{h1jk} + r_m + (sr)_{h\tilde{m}} + \varepsilon_{h1jkm} \qquad (19.2.12)$$

and

$$y_{h2jkm} = s_h + \eta_{h2jk} + r_m + (sr)_{h\tilde{m}} + \varepsilon_{h2jkm}. \qquad (19.2.13)$$

Parameter estimates can be obtained by least squares, i.e., by fitting the models ignoring the η errors. The fitted values are given in Table 19.13. Note that the rows and columns have been rearranged from those used for the data in Table 19.6.

For Fill A, either surface treatment and any level of rotation, estimated weight loss increases by 31.08 as the proportion goes up.

For Fill A, the effect of going from 1000 to 2000 rotations is an estimated decrease in weight loss of 12.0 units with a surface treatment but an estimated increase in weight loss of 8.0 units without the surface treatment. The estimated effect of going from 2000 to 3000 rotations is a drop of 41.1 units in weight loss regardless of the surface treatment.

For Fill A, the estimated effect of the surface treatment is an additional 20.6 units in weight loss at either 2000 or 3000 rotations but it is an additional 40.5 units at 1000 rotations.

For Fill B, the estimated weight loss does not depend on proportions.

For Fill B, the effect of going from 1000 to 2000 rotations is an estimated decrease in weight loss of 111.0 units with a surface treatment but only an estimated decrease in weight loss of 43.2

Table 19.13: *Abrasion resistance data fitted values: Final model.*

Fill	Surf. treat.	1000			2000			3000		
		25%	50%	75%	25%	50%	75%	25%	50%	75%
A	Yes	203.9	235.0	266.1	192.0	223.0	254.1	150.9	182.0	213.0
	Yes	203.9	235.0	266.1	192.0	223.0	254.1	150.9	182.0	213.0
	No	163.4	194.5	225.6	171.4	202.5	233.5	130.3	161.4	192.5
	No	163.4	194.5	225.6	171.4	202.5	233.5	130.3	161.4	192.5
B	Yes	227.0	227.0	227.0	116.0	116.0	116.0	93.9	93.9	93.9
	Yes	227.0	227.0	227.0	116.0	116.0	116.0	93.9	93.9	93.9
	No	135.2	135.2	135.2	92.0	92.0	92.0	70.0	70.0	70.0
	No	135.2	135.2	135.2	92.0	92.0	92.0	70.0	70.0	70.0

without the surface treatment. The estimated effect of going from 2000 to 3000 rotations is a drop of 22.1 units in weight loss regardless of the surface treatment.

For Fill B, the estimated effect of the surface treatment is an additional 24.0 units in weight loss at either 2000 or 3000 rotations but it is an additional 91.8 units at 1000 rotations.

Most of these estimates are identical to estimates based on the balanced analysis presented in Christensen (1996, Section 12.2). The exceptions are the estimates that compare results for 1000 and 2000 rotations for fixed levels of surface treatment, and fill (proportions being irrelevant). The estimates in Christensen (1996, Section 12.2) were somewhat more naive in that they did not incorporate the lack of $\mathbf{S} * \mathbf{F}$ interaction at 2000 and 3000 rotations.

The same information can be obtained from the tables of coefficients for models (19.2.12) and (19.2.13) but it is much more straightforward to get the estimates from the table of fitted values. In particular, fitting the models (with intercepts but without the whole-plot error terms) in R gives

Table of Coefficients

Fill A: Model (19.2.12)		Fill B: Model (19.2.13)	
Predictor	*Est*	Predictor	*Est*
(Intercept)	172.833	(Intercept)	227.000
Sa2	−40.500	Sb2	−91.833
pa	31.083		
RTa2	7.958	RTb2	−43.125
RTa3	−33.125	RTb3	−65.208
Sa1:Mtildea2	−19.917	Sb1:Mtildeb2	−67.917
Sa2:Mtildea2	NA	Sb2:Mtildeb2	NA

The whole-plot errors were not included in the fitted models so the standard errors, t statistics, and P values are all invalid and not reported.

The estimates for Fill A that we obtained from Table 19.13 can now be found as

$$
\begin{aligned}
31.083 &= \text{pa} \\
-11.959 &= -19.917 + 7.958 = \text{Sa1:Mtildea2} + \text{RTa2} \\
7.958 &= \text{RTa2} \\
-41.083 &= -33.125 - 7.958 = \text{RTa3} - \text{RTa2} \\
20.583 &= 40.500 - 19.917 = 20.583 = -\text{Sa2} + \text{Sa1:Mtildea2} \\
40.500 &= -\text{Sa2}
\end{aligned}
$$

and the estimates for Fill B are

$$
-111.042 \;=\; -67.917 - 43.125 = \text{Sb1:Mtildeb2} + \text{RTb2}
$$

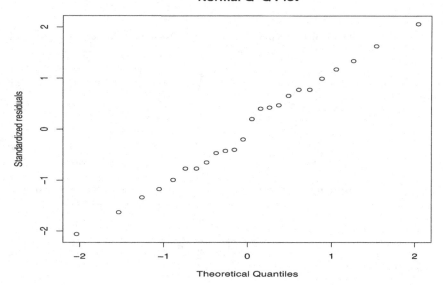

Figure 19.6: *Normal plot of whole-plot residuals, $W' = 0.98$, Box data.*

$$
\begin{aligned}
-43.125 &= \text{RTb2} \\
-22.083 &= -65.208 + 43.125 = \text{RTb3} - \text{RTb2} \\
23.916 &= 91.833 - 67.917 = -\text{Sb2} + \text{Sb1:Mtildeb2} \\
91.833 &= -\text{Sb2}.
\end{aligned}
$$

The down side of looking at the coefficients is that it is by no means clear how to figure out that these parameter estimates, and linear combinations of parameter estimates, are what one wants to look at. It is much easier to look at the table of fitted values to isolate features of interest corresponding to the fitted model.

If you were to include fixed η effects when fitting models (19.2.12) and (19.2.13), you would get the same estimates of any terms that involve the subplot treatments. In this example those would be RTa2, RTb2, RTa3, RTb3, Sa1:Mtildea2, and Sb1:Mtildeb2. Moreover, the reported standard errors for these parameter estimates would be appropriate. (Although even better standard errors could be constructed by pooling the error estimates from models (19.2.12) and (19.2.13).) With fixed η effects in the models, estimates of any whole-plot terms (any terms not previously listed) depend entirely on the side conditions used to fit the model.

Finally, we examine residual plots for Model (19.2.1). The Error 1 plots are based on a model for the whole plots that averages observations in subplots. Figures 19.6 and 19.7 contain residual plots for the Error 1 residuals. The Error 2 plots are based on Model (19.2.2). Figures 19.8 and 19.9 contain residual plots for the Error 2 residuals. We see no serious problems in any of the plots. □

19.3 Multivariate analysis of variance

The multivariate approach to analyzing data that contain multiple measurements on each subject involves using the multiple measures as separate dependent variables in a collection of standard analyses. The method of analysis, known as *multivariate analysis of variance (MANOVA)* or more generally as *multivariate linear models*, then combines results from the several linear models. A detailed discussion of MANOVA is beyond the scope of this book, but we present a short introduction to some of the underlying ideas.

For simplicity, we focus on a balanced analysis but there is nothing in the general theory that

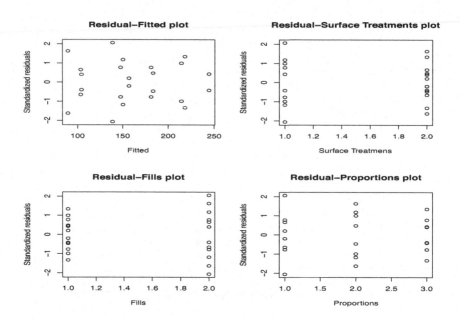

Figure 19.7: *Whole-plot residuals plots, Box data.*

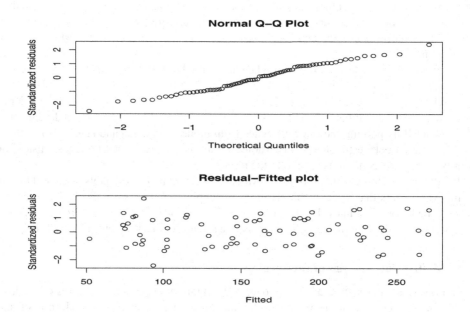

Figure 19.8 *Normal plot of subplot residuals, $W' = 0.98$ and subplot residuals versus predicted values, Box data.*

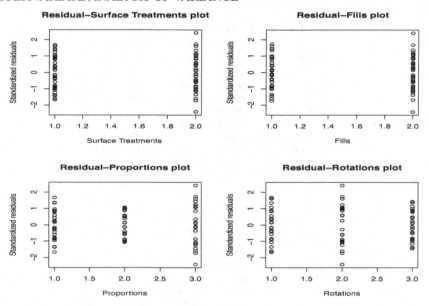

Figure 19.9: *Subplot residuals versus factor values, Box data.*

requires balance except that there be no missing observations among the multiple measures on a subject. Entirely missing a subject causes few problems. The discussion in Christensen (2001, Section 1.5) is quite general and particularly relevant in that it makes extensive comparisons to split-plot analyses. Unfortunately, the mathematical level of Christensen (2001) is much higher than the level of this book. Almost all Statistics books on multivariate analysis deal with MANOVA. Johnson and Wichern (2007) or Johnson (1998) are reasonable places to look for more information on the subject.

The discussion in this section makes some use of matrices. Matrices are reviewed in Appendix A.

EXAMPLE 19.3.1. Consider again the Box (1950) data on the abrasion resistance of a fabric. We began in Section 16.2 by analyzing the weight losses obtained after 1000 revolutions of the testing machine. In the split-plot analysis we combined these data for 1000 rotations with the data for 2000 and 3000 rotations. In the multivariate approach, we revert to the earlier analysis and fit separate ANOVA models for the data from 1000 rotations, 2000 rotations, and 3000 rotations. Again, the three factors are referred to as $\mathbf{S}$, $\mathbf{F}$, and $\mathbf{P}$, respectively. The variables $y_{hijk,1}$, $y_{hijk,2}$, and $y_{hijk,3}$ denote the data from 1000, 2000, and 3000 rotations, respectively. We fit the models

$$
\begin{aligned}
y_{hijk,1} &= \mu_{hijk,1} + \varepsilon_{hijk,1} \\
&= \mu_1 + s_{h,1} + f_{i,1} + p_{j,1} \\
&\quad + (sf)_{hi,1} + (sp)_{hj,1} + (fp)_{ij,1} + (sfp)_{hij,1} + \varepsilon_{hijk,1},
\end{aligned}
$$

$$
\begin{aligned}
y_{hijk,2} &= \mu_{hijk,2} + \varepsilon_{hijk,2} \\
&= \mu_2 + s_{h,2} + f_{i,2} + p_{j,2} \\
&\quad + (sf)_{hi,2} + (sp)_{hj,2} + (fp)_{ij,2} + (sfp)_{hij,2} + \varepsilon_{hijk,2},
\end{aligned}
$$

and

$$
\begin{aligned}
y_{hijk,3} &= \mu_{hijk,3} + \varepsilon_{hijk,3} \\
&= \mu_3 + s_{h,3} + f_{i,3} + p_{j,3} \\
&\quad + (sf)_{hi,3} + (sp)_{hj,3} + (fp)_{ij,3} + (sfp)_{hij,3} + \varepsilon_{hijk,3}
\end{aligned}
$$

Table 19.14: *Analysis of variance for* y_1.

Source	df	SS	MS	F	P
S	1	26268.2	26268.2	97.74	0.000
F	1	6800.7	6800.7	25.30	0.000
P	2	5967.6	2983.8	11.10	0.002
S*F	1	3952.7	3952.7	14.71	0.002
S*P	2	1186.1	593.0	2.21	0.153
F*P	2	3529.1	1764.5	6.57	0.012
S*F*P	2	478.6	239.3	0.89	0.436
Error	12	3225.0	268.8		
Total	23	51407.8			

Table 19.15: *Analysis of variance for* y_2.

Source	df	SS	MS	F	P
S	1	5017.0	5017.0	25.03	0.000
F	1	70959.4	70959.4	353.99	0.000
P	2	7969.0	3984.5	19.88	0.000
S*F	1	57.0	57.0	0.28	0.603
S*P	2	44.3	22.2	0.11	0.896
F*P	2	6031.0	3015.5	15.04	0.001
S*F*P	2	14.3	7.2	0.04	0.965
Error	12	2405.5	200.5		
Total	23	92497.6			

$h = 1,2, i = 1,2, j = 1,2,3, k = 1,2$.

As in standard ANOVA models, we assume that the individuals (on which the repeated measures were taken) are independent. Thus, for fixed $m = 1,2,3$, the $\varepsilon_{hijk,m}$s are independent $N(0, \sigma_{mm})$ random variables. We are now using a double subscript in σ_{mm} to denote a variance rather than writing σ_m^2. As usual, the errors on a common dependent variable, say $\varepsilon_{hijk,m}$ and $\varepsilon_{h'i'j'k',m}$, are independent when $(h,i,j,k) \neq (h',i',j',k')$, but we also assume that the errors on different dependent variables, say $\varepsilon_{hijk,m}$ and $\varepsilon_{h'i'j'k',m'}$, are independent when $(h,i,j,k) \neq (h',i',j',k')$. However, not all of the errors for all the variables are assumed independent. Two observations (or errors) on the same subject are *not* assumed to be independent. For fixed h, i, j, k the errors for any two variables are possibly correlated with, say, $\text{Cov}(\varepsilon_{hijk,m}, \varepsilon_{hijk,m'}) = \sigma_{mm'}$.

The models for each variable are of the same form but the parameters differ for the different dependent variables $y_{hijk,m}$. All the parameters have an additional subscript to indicate which dependent variable they belong to. The essence of the procedure is simply to fit each of the models individually and then to combine results. Fitting individually gives three separate sets of residuals, $\hat{\varepsilon}_{hijk,m} = y_{hijk,m} - \bar{y}_{hij\cdot,m}$ for $m = 1,2,3$, so three separate sets of residual plots and three separate ANOVA tables. The three ANOVA tables are given as Tables 19.14, 19.15, and 19.16. (Table 19.14 reproduces Table 16.10.) Each variable can be analyzed in detail using the ordinary methods for multifactor models illustrated in Section 16.2. Residual plots for y_1 were previously given in Section 16.2 as Figures 16.3 and 16.4 with additional plots given here. The top left residual plot for y_1 in Figure 19.10 was given as Figure 16.3. Residual plots for the analyses on y_2 and y_3 are given in Figures 19.11 through 19.14.

The key to multivariate analysis of variance is to combine results *across* the three variables y_1, y_2, and y_3. Recall that the mean squared errors are just the sums of the squared residuals divided by

Table 19.16: *Analysis of variance for y_3.*

Source	df	SS	MS	F	P
S	1	1457.0	1457.0	6.57	0.025
F	1	48330.4	48330.4	217.83	0.000
P	2	1396.6	698.3	3.15	0.080
S $*$ F	1	0.4	0.4	0.00	0.968
S $*$ P	2	250.6	125.3	0.56	0.583
F $*$ P	2	1740.3	870.1	3.92	0.049
S $*$ F $*$ P	2	272.2	136.1	0.61	0.558
Error	12	2662.5	221.9		
Total	23	56110.0			

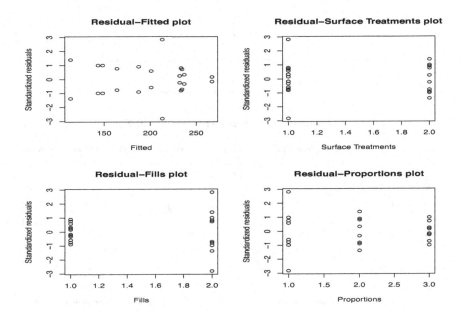

Figure 19.10: *Residual plots for y_1.*

the error degrees of freedom, i.e.,

$$MSE_{mm} \equiv s_{mm} = \frac{1}{dfE} \sum_{hijk} \hat{\varepsilon}_{hijk,m}^2.$$

This provides an estimate of σ_{mm}. We can also use the residuals to estimate covariances between the three variables. The estimate of $\sigma_{mm'}$ is

$$MSE_{mm'} \equiv s_{mm'} = \frac{1}{dfE} \sum_{hijk} \hat{\varepsilon}_{hijk,m} \hat{\varepsilon}_{hijk,m'}.$$

We now form the estimates into a matrix of estimated covariances

$$S = \begin{bmatrix} s_{11} & s_{12} & s_{13} \\ s_{21} & s_{22} & s_{23} \\ s_{31} & s_{32} & s_{33} \end{bmatrix}.$$

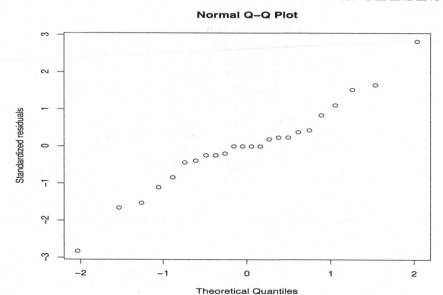

Figure 19.11: *Normal plot for y_2, $W' = 0.97$.*

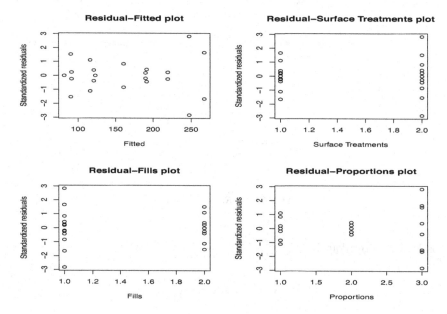

Figure 19.12: *Residual plots for y_2.*

Note that $s_{mm'} = s_{m'm}$, e.g., $s_{12} = s_{21}$. The matrix S provides an estimate of the covariance matrix

$$\Sigma \equiv \begin{bmatrix} \sigma_{11} & \sigma_{12} & \sigma_{13} \\ \sigma_{21} & \sigma_{22} & \sigma_{23} \\ \sigma_{31} & \sigma_{32} & \sigma_{33} \end{bmatrix}.$$

The key difference between this analysis and the split-plot analysis is that this analysis makes no assumptions about the variances and covariances in Σ. The split-plot analysis assumes that

$$\sigma_{11} = \sigma_{22} = \sigma_{33} = \sigma_w^2 + \sigma_s^2$$

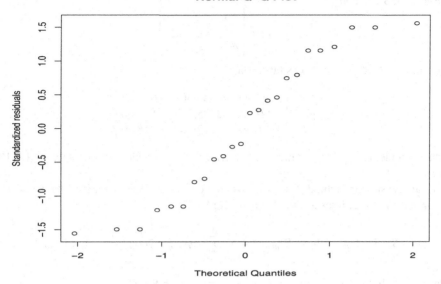

Figure 19.13: *Normal plot for y_3, $W' = 0.94$.*

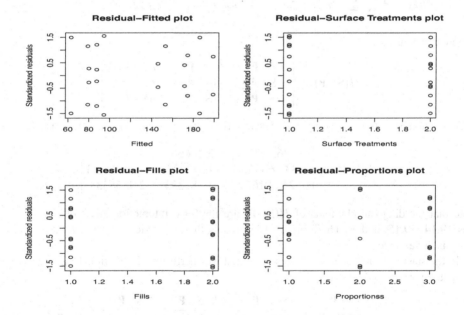

Figure 19.14: *Residual–prediction plot for y_3.*

and that for $m \neq m'$,

$$\sigma_{mm'} = \sigma_w^2.$$

Similarly, we can construct a matrix that contains sums of squares error and sums of cross products error. Write

$$e_{mm'} \equiv \sum_{hijk} \hat{\varepsilon}_{hijk,m} \hat{\varepsilon}_{hijk,m'}$$

where $e_{mm} = SSE_{mm}$ and

$$E \equiv \begin{bmatrix} e_{11} & e_{12} & e_{13} \\ e_{21} & e_{22} & e_{23} \\ e_{31} & e_{32} & e_{33} \end{bmatrix}.$$

Obviously, $E = (dfE)S$. For Box's fabric data,

$$E = \begin{bmatrix} 3225.00 & -80.50 & 1656.50 \\ -80.50 & 2405.50 & -112.00 \\ 1656.50 & -112.00 & 2662.50 \end{bmatrix}.$$

The diagonal elements of this matrix are the error sums of squares from Tables 19.14, 19.15, and 19.16.

We can use similar methods for every line in the three analysis of variance tables. For example, each variable $m = 1, 2, 3$ has a sum of squares for $\mathbf{S} * \mathbf{P}$, say,

$$SS(\mathbf{S} * \mathbf{P})_{mm} \equiv h(\mathbf{S} * \mathbf{P})_{mm} = 4 \sum_{h=1}^{2} \sum_{j=1}^{3} \left(\bar{y}_{h \cdot j \cdot, m} - \bar{y}_{h \cdots, m} - \bar{y}_{\cdot \cdot j \cdot, m} + \bar{y}_{\cdots, m} \right)^2.$$

We can also include cross products using $SS(\mathbf{S} * \mathbf{P})_{mm'} \equiv h(\mathbf{S} * \mathbf{P})_{mm'}$, where

$$h(\mathbf{S} * \mathbf{P})_{mm'} = 4 \sum_{h=1}^{2} \sum_{j=1}^{3} \left(\bar{y}_{h \cdot j \cdot, m} - \bar{y}_{h \cdots, m} - \bar{y}_{\cdot \cdot j \cdot, m} + \bar{y}_{\cdots, m} \right) \left(\bar{y}_{h \cdot j \cdot, m'} - \bar{y}_{h \cdots, m'} - \bar{y}_{\cdot \cdot j \cdot, m'} + \bar{y}_{\cdots, m'} \right)$$

and create a matrix

$$H(\mathbf{S} * \mathbf{P}) \equiv \begin{bmatrix} h(\mathbf{S} * \mathbf{P})_{11} & h(\mathbf{S} * \mathbf{P})_{12} & h(\mathbf{S} * \mathbf{P})_{13} \\ h(\mathbf{S} * \mathbf{P})_{21} & h(\mathbf{S} * \mathbf{P})_{22} & h(\mathbf{S} * \mathbf{P})_{23} \\ h(\mathbf{S} * \mathbf{P})_{31} & h(\mathbf{S} * \mathbf{P})_{32} & h(\mathbf{S} * \mathbf{P})_{33} \end{bmatrix}.$$

(The nice algebraic formulae only exist because the entire model is balanced.) For the fabric data

$$H(\mathbf{S} * \mathbf{P}) = \begin{bmatrix} 1186.0833 & -33.166667 & 526.79167 \\ -33.166667 & 44.333333 & -41.583333 \\ 526.79167 & -41.583333 & 250.58333 \end{bmatrix}.$$

Note that the diagonal elements of $H(\mathbf{S} * \mathbf{P})$ are the $\mathbf{S} * \mathbf{P}$ interaction sums of squares from Tables 19.14, 19.15, and 19.16. Table 19.17 contains the H matrices for all of the sources in the analysis of variance.

In the standard (univariate) analysis of y_1 that was performed in Section 16.2, the test for $\mathbf{S} * \mathbf{P}$ interactions was based on

$$F = \frac{MS(\mathbf{S} * \mathbf{P})_{11}}{MSE_{11}} = \frac{SS(\mathbf{S} * \mathbf{P})_{11}}{SSE_{11}} \frac{1/df(\mathbf{S} * \mathbf{P})}{1/dfE} = \frac{h(\mathbf{S} * \mathbf{P})_{11}}{e_{11}} \frac{dfE}{df(\mathbf{S} * \mathbf{P})}.$$

The last two equalities are given to emphasize that the test depends on the $y_{hijk,1}$s only through $h(\mathbf{S} * \mathbf{P})_{11} [e_{11}]^{-1}$. Similarly, a multivariate test of $\mathbf{S} * \mathbf{P}$ is a function of the matrices

$$H(\mathbf{S} * \mathbf{P})E^{-1},$$

where E^{-1} is the matrix inverse of E. A major difference between the univariate and multivariate procedures is that there is no uniform agreement on how to use $H(\mathbf{S} * \mathbf{P})E^{-1}$ to construct a test. The *generalized likelihood ratio* test, also known as *Wilks' lambda*, is

$$\Lambda(\mathbf{S} * \mathbf{P}) \equiv \frac{1}{|I + H(\mathbf{S} * \mathbf{P})E^{-1}|}$$

Table 19.17: *MANOVA statistics.*

$$H(GRANDMEAN) = \begin{bmatrix} 940104.17 & 752281.25 & 602260.42 \\ 752281.25 & 601983.37 & 481935.13 \\ 602260.42 & 481935.13 & 385827.04 \end{bmatrix}$$

$$H(S) = \begin{bmatrix} 26268.167 & 11479.917 & 6186.5833 \\ 11479.917 & 5017.0417 & 2703.7083 \\ 6186.5833 & 2703.7083 & 1457.0417 \end{bmatrix}$$

$$H(F) = \begin{bmatrix} 6800.6667 & 21967.500 & 18129.500 \\ 21967.500 & 70959.375 & 58561.875 \\ 18129.500 & 58561.875 & 48330.375 \end{bmatrix}$$

$$H(P) = \begin{bmatrix} 5967.5833 & 6818.2500 & 2646.9583 \\ 6818.2500 & 7969.0000 & 3223.7500 \\ 2646.9583 & 3223.7500 & 1396.5833 \end{bmatrix}$$

$$H(S*F) = \begin{bmatrix} 3952.6667 & 474.83333 & 38.500000 \\ 474.83333 & 57.041667 & 4.6250000 \\ 38.500000 & 4.6250000 & 0.37500000 \end{bmatrix}$$

$$H(S*P) = \begin{bmatrix} 1186.0833 & -33.166667 & 526.79167 \\ -33.166667 & 44.333333 & -41.583333 \\ 526.79167 & -41.583333 & 250.58333 \end{bmatrix}$$

$$H(F*P) = \begin{bmatrix} 3529.0833 & 4275.5000 & 2374.1250 \\ 4275.5000 & 6031.0000 & 2527.2500 \\ 2374.1250 & 2527.2500 & 1740.2500 \end{bmatrix}$$

$$H(S*F*P) = \begin{bmatrix} 478.58333 & 4.4166667 & 119.62500 \\ 4.4166667 & 14.333333 & -57.750000 \\ 119.62500 & -57.750000 & 272.25000 \end{bmatrix}$$

$$E = \begin{bmatrix} 3225.00 & -80.50 & 1656.50 \\ -80.50 & 2405.50 & -112.00 \\ 1656.50 & -112.00 & 2662.50 \end{bmatrix}$$

where I indicates a 3×3 identity matrix and $|A|$ denotes the determinant of a matrix A. *Roy's maximum root statistic* is the maximum eigenvalue of $H(S*P)E^{-1}$, say, $\phi_{max}(S*P)$. On occasion, Roy's statistic is taken as

$$\theta_{max}(S*P) \equiv \frac{\phi_{max}(S*P)}{1+\phi_{max}(S*P)}.$$

A third statistic is the *Lawley–Hotelling trace*,

$$T^2(S*P) \equiv dfE \ \text{tr}\left[H(S*P)E^{-1}\right],$$

and a final statistic is *Pillai's trace*,

$$V(S*P) \equiv \text{tr}\left[H(S*P)(E+H(S*P))^{-1}\right].$$

Similar test statistics Λ, ϕ, θ, T^2 and V can be constructed for all of the other main effects and interactions. It can be shown that for H terms with only one degree of freedom, these test statistics are equivalent to each other and to an F statistic. In such cases, we only present T^2 and the F value.

Table 19.18 presents the test statistics for each term. When the F statistic is exactly correct, it is given in the table. In other cases, the table presents F statistic approximations. The approximations

Table 19.18: *Multivariate statistics.*

Effect	Statistics	F	df	P
GRAND MEAN	$T^2 = 6836.64$	1899.07	3, 10	0.000
S	$T^2 = 137.92488$	38.31	3, 10	0.000
F	$T^2 = 612.96228$	170.27	3, 10	0.000
P	$\Lambda = 0.13732$	5.66	6, 20	0.001
	$T^2 = 65.31504$	8.16	6, 18	0.000
	$V = 0.97796$	3.51	6, 22	0.014
	$\phi_{max} = 5.28405$			
S * F	$T^2 = 21.66648$	6.02	3, 10	0.013
S * P	$\Lambda = 0.71068$	0.62	6, 20	0.712
	$T^2 = 4.76808$	0.60	6, 18	0.730
	$V = 0.29626$	0.64	6, 22	0.699
	$\phi_{max} = 0.37102$			
F * P	$\Lambda = 0.17843$	4.56	6, 20	0.005
	$T^2 = 46.03092$	5.75	6, 18	0.002
	$V = 0.95870$	3.38	6, 22	0.016
	$\phi_{max} = 3.62383$			
S * F * P	$\Lambda = 0.75452$	0.50	6, 20	0.798
	$T^2 = 3.65820$	0.46	6, 18	0.831
	$V = 0.26095$	0.55	6, 22	0.765
	$\phi_{max} = 0.20472$			

are commonly used and discussed; see, for example, Rao (1973, chapter 8) or Christensen (2001, Section 1.2). Degrees of freedom for the F approximations and P values are also given.

Each effect in Table 19.18 corresponds to a combination of a whole-plot effect and a whole-plot-by-subplot interaction from the split-plot analysis Table 19.7. For example, the multivariate effect **S** corresponds to combining the effects **S** and **S** * **R** from the univariate analysis. The highest-order terms in the table that are significant are the **F** * **P** and the **S** * **F** terms. Relative to the split-plot analysis, these suggest the presence of **F** * **P** interaction or **F** * **P** * **R** interaction and **S** * **F** interaction or **S** * **F** * **R** interaction. In Section 19.2, we found the merest suggestion of an **F** * **P** * **R** interaction but clear evidence of an **F** * **P** interaction; we also found clear evidence of an **S** * **F** * **R** interaction. However, the split-plot results were obtained under different, and perhaps less appropriate, assumptions.

□

To complete a multivariate analysis, additional modeling is needed (or MANOVA contrasts for balanced data). The MANOVA assumptions also suggest some alternative residual analysis. We will not discuss either of these subjects. Moreover, our analysis has exploited the balance in **S**, **F**, and **P** so that we have not needed to examine various sequences of models that would, in general, determine different H matrices for the effects. (Balance in **R** is required for the MANOVA).

Finally, a personal warning. One should not underestimate how much one can learn from simply doing the analyses for the individual variables. Personally, I would look thoroughly at each individual variable (number of rotations in our example) before worrying about what a multivariate analysis can add.

19.4   Random effects models

In this section we consider two special cases of split-plot models. First we consider a model in which several "identical" measurements are taken on the same subject. These measurements on a unit involve some random errors but they do not involve the error associated with unit-to-unit variation. (Generally, these measurements are taken at essentially the same time so that trends over time are irrelevant.) Such models are called subsampling models. The second class of models are those in which some treatment effects in an ANOVA can actually be considered as random. For simplicity, both discussions are restricted to balanced models. Unbalanced models are much more difficult to deal with and typically require a knowledge of linear model theory, cf. Christensen (2011, especially Chapter 12).

19.4.1 Subsampling

It is my impression that many of the disasters that occur in planning and analyzing studies occur because people misunderstand subsampling. The following is both a true story and part of the folklore of the Statistics program at the University of Minnesota. A graduate student wanted to study the effects of two drugs on mice. The student collected 200 observations in the following way. Two mice were randomly assigned to each drug. From each mouse, tissue samples were collected at 50 sites. The subjects were the mice because the drugs were applied to the mice, not to the tissue sites. There are two sources of variation: mouse-to-mouse variation and within-mouse variation. The 50 observations (*subsamples*) on each mouse are very useful in reducing the within-mouse variation but do nothing to reduce mouse-to-mouse variation. Relative to the mouse-to-mouse variation, which is likely to be larger than the within-mouse variation, there are only two observations that have the same treatment. As a result, each of the two treatment groups provides only one degree of freedom for estimating the variance that applies to treatment comparisons. In other words, the experiment provides two degrees of freedom for (the appropriate) error. Obviously a lot of work went into collecting the 200 observations. The work was wasted! Moreover, the problem in the design of this experiment could easily have been compounded by an analysis that ignored the subsampling problem. If subsampling is ignored in the analysis of such data, the *MSE* is inappropriately small and effects look more significant than they really are. (Fortunately, none of the many Statistics students that were approached to analyze these data were willing to do it incorrectly.)

Another example comes from Montana State University. A Range science graduate student wanted to compare two types of mountain meadows. He had located two such meadows and was planning to take extensive measurements on each. It had not occurred to him that this procedure would look only at within-meadow variation and that there was variation between meadows that he was ignoring.

Consider the subsampling model

$$y_{ijk} = \mu_i + \eta_{ij} + \varepsilon_{ijk} \tag{19.4.1}$$

where $i = 1, \ldots, a$ is the number of treatments, $j = 1, \ldots, n_i$ is the number of replications on different subjects, and $k = 1, \ldots, N$ is the number of subsamples on each subject. We assume that the ε_{ijk}s are independent $N(0, \sigma_s^2)$ random variables, that the η_{ij}s are independent $N(0, \sigma_w^2)$, and that the η_{ij}s and ε_{ijk}s are independent. The ηs indicate errors (variability) that occur from subject to subject, whereas the εs indicate errors (variability) that occur in measurements taken on a given subject. Model (19.4.1) can be viewed as a special case of a split-plot model in which there are no subplot treatments. If there are no subplot treatments, interest lies exclusively in the whole-plot analysis. The whole-plot analysis can be conducted in the usual way by taking the data to be the averages over the subsamples (subplots).

We can be more formal by using Model (19.4.1) to obtain

$$\bar{y}_{ij\cdot} = \mu_i + e_{ij} \tag{19.4.2}$$

where we define

$$e_{ij} \equiv \eta_{ij} + \bar{\varepsilon}_{ij\cdot}$$

and have $i = 1, \ldots, a$, $j = 1, \ldots, n_i$. Using Proposition 1.2.11, it is not difficult to see that the e_{ij}s are independent $N(0, \sigma_w^2 + \sigma_s^2/N)$, so that Model (19.4.2) is just an unbalanced one-way ANOVA model and can be analyzed as such. If desired, the methods of the next subsection can be used to estimate the between-unit (whole-plot) variance σ_w^2 and the within-unit (subplot) variance σ_s^2. Note that our analysis in Example 19.1.1 was actually on a model similar to (19.4.2). The data analyzed were averages of two repeat measurements of dynamic absorption.

Model (19.4.2) also helps to formalize the benefits of subsampling. We have N subsamples that lead to $\text{Var}(e_{ij}) = \sigma_w^2 + \sigma_s^2/N$. If we did not take subsamples, the variance would be $\sigma_w^2 + \sigma_s^2$, so we have reduced one of the terms in the variance by subsampling. If the within-unit variance σ_s^2 is large relative to the between-unit variance σ_w^2, subsampling can be very beneficial. If the between-unit variance σ_w^2 is substantial when compared to the within-unit variance σ_s^2, subsampling has very limited benefits. In this latter case, it is important to obtain a substantial number of true replications involving the between-unit variability with subsampling based on convenience (rather than importance).

Model (19.4.1) was chosen to have unequal numbers of units on each treatment but a balanced number of subsamples. This was done to suggest the generality of the procedure. Subsamples can be incorporated into any linear model and, as long as the number of subsamples is constant for each unit, a simple analysis can be obtained by averaging the subsamples for each unit and using the averages as data. Christensen (2011, Section 11.4) provides a closely related discussion that is not too mathematical.

19.4.2   Random effects

We begin with an example.

EXAMPLE 19.4.1.   Ott (1949) presented data on an electrical characteristic associated with ceramic components for a phonograph (one of those ancient machines that played vinyl records). Ott and Schilling (1990) and Ryan (1989) have also considered these data. Ceramic pieces were cut from strips, each of which could provide 25 pieces. It was decided to take 7 pieces from each strip, manufacture the 7 ceramic phonograph components, and measure the electrical characteristic on each. The data from 4 strips are given below. (These are actually the third through sixth of the strips reported by Ott.)

Strip	Observations						
1	17.3	15.8	16.8	17.2	16.2	16.9	14.9
2	16.9	15.8	16.9	16.8	16.6	16.0	16.6
3	15.5	16.6	15.9	16.5	16.1	16.2	15.7
4	13.5	14.5	16.0	15.9	13.7	15.2	15.9

The standard analysis looks for differences between the means of these four specific ceramic strips. An alternative approach to these data is to think of the four ceramic strips as being a random sample from the population of ceramic strips that are involved in making the assemblies. If we do that, we have two sources of variability, variability among the observations on a given strip and variability between different ceramic strips. Our goal in this subsection is to estimate the variances and test whether there is any variability between strips.   □

Consider a balanced one-way ANOVA model

$$y_{ij} = \mu + \alpha_i + \varepsilon_{ij}$$

where $i = 1,\ldots,a$ and $j = 1,\ldots,N$. As usual, we assume that the ε_{ij}s are independent $N(0,\sigma^2)$ random variables, but now, rather than assuming that the α_is are fixed treatment effects, we assume that they are *random treatment effects*. In particular, assume that the α_is are independent $N(0,\sigma_A^2)$ random variables that are also independent of the ε_{ij}s. This model can be viewed as a split-plot model in which there are no whole-plot factors or subplot factors.

The analysis revolves around the analysis of variance table and the use of Proposition 1.2.11. As usual in a one-way ANOVA, begin with the summary statistics $\bar{y}_{i\cdot}$ and s_i^2, $i = 1,\ldots,a$. In comparing the observations within a single strip, there is no strip-to-strip variability. The sample variances s_i^2

each involve comparisons only within a given strip, so each provides an estimate of the within-strip variance, σ^2. In particular, $E(s_i^2) = \sigma^2$. Clearly, if we pool these estimates we continue to get an estimate of σ^2. In particular,

$$E(MSE) = \sigma^2.$$

We now examine $MSGrps$. Before proceeding note that by independence of the α_is and the ε_{ij}s,

$$
\begin{aligned}
\text{Var}(y_{ij}) &= \text{Var}(\mu + \alpha_i + \varepsilon_{ij}) \\
&= \text{Var}(\alpha_i) + \text{Var}(\varepsilon_{ij}) \\
&= \sigma_A^2 + \sigma^2.
\end{aligned}
$$

Thus $\text{Var}(y_{ij})$ is the sum of two *variance components* σ_A^2 and σ^2. Moreover,

$$
\begin{aligned}
\text{Var}(\bar{y}_{i\cdot}) &= \text{Var}(\mu + \alpha_i + \bar{\varepsilon}_{i\cdot}) \\
&= \text{Var}(\alpha_i) + \text{Var}(\bar{\varepsilon}_{i\cdot}) \\
&= \sigma_A^2 + \frac{\sigma^2}{N}
\end{aligned}
$$

because $\bar{\varepsilon}_{i\cdot}$ is the sample mean of N independent random variables that have variance σ^2. It is easily seen that $E(\bar{y}_{i\cdot}) = \mu$. The $\bar{y}_{i\cdot}$s form a random sample of size a, cf. Christensen (1996, Chapter 5). The population that they are sampled from is $N(\mu, \sigma_A^2 + \sigma^2/N)$. Clearly, the sample variance of the $\bar{y}_{i\cdot}$s provides an estimate of $\sigma_A^2 + \sigma^2/N$. The $MSGrps$ is N times the sample variance of the $\bar{y}_{i\cdot}$s, so $MSGrps$ provides an unbiased estimate of $N\sigma_A^2 + \sigma^2$.

We already have an estimate of σ^2. To obtain an estimate of σ_A^2 use the results of the previous paragraph and take

$$\hat{\sigma}_A^2 = \frac{MSGrps - MSE}{N}.$$

It is a simple exercise to show that

$$E(\hat{\sigma}_A^2) = \sigma_A^2.$$

Note, however, that the quality of this estimate depends crucially on a, the number of groups. To see this, note that the best estimate we could get for σ_A^2 would be if we actually got to see the α_is. In that case, $\sum_{i=1}^{a} \alpha_i^2/a$ is the best estimate we could get of σ_A^2; an estimate that has a degrees of freedom. If a is small, the estimate will be lousy. But we cannot even do as well as this. We don't get to see the α_is, we have to estimate them with $\bar{y}_{i\cdot} - \bar{y}_{\cdot\cdot}$ and then $\sum_{i=1}^{a}(\bar{y}_{i\cdot} - \bar{y}_{\cdot\cdot})^2/(a-1)$ gives us an $a-1$ degree-of-freedom estimate of $\sigma_A^2 + \sigma^2/N$; not even an estimate of σ_A^2. To get a good estimate of σ_A^2, we need a large; not N.

The usual F statistic is $MSGrps/MSE$. Clearly, it is a (biased) estimate of

$$\frac{N\sigma_A^2 + \sigma^2}{\sigma^2} = 1 + \frac{N\sigma_A^2}{\sigma^2}.$$

If $H_0 : \sigma_A^2 = 0$ holds, the F statistic should be about 1. In general, if H_0 holds,

$$\frac{MSGrps}{MSE} \sim F(a-1, dfE),$$

and the usual F test can be interpreted as a test of $H_0 : \sigma_A^2 = 0$. Interestingly, however, for this test to be good, we need N large; not a. Typically, it is easier to get N large than it is to get a large, so typically it is easier to tell whether $\sigma_A^2 \neq 0$ than it is to tell what σ_A^2 actually is.

EXAMPLE 19.4.1 CONTINUED.   For the electrical characteristic data, the analysis of variance table is given below.

Analysis of Variance: Electrical characteristic data

Source	df	SS	MS	F	P
Treatments	3	10.873	3.624	6.45	0.002
Error	24	13.477	0.562		
Total	27	24.350			

The F statistic shows strong evidence that variability exists between ceramic strips. The estimate of within-strip variability is $MSE = 0.562$. With 7 observations on each ceramic strip, the estimate of between-strip variability is

$$\hat{\sigma}_A^2 = \frac{MSGrps - MSE}{N} = \frac{3.624 - 0.562}{7} = 0.437,$$

but it is not a very good estimate, being worse than an estimate based on 3 degrees of freedom.

While in many ways this random effects analysis seems more appropriate for the relatively undifferentiated strips being considered, this analysis also seems less informative for these data than the fixed effects analysis. It is easy to see that most of the between-strip "variation" is due to a single strip, number 4, being substantially different from the others. Are we to consider this strip an outlier in the population of ceramic strips? Having three sample means that are quite close and one that is substantially different certainly calls into question the assumption that the random treatment effects are normally distributed. Most importantly, some kind of analysis that looks at individual sample means is necessary to have any chance of identifying an odd strip.   □

While the argument given here works only for balanced data, the corresponding model fitting ideas give similar results for unbalanced one-way ANOVA data. In particular,

$$\hat{\sigma}_A^2 = \frac{SSGrps - MSE(a-1)}{n - \sum_{i=1}^{a} N_i^2/n},$$

and the usual F test gives an appropriate test of $H_0 : \sigma_A^2 = 0$. Christensen (2011, Section 12.9, Subsection 12.10.11) provides theoretical justification for these claims but does not treat this particular example.

The ideas behind the analysis of the balanced one-way ANOVA model generalize nicely to other balanced models. Consider the balanced two-way with replication,

$$y_{ijk} = \mu + \alpha_i + \beta_j + \gamma_{ij} + \varepsilon_{ijk},$$

where $i = 1, \ldots, a$, $j = 1, \ldots, b$, and $k = 1, \ldots, N$. Assume that the ε_{ijk}s are independent $N(0, \sigma^2)$ random variables, that the γ_{ij}s are independent $N(0, \sigma_\gamma^2)$, and that the ε_{ijk}s and γ_{ij}s are independent. This model involves two variance components, σ_γ^2 and σ^2.

The theory alluded to earlier leads to the following results. MSE still estimates σ^2. $MS(\gamma)$ estimates

$$E[MS(\gamma)] = \sigma^2 + N\sigma_\gamma^2.$$

The usual interaction test is a test of $H_0 : \sigma_\gamma^2 = 0$.

In addition, for main effects, $MS(\beta)$ estimates

$$E[MS(\beta)] = \sigma^2 + N\sigma_\gamma^2 + \frac{aN}{b-1} \sum_{j=1}^{b} (\beta_j - \bar{\beta}_.)^2.$$

When the β_js are all equal, $MS(\beta)$ estimates $\sigma^2 + N\sigma_\gamma^2$. It follows that to obtain an F test for equality of the β_js, the test must reject when $MS(\beta)$ is much larger than $MS(\gamma)$. In particular, an α-level test rejects if

$$\frac{MS(\beta)}{MS(\gamma)} > F(1 - \alpha, a - 1, [a - 1][b - 1]).$$

Table 19.19: *Cornell's scaled vinyl thickness values.*

		Replication 1				Replication 2			
	Rate	High	Low	Low	High	High	Low	Low	High
	Temp	Low	High	Low	High	Low	High	Low	High
	1	8	12	7	12	7	10	8	11
	2	6	9	7	10	5	8	6	9
Blend	3	10	13	9	14	11	12	10	12
	4	4	6	5	6	5	3	4	5
	5	11	15	9	13	10	11	7	9

This is just the usual result *except that the MSE has been replaced by the MS(γ)*. The analysis of effects, i.e., further modeling or contrasts, involving the β_js also follows the standard pattern but with $MS(\gamma)$ used in place of *MSE*. Similar results hold for investigating the α_is. Basically, you can think of the ε_{ijk}s as subsampling errors and do the analysis on the $\bar{y}_{ij}$.s.

The moral of this analysis is that one needs to think very carefully about whether to model interactions as fixed effects or random effects. It would seem that if you do not care about interactions, if they are just an annoyance in evaluating the main effects, you probably should treat them as random and use the interaction mean square as the appropriate estimate of variability. A related way of thinking is to stipulate that you do not care about any main effects unless they are large enough to show up above any interaction. In particular, that is essentially what is done in a randomized complete block design. An RCB takes the block-by-treatment interaction as the error and only treatment effects that are strong enough to show up over and above any block-by-treatment interaction are deemed significant. On the other hand, if interactions are something of direct interest, they should typically be treated as fixed effects.

19.5 Exercises

EXERCISE 19.5.1.   In Exercises 17.11.3, 17.11.4, and 18.7.1, we considered data from Cornell (1988) on scaled vinyl thicknesses. Exercise 17.11.3 involved five blends of vinyl and we discussed the fact that the production process was set up eight times with a group of five blends run on each setting. The eight production settings were those in Exercise 17.11.4. The complete data are displayed in Table 19.19.

(a) Identify the design for this experiment and give an appropriate model. List all the assumptions made in the model.

(b) Analyze the data. Give an appropriate analysis of variance table. Examine appropriate contrasts using the LSD method with an α of .05.

(c) Check the assumptions of the model and adjust the analysis appropriately.

(d) Discuss the relationship between the current analysis and those conducted earlier.

EXERCISE 19.5.2.   Wilm (1945) presented data involving the effect of logging on soil moisture deficits under a forest. Treatments consist of five intensities of logging. Treatments were identified as the volume of the trees left standing after logging that were larger than 9.6 inches in diameter. The logging treatments were uncut, 6000 board-feet, 4000 board-feet, 2000 board-feet, 0 board-feet. The experiment was conducted by selecting four blocks (A,B,C,D) of forest. These were subdivided into five plots. Within each block each of the treatments were randomly assigned to a plot. Soil moisture deficits were measured in each of three consecutive years, 1941, 1942, and 1943. The data are presented in Table 19.20.

Table 19.20: *Soil moisture deficits as affected by logging.*

Treatment	Year	Block			
		A	B	C	D
Uncut	41	2.40	0.98	1.38	1.37
	42	3.32	1.91	2.36	1.62
	43	2.59	1.44	1.66	1.75
6000	41	1.76	1.65	1.69	1.11
	42	2.78	2.07	2.98	2.50
	43	2.27	2.28	2.16	2.06
4000	41	1.43	1.30	0.18	1.66
	42	2.51	1.48	1.83	2.36
	43	1.54	1.46	0.16	1.84
2000	41	1.24	0.70	0.69	0.82
	42	3.29	2.00	1.38	1.98
	43	2.67	1.44	1.75	1.56
None	41	0.79	0.21	0.01	0.16
	42	1.70	1.44	2.65	2.15
	43	1.62	1.26	1.36	1.87

Treatments are volumes of timber left standing
in trees withdiameters greater than 9.6 inches.
Volumes are measured in board-feet.

(a) Identify the design for this experiment and give an appropriate model. List all the assumptions made in the model.

(b) Analyze the data. Give an appropriate analysis of variance table. Examine appropriate contrasts. In particular, compare the uncut plots to the average of the other plots and use polynomials to examine differences among the other four treatments. Discuss the reasonableness of this procedure in which the 'uncut' treatment is excluded when fitting the polynomials.

(c) Check the assumptions of the model and adjust the analysis appropriately. What assumptions are difficult to check? Identify any such assumptions that are particularly suspicious.

EXERCISE 19.5.3.   Day and del Priore (1953) report data from an experiment on the noise generated by various reduction gear designs. The data were collected because of the Navy's interest in building quiet submarines. Primary interest focused on the direction of lubricant application. Lubricants were applied either inmesh (I) or tangent (T) and either at the top (T) or the bottom (B). Thus the direction TB indicates tangent, bottom while IT is inmesh, top.

Four additional factors were considered. Load was 25%, 100%, or 125%. The temperature of the input lubricant was 90, 120, or 160 degrees F. The volume of lubricant flow was 0.5 gpm, 1 gpm, or 2 gpm. The speed was either 300 rpm or 1200 rpm. Temperature and volume were of less interest than direction; speed and load were of even less interest. It was considered that load, temperature, and volume would not interact but that speed might interact with the other factors. There was little idea whether direction would interact with other factors. As a result, a split-plot design with whole plots in a 3 × 3 Latin square was used. The factors used in defining the whole-plot Latin square were load, temperature, and volume. The subplot factors were speed and the direction factors.

The data are presented in Table 19.21. The four observations with 100% load, 90-degree temperature, 0.5-gpm volume, and lubricant applied tangentially were not made. Substitutes for these values were used. As an approximate analysis, treat the substitute values as real values but subtract four degrees of freedom from the subplot error. Analyze the data.

EXERCISE 19.5.4.   In Exercise 16.4.1 and Table 16.19 we presented Baten's (1956) data on lengths of steel bars. The bars were made with one of two heat treatments (W, L) and cut on one of four screw machines (A, B, C, D) at one of three times of day (8 am, 11 am, 3 pm). There are

Table 19.21: *Gear test data.*

Load	Direction	.5 gpm		1 gpm		2 gpm	
		Temp 120		Temp 90		Temp 160	
	TB	92.7	81.4	91.3	68.0	86.9	78.2
25%	IT	95.9	79.2	87.7	77.7	90.7	97.9
	IB	92.7	85.5	93.6	76.2	92.1	80.2
	TT	92.2	81.4	92.9	72.2	90.6	85.8
		Temp 90		Temp 160		Temp 120	
	TB	94.2*	80.2*	89.7	86.0	91.9	84.8
100%	IT	88.6	83.7	87.8	86.6	85.4	79.5
	IB	89.8	83.9	90.4	79.3	85.7	86.9
	TT	89.8*	75.4*	90.4	85.0	82.6	79.0
		Temp 160		Temp 120		Temp 90	
	TB	88.7	94.2	90.3	86.7	88.4	75.8
125%	IT	92.1	91.1	90.3	83.5	86.3	71.2
	IB	91.7	89.2	90.4	86.6	88.3	87.9
	TT	93.4	86.2	89.7	83.0	88.6	84.5
		300	1200	300	1200	300	1200
		Speed		Speed		Speed	
* indicates a replacement for missing data.							

distressing aspects to Baten's article. First, he never mentions what the heat treatments are. Second, he does not discuss how the four screw machines differ or whether the same person operates the same machine all the time. If the machines were largely the same and one specific person always operates the same machine all the time, then machine differences would be due to operators rather than machines. If the machines were different and one person operates the same machine all the time, it becomes impossible to tell whether machine differences are due to machines or operators. Most importantly, Baten does not discuss how the replications were obtained. In particular, consider the role of day-to-day variation in the analysis.

If the 12 observations on a heat treatment–machine combination are all taken on the same day, there is no replication in the experiment that accounts for day-to-day variation. In that case the average of the four numbers for each heat treatment–machine–time combination gives essentially one observation and for each heat treatment–machine combination the three time means are correlated. To obtain an analysis, the heat–machine interaction and the heat–machine–time interaction would have to be used as the two error terms.

Suppose the 12 observations on a heat treatment–machine combination are taken on four different days with one observation obtained on each day for each time period. Then the three observations on a given day are correlated but the observations on different days are independent. This leads to a traditional split-plot analysis.

Finally, suppose that the 12 observations on a heat treatment–machine combination are all taken on 12 different days. Yet another analysis is appropriate.

Compare the results of these three different methods of analyzing the experiment. If the day-to-day variability is no larger than the within-day variability, there should be little difference. When considering the analysis that assumes 12 observations taken on four different days, treat the order of the four heat treatment–machine–time observations as indicating the day. For example, with heat treatment W and machine A, take 9, 3, and 4 as the three time observations on the second day.

EXERCISE 19.5.5.   Reanalyze Mandel's (1972) data from Example 12.4.1 and Table 12.4 assuming that the five laboratories are a random sample from a population of laboratories. Include estimates of both variance components.

EXERCISE 19.5.6.   Reanalyze the data of Example 17.4.1 assuming that the Disk-by-Window interaction is a random effect. Include estimates of both variance components.

Table 19.22: *Snedecor and Haber (1946) cutting dates on asparagus.*

Year	a	b	c	a	b	c	a	b	c
					Treatments				
29	201	301	362	185	236	341	209	226	357
30	230	296	353	216	256	328	219	212	354
31	324	543	594	317	397	487	357	358	560
32	512	778	755	448	639	622	496	545	685
33	399	644	580	361	483	445	344	415	520
34	891	1147	961	783	998	802	841	833	871
35	449	585	535	409	525	478	418	451	538
36	595	807	548	566	843	510	622	719	578
37	632	804	565	629	841	576	636	735	634
38	527	749	353	527	823	299	530	731	413
29	219	330	427	225	307	382	219	342	464
30	222	301	391	239	297	321	216	287	364
31	348	521	599	347	463	502	356	557	584
32	487	742	802	512	711	684	508	768	819
33	372	534	573	405	577	467	377	529	612
34	773	1051	880	786	1066	763	780	969	1028
35	382	570	540	415	610	468	407	526	651
36	505	737	577	549	779	548	595	772	660
37	534	791	524	559	741	621	626	826	673
38	434	614	343	433	706	352	518	722	424

Blocks are indicated by vertical and horizontal lines.

EXERCISE 19.5.7.   People who really want to test their skill may wish to examine the data presented in Snedecor and Haber (1946) and repeated in Table 19.22. The experiment was to examine the effects of three cutting dates on asparagus. Six blocks were used. One plot was assigned a cutting date of June 1 (a), one a cutting date of June 15 (b), and the last a cutting date of July 1 (c). Data were collected on these plots for 10 years.

Try to come up with an intelligible summary of the data that would be of use to someone growing asparagus. In particular, the experiment was planned to run for the effective lifetime of the planting, normally 20 years or longer. The experiment was cut short due to lack of labor but interest remained in predicting behavior ten years after the termination of data collection. As most effects seem to be significant, I would be inclined to focus on effects that seem relatively large rather than on statistically significant effects.

EXERCISE 19.5.8.   Reconsider the data of Exercises 15.5.2, 15.5.3, and Table 15.10 from Smith, Gnanadesikan, and Hughes (1962). Perform a multivariate ACOVA on the data. Are the data repeated measures data? Is it reasonable to apply a split-plot model to the data? If so, do so.

Chapter 20

Logistic Regression: Predicting Counts

For the most part, this book concerns itself with measurement data and the corresponding analyses based on normal distributions. In this chapter and the next we consider data that consist of counts. Elementary count data were introduced in Chapter 5.

Frequently, data are collected on whether or not a certain event occurs. A mouse dies when exposed to a dose of chloracetic acid or it does not. In the past, O-rings failed during a space shuttle launch or they did not. Men have coronary incidents or they do not. These are modeled as random events and we collect data on how often the event occurs. We also collect data on potential predictor (explanatory) variables. For example, we use the size of dose to estimate the probability that a mouse will die when exposed. We use the atmospheric temperature at launch time to estimate the probability that O-rings fail. We may use weight, cholesterol, and blood pressure to estimate the probability that men have coronary incidents. Once we have estimated the probability that these events will occur, we are ready to make predictions. In this chapter we investigate the use of logistic models to estimate probabilities. Logistic models (also known as logit models) are linear models for the log-odds that an event will occur. For a more complete discussion of logistic and logit models see Christensen (1997).

Section 20.1 introduces models for predicting count data. Section 20.2 presents a simple model with one predictor variable where the data are the proportions of trials that display the event. It also discusses the output one typically obtains from running a logistic regression program. Section 20.3 discusses how to perform model tests with count data. Section 20.4 discusses how logistic models are fitted. Section 20.5 introduces the important special case in which each observation is a separate trial that either displays the event or does not. Section 20.6 explores the use of multiple continuous predictors. Section 20.7 examines ANOVA type models with Section 20.8 examining ACOVA type models.

20.1 Models for binomial data

Logistic regression is a method of modeling the relationships between probabilities and predictor variables. We begin with an example.

EXAMPLE 20.1.1. Woodward et al. (1941) reported data on 120 mice divided into 12 groups of 10. The mice in each group were exposed to a specific dose of chloracetic acid and the observations consist of the number in each group that lived and died. Doses were measured in grams of acid per kilogram of body weight. The data are given in Table 20.1, along with the proportions y_h of mice who died at each dose x_h.

We could analyze these data using the methods discussed earlier in Chapter 5. We have samples from twelve populations. We could test to see if the populations are the same. We don't think they are because we think survival depends on dose. More importantly, we want to try to model the relationship between dose level and the probability of dying, because that allows us to make predictions about the probability of dying for any dose level that is similar to the doses in the original data. □

In Section 3.1 we talked about models for measurement data y_h, $h = 1, \ldots, n$ with $E(y_h) \equiv \mu_h$

Table 20.1: *Lethality of chloracetic acid.*

Dose (x_h)	Group (h)	Died	Survived	Total	Proportion (y_h)
.0794	1	1	9	10	0.1
.1000	2	2	8	10	0.2
.1259	3	1	9	10	0.1
.1413	4	0	10	10	0.0
.1500	5	1	9	10	0.1
.1588	6	2	8	10	0.2
.1778	7	4	6	10	0.4
.1995	8	6	4	10	0.6
.2239	9	4	6	10	0.4
.2512	10	5	5	10	0.5
.2818	11	5	5	10	0.5
.3162	12	8	2	10	0.8

and $\mathrm{Var}(y_h) = \sigma^2$. For testing models, we eventually assumed

$$y_h\text{s} \quad \text{independent} \quad N(\mu_h, \sigma^2),$$

with some model for the μ_hs. In Section 3.9 we got more specific about models, writing

$$y_h\text{s} \quad \text{independent} \quad N[m(x_h), \sigma^2],$$

where x_h is the value of some predictor variable or vector and $m(\cdot)$ is the model for the means, i.e.,

$$\mu_h \equiv m(x_h).$$

We then discussed a variety of models $m(\cdot)$ that could be used for various types of predictor variables and exploited those models in subsequent chapters.

In this chapter, we discuss similar models for data that are binomial proportions. In Section 1.4 we discussed binomial sampling. In particular, if we have N independent trials of whether some event occurs (e.g., flipping a coin and seeing heads) and if each trial has the same probability p that the event occurs, then the number of occurrences is a binomial random variable W, say

$$W \quad \sim \quad \mathrm{Bin}(N, p),$$

with

$$E(W) = Np \quad \text{and} \quad \mathrm{Var}(W) = Np(1-p).$$

We will be interested in binomial proportions

$$y \equiv \frac{W}{N},$$

with

$$E(y) = p$$

and

$$\mathrm{Var}(y) = \frac{p(1-p)}{N},$$

see Proposition 1.2.11. In applications, N is known and p is an unknown parameter to be modeled and estimated.

In general, we assume n independent binomial proportions y_h for which we know the number of trials N_h, i.e.,

$$N_h y_h \quad \text{independent} \quad \mathrm{Bin}(N_h, p_h), \qquad h = 1, \ldots, n.$$

With $E(y_h) = p_h$, much like we did for measurement data, we want to create a model for the p_hs that depends on a predictor x_h. In fact, we would like to use the same models, simple linear regression, multiple regression, one-way ANOVA and multifactor ANOVA, that we used for measurement data. But before we can do that, we need to deal with a problem.

We want to create models for $p_h = E(y_h)$, but with binomial proportions this mean value is always a probability and probabilities are required to be between 0 and 1. If we wrote a simple linear regression model such as $p_h = \beta_0 + \beta_1 x_h$ for some predictor variable x, nothing forces the probabilities to be between 0 and 1. When modeling probabilities, it seems reasonable to ask that they be between 0 and 1.

Rather than modeling the probabilities directly, we model a function of the probabilities that is not restricted between 0 and 1. In particular, we model the log of the odds, rather than the actual probabilities. The odds O_h are defined to be the probability that the event occurs, divided by the probability that it does not occur, thus

$$O_h \equiv \frac{p_h}{1 - p_h}.$$

Probabilities must be between 0 and 1, so the odds can take any values between 0 and $+\infty$. Taking the log of the odds permits any values between $-\infty$ and $+\infty$, so we consider models

$$\log\left(\frac{p_h}{1 - p_h}\right) = m(x_h), \qquad (20.1.1)$$

where $m(\cdot)$ is any of the models that we considered earlier.

Two different names have been used for such models. If $m(x_h)$ corresponds to a one-sample, two-sample, one-way ANOVA, or multifactor ANOVA, these models have often been called *logit models*. The name stems from using the transformation

$$\eta = f(p) \equiv \log\left(\frac{p}{1 - p}\right),$$

which is known as the *logit transform*. It maps the unit interval into the real line. On the other hand, if the model $m(x_h)$ corresponds to any sort of regression model, models like (20.1.1) are called *logistic regression* models. These models are named after the *logistic transform*, which is the inverse of the logit transform,

$$p = g(\eta) \equiv \frac{e^\eta}{1 + e^\eta}.$$

The functions are inverses in the sense that $g(f(p)) = p$ and $f(g(\eta)) = \eta$. To perform any worthwhile data analysis requires using both the logit transform and the logistic transform, so it really does not matter what you call the models. These days, any model of the form (20.1.1) is often called logistic regression, regardless of whether $m(x_h)$ corresponds to a regression model.

In Chapter 3, to perform tests and construct confidence intervals, we assumed that the y_h observations were independent, with a common variance σ^2, and normally distributed. In this chapter, to perform tests and construct confidence intervals similar to those used earlier, we need to rely on having large amounts of data. That can happen in two different ways. The best way is to have the N_h values large for every value of h. In the chloracetic acid data, each N_h is 10, which is probably large enough. Unfortunately, this best way to have the data may be the least common way of actually obtaining data. The other and more common way to get a lot of data is to have the number of proportions n reasonably large but the N_hs possibly small. Frequently, the N_hs all equal 1. When worrying about O-ring failure, each shuttle launch is a separate trial, $N_h = 1$, but we have $n = 23$ launches to examine. When examining coronary incidents, each man is a separate trial, $N_h = 1$, but we have $n = 200$ men to examine. In other words, if the N_hs are all large, we don't really care if n is large or not. If the N_hs are not all large, we need n to be large. A key point is that n needs to be large relative to the number of parameters we fit in our model. For the O-ring data, we will only fit two parameters, so $n = 23$ is probably reasonable. For the coronary incident data, we have many more predictors, so we need many more subjects. In fact, we will need to resist the temptation to fit too many parameters to the data.

20.2   Simple linear logistic regression

In simple linear logistic regression we use a single measurement variable to predict probabilities.

EXAMPLE 20.2.1.   In Example 20.1.1 and Table 20.1 we presented the data of Woodward et al. (1941) on the slaughter of mice. These data are extremely well behaved in that they all have the same reasonably large number of trials $N_h = 10$, $h = 1, \ldots, 12$, and there is only one measurement predictor variable, the dose x_h.

A simple linear logistic regression model has

$$\log\left(\frac{p_h}{1 - p_h}\right) = \beta_0 + \beta_1 x_h, \tag{20.2.1}$$

so our model fits a straight line in dose to the log-odds. Alternatively,

$$p_h = \frac{e^{\beta_0 + \beta_1 x_h}}{1 + e^{\beta_0 + \beta_1 x_h}}.$$

Indeed, for an arbitrary dose x we can write

$$p(x) = \frac{e^{\beta_0 + \beta_1 x}}{1 + e^{\beta_0 + \beta_1 x}}. \tag{20.2.2}$$

Standard computer output involves a table of coefficients:

Table of Coefficients: Model (20.2.1).

Predictor	$\hat{\beta}_k$	SE($\hat{\beta}_k$)	t	P
Constant	-3.56974	0.705330	-5.06	0.000
Dose	14.6369	3.33248	4.39	0.000

The validity of everything but the point estimates relies on having large amounts of data. Using the point estimates gives the *linear predictor*

$$\hat{\eta}(x) = \hat{\beta}_0 + \hat{\beta}_1 x = -3.56974 + 14.6369x.$$

Applying the logistic transformation to the linear predictor gives the estimated probability for any x,

$$\hat{p}(x) = \frac{e^{\hat{\eta}(x)}}{1 + e^{\hat{\eta}(x)}}.$$

This function is plotted in Figure 20.1. The approximate model is unlikely to fit well outside the range of the x_h values that actually occurred in Table 20.1, although since this range of x_h values gets the fitted values reasonably close to both zero and one, predicting outside the range of the observed doses may be less of a problem than in regression for measurement data.

The table of coefficients is used exactly like previous tables of coefficients, e.g., $\hat{\beta}_1 = 14.64$ is the estimated slope parameter and SE($\hat{\beta}_1$) = 3.326 is its standard error. The t values are simply the estimates divided by their standard errors, so they provide statistics for testing whether the regression coefficient equals 0. The P values are based on large sample normal approximations, i.e., the t statistics are compared to a $t(\infty)$ distribution. Clearly, there is a significant effect for fitting the dose, so we reject the hypothesis that $\beta_1 = 0$. The dose helps explain the data.

Many computer programs expand the table of coefficients to include odds ratios, defined as

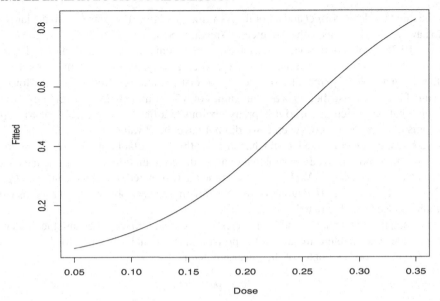

Figure 20.1: *Fitted probabilities as a function of dose.*

$\xi_k \equiv e^{\beta_k}$, and a confidence interval for the odds ratio. The $(1 - \alpha)$ confidence interval for ξ_k is typically found by exponentiating the limits of the confidence interval for β_k, i.e., it is (e^{L_k}, e^{U_k}) where $L_k \equiv \hat{\beta}_k - t(1 - \alpha/2, \infty)\mathrm{SE}(\hat{\beta}_k)$ and $U_k \equiv \hat{\beta}_k + t(1 - \alpha/2, \infty)\mathrm{SE}(\hat{\beta}_k)$ provide the $(1 - \alpha)100\%$ confidence limits for β_k.

Additional standard output includes the Log-Likelihood $= -63.945$ (explained in Section 20.4) and a model-based χ^2 test for $\beta_1 = 0$ that is explained in Section 20.3. The model-based test for $\beta_1 = 0$ has $G^2 = 23.450$ with $df = 1$ and a P value of 0.000, obtained by comparing 23.450 to a $\chi^2(1)$ distribution. This test provides substantial evidence that death is related to dose. □

20.2.1 Goodness-of-fit tests

Computer programs written specifically for logistic regression frequently report goodness-of-fit tests. If a valid goodness-of-fit test is rejected, it suggests that the fitted model is wrong. Typical output is

Goodness-of-Fit Tests.

Method	Chi-Square	df	P
Pearson (X^2)	8.7421	10	0.557
Deviance (G^2)	10.2537	10	0.419
Hosmer–Lemeshow	6.7203	4	0.151

There are problems with this output. First of all, as Hosmer, Lemeshow, and colleagues established in Hosmer et al. (1997), their χ^2 test isn't worth the toner it takes to print it. It amazes me that so many programs persist in computing it. (It is not a bad idea for a test statistic, but there was never any reason to think the statistic had a χ^2 distribution.) Indeed, for many data (perhaps most), even the Pearson and deviance statistics do not have χ^2 distributions, because the N_hs are not all large. The mouse data do not have this problem. Finally, the deviance reported in the table is often problematic in many specialized programs for doing logistic regression. The deviance is well-defined for the mouse data because all the x_hs are distinct, but as we will see later, specialized programs for logistic regression frequently pool cases together to increase the size of the N_hs, which can destroy the usefulness of the numbers reported as the deviance.

If the fitted model is correct and all of the N_h values are large, the Pearson and deviance statistics should have χ^2 distributions with df degrees of freedom. So the question becomes, "Do $X^2 = 8.7421$ and $G^2 = 10.254$ look like they could reasonably come from a $\chi^2(10)$ distribution?" To answer that question, check whether the P values are small. Alternatively, we could compare the test statistics to values from a table of percentiles for the $\chi^2(10)$ distribution; see Appendix B.2. However, since the mean of a $\chi^2(df)$ distribution is df, our values of 8.7421 and 10.254 are very close to the mean of the distribution, which is 10, so it is pretty obvious that the data are consistent with the simple linear logistic regression model even if we did not have the P values given to us. The Pearson and deviance statistics are computed as in Chapter 5 for the 12×2 table of 12 rows (dose groups) and 2 columns (Died and Survived), except that to make the computations one must define the observed counts as $O_{h1} = N_h y_h$, $O_{h2} = N_h(1 - y_h)$ and define the (estimated) expected counts as $\hat{E}_{h1} = N_h \hat{p}_h$, and $\hat{E}_{h2} = N_h(1 - \hat{p}_h)$. The 10 degrees of freedom are the number of rows $n = 12$ minus the number of parameters we fit in the model, $p = 2$.

The reason that the deviance and Pearson tests work as advertised is because the fitted regression model provides reasonable estimates of the probabilities for each case, i.e., for $h = 1, \ldots, n$ Model (20.2.1) provides good estimates of the linear predictor

$$\hat{\eta}_h \equiv \hat{\beta}_0 + \hat{\beta}_1 x_h$$

and

$$\hat{p}_h \equiv \frac{e^{\hat{\eta}_h}}{1 + e^{\hat{\eta}_h}} = \hat{p}(x_h),$$

but in addition with large N_hs the values y_h from Table 20.1 provide reasonable estimates for the twelve death probabilities without fitting any obvious model. The problem with the Pearson and deviance goodness-of-fit tests is that when some of the N_hs are small, the y_hs no longer provide good estimates of the case probabilities, whence the $\chi^2(df)$ is no longer an appropriate reference distribution for the Pearson and deviance statistics.

As we will see in Section 20.5, in an attempt to get valid goodness-of-fit tests, many computer programs for logistic regression redefine the N_hs to make them larger (and n smaller). They do this by pooling together any cases that have exactly the same predictor variables x. With continuous predictors I have never seen this pooling procedure get the N_hs large enough to validate a χ^2 distribution but we will see that it is certainly possible.

Although the deviance G^2 may or may not provide a valid goodness-of-fit test, ideally the deviance is extremely useful for constructing model tests. Unfortunately, different models with different predictors typically have different poolings, which destroys the usefulness of the deviance as a tool for comparing models. When using logistic regression programs, one must compare models by constructing the likelihood ratio test statistic from the reported log-likelihoods, rather than the deviance. It is also possible to fit logistic regression models by using programs for fitting *generalized linear models*. ("Generalized linear models" are something distinct from "general linear models.") Generalized linear model programs rarely indulge in the pooling silliness that logistic regression programs often display, so their reported deviance values can be used to compare models.

20.2.2 Assessing predictive ability

We can measure the predictive ability of the model through R^2, which is the squared correlation between the y_h values and the $\hat{p}_h$ values. For these data $R^2 = 0.759$, which is quite high for a logistic regression. The high value is related to the fact that we have 10 observations in each binomial proportion. We are evaluating the model on its ability to predict the outcome of 10 trials, not predicting the outcome of one trial.

Frequently, with *dose-response* data like the Woodward data, one uses the log-dose as a predictor, i.e., the model becomes

$$\log \left(\frac{p_h}{1 - p_h} \right) = \beta_0 + \beta_1 \log(x_h).$$

For these data we get $R^2 = 0.760$ based on the log dose, which indicates that log-dose is not much of an improvement over dose.

Suppose we want to predict a new value y_0 that is the binomial proportion from N_0 trials observed at x_0. The predictive ability of a model depends a great deal on where the predictor variables are located. At $x_0 = -\beta_0/\beta_1$, from Equation (20.2.2) the probability is $p(x_0) = 0.5$. If $N_0 = 1$, nobody can predict well an individual 50:50 outcome like a coin toss, but, as N_0 increases, the binomial proportion y_0 has less variability, so ideally it can be predicted better. As x_0 gets further from $-\beta_0/\beta_1$, $p(x_0)$ gets further from 0.5, so closer to 0 or 1. When the probability is close to 0 or 1, predicting the outcome is relatively easy for any value of N_0.

For these data, $-\beta_0/\beta_1$ is called the LD_{50}, which denotes the *lethal dose 50* and is defined to be the dose at which lethality is 50%. In other contexts this number might be called the *effective dose 50*, denoted ED_{50}.

Any overall measure of the predictive ability of a model, like R^2, compares the y_h values to the $\hat{p}(x_h)$ values, $h = 1, \ldots, n$. Any such measure depends a great deal on the extent to which the x_h values have true probabilities close to 0 or 1, which can always be predicted well, or x_h values with true probabilities close to 0.5, where predictability depends crucially on N_h. For the mouse data, we defined $n = 12$ groups with $N_h = 10$, $h = 1, \ldots, 12$, but at the other extreme we could redefine the data based on the $n = 120$ individual mice with $N_h = 1$, $h = 1, \ldots, 120$. These give the same $\hat{p}(x)$ but lead to different prediction problems with the former predicting group proportions and the latter predicting individual outcomes. It is easier to do the former problem well. R^2 differs for the two problems with $R^2 = 0.191$ in the $n = 120$, 0-1 data. A similar phenomenon was explored in Subsection 12.5.2 and those results are extended in Exercise 20.9.4.

To use R^2 in this context, the (x_h, y_h) pairs should be a random sample from some population and we should be thinking about predicting a new value y_0 after observing x_0 from that population. The criterion seems plausible as long as the N_hs are all the same for $h = 0, 1, \ldots, n$, even if some creativity might be needed to imagine the distribution on x_0.

Many programs for fitting logistic regression report other values that can be used to assess the predictive ability of the model. Typical output includes:

Measures of Association
between the Response Variable and Predicted Probabilities

Pairs	Number	Percent	Summary Measures	
Concordant	2326	73.6	Somers' D	0.53
Discordant	636	20.1	Goodman–Kruskal Gamma	0.57
Ties	197	6.2	Kendall's Tau-a	0.24
Total	3159	100.0		

This table is based on looking at pairs of 0-1 observations. For these mouse data, that involves looking at the 0-1 scores for each of the $n = 120$ mice. Here 1 indicates that a mouse died. With n the total number of subjects, let n_1 and n_0 be the number of subjects scoring 1 and 0, respectively. For each pair i, j with $i \neq j$, we are going to look at the sign of $(y_i - y_j)(\hat{p}_i - \hat{p}_j)$. The total number of pairs is $\binom{n}{2}$, of these we will ignore the $\binom{n_1}{2}$ pairs that have $(y_i - y_j) = (1 - 1) = 0$ and the $\binom{n_2}{2}$ pairs that have $(y_i - y_j) = (0 - 0) = 0$. For the mouse data with 39 mice having died, the total number of pairs that we will consider are

$$\binom{120}{2} - \binom{39}{2} - \binom{81}{2} = 7140 - 741 - 3240 = 3159.$$

- The number of concordant pairs is $\mathcal{C}$; the number of pairs with $(y_i - y_j)(\hat{p}_i - \hat{p}_j) > 0$.
- The number of discordant pairs is $\mathcal{D}$; the number of pairs with $(y_i - y_j)(\hat{p}_i - \hat{p}_j) < 0$.
- The number of ties is $\mathcal{T}$; the number of pairs with $(\hat{p}_i - \hat{p}_j) = 0$.

The idea is that the higher the percentage of concordant pairs, the better the predictive ability of the model.

The 197 ties occur because many of the 120 cases have the same predictor variable. In particular, the number of ties in these data is the sum over the 12 categories of the number died times the number survived.

The three summary measures of association commonly given are

$$\text{Somers' D} \equiv \frac{C - D}{C + D + T} = \frac{C - D}{\binom{n}{2} - \binom{n_1}{2} - \binom{n_2}{2}},$$

$$\text{Goodman–Kruskal Gamma} \equiv \frac{C - D}{C + D},$$

and

$$\text{Kendall's Tau-a} \equiv \frac{C - D}{\binom{n}{2}}.$$

Other versions of Kendall's Tau make adjustments for the number of ties. It is pretty obvious that

$$\text{Goodman–Kruskal Gamma} \geq \text{Somers' D} \geq \text{Kendall's Tau-a}.$$

For the same y_hs, increasing the N_hs by a constant multiple does not affect any of the measures of predictive ability but it does increase the goodness-of-fit statistics and also makes them more valid.

20.2.3   Case diagnostics

Diagnostic quantities that are similar to those for standard regression can be computed. Raw residuals, $y_h - \hat{p}_h$, are not of much interest. The *Pearson residuals* are just the observations minus their estimated probability divided by the standard error of the observation, i.e.,

$$r_h = \frac{y_h - \hat{p}_h}{\sqrt{\hat{p}_h(1 - \hat{p}_h)/N_h}}.$$

This SE does not really account for the process of fitting the model, i.e., estimating p_h. We can incorporate the fitting process by incorporating the leverage, say, a_h. A *standardized Pearson residual* is

$$\tilde{r}_h = \frac{y_h - \hat{p}_h}{\sqrt{\hat{p}_h(1 - \hat{p}_h)(1 - a_h)/N_h}}.$$

Leverages for logistic regression are similar in spirit to those discussed in Chapters 7 and 11, but rather more complicated to compute. Values near 1 are still high-leverage points and the $2r/n$ and $3r/n$ rules of thumb can be applied where r is the number of (functionally distinct) parameters in the model. Table 20.2 contains diagnostics for the mouse data. Nothing seems overly disturbing.

I prefer using the standardized Pearson residuals, but the Pearson residuals often get used because of their simplicity. When all N_hs are large, both residuals can be compared to a $N(0,1)$ distribution to assess whether they are consistent with the model and the other data. In this large N_h case, much like the spirit of Chapter 5, we use the residuals to identify cases that cause problems in the goodness-of-fit test. Even with small N_hs, where no valid goodness-of-fit test is present, the residuals are used to identify potential problems.

With measurement data, residuals are used to check for outliers in the dependent variable, i.e., values of the dependent variable that do not seem to belong with the rest of the data. With count data it is uncommon to get anything that is really an outlier in the counts. The y_hs are proportions, so outliers would be values that are not between 0 and 1. With count data, large residuals really highlight areas where the model is not fitting the data very well. If you have a high dose of poison but very few mice die, something is wrong. The problem is often something that we have left out of the model.

Table 20.2: *Diagnostics for mouse data.*

Group	y_h	$\hat{p}_h$	r_h	$\tilde{r}_h$	Leverage	Cook
1	0.1	0.083	0.200	0.218	0.161	0.005
2	0.2	0.109	0.930	1.012	0.156	0.095
3	0.1	0.151	−0.450	−0.486	0.141	0.019
4	0.0	0.182	−1.493	−1.600	0.130	0.191
5	0.1	0.202	−0.803	−0.858	0.123	0.052
6	0.2	0.223	−0.178	−0.190	0.117	0.002
7	0.4	0.275	0.882	0.933	0.106	0.052
8	0.6	0.343	1.712	1.810	0.106	0.194
9	0.4	0.427	−0.175	−0.188	0.129	0.003
10	0.5	0.527	−0.169	−0.188	0.186	0.004
11	0.5	0.635	−0.889	−1.044	0.275	0.207
12	0.8	0.742	0.417	0.525	0.370	0.081

20.3   Model testing

Based on the results of a valid goodness-of-fit test, we already have reason to believe that a simple linear logistic regression fits the chloracetic acid data reasonably well, but for the purpose of illustrating the procedure for testing models, we will test the simple linear logistic model against a cubic polynomial logistic model. This section demonstrates the test. In the next section we discuss the motivation for it.

In Section 20.2 we gave the table of coefficients and the table of goodness-of-fit tests for the simple linear logistic regression model

$$\log\left(\frac{p_h}{1-p_h}\right) = \beta_0 + \beta_1 x_h. \tag{20.3.1}$$

The table of coefficients along with the deviance information follows.

Table of Coefficients: Model (20.3.1).

Predictor	$\hat{\beta}_k$	SE($\hat{\beta}_k$)	t	P
Constant	−3.56974	0.705330	−5.06	0.000
Dose	14.6369	3.33248	4.39	0.000

Deviance: $G^2 = 10.254$ $df = 10$

Additional standard output includes the Log-Likelihood $= -63.945$ (explained in Section 20.4) and a model-based test for $\beta_1 = 0$ (that is also discussed in Section 20.4), for which the test statistic is $G^2 = 23.450$ with $df = 1$ and a P value of 0.000.

The cubic polynomial logistic regression is

$$\log\left(\frac{p_h}{1-p_h}\right) = \gamma_0 + \gamma_1 x_h + \gamma_2 x_h^2 + \gamma_3 x_h^3. \tag{20.3.2}$$

with

Table of Coefficients: Model (20.3.2).

Predictor	$\hat{\gamma}_k$	SE($\hat{\gamma}_k$)	t	P
Constant	−2.47396	4.99096	−0.50	0.620
dose	−5.76314	83.1709	−0.07	0.945
x^2	114.558	434.717	0.26	0.792
x^3	−196.844	714.422	−0.28	0.783

and goodness-of-fit tests

Goodness-of-Fit Tests: Model (20.3.2).

Method	Chi-Square	df	P
Pearson	8.7367	8	0.365
Deviance	10.1700	8	0.253
Hosmer–Lemeshow	6.3389	4	0.175

Additional standard output includes the Log-Likelihood $= -63.903$ and a model-based test that all slopes are zero, i.e., $0 = \gamma_1 = \gamma_2 = \gamma_3$, that has $G^2 = 23.534$ with $df = 3$, and a P value of 0.000.

To test the full cubic model against the reduced simple linear model, we compute the likelihood ratio test statistic from the log-likelihoods,

$$G^2 = -2[(-63.903) - (-63.945)] = 0.084.$$

There are 4 parameters in Model (20.3.2) and only 2 parameters in Model (20.3.1) so there are $4 - 2 = 2$ degrees of freedom associated with this test. When the total number of cases n is large compared to the number of parameters in the full model, we can compare $G^2 = 0.084$ to a $\chi^2(4-2)$ distribution. This provides no evidence that the cubic model fits better than the simple linear model. Note that the validity of this test does not depend on having the N_hs large.

For these data, we can also obtain G^2 by the difference in deviances reported for the two models,

$$G^2 = 10.254 - 10.1700 = 0.084.$$

The difference in the deviance degrees of freedom is $10 - 8 = 2$, which is also the correct degrees of freedom.

Although finding likelihood ratio tests by subtracting deviances and deviance degrees of freedom is our preferred computational tool, unfortunately, *subtracting the deviances and the deviance degrees of freedom cannot be trusted to give the correct G^2 and degrees of freedom* when using programs designed for fitting logistic models (as opposed to programs for fitting generalized linear models). As discussed in Section 20.5, many logistic regression programs pool cases with identical predictor variables prior to computing the deviance and when models use different predictors, the pooling often changes, which screws up the test. Subtracting the deviances and deviance degrees of freedom does typically give the correct result when using programs for generalized linear models.

The standard output for Model (20.3.1) also included a model-based test for $\beta_1 = 0$ with $G^2 = 23.450, df = 1$, and a P value of 0.000. This is the likelihood ratio test for comparing the full model (20.3.1) with the intercept-only model

$$\log\left(\frac{p_h}{1 - p_h}\right) = \delta_0. \tag{20.3.3}$$

Alas, many logistic regression programs do not like to fit Model (20.3.3), so we take the program's word for the result of the test. (Programs for generalized linear models are more willing to fit Model (20.3.3).) Finding the test statistic is discussed in Section 20.5.

The usual output for fitting Model (20.3.2) has a model-based test that all slopes are zero, i.e., that $0 = \gamma_1 = \gamma_2 = \gamma_3$, for which $G^2 = 23.534$ with $df = 3$ and a P value of 0.000. This is the likelihood ratio test for the full model (20.3.2) against the reduced "intercept-only" model (20.3.3). Generally, when fitting a model these additional reported G^2 tests are for comparing the current model to the intercept-only model (20.3.3).

20.4   Fitting logistic models

In this section we discuss the ideas behind our methods for estimating parameters and for testing models. First we define the likelihood function. Our point estimates are *maximum likelihood estimates (MLEs)*, which are the parameter values that maximize the likelihood function. We compare models by comparing the maximum value that the likelihood function achieves under each model.

Such tests are *(generalized) likelihood ratio tests* for binomial count data. While we did not present the likelihood function for normal data, least squares estimates are also MLEs and F tests are also equivalent to (generalized) likelihood ratio tests.

Our logistic models take the form

$$\log\left(\frac{p_h}{1 - p_h}\right) = m(x_h), \tag{20.4.1}$$

where x_h is a vector of measurement or categorical variables and $m(\cdot)$ is any of the models that we have considered earlier for such predictor variables. The model $m(x_h)$ can correspond to a one-sample, two-sample, one-way ANOVA, or multifactor ANOVA model or any sort of regression model. We can solve (20.4.1) for p_h by writing

$$p_h = \frac{e^{m(x_h)}}{1 + e^{m(x_h)}}. \tag{20.4.2}$$

Given the estimate $\hat{m}(x)$ we get

$$\hat{p}(x) = \frac{e^{\hat{m}(x)}}{1 + e^{\hat{m}(x)}}.$$

For example, given the estimates $\hat{\beta}_0$ and $\hat{\beta}_1$ for a simple linear logistic regression, we get

$$\hat{p}(x) = \frac{\exp(\hat{\beta}_0 + \hat{\beta}_1 x)}{1 + \exp(\hat{\beta}_0 + \hat{\beta}_1 x)}. \tag{20.4.3}$$

In particular, this formula provides the $\hat{p}_h$s when doing predictions at the x_hs.

Estimates of coefficients are found by maximizing the likelihood function. The likelihood function is the probability of getting the data that were actually observed. It is a function of the unknown model parameters contained in $m(\cdot)$. Because the $N_h y_h$s are independent binomials, the likelihood function is

$$L(p_1, \ldots, p_n) = \prod_{h=1}^{n} \binom{N_h}{N_h y_h} p_h^{N_h y_h} (1 - p_h)^{N_h - N_h y_h}. \tag{20.4.4}$$

For a particular proportion y_h, $N_h y_y$ is $\text{Bin}(N_h, p_h)$ and the probability from Section 1.4 is an individual term on the right. We multiply the individual terms because the $N_h y_h$s are independent.

If we substitute for the p_hs using (20.4.2) into the likelihood function (20.4.4), the likelihood becomes a function of the model parameters. For example, if $m(x_h) = \beta_0 + \beta_1 x_h$ the likelihood becomes a function of the model parameters β_0 and β_1 for known values of (x_h, y_h, N_h), $h = 1, \ldots, n$. Computer programs maximize this function of β_0 and β_1 to give maximum likelihood estimates $\hat{\beta}_0$ and $\hat{\beta}_1$ along with approximate standard errors. The estimates have approximate normal distributions for large sample sizes. For the large sample approximations to be valid, it is typically enough that the total number of trials in the entire data n be large relative to the number of model parameters; the individual sample sizes N_h need not be large. The normal approximations also hold if all the N_hs are large regardless of the size of n.

In Section 11.3 we found the least squares estimates for linear regression models. Although we did not explicitly give the likelihood function for regression models with normally distributed data, we mentioned that the least squares estimates were also maximum likelihood estimates. Unfortunately, for logistic regression there are no closed-form solutions for the estimates and standard errors like those presented for measurement data in Chapter 11. For logistic regression, different computer programs may give *slightly* different results because the computations are more complex.

Maximum likelihood theory also provides a (generalized) likelihood ratio (LR) test for a full model versus a reduced model. Suppose the full model is

$$\log\left(\frac{p_{Fh}}{1 - p_{Fh}}\right) = m_F(x_h).$$

Fitting the model leads to estimated probabilities $\hat{p}_{Fh}$. The reduced model must be a special case of the full model, say,

$$\log\left(\frac{p_{Rh}}{1 - p_{Rh}}\right) = m_R(x_h),$$

with fitted probabilities $\hat{p}_{Rh}$. The commonly used form of the likelihood ratio test statistic is,

$$
\begin{aligned}
G^2 &= -2\log\left(\frac{L(\hat{p}_{R1},\ldots,\hat{p}_{Rn})}{L(\hat{p}_{F1},\ldots,\hat{p}_{Fn})}\right) \\
&= 2\sum_{h=1}^{n}\left\{N_h y_h \log(\hat{p}_{Fh}/\hat{p}_{Rh}) + N_h(1 - y_h)\log[(1 - \hat{p}_{Fh})/(1 - \hat{p}_{Rh})]\right\},
\end{aligned}
$$

where the second equality is based on Equation (20.4.4). An alternative to the LR test statistic is the Pearson test statistic, which is

$$X^2 = \sum_{h=1}^{n}\frac{(N_h\hat{p}_{Fh} - N_h\hat{p}_{Rh})^2}{N_h\hat{p}_{Rn}(1 - \hat{p}_{Rn})} = \sum_{h=1}^{n}\left[\frac{\hat{p}_{Fh} - \hat{p}_{Rh}}{\sqrt{\hat{p}_{Rn}(1 - \hat{p}_{Rn})/N_h}}\right]^2.$$

We make minimal use of X^2 in our discussions.

If the reduced model is true and the sample size n is large relative to the number of parameters in the full model, G^2 and X^2 have asymptotic χ^2 distributions where the degrees of freedom is the difference in the number of (functionally distinct) parameters between the two models. The same χ^2 distribution holds even if n is not large when the N_hs are all large.

Many computer programs for fitting a model report the value of the log-likelihood,

$$\ell(\hat{p}_1,\ldots,\hat{p}_n) \equiv \log[L(\hat{p}_1,\ldots,\hat{p}_n)].$$

To compare a full and reduced model, G^2 is twice the absolute value of the difference between these values. When using logistic regression programs (as opposed to generalized linear model programs), this is how one needs to compute G^2.

The smallest interesting logistic model that we can fit to the data is the *intercept-only model*

$$\log\left(\frac{p_h}{1 - p_h}\right) = \beta_0. \tag{20.4.5}$$

The largest logistic model that we can fit to the data is the *saturated model* that has a separate parameter for each case,

$$\log\left(\frac{p_h}{1 - p_h}\right) = \gamma_h. \tag{20.4.6}$$

Interesting models tend to be somewhere between these two. Many computer programs automatically report the results of testing the fitted model against both of these.

For standard simple linear regression, we have two tests for $H_0 : \beta_1 = 0$, a t test and an F test, and the two tests are equivalent, e.g., always give the same P value, cf. Section 6.1. For simple linear logistic regression we have a t test for $H_0 : \beta_1 = 0$, and testing against the intercept-only model provides a G^2 test for $H_0 : \beta_1 = 0$. As will be seen in Section 20.5 for the O-ring data, these tests typically do not give the same P values. The two tests are not equivalent. For the mouse data, both P values were reported as 0.000, so one could not see that the two P values were different beyond the three decimal points reported.

Testing a fitted model $m(\cdot)$ against the saturated model (20.4.6) is called a *goodness-of-fit test*. The fitted probabilities under Model (20.4.6) are just the observed proportions for each case, the y_hs. The *deviance* for a fitted model is defined as G^2 for testing the fitted model against the saturated model (20.4.6),

$$
\begin{aligned}
G^2 &= -2\log\left(\frac{L(\hat{p}_1,\ldots,\hat{p}_n)}{L(y_1,\ldots,y_n)}\right) \\
&= 2\sum_{h=1}^{n}[N_h y_h \log(y_h/\hat{p}_h) + (N_h - N_h y_h)\log((1 - y_h)/(1 - \hat{p}_h))].
\end{aligned}
\tag{20.4.7}
$$

In this formula, if $a = 0$, then $a\log(a)$ is taken as zero. The degrees of freedom for the deviance are n (the number of parameters in Model (20.4.6)) minus the number of (functionally distinct) parameters in the fitted model.

The problem with the goodness-of-fit test is that the number of parameters in Model (20.4.6) is the sample size n, so the only way for G^2 to have an asymptotic χ^2 distribution is if all the N_hs are large. For the mouse death data, the N_hs are all 10, which is probably fine, but for a great many data sets, all the N_hs are 1, so a χ^2 test of the goodness-of-fit statistic is not appropriate. A similar conclusion holds for the Pearson statistic.

As also discussed in the next section, in an effort to increase the size of the N_hs, many logistic regression computer programs pool together any cases for which $x_h = x_i$. Thus, instead of having two cases with $N_h y_h \sim \text{Bin}(N_h, p_h)$ and $N_i y_i \sim \text{Bin}(N_i, p_i)$, the two cases get pooled into a single case with $(N_h y_h + N_i y_i) \sim \text{Bin}(N_h + N_i, p_h)$. Note that if $x_h = x_i$, it follows that $p_h = p_i$ and the new proportion would be $(N_h y_h + N_i y_i)/(N_h + N_i)$. I have never encountered regression data with so few distinct x_h values that this pooling procedure actually accomplished its purpose of making all the group sizes reasonably large, but if the mouse data were presented as 120 mice that either died or not along with their dose, such pooling would work fine.

Ideally, the deviance G^2 of (20.4.7) could be used analogously to the SSE in normal theory and the degrees of freedom for the deviance of (20.4.7) would be analogous to the dfE. To compare a full and reduced model you just subtract their deviances (rather than their SSEs) and compare the test statistic to a χ^2 with degrees of freedom equal to the difference in the deviance degrees of freedom (rather than differencing the dfEs). This procedure works just fine when fitting the models using programs for fitting generalized linear models. The invidious thing about the pooling procedure of the previous paragraph is that when you change the model from reduced to full, you often change the predictor vector x_h in such a way that it changes which cases have $x_h = x_i$. When comparing a full and a reduced model, the models may well have different cases pooled together, which means that the difference in deviances no longer provides the appropriate G^2 for testing the models. In such cases G^2 needs to be computed directly from the log-likelihood.

After discussing the commonly reported goodness-of-fit statistics in the next section, we will no longer discuss any deviance values that are obtained by pooling. After Subsection 20.5.1, the deviances we discuss may not be those reported by a logistic regression program but they should be those obtained by a generalized linear models program.

20.5 Binary data

Logistic regression is often used when the binomial sample sizes are all 1. The resulting binary data consist entirely of 0s and 1s.

EXAMPLE 20.5.1. *O-ring Data.*
Table 20.3 presents data from Dalal, Fowlkes, and Hoadley (1989) on field O-ring failures in the 23 pre-*Challenger* space shuttle launches. *Challenger* was the shuttle that blew up on take-off. Atmospheric temperature is the predictor variable. The *Challenger* explosion occurred during a takeoff at 31 degrees Fahrenheit. Each flight is viewed as an independent trial. The result of a trial is 1 if any field O-rings failed on the flight and 0 if all the O-rings functioned properly. A simple linear logistic regression uses temperature to model the probability that any O-ring failed. Such a model allows us to predict O-ring failure from temperature.

Table 20.3: *O-ring failure data*.

Case	Flight	Failure	Temperature	Case	Flight	Failure	Temperature
1	14	1	53	13	2	1	70
2	9	1	57	14	11	1	70
3	23	1	58	15	6	0	72
4	10	1	63	16	7	0	73
5	1	0	66	17	16	0	75
6	5	0	67	18	21	1	75
7	13	0	67	19	19	0	76
8	15	0	67	20	22	0	76
9	4	0	68	21	12	0	78
10	3	0	69	22	20	0	79
11	8	0	70	23	18	0	81
12	17	0	70				

Let p_i be the probability that any O-ring fails in case i. The simple linear logistic regression model is

$$\text{logit}(p_i) \equiv \log\left(\frac{p_i}{1-p_i}\right) = \beta_0 + \beta_1 x_i,$$

where x_i is the known temperature and β_0 and β_1 are unknown intercept and slope parameters (coefficients).

Maximum likelihood theory gives the coefficient estimates, standard errors, and t values as

Table of Coefficients: O-rings

Predictor	$\hat{\beta}_k$	$SE(\hat{\beta}_k)$	t	P
Constant	15.0429	7.37862	2.04	0.041
Temperature	-0.232163	0.108236	-2.14	0.032

The t values are the estimate divided by the standard error. For testing $H_0 : \beta_1 = 0$, the value $t = -2.14$ yields a P value that is approximately 0.03, so there is evidence that temperature does help predict O-ring failure. Alternatively, a model-based test of $\beta_1 = 0$ compares the simple linear logistic model to an intercept-only model and gives $G^2 = 7.952$ with $df = 1$ and $P = 0.005$. These tests should be reasonably valid because $n = 23$ is reasonably large relative to the 2 parameters in the fitted model. The log-likelihood is $\ell = -10.158$.

Figure 20.2 gives a plot of the estimated probabilities as a function of temperature,

$$\hat{p}(x) = \frac{e^{15.0429-0.232163x}}{1+e^{15.0429-0.232163x}}.$$

The *Challenger* was launched at $x = 31$ degrees, so the predicted log odds are $15.04 - 0.2321(31) = 7.8449$ and the predicted probability of an O-ring failure is $e^{7.8449}/(1 + e^{7.8449}) = 0.9996$. Actually, there are problems with this prediction because we are predicting very far from the observed data. The lowest temperature at which a shuttle had previously been launched was 53 degrees, very far from 31 degrees. According to the fitted model, a launch at 53 degrees has probability 0.939 of O-ring failure, so even with the caveat about predicting beyond the range of the data, the model indicates an overwhelming probability of failure.

20.5.1 *Goodness-of-fit tests*

Many specialized logistic regression computer programs report the following goodness-of-fit statistics for the O-ring data.

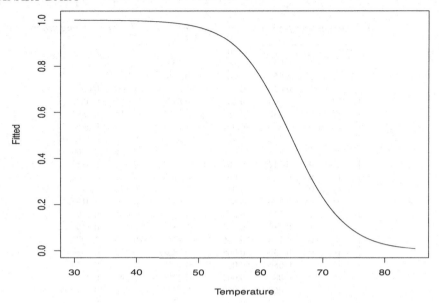

Figure 20.2: *O-ring estimated failure probabilities.*

Goodness-of-Fit Tests			
Method	Chi-Square	df	P
Pearson	11.1303	14	0.676
Deviance	11.9974	14	0.607
Hosmer–Lemeshow	9.7119	8	0.286

For 0-1 data, these are all useless. The Hosmer–Lemeshow statistic does not have a χ^2 distribution. For computing the Pearson and deviance statistics the 23 original cases have been pooled into $\tilde{n} = 16$ new cases based on duplicate temperatures. This gives binomial sample sizes of $\tilde{N}_6 = 3$, $\tilde{N}_9 = 4$, $\tilde{N}_{12} = \tilde{N}_{13} = 2$, and $\tilde{N}_h = 1$ for all other cases. With two parameters in the fitted model, the reported degrees of freedom are $14 = 16 - 2$. To have a valid $\chi^2(14)$ test, all the $\tilde{N}_h$s would need to be large, but none of them are. Pooling does not give a valid χ^2 test and it also eliminates the deviance as a useful tool in model testing.

Henceforward, we only report deviances that are not obtained by pooling. These are the likelihood ratio test statistics for the fitted model against the saturated model with the corresponding degrees of freedom. Test statistics for any full and reduced models can then be obtained by subtracting the corresponding deviances from each other just as the degrees of freedom for the test can be obtained by subtraction. These deviances can generally be found by fitting logistic models as special cases in programs for fitting generalized linear models. When using specialized logistic regression software, great care must be taken and the safest bet is to always use log-likelihoods to obtain test statistics.

EXAMPLE 20.5.1 CONTINUED.   For the simple linear logistic regression model

$$\log\left(\frac{p_i}{1-p_i}\right) = \beta_0 + \beta_1 x_i. \tag{20.5.1}$$

Without pooling, the deviance is $G^2 = 20.315$ with $21 = 23 - 2 = n - 2$ degrees of freedom. For the intercept-only model

$$\log\left(\frac{p_i}{1-p_i}\right) = \beta_0 \tag{20.5.2}$$

Table 20.4: *Diagnostics for Challenger data: Generalized linear modeling program.*

Case	y_h	$\hat{p}_h$	Leverage	r_h	$\tilde{r}_h$	Cook
1	1	0.939	0.167	0.254	0.279	0.0078
2	1	0.859	0.208	0.405	0.455	0.0271
3	1	0.829	0.209	0.454	0.511	0.0345
4	1	0.603	0.143	0.812	0.877	0.0641
5	0	0.430	0.086	−0.8694	−0.910	0.0391
6	0	0.375	0.074	−0.7741	−0.805	0.0260
7	0	0.375	0.074	−0.7741	−0.805	0.0260
8	0	0.375	0.074	−0.7741	−0.805	0.0260
9	0	0.322	0.067	−0.6893	−0.714	0.0183
10	0	0.274	0.063	−0.6138	−0.634	0.0136
11	0	0.230	0.063	−0.5465	−0.564	0.0107
12	0	0.230	0.063	−0.5465	−0.564	0.0107
13	1	0.230	0.063	1.830	1.890	0.1196
14	1	0.230	0.063	1.830	1.890	0.1196
15	0	0.158	0.066	−0.4333	−0.448	0.0071
16	0	0.130	0.068	−0.3858	−0.400	0.0058
17	0	0.086	0.069	−0.3059	−0.317	0.0037
18	1	0.086	0.069	3.270	3.389	0.4265
19	0	0.069	0.068	−0.2723	−0.282	0.0029
20	0	0.069	0.068	−0.2723	−0.282	0.0029
21	0	0.045	0.063	−0.2159	−0.223	0.0017
22	0	0.036	0.059	−0.1922	−0.198	0.0012
23	0	0.023	0.051	−0.1524	−0.156	0.0007

the deviance is $G^2 = 28.267$ with $22 = 23 - 1 = n - 1$ degrees of freedom. Since $N_i = 1$ for all i, neither of these G^2s is compared directly to a chi-squared distribution. However, the model-based test for $H_0 : \beta_1 = 0$ has $G^2 = 28.267 - 20.315 = 7.952$ on $df = 22 - 21 = 1$, which agrees with the test reported earlier even though the deviance for Model (20.5.1) is different from that reported earlier. Comparing $G^2 = 7.952$ to a $\chi^2(1)$ distribution, the P value for the test is approximately 0.005. It is considerably smaller than the P value for the t test of H_0.   □

It can be difficult to get even generalized linear model programs to fit the intercept-only model but the deviance G^2 can be obtained from the formula in Section 20.4. Given the estimate $\hat{\beta}_0$ for Model (20.5.2), we get $\hat{p}_i = e^{\hat{\beta}_0}/(1 + e^{\hat{\beta}_0})$ for all i, and apply the formula. In general, for the intercept-only model $\hat{p}_i = \sum_{i=1}^n N_i y_i / \sum_{i=1}^n N_i$, which, for binary data, reduces to $\hat{p}_i = \sum_{i=1}^n y_i / n$. The degrees of freedom are the number of cases minus the number of fitted parameters, $n - 1$.

20.5.2 Case diagnostics

The residuals and leverages in Table 20.4 have been computed from a program for generalized linear modeling. The residuals and leverages have also been computed in Table 20.5 using pooling, which is why some values are missing. Cases 6, 7, 8, cases 11, 12, 13, 14, cases 17, 18, and cases 19, 20 all have duplicated temperatures with residuals and leverages reported only for the first case. In Table 20.5 the reported leverage for case 6 is 0.22. Without pooling in Table 20.4 this leverage is distributed as $0.22/3 = 0.074$ for each of cases 6, 7, and 8.

20.5.3 Assessing predictive ability

In any predictive model, a reasonable measure of the predictive ability of a model is the squared correlation between the actual observations and the predicted (fitted) values. For the challenger data, this number is

$$R^2 = 0.346.$$

Table 20.5: *Diagnostics for* Challenger *data: Logistic regression program.*

Case	y_h	$\hat{p}_h$	Leverage	r_h	$\tilde{r}_h$
1	1	0.939	0.167	0.254	0.279
2	1	0.859	0.208	0.405	0.455
3	1	0.829	0.209	0.454	0.511
4	1	0.603	0.143	0.812	0.877
5	0	0.430	0.086	−0.869	−0.910
6	0	0.375	0.223	−1.341	−1.521
7	0	0.375	*	*	*
8	0	0.375	*	*	*
9	0	0.322	0.067	−0.689	−0.714
10	0	0.274	0.063	−0.614	−0.634
11	0	0.230	0.251	1.283	1.483
12	0	0.230	*	*	*
13	1	0.230	*	*	*
14	1	0.230	*	*	*
15	0	0.158	0.066	−0.433	−0.448
16	0	0.130	0.068	−0.386	−0.400
17	0	0.086	0.138	2.096	2.258
18	1	0.086	*	*	*
19	0	0.069	0.136	−0.385	−0.414
20	0	0.069	*	*	*
21	0	0.045	0.063	−0.216	−0.223
22	0	0.036	0.059	−0.192	−0.198
23	0	0.023	0.051	−0.152	−0.156

This may seem like a small number, but it is difficult to predict well in 0-1 logistic regression even when you know the perfect model $p(x)$. The predictive ability of the model depends on the x values one is likely to see. For x values that correspond to $p(x)$ close to 0 or 1, the model will make very good predictions. But for x values with $p(x) \doteq 0.5$, we will never be able to make reliable predictions. Those predictions will be no better than predictions for the flip of a coin.

The other commonly used predictive measures for these data are given below.

Measures of Association
Between the Response Variable and Predicted Probabilities

Pairs	Number	Percent	Summary Measures	
Concordant	85	75.9	Somers' D	0.56
Discordant	22	19.6	Goodman–Kruskal Gamma	0.59
Ties	5	4.5	Kendall's Tau-a	0.25
Total	112	100.0		

20.6 Multiple logistic regression

This section examines regression models for the log-odds of a two-category response variable in which we use more than one predictor variable. The discussion is centered around an example.

EXAMPLE 20.6.1. *Chapman Data.*
Dixon and Massey (1983) and Christensen (1997) present data on 200 men taken from the Los Angeles Heart Study conducted under the supervision of John M. Chapman, UCLA. The data consist of seven variables:

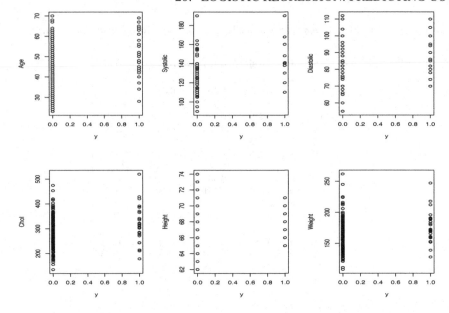

Figure 20.3: *Coronary incident scatterplot matrix.*

Abbreviation	Variable	Units
Ag	Age:	in years
S	Systolic Blood Pressure:	millimeters of mercury
D	Diastolic Blood Pressure:	millimeters of mercury
Ch	Cholesterol:	milligrams per DL
H	Height:	inches
W	Weight:	pounds
CN	Coronary incident:	1 if an incident had occurred in the previous ten years; 0 otherwise

Of the 200 cases, 26 had coronary incidents. The data are available on the Internet, like all the data in this book, through the webpage:

http://stat.unm.edu/~fletcher.

The data are part of the data that go along with both this book and the book *Log-Linear Models and Logistic Regression*. They are also available on the Internet via STATLIB. Figure 20.3 plots each variable against $y = CN$. Figures 20.4 through 20.7 provide a scatterplot matrix of the predictor variables.

Let p_i be the probability of a coronary incident for the ith man. We begin with the logistic regression model

$$\log[p_i/(1 - p_i)] = \beta_0 + \beta_1 Ag_i + \beta_2 S_i + \beta_3 D_i + \beta_4 Ch_i + \beta_5 H_i + \beta_6 W_i \qquad (20.6.1)$$

$i = 1, \ldots, 200$. The maximum likelihood fit of this model is given in Table 20.6. The deviance df is the number of cases, 200, minus the number of fitted parameters, 7. Based on the t values, none of the variables really stand out. There are suggestions of age, cholesterol, and weight effects. The (unpooled) deviance G^2 *would* look good except that, as discussed earlier, with $N_i = 1$ for all i there is no basis for comparing it to a $\chi^2(193)$ distribution.

Prediction follows as usual,

$$\log[\hat{p}_i/(1 - \hat{p}_i)] = \hat{\beta}_0 + \hat{\beta}_1 Ag_i + \hat{\beta}_2 S_i + \hat{\beta}_3 D_i + \hat{\beta}_4 Ch_i + \hat{\beta}_5 H_i + \hat{\beta}_6 W_i.$$

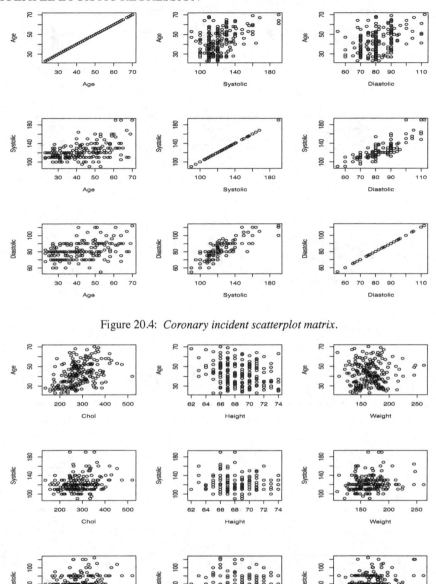

Figure 20.4: *Coronary incident scatterplot matrix.*

Figure 20.5: *Coronary incident scatterplot matrix.*

Table 20.6: *Table of Coefficients: Model (20.6.1).*

Predictor	$\hat{\beta}_k$	SE($\hat{\beta}_k$)	t
Constant	−4.5173	7.481	−0.60
Ag	0.04590	0.02354	1.95
S	0.00686	0.02020	0.34
D	−0.00694	0.03835	−0.18
Ch	0.00631	0.00363	1.74
H	−0.07400	0.10622	−0.70
W	0.02014	0.00987	2.04
Deviance:	$G^2 = 134.9$	$df = 193$	

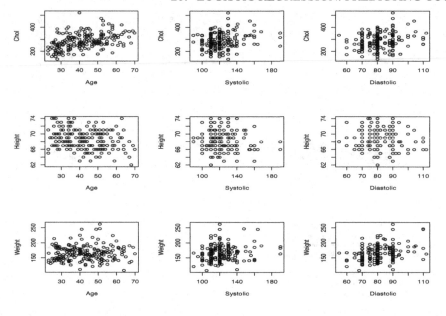

Figure 20.6: *Coronary incident scatterplot matrix.*

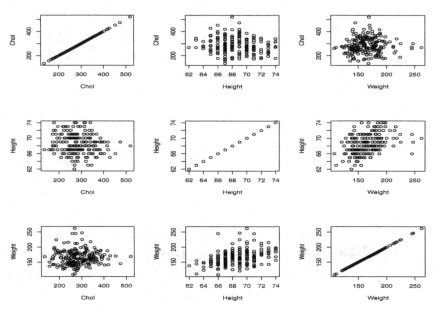

Figure 20.7: *Coronary incident scatterplot matrix.*

For a 60-year-old man with blood pressure of 140 over 90, a cholesterol reading of 200, who is 69 inches tall and weighs 200 pounds, the estimated log odds of a coronary incident are

$$\log[\hat{p}/(1-\hat{p})] = -4.5173 + 0.04590(60) + 0.00686(140) - 0.00694(90)$$
$$+ 0.00631(200) - 0.07400(69) + 0.02014(200) = -1.2435.$$

The probability of a coronary incident is estimated as

$$\hat{p} = \frac{e^{-1.2435}}{1 + e^{-1.2435}} = 0.224.$$

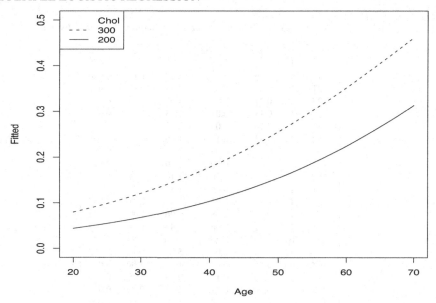

Figure 20.8 *Coronary incident probabilities as a function of age for* $S = 140, D = 90, H = 69, W = 200$. *Solid* $Ch = 200$, *dashed* $Ch = 300$.

Figure 20.8 plots the estimated probability of a coronary incident as a function of age for people with $S = 140, D = 90, H = 69, W = 200$ and either $Ch = 200$ (solid line) or $Ch = 300$ (dashed line).

Diagnostic quantities for the cases with the largest Cook's distances are given in Table 20.7. They include 19 of the 26 cases that had coronary incidents. The large residuals are for people who had low probabilities for a coronary incident but had one nonetheless. High leverages correspond to unusual data. For example, case 41 has the highest cholesterol. Case 108 is the heaviest man in the data.

We now consider fitting some reduced models. Simple linear logistic regressions were fitted for each of the variables with high t values, i.e., Ag, Ch, and W. Regressions with variables that seem naturally paired were also fitted, i.e., S,D and H,W. Table 20.8 contains the models along with df, G^2, $A - q$, and A^*. The first two of these are the deviance degrees of freedom and the deviance. No P values are given because the asymptotic χ^2 approximation does not hold. Also given are two analogues of Mallow's C_p statistic, $A - q$ and A^*. $A - q \equiv G^2 - 2(df)$ is the *Akaike information criterion (AIC)* less twice the number of trials ($q \equiv 2n$). A^* is a version of the Akaike information criterion defined for comparing Model (20.6.1) to various submodels. It gives numerical values similar to the C_p statistic and is defined by

$$A^* = (G^2 - 134.9) - (7 - 2p).$$

Here 134.9 is the deviance G^2 for the full model (20.6.1), 7 comes from the degrees of freedom for the full model (6 explanatory variables plus an intercept), and p comes from the degrees of freedom for the submodel ($p = 1 + $ number of explanatory variables). The information in $A - q$ and A^* is identical, $A^* = 258.1 + (A - q)$. (The value $258.1 = 2n - G^2$[full model] $- p$[full model] $= 400 - 134.9 - 7$ does not depend on the reduced model.) A^* is listed because it is a little easier to look at since it takes values similar to C_p. Computer programs rarely report $A - q$ or A^*. (The glm procedure in the R language provides a version of the AIC.) $A - q$ is very easy to compute from the deviance and its degrees of freedom.

Of the models listed in Table 20.8,

$$\log[p_i/(1 - p_i)] = \gamma_0 + \gamma_1 Ag_i \tag{20.6.2}$$

Table 20.7: *Diagnostics for Chapman data. Cases with high Cook's distances.*

Case	y_h	$\hat{p}_h$	Leverage	r_h	$\tilde{r}_h$	Cook
5	1	0.36	0.13	1.32	1.42	0.043
19	1	0.46	0.15	1.08	1.17	0.036
21	1	0.08	0.02	3.34	3.37	0.028
27	1	0.21	0.03	1.97	1.99	0.016
29	1	0.11	0.01	2.73	2.75	0.016
39	1	0.16	0.03	2.33	2.36	0.022
41	1	0.31	0.15	1.46	1.59	0.065
42	1	0.12	0.03	2.60	2.63	0.027
44	1	0.41	0.09	1.19	1.24	0.021
48	1	0.18	0.06	2.14	2.21	0.045
51	1	0.34	0.06	1.39	1.44	0.019
54	1	0.19	0.03	2.07	2.09	0.017
55	1	0.52	0.08	0.96	1.00	0.012
81	1	0.32	0.06	1.44	1.49	0.021
84	0	0.36	0.20	-0.74	-0.83	0.026
86	1	0.03	0.01	5.95	5.98	0.052
108	0	0.45	0.17	-0.91	-1.00	0.029
111	1	0.56	0.11	0.89	0.95	0.015
113	0	0.37	0.21	-0.76	-0.85	0.027
114	0	0.46	0.14	-0.93	-1.00	0.024
116	0	0.41	0.10	-0.84	-0.89	0.013
123	1	0.36	0.07	1.35	1.40	0.022
124	1	0.12	0.02	2.70	2.72	0.019
126	1	0.13	0.04	2.64	2.70	0.047

Table 20.8: *Models for Chapman data.*

Variables	df	G^2	$A - q$	A^*
Ag,S,D,Ch,H,W	193	134.9	−251.1	7
Ag	198	142.7	−253.3	4.8
W	198	150.1	−245.9	12.2
H,W	197	146.8	−247.2	10.9
Ch	198	146.9	−249.1	9.0
S,D	197	147.9	−246.1	12.0
Intercept	199	154.6	−243.4	14.7

is the only model that is better than the full model based on the information criterion, i.e., A^* is 4.8 for this model, less than the 7 for Model (20.6.1).

Asymptotically valid tests of submodels against Model (20.6.1) are available. These are performed in the usual way, i.e., the differences in deviance degrees of freedom and deviance G^2s give the appropriate values for the tests. For example, the test of Model (20.6.2) versus Model (20.6.1) has $G^2 = 142.7 - 134.9 = 7.8$ with $df = 198 - 193 = 5$. This and other tests are given below.

Tests against Model (20.6.1)

Model	df	G^2
Ag	5	7.8
W	5	15.2**
H,W	4	11.9*
Ch	5	12.0*
S,D	4	13.0*
Intercept	6	19.7**

All of the test statistics are significant at the 0.05 level, except for that associated with Model (20.6.2). This indicates that none of the models other than (20.6.2) is an adequate substitute for

Table 20.9: *Chapman data models that include Age.*

Variables	df	G^2	A^*
Ag,S,D,Ch,H,W	193	134.9	7.0
Ag,S,D	196	141.4	7.5
Ag,S,Ch	196	139.3	5.4
Ag,S,H	196	141.9	8.0
Ag,S,W	196	138.4	4.5
Ag,D,Ch	196	139.0	5.1
Ag,D,H	196	141.4	7.5
Ag,D,W	196	138.5	4.6
Ag,Ch,H	196	139.9	6.0
Ag,Ch,W	196	135.5	1.6
Ag,H,W	196	138.1	4.2
Ag,S	197	141.9	6.0
Ag,D	197	141.4	5.5
Ag,Ch	197	139.9	4.0
Ag,H	197	142.7	6.8
Ag,W	197	138.8	2.9
Ag	198	142.7	4.8

the full model (20.6.1). In this table, one asterisk indicates significance at the 0.05 level and two asterisks denotes significance at the 0.01 level.

Our next step is to investigate models that include Ag and some other variables. If we can find one or two variables that account for most of the value $G^2 = 7.8$, we may have an improvement over Model (20.6.2). If it takes three or more variables to explain the 7.8, Model (20.6.2) will continue to be the best-looking model. (Note that $\chi^2(.95, 3) = 7.81$, so a model with three more variables than Model (20.6.2) and the same G^2 fit as Model (20.6.1) would still not demonstrate a significant lack of fit in Model (20.6.2).)

Fits for all models that involve Ag and either one or two other explanatory variables are given in Table 20.9. Based on the A^* values, two models stand out:

$$\log[p_i/(1 - p_i)] = \gamma_0 + \gamma_1 Ag_i + \gamma_2 W_i \qquad (20.6.3)$$

with $A^* = 2.9$ and

$$\log[p_i/(1 - p_i)] = \eta_0 + \eta_1 Ag_i + \eta_2 W_i + \eta_3 Ch_i \qquad (20.6.4)$$

with $A^* = 1.6$.

The estimated parameters and standard errors for model (20.6.3) are

Table of Coefficients: Model (20.6.3).

Variable	$\hat{\gamma}_k$	SE($\hat{\gamma}_k$)
Intercept	−7.513	1.706
Ag	0.06358	0.01963
W	0.01600	0.00794

For Model (20.6.4), these are

Table of Coefficients: Model (20.6.4).

Variable	$\hat{\eta}_k$	SE($\hat{\eta}_k$)
Intercept	−9.255	2.061
Ag	0.05300	0.02074
W	0.01754	0.003575
Ch	0.006517	0.008243

The coefficients for Ag and W are quite stable in the two models. The coefficients of Ag, W, and Ch are all positive, so that a small increase in age, weight, or cholesterol is associated with a small increase in the odds of having a coronary incident. Note that we are establishing association, not causation. The data tell us that higher cholesterol is related to higher probabilities, not that it causes higher probabilities.

As in standard regression, interpreting regression coefficients can be very tricky. The fact that the regression coefficients are all positive conforms with the conventional wisdom that high values for any of these factors is associated with increased chance of heart trouble. However, as in standard regression analysis, correlations between predictor variables can make interpretations of individual regression coefficients almost impossible.

It is interesting to note that from fitting Model (20.6.1) the estimated regression coefficient for D, diastolic blood pressure, is negative, cf. Table 20.6. A naive interpretation would be that as diastolic blood pressure goes up, the probability of a coronary incident goes down. (If the log odds go down, the probability goes down.) This is contrary to common opinion about how these things work. Actually, this is really just an example of the fallacy of trying to interpret regression coefficients. The regression coefficients have been determined so that the fitted model explains these particular data as well as possible. As mentioned, correlations between the predictor variables can have a huge effect on the estimated regression coefficients. The sample correlation between S and D is 0.802, so estimated regression coefficients for these variables are unreliable. Moreover, it is not even enough just to check pairwise correlations between variables; any large partial correlations will also adversely affect coefficient interpretations. Fortunately, such correlations should not normally have an adverse affect on the predictive ability of the model; they only adversely affect attempts to interpret regression coefficients. Another excuse for the D coefficient $\hat{\beta}_3$ being negative is that, from the t value, β_3 is not significantly different from 0.

The estimated blood pressure coefficients from Model (20.6.1) also suggest an interesting hypothesis. (The hypothesis would be more interesting if the individual coefficients were significant, but we wish to demonstrate a modeling technique.) The coefficient for D is -0.00694, which is approximately the negative of the coefficient for S, 0.00686. This suggests that perhaps $\beta_3 = -\beta_2$ in Model (20.6.1). If we incorporate this hypothesis into Model (20.6.1) we get

$$
\begin{aligned}
&\log[p_i/(1-p_i)] \\
&= \quad \beta_0 + \beta_1 Ag_i + \beta_2 S_i + (-\beta_2)D_i + \beta_4 Ch_i + \beta_5 H_i + \beta_6 W_i \quad\quad (20.6.5) \\
&= \quad \beta_0 + \beta_1 Ag_i + \beta_2 (S_i - D_i) + \beta_4 Ch_i + \beta_5 H_i + \beta_6 W_i,
\end{aligned}
$$

which gives deviance $G^2 = 134.9$ on $df = 194$. This model is a reduced model relative to Model (20.6.1), so from Table 20.9 a test of it against Model (20.6.1) has

$$
G^2 = 134.9 - 134.9 = 0.0,
$$

with $df = 194 - 193 = 1$. The G^2 is essentially 0, so the data are consistent with the reduced model. Of course this reduced model was suggested by the fitted full model, so any formal test would be biased—but then one does not accept null hypotheses anyway, and the whole point of choosing this reduced model was that it seemed likely to give a G^2 close to that of Model (20.6.1). We note that the new variable $S - D$ is still not significant in Model (20.6.5); it only has a t value of $0.006834/0.01877 = 0.36$.

If we wanted to test something like $H_0 : \beta_3 = -0.005$, the reduced model is

$$
\log[p_i/(1-p_i)] = \beta_0 + \beta_1 Ag_i + \beta_2 S_i + (-0.005)D_i + \beta_4 Ch_i + \beta_5 H_i + \beta_6 W_i
$$

and involves a known term $(-0.005)D_i$ in the linear predictor. This known term is called an *offset*. To fit a model with an offset, most computer programs require that the offset be specified separately and that the model be specified without it, i.e., as

$$
\log[p_i/(1-p_i)] = \beta_0 + \beta_1 Ag_i + \beta_2 S_i + \beta_4 Ch_i + \beta_5 H_i + \beta_6 W_i.
$$

The use of an offset is illustrated in Section 21.6.

Table 20.10: *Muscle tension change data.*

Tension (h)	Weight (i)	Muscle (j)	Drug (k) Drug 1	Drug 2
	High	Type 1	3	21
		Type 2	23	11
High				
	Low	Type 1	22	32
		Type 2	4	12
	High	Type 1	3	10
		Type 2	41	21
Low				
	Low	Type 1	45	23
		Type 2	6	22

We learned earlier that, relative to Model (20.6.1), either model (20.6.3) or (20.6.4) does an adequate job of explaining the data. This conclusion was based on looking at A^* values, but would also be obtained by doing formal tests of models.

Christensen (1997, Section 4.4) discusses how to perform best subset selection, similar to Section 10.2, for logistic regression. His preferred method requires access to a standard best subset selection program that allows weighted regression. He does not recommend the score test procedure used by SAS in PROC LOGISTIC. Stepwise methods, like backward elimination and forward selection, are relatively easy to apply.

20.7 ANOVA type logit models

In this section, analysis-of-variance-type models for the log odds of a two-category response variable are discussed. For ANOVA models, binary data can often be pooled to obtain reasonably large group sizes. More often, the data come presented in groups. We begin with a standard example.

EXAMPLE 20.7.1.   A study on mice examined the relationship between two drugs and muscle tension. Each mouse had a muscle identified and its tension measured. A randomly selected drug was administered to the mouse and the change in muscle tension was evaluated. Muscles of two types were used. The weight of the muscle was also measured. Factors and levels are as follow.

Factor	Abbreviation	Levels
Change in muscle tension	T	High, Low
Weight of muscle	W	High, Low
Muscle type	M	Type 1, Type 2
Drug	D	Drug 1, Drug 2

The data in Table 20.10 are counts (rather than proportions) for every combination of the factors. Probabilities p_{hijk} can be defined for every factor combination with $p_{1ijk} + p_{2ijk} = 1$, so the odds of a high tension change are p_{1ijk}/p_{2ijk}.

Change in muscle tension is a response factor. Weight, muscle type, and drug are all predictor variables. We model the log odds of having a high change in muscle tension (given the levels of the explanatory factors), so the observed proportion of, say, high change for Weight = Low, Muscle = 2, Drug = 1 is, from Table 20.10, $4/(4+6)$. The most general ANOVA model (saturated model) includes all main effects and all interactions between the explanatory factors, i.e.,

$$\log(p_{1ijk}/p_{2ijk}) = G + W_i + M_j + D_k \qquad (20.7.1)$$
$$+ (WM)_{ij} + (WD)_{ik} + (MD)_{jk}$$
$$+ (WMD)_{ijk}.$$

Table 20.11: *Statistics for muscle tension logit models.*

Logit Model	df	G^2	P	$A-q$
$[WM][WD][MD]$	1	0.111	0.7389	-1.889
$[WM][WD]$	2	2.810	0.2440	-1.190
$[WM][MD]$	2	0.1195	0.9417	-3.8805
$[WD][MD]$	2	1.059	0.5948	-2.941
$[WM][D]$	3	4.669	0.1966	-1.331
$[WD][M]$	3	3.726	0.2919	-2.274
$[MD][W]$	3	1.060	0.7898	-4.940
$[W][M][D]$	4	5.311	0.2559	-2.689
$[W][M]$	5	11.35	0.0443	1.35
$[W][D]$	5	12.29	0.0307	2.29
$[M][D]$	5	7.698	0.1727	-2.302

Table 20.12: *Estimated odds of high tension change for [MD][W].*

Weight	Muscle	Drug	
		Drug 1	Drug 2
High	Type 1	0.625	1.827
	Type 2	0.590	.592
Low	Type 1	0.512	1.496
	Type 2	0.483	.485

As usual, this is equivalent to a model with just the highest-order effects,

$$\log(p_{1ijk}/p_{2ijk}) = (WMD)_{ijk}.$$

As introduced in earlier chapters, we denote this model [WMD] with similar notations for other models that focus on the highest-order effects.

Models can be fitted by maximum likelihood. Reduced models can be tested. Estimates and asymptotic standard errors can be obtained. The analysis of model (20.7.1) is similar to that of an unbalanced three-factor ANOVA model as illustrated in Chapter 16.

Table 20.11 gives a list of ANOVA type logit models, deviance df, deviance G^2, P values for testing the fitted model against Model (20.7.1), and $A-q$ values. Clearly, the best-fitting logit models are the models [MD][W] and [WM][MD]. Both involve the muscle-type-by-drug interaction and a main effect for weight. One of the models includes the muscle-type-by-weight interaction. Note that P values associated with saturated model goodness-of-fit tests are appropriate here because we are not dealing with 0-1 data. (The smallest group size is $3+3=6$.)

The estimated odds for a high tension change using [MD][W] are given in Table 20.12. The estimated odds are 1.22 times greater for high-weight muscles than for low-weight muscles. For example, in Table 20.12, $0.625/0.512 = 1.22$ but also $1.22 = 0.590/0.483 = 1.827/1.495 = 0.592/0.485$. This corresponds to the main effect for weight in the logit model. The odds also involve a muscle-type-by-drug interaction. To establish the nature of this interaction, consider the four estimated odds for high weights with various muscles and drugs. These are the four values at the top of Table 20.12, e.g., for muscle type 1, drug 1 this is 0.625. In every muscle type–drug combination other than type 1, drug 2, the estimated odds of having a high tension change are about 0.6. The estimated probability of having a high tension change is about $0.6/(1+0.6) = 0.375$. However, for type 1, drug 2, the estimated odds are 1.827 and the estimated probability of a high tension change is $1.827/(1+1.827) = 0.646$. The chance of having a high tension change is much greater for the combination muscle type 1, drug 2 than for any other muscle type–drug combination. A similar analysis holds for the low-weight odds $\hat{p}_{12jk}/(1-\hat{p}_{12jk})$ but the actual values of the odds are smaller by a multiplicative factor of 1.22 because of the main effect for weight.

The other logit model that fits quite well is [WM][MD]. Tables 20.13 and 20.14 both contain

Table 20.13: *Estimated odds for [WM][MD].*

		Drug	
Weight	Muscle	Drug 1	Drug 2
High	Type 1	0.809	2.202
	Type 2	0.569	0.512
Low	Type 1	0.499	1.358
	Type 2	0.619	0.557

Table 20.14: *Estimated odds for [WM][MD].*

		Drug	
Muscle	Weight	Drug 1	Drug 2
Type 1	High	0.809	2.202
	Low	0.499	1.358
Type 2	High	0.569	0.512
	Low	0.619	0.557

the estimated odds of high tension change for this model. The difference between Tables 20.13 and 20.14 is that the rows of Table 20.13 have been rearranged in Table 20.14. This sounds like a trivial change, but examination of the tables shows that Table 20.14 is easier to interpret. The reason for changing from Table 20.13 to Table 20.14 is the nature of the logit model. The model [WM][MD] has M in both terms, so it is easiest to interpret the fitted model when fixing the level of M. For a fixed level of M, the effects of W and D are additive in the log odds, although the size of those effects change with the level of M.

Looking at the type 2 muscles in Table 20.14, the high-weight odds are 0.919 times the low-weight odds. Also, the drug 1 odds are 1.111 times the drug 2 odds. Neither of these are really very striking differences. For muscle type 2, the odds of a high tension change are about the same regardless of weight and drug. Contrary to our previous model, they do not seem to depend much on weight and to the extent that they do depend on weight, the odds go down rather than up for higher weights.

Looking at the type 1 muscles, we see the dominant features of the previous model reproduced. The odds of high tension change are 1.622 times greater for high weights than for low weights. The odds of high tension change are 2.722 times greater for drug 2 than for drug 1.

Both models indicate that for type 1 muscles, high weight increases the odds and drug 2 increases the odds. Both models indicate that for type 2 muscles, drug 2 does not substantially change the odds. The difference between the models [MD][W] and [WM][MD] is that [MD][W] indicates that for type 2 muscles, high weight should increase the odds, but [WM][MD] indicates little change for high weight and, in fact, what change there is indicates a decrease in the odds.

20.8 Ordered categories

In dealing with ANOVA models for measurement data, when one or more factors had quantitative levels, it was useful to model effects with polynomials. Similar results apply to logit models.

EXAMPLE 20.8.1.   Consider data in which there are four factors defining a $2 \times 2 \times 2 \times 6$ table. The factors are

Table 20.15: *Abortion opinion data.*

RACE	SEX	OPINION	AGE 18–25	26–35	36–45	46–55	56–65	66+
White	Male	Yes	96	138	117	75	72	83
		No	44	64	56	48	49	60
	Female	Yes	140	171	152	101	102	111
		No	43	65	58	51	58	67
Nonwhite	Male	Yes	24	18	16	12	6	4
		No	5	7	7	6	8	10
	Female	Yes	21	25	20	17	14	13
		No	4	6	5	5	5	5

Factor	Abbrev-iation	Levels
Race (h)	R	White, Nonwhite
Sex (i)	S	Male, Female
Opinion (j)	O	Yes = Supports Legalized Abortion
		No = Opposed to Legalized Abortion
Age (k)	A	18–25, 26–35, 36–45, 46–55, 56–65, 66+ years

Opinion is the response factor. Age has ordered categories. The data are given in Table 20.15. The probability of a Yes opinion for Race h, Sex i, Age k is $p_{hik} \equiv p_{hi1k}$. The corresponding No probability has $1 - p_{hik} \equiv p_{hi2k}$.

As in the previous section, we could fit a three-factor ANOVA type logit model to these data. Table 20.16 contains fitting information for standard three-factor models wherein [] indicates the intercept (grand mean) only model. From the deviances and $A - q$ in Table 20.16 a good-fitting logit model is

$$\log[p_{hik}/(1 - p_{hik})] = (RS)_{hi} + A_k. \tag{20.8.1}$$

Fitting this model gives the estimated odds of supporting relative to opposing legalized abortion that follow.

Odds of Support versus Opposed: Model (20.8.1)

Race	Sex	Age 18–25	26–35	36–45	46–55	56–65	65+
White	Male	2.52	2.14	2.09	1.60	1.38	1.28
	Female	3.18	2.70	2.64	2.01	1.75	1.62
Nonwhite	Male	2.48	2.11	2.06	1.57	1.36	1.26
	Female	5.08	4.31	4.22	3.22	2.79	2.58

The deviance G^2 is 9.104 with 15 df. The G^2 for fitting [R][S][A] is 11.77 on 16 df. The difference in G^2s is not large, so the reduced logit model $\log[p_{hik}/(1 - p_{hik})] = R_h + S_i + A_k$ may fit adequately, but we continue to examine Model (20.8.1).

The odds suggest two things: 1) odds decrease as age increases, and 2) the odds for males are about the same, regardless of race. We fit models that incorporate these suggestions. Of course, because the data are suggesting the models, formal tests of significance will be even less appropriate than usual but G^2s still give a reasonable measure of the quality of model fit.

To model odds that are decreasing with age we incorporate a linear trend in ages. In the absence of specific ages to associate with the age categories we simply use the scores $k = 1, 2, \ldots, 6$. These

Table 20.16: *Logit models for the abortion opinion data.*

Model	df	G^2	$A-q$
[RS][RA][SA]	5	4.161	−5.839
[RS][RA]	10	4.434	−15.566
[RS][SA]	10	8.903	−11.097
[RA][SA]	6	7.443	−4.557
[RS][A]	15	9.104	−20.896
[RA][S]	11	7.707	−14.23
[SA][R]	11	11.564	−10.436
[R][S][A]	16	11.772	−20.228
[R][S]	21	40.521	−1.479
[R][A]	17	21.605	−12.395
[S][A]	17	14.084	−19.916
[R]	22	49.856	5.856
[S]	22	43.451	−0.549
[A]	18	23.799	−12.201
[]	23	52.636	6.636

quantitative levels suggest fitting the ACOVA model

$$\log[p_{hik}/(1 - p_{hik})] = (RS)_{hi} + \gamma k. \tag{20.8.2}$$

The deviance G^2 is 10.18 on 19 df, so the linear trend in coded ages fits very well. Recall that Model (20.8.1) has $G^2 = 9.104$ on 15 df, so a test of Model (20.8.2) versus Model (20.8.1) has $G^2 = 10.18 - 9.104 = 1.08$ on $19 - 15 = 4$ df.

To incorporate the idea that males have the same odds of support, recode the indices for races and sexes. The indices for the $(RS)_{hi}$ terms are $(h, i) = (1, 1), (1, 2), (2, 1), (2, 2)$. We recode with new indexes (f, e) having the correspondence

$$\begin{array}{ccccc} (h, i) & (1,1) & (1,2) & (2,1) & (2,2) \\ (f, e) & (1,1) & (2,1) & (1,2) & (3,1). \end{array}$$

The model

$$\log[p_{fek}/(1 - p_{fek})] = (RS)_{fe} + A_k$$

gives exactly the same fit as Model (20.8.1). Together, the subscripts f, e, and k still distinguish all of the cases in the data. The point of this recoding is that the single subscript f distinguishes between males and the two female groups but does not distinguish between white and nonwhite males, so now if we fit the model

$$\log[p_{fek}/(1 - p_{fek})] = (RS)_f + A_k, \tag{20.8.3}$$

we have a model that treats the two male groups the same. To fit this, you generally do not need to define the index e in your data file, even though it will implicitly exist in the model.

Of course, Model (20.8.3) is a reduced model relative to Model (20.8.1). Model (20.8.3) has deviance $G^2 = 9.110$ on 16 df, so the comparison between models has $G^2 = 9.110 - 9.104 = 0.006$ on $16 - 15 = 1$ df. We have lost almost nothing by going from Model (20.8.1) to Model (20.8.3).

Finally, we can write a model that incorporates both the trend in ages and the equality for males

$$\log[p_{fek}/(1 - p_{fek})] = (RS)_f + \gamma k. \tag{20.8.4}$$

This has $G^2 = 10.19$ on 20 df. Thus, relative to Model (20.8.1), we have dropped 5 df from the model, yet only increased the G^2 by $10.19 - 9.10 = 1.09$. Rather than fitting Model (20.8.4), we fit the equivalent model that includes an intercept (grand mean) μ. The estimates and standard errors for this model, using the side condition $(RS)_1 = 0$, are

Table 20.17: *French convictions.*

Year	Convictions	Accusations
1825	4594	7234
1826	4348	6988
1827	4236	6929
1828	4551	7396
1829	4475	7373
1830	4130	6962

Table of Coefficients: Model related to (20.8.4)

Parameter	*Est*	SE	*t*
μ	1.071	0.1126	9.51
$(RS)_1$	0	—	—
$(RS)_2$	0.2344	0.09265	2.53
$(RS)_3$	0.6998	0.2166	3.23
γ	−0.1410	0.02674	−5.27

All of the terms seem important. With this side condition, $\widehat{(RS)}_2$ is actually an estimate of $(RS)_2 - (RS)_1$, so the t score 2.53 is an indication that white females have an effect on the odds of support that is different from males. Similarly, $\widehat{(RS)}_3$ is an estimate of the difference in effect between nonwhite females and males.

The estimated odds of support are

Odds of Support: Model (20.8.4).

Race-Sex	Age					
	18–25	26–35	36–45	46–55	56–65	65+
Male	2.535	2.201	1.912	1.661	1.442	1.253
White female	3.204	2.783	2.417	2.099	1.823	1.583
Nonwhite female	5.103	4.432	3.850	3.343	2.904	2.522

The odds can be transformed into probabilities of support. To most people, probabilities are easier to interpret than odds. The estimated probability that a white female between 46 and 55 years of age supports legalized abortion is $2.099/(1 + 2.099) = 0.677$. The odds are about 2, so the probability is about twice as great that such a person will support legalized abortion rather than oppose it.

20.9   Exercises

EXERCISE 20.9.1.   Fit a logistic model to the data of Table 20.17 that relates probability of conviction to year. Is there evidence of a trend in the conviction rates over time? Is there evidence for a lack of fit?

EXERCISE 20.9.2.   Stigler (1986, p. 208) reports data from the *Edinburgh Medical and Surgical Journal* (1817) on the relationship between heights and chest circumferences for Scottish militia men. Measurements were made in inches. We concern ourselves with two groups of men, those with 39-inch chests and those with 40-inch chests. The data are given in Table 20.18. Test whether the distribution of heights is the same for these two groups, cf. Chapter 5.

Is it reasonable to fit a logistic regression to the data of Table 20.18? Why or why not? Explain what such a model would be doing. Whether reasonable or not, fitting such a model can be done. Fit a logistic model and discuss the results. Is there evidence for a lack of fit?

EXERCISE 20.9.3.   Chapman, Masinda, and Strong (1995) give the data in Table 20.19. These

Table 20.18: *Heights and chest circumferences.*

Chest	Heights					Total
	64–65	66–67	68–69	70–71	71–73	
39	142	442	341	117	20	1062
40	118	337	436	153	38	1082
Total	260	779	777	270	58	2144

Table 20.19: *Unpopped kernels.*

Time	Trials		
	1	2	3
30	144	145	141
45	125	125	118
60	101	138	119
120	197	112	92
150	109	101	61
165	64	54	78
180	34	23	50
210	25	31	36
225	25	27	8
240	11	12	27
255	3	0	2

are the number out of 150 popcorn kernels that fail to pop when microwaved for a given amount of time. There are three replicates. Fit a logistic regression with time as the predictor.

EXERCISE 20.9.4.   Use the results of Subsection 12.5.2 and the not-so-obvious fact that $\sum_{i=1}^{n} \hat{p}_i/n = \sum_{i=1}^{n} y_i/n \equiv \bar{y}$ to show that, for the mouse data of Section 20.1, the formula for R^2 when $n = 12, N_h \equiv N = 10$ is

$$R^2 = \frac{\left[\sum_{i=1}^{n}(\hat{p}_i - \bar{y})(y_i - \bar{y})\right]^2}{\left[\sum_{i=1}^{n}(\hat{p}_i - \bar{y})^2\right]\left[\sum_{i=1}^{n}(y_i - \bar{y})^2\right]}$$

but that for 0-1 data we get the smaller value

$$R^2 = \frac{\left[N\sum_{i=1}^{n}(\hat{p}_i - \bar{y})(y_i - \bar{y})\right]^2}{\left[N\sum_{i=1}^{n}(\hat{p}_i - \bar{y})^2\right]\left[N\sum_{i=1}^{n}(y_i - \bar{y})^2 + N\sum_{i=1}^{n}y_i(1 - y_i)\right]}.$$

EXERCISE 20.9.5.   For the $n = 12$ version of the mouse data, reanalyze it using the log of the dose as a predictor variable. Create a version of the data with $n = 120$ and reanalyze both the x and $\log(x)$ versions. Compare. Among other things compare the deviances and R^2 values.

Chapter 21

Log-Linear Models: Describing Count Data

In a longitudinal study discussed by Christensen (1997), 2121 people neither exercised regularly nor developed cardiovascular disease during the study. These subjects were cross-classified by three factors: Personality type (A,B), Cholesterol level (normal, high), and Diastolic Blood Pressure (normal, high). The data are given in Table 21.1.

Table 21.1 is a three-way table of counts. The three factors are Personality, Cholesterol level, and Diastolic Blood Pressure. Each factor happens to be at two levels, but that is of no particular consequence for our modeling of the data. We can analyze the data by fitting three-way ANOVA type models to it. However, count data are not normally distributed, so standard ANOVA methods are inappropriate. In particular, random variables for counts tend to have variances that depend on their mean values. Standard sampling schemes for count data are multinomial sampling and Poisson sampling. In this case, we can think of the data as being a sample of 2121 from a multinomial distribution, cf. Section 1.5.

In general we assume that our data are independent Poisson random variables, say,

$$y_h \sim \text{Pois}(\mu_h), \qquad h = 1, \ldots, n$$

and create models

$$\log(\mu_h) = m(x_h)$$

where x_h is a predictor variable or vector associated with case h. Most often x_h contains factor variables. Our standard models $m(\cdot)$ are all linear in their parameters, so these are called *log-linear models*. Rather than Poisson sampling, most of our data are multinomial or a combination of independent multinomials. For models with an intercept (or ANOVA models), the analysis for Poisson data is the same as the analysis for multinomial data. A similar statement holds for independent multinomial data provided appropriate group effects are included in the model. For a complete discussion of fitting log-linear models with various sampling schemes and the relationship between this approach and ANOVA modeling for normal data; see Christensen (1997).

Log-linear models are more often used for exploring relationships between factors than for prediction. In prediction, one tends to focus on a particular factor (variable) of interest, called the response factor, and use the other factors (variables) to predict or explain the response. If a factor of interest has only two possible outcomes, the logistic models of Chapter 21 can be used for prediction. Log-linear models are more often used to model independence relationships between factors.

Table 21.1: *Personality, cholesterol, blood pressure data.*

y_{ijk} Personality	Cholesterol	Diastolic Blood Pressure Normal	High
A	Normal	716	79
	High	207	25
B	Normal	819	67
	High	186	22

Table 21.2: *Religion and occupations.*

Religion	Occupation A	B	C	D	Total
Protestant	210	277	254	394	1135
Roman Catholic	102	140	127	279	648
Jewish	36	60	30	17	143
Total	348	477	411	690	1926

In this chapter we examine log-linear models, we also relate them to logistic regression models and show how to use log-linear models to develop prediction models for factors that have more than two possible outcomes.

21.1 Models for two-factor tables

Consider a 3×4 table such as that given in Table 21.2. These are data extracted from Lazerwitz (1961) and were considered in Chapter 5. The data are from three religious groups and give the numbers of people who practice various occupations. The occupations are A, professions; B, owners, managers, and officials; C, clerical and sales; and D, skilled workers.

Consider fitting ANOVA type models to the logs of the expected cell counts. Let y_{ij} denote the observed count in row i and column j of the table and let μ_{ij} denote the expected count in the i, j cell. The two-way model with interaction can be written

$$\log(\mu_{ij}) = \mu + \alpha_i + \eta_j + \gamma_{ij}, \qquad i = 1, \ldots 3 \,; \, j = 1, \ldots, 4.$$

An alternative notation is often used for log-linear models that merely changes the names of the parameters,

$$\log(\mu_{ij}) = u + u_{1(i)} + u_{2(j)} + u_{12(ij)}. \tag{21.1.1}$$

This log-linear model imposes no constraints on the table of cell means because it includes a separate parameter for every cell in the table. Actually, the u, $u_{1(i)}$, and $u_{2(j)}$ terms are all redundant because the $u_{12(ij)}$ terms alone provide a parameter for explaining every expected cell count in the table. This model will fit the data for any two-factor table perfectly! In other words, it will lead to $\hat{\mu}_{ij} = y_{ij}$. Because it has a parameter for every cell, this model is referred to as the *saturated model*.

The log-linear model that includes only main effects is

$$\log(\mu_{ij}) = u + u_{1(i)} + u_{2(j)}. \tag{21.1.2}$$

In terms of Table 21.2, if this model fits the data, it says the data are explained adequately by a model in which religion and occupation are independent, cf. Christensen (1997). If religion and occupation are independent, knowing one's religion gives no new information about a person's occupation. That makes sense relative to the model involving only main effects, because then religion affects only the terms $u_{1(i)}$ and has no effect on the contribution from occupation, which is the additive term $u_{2(j)}$. On the other hand, if Model (21.1.1) applies to the data, the interaction terms $u_{12(ij)}$ allow the possibility of different occupation effects for every religious group. (It turns out that the model of independence does not fit these data well.)

21.1.1 Lancaster–Irwin partitioning

Lancaster–Irwin partitioning was illustrated in Section 5.6. We now demonstrate how to accomplish such partitions using log-linear models. In particular, we fit the reduced and collapsed tables given in Table 5.10. The religion and occupations data of Table 21.2 are rewritten in Table 21.3 in a form suitable for computing the fits to models (21.1.1) and (21.1.2). The table also includes new

Table 21.3 *Religion and occupations: i is Religion, j is Occupation, k collapses Roman Catholic and Protestant, k and m together uniquely define religions.*

y	i	j	k	m
210	1	1	0	1
277	1	2	0	1
254	1	3	0	1
394	1	4	0	1
102	2	1	0	2
140	2	2	0	2
127	2	3	0	2
279	2	4	0	2
36	3	1	3	0
60	3	2	3	0
30	3	3	3	0
17	3	4	3	0

subscripts k and m that replace i and can be used to fit the reduced table. The full interaction model (21.1.1) can be written as

$$\log(\mu_{kmj}) = u + u_{1(km)} + u_{2(j)} + u_{12(kmj)}, \tag{21.1.3}$$

whereas the main-effects model (21.1.2) can be rewritten as

$$\log(\mu_{kmj}) = u + u_{1(km)} + u_{2(j)}. \tag{21.1.4}$$

The reduced table of Tables 5.10 and 5.11 can be fitted using the model

$$\log(\mu_{kmj}) = u + u_{1(km)} + u_{2(kj)}. \tag{21.1.5}$$

This model has a separate parameter for each Jewish occupation, so it effectively leaves them alone, and fits an independence model to the Roman Catholics and Protestants, cf. Christensen (1997, Exercise 8.4.3). The pair of subscripts km uniquely define the three religious groups, so in Model (21.1.5) the term $u_{1(km)}$ is really a main effect for religions. The terms $u_{2(kj)}$ define main effects for occupations when $k = 0$, i.e., for Roman Catholics and Protestants, but the terms $u_{2(kj)}$ define separate effects for each Jewish occupation when $k = 3$. *The key ideas are that k has a unique value for each religious category except it does not distinguish between the categories that are to be collapsed, and that k and m uniquely define the religions.* Thus m needs to have different values for each religion that is to be collapsed. In particular, assuming i never takes on the value 0, we can define the new variables as

$$k = \begin{cases} 0 & \text{if row } i \text{ is collapsed} \\ i & \text{if row } i \text{ is not collapsed} \end{cases} \qquad m = \begin{cases} i & \text{if row } i \text{ is collapsed} \\ 0 & \text{if row } i \text{ is not collapsed.} \end{cases}$$

Note that Model (21.1.5) is a special case of the full interaction model (21.1.3) but is more general than the independence model (21.1.4).

For fitting Model (21.1.5), $G^2 = 12.206$ on $df = 3$, which are the deviance and degrees of freedom for the reduced table. For fitting models (21.1.2) and (21.1.4) the deviance is $G^2 = 64.342$ on $df = 6$. The deviance and degrees of freedom for the collapsed table are obtained by subtraction, $G^2 = 64.342 - 12.206 = 52.136$ on $df = 6 - 3 = 3$. These are roughly similar to the corresponding Pearson χ^2 statistics discussed in Chapter 5.

21.2 Models for three-factor tables

Consider a three-factor table of counts such as Table 21.1. The *saturated* model is the model with three-factor interaction terms,

$$\log(\mu_{ijk}) = u + u_{1(i)} + u_{2(j)} + u_{3(k)} + u_{12(ij)} + u_{13(ik)} + u_{23(jk)} + u_{123(ijk)}.$$

This model has a separate $u_{123(ijk)}$ parameter for each cell in the table, so this model will fit any three-factor table perfectly. Note that the model is grossly overparameterized; an equivalent model is simply

$$\log(\mu_{ijk}) = u_{123(ijk)}.$$

We abbreviate this model as [123]. The only reason for using the overparameterized model is that it is suggestive of some interesting submodels. In the following discussion, statements are made about log-linear models implying independence relationships. See Christensen (1997, Chapter 3) for the validity of these claims.

The model with main effects only is

$$\log(\mu_{ijk}) = u + u_{1(i)} + u_{2(j)} + u_{3(k)}. \tag{21.2.1}$$

This model implies complete independence of the three factors in the table. Note that a less overparameterized version of Model (21.2.1) is

$$\log(\mu_{ijk}) = u_{1(i)} + u_{2(j)} + u_{3(k)}.$$

We can abbreviate the model as [1][2][3].

We can also look at models that include main effects and only one two-factor interaction, for example,

$$\log(\mu_{ijk}) = u + u_{1(i)} + u_{2(j)} + u_{3(k)} + u_{23(jk)}. \tag{21.2.2}$$

This model implies that factor 1 is independent of factors 2 and 3. We can also get two other models with similar interpretations by including all the main effects but only the $u_{12(ij)}$ interaction (factor 3 independent of factors 1 and 2) or the main effects and only the $u_{13(ik)}$ interaction (factor 2 independent of factors 1 and 3). Note that a less overparameterized version of Model (21.2.2) is

$$\log(m_{ijk}) = u_{1(i)} + u_{23(jk)}.$$

We can abbreviate the model as [1][23].

Now consider models that include two two-factor interactions, for example,

$$\log(\mu_{ijk}) = u + u_{1(i)} + u_{2(j)} + u_{3(k)} + u_{13(ik)} + u_{23(jk)}. \tag{21.2.3}$$

This model implies that factor 1 is independent of factor 2 given factor 3. Note that factor 3 is in both interaction terms, which is why the model has 1 and 2 independent given 3. We can also get two other models with similar interpretations by including only the $u_{12(ij)}$ and $u_{23(jk)}$ interactions (factors 1 and 3 independent given factor 2) or only the $u_{12(ij)}$ and $u_{13(ik)}$ interactions (factors 2 and 3 independent given factor 1). Note that a less overparameterized version of Model (21.2.3) is

$$\log(\mu_{ijk}) = u_{13(ik)} + u_{23(jk)}.$$

We abbreviate the model as [13][23].

The no three-factor interaction model is

$$\log(\mu_{ijk}) = u + u_{1(i)} + u_{2(j)} + u_{3(k)} + u_{12(ij)} + u_{13(ik)} + u_{23(jk)}.$$

A less overparameterized version is

$$\log(\mu_{ijk}) = u_{12(ij)} + u_{13(ik)} + u_{23(jk)},$$

which we can abbreviate as [12][13][23]. This model has no nice interpretation in terms of independence.

21.2.1 Testing models

Testing models works much the same as it does for linear models except with Poisson or multinomial data there is no need to estimate a variance. Thus, the tests are similar to looking at only the numerator sums of squares for an analysis of variance or regression test. (This would be totally appropriate in ANOVA and regression if we ever actually knew the value of σ^2.) The *deviance* of a model is

$$G^2 = 2 \sum_{\text{all cells}} y_{ijk} \log\left(y_{ijk}/\hat{\mu}_{ijk}\right),$$

where $\hat{\mu}_{ijk}$ is the (maximum likelihood) estimated expected cell count based on the model we are testing. The subscripts in G^2 are written for a three-dimensional table, but the subscripts are irrelevant. What is relevant is summing over all cells.

G^2 gives a test of whether the model gives an adequate explanation of the data relative to the saturated model. (Recall that the saturated model always fits the data perfectly because it has a separate parameter for every cell in the table. Thus the estimated cell counts for a saturated model are always just the observed cell counts.) The *deviance degrees of freedom* (df) for G^2 are the degrees of freedom for the (interaction) terms that have been left out of the saturated model. So in an $I \times J \times K$ table, the model that drops out the three-factor interaction has $(I-1)(J-1)(K-1)$ degrees of freedom for G^2. If we drop out the [12] interaction as well as the three-factor interaction, G^2 has $(I-1)(J-1)(K-1) + (I-1)(J-1)$ degrees of freedom. If the observations y in each cell are reasonably large, the G^2 statistics can be compared to a χ^2 distribution with the appropriate number of degrees of freedom. A large G^2 indicates that the model fits poorly as compared to the saturated model.

Alternatively, one can use the Pearson statistic,

$$X^2 = \sum_{\text{all cells}} \frac{(y_{ijk} - \hat{\mu}_{ijk})^2}{\hat{\mu}_{ijk}}$$

for testing the model. It has the same degrees of freedom as G^2.

To test a full model against a reduced model (Red.), we compare deviance G^2s. In particular, the test statistic is

$$
\begin{aligned}
G^2(Red.\ versus\ Full) &\equiv G^2(Red.) - G^2(Full) \\
&= 2 \sum_{\text{all cells}} y_{ijk} \log\left(y_{ijk}/\hat{\mu}_{Rijk}\right) - 2 \sum_{\text{all cells}} y_{ijk} \log\left(y_{ijk}/\hat{\mu}_{Fijk}\right) \\
&= 2 \sum_{\text{all cells}} y_{ijk} \log\left(\hat{\mu}_{Fijk}/\hat{\mu}_{Rijk}\right) \\
&= 2 \sum_{\text{all cells}} \hat{\mu}_{Fijk} \log\left(\hat{\mu}_{Fijk}/\hat{\mu}_{Rijk}\right)
\end{aligned}
$$

where $\hat{\mu}_{Rijk}$ and $\hat{\mu}_{Fijk}$ are the maximum likelihood estimates from the reduced and full models, respectively, and showing the last equality is beyond the scope of this book. The degrees of freedom for the test is the difference in deviance degrees of freedom, $df(Red.) - df(Full)$.

In Chapter 20 we introduced AIC as a tool for model selection. For log-linear models, maximizing Akaike's information criterion amounts to choosing the model "M" that *minimizes*

$$A_M = G^2(M) - [q - 2r],$$

where $G^2(M)$ is the deviance for testing the M model against the saturated model, r is the number of degrees of freedom for the M model (not the degrees of freedom for the model's deviance), and there are $q \equiv n$ degrees of freedom for the saturated model, i.e., $q \equiv n$ cells in the table.

Given a list of models to be compared along with their G^2 statistics and the degrees of freedom for the tests, a slight modification of A_M is easier to compute,

$$
\begin{aligned}
A_M - q &= G^2(M) - 2[q - r] \\
&= G^2(M) - 2df(M).
\end{aligned}
$$

Because q does not depend on the model M, minimizing $A_M - q$ is equivalent to minimizing A_M. Note that for the saturated model, $A - q \equiv 0$.

EXAMPLE 21.2.1.   For the personality (1), cholesterol (2), blood pressure (3) data of Table 21.1, testing models against the saturated model gives deviance and AIC values

Model	df	G^2	$A - q$
[12][13][23]	1	0.613	−1.387
[12][13]	2	2.062	−1.938
[12][23]	2	2.980	−1.020
[13][23]	2	4.563	0.563
[1][23]	3	7.101	1.101
[2][13]	3	6.184	0.184
[3][12]	3	4.602	−1.398
[1][2][3]	4	8.723	0.723

Comparing the G^2 values to 95th percentiles of χ^2 distributions with the appropriate number of degrees of freedom, all of the models seem to explain the data adequately.

To test, for example, the reduced model [1][2][3] against a full model [12][13], the test statistic is

$$
G^2([1][2][3] \text{ versus } [12][13]) = G^2([1][2][3]) - G^2([12][13]) = 8.723 - 2.062 = 6.661,
$$

on $df([1][2][3]) - df([12][13]) = 4 - 2 = 2$. The test statistic is greater than $\chi^2(.95, 2) = 5.991$, so we can reject the model of complete independence [1][2][3]. Complete independence fits the data significantly worse than the model [12][13] in which cholesterol and blood pressure are independent given personality type. So even though [1][2][3] fits adequately relative to the saturated model [123], it fits the data inadequately relative to [12][13].

For the personality data, AIC suggests three attractive models: [3][12], [12][13], and [12][13][23]. Model [12][23] is also not bad.   □

With only three factors it is easy to look at all possible models. Model selection procedures become more important when dealing with tables having more than three factors, cf. Christensen (1997) and Section 21.4.

21.3   Estimation and odds ratios

For log-linear models, the primary goal of estimation is to obtain the estimated expected cell counts $\hat{\mu}_h$. Iterative computing methods are needed to find these. The Newton–Raphson method (iteratively reweighted least squares) provides an estimate of the model parameters. The other estimation method is iterative proportional fitting (the Deming–Stephans algorithm) most often used only for ANOVA type models. This directly provides estimates $\hat{\mu}_h$ from which parameter estimates can be back calculated; see Christensen (1997) for details. Programs for fitting generalized linear models typically use Newton–Raphson. The BMDP procedure 4F is restricted to ANOVA type models but uses iterative proportional fitting, and as a result it provides some nice features not available in many other programs. (Over the years, BMDP 4F has become an obscure piece of software, but I know of nothing that approximates its capabilities. It is now available through z/OS mainframe versions of SAS.)

EXAMPLE 21.3.1. Kihlberg, Narragon, and Campbell (1964), Fienberg (1980), and Christensen (1997) present data on automobile injuries. The model of no three-factor interaction

$$\log(\mu_{ijk}) = u + u_{1(i)} + u_{2(j)} + u_{3(k)} + u_{12(ij)} + u_{13(ik)} + u_{23(jk)}$$

fits the data very well. Below are given the data and the estimated expected cell counts based on the model.

$y_{ijk}(\hat{\mu}_{ijk})$		Accident Type (k)			
		Collision		Rollover	
Injury (j)		Not Severe	Severe	Not Severe	Severe
Driver	No	350 (350.49)	150 (149.51)	60 (59.51)	112 (112.49)
Ejected (i)	Yes	26 (25.51)	23 (23.49)	19 (19.49)	80 (79.51)

One way to examine a fitted model is to look at the estimated odds ratios. For multinomial sampling we first use the unrestricted estimates of the probabilities $p_{ijk} \equiv \mu_{ijk}/\mu_{...} = \mu_{ijk}/y_{...}$, which are $\hat{p}_{ijk} = y_{ijk}/y_{...}$. The estimated odds of a not-severe injury when the driver is not ejected in a collision are

$$\frac{y_{111}/y_{...}}{y_{121}/y_{...}} = \frac{y_{111}}{y_{121}} = \frac{y_{111}}{y_{121}} = \frac{350}{150} = 2.333.$$

The estimated odds of a not-severe injury when the driver is ejected in a collision are

$$\frac{26}{23} = 1.130$$

In a collision, the odds of a not-severe injury are

$$2.064 = \frac{2.333}{1.130} = \frac{350(23)}{26(150)}$$

times greater if one is not ejected from the car. This is known as an *odds ratio*. Similarly, in a rollover, the odds of a not-severe injury are

$$2.256 = \frac{60(80)}{19(112)}$$

times greater if one is not ejected from the car. These odds ratios are quite close to one another. In the no-three-way-interaction model, these odds ratios are forced to be the same. If we make the same computations using the estimated expected cell counts, we get

$$2.158 = \frac{350.49(23.49)}{25.51(149.51)} = \frac{59.51(79.51)}{14.49(112.49)}.$$

For both collisions and rollovers, the odds of a severe injury are about twice as large if the driver is ejected from the vehicle than if not. Equivalently, the odds of having a not-severe injury are about twice as great if the driver is not ejected from the vehicle than if the driver is ejected. It should be noted that the odds of being severely injured in a rollover are consistently much higher than in a collision. What we have concluded in our analysis is that the *relative* effect of the driver being ejected is the same for both types of accident and that being ejected substantially increases one's chances of being severely injured. □

All of the models discussed in Section 21.2 have interpretations in terms of odds ratios. An odds ratio keeps one of the three indexes fixed, say, k, and looks at quantities like

$$\frac{\hat{p}_{ijk}\hat{p}_{i'j'k}}{\hat{p}_{ij'k}\hat{p}_{i'jk}} = \frac{\hat{\mu}_{ijk}\hat{\mu}_{i'j'k}}{\hat{\mu}_{ij'k}\hat{\mu}_{i'jk}}.$$

In the model of complete independence (21.2.1), these values will always be 1, no matter how they are constructed. In Model (21.2.2) where factor 1 is independent of factors 2 and 3, any estimated odds ratio that fixes either the level k or the level j will equal one, and all odds ratios that fix i will be the same regardless of the value of i. In Model (21.2.3) where factors 1 and 2 are independent given 3, any estimated odds ratio that fixes the level k will equal one and all odds ratios that fix i will be the same regardless of the value of i; similarly all odds ratios that fix j will be the same regardless of the value of j. Odds ratios that equal one are directly related to certain interaction contrasts equaling zero, cf. Christensen (1997)

EXAMPLE 21.3.2.   Consider data from Everitt (1977) and Christensen (1997) on classroom behavior. The three factors are Classroom Behavior (Deviant or Nondeviant), Risk of the home situation: not at risk (N) or at risk (R), and Adversity of the school situation (Low, Medium, or High). The data and estimated expected cell counts for the model $\log(\mu_{ijk}) = u + u_{1(i)} + u_{2(j)} + u_{3(k)} + u_{23(jk)}$, in which behavior is independent of risk and adversity, are given below.

y_{ijk} ($\hat{\mu}_{ijk}$)		Adversity (k)					
		Low		Medium		High	
Risk (j)		N	R	N	R	N	R
	Non.	16	7	15	34	5	3
Classroom		(14.02)	(6.60)	(14.85)	(34.64)	(4.95)	(4.95)
Behavior (i)	Dev.	1	1	3	8	1	3
		(2.98)	(1.40)	(3.15)	(7.36)	(1.05)	(1.05)

Subject to round-off error, the estimate of the odds of nondeviant behavior is

$$\frac{\hat{\mu}_{1jk}}{\hat{\mu}_{2jk}} = \frac{14.02}{2.98} = \frac{6.60}{1.40} = \cdots = \frac{4.95}{1.05} = 4.702 = e^{1.548}.$$

Thus, any odds ratio in which either j or k is held fixed always equals 1. The estimate of the log odds of nondeviant behavior is

$$\begin{aligned} \log(\hat{\mu}_{1jk}/\hat{\mu}_{2jk}) &= \log(\hat{\mu}_{1jk}) - \log(\hat{\mu}_{2jk}) \\ &= \hat{u}_{1(1)} - \hat{u}_{1(2)} = 1.548. \end{aligned}$$

The odds of having a home situation that is not at risk depend on the adversity level. Up to round-off error, the odds satisfy

$$\frac{\hat{\mu}_{i1k}}{\hat{\mu}_{i2k}} = \begin{cases} 14.02/6.60 = 2.98/1.40 = 2.125 & k=1 \\ 14.85/34.64 = 3.15/7.36 = 0.428 & k=2 \\ 4.95/4.95 = 1.05/1.05 = 1 & k=3 \end{cases}.$$

The odds ratios do not depend on i, so any odds ratios that fix i (but change k) will equal each other, but will not necessarily equal 1; whereas odds ratios that fix k (but change i) will always equal 1.

21.4   Higher-dimensional tables

Tables with four or more factors can also be modeled. The basic ideas from three-dimensional tables continue to apply, but the models become more complicated.

EXAMPLE 21.4.1.   Consider again the data of Example 20.7.1 and Table 20.10 on muscle tension change. Here we examine models for expected cell counts but do not identify tension as a response.

The log-linear model of all main effects is

$$\log(\mu_{hijk}) = \gamma + \tau_h + \omega_i + \mu_j + \delta_k. \qquad (21.4.1)$$

The model of all two-factor interactions is

$$
\begin{aligned}
\log(\mu_{hijk}) &= \gamma + \tau_h + \omega_i + \mu_j + \delta_k + (\tau\omega)_{hi} + (\tau\mu)_{hj} + (\tau\delta)_{hk} \\
&\quad + (\omega\mu)_{ij} + (\omega\delta)_{ik} + (\mu\delta)_{jk}.
\end{aligned}
\tag{21.4.2}
$$

The model of all three-factor interactions is

$$
\begin{aligned}
\log(\mu_{hijk}) &= \gamma + \tau_h + \omega_i + \mu_j + \delta_k + (\tau\omega)_{hi} + (\tau\mu)_{hj} + (\tau\delta)_{hk} \\
&\quad + (\omega\mu)_{ij} + (\omega\delta)_{ik} + (\mu\delta)_{jk} \\
&\quad + (\tau\omega\mu)_{hij} + (\tau\omega\delta)_{hik} + (\tau\mu\delta)_{hjk} + (\omega\mu\delta)_{ijk}.
\end{aligned}
\tag{21.4.3}
$$

Removing some redundant parameters gives

$$
\log(\mu_{hijk}) = \tau_h + \omega_i + \mu_j + \delta_k,
\tag{21.4.1}
$$

$$
\log(\mu_{hijk}) = (\tau\omega)_{hi} + (\tau\mu)_{hj} + (\tau\delta)_{hk} + (\omega\mu)_{ij} + (\omega\delta)_{ik} + (\mu\delta)_{jk},
\tag{21.4.2}
$$

and

$$
\log(\mu_{hijk}) = (\tau\omega\mu)_{hij} + (\tau\omega\delta)_{hik} + (\tau\mu\delta)_{hjk} + (\omega\mu\delta)_{ijk}.
\tag{21.4.3}
$$

Corresponding shorthand notations are

$$
[T][W][M][D]
\tag{21.4.1}
$$

$$
[TW][TM][WM][TD][WD][MD]
\tag{21.4.2}
$$

$$
[TWM][TWD][TMD][WMD].
\tag{21.4.3}
$$

Statistics for testing models against the saturated model are given below. The only model considered that fits these data has all three-factor interactions.

Model	df	G^2	P	$A-q$
[TWM][TWD][TMD][WMD]	1	0.11	0.74	-1.89
[TW][TM][WM][TD][WD][MD]	5	47.67	0.00	37.67
[T][W][M][D]	11	127.4	0.00	105.4

To test reduced models, say, [TWM][TWD][TMD][WMD] against the reduced model of all two-factor terms [TW][TM][WM][TD][WD][MD], compute $G^2 = 47.67 - 0.11 = 47.56$ with $df = 5 - 1 = 4$. Clearly, the reduced model does not fit. A reasonable beginning for modelling these data would be to find which, if any, three-factor interactions can be removed without harming the model significantly. □

EXAMPLE 21.4.2. In later sections we will consider an expanded version of the abortion opinion data of Example 20.8.1 and Table 20.15 that includes another opinion category that was ignored in Chapter 20. The four factors now define a $2 \times 2 \times 3 \times 6$ table. The factors and levels are

Factor	Abbreviation	Levels
Race	R	White, Nonwhite
Sex	S	Male, Female
Opinion	O	Yes = Supports Legalized Abortion
		No = Opposed to Legalized Abortion
		Und = Undecided
Age	A	18–25, 26–35, 36–45, 46–55, 56–65, 66+ years

The data are given in Table 21.4. □

Table 21.4: *Abortion opinion data.*

RACE (h)	SEX (i)	OPINION (j)	18–25	26–35	36–45	46–55	56–65	66+
					AGE (k)			
White	Male	Yes	96	138	117	75	72	83
		No	44	64	56	48	49	60
		Und	1	2	6	5	6	8
	Female	Yes	140	171	152	101	102	111
		No	43	65	58	51	58	67
		Und	1	4	9	9	10	16
Nonwhite	Male	Yes	24	18	16	12	6	4
		No	5	7	7	6	8	10
		Und	2	1	3	4	3	4
	Female	Yes	21	25	20	17	14	13
		No	4	6	5	5	5	5
		Und	1	2	1	1	1	1

21.5   Ordered categories

In ANOVA models with one or more factors having quantitative levels, it is useful to model effects with polynomials. Similar results apply to log-linear models.

EXAMPLE 21.5.1.   Men from Framingham, Massachusetts were categorized by their serum cholesterol and systolic blood pressure. Consider the subsample that did not develop coronary heart disease during the follow-up period. The data are as follows.

Cholesterol	Blood Pressure (in mm Hg)				
(in mg/100 cc)	<127	127–146	147–166	167+	Totals
<200	117	121	47	22	307
200–219	85	98	43	20	246
220–259	119	209	68	43	439
≥260	67	99	46	33	245
Totals	388	527	204	118	1237

Both factors have ordered levels, although there is no one number associated with each level. We consider quantitative levels $1, 2, 3, 4$ for both factors. Obviously, this is somewhat arbitrary. An alternative approach that involves nonlinear modeling is to estimate the quantitative levels for each factor, cf. Christensen (1997) for more information.

Now consider four models that involve the quantitative levels:

Abbreviation	Model
[C][P][C_1]	$\log(\mu_{ij}) = u + u_{C(i)} + u_{P(j)} + C_{1i} \cdot j$
[C][P][P_1]	$\log(\mu_{ij}) = u + u_{C(i)} + u_{P(j)} + P_{1j} \cdot i$
[C][P][γ]	$\log(\mu_{ij}) = u + u_{C(i)} + u_{P(j)} + \gamma \cdot i \cdot j$
[C][P]	$\log(\mu_{ij}) = u + u_{C(i)} + u_{P(j)}.$

These are called the *row effects, column effects, uniform association,* and *independence* models, respectively. The fits for the models relative to the saturated model are

Model	df	G^2	$A-q$
$[C][P][C_1]$	6	7.404	−4.596
$[C][P][P_1]$	6	5.534	−6.466
$[C][P][\gamma]$	8	7.429	−8.571
$[C][P]$	9	20.38	2.38

The best-fitting model is the uniform association model

$$\log(\mu_{ij}) = u + u_{C(i)} + u_{P(j)} + \gamma \cdot i \cdot j.$$

Using the side conditions $u_{C(1)} = u_{P(1)} = 0$, the parameter estimates and large sample standard errors are

Parameter	Estimate	Standard Error
u	4.614	0.0699
$u_{C(1)}$	0	—
$u_{C(2)}$	−0.4253	0.1015
$u_{C(3)}$	−0.0589	0.1363
$u_{C(4)}$	−0.8645	0.1985
$u_{P(1)}$	0	—
$u_{P(2)}$	0.0516	0.0965
$u_{P(3)}$	−1.164	0.1698
$u_{P(4)}$	−1.991	0.2522
γ	0.1044	0.0293

The estimated cell counts are

Estimated Cell Counts $\hat{\mu}_{ij}$: Uniform Association

Cholesterol	<127	127–146	147–166	167+
<200	112.0	131.0	43.1	20.9
200–219	81.3	105.4	38.5	20.8
220–259	130.1	187.4	76.0	45.5
≥260	64.5	103.2	46.4	30.8

Column header group: Blood Pressure (spanning <127, 127–146, 147–166, 167+).

Because these are obtained from the uniform association model, the odds ratios for consecutive table entries are identical. For example, the odds of blood pressure < 127 relative to blood pressure 127–146 for men with cholesterol < 200 are 1.11 times the similar odds for men with cholesterol of 200–219; up to round-off error

$$\frac{112.0/131.0}{81.3/105.4} = \frac{112.0(105.4)}{81.3(131.0)} = e^{0.1044} = 1.11$$

where $0.1044 = \hat{\gamma}$. Similarly, the odds of blood pressure 127–146 relative to blood pressure 147–166 for men with cholesterol < 200 are 1.11 times the odds for men with cholesterol of 200–219:

$$\frac{131.0(38.5)}{105.4(43.1)} = e^{0.1044} = 1.11.$$

Also the odds of blood pressure < 127 relative to blood pressure 127–146 for men with cholesterol 200–219 are 1.11 times the odds for men with cholesterol of 220–259:

$$\frac{81.3(187.4)}{130.1(105.4)} = e^{0.1044} = 1.11.$$

For consecutive categories, the odds of lower blood pressure are 1.11 times greater with lower blood cholesterol than with higher blood cholesterol.

Of course, we can also compare nonconsecutive categories. For categories that are one step away from consecutive, the odds of lower blood pressure are $1.23 = e^{2(0.1044)}$ times greater with lower cholesterol than with higher cholesterol. For example, the odds of having blood pressure <127 compared to having blood pressure of 147–166 with cholesterol <200 are $1.23 = e^{2(0.1044)}$ times those for cholesterol 200–219. To check this observe that

$$\frac{112.0(38.5)}{81.3(43.1)} = 1.23.$$

Similarly, the odds of having blood pressure <127 compared to having blood pressure of 127–146 with cholesterol <200 are 1.23 times those for cholesterol 220–259. Extending this leads to observing that the odds of having blood pressure <127 compared to having blood pressure of 167+ with cholesterol <200 are $2.559 = e^{9(0.1044)}$ times those for cholesterol ≥ 260.

It is of interest to compare the estimated cell counts obtained under uniform association with the estimated cell counts under independence. The estimated cell counts under independence are

Estimated Cell Counts $\hat{\mu}_{ij}$: Independence

Cholesterol	Blood Pressure			
	<127	127–146	147–166	167+
<200	96.3	130.8	50.6	29.3
200–219	77.2	104.8	40.6	23.7
220–259	137.7	187.0	72.4	41.9
≥ 260	76.85	104.4	40.4	23.4

All of the estimated odds ratios from the independence model are 1. Relative to independence, with $\gamma > 0$, the uniform association model increases the estimated cell counts for cells with 1) high cholesterol and high blood pressure and 2) low cholesterol and low blood pressure. Also, the uniform association model decreases the estimated cell counts for cells with 1) high cholesterol and low blood pressure and 2) low cholesterol and high blood pressure.

EXAMPLE 21.5.2. For the Abortion Opinion data of Table 21.4, the model [RSO][OA] fits well. The ages are quantitative levels. We consider whether using the quantitative nature of this factor leads to a more succinct model. The age categories are 18–25, 26–35, 36–45, 46–55, 56–65, and 66+. For lack of a better idea, the category scores were taken as 1, 2, 3, 4, 5, and 6. Since the first and last age categories are different from the other four, the use of the scores 1 and 6 are particularly open to question. Two new models were considered:

Abbreviation	Model
[RSO][OA]	$\log(\mu_{hijk}) = u_{RSO(hij)} + u_{OA(jk)}$
[RSO][A][O_1]	$\log(\mu_{hijk}) = u_{RSO(hij)} + u_{A(k)} + O_1 jk$
[RSO][A][O_1][O_2]	$\log(\mu_{hijk}) = u_{RSO(hij)} + u_{A(k)} + O_1 jk + O_2 jk^2$

Both of these are reduced models relative to [RSO][OA]. To compare models, we need the following statistics

Model	df	G^2	$A-q$
[RSO][OA]	45	24.77	−65.23
[RSO][A][O_1][O_2]	51	26.99	−75.01
[RSO][A][O_1]	53	29.33	−76.67

Comparing [RSO][A][O_1] versus [RSO][OA] gives $G^2 = 29.33 - 24.77 = 4.56$ with degrees of

Table 21.5: *Textile faults.*

Roll	Length (l)	Faults (y)	Roll	Length (l)	Faults (y)
1	551	6	17	543	8
2	651	4	18	842	9
3	832	17	19	905	23
4	375	9	20	542	9
5	715	14	21	522	6
6	868	8	22	122	1
7	271	5	23	657	9
8	630	7	24	170	4
9	491	7	25	738	9
10	372	7	26	371	14
11	645	6	27	735	17
12	441	8	28	749	10
13	895	28	29	495	7
14	458	4	30	716	3
15	642	10	31	952	9
16	492	4	32	417	2

freedom $53 - 45 = 8$. The G^2 value is not significant. Similarly, [RSO][A][O_1][O_2] is an adequate model relative to [RSO[OA]]. The test for [O_2] has $G^2 = 29.33 - 27.99 = 1.34$ on 2 df, which is not significant. The model with only [O_1] fits the data well.

I should perhaps mention that, although it fits very well, [RSO][OA] is a strange model for these data. The model suggests that race–sex combinations are independent of age given one's opinions. One would not expect something as ephemeral as an opinion to affect things as concrete as race, sex, and age.

21.6   Offsets

Most of our examples have involved multinomial data with ANOVA type models. Now we consider a regression with Poisson data. This example also involves a term in the linear predictor that is known.

EXAMPLE 21.6.1.   Consider the data in Table 21.5. This is data from Bissell (1972) on the number of faults in pieces of fabric. It is reasonable to model the number of faults in any piece of fabric as Poisson and to use the length of the fabric as a predictor variable for the number of faults.

In general, we assume the existence of n independent random variables y_h with $y_h \sim Pois(\mu_h)$. In this example, a reasonable model might be that the expected number of faults μ_h is some number λ times the length of the piece of fabric, say l_h, i.e.,

$$\mu_h = \lambda l_h. \qquad (21.6.1)$$

Such a model assumes that the faults are being generated at a constant rate, and therefore the expected number of faults is proportional to the length. We can rewrite Model (21.6.1) as a log-linear model

$$\log(\mu_h) = \log(\lambda) + \log(l_h)$$

or

$$\log(\mu_h) = \beta_0 + (1)\log(l_h), \qquad (21.6.2)$$

where $\beta_0 \equiv \log(\lambda)$. If we generalize Model (21.6.2) it might look a bit more familiar. Using a simple linear regression structure with $\log(l_h)$ as a predictor variable, we have the more general model

$$\log(\mu_h) = \beta_0 + \beta_1 \log(l_h). \qquad (21.6.3)$$

Table 21.6: *Model fits: Textile fault data.*

Model	df	Deviance
(21.6.2)	31	64.538
(21.6.3)	30	64.537

Table 21.7: *Tables of Coefficients: Models (21.6.1) and (21.6.2).*

Model (21.6.2)		
Parameter	Estimate	SE
β_0	−4.193	0.05934
Model (21.6.3)		
Parameter	Estimate	SE
β_0	−4.173	1.135
β_1	0.9969	0.1759

On the original scale, this model has

$$\mu_h = \lambda l_h^{\beta_1}.$$

Model (21.6.2) is the special case of Model (21.6.3) with $\beta_1 = 1$, and we would like to test the models. Model (21.6.2) includes a term $\log(l_h)$ that is not multiplied by an unknown parameter. We encountered such terms in normal theory linear models, and dealt with them by constructing a new dependent variable. Dealing with such terms in nonnormal models is more complicated but many computer programs handle them easily. A term like $\log(l_h)$ that is used in the linear predictor of a model like (21.6.2) that *is not* multiplied by an unknown parameter is often called an *offset*. Different computer programs have different ways of specifying an offset.

Fitting models (21.6.2) and (21.6.3) gives the results in Table 21.6. The test statistic for the adequacy of Model (21.6.2) relative to Model (21.6.3) is

$$G^2 = 64.538 - 64.537 = 0.001$$

on $31 - 30 = 1$ df. Clearly, Model (21.6.3) adds little to (21.6.2). The parameter estimates for both models are given in Table 21.7. The estimated slope in Model (21.6.3) is remarkably close to 1.

Of course it is not clear that either of these models fit well. The squared correlation between the observations and the predicted values is only $R^2 = 0.34$ and if we think the sample sizes are large in each of the 32 cells, the deviance would give a highly significate result when testing for lack of fit.

$\square$

21.7   Relation to logistic models

All of the logistic models used in Chapter 20 are equivalent to log-linear models. This is discussed in detail in Christensen (1997). Here, we merely show that log-linear models imply standard logistic models. (It is considerably harder to show that a logistic model implies a particular log-linear model.) A logistic regression involves I independent binomial random variables. We can think of this as determining an $I \times 2$ table, where the 2 columns indicate successes and failures.

Note that in this chapter, n denotes the entire number of cells being considered, whereas in Chapter 20, n denoted the number of independent binomials being considered. Thus, in Chapter 20, $n \equiv I$ but in this chapter $n = 2I$, and later, when we consider more than two possible responses, n will be a larger multiple of I.

Consider a log-linear model for a two-dimensional table that involves the use of a continuous predictor variable x to model interaction. The row effects model of Section 21.5 for the ith individual (or group) having the jth response is

$$\log(\mu_{ij}) = u_{1(i)} + u_{2(j)} + \eta_j x_i$$

where the usual interaction term $u_{12(ij)}$ is replaced in the model by a more specific interaction term, $\eta_j x_i$. Of course, x_i is the known predictor variable and η_j is an unknown parameter. This is an interaction term because it involves both the i and j subscripts, just like $u_{12(ij)}$. The relationship between the logistic model and the log-linear model is that, with $j = 1$ denoting "success" and p_i denoting the probability of success,

$$
\begin{aligned}
\log\left(\frac{p_i}{1 - p_i}\right) &= \log\left(\frac{\mu_{i1}}{\mu_{i2}}\right) \\
&= \log(\mu_{i1}) - \log(\mu_{i2}) \\
&= \left[u_{1(i)} + u_{2(1)} + \eta_1 x_i\right] - \left[u_{1(i)} + u_{2(2)} + \eta_2 x_i\right] \\
&= \left[u_{2(1)} - u_{2(2)}\right] + \left[\eta_1 x_i - \eta_2 x_i\right] \\
&\equiv \beta_0 + \beta_1 x_i
\end{aligned}
$$

where $\beta_0 \equiv \left[u_{2(1)} - u_{2(2)}\right]$ and $\beta_1 \equiv [\eta_1 - \eta_2]$.

Similarly, the log-linear model with two predictor variables

$$
\log(\mu_{ij}) = u_{1(i)} + u_{2(j)} + \eta_{1j} x_{i1} + \eta_{2j} x_{i2}
$$

implies

$$
\begin{aligned}
\log\left(\frac{p_i}{1 - p_i}\right) &= \log\left(\frac{\mu_{i1}}{\mu_{i2}}\right) \\
&= \log(\mu_{i1}) - \log(\mu_{i2}) \\
&= \left[u_{1(i)} + u_{2(1)} + \eta_{11} x_{i1} + \eta_{21} x_{i2}\right] \\
&\quad - \left[u_{1(i)} + u_{2(2)} + \eta_{12} x_{i1} + \eta_{22} x_{i2}\right] \\
&= \left[u_{2(1)} - u_{2(2)}\right] + \left[\eta_{11} x_{i1} - \eta_{12} x_{i1}\right] + \left[\eta_{21} x_{i2} - \eta_{22} x_{i2}\right] \\
&\equiv \beta_0 + \beta_1 x_{i1} + \beta_2 x_{i2}
\end{aligned}
$$

where $\beta_0 \equiv \left[u_{2(1)} - u_{2(2)}\right]$, $\beta_1 \equiv [\eta_{11} - \eta_{12}]$, and $\beta_2 \equiv [\eta_{21} - \eta_{22}]$.

Similar results apply to ANOVA type models. The logit model from Equation (20.8.1)

$$
\log[p_{hik}/(1 - p_{hik})] = (RS)_{hi} + A_k.
$$

is equivalent to the log-linear model

$$
\log(\mu_{hijk}) = [RSA]_{hik} + [RSO]_{hij} + [OA]_{jk}
$$

in which the terms of the logit model become terms in the log-linear model except that the response factor O with subscript j is incorporated into the terms, e.g., $(RS)_{hi}$ becomes $[RSO]_{hij}$ and A_k becomes $[OA]_{jk}$. Also, the log-linear model has an effect for every combination of the explanatory factors, e.g., it includes $[RSA]_{hik}$.

The logit model

$$
\log(\mu_{hi1k}/\mu_{hi2k}) = (RS)_{hi} + \gamma k
$$

corresponds to the log-linear model

$$
\log(\mu_{hijk}) = [RSA]_{hik} + [RSO]_{hij} + \gamma_j k
$$

where we have added the highest-order interaction term not involving O, [RSA], and made the (RS) and γ terms depend on the opinion level j.

Table 21.8: *Log-linear models for the abortion opinion data.*

Model	df	G^2	$A - q$
[RSA][RSO][ROA][SOA]	10	6.12	−13.88
[RSA][RSO][ROA]	20	7.55	−32.45
[RSA][RSO][SOA]	20	13.29	−26.71
[RSA][ROA][SOA]	12	16.62	−7.38
[RSA][RSO][OA]	30	14.43	−45.57
[RSA][ROA][SO]	22	17.79	−26.21
[RSA][SOA][RO]	22	23.09	−20.91
[RSA][RO][SO][OA]	32	24.39	−39.61
[RSA][RO][SO]	42	87.54	3.54
[RSA][RO][OA]	34	34.41	−33.59
[RSA][SO][OA]	34	39.63	−28.37
[RSA][RO]	44	97.06	9.06
[RSA][SO]	44	101.9	13.9
[RSA][OA]	36	49.37	−22.63
[RSA][O]	46	111.1	19.1

21.8 Multinomial responses

Logistic regression is used to predict a dependent variable with binary responses. Binary responses involve only two outcomes, either "success" or "failure." On occasion, we need to predict a dependent variable that has more than two possible outcomes. This can be done by performing a sequence of logistic regressions, although there are several different ways to define such sequences. Another approach is to realize, as in the previous section, that logistic models are equivalent to log-linear models and to use log-linear models to deal with multiple response categories. Christensen (1997) presents a more systematic approach to these problems. Here, we restrict ourselves to presenting some examples.

EXAMPLE 21.8.1.   We now examine fitting models to the data from Table 21.4 on race, sex, opinions on abortion, and age. We treat opinions as a three-category response variable. In a log-linear model, the variables are treated symmetrically. The analysis looks for relationships among any of the variables. Here we consider opinions as a response variable. This changes the analysis in that we think of having separate independent samples from every combination of the predictor variables. Under this sampling scheme, the interaction among all of the predictors, [RSA], must be included in all models. Table 21.8 presents fits for all the models that include [RSA] and correspond to ANOVA-type logit models.

Using AIC, the best-fitting model is clearly [RSA][RSO][OA]. The fitted values for [RSA][RSO][OA] are given in Table 21.9.

This log-linear model can be used directly to fit multiple logit models that address specific issues related to the multinomial responses. The method of identifying these logit models is precisely as illustrated in the previous section. For these data we might consider two logit models, one that examines the odds of the first category (supporting legalized abortion) to the second category (opposing legalized abortion) and another that examines the second category (odds of opposing legalized abortion) to the third (being undecided),

$$\log(\mu_{hi1k}/\mu_{hi2k}) = w^1_{RS(hi)} + w^1_{A(R)}$$
$$\log(\mu_{hi2k}/\mu_{hi3k}) = w^2_{RS(hi)} + w^2_{A(k)}.$$

Alternatively, we might examine the odds of each of the first two categories relative to the third, i.e., the odds of supporting legalized abortion to undecided and the odds of opposing to undecided:

$$\log(\mu_{hi1k}/\mu_{hi3k}) = v^1_{RS(hi)} + v^1_{A(k)}$$
$$\log(\mu_{hi2k}/\mu_{hi3k}) = v^2_{RS(hi)} + v^2_{A(k)}.$$

Table 21.9: *Fitted values for [RSA][RSO][OA].*

Race	Sex	Opinion	18-25	26-35	36-45	46-55	56-65	65+
					Age			
White	Male	Support	100.1	137.2	117.5	75.62	70.58	80.10
		Oppose	39.73	64.23	56.17	47.33	50.99	62.55
		Undec.	1.21	2.59	5.36	5.05	5.43	8.35
	Female	Support	138.4	172.0	152.4	101.8	101.7	110.7
		Oppose	43.49	63.77	57.68	50.44	58.19	68.43
		Undec.	2.16	4.18	8.96	8.76	10.08	14.86
Nonwhite	Male	Support	21.19	16.57	15.20	11.20	8.04	7.80
		Oppose	8.54	7.88	7.38	7.11	5.90	6.18
		Undec.	1.27	1.54	3.42	3.69	3.06	4.02
	Female	Support	21.40	26.20	19.98	16.38	13.64	12.40
		Oppose	4.24	6.12	4.77	5.12	4.92	4.83
		Undec.	0.36	0.68	1.25	1.50	1.44	1.77

Table 21.10: *Estimated odds of Support versus Oppose.*

		Legalized Abortion					
		(Based on the log-linear model [RSA][RSO][OA])					
				Age			
Race	Sex	18–25	26–35	36–45	46–55	56–65	65+
White	Male	2.52	2.14	2.09	1.60	1.38	1.28
	Female	3.18	2.70	2.64	2.02	1.75	1.62
Nonwhite	Male	2.48	2.10	2.06	1.57	1.36	1.26
	Female	5.05	4.28	4.19	3.20	2.77	2.57

Other possibilities exist. Of those mentioned, the only odds that seem particularly interesting to the author are the odds of supporting to opposing. In the second pair of models, the category "undecided" is being used as a standard level to which other levels are compared. This seems a particularly bad choice in the context of these data. The fact that undecided happens to be the last opinion category listed in the table is no reason for it to be chosen as the standard of comparison. Either of the other categories would make a better standard.

Neither of these pairs of models are particularly appealing for these data, so we only illustrate a few salient points before moving on. Consider the odds of support relative to opposed. The odds can be obtained from the fitted values in Table 21.9. For example, the odds for young white males are $100.1/39.73 = 2.52$. The full table of odds is given in Table 21.10. Except for nonwhite females, the odds of support are essentially identical to those obtained in Section 20.8 in which undecideds were excluded. The four values vary from age to age by a constant multiple depending on the ages involved. The odds of support decrease steadily with age. The model has no inherent structure among the four race–sex categories; however, the odds for white males and nonwhite males are surprisingly similar. Nonwhite females are most likely to support legalized abortion, white females are next, and males are least likely to support legalized abortion. Confidence intervals for log odds or log odds ratios can be found using methods in Christensen (1997).

Another approach to modeling is to examine the set of three models that consists of the odds of supporting, the odds of opposing, and the odds of undecided (in each case, the odds are defined relative to the union of the other categories).

Finally, we could examine two models, one for the odds of supporting to opposing and one for the odds of undecided to having an opinion. (Note the similarity to Lancaster–Irwin partitioning.) Fitting these two models involves fitting log-linear models to two sets of data. Eliminating all undecideds from the data, we fit [RSA][RSO][OA] to the $2 \times 2 \times 2 \times 6$ table containing only the "support" and "oppose" categories. We essentially did this already in Section 20.8.

We now pool the support and oppose categories to get a $2 \times 2 \times 2 \times 6$ table in which the opinions are "support or oppose" and "undecided." Again, the model [RSA][RSO][OA] is fitted to the data. For this model, we report only the estimated odds.

Odds of Being Decided on Abortion

Race	Sex	Age					
		18–25	26–35	36–45	46–55	56–65	65+
White	Male	116.79	78.52	32.67	24.34	22.26	16.95
	Female	83.43	56.08	23.34	17.38	15.90	12.11
Nonwhite	Male	23.76	15.97	6.65	4.95	4.53	3.45
	Female	68.82	46.26	19.25	14.34	13.12	9.99

The estimated odds vary from age to age by a constant multiple. The odds decrease with age, so older people are less likely to take a position. White males are most likely to state a position. Nonwhite males are least likely to state a position. (Recall from Section 20.8 that white and non-white males take nearly the same positions but now we see that they state positions very differently.) White and nonwhite females have odds of being decided that are somewhat similar.

With support and opposed collapsed, the G^2 for [RSA][RSO][OA] turns out to be 5.176 on 15 df. The G^2 for the smaller model [RSA][RO][SO][OA] is 12.71 on 16 df. The difference is very large. Although, as seen in Section 20.8 and specifically Table 20.15, a main-effects-only logit model fits the support–opposition data quite well, to deal with the undecided category requires a race–sex interaction.

Additional modeling similar to that in Section 20.8 can be applied to the odds of having made a decision on legalized abortion.

21.9 Logistic discrimination and allocation

Consider four populations of people determined by age: adult, adolescent, child, infant. These are common distinctions, but the populations are not clearly defined. It is not clear when infants become children, when children become adolescents, nor when adolescents become adults. Nonetheless, most people can clearly be identified as members of one of these four groups. It might be of interest to see whether one can *discriminate* between these populations on the basis of various measurements like height and weight. Another interesting problem is predicting the population of a new individual given only the measurements like height and weight. The problem of predicting the population of a new case is referred to as *allocation*. In a standard discrimination–allocation problem, independent samples are taken from each population. The factor of interest in these problems is the population, but it is not a response factor in the sense used elsewhere in this book. The logistic regression approach (or as presented here, the log-linear model approach so as to handle more than two populations) to discrimination treats the distribution for each population as a multinomial. While the procedures illustrated are quite straightforward, their philosophical justification is more complex, cf. Christensen (1997).

EXAMPLE 21.9.1.   Aitchison and Dunsmore (1975, p. 212) consider 21 individuals with one of 3 types of Cushing's syndrome. Cushing's syndrome involves overproduction of cortisol. The three types considered are

A – adenoma
B – bilateral hyperplasia
C – carcinoma

The case variables considered are the rates at which two steroid metabolites are excreted in the

Table 21.11: *Cushing's syndrome data.*

Case	Type	TETRA	PREG	Case	Type	TETRA	PREG
1	A	3.1	11.70	12	B	15.4	3.60
2	A	3.0	1.30	13	B	7.7	1.60
3	A	1.9	0.10	14	B	6.5	0.40
4	A	3.8	0.04	15	B	5.7	0.40
5	A	4.1	1.10	16	B	13.6	1.60
6	A	1.9	0.40	17	C	10.2	6.40
7	B	8.3	1.00	18	C	9.2	7.90
8	B	3.8	0.20	19	C	9.6	3.10
9	B	3.9	0.60	20	C	53.8	2.50
10	B	7.8	1.20	21	C	15.8	7.60
11	B	9.1	0.60				

urine. (These are measured in milligrams per day.) The two steroids are

TETRA —Tetrahydrocortisone

and

PREG —Pregnanetriol.

The data are listed in Table 21.11. Note the strange PREG value for Case 4.

The data determine the 3×21 table

	Case(j)														
Type(i)	1	2	3	4	5	6	7	8	⋯	16	17	18	19	20	21
A	1	1	1	1	1	1	0	0	⋯	0	0	0	0	0	0
B	0	0	0	0	0	0	1	1	⋯	1	0	0	0	0	0
C	0	0	0	0	0	0	0	0	⋯	0	1	1	1	1	1

to which we fit a log-linear model.

The case variables TETRA and PREG are used to model the interaction in this table. The case variables are highly skewed, so, following Aitchison and Dunsmore, we analyze the transformed variables $TL \equiv \log(\text{TETRA})$ and $PL \equiv \log(\text{PREG})$. The transformed data are plotted in Figure 21.1.

The evaluation of the relationship is based on the relative likelihoods of the three syndrome types. Thus with i denoting the population for any case j, our interest is in the relative sizes of p_{1j}, p_{2j}, and p_{3j}. Estimates of these quantities are easily obtained from the $\hat{\mu}_{ij}$s. Simply take the fitted mean value $\hat{\mu}_{ij}$ and divide by the number of observations from population i,

$$\hat{p}_{ij} = \frac{\hat{\mu}_{ij}}{y_{i\cdot}}. \qquad (21.9.1)$$

For a new patient of unknown syndrome type but whose values of TL and PL place them in category j, the most likely type of Cushing's syndrome is that with the largest value among p_{1j}, p_{2j}, and p_{3j}. In practice, new patients are unlikely to fall into one of the 21 previously observed categories but the modeling procedure is flexible enough to allow allocation of individuals having any values of TL and PL.

Discrimination

The main effects model is

$$\log(\mu_{ij}) = \alpha_i + \beta_j \qquad i = 1,2,3 \qquad j = 1,\ldots,21.$$

We want to use TL and PL to help model the interaction, so fit

$$\log(\mu_{ij}) = \alpha_i + \beta_j + \gamma_{1i}(TL)_j + \gamma_{2i}(PL)_j, \qquad (21.9.2)$$

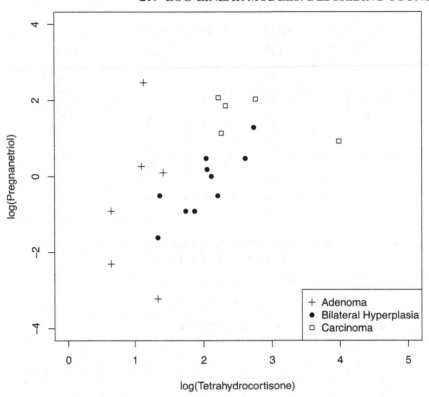

Figure 21.1: *Cushing's syndrome data.*

$i = 1, 2, 3$, $j = 1, \ldots, 21$. Taking differences gives, for example,

$$\log(\mu_{1j}/\mu_{2j}) = (\alpha_1 - \alpha_2) + (\gamma_{11} - \gamma_{12})(TL)_j + (\gamma_{21} - \gamma_{22})(PL)_j,$$

which can be written as

$$\log(\mu_{1j}/\mu_{2j}) = \delta_0 + \delta_1(TL)_j + \delta_2(PL)_j.$$

Although this looks like a logistic regression model, *it has a fundamentally different interpretation*. A value p_{ij} is the likelihood within population i of observing category j. Having fitted Model (21.9.2), the estimate of the log of the likelihood ratio is

$$\log\left(\frac{\hat{p}_{1j}}{\hat{p}_{2j}}\right) = \log\left(\frac{\hat{\mu}_{1j}/y_{1\cdot}}{\hat{\mu}_{2j}/y_{2\cdot}}\right) = \log\left(\frac{\hat{\mu}_{1j}}{\hat{\mu}_{2j}}\right) - \log\left(\frac{y_{1\cdot}}{y_{2\cdot}}\right).$$

The ratio p_{1j}/p_{2j} is *not* an odds of Type *A* relative to Type *B*. Both numbers are probabilities but they are probabilities from different populations. The correct interpretation of p_{1j}/p_{2j} is as a likelihood ratio, specifically the likelihood of Type *A* relative to Type *B*.

The G^2 for Model (21.9.2) is 12.30 on 36 degrees of freedom. As in logistic regression, although G^2 is a valid measure of goodness of fit, G^2 cannot legitimately be compared to a χ^2 distribution. However, we can test reduced models. The model

$$\log(\mu_{ij}) = \alpha_i + \beta_j + \gamma_{1i}(TL)_j$$

has $G^2 = 21.34$ on 38 degrees of freedom, and

$$\log(\mu_{ij}) = \alpha_i + \beta_j + \gamma_{2i}(PL)_j$$

Table 21.12: *Estimated probabilities for the three multinomials.*

	Group				Group		
Case	A	B	C	Case	A	B	C
1	0.1485	0.0012	0.0195	12	0.0000	0.0295	0.1411
2	0.1644	0.0014	0.0000	13	0.0000	0.0966	0.0068
3	0.1667	0.0000	0.0000	14	0.0001	0.0999	0.0000
4	0.0842	0.0495	0.0000	15	0.0009	0.0995	0.0000
5	0.0722	0.0565	0.0003	16	0.0000	0.0907	0.0185
6	0.1667	0.0000	0.0000	17	0.0000	0.0102	0.1797
7	0.0000	0.0993	0.0015	18	0.0000	0.0060	0.1879
8	0.1003	0.0398	0.0000	19	0.0000	0.0634	0.0733
9	0.0960	0.0424	0.0000	20	0.0000	0.0131	0.1738
10	0.0000	0.0987	0.0025	21	0.0000	0.0026	0.1948
11	0.0000	0.0999	0.0003				

has $G^2 = 37.23$ on 38 degrees of freedom. Neither of the reduced models provides an adequate fit. (Recall that χ^2 tests of model comparisons like these were also valid for logistic regression.)

Table 21.12 contains estimated probabilities for the three populations. The probabilities are computed using Equation (21.9.1) and Model (21.9.2).

Table 21.13 illustrates two Bayesian analyses. For each case j, it gives the estimated posterior probability that the case belongs to each of the three syndrome types. The data consist of the observed TL and PL values in category j. Given that the syndrome type is i, the estimated probability of observing data in category j is $\hat{p}_{ij}$. Let $\pi(i)$ be the prior probability that the case is of syndrome type i. Bayes theorem gives

$$\hat{\pi}(i|Data) = \frac{\hat{p}_{ij}\pi(i)}{\sum_{i=1}^{3} \hat{p}_{ij}\pi(i)} .$$

Two choices of prior probabilities are used in Table 21.13: probabilities proportional to sample sizes, i.e., $\pi(i) = y_{i.}/y_{..}$, and equal probabilities $\pi(i) = \frac{1}{3}$. Prior probabilities proportional to sample sizes are *rarely appropriate*, but they relate in simple ways to standard output, so they are often given more prominence than they deserve. Both of the sets of posterior probabilities are easily obtained. The table entries for proportional probabilities are just the $\hat{\mu}_{ij}$ values from fitting the log-linear model in the usual way. This follows from two facts: first, $\hat{\mu}_{ij} = y_{i.}\hat{p}_{ij}$, and second, the model fixes the column totals so $\hat{\mu}_{.j} = 1 = y_{.j}$. To obtain the equal probabilities values, simply divide the entries in Table 21.12 by the sum of the three probabilities for each case. Cases that are misclassified by either procedure are indicated with a double asterisk in Table 21.13.

Allocation

Model (21.9.2) includes a separate term β_j for each case so it is not clear how Model (21.9.2) can be used to allocate future cases. We begin with logit models and work back to an allocation model. From (21.9.2), we can model the probability ratio of type A relative to type B

$$\begin{aligned}
\log(p_{1j}/p_{2j}) \\
= \log(\mu_{1j}/\mu_{2j}) - \log(y_{1.}/y_{2.}) \quad (21.9.3) \\
= (\alpha_1 - \alpha_2) + (\gamma_{11} - \gamma_{12})(TL)_j + (\gamma_{21} - \gamma_{22})(PL)_j - \log(y_{1.}/y_{2.}).
\end{aligned}$$

The log-likelihoods of A relative to C are

$$\begin{aligned}
\log(p_{1j}/p_{3j}) \\
= \log(\mu_{1j}/\mu_{3j}) - \log(y_{1.}/y_{3.}) \quad (21.9.4) \\
= (\alpha_1 - \alpha_3) + (\gamma_{11} - \gamma_{13})(TL)_j + (\gamma_{21} - \gamma_{23})(PL)_j - \log(y_{1.}/y_{3.}).
\end{aligned}$$

Fitting Model (21.9.2) gives the estimated parameters.

Table 21.13: *Estimated posterior probabilities of classification.*

Case		Group	Proportional Prior Probabilities A	B	C	Equal Prior Probabilities A	B	C
1		A	.89	.01	.10	.88	.01	.12
2		A	.99	.01	.00	.99	.01	.00
3		A	1.00	.00	.00	1.00	.00	.00
4		A	.50	.50	.00	.63	.37	.00
5	**	A	.43	.57	.00	.56	.44	.00
6		A	1.00	.00	.00	1.00	.00	.00
7		B	.00	.99	.01	.00	.99	.01
8	**	B	.60	.40	.00	.72	.28	.00
9	**	B	.58	.42	.00	.69	.31	.00
10		B	.00	.99	.01	.00	.97	.03
11		B	.00	1.00	.00	.00	1.00	.00
12	**	B	.00	.29	.71	.00	.17	.83
13		B	.00	.97	.03	.00	.93	.07
14		B	.00	1.00	.00	.00	1.00	.00
15		B	.01	.99	.00	.01	.99	.00
16		B	.00	.91	.09	.00	.83	.17
17		C	.00	.10	.90	.00	.05	.95
18		C	.00	.06	.94	.00	.03	.97
19	**	C	.00	.63	.37	.00	.46	.54
20		C	.00	.13	.87	.00	.07	.93
21		C	.00	.03	.97	.00	.01	.99

Par.	Est.	Par.	Est.	Par.	Est.
α_1	0.0	γ_{11}	−16.29	γ_{21}	−3.359
α_2	−20.06	γ_{12}	−1.865	γ_{22}	−3.604
α_3	−28.91	γ_{13}	0.0	γ_{23}	0.0

where the estimates with values of 0 are really side conditions imposed on the collection of estimates to make it unique.

For a new case with values TL and PL, we plug estimates into equations (21.9.3) and (21.9.4) to get

$$\log(\hat{p}_1/\hat{p}_2) = 20.06 + (-16.29 + 1.865)TL + (-3.359 + 3.604)PL - \log(6/10)$$

and

$$\log(\hat{p}_1/\hat{p}_3) = 28.91 - 16.29(TL) - 3.359(PL) - \log(6/5).$$

For example, if the new case has a tetrahydrocortisone reading of 4.1 and a pregnanetriol reading of 1.10 then $\log(\hat{p}_1/\hat{p}_2) = 0.24069$ and $\log(\hat{p}_1/\hat{p}_3) = 5.4226$. The likelihood ratios are

$$\hat{p}_1/\hat{p}_2 = 1.2721$$
$$\hat{p}_1/\hat{p}_3 = 226.45$$

and by division

$$\hat{p}_2/\hat{p}_3 = 226.45/1.2721 = 178.01.$$

It follows that Type A is a little more likely than Type B and that both are much more likely than Type C.

One can also obtain estimated posterior probabilities for a new case. The posterior odds are

$$\frac{\hat{\pi}(1|Data)}{\hat{\pi}(2|Data)} = \frac{\hat{p}_1}{\hat{p}_2}\frac{\pi(1)}{\pi(2)} \equiv \hat{O}_2$$

and

$$\frac{\hat{\pi}(1|Data)}{\hat{\pi}(3|Data)} = \frac{\hat{p}_1}{\hat{p}_3}\frac{\pi(1)}{\pi(3)} \equiv \hat{O}_3.$$

Using the fact that $\hat{\pi}(1|Data) + \hat{\pi}(2|Data) + \hat{\pi}(3|Data) = 1$, we can solve for $\hat{\pi}(i|Data)$, $i = 1,2,3$,

$$\hat{\pi}(1|Data) = \left[1 + \frac{1}{\hat{O}_2} + \frac{1}{\hat{O}_3}\right]^{-1} = \frac{\hat{O}_2\hat{O}_3}{\hat{O}_2\hat{O}_3 + \hat{O}_3 + \hat{O}_2}$$

$$\hat{\pi}(2|Data) = \frac{1}{\hat{O}_2}\left[1 + \frac{1}{\hat{O}_2} + \frac{1}{\hat{O}_3}\right]^{-1} = \frac{\hat{O}_3}{\hat{O}_2\hat{O}_3 + \hat{O}_3 + \hat{O}_2}$$

$$\hat{\pi}(3|Data) = \frac{1}{\hat{O}_3}\left[1 + \frac{1}{\hat{O}_2} + \frac{1}{\hat{O}_3}\right]^{-1} = \frac{\hat{O}_2}{\hat{O}_2\hat{O}_3 + \hat{O}_3 + \hat{O}_2}.$$

Using TETRA $= 4.10$ and PREG $= 1.10$, the assumption $\pi(i) = y_i./y_{..}$ and more numerical accuracy in the parameter estimates than was reported earlier,

$$\hat{\pi}(1|Data) = 0.433$$
$$\hat{\pi}(2|Data) = 0.565$$
$$\hat{\pi}(3|Data) = 0.002.$$

Assuming $\pi(i) = 1/3$ gives

$$\hat{\pi}(1|Data) = 0.560$$
$$\hat{\pi}(2|Data) = 0.438$$
$$\hat{\pi}(3|Data) = 0.002.$$

Note that the values of tetrahydrocortisone and pregnanetriol used are identical to those for case 5; thus the $\hat{\pi}(i|Data)$s are identical to those listed in Table 21.13 for case 5.

To use the log-linear model approach illustrated here, one needs to fit a 3×21 table. Typically, a data file of 63 entries is needed. Three rows of the data file are associated with each of the 21 cases. Each data entry has to be identified by case and by type. In addition, the case variables should be included in the file in such a way that all three rows for a case include the corresponding case variables, TL and PL. Model (21.9.2) is easily fitted using R or SAS PROC GENMOD.

It is easy just to fit log-linear models to data such as that in Table 21.11 and get $\hat{\mu}_{ij}$s, or, when there are only two populations, fit logistic models and get $\hat{p}_{ij}$s. If you treat these values as estimated probabilities for being in the various populations, you are doing a Bayesian analysis with prior probabilities proportional to sample sizes. This is rarely an appropriate methodology.

21.10 Exercises

EXERCISE 21.10.1.   Watkins, Bergman, and Horton (1994) presented data on a complicated designed experiment that generated counts. The dependent variable is the number of ends cut by a tool. The experiment was a half replication of an experiment with five factors each at two levels, i.e., a half rep. of a 2^5 or a 2^{5-1}. The factors in the design of the experiment were the first five factors listed in Table 21.14. There are two different chasers, two different coolants, the two speeds were coded as intermediate (1) and high (2), two different pipes, and two different rake angles. In addition, two covariates were observed. On each run it was noted whether the spindle was left (1) or right (2). In the course of the experiment, two new heads were installed. A new head was installed prior to run number 8 and also prior to the second observation of run 15. The data are given in Table 21.15. It seems reasonable to treat all of the observations as independent Poisson random variables. Analysis of variance type models on seven factors can be performed. Analyze the data.

EXERCISE 21.10.2.   Bisgaard and Fuller (1995) give the data in Table 21.16 on the numbers of defectives per grille in a process examined by Chrysler Motors Engineering. The data are from a fractional factorial design. The factors are: A, Mold Cycle; B, Viscosity; C, Mold Temp; D, Mold

Table 21.14: *Tool life factors.*

Run	Chaser	Coolant	Speed	Pipe	Rak e Angle	Spindle	Head
Run	Ch	Cl	Spd	P	RA	Spn	H

Table 21.15: *Poisson data.*

Run	Ch	Cl	Spd	P	RA	Spn	H	Ends cut		
1	1	1	1	1	1	1	1	137	24	58
2	1	1	1	2	2	2	1	89	41	26
3	1	1	2	1	2	1	1	56	34	199
4	1	1	2	2	1	1	1	545	105	106
4	1	1	2	2	1	1	1	122	132	168
4	1	1	2	2	1	2	1	74	66	49
5	1	2	1	1	2	1	1	428	157	188
6	1	2	1	2	1	2	1	352	68	97
7	1	2	2	1	1	1	1	320	750	988
8	1	2	2	2	2	2	2	632	73	529
9	2	1	1	1	2	2	2	44	41	7
10	2	1	1	2	1	2	2	2	3	4
11	2	1	2	1	1	2	2	112	64	59
12	2	1	2	2	2	2	2	59	45	9
13	2	2	1	1	1	2	2	40	7	32
14	2	2	1	2	2	2	2	21	42	120
15	2	2	2	1	2	2	2	14	19	28
15	2	2	2	1	2	2	3	168	34	58
16	2	2	2	2	1	2	3	70	57	18

Pressure; E, Weight; F, Priming; G, Thickening Process; H, Glass Type; J, Cutting Pattern. Fit a main-effects-only model. Try to fit a model with all main effects and all two-factor interactions.

EXERCISE 21.10.3.   Reanalyze the data of Table 21.11 using a model based on TETRA and PREG rather than *TL* and *PL*. How much does deleting Case 4 affect the conclusions in either analysis?

EXERCISE 21.10.4.   Find two data sets from earlier chapters that are good candidates for analysis by log-linear models and reanalyze them. (There are many data sets that consist of counts.)

Table 21.16: *Grille defectives.*

Case	Treatment subscripts									Defectives
	a	b	c	d	e	f	g	h	j	
1	0	0	0	0	1	0	1	0	1	56
2	1	0	0	0	1	0	0	1	0	17
3	0	1	0	0	0	1	1	0	0	2
4	1	1	0	0	0	1	0	1	1	4
5	0	0	1	0	1	1	0	1	1	3
6	1	0	1	0	1	1	1	0	0	4
7	0	1	1	0	0	0	0	1	0	50
8	1	1	1	0	0	0	1	0	1	2
9	0	0	0	1	0	1	1	1	1	1
10	1	0	0	1	0	1	0	0	0	0
11	0	1	0	1	1	0	1	1	0	3
12	1	1	0	1	1	0	0	0	1	12
13	0	0	1	1	0	0	0	0	1	3
14	1	0	1	1	0	0	1	1	0	4
15	0	1	1	1	1	1	0	0	0	0
16	1	1	1	1	1	1	1	1	1	0

Chapter 22

Exponential and Gamma Regression: Time-to-Event Data

Time-to-event data is just that: measurements of how long it takes before some event occurs. If that event involves a machine or a machine component, analyzing such data has traditionally been called *reliability* analysis. If that event is the death of a medical patient, the analysis is called *survival analysis*. More generally, survival analysis is used to describe any analysis of time-to-event data in the biological or medical fields and reliability analysis is used for applications in the physical and engineering sciences.

Traditionally, a major distinction between reliability and survival analysis was that survival analysis dealt with lost (partially observed) observations and reliability did not. By lost observations we mean, say, patients who began the study but then were lost to the study. For such patients, the exact time of survival is unknown; one only knows at what time the patient was last contacted alive. This form of partial information is known as *censoring*. These days censoring seems to come up often in reliability also. *The examples presented in this chapter do not involve censoring*, but the use of linear structures for modeling the data depends little on whether the data are censored or not. To introduce a detailed discussion of censoring would take us too far afield.

Another traditional difference between reliability and survival analysis has been that people doing reliability have been happy to assume parametric models for the distribution of the data, while survival analysis has focused strongly on nonparametric models, in which no specific distributional assumptions are made. Personally, I think that the least important assumption one typically makes in a data analysis is the distributional assumption. I think that the assumptions of independence, having an appropriate mean structure (no lack of fit), and the assumption of an appropriate variability model (e.g., equal variances for normal data), are all more important than the distributional assumption. Moreover, if residuals are available, it is pretty easy to check the distributional assumption. For what little it is worth, my personal opinion is that the emphasis on nonparametric methods in survival analysis is often misplaced. (This is not to be confused with nonparametric regression methods in which one makes no strong assumption about the functional form of the mean structure.)

In this chapter we examine two parametric approaches that are specific to the analysis of time-to-event data. (Another parametric approach is just to take logarithms of the times and go on your merry way.) The first parametric approach is probably the oldest, *exponential regression*. In standard regression, we assume that each observation has a normal distribution but that the expected value of an observation follows some linear model. In exponential regression we assume that each observation follows an exponential distribution but that the log of the expected value follows a linear model. The second method is a generalization of the first. The exponential distribution is a special case of the gamma distribution. We can assume that each observation follows a gamma distribution but that the log of the expected value follows a linear model. A nonparametric method that involves the use of linear structures is the *Cox proportional hazards model*, but that will not be discussed. As always, this book is less concerned with data analysis, or even the basis for these models, than with the fact that the same linear structures illustrated in earlier chapters can still be used to analyze such

Table 22.1: *Leukemia survival data.*

y	WBC	AG	y	WBC	AG	y	WBC	AG	y	WBC	AG
65	2300	1	143	7000	1	56	4400	2	2	27000	2
156	750	1	56	9400	1	65	3000	2	3	28000	2
100	4300	1	26	32000	1	17	4000	2	8	31000	2
134	2600	1	22	35000	1	7	1500	2	4	26000	2
16	6000	1	1	100000	1	16	9000	2	3	21000	2
108	10500	1	1	100000	1	22	5300	2	30	79000	2
121	10000	1	5	52000	1	3	10000	2	4	100000	2
4	17000	1	65	100000	1	4	19000	2	43	100000	2
39	5400	1									

data. Censoring and Cox models are discussed in a wide variety of places including Christensen et al. (2011).

22.1   Exponential regression

Feigl and Zelen (1965) examined data on y, the number of weeks a patient survived after diagnosis of acute myelogenous leukemia. A predictor variable is the white blood cell count at diagnosis (WBC) and a grouping variable is whether the patient was AG positive (coded here as 1) or AG negative (2). The data are given in Table 22.1.

We assume that times to survival have exponential distributions. The exponential distribution depends on only one parameter. Often, if $y \sim Exp(\lambda)$, the density of y is written

$$f(y|\lambda) = \lambda e^{-\lambda y}$$

for $y > 0$ and $\lambda > 0$. It can be shown that $E(y) \equiv \mu = 1/\lambda$ and that $Var(y) = 1/\lambda^2$. Sometimes the density is written in terms of the parameter μ.

In exponential regression, $y_1, \ldots, y_n$ are independent $Exp(\lambda_i)$. Each y_i has an associated vector of predictor variables, x_i. We assume a linear structure

$$\log\left(\frac{1}{\lambda_i}\right) \equiv \log(\mu_i) = x_i'\beta.$$

Taking exponents gives

$$\frac{1}{\lambda_i} = \mu_i = \exp(x_i'\beta)$$

or

$$\lambda_i = e^{-x_i'\beta}.$$

Substituting into the density, the y_is are independent with densities

$$f(y_i|\beta) = \exp\left(-x_i'\beta - ye^{-x_i'\beta}\right).$$

As in previous chapters, for any model the deviance statistic is used as the basis for checking model fits.

EXAMPLE 22.1.1   The structure of the Feigl and Zelen data is exactly similar to an analysis of covariance. We use similar models except that the y_is have exponential distributions rather than normal distributions. In other words, we use exactly the same kinds of linear structures to model these data as were used in Chapter 15. Let i indicate the AG status and let j denote the observations within each AG status. Also, we use the base 10 log of the white blood cell count (lw) as a predictor variable. Begin by fitting a model that includes just an overall mean,

$$\log(\mu_{ij}) = v. \tag{22.1.1}$$

Table 22.2: *Model fits: Exponential regression.*

Model #	Model	Deviance	df
(22.1.1)	v	58.138	32
(22.1.2)	v_i	46.198	31
(22.1.3)	$v + \gamma(lw)_{ij}$	47.808	31
(22.1.4)	$v_i + \gamma(lw)_{ij}$	40.319	30
(22.1.5)	$v_i + \gamma_i(lw)_{ij}$	38.555	29

Table 22.3: *Tables of Coefficients: Exponential regression.*

Model (22.1.4)				Model (22.1.5)			
Parameter	Parameter	Estimate	SE	Parameter	Parameter	Estimate	SE
A(1)	v_1	6.8331	1.2671	A(1)	v_1	8.4782	1.7120
A(2)	v_2	5.8154	1.2932	A(2)	v_2	4.3433	1.6382
LW	γ	-0.7009	0.3036	A(1).LW	γ_1	-1.1095	0.4138
				A(2).LW	γ_2	-0.3546	0.3874

Next fit a model that includes effects for AG group,

$$\log(\mu_{ij}) = v_i . \tag{22.1.2}$$

Alternatively, fit a simple linear regression model in the predictor variable,

$$\log(\mu_{ij}) = v + \gamma(lw)_{ij} . \tag{22.1.3}$$

Next fit an analysis of covariance model,

$$\log(\mu_{ij}) = v_i + \gamma(lw)_{ij} . \tag{22.1.4}$$

Finally, fit a model with separate regressions for each group,

$$\log(\mu_{ij}) = v_i + \gamma_i(lw)_{ij} . \tag{22.1.5}$$

Recall that models (22.1.2) and (22.1.3) are not comparable, so this hierarchy defines two sequences of models. Table 22.2 gives the fits for all the models.

As with (most) logistic regression and unlike (most) log-linear models for count data, *the deviances cannot be used directly as lack-of-fit tests for the models.* However, as with both logistic regression and log-linear models, we can use differences in deviances to compare models. For large samples, these differences are compared to a χ^2 distribution with degrees of freedom equal to the difference in deviance degrees of freedom. For example, testing Model (22.1.1) against Model (22.1.3) gives

$$D = 58.138 - 47.808 = 10.33,$$

on $32 - 31 = 1$ df. This is highly significant when compared to a $\chi^2(1)$ distribution suggesting that the slope of the $\log_{10}$ WBC count is important when ignoring AG. Similarly, testing Model (22.1.2) versus Model (22.1.4) yields a P value just larger than 0.01, which is evidence for the importance of WBC after fitting AG. Model (22.1.4) is significantly better than all of the smaller models and Model (22.1.5) is not significantly better than Model (22.1.4). We present parameter estimates for models (22.1.4) and (22.1.5) in Table 22.3.

The methods illustrated in earlier chapters can all be used to continue the modeling process. For example, in Model (22.1.5), to test $v_1 = 2v_2$ we need to replace the factor AG by a regression variable that will take the value 2 in the first AG group and the value 1 in the second AG group. (Relative to Table 22.1, such a variable can be created as $NAG = 3 - AG$.) The model is not allowed to have an intercept, so it is

$$\log(\mu_{ij}) = \beta (nag)_{ij} + \gamma_i (lw)_{ij} .$$

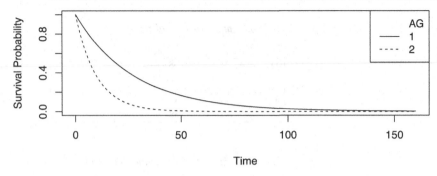

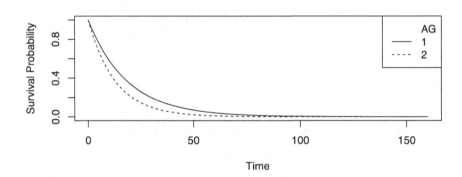

Figure 22.1 *Exponential regression estimated survival functions for lw = 5. Top, Model (22.1.4); bottom, Model (22.1.5).*

Fitting gives a deviance of 38.5578 on 30 df. Comparing this to Model (22.1.5) gives

$$D = 38.558 - 38.555 = 0.003$$

on $30 - 29 = 1$ df. It fits remarkably well, as it should since we used the data to suggest the reduced model! □

Perhaps the most useful end result from an exponential regression is a plotted cumulative distribution function, $F(y)$, or plotted survival function, $S(y) \equiv 1 - F(y)$. Such plots are made for specific values of the vector x using the estimate of β. For exponential regression, the survival function for given x and β is

$$S(y|x,\beta) = \exp\left(-ye^{-x'\beta}\right). \tag{22.1.6}$$

Figure 22.1 plots the maximum likelihood estimated survival curves for $lw = 5$ and both $AG = 1, 2$. The top panel is from Model (22.1.4) and the bottom is Model (22.1.5). They simply substitute into Equation (22.1.6) the appropriate x vectors and estimate of β from Table 22.3. The point estimates seem to give very different pictures of the importance of AG but remember that there is no significant difference between the models. These pictures ignore the variability in the estimates.

22.1.1 Computing issues

Unfortunately, different computer programs can give different results for exponential regression. Point estimates typically agree, but programs can use different procedures for finding standard errors. Minitab's "Regression with Life Data" and SAS's PROC GENMOD give results that agree.

Table 22.4: *Table of Coefficients: Model (22.1.7).*

Variable	$\hat{\beta}_k$	Minitab/SAS $SE(\hat{\beta}_k)$	R $SE(\hat{\beta}_k)$
Intercept	8.4782	1.7120	1.6555
a	−4.1349	2.3695	2.5703
lw	−1.1095	0.4138	0.3998
a*lw	0.7548	0.5668	0.6145

R's "glm" procedure gives different results. Because Minitab will not fit Model (22.1.5), to demonstrate the issue I created a regression version of Model (22.1.5) by defining a 0-1 indicator variable $a \equiv AG - 1$ that identifies individuals with $AG = 2$. I fitted the model

$$\log(\mu_{ij}) = \beta_0 + \beta_1 a_{ij} + \beta_2 (lw)_{ij} + \beta_3 [a_{ij} \times (lw)_{ij}] . \tag{22.1.7}$$

The correspondence between the parameters in models (22.1.7) and (22.1.5) is: $\beta_0 = v_1$, $\beta_1 = v_2 - v_1$, $\beta_2 = \gamma_1$, and $\beta_3 = \gamma_2 - \gamma_1$. Table 22.4 gives the results. The standard errors for $\hat{\beta}_0$ and $\hat{\beta}_2$ are larger in GENMOD and Minitab than in R but the reverse is true for $\hat{\beta}_1$ and $\hat{\beta}_3$. Note that the relatively small t values for β_3, $0.7548/0.5668$ and $0.7548/0.6145$, confirm the earlier deviance test that there is not much need to use Model (22.1.5) rather than Model (22.1.4).

Actually, neither SAS nor R actually fits exponential regression. They both fit gamma regression (as in the next section), but they let one specify the scale parameter, which allows one to fit exponential regression. Minitab does exponential regression but not gamma regression.

22.2 Gamma regression

A generalization of exponential regression involves using the gamma distribution. A random variable y has a gamma distribution with parameters α and λ, written $y \sim \text{Gamma}(\alpha, \lambda)$, if it has the probability density function

$$f(y|\alpha, \lambda) = \frac{\lambda^\alpha}{\Gamma(\alpha)} e^{-\lambda y} y^{\alpha-1}$$

for $y > 0$, $\lambda > 0$, $\alpha > 0$. The expected value is

$$\text{E}(y) \equiv \mu = \frac{\alpha}{\lambda}$$

and the variance is

$$\text{Var}(y) = \frac{\alpha}{\lambda^2}.$$

The special case of $\alpha = 1$ gives the exponential distribution.

In gamma regression we assume n independent observations y_i with

$$y_i \sim \text{Gamma}(\alpha, \lambda_i), \qquad \text{E}(y_i) \equiv \mu_i = \frac{\alpha}{\lambda_i}, \qquad \log(\mu_i) = x_i'\beta.$$

EXAMPLE 22.2.1   We can also use gamma regression to model the Feigl and Zelen data. The linear models we consider are exactly the same as those given in the previous section. Moreover, fitting these models gives exactly the same deviances, degrees of freedom, and parameter estimates as in exponential regression. What differs from exponential regression are the standard errors of parameter estimates and how the deviances are used. For example, Table 22.5 gives parameter estimates and standard errors for fitting Model (22.1.5) using gamma regression. The parameter estimates are identical to those from Section 22.1 but the standard errors are different. In exponential regression,

Table 22.5: *Table of Coefficients: Model (22.1.5).*

Parameter	Gamma regression Estimate	SE
A(1)	8.4782	1.7212
A(2)	4.3433	1.6470
A(1).LW	−1.1095	0.4160
A(2).LW	−0.3546	0.3895
Scale	0.9894	0.2143
The scale parameter was estimated by maximum likelihood.		

the variance of an observation is a direct function of the mean. Thus in exponential regression, deviances are used in ways similar to those used for binomial, multinomial, and Poisson data as illustrated in Chapters 20 and 21. The gamma distribution has two parameters, like a normal distribution, and deviances in gamma regression are used like sums of squares error in normal theory models.

In gamma regression, as in normal theory regression, when testing models, we must adjust for the scale parameter. As in normal theory, the largest model fitted must be assumed to fit the data. Thus, to test Model (22.1.4) against the larger model (22.1.5), we construct a pseudo F statistic,

$$F_{obs} = \frac{40.319 - 38.555}{30 - 29} \Bigg/ \frac{38.555}{29} = 1.33$$

and compare the statistic to an $F(1, 29)$ distribution. Unlike normal theory, the F distribution is merely an *approximate distribution* that is valid for large samples. Clearly, the test provides no evidence that we need separate regressions over and above the ACOVA model.

As always, we can use advanced ideas of linear modeling. Suppose that in Model (22.1.4) we want to incorporate the hypothesis that the slope of *lw* is −1. That gives us the model

$$\log(\mu_{ij}) = v_i + (-1)(lw)_{ij} .$$

In normal theory, we would just use $(-1)(lw)_{ij}$ to alter the dependent variable. Here, the procedure is a bit more complex, but standard computer programs accommodate such models by using an *offset*. An offset is just a term in a linear model that *is not* multiplied by an unknown parameter. Computer commands are illustrated on the website. For now, merely note that the deviance of the model is 41.407 on 31 *df*. Testing the model against (22.1.4) gives

$$F_{obs} = \frac{41.407 - 40.319}{31 - 30} \Bigg/ \frac{40.319}{30} = 0.81$$

and no evidence against $H_0 : \gamma = -1$. Alternatively, for this simple hypothesis we could compute the "Wald" statistic, which is $[\hat{\gamma} - (-1)]/\text{SE}(\hat{\gamma})$, from the table of coefficients for gamma regression with Model (22.1.4), cf. Exercise 22.3.1.   □

Rather than using a model with $\log(\mu_i) = x_i'\beta$, some people prefer a model that involves assuming $-1/\mu_i = x_i'\beta$. This is the *canonical link function* and is the default link in some programs. However, in such models not all β vectors are permissible, because β must be restricted so that $x_i'\beta < 0$ for all i.

In addition to being an approach to modeling time-to-event data, gamma regression is often used to model situations in which the data have a constant coefficient of variation. The coefficient of variation is

$$\frac{\sqrt{\text{Var}(y_i)}}{\text{E}(y_i)} = \frac{\sqrt{\alpha/\lambda_i^2}}{\alpha/\lambda_i} = \frac{1}{\sqrt{\alpha}}.$$

In such cases, gamma regression is an alternative to doing a standard linear model analysis on the logs of the data, cf. Section 7.3.

Table 22.6: *Table of Coefficients: Model (22.1.7).*

| | Gamma regression | | |
| | | SAS | R |
Variable	$\hat{\beta}_k$	$SE(\hat{\beta}_k)$	$SE(\hat{\beta}_k)$
Intercept	8.4782	1.7212	1.7490
a	−4.1349	2.3822	2.7155
lw	−1.1095	0.4160	0.4224
a*lw	0.7548	0.5699	0.6492
SAS Scale	0.9894	0.2143	
R Scale	1.116173	—-	

22.2.1 Computing issues

As with exponential regression, different computer programs give different results for gamma regression. Again, point estimates of regression parameters typically agree, but programs can use different procedures for finding standard errors and estimating the scale parameter. Again, we demonstrate by fitting (22.1.7) with results now in Table 22.6. As suggested, the regression point estimates are fine; they agree with each other and with exponential regression. The SAS GENMOD standard errors are obtained by dividing the exponential regression standard errors by the square root of the SAS scale parameter, e.g., $1.7212 = 1.7120/\sqrt{0.9894}$. The R "glm" standard errors are obtained by multiplying the R exponential regression standard errors by the square root of the R scale parameter, e.g., $1.7490 = 1.6555 \times \sqrt{1.116173}$. Unfortunately, $1/\sqrt{0.9894} \neq \sqrt{1.116173}$, so the programs not only have different definitions of the scale parameter but actually have different estimates of the scale parameter even if we made the definitions agree. And even if the scale estimates agreed, the standard errors would not because they do not agree for exponential regression. Some programs, like GLIM, use the deviance divided by its degrees of freedom as a "scale" parameter.

22.3 Exercises

EXERCISE 22.3.1. Fit Model (22.1.4) to the Feigl-Zelen data using gamma regression and compare the Wald test of $H_0 : \gamma = -1$ to the deviance (generalized likelihood ratio) test.

EXERCISE 22.3.2. Reanalyze the Feigl-Zelen data by taking a log transform and using the methods of Chapter 15. How do the results change? Do you have any way to decide which analysis is superior?

EXERCISE 22.3.3. The time to an event is a measurement of time. Can you think of any reasons why time measurements should be treated differently from other measurements?

EXERCISE 22.3.4. How would you define fitted values, residuals, and "crude" standardized residuals (ones that do not account for variability associated with fitting the model) in gamma regression?

EXERCISE 22.3.5. One way to compare the predictive ability of models is to compute R^2 as the squared sample correlation betweens the values $(y_h, \hat{y}_h)$. Based on this criterion, will gamma regression always look better than its special case exponential regression?

## Chapter 23

Nonlinear Regression

Most relationships between predictor variables and the mean values of observations are nonlinear. Fortunately, the "linear" in linear models refers to how the coefficients are incorporated into the model, not to having a linear relationship between the predictor variables and the mean values of observations. In Chapter 8 we discussed methods for fitting nonlinear relationships using models that are linear in the parameters. Moreover, Taylor's theorem from calculus indicates that even simple linear models and low-order polynomial models can make good approximate models to nonlinear relationships. Nonetheless, when we have special knowledge about the relationship between mean values and predictor variables, nonlinear regression provides a way to use that knowledge and thus can provide much better models. The biggest difficulty with nonlinear regression is that to use it you need detailed knowledge about the process generating the data, i.e., you need a good idea about the appropriate nonlinear relationship between the parameters associated with the predictor variables and the mean values of the observations. Nonlinear regression is a technique with wide applicability in the biological and physical sciences.

From a statistical point of view, nonlinear regression models are much more difficult to work with than linear regression models. It is harder to obtain estimates of the parameters. It is harder to do good statistical inference once those parameter estimates are obtained. Section 1 introduces nonlinear regression models. In section 2 we discuss parameter estimation. Section 3 examines methods for statistical inference. Section 4 considers the choice that is sometimes available between doing nonlinear regression and doing linear regression on transformed data. For a much more extensive treatment of nonlinear regression; see Seber and Wild (1989).

23.1 Introduction and examples

We have considered linear regression models

$$y_i = \beta_0 + \beta_1 x_{i1} + \cdots + \beta_{p-1} x_{i p-1} + \varepsilon_i \tag{23.1.1}$$

$i = 1, \ldots, n$ that we can write with vectors as

$$y_i = x_i' \beta + \varepsilon_i \tag{23.1.2}$$

where $x_i' = (1, x_{i1}, \ldots, x_{i p-1})$ and $\beta = (\beta_0, \ldots, \beta_{p-1})'$ These models are linear in the sense that $\mathrm{E}(y_i) = x_i' \beta$ where the unknown parameters, the β_js, are multiplied by known constants, the x_{ij}s, and added together. In this chapter we consider an important generalization of this model, *nonlinear regression*. A nonlinear regression model is simply a model for $\mathrm{E}(y_i)$ that does not combine the parameters of the model in a linear fashion.

EXAMPLE 23.1.1. *Some nonlinear regression models*
Almost any nonlinear function can be made into a nonlinear regression model. Consider the following four nonlinear functions of parameters β_j and a single predictor variable x:

$$f_1(x; \beta_0, \beta_1, \beta_2) = \beta_0 + \beta_1 \sin(\beta_2 x)$$

$$f_2(x;\beta_0,\beta_1,\beta_2) = \beta_0 + \beta_1 e^{\beta_2 x}$$
$$f_3(x;\beta_0,\beta_1,\beta_2) = \beta_0/[1 + \beta_1 e^{\beta_2 x}]$$
$$f_4(x;\beta_0,\beta_1,\beta_2,\beta_3) = \beta_0 + \beta_1[e^{\beta_2 x} - e^{\beta_3 x}].$$

Each of these can be made into a nonlinear regression model. Using f_4, we can write a model for data pairs (y_i, x_i), $i = 1, \ldots, n$:

$$y_i = \beta_0 + \beta_1[e^{\beta_2 x_i} - e^{\beta_3 x_i}] + \varepsilon_i$$
$$\equiv f_4(x_i;\beta_0,\beta_1,\beta_2,\beta_3) + \varepsilon_i.$$

Similarly, for $k = 1, 2, 3$ we can write models

$$y_i = f_k(x_i;\beta_0,\beta_1,\beta_2) + \varepsilon_i.$$

As usual, we assume that the ε_is are independent $N(0,\sigma^2)$ random variables. As alluded to earlier, the problem is to find an appropriate function $f(\cdot)$ for the data at hand.   □

In general, for s predictor variables and p regression parameters we can write a nonlinear regression model that generalizes Model (23.1.1) as

$$y_i = f(x_{i1}, \ldots, x_{is}; \beta_0, \beta_1, \ldots, \beta_{p-1}) + \varepsilon_i, \quad \varepsilon_i \text{s indep. } N(0,\sigma^2)$$

$i = 1, \ldots, n$. This is quite an awkward way to write $f(\cdot)$, so we write the model in vector form as

$$y_i = f(x_i;\beta) + \varepsilon_i, \quad \varepsilon_i \text{s indep. } N(0,\sigma^2) \tag{23.1.3}$$

where $x_i = (x_{i1}, \ldots, x_{is})'$ and $\beta = (\beta_0, \beta_1, \ldots, \beta_{p-1})'$ are vectors defined similarly to Model (23.1.2). Note that

$$E(y_i) = f(x_i;\beta).$$

EXAMPLE 23.1.2.   Pritchard, Downie, and Bacon (1977) reported data from Jaswal et al. (1969) on the initial rate r of benzene oxidation over a vanadium pentoxide catalyst. The predictor variables involve three levels of the temperature, T, for the reactions, different oxygen and benzene concentrations, x_1 and x_2, and the observed number of moles of oxygen consumed per mole of benzene, x_4. Based on chemical theory, a steady state adsorption model was proposed. One algebraically simple form of this model is

$$y_i = \exp[\beta_0 + \beta_1 x_{i3}]\frac{1}{x_{i2}} + \exp[\beta_2 + \beta_3 x_{i3}]\frac{x_{i4}}{x_{i1}} + \varepsilon_i, \tag{23.1.4}$$

where $y = 100/r$ and the temperature is involved through $x_3 = 1/T - 1/648$. The data are given in Table 23.1.

The function giving the mean structure for Model (23.1.4) is

$$f(x;\beta) \equiv f(x_1, x_2, x_3, x_4; \beta_0, \beta_2, \beta_3, \beta_4) = \exp[\beta_0 + \beta_1 x_3]\frac{1}{x_2} + \exp[\beta_2 + \beta_3 x_3]\frac{x_4}{x_1}. \tag{23.1.5}$$

□

23.2 Estimation

We used least squares estimation to obtain the $\hat{\beta}_j$s in linear regression; we will continue to use least squares estimation in nonlinear regression. For the linear regression Model (23.1.2), least squares estimates minimize

$$SSE(\beta) \equiv \sum_{i=1}^{n} [y_i - E(y_i)]^2 = \sum_{i=1}^{n} [y_i - x_i'\beta]^2.$$

Table 23.1: *Benzene oxidation data.*

Obs.	x_1	x_2	T	x_4	$r = 100/y$	Obs.	x_1	x_2	T	x_4	$r = 100/y$
1	134.5	19.1	623	5.74	218	28	30.0	20.0	648	5.64	294
2	108.0	20.0	623	5.50	189	29	16.3	20.0	648	5.61	233
3	68.6	19.9	623	5.44	192	30	16.5	20.0	648	5.63	222
4	49.5	20.0	623	5.55	174	31	20.4	12.5	648	5.70	188
5	41.7	20.0	623	5.45	152	32	20.5	16.6	648	5.67	231
6	29.4	19.9	623	6.31	139	33	20.8	20.0	648	5.63	239
7	22.5	20.0	623	5.39	118	34	21.3	30.0	648	5.63	301
8	17.2	19.9	623	5.60	120	35	19.6	43.3	648	5.62	252
9	17.0	19.7	623	5.61	122	36	20.6	20.0	648	5.72	217
10	22.8	20.0	623	5.54	132	37	20.5	30.0	648	5.43	276
11	41.3	20.0	623	5.52	167	38	20.3	42.7	648	5.60	467
12	59.6	20.0	623	5.53	208	39	16.0	19.1	673	5.88	429
13	119.7	20.0	623	5.50	216	40	23.5	20.0	673	6.01	475
14	158.2	20.0	623	5.48	294	41	132.8	20.0	673	6.48	1129
15	23.3	20.0	648	5.65	229	42	107.7	20.0	673	6.26	957
16	40.8	20.0	648	5.95	296	43	68.5	20.0	673	6.40	745
17	140.3	20.0	648	5.98	547	44	47.2	19.7	673	5.82	649
18	140.8	19.9	648	5.96	582	45	42.5	20.3	673	5.86	742
19	141.2	20.0	648	5.64	480	46	30.1	20.0	673	5.87	662
20	140.0	19.7	648	5.56	493	47	11.2	20.0	673	5.87	373
21	121.2	19.96	648	6.06	513	48	17.1	20.0	673	5.84	440
22	104.7	19.7	648	5.63	411	49	65.8	20.0	673	5.85	662
23	40.8	20.0	648	6.09	349	50	108.2	20.0	673	5.86	724
24	22.6	20.0	648	5.88	226	51	123.5	20.0	673	5.85	915
25	55.2	20.0	648	5.64	338	52	160.0	20.0	673	5.81	944
26	55.4	20.0	648	5.64	351	53	66.4	20.0	673	5.87	713
27	29.5	20.0	648	5.63	295	54	66.5	20.0	673	5.88	736

For the nonlinear regression Model (23.1.3), least squares estimates minimize

$$SSE(\beta) \equiv \sum_{i=1}^{n} [y_i - E(y_i)]^2 = \sum_{i=1}^{n} [y_i - f(x_i;\beta)]^2. \tag{23.2.1}$$

As shown below, in nonlinear regression with independent $N(0,\sigma^2)$ errors, the least squares estimates are also maximum likelihood estimates. Not surprisingly, finding the minimum of a function like (23.2.1) involves extensive use of calculus. We present in detail the Gauss–Newton algorithm for finding the least squares estimates and briefly mention an alternative method for finding the estimates.

23.2.1   The Gauss–Newton algorithm

The Gauss–Newton algorithm produces a series of vectors β^r that we hope converge to the least squares estimate $\hat{\beta}$. The algorithm requires an initial value for the vector β, say β^0. This can be thought of as a guess for $\hat{\beta}$. We use matrix methods similar to those in Chapter 11 to present the algorithm.

In matrix notation write $Y = (y_1, \ldots, y_n)'$, $e = (\varepsilon_1, \ldots, \varepsilon_n)'$, and

$$F(X;\beta) \equiv \begin{bmatrix} f(x_1;\beta) \\ \vdots \\ f(x_n;\beta) \end{bmatrix}.$$

We can now write Model (23.1.3) as

$$Y = F(X;\beta) + e, \quad \varepsilon_i\text{s indep. } N(0,\sigma^2). \tag{23.2.2}$$

Given β^r, the algorithm defines β^{r+1}. Define the matrix Z_r as the $n \times p$ matrix of partial derivatives $\partial f(x_i; \beta)/\partial \beta_j$ evaluated at β^r. Note that to find the ith row of Z_r, we need only differentiate to find the p partial derivatives $\partial f(x; \beta)/\partial \beta_j$ and evaluate these p functions at $x = x_i$ and $\beta = \beta^r$. For β values that are sufficiently close to β^r, a vector version of Taylor's theorem from calculus gives the approximation

$$F(X; \beta) \doteq F(X; \beta^r) + Z_r(\beta - \beta^r). \tag{23.2.3}$$

Here, because β^r is known, $F(X; \beta^r)$ and Z_r are known. Substituting the approximation (23.2.3) into Equation (23.2.2), we get the *approximate* model

$$
\begin{aligned}
Y &= F(X; \beta^r) + Z_r(\beta - \beta^r) + e \\
  &= F(X; \beta^r) + Z_r \beta - Z_r \beta^r + e.
\end{aligned}
$$

Rearranging terms gives

$$[Y - F(X; \beta^r) + Z_r \beta^r] = Z_r \beta + e. \tag{23.2.4}$$

If Z_r has full column rank, this is simply a linear regression model. The dependent variable vector is $Y - F(X; \beta^r) + Z_r \beta^r$, the matrix of predictor variables (model matrix) is Z_r, the parameter vector is β, and the error vector is e. Using least squares to estimate β gives us

$$
\begin{aligned}
\beta^{r+1} &= (Z_r'Z_r)^{-1}Z_r'[Y - F(X; \beta^r) + Z_r \beta^r] \\
&= (Z_r'Z_r)^{-1}Z_r'[Y - F(X; \beta^r)] + (Z_r'Z_r)^{-1}Z_r'Z_r \beta^r \\
&= (Z_r'Z_r)^{-1}Z_r'[Y - F(X; \beta^r)] + \beta^r. \tag{23.2.5}
\end{aligned}
$$

From linear regression theory, the value β^{r+1} minimizes the function

$$SSE_r(\beta) \equiv [\{Y - F(X; \beta^r) + Z_r \beta^r\} - Z_r \beta]' [\{Y - F(X; \beta^r) + Z_r \beta^r\} - Z_r \beta].$$

Actually, we wish to minimize the function defined in (23.2.1). In matrix form, (23.2.1) is

$$SSE(\beta) = [Y - F(X; \beta)]' [Y - F(X; \beta)].$$

From (23.2.3), we have $SSE_r(\beta) \doteq SSE(\beta)$ for βs near β^r. If β^r is near the least squares estimate $\hat{\beta}$, the minimum of $SSE_r(\beta)$ should be close to the minimum of $SSE(\beta)$. While β^{r+1} minimizes $SSE_r(\beta)$ exactly, β^{r+1} is merely an approximation to the estimate $\hat{\beta}$ that minimizes $SSE(\beta)$. However, when β^r is close to $\hat{\beta}$, the approximation (23.2.3) is good. At the end of this subsection, we give a geometric argument that β^r converges to the least squares estimate.

EXAMPLE 23.2.1.   *Multiple linear regression*
Suppose we treat Model (23.1.2) as a nonlinear regression model. Then $f(x_i; \beta) = x_i'\beta$, $F(X; \beta) = X\beta$, $\partial f(x_i; \beta)/\partial \beta_j = x_{ij}$, where $x_{i0} = 1$, and $Z_r = X$. From standard linear regression theory we know that $\hat{\beta} = (X'X)^{-1}X'Y$. Using the Gauss–Newton algorithm (23.2.5) with any β^0,

$$
\begin{aligned}
\beta^1 &= (Z_r'Z_r)^{-1}Z_r'[Y - F(X; \beta^0) + Z_0 \beta^0] \\
&= (X'X)^{-1}X'[Y - X\beta^0 + X\beta^0] \\
&= (X'X)^{-1}X'Y \\
&= \hat{\beta}.
\end{aligned}
$$

Thus, for a linear regression problem, the Gauss–Newton algorithm arrives at $\hat{\beta}$ in only one iteration.
□

EXAMPLE 23.2.2.   To perform the analysis on the benzene oxidation data, we need the partial derivatives of the function (23.1.5):

$$\frac{\partial f(x;\beta)}{\partial \beta_0} = \exp[\beta_0 + \beta_1 x_3]\frac{1}{x_2}$$

$$\frac{\partial f(x;\beta)}{\partial \beta_1} = \exp[\beta_0 + \beta_1 x_3]\frac{x_3}{x_2}$$

$$\frac{\partial f(x;\beta)}{\partial \beta_2} = \exp[\beta_2 + \beta_3 x_3]\frac{x_4}{x_1}$$

$$\frac{\partial f(x;\beta)}{\partial \beta_3} = \exp[\beta_2 + \beta_3 x_3]\frac{x_3 x_4}{x_1}.$$

With $\beta^1 = (0.843092, 11427.598, 0.039828, 2018.7689)'$, we illustrate one step of the algorithm. The dependent variable in Model (23.2.4) is

$$Y - F(X;\beta^1) + Z_1\beta^1 = \begin{bmatrix} 0.458716 \\ 0.529101 \\ 0.520833 \\ 0.574713 \\ 0.657895 \\ \vdots \\ 0.140252 \\ 0.135870 \end{bmatrix} - \begin{bmatrix} 0.297187 \\ 0.295806 \\ 0.330450 \\ 0.367968 \\ 0.389871 \\ \vdots \\ 0.142284 \\ 0.142300 \end{bmatrix} + \begin{bmatrix} 0.391121 \\ 0.375497 \\ 0.382850 \\ 0.387393 \\ 0.391003 \\ \vdots \\ 0.005125 \\ 0.005123 \end{bmatrix} = \begin{bmatrix} 0.552649 \\ 0.608792 \\ 0.573233 \\ 0.594137 \\ 0.659027 \\ \vdots \\ 0.003093 \\ -0.001307 \end{bmatrix}.$$

The model matrix in Model (23.2.4) is

$$Z_1 = \begin{bmatrix} 0.246862 & 0.0000153 & 0.050325 & 0.0000031 \\ 0.235753 & 0.0000146 & 0.060052 & 0.0000037 \\ 0.236938 & 0.0000147 & 0.093512 & 0.0000058 \\ 0.235753 & 0.0000146 & 0.132214 & 0.0000082 \\ 0.235753 & 0.0000146 & 0.154117 & 0.0000095 \\ \vdots & \vdots & \vdots & \vdots \\ 0.060341 & -0.0000035 & 0.081942 & -0.0000047 \\ 0.060341 & -0.0000035 & 0.081958 & -0.0000047 \end{bmatrix}.$$

Fitting Model (23.2.4) gives the estimate $\beta^2 = (1.42986, 12717, -0.15060, 9087.3)'$. Eventually, the sequence converges to $\hat{\beta}' = (1.3130, 11908, -0.23463, 10559.5)$. □

In practice, methods related to Marquardt (1963) are often used to find the least squares estimates. These involve use of a statistical procedure known as ridge regression, cf. Seber and Wild (1989, p. 624). Marquardt's method involves modifying Model (23.2.4) to estimate $\beta - \beta^r$ by subtracting $Z_r\beta^r$ from both sides of the equality. Now, rather than using the least squares estimate $\beta^{r+1} - \beta^r = (Z_r'Z_r)^{-1}Z_r'[Y - F(X;\beta^r)]$, the simplest form of ridge regression (cf. Christensen, 2011) uses the estimate

$$\beta^{r+1} - \beta^r = (Z_r'Z_r + kI_p)^{-1}Z_r'[Y - F(X;\beta^r)],$$

where I_p is a $p \times p$ identity matrix and k is a number that needs to be determined. More complicated forms of ridge regression involve replacing I_p with a diagonal matrix.

When the sequence of values β^r stops changing (converges), β^r is the least squares estimate. We will use a geometric argument to justify this statement. The argument applies to both the Gauss–Newton algorithm and the Marquardt method. By definition, $SSE(\beta)$ is the squared length of the vector $Y - F(X;\beta)$, i.e., it is the square of the distance between Y and $F(X;\beta)$. Geometrically, $\hat{\beta}$ is the value of β that makes $Y - F(X;\beta)$ as short a vector as possible. Y can be viewed as either a

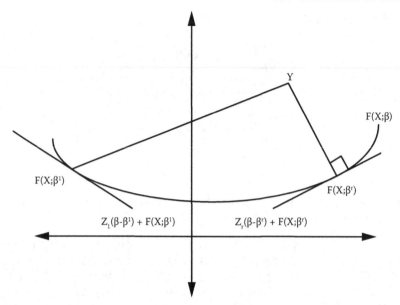

Figure 23.1: *The geometry of nonlinear least squares estimation.*

point in $\mathbf{R}^n$ or as a vector in $\mathbf{R}^n$. For now, think of it as a point. $Y - F(X;\beta)$ is as short as possible when the line connecting Y and $F(X;\beta)$ is perpendicular to the surface $F(X;\beta)$. By definition, a line is perpendicular to a surface if it is perpendicular to the tangent plane of the surface at the point of intersection between the line and the surface. Thus in Figure 23.1, β^r has $Y - F(X;\beta^r)$ as short as possible but β^1 does not have $Y - F(X;\beta^1)$ as short as possible. We will show that when β^r converges, the line connecting Y and $F(X;\beta^r)$ is perpendicular to the tangent plane at β^r and thus $Y - F(X;\beta^r)$ is as short as possible. To do this technically, i.e., using vectors, we need to subtract $F(X;\beta^r)$ from everything. Thus we want to show that $Y - F(X;\beta^r)$ is a vector that is perpendicular to the surface $F(X;\beta) - F(X;\beta^r)$. From (23.2.3), the tangent plane to the surface $F(X;\beta)$ at β^r is $F(X;\beta^r) + Z_r(\beta - \beta^r)$, so the tangent plane to the surface $F(X;\beta) - F(X;\beta^r)$ is just $Z_r(\beta - \beta^r)$. Thus we need to show that when β^r converges, $Y - F(X;\beta^r)$ is perpendicular to the plane defined by Z_r. Algebraically, this means showing that

$$0 = Z_r'[Y - F(X;\beta^r)].$$

From the Gauss–Newton algorithm, at convergence we have $\beta^{r+1} = \beta^r$ and by (23.2.5) $\beta^{r+1} = (Z_r'Z_r)^{-1}Z_r'[Y - F(X;\beta^r)] + \beta^r$, so we must have

$$0 = (Z_r'Z_r)^{-1}Z_r'[Y - F(X;\beta^r)]. \tag{23.2.6}$$

This occurs precisely when $0 = Z_r'[Y - F(X;\beta^r)]$ because you can go back and forth between the two equations by multiplying with $(Z_r'Z_r)$ and $(Z_r'Z_r)^{-1}$, respectively. Thus β^r is the value that makes $Y - F(X;\beta)$ as short a vector as possible and $\beta^r = \hat{\beta}$. Essentially the same argument applies to the Marquardt method except Equation (23.2.6) is replaced by $0 = (Z_r'Z_r + kI_p)^{-1}Z_r'[Y - F(X;\beta^r)]$.

The problem with this geometric argument—and indeed with the algorithms themselves—is that sometimes there is more than one β for which $Y - F(X;\beta)$ is perpendicular to the surface $F(X;\beta)$. If you start with an unfortunate choice of β^0, the sequence might converge to a value that does not minimize $SSE(\beta)$ over *all* β but only in a region around β^0. In fact, sometimes the sequence β^r might not even converge.

23.2.2 Maximum likelihood estimation

Nonlinear regression is a problem in which least squares estimates are maximum likelihood estimates. We now show this. The density of a random variable y with distribution $N(\mu, \sigma^2)$ is

$$\phi(y) = \frac{1}{\sqrt{2\pi}\sqrt{\sigma^2}} \exp[-(y-\mu)^2/2\sigma^2].$$

The joint density of independent random variables is obtained by multiplying the densities of the individual random variables. From Model (23.1.3), the y_is are independent $N[f(x_i;\beta), \sigma^2]$ random variables, so

$$\phi(Y) \equiv \phi(y_1, \ldots, y_n) = \prod_{i=1}^{n} \phi(y_i)$$

$$= \prod_{i=1}^{n} \frac{1}{\sqrt{2\pi}\sqrt{\sigma^2}} \exp[-\{y_i - f(x_i;\beta)\}^2/2\sigma^2]$$

$$= \left[\frac{1}{\sqrt{2\pi}}\right]^n \left[\sqrt{\sigma^2}\right]^{-n} \exp\left[-\frac{1}{2\sigma^2}\sum_{i=1}^{n}\{y_i - f(x_i;\beta)\}^2\right]$$

$$= \left[\frac{1}{\sqrt{2\pi}}\right]^n \left[\sqrt{\sigma^2}\right]^{-n} \exp\left[-\frac{1}{2\sigma^2}SSE(\beta)\right].$$

The density is a function of Y for fixed values of β and σ^2. The likelihood is exactly the same function except that the likelihood is a function of β and σ^2 for fixed values of the observations y_i. Thus, the likelihood function is

$$L(\beta, \sigma^2) = \left[\frac{1}{\sqrt{2\pi}}\right]^n \left[\sqrt{\sigma^2}\right]^{-n} \exp\left[-\frac{1}{2\sigma^2}SSE(\beta)\right].$$

The maximum likelihood estimates of β and σ^2 are those values that maximize $L(\beta, \sigma^2)$. For *any* given value of σ^2, the likelihood is a simple function of $SSE(\beta)$. In fact, the likelihood is maximized by whatever value of β that minimizes $SSE(\beta)$, i.e., the least squares estimate $\hat{\beta}$. Moreover, the function $SSE(\beta)$ does not involve σ^2, so $\hat{\beta}$ does not involve σ^2 and the maximum of $L(\beta, \sigma^2)$ occurs wherever the maximum of $L(\hat{\beta}, \sigma^2)$ occurs. This is now a function of σ^2 alone. Differentiating with respect to σ^2, it is not difficult to see that the maximum likelihood estimate of σ^2 is

$$\hat{\sigma}^2 = \frac{SSE(\hat{\beta})}{n} = \frac{1}{n}\sum_{i=1}^{n}\left[y_i - f(x_i;\hat{\beta})\right]^2.$$

Alternatively, by analogy to linear regression, an estimate of σ^2 is

$$MSE = \frac{SSE(\hat{\beta})}{n-p} = \frac{1}{n-p}\sum_{i=1}^{n}\left[y_i - f(x_i;\hat{\beta})\right]^2.$$

Incidentally, *these exact same arguments apply to linear regression, showing that least squares estimates are also maximum likelihood estimates in linear regression.*

23.3 Statistical inference

Statistical inference for nonlinear regression is based entirely on versions of the central limit theorem. It requires a large sample size for the procedures to be approximately valid. The entire analysis can be conducted as if the multiple linear regression model

$$\left[Y - F(X;\hat{\beta}) + Z_*\hat{\beta}\right] = Z_*\beta + e, \quad \varepsilon_i\text{s indep. } N(0, \sigma^2) \tag{23.3.1}$$

were valid. Here Z_* is just like Z_r from the previous section except that the partial derivatives are evaluated at $\hat{\beta}$ rather than β^r. In other words, Z_* is the $n \times p$ matrix of partial derivatives $\partial f(x_i;\beta)/\partial \beta_j$ evaluated at $\hat{\beta}$. Actually, Model (23.3.1) is simply the linear model (23.2.4) from the Gauss–Newton algorithm evaluated when β^r has converged to $\hat{\beta}$.

EXAMPLE 23.3.1.   *Inference on regression parameters*
For the benzene oxidation data, $\hat{\beta}' = (1.3130, 11908, -.23463, 10559.5)$. It follows that the dependent variable for Model (23.3.1) is

$$
Y - F(X;\hat{\beta}) + Z_*\hat{\beta} =
\begin{bmatrix}
0.458716 \\
0.529101 \\
0.520833 \\
0.574713 \\
0.657895 \\
\vdots \\
0.140252 \\
0.135870
\end{bmatrix}
-
\begin{bmatrix}
0.471786 \\
0.466022 \\
0.511129 \\
0.559092 \\
0.587341 \\
\vdots \\
0.132084 \\
0.132091
\end{bmatrix}
+
\begin{bmatrix}
0.86150 \\
0.82922 \\
0.85132 \\
0.86824 \\
0.88009 \\
\vdots \\
0.02715 \\
0.02714
\end{bmatrix}
=
\begin{bmatrix}
0.84843 \\
0.89230 \\
0.86102 \\
0.88386 \\
0.95064 \\
\vdots \\
0.03532 \\
0.03092
\end{bmatrix}
$$

and the model matrix for Model (23.3.1) is

$$
Z_* =
\begin{bmatrix}
0.406880 & 0.0000252 & 0.064906 & 0.0000040 \\
0.388570 & 0.0000241 & 0.077452 & 0.0000048 \\
0.390523 & 0.0000242 & 0.120606 & 0.0000075 \\
0.388570 & 0.0000241 & 0.170522 & 0.0000106 \\
0.388570 & 0.0000241 & 0.198771 & 0.0000123 \\
\vdots & \vdots & \vdots & \vdots \\
0.093918 & -0.0000054 & 0.038166 & -0.0000022 \\
0.093918 & -0.0000054 & 0.038174 & -0.0000022
\end{bmatrix}.
$$

The size of the values in the second and fourth columns could easily cause numerical instability, but there were no signs of such problems in this analysis. Note also that the two small columns of Z_* correspond to the large values of $\hat{\beta}$. Fitting this model gives $SSE = 0.0810169059$ with $dfE = 54 - 4 = 50$, so $MSE = 0.0016203381$. The parameters, estimates, large sample standard errors, t statistics, P values, and 95% confidence intervals for the parameters are given below.

Table of Coefficients

Par	Est	Asymptotic SE(Est)	t	P	95% Confidence interval
β_0	1.3130	0.0600724	21.86	0.000	$(1.1923696, 1.433687)$
β_1	11908	1118.1335	10.65	0.000	$(9662.1654177, 14153.831076)$
β_2	$-.23463$	0.0645778	-3.63	0.001	$(-0.3643371, -0.104921)$
β_3	10559.5	1311.4420	8.05	0.000	$(7925.4156062, 13193.622791)$

Generally, $\text{Cov}(\hat{\beta})$ is estimated with

$$
MSE(Z_*'Z_*)^{-1} = 0.0016203381
\begin{bmatrix}
2 & -30407 & -2 & 24296 \\
-30407 & 771578688 & 24805 & -720666240 \\
-2 & 24805 & 3 & -30620 \\
24296 & -720666240 & -30620 & 1061435136
\end{bmatrix};
$$

however, here we begin to see some numerical instability, at least in the reporting of this matrix. For example, using this matrix, $\text{SE}(\hat{\beta}_0) \doteq .0569 = \sqrt{0.0016203381(2)}$. The 2 in the matrix has been rounded off because of the large numbers in other entries of the matrix. In reality, $\text{SE}(\hat{\beta}_0) = .0600724 = \sqrt{0.0016203381(2.22712375)}$. $\square$

The primary complication from using Model (23.3.1) involves forming confidence intervals for points on the regression surface and prediction intervals. Suppose we want to predict a new value y_0 for a given vector of predictor variable values, say x_0. Unfortunately, Model (23.3.1) is not set up to predict y_0 but rather to provide a prediction of $y_0 - f(x_0; \hat{\beta}) + z'_{*0}\hat{\beta}$, where z'_{*0} is $(\partial f(x_0; \beta)/\partial \beta_0, \ldots, \partial f(x_0; \beta)/\partial \beta_{p-1})$ evaluated at $\hat{\beta}$. Happily, a simple modification of the prediction interval for $y_0 - f(x_0; \hat{\beta}) + z'_{*0}\hat{\beta}$ produces a prediction interval for y_0. As in Section 11.4, the $(1 - \alpha)100\%$ prediction interval has endpoints $z'_{*0}\hat{\beta} \pm W_p$, where

$$W_p \equiv t\left(1 - \frac{\alpha}{2}, n - p\right) \sqrt{MSE\left[1 + z'_{*0}(Z'_* Z_*)^{-1} z_{*0}\right]}.$$

In other words, the prediction interval is

$$z'_{*0}\hat{\beta} - W_p < y_0 - f(x_0; \hat{\beta}) + z'_{*0}\hat{\beta} < z'_{*0}\hat{\beta} + W_p.$$

To make this into an interval for y_0, simply add $f(x_0; \hat{\beta}) - z'_{*0}\hat{\beta}$ to each term, giving the interval

$$f(x_0; \hat{\beta}) - W_p < y_0 < f(x_0; \hat{\beta}) + W_p.$$

Similarly, the $(1 - \alpha)100\%$ confidence interval from Model (23.3.1) for a point on the surface gives a confidence interval for $z'_{*0}\beta$ rather than for $f(x_0; \beta)$. Defining

$$W_s \equiv t\left(1 - \frac{\alpha}{2}, n - p\right) \sqrt{MSE\, z'_{*0}(Z'_* Z_*)^{-1} z'_{*0}},$$

the confidence interval for $z'_{*0}\beta$ is

$$z'_{*0}\hat{\beta} - W_s < z'_{*0}\beta < z'_{*0}\hat{\beta} + W_s.$$

As in (23.2.3), $f(x_0; \beta) \doteq f(x_0; \hat{\beta}) + z'_{*0}(\beta - \hat{\beta})$, or equivalently,

$$f(x_0; \beta) - f(x_0; \hat{\beta}) + z'_{*0}\hat{\beta} \doteq z'_{*0}\beta.$$

We can substitute into the confidence interval to get

$$z'_{*0}\hat{\beta} - W_s < f(x_0; \beta) - f(x_0; \hat{\beta}) + z'_{*0}\hat{\beta} < z'_{*0}\hat{\beta} + W_s$$

and again, adding $f(x_0; \hat{\beta}) - z'_{*0}\hat{\beta}$ to each term gives

$$f(x_0; \hat{\beta}) - W_s < f(x_0; \beta) < f(x_0; \hat{\beta}) + W_s.$$

EXAMPLE 23.3.2.   *Prediction*
For the benzene oxidation data, we choose to make a prediction at $x'_0 = (x_{01}, x_{02}, x_{03}, x_{04}) = (100, 20, 0, 5.7)$. Using x_0 and $\hat{\beta}$ to evaluate the partial derivatives, the vector used for making predictions in Model (23.3.1) is $z'_{*0} = (0.185871, 0, 0.0450792, 0)$ and the prediction, i.e. the estimate of the value on the surface at z_{*0}, for Model (23.3.1) is $z'_{*0}\hat{\beta} = 0.233477$. The standard error of the surface is 0.00897 and the standard error for prediction is $\sqrt{0.0016203381 + 0.00897^2}$. Model (23.3.1) gives the 95% confidence interval for the surface as (0.21545,0.25150) and the 95% prediction interval as (0.15062,0.31633). The actual prediction (estimate of the value on the surface at x_0) is $f(x_0; \hat{\beta}) = 0.230950$. The confidence interval and prediction interval need to be adjusted by $f(x_0; \hat{\beta}) - z'_{*0}\hat{\beta} = 0.230950 - 0.233477 = -0.002527$. This term needs to be added to the endpoints of the intervals, giving a 95% confidence interval for the surface of (0.21292,0.24897) and a 95%

prediction interval of $(0.14809, 0.31380)$. Actually, our interest is in $r = 100/y$ rather than y, so a 95% prediction interval for r is $(100/.31380, 100/.14809)$, which is $(318.7, 675.3)$. $\qquad \square$

We can also test full models against reduced models. Again, write the full model as

$$y_i = f(x_i; \beta) + \varepsilon_i, \quad \varepsilon_i\text{s indep. } N(0, \sigma^2), \tag{23.3.2}$$

which, when fitted, gives $SSE(\hat{\beta})$, and write the reduced model as

$$y_i = f_0(x_i; \gamma) + \varepsilon_i \tag{23.3.3}$$

with $\gamma' = (\gamma_0, \ldots, \gamma_{q-1})$. When fitted, Model (23.3.3) gives $SSE(\hat{\gamma})$. The simplest way of ensuring that Model (23.3.3) is a reduced model relative to Model (23.3.2) is by specifying constraints on the parameters.

EXAMPLE 23.3.3.   In Section 1 we considered the model $y_i = \beta_0 + \beta_1[e^{\beta_2 x_i} - e^{\beta_3 x_i}] + \varepsilon_i$ with $p = 4$. If we specify $H_0 : \beta_1 = 4; 2\beta_2 = \beta_3$, the reduced model is $y_i = \beta_0 + 4[e^{\beta_2 x_i} - e^{2\beta_2 x_i}] + \varepsilon_i$. The parameters do not mean the same things in the reduced model as in the original model, so we can rewrite the reduced model as $y_i = \gamma_0 + 4[e^{\gamma_1 x_i} - e^{2\gamma_1 x_i}] + \varepsilon_i$ with $q = 2$. This particular reduced model can also be rewritten as $y_i = \gamma_0 + 4[e^{\gamma_1 x_i}(1 - e^{\gamma_1 x_i})] + \varepsilon_i$, which is beginning to look quite different from the full model. $\qquad \square$

Corresponding to Model (23.3.3), there is a linear model similar to Model (23.3.1),

$$[Y - F_0(X; \hat{\gamma}) + Z_{*0}\hat{\gamma}] = Z_{*0}\gamma + e.$$

Alas, this model will typically *not* be a reduced model relative to Model (23.3.1). In fact, the dependent variables (left-hand sides of the equations) are not even the same. Nonetheless, because Model (23.3.3) is a reduced version of Model (23.3.2), we can test the models in the usual way by using sums of squares error. Reject the reduced model with an α-level test if

$$\frac{[SSE(\hat{\gamma}) - SSE(\hat{\beta})]/(p-q)}{SSE(\hat{\beta})/(n-p)} > F(1 - \alpha, p - q, n - p). \tag{23.3.4}$$

Of course, as in all of inference for nonlinear regression, the test is only a large sample approximation. The test statistic does not have exactly an F distribution when the reduced model is true.

EXAMPLE 23.3.4.   *Testing a reduced model*
Consider the reduced model obtained from (23.1.4) by setting $\beta_0 = \beta_2$ and $\beta_1 = \beta_3$. We can rewrite the model as

$$y_i = \exp[\gamma_0 + \gamma_1 x_{i3}]\frac{1}{x_{i2}} + \exp[\gamma_0 + \gamma_1 x_{i3}]\frac{x_{i4}}{x_{i1}} + \varepsilon_i.$$

This model has $q = 2$ parameters. The partial derivatives of the function

$$f_0(x; \gamma) \equiv f(x_1, x_2, x_3, x_4; \gamma_0, \gamma_1) = \exp[\gamma_0 + \gamma_1 x_3]\frac{1}{x_2} + \exp[\gamma_0 + \gamma_1 x_3]\frac{x_4}{x_1}$$

are

$$\frac{\partial f_0(x; \gamma)}{\partial \gamma_0} = \exp[\gamma_0 + \gamma_1 x_3]\frac{1}{x_2} + \exp[\gamma_0 + \gamma_1 x_3]\frac{x_4}{x_1}$$

$$\frac{\partial f_0(x; \gamma)}{\partial \gamma_1} = \exp[\gamma_0 + \gamma_1 x_3]\frac{x_3}{x_2} + \exp[\gamma_0 + \gamma_1 x_3]\frac{x_3 x_4}{x_1}.$$

Fitting the model gives estimated parameters $\hat{\gamma}' = (9.267172, 12155.54478)$ with $SSE =$

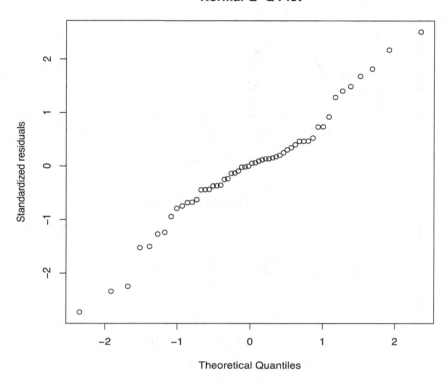

Figure 23.2: *Rankit plot of standardized residuals, $W' = .965$.*

0.4919048545 on $dfE = 54 - 2 = 52$ for $MSE = 0.0094597087$. From Inequality (23.3.4) and Example 23.3.1, the test statistic is

$$\frac{[0.4919048545 - 0.0810169059]/[4 - 2]}{0.0016203381} = 126.79.$$

With an F statistic this large, the test will be rejected for any reasonable α level.   □

EXAMPLE 23.3.5.   *Diagnostics*
Model (23.3.1) is a linear regression model so we can do the usual things to it. Figure 23.2 contains a normal plot of the standardized residuals. This does not look too bad to me and the Wilk–Francia statistic of $W' = .965$ is not significantly low.

The fitted values from Model (23.3.1) are denoted $\hat{d}_i$ whereas more intuitive fitted values are $\hat{y}_i \equiv f(x_i, \hat{\beta})$. The residuals from fitting Model (23.3.1) equal $y_i - \hat{y}_i$. Figure 23.3 contains plots of the standardized residuals versus both types of fitted values. The plot versus $\hat{d}$ is most notable for the large empty space in the middle. The plot against $\hat{y}$ has some suggestion of increasing variance especially as there are fewer points on the right but they are more spread out than the many points on the left.

Figure 23.4 contains plots of the standardized residuals versus the predictor variables x_1, x_2, T, x_4. To some extent in the plot versus x_1 and very clearly in the plots versus T and x_4, we see a traditional horn shape associated with heteroscedastic variances. However, in the plot of x_4, there are many points on the left and only four points on the right, so some additional spread on the left is to be expected. On the other hand, if not for the one point in the bottom right, the plot versus x_1 would be a classic horn shape. These plots call in question all of the inferential procedures that we have illustrated because the analysis *assumes* that the variance is the same for each observation.

Residual–dhat plot

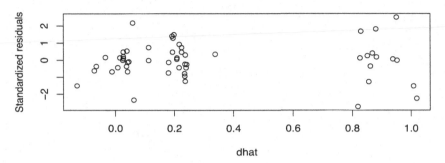

Residual–yhat plot

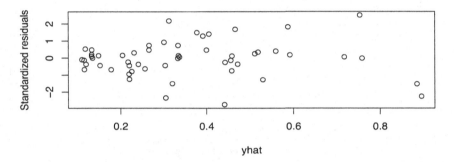

Figure 23.3: *Standardized residuals versus fitted values.*

Table 23.2 contains standard diagnostic quantities from fitting Model (23.3.1). We use these quantities in the usual way but possible problems are discussed at the end of the section. Given that there are 54 cases, none of the standardized residuals r or standardized deleted residuals t look exceptionally large.

Figure 23.5 contains index plots of the leverages and Cook's distances. They simply plot the value against the observation number for each case. Neither plot looks too bad to me (at least at 12:15 a.m. while I am doing this). However, there are some leverages that exceed the $3p/n = 3(4)/54 = 0.222$ rule.

For more on how to analyze these data, see Pritchard et al. (1977) and Carroll and Ruppert (1984).   □

Unlike linear regression, where the procedure is dominated by the predictor variables, nonlinear regression is very parameter oriented. This is perhaps excusable because in nonlinear regression there is usually some specific theory suggesting the regression model and that theory may give meaning to the parameters. Nonetheless, one can create big statistical problems or remove statistical problems simply by the choice of the parameterization. For example, Model (23.1.4) can be rewritten as

$$y_i = \gamma_0 \exp[\gamma_1 x_{i3}] \frac{1}{x_2} + \gamma_2 \exp[\gamma_3 x_{i3}] \frac{x_4}{x_1} + \varepsilon_i. \tag{23.3.5}$$

If, say, $\gamma_0 = 0$, the entire term $\gamma_0 \exp[\gamma_1 x_{i3}]/x_2$ vanishes. This term is the only place in which the parameter γ_1 appears. So if $\gamma_0 = 0$, it will be impossible to learn about γ_1. More to the point, if γ_0 is near zero, it will be very difficult to learn about γ_1. (Of course, one could argue that from the viewpoint of prediction, one may not care much what γ_1 is if γ_0 is very near zero and x_3 is of moderate size.) In any case, unlike linear regression, the value of one parameter can affect what we learn

Table 23.2: *Diagnostic statistics.*

Obs.	$\hat{e}$	$\hat{d}$	Leverage	r	t	C
1	−0.01307	0.86148	0.15936	−0.35412	−0.35103	0.005944
2	0.06308	0.82920	0.13152	1.68152	1.71375	0.107045
3	0.00971	0.85129	0.10156	0.25436	0.25195	0.001828
4	0.01562	0.86822	0.07718	0.40396	0.40054	0.003412
5	0.07055	0.88006	0.07046	1.81795	1.86227	0.062624
6	0.00248	0.93759	0.09688	0.06491	0.06427	0.000113
7	0.09455	0.94948	0.12257	2.50767	2.65506	0.219614
8	−0.05236	1.00834	0.27329	−1.52579	−1.54686	0.218863
9	−0.07670	1.01929	0.28149	−2.24800	−2.34713	0.494925
10	−0.00054	0.95166	0.12675	−0.01435	−0.01419	0.000007
11	0.00696	0.88195	0.06980	0.17923	0.17748	0.000603
12	−0.04892	0.85589	0.08901	−1.27316	−1.28133	0.039595
13	0.00451	0.82602	0.13811	0.12073	0.11951	0.000584
14	−0.10112	0.81881	0.15431	−2.73157	−2.93174	0.340371
15	0.05903	0.19905	0.03341	1.49170	1.51071	0.019231
16	0.03663	0.21699	0.02807	0.92310	0.92172	0.006153
17	−0.03676	0.23614	0.05558	−0.93981	−0.93868	0.012994
18	−0.04846	0.23742	0.05636	−1.23931	−1.24613	0.022932
19	−0.00913	0.23663	0.05675	−0.23346	−0.23122	0.000820
20	−0.01727	0.24039	0.05887	−0.44225	−0.43864	0.003058
21	−0.03085	0.23526	0.05273	−0.78755	−0.78450	0.008630
22	0.01208	0.23778	0.05285	0.30838	0.30559	0.001327
23	−0.01739	0.21635	0.02775	−0.43803	−0.43446	0.001369
24	0.05084	0.19577	0.03765	1.28753	1.29625	0.016213
25	0.02918	0.22509	0.03552	0.73818	0.73478	0.005018
26	0.01852	0.22516	0.03561	0.46839	0.46472	0.002025
27	0.00218	0.20863	0.02682	0.05485	0.05430	0.000021
28	0.00558	0.20916	0.02671	0.14058	0.13920	0.000136
29	−0.02888	0.18018	0.07150	−0.74455	−0.74120	0.010672
30	−0.00527	0.18073	0.06992	−0.13582	−0.13450	0.000347
31	0.01355	0.33863	0.06789	0.34853	0.34546	0.002212
32	−0.00978	0.24271	0.04494	−0.24865	−0.24631	0.000727
33	0.01847	0.19382	0.04064	0.46855	0.46486	0.002325
34	−0.00073	0.11365	0.04625	−0.01853	−0.01836	0.000004
35	0.08420	0.05952	0.07329	2.17301	2.26054	0.093364
36	0.05536	0.19252	0.04282	1.40571	1.41991	0.022102
37	0.02892	0.11355	0.04651	0.73583	0.73239	0.006603
38	−0.09110	0.06312	0.06575	−2.34133	−2.45639	0.096455
39	−0.02390	-0.07127	0.09852	−0.62540	−0.62158	0.010687
40	0.00620	-0.03354	0.04176	0.15727	0.15571	0.000269
41	−0.02641	0.04151	0.04883	−0.67271	−0.66896	0.005807
42	−0.01452	0.03812	0.04522	−0.36911	−0.36588	0.001613
43	−0.00003	0.02532	0.03438	−0.00065	−0.00063	0.000000
44	0.00550	0.01539	0.02983	0.13875	0.13740	0.000148
45	−0.01729	0.00833	0.02610	−0.43515	−0.43159	0.001268
46	−0.02705	-0.01152	0.02799	−0.68170	−0.67801	0.003346
47	−0.05209	-0.13085	0.26019	−1.50456	−1.52442	0.199053
48	−0.01409	-0.06464	0.08364	−0.36561	−0.36245	0.003051
49	0.01876	0.02696	0.03552	0.47447	0.47078	0.002073
50	0.02082	0.03956	0.04672	0.52980	0.52598	0.003439
51	−0.00508	0.04202	0.04941	−0.12939	−0.12809	0.000217
52	−0.00366	0.04603	0.05414	−0.09355	−0.09259	0.000125
53	0.00817	0.02714	0.03565	0.20665	0.20468	0.000395
54	0.00378	0.02714	0.03565	0.09558	0.09465	0.000084

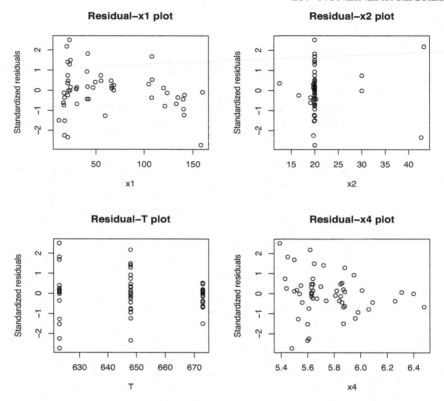

Figure 23.4: *Standardized residuals versus predictor variables.*

about other parameters. (In linear regression, the values of some predictor variables affect what we can learn about the parameters for other predictor variables, but it is not the parameters themselves that create the problem. In fact, in nonlinear regression, as the benzene example indicates, the predictor variables are not necessarily associated with any particular parameter.) In Model (23.1.4) we have ameliorated the problem of γ_0 near 0 by using the parameter β_0. When γ_0 approaches zero, β_0 approaches negative infinity, so this problem with the coefficient of x_{i3}, i.e. γ_1 or β_1, will not arise for finite β_0. However, unlike (23.3.5), Model (23.1.4) cannot deal with the possibility of $\gamma_0 < 0$. Similar problems can occur with γ_2.

All of the methods in this section depend crucially on the quality of the approximation in (23.2.3) when $\beta^r = \hat{\beta}$. If this approximation is poor, these methods can be very misleading. In particular, Cook and Tsai (1985, 1990) discuss problems with residual analysis when the approximation is poor and discuss diagnostics for the quality of the normal approximation. St. Laurent and Cook (1992) discuss concepts of leverage for nonlinear regression. For large samples, the true value of β should be close to $\hat{\beta}$ and the approximation should be good. (This conclusion also depends on having the standard errors for functions of $\hat{\beta}$ small in large samples.) But it is very difficult to tell what constitutes a 'large sample.' As a practical matter, the quality of the approximation depends a great deal on the amount of curvature found in $f(x; \beta)$ near $\beta = \hat{\beta}$. This curvature is conveniently measured by the second partial derivatives $\partial^2 f(x; \beta) / \partial \beta_j \partial \beta_k$ evaluated at $\hat{\beta}$. A good analysis of nonlinear regression data should include an examination of curvature, but such an examination is beyond the scope of this book, cf. Seber and Wild (1989, Chapter 4).

Leverage plot

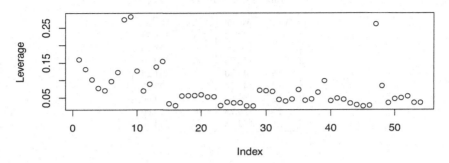

Cook's distance plot

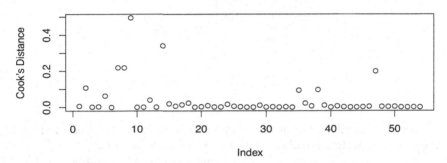

Figure 23.5: *Index plots of diagnostics.*

23.4 Linearizable models

Some nonlinear relationships can be changed into linear relationships. The nonlinear regression model (23.1.3) indicates that

$$E(y_i) = f(x_i; \beta).$$

Sometimes $f(x_i; \beta)$ can be written as

$$f(x_i; \beta) = f(x_i' \beta).$$

If f is invertible, we get

$$f^{-1}[E(y_i)] = x_i' \beta.$$

Often it is not too clear whether we should be modeling $f^{-1}[E(y_i)] = x_i' \beta$ or $E[f^{-1}(y_i)] = x_i' \beta$. As we saw, the first of these comes from nonlinear regression. The second equality suggests the *linear* regression model,

$$f^{-1}(y_i) = x_i' \beta + \varepsilon_i. \qquad (23.4.1)$$

It can be very difficult to choose between analyzing the nonlinear model (23.1.3) and the linear model (23.4.1). The decision is often based on which model gives better approximations to the assumption of independent identically distributed mean zero normal errors.

EXAMPLE 23.4.1.   In Section 7.3 we analyzed the Hooker data using a linear model $\log(y_i) = \beta_0 + \beta_1 x_i + \varepsilon_i$. Exponentiating both sides gives $y_i = \exp[\beta_0 + \beta_1 x_i + \varepsilon_i]$, which we can rewrite as $y_i = \exp[\beta_0 + \beta_1 x_i]\xi_i$, where ξ_i is a multiplicative error term with $\xi_i = \exp(\varepsilon_i)$. Alternatively, we could fit a nonlinear regression model

$$y_i = \exp[\beta_0 + \beta_1 x_i] + \varepsilon_i. \qquad (23.4.2)$$

Table 23.3: *Day (1966) data.*

Week	Weight	Week	Weight
8	140.50	34	163.75
12	139.25	35	168.75
14	138.75	36	170.00
15	140.00	37	171.25
19	147.25	38	173.00
23	150.50	39	174.00
27	156.75	40	174.00
31	162.75	42	174.50

Table 23.4: *Bliss and James (1966) data.*

x	z	x	z
.200	99	.0150	172
.100	115	.0100	188
.075	119	.0075	284
.050	112	.0070	227
.0375	126	.0060	275
.0250	149	.0050	525
.0200	152	.0025	948

The difference between these two models is that in the first model (the linearized model) the errors on the original scale are multiplied by the regression structure $\exp[\beta_0 + \beta_1 x_i]$, whereas in the nonlinear model the errors are additive, i.e., are added to the regression structure. To fit the nonlinear model (23.4.2), we need the partial derivatives of $f(x; \beta_0, \beta_1) \equiv \exp[\beta_0 + \beta_1 x]$, namely $\partial f(x; \beta_0, \beta_1)/\partial \beta_0 = \exp[\beta_0 + \beta_1 x]$ and $\partial f(x; \beta_0, \beta_1)/\partial \beta_1 = \exp[\beta_0 + \beta_1 x]x$. As mentioned earlier, the choice between using the linearized model from Section 7.3 or the nonlinear regression model (23.4.2) is often based on which model seems to have better residual plots, etc. Exercise 23.5.1 asks for this comparison. □

23.5 Exercises

EXERCISE 23.5.1.   Fit the nonlinear regression (23.4.2) to the Hooker data and compare the fit of this model to the fit of the linearized model described in Section 7.3.

EXERCISE 23.5.2.   For pregnant women, Day (1966) modeled the relationship between weight z and week of gestation x with

$$E(y) = \beta_0 + \exp[\beta_1 + \beta_2 x]$$

where $y = 1/\sqrt{z - z_0}$ and z_0 is the initial weight of the woman. For a woman with initial weight of 138 pounds, the data in Table 23.3 were recorded.

Fit the model $y_i = \beta_0 + \exp[\beta_1 + \beta_2 x_i] + \varepsilon_i$. Test whether each parameter is equal to zero, give 95% confidence intervals for each parameter, give 95% prediction intervals and surface confidence intervals for $x = 21$ weeks, and check the diagnostic quantities. Test the reduced model defined by $H_0 : \beta_0 = 0; \beta_1 = 0$.

EXERCISE 23.5.3.   Following Bliss and James (1966), fit the model $y_i = (x_i \beta_0)/(x_i + \beta_1) + \varepsilon_i$ to the following data on the relationship between reaction velocity y and concentration of substrate x.

x	.138	.220	.291	.560	.766	1.460
y	.148	.171	.234	.324	.390	.493

Test whether each parameter is equal to zero, give 99% confidence intervals for each parameter, give 99% prediction intervals and surface confidence intervals for $x = .5$, and check the diagnostic quantities.

EXERCISE 23.5.4.   Bliss and James (1966) give data on the median survival time z of house flies following application of the pesticide DDT at a level of molar concentration x. Letting $y = 100/z$, fit the model $y_i = \beta_0 + \beta_1 x_i / (x_i + \beta_2) + \varepsilon_i$ to the data given in Table 23.4.

Test whether each parameter is equal to zero, give 99% confidence intervals for each parameter, give 95% prediction intervals and surface confidence intervals for a concentration of $x = .03$, and check the diagnostic quantities. Find the SSE and test the reduced model defined by $H_0 : \beta_0 = 0, \beta_2 = .0125$. Test $H_0 : \beta_2 = .0125$.

Appendix A: Matrices

A matrix is a rectangular array of numbers. Such arrays have *rows* and *columns*. The numbers of rows and columns are referred to as the *dimensions* of a matrix. A matrix with, say, 5 rows and 3 columns is referred to as a 5×3 matrix.

EXAMPLE A.0.1.   Three matrices are given below along with their dimensions.

$$\begin{bmatrix} 1 & 4 \\ 2 & 5 \\ 3 & 6 \end{bmatrix}, \quad \begin{bmatrix} 20 & 80 \\ 90 & 140 \end{bmatrix}, \quad \begin{bmatrix} 6 \\ 180 \\ -3 \\ 0 \end{bmatrix}.$$

$$3 \times 2 \qquad\qquad 2 \times 2 \qquad\qquad 4 \times 1$$

□

Let r be an arbitrary positive integer. A matrix with r rows and r columns, i.e., an $r \times r$ matrix, is called a *square matrix*. The second matrix in Example A.0.1 is square. A matrix with only one column, i.e., an $r \times 1$ matrix, is a *vector*, sometimes called a *column vector*. The third matrix in Example A.0.1 is a vector. A $1 \times r$ matrix is sometimes called a *row vector*.

An arbitrary matrix A is often written

$$A = [a_{ij}]$$

where a_{ij} denotes the element of A in the ith row and jth column. Two matrices are equal if they have the same dimensions and all of their elements (entries) are equal. Thus for $r \times c$ matrices $A = [a_{ij}]$ and $B = [b_{ij}]$, $A = B$ if and only if $a_{ij} = b_{ij}$ for every $i = 1, \ldots, r$ and $j = 1, \ldots, c$.

EXAMPLE A.0.2.   Let

$$A = \begin{bmatrix} 20 & 80 \\ 90 & 140 \end{bmatrix} \text{ and } B = \begin{bmatrix} b_{11} & b_{12} \\ b_{21} & b_{22} \end{bmatrix}.$$

If $B = A$, then $b_{11} = 20, b_{12} = 80, b_{21} = 90$, and $b_{22} = 140$.

□

The *transpose* of a matrix A, denoted A', changes the rows of A into columns of a new matrix A'. If A is an $r \times c$ matrix, the transpose A' is a $c \times r$ matrix. In particular, if we write $A' = [\tilde{a}_{ij}]$, then the element in row i and column j of A' is defined to be $\tilde{a}_{ij} = a_{ji}$.

EXAMPLE A.0.3.

$$\begin{bmatrix} 1 & 4 \\ 2 & 5 \\ 3 & 6 \end{bmatrix}' = \begin{bmatrix} 1 & 2 & 3 \\ 4 & 5 & 6 \end{bmatrix}$$

and

$$\begin{bmatrix} 20 & 80 \\ 90 & 140 \end{bmatrix}' = \begin{bmatrix} 20 & 90 \\ 80 & 140 \end{bmatrix}.$$

The transpose of a column vector is a row vector,

$$\begin{bmatrix} 6 \\ 180 \\ -3 \\ 0 \end{bmatrix}' = [6 \quad 180 \quad -3 \quad 0].$$ □

A.1   Matrix addition and subtraction

Two matrices can be added (or subtracted) if they have the same dimensions, that is, if they have the same number of rows and columns. Addition and subtraction is performed elementwise.

EXAMPLE A.1.1.

$$\begin{bmatrix} 1 & 4 \\ 2 & 5 \\ 3 & 6 \end{bmatrix} + \begin{bmatrix} 2 & 8 \\ 4 & 10 \\ 6 & 12 \end{bmatrix} = \begin{bmatrix} 1+2 & 4+8 \\ 2+4 & 5+10 \\ 3+6 & 6+12 \end{bmatrix} = \begin{bmatrix} 3 & 12 \\ 6 & 15 \\ 9 & 18 \end{bmatrix}.$$

$$\begin{bmatrix} 20 & 80 \\ 90 & 140 \end{bmatrix} - \begin{bmatrix} -15 & -75 \\ 80 & 130 \end{bmatrix} = \begin{bmatrix} 35 & 155 \\ 10 & 10 \end{bmatrix}.$$ □

In general, if A and B are $r \times c$ matrices with $A = [a_{ij}]$ and $B = [b_{ij}]$, then

$$A + B = [a_{ij} + b_{ij}] \text{ and } A - B = [a_{ij} - b_{ij}].$$

A.2   Scalar multiplication

Any matrix can be multiplied by a scalar. Multiplication by a scalar (a *real number*) is elementwise.

EXAMPLE A.2.1.   Scalar multiplication gives

$$\frac{1}{10} \begin{bmatrix} 20 & 80 \\ 90 & 140 \end{bmatrix} = \begin{bmatrix} 20/10 & 80/10 \\ 90/10 & 140/10 \end{bmatrix} = \begin{bmatrix} 2 & 8 \\ 9 & 14 \end{bmatrix}.$$

$$2[6 \quad 180 \quad -3 \quad 0] = [12 \quad 360 \quad -6 \quad 0].$$ □

In general, if λ is any number and $A = [a_{ij}]$, then

$$\lambda A = [\lambda a_{ij}].$$

A.3   Matrix multiplication

Two matrices can be multiplied together if the number of columns in the first matrix is the same as the number of rows in the second matrix. In the process of multiplication, the rows of the first matrix are matched up with the columns of the second matrix.

EXAMPLE A.3.1.

$$\begin{bmatrix} 1 & 4 \\ 2 & 5 \\ 3 & 6 \end{bmatrix} \begin{bmatrix} 20 & 80 \\ 90 & 140 \end{bmatrix} = \begin{bmatrix} (1)(20)+(4)(90) & (1)(80)+(4)(140) \\ (2)(20)+(5)(90) & (2)(80)+(5)(140) \\ (3)(20)+(6)(90) & (3)(80)+(6)(140) \end{bmatrix}$$

$$= \begin{bmatrix} 380 & 640 \\ 490 & 860 \\ 600 & 1080 \end{bmatrix}.$$

The entry in the first row and column of the product matrix, $(1)(20)+(4)(90)$, matches the elements in the first row of the first matrix, $(1\ 4)$, with the elements in the first column of the second matrix, $\begin{pmatrix} 20 \\ 90 \end{pmatrix}$. The 1 in $(1\ 4)$ is matched up with the 20 in $\begin{pmatrix} 20 \\ 90 \end{pmatrix}$ and these numbers are multiplied. Similarly, the 4 in $(1\ 4)$ is matched up with the 90 in $\begin{pmatrix} 20 \\ 90 \end{pmatrix}$ and the numbers are multiplied. Finally, the two products are added to obtain the entry $(1)(20)+(4)(90)$. Similarly, the entry in the third row, second column of the product, $(3)(80)+(6)(140)$, matches the elements in the third row of the first matrix, $(3\ 6)$, with the elements in the second column of the second matrix, $\begin{pmatrix} 80 \\ 140 \end{pmatrix}$. After multiplying and adding we get the entry $(3)(80)+(6)(140)$. To carry out this matching, the number of columns in the first matrix must equal the number of rows in the second matrix. The matrix product has the same number of rows as the first matrix and the same number of columns as the second because each row of the first matrix can be matched with each column of the second.

□

EXAMPLE A.3.2. We illustrate another matrix multiplication commonly performed in Statistics, multiplying a matrix on its left by the transpose of that matrix, i.e., computing $A'A$.

$$\begin{bmatrix} 1 & 4 \\ 2 & 5 \\ 3 & 6 \end{bmatrix}' \begin{bmatrix} 1 & 4 \\ 2 & 5 \\ 3 & 6 \end{bmatrix} = \begin{bmatrix} 1 & 2 & 3 \\ 4 & 5 & 6 \end{bmatrix} \begin{bmatrix} 1 & 4 \\ 2 & 5 \\ 3 & 6 \end{bmatrix}$$

$$= \begin{bmatrix} 1+4+9 & 4+10+18 \\ 4+10+18 & 16+25+36 \end{bmatrix}$$

$$= \begin{bmatrix} 14 & 32 \\ 32 & 77 \end{bmatrix}.$$

□

Notice that in matrix multiplication the roles of the first matrix and the second matrix are *not* interchangeable. In particular, if we reverse the order of the matrices in Example A.3.1, the matrix product

$$\begin{bmatrix} 20 & 80 \\ 90 & 140 \end{bmatrix} \begin{bmatrix} 1 & 4 \\ 2 & 5 \\ 3 & 6 \end{bmatrix}$$

is undefined because the first matrix has two columns while the second matrix has three rows. Even when the matrix products are defined for both AB and BA, the results of the multiplication typically differ. If A is $r \times s$ and B is $s \times r$, then AB is an $r \times r$ matrix and BA is and $s \times s$ matrix. When $r \neq s$, clearly $AB \neq BA$, but even when $r = s$ we still can not expect AB to equal BA.

EXAMPLE A.3.3. Consider two square matrices, say,

$$A = \begin{bmatrix} 1 & 2 \\ 3 & 4 \end{bmatrix} \quad B = \begin{bmatrix} 0 & 2 \\ 1 & 2 \end{bmatrix}.$$

Multiplication gives

$$AB = \begin{bmatrix} 2 & 6 \\ 4 & 14 \end{bmatrix}$$

and

$$BA = \begin{bmatrix} 6 & 8 \\ 7 & 10 \end{bmatrix},$$

so $AB \neq BA$.

□

In general if $A = [a_{ij}]$ is an $r \times s$ matrix and $B = [b_{ij}]$ is a $s \times c$ matrix, then

$$AB = [d_{ij}]$$

is the $r \times c$ matrix with

$$d_{ij} = \sum_{\ell=1}^{s} a_{i\ell} b_{\ell j}.$$

A useful result is that the transpose of the product AB is the product, in reverse order, of the transposed matrices, i.e. $(AB)' = B'A'$.

EXAMPLE A.3.4.   As seen in Example A.3.1,

$$AB \equiv \begin{bmatrix} 1 & 4 \\ 2 & 5 \\ 3 & 6 \end{bmatrix} \begin{bmatrix} 20 & 80 \\ 90 & 140 \end{bmatrix} = \begin{bmatrix} 380 & 640 \\ 490 & 860 \\ 600 & 1080 \end{bmatrix} \equiv C.$$

The transpose of this matrix is

$$C' = \begin{bmatrix} 380 & 490 & 600 \\ 640 & 860 & 1080 \end{bmatrix} = \begin{bmatrix} 20 & 90 \\ 80 & 140 \end{bmatrix} \begin{bmatrix} 1 & 2 & 3 \\ 4 & 5 & 6 \end{bmatrix} = B'A'.$$

□

Let $a = (a_1, \ldots, a_n)'$ be a vector. A very useful property of vectors is that

$$a'a = \sum_{i=1}^{n} a_i^2 \geq 0.$$

A.4   Special matrices

If $A = A'$, then A is said to be *symmetric*. If $A = [a_{ij}]$ and $A = A'$, then $a_{ij} = a_{ji}$. The entry in row i and column j is the same as the entry in row j and column i. Only square matrices can be symmetric.

EXAMPLE A.4.1.   The matrix

$$A = \begin{bmatrix} 4 & 3 & 1 \\ 3 & 2 & 6 \\ 1 & 6 & 5 \end{bmatrix}$$

has $A = A'$. A is symmetric about the diagonal that runs from the upper left to the lower right.   □

For any $r \times c$ matrix A, the product $A'A$ is always symmetric. This was illustrated in Example 23.3.2. More generally, write $A = [a_{ij}]$, $A' = [\tilde{a}_{ij}]$ with $\tilde{a}_{ij} = a_{ji}$, and

$$A'A = [d_{ij}] = \left[\sum_{\ell=1}^{c} \tilde{a}_{i\ell} a_{\ell j} \right].$$

Note that

$$d_{ij} = \sum_{\ell=1}^{c} \tilde{a}_{i\ell} a_{\ell j} = \sum_{\ell=1}^{c} a_{\ell i} a_{\ell j} = \sum_{\ell=1}^{c} \tilde{a}_{j\ell} a_{\ell i} = d_{ji}$$

so the matrix is symmetric.

Diagonal matrices are square matrices with all off-diagonal elements equal to zero.

EXAMPLE A.4.2.   The matrices

$$\begin{bmatrix} 1 & 0 & 0 \\ 0 & 2 & 0 \\ 0 & 0 & 3 \end{bmatrix}, \quad \begin{bmatrix} 20 & 0 \\ 0 & -3 \end{bmatrix}, \text{ and } \begin{bmatrix} 1 & 0 & 0 \\ 0 & 1 & 0 \\ 0 & 0 & 1 \end{bmatrix}$$

are diagonal. □

In general, a diagonal matrix is a square matrix $A = [a_{ij}]$ with $a_{ij} = 0$ for $i \neq j$. Obviously, diagonally matrices are symmetric.

An *identity matrix* is a diagonal matrix with all 1s along the diagonal, i.e., $a_{ii} = 1$ for all i. The third matrix in Example A.4.2 above is a 3×3 identity matrix. The identity matrix gets it name because any matrix multiplied by an identity matrix remains unchanged.

EXAMPLE A.4.3.

$$\begin{bmatrix} 1 & 4 \\ 2 & 5 \\ 3 & 6 \end{bmatrix} \begin{bmatrix} 1 & 0 \\ 0 & 1 \end{bmatrix} = \begin{bmatrix} 1 & 4 \\ 2 & 5 \\ 3 & 6 \end{bmatrix}.$$

$$\begin{bmatrix} 1 & 0 & 0 \\ 0 & 1 & 0 \\ 0 & 0 & 1 \end{bmatrix} \begin{bmatrix} 1 & 4 \\ 2 & 5 \\ 3 & 6 \end{bmatrix} = \begin{bmatrix} 1 & 4 \\ 2 & 5 \\ 3 & 6 \end{bmatrix}.$$

□

An $r \times r$ identity matrix is denoted I_r with the subscript deleted if the dimension is clear.

A *zero matrix* is a matrix that consists entirely of zeros. Obviously, the product of any matrix multiplied by a zero matrix is zero.

EXAMPLE A.4.4.

$$\begin{bmatrix} 0 & 0 \\ 0 & 0 \\ 0 & 0 \end{bmatrix}, \quad \begin{bmatrix} 0 \\ 0 \\ 0 \\ 0 \end{bmatrix}.$$

□

Often a zero matrix is denoted by 0 where the dimension of the matrix, and the fact that it is a matrix rather than a scalar, must be inferred from the context.

A matrix M that has the property $MM = M$ is called *idempotent*. A symmetric idempotent matrix is a *perpendicular projection operator*.

EXAMPLE A.4.5.   The following matrices are both symmetric and idempotent:

$$\begin{bmatrix} 1 & 0 & 0 \\ 0 & 1 & 0 \\ 0 & 0 & 1 \end{bmatrix}, \quad \begin{bmatrix} 1/3 & 1/3 & 1/3 \\ 1/3 & 1/3 & 1/3 \\ 1/3 & 1/3 & 1/3 \end{bmatrix}, \quad \begin{bmatrix} .5 & .5 & 0 \\ .5 & .5 & 0 \\ 0 & 0 & 1 \end{bmatrix}.$$

□

A.5   Linear dependence and rank

Consider the matrix

$$A = \begin{bmatrix} 1 & 2 & 5 & 1 \\ 2 & 2 & 10 & 6 \\ 3 & 4 & 15 & 1 \end{bmatrix}.$$

Note that each column of A can be viewed as a vector. The *column space* of A, denoted $C(A)$, is the collection of all vectors that can be written as *a linear combination of the columns of A*. In other words, $C(A)$ is the set of all vectors that can be written as

$$\lambda_1 \begin{bmatrix} 1 \\ 2 \\ 3 \end{bmatrix} + \lambda_2 \begin{bmatrix} 2 \\ 2 \\ 4 \end{bmatrix} + \lambda_3 \begin{bmatrix} 5 \\ 10 \\ 15 \end{bmatrix} + \lambda_4 \begin{bmatrix} 1 \\ 6 \\ 1 \end{bmatrix} = A \begin{bmatrix} \lambda_1 \\ \lambda_2 \\ \lambda_3 \\ \lambda_4 \end{bmatrix} = A\lambda$$

for some vector $\lambda = (\lambda_1, \lambda_2, \lambda_3, \lambda_4)'$.

The columns of any matrix A are *linearly dependent* if they contain redundant information. Specifically, let x be some vector in $C(A)$. The columns of A are linearly dependent if we can find two distinct vectors λ and γ such that $x = A\lambda$ and $x = A\gamma$. Thus two distinct linear combinations of the columns of A give rise to the same vector x. Note that $\lambda \neq \gamma$ because λ and γ are distinct. Note also that, using a distributive property of matrix multiplication, $A(\lambda - \gamma) = A\lambda - A\gamma = 0$, where $\lambda - \gamma \neq 0$. This condition is frequently used as an alternative definition for linear dependence, i.e., the columns of A are linearly dependent if there exists a vector $\delta \neq 0$ such that $A\delta = 0$. If the columns of A are not linearly dependent, they are *linearly independent*.

EXAMPLE A.5.1.   Observe that the example matrix A given at the beginning of the section has

$$\begin{bmatrix} 1 & 2 & 5 & 1 \\ 2 & 2 & 10 & 6 \\ 3 & 4 & 15 & 1 \end{bmatrix} \begin{bmatrix} 5 \\ 0 \\ -1 \\ 0 \end{bmatrix} = \begin{bmatrix} 0 \\ 0 \\ 0 \end{bmatrix},$$

so the columns of A are linearly dependent.   □

The *rank* of A is the smallest number of columns of A that can generate $C(A)$. It is also the maximum number of linearly independent columns in A.

EXAMPLE A.5.2.   The matrix

$$A = \begin{bmatrix} 1 & 2 & 5 & 1 \\ 2 & 2 & 10 & 6 \\ 3 & 4 & 15 & 1 \end{bmatrix}$$

has rank 3 because the columns

$$\begin{bmatrix} 1 \\ 2 \\ 3 \end{bmatrix}, \begin{bmatrix} 2 \\ 2 \\ 4 \end{bmatrix}, \begin{bmatrix} 1 \\ 6 \\ 1 \end{bmatrix}$$

generate $C(A)$. We saw in Example A.5.1 that the column $(5, 10, 15)'$ was redundant. None of the other three columns are redundant; they are linearly independent. In other words, the only way to get

$$\begin{bmatrix} 1 & 2 & 1 \\ 2 & 2 & 6 \\ 3 & 4 & 1 \end{bmatrix} \delta = \begin{bmatrix} 0 \\ 0 \\ 0 \end{bmatrix}$$

is to take $\delta = (0,0,0)'$.   □

A.6   Inverse matrices

The *inverse* of a square matrix A is the matrix A^{-1} such that

$$AA^{-1} = A^{-1}A = I.$$

The inverse of A exists only if the columns of A are linearly independent. Typically, it is difficult to find inverses without the aid of a computer. For a 2×2 matrix

$$A = \begin{bmatrix} a_{11} & a_{12} \\ a_{21} & a_{22} \end{bmatrix},$$

the inverse is given by

$$A^{-1} = \frac{1}{a_{11}a_{22} - a_{12}a_{21}} \begin{bmatrix} a_{22} & -a_{12} \\ -a_{21} & a_{11} \end{bmatrix}. \tag{A.6.1}$$

To confirm that this is correct, multiply AA^{-1} to see that it gives the identity matrix. Moderately complicated formulae exist for computing the inverse of 3×3 matrices. Inverses of larger matrices become very difficult to compute by hand. Of course computers are ideally suited for finding such things.

One use for inverse matrices is in solving systems of equations.

EXAMPLE A.6.1. Consider the system of equations

$$
\begin{aligned}
2x + 4y &= 20 \\
3x + 4y &= 10.
\end{aligned}
$$

We can write this in matrix form as

$$
\begin{bmatrix} 2 & 4 \\ 3 & 4 \end{bmatrix} \begin{bmatrix} x \\ y \end{bmatrix} = \begin{bmatrix} 20 \\ 10 \end{bmatrix}.
$$

Multiplying on the left by the inverse of the coefficient matrix gives

$$
\begin{bmatrix} 2 & 4 \\ 3 & 4 \end{bmatrix}^{-1} \begin{bmatrix} 2 & 4 \\ 3 & 4 \end{bmatrix} \begin{bmatrix} x \\ y \end{bmatrix} = \begin{bmatrix} 2 & 4 \\ 3 & 4 \end{bmatrix}^{-1} \begin{bmatrix} 20 \\ 10 \end{bmatrix}.
$$

Using the definition of the inverse on the left-hand side of the equality and the formula in (A.6.1) on the right-hand side gives

$$
\begin{bmatrix} 1 & 0 \\ 0 & 1 \end{bmatrix} \begin{bmatrix} x \\ y \end{bmatrix} = \begin{bmatrix} -1 & 1 \\ 3/4 & -1/2 \end{bmatrix} \begin{bmatrix} 20 \\ 10 \end{bmatrix}
$$

or

$$
\begin{bmatrix} x \\ y \end{bmatrix} = \begin{bmatrix} -10 \\ 10 \end{bmatrix}.
$$

Thus $(x,y) = (-10, 10)$ is the solution for the two equations, i.e., $2(-10) + 4(10) = 20$ and $3(-10) + 4(10) = 10$. □

More generally, a system of equations, say,

$$
\begin{aligned}
a_{11} y_1 + a_{12} y_2 + a_{13} y_3 &= c_1 \\
a_{21} y_1 + a_{22} y_2 + a_{23} y_3 &= c_2 \\
a_{31} y_1 + a_{32} y_2 + a_{33} y_3 &= c_3
\end{aligned}
$$

in which the a_{ij}s and c_is are known and the y_is are variables, can be written in matrix form as

$$
\begin{bmatrix} a_{11} & a_{12} & a_{13} \\ a_{21} & a_{22} & a_{23} \\ a_{31} & a_{32} & a_{33} \end{bmatrix} \begin{bmatrix} y_1 \\ y_2 \\ y_3 \end{bmatrix} = \begin{bmatrix} c_1 \\ c_2 \\ c_3 \end{bmatrix}
$$

or

$$
AY = C.
$$

To find Y simply observe that $AY = C$ implies $A^{-1}AY = A^{-1}C$ and $Y = A^{-1}C$. Of course this argument assumes that A^{-1} exists, which is not always the case. Moreover, the procedure obviously extends to larger sets of equations.

On a computer, there are better ways of finding solutions to systems of equations than finding the inverse of a matrix. In fact, inverses are often found by solving systems of equations. For example, in a 3×3 case the first column of A^{-1} can be found as the solution to

$$\begin{bmatrix} a_{11} & a_{12} & a_{13} \\ a_{21} & a_{22} & a_{23} \\ a_{31} & a_{32} & a_{33} \end{bmatrix} \begin{bmatrix} y_1 \\ y_2 \\ y_3 \end{bmatrix} = \begin{bmatrix} 1 \\ 0 \\ 0 \end{bmatrix}.$$

For a special type of square matrix, called an *orthogonal matrix*, the transpose is also the inverse. In other words, a square matrix P is an orthogonal matrix if

$$P'P = I = PP'.$$

To establish that P is orthogonal, it is enough to show either that $P'P = I$ or that $PP' = I$. Orthogonal matrices are particularly useful in discussions of eigenvalues and principal component regression.

A.7   A list of useful properties

The following proposition summarizes many of the key properties of matrices and the operations performed on them.

Proposition A.7.1.   Let A, B, and C be matrices of appropriate dimensions and let λ be a scalar.

$$\begin{aligned} A + B &= B + A \\ (A + B) + C &= A + (B + C) \\ (AB)C &= A(BC) \\ C(A + B) &= CA + CB \\ \lambda(A + B) &= \lambda A + \lambda B \\ (A')' &= A \\ (A + B)' &= A' + B' \\ (AB)' &= B'A' \\ \left(A^{-1}\right)^{-1} &= A \\ (A')^{-1} &= \left(A^{-1}\right)' \\ (AB)^{-1} &= B^{-1}A^{-1}. \end{aligned}$$

The last equality only holds when A and B both have inverses. The second-to-last property implies that the inverse of a symmetric matrix is symmetric because then $A^{-1} = (A')^{-1} = (A^{-1})'$. This is a very important property.

A.8   Eigenvalues and eigenvectors

Let A be a square matrix. A scalar ϕ is an eigenvalue of A and $x \neq 0$ is an eigenvector for A corresponding to ϕ if

$$Ax = \phi x.$$

EXAMPLE A.8.1.   Consider the matrix

$$A = \begin{bmatrix} 3 & 1 & -1 \\ 1 & 3 & -1 \\ -1 & -1 & 5 \end{bmatrix}.$$

The value 3 is an eigenvalue and any nonzero multiple of the vector $(1,1,1)'$ is a corresponding eigenvector. For example,

$$
\begin{bmatrix} 3 & 1 & -1 \\ 1 & 3 & -1 \\ -1 & -1 & 5 \end{bmatrix} \begin{bmatrix} 1 \\ 1 \\ 1 \end{bmatrix} = \begin{bmatrix} 3 \\ 3 \\ 3 \end{bmatrix} = 3 \begin{bmatrix} 1 \\ 1 \\ 1 \end{bmatrix}.
$$

Similarly, if we consider a multiple, say, $4(1,1,1)'$,

$$
\begin{bmatrix} 3 & 1 & -1 \\ 1 & 3 & -1 \\ -1 & -1 & 5 \end{bmatrix} \begin{bmatrix} 4 \\ 4 \\ 4 \end{bmatrix} = \begin{bmatrix} 12 \\ 12 \\ 12 \end{bmatrix} = 3 \begin{bmatrix} 4 \\ 4 \\ 4 \end{bmatrix}.
$$

The value 2 is also an eigenvalue with eigenvectors that are nonzero multiples of $(1,-1,0)'$.

$$
\begin{bmatrix} 3 & 1 & -1 \\ 1 & 3 & -1 \\ -1 & -1 & 5 \end{bmatrix} \begin{bmatrix} 1 \\ -1 \\ 0 \end{bmatrix} = \begin{bmatrix} 2 \\ -2 \\ 0 \end{bmatrix} = 2 \begin{bmatrix} 1 \\ -1 \\ 0 \end{bmatrix}.
$$

Finally, 6 is an eigenvalue with eigenvectors that are nonzero multiples of $(1,1,-2)'$. $\square$

Proposition A.8.2. Let A be a symmetric matrix, then for a diagonal matrix $D(\phi_i)$ consisting of eigenvalues there exists an orthogonal matrix P whose columns are corresponding eigenvectors such that

$$
A = PD(\phi_i)P'.
$$

EXAMPLE A.8.3. Consider again the matrix

$$
A = \begin{bmatrix} 3 & 1 & -1 \\ 1 & 3 & -1 \\ -1 & -1 & 5 \end{bmatrix}.
$$

In writing $A = PD(\phi_i)P'$, the diagonal matrix is

$$
D(\phi_i) = \begin{bmatrix} 3 & 0 & 0 \\ 0 & 2 & 0 \\ 0 & 0 & 6 \end{bmatrix}.
$$

The orthogonal matrix is

$$
P = \begin{bmatrix} \frac{1}{\sqrt{3}} & \frac{1}{\sqrt{2}} & \frac{1}{\sqrt{6}} \\ \frac{1}{\sqrt{3}} & \frac{-1}{\sqrt{2}} & \frac{1}{\sqrt{6}} \\ \frac{1}{\sqrt{3}} & 0 & \frac{-2}{\sqrt{6}} \end{bmatrix}.
$$

We leave it to the reader to verify that $PD(\phi_i)P' = A$ and that $P'P = I$.

Note that the columns of P are multiples of the vectors identified as eigenvectors in Example A.8.1; hence the columns of P are also eigenvectors. The multiples of the eigenvectors were chosen so that $PP' = I$ and $P'P = I$. Moreover, the first column of P is an eigenvector corresponding to 3, which is the first eigenvalue listed in $D(\phi_i)$. Similarly, the second column of P is an eigenvector corresponding to 2 and the third column corresponds to the third listed eigenvalue, 6.

With a 3×3 matrix A having three *distinct* eigenvalues, any matrix P with eigenvectors for columns would have $P'P$ a diagonal matrix, but the multiples of the eigenvectors must be chosen so that the diagonal entries of $P'P$ are all 1. $\square$

EXAMPLE A.8.4.   Consider the matrix

$$B = \begin{bmatrix} 5 & -1 & -1 \\ -1 & 5 & -1 \\ -1 & -1 & 5 \end{bmatrix}.$$

This matrix is closely related to the matrix in Example A.8.1. The matrix B has 3 as an eigenvalue with corresponding eigenvectors that are multiples of $(1,1,1)'$, just like the matrix A. Once again 6 is an eigenvalue with corresponding eigenvector $(1,1,-2)'$ and once again $(1,-1,0)'$ is an eigenvector, but now, unlike A, $(1,-1,0)$ also corresponds to the eigenvalue 6. We leave it to the reader to verify these facts. The point is that in this matrix, 6 is an eigenvalue that has two linearly independent eigenvectors. In such cases, any nonzero linear combination of the two eigenvectors is also an eigenvector. For example, it is easy to see that

$$3 \begin{bmatrix} 1 \\ -1 \\ 0 \end{bmatrix} + 2 \begin{bmatrix} 1 \\ 1 \\ -2 \end{bmatrix} = \begin{bmatrix} 5 \\ -1 \\ -4 \end{bmatrix}$$

is an eigenvector corresponding to the eigenvalue 6.

To write $B = PD(\phi)P'$ as in Proposition A.8.2, $D(\phi)$ has 3, 6, and 6 down the diagonal and one choice of P is that given in Example A.8.3. However, because one of the eigenvalues occurs more than once in the diagonal matrix, there are many choices for P.   □

Generally, if we need eigenvalues or eigenvectors we get a computer to find them for us.
Two frequently used functions of a square matrix are the determinant and the trace.

Definition A.8.5.
   a) The determinant of a square matrix is the product of the eigenvalues of the matrix.
   b) The trace of a square matrix is the sum of the eigenvalues of the matrix.

In fact, one can show that the trace of a square matrix also equals the sum of the diagonal elements of that matrix.

Appendix B: Tables

B.1 Tables of the *t* distribution

Table B.1: *Percentage points of the t distribution.*

				α levels				
Two-sided	.20	.10	.05	.04	.02	.01	.002	.001
One-sided	.10	.05	.025	.02	.01	.005	.001	.0005
					Percentiles			
df	0.90	0.95	0.975	0.98	0.99	0.995	0.999	0.9995
1	3.078	6.314	12.7062	15.8946	31.8206	63.6570	318.317	636.607
2	1.886	2.920	4.3027	4.8487	6.9646	9.9248	22.327	31.598
3	1.638	2.353	3.1824	3.4819	4.5407	5.8409	10.215	12.924
4	1.533	2.132	2.7764	2.9985	3.7470	4.6041	7.173	8.610
5	1.476	2.015	2.5706	2.7565	3.3649	4.0322	5.893	6.869
6	1.440	1.943	2.4469	2.6122	3.1427	3.7075	5.208	5.959
7	1.415	1.895	2.3646	2.5168	2.9980	3.4995	4.785	5.408
8	1.397	1.860	2.3060	2.4490	2.8965	3.3554	4.501	5.041
9	1.383	1.833	2.2622	2.3984	2.8214	3.2499	4.297	4.781
10	1.372	1.812	2.2281	2.3593	2.7638	3.1693	4.144	4.587
11	1.363	1.796	2.2010	2.3281	2.7181	3.1058	4.025	4.437
12	1.356	1.782	2.1788	2.3027	2.6810	3.0546	3.930	4.318
13	1.350	1.771	2.1604	2.2816	2.6503	3.0123	3.852	4.221
14	1.345	1.761	2.1448	2.2638	2.6245	2.9769	3.787	4.140
15	1.341	1.753	2.1315	2.2485	2.6025	2.9467	3.733	4.073
16	1.337	1.746	2.1199	2.2354	2.5835	2.9208	3.686	4.015
17	1.333	1.740	2.1098	2.2239	2.5669	2.8982	3.646	3.965
18	1.330	1.734	2.1009	2.2137	2.5524	2.8784	3.611	3.922
19	1.328	1.729	2.0930	2.2047	2.5395	2.8610	3.579	3.883
20	1.325	1.725	2.0860	2.1967	2.5280	2.8453	3.552	3.850
21	1.323	1.721	2.0796	2.1894	2.5176	2.8314	3.527	3.819
22	1.321	1.717	2.0739	2.1829	2.5083	2.8188	3.505	3.792
23	1.319	1.714	2.0687	2.1769	2.4999	2.8073	3.485	3.768
24	1.318	1.711	2.0639	2.1716	2.4922	2.7969	3.467	3.745
25	1.316	1.708	2.0595	2.1666	2.4851	2.7874	3.450	3.725
26	1.315	1.706	2.0555	2.1620	2.4786	2.7787	3.435	3.707
27	1.314	1.703	2.0518	2.1578	2.4727	2.7707	3.421	3.690
28	1.313	1.701	2.0484	2.1539	2.4671	2.7633	3.408	3.674
29	1.311	1.699	2.0452	2.1503	2.4620	2.7564	3.396	3.659
30	1.310	1.697	2.0423	2.1470	2.4573	2.7500	3.385	3.646
31	1.309	1.696	2.0395	2.1438	2.4528	2.7441	3.375	3.633
32	1.309	1.694	2.0369	2.1409	2.4487	2.7385	3.365	3.622
33	1.308	1.692	2.0345	2.1382	2.4448	2.7333	3.356	3.611
34	1.307	1.691	2.0323	2.1356	2.4412	2.7284	3.348	3.601
35	1.306	1.690	2.0301	2.1332	2.4377	2.7238	3.340	3.591

Table B.2: *Percentage points of the t distribution.*

				α levels				
Two-sided	.20	.10	.05	.04	.02	.01	.002	.001
One-sided	.10	.05	.025	.02	.01	.005	.001	.0005
					Percentiles			
df	0.90	0.95	0.975	0.98	0.99	0.995	0.999	0.9995
36	1.306	1.688	2.0281	2.1309	2.4345	2.7195	3.333	3.582
37	1.305	1.687	2.0262	2.1287	2.4314	2.7154	3.326	3.574
38	1.304	1.686	2.0244	2.1267	2.4286	2.7116	3.319	3.566
39	1.304	1.685	2.0227	2.1247	2.4258	2.7079	3.313	3.558
40	1.303	1.684	2.0211	2.1229	2.4233	2.7045	3.307	3.551
41	1.303	1.683	2.0196	2.1212	2.4208	2.7012	3.301	3.544
42	1.302	1.682	2.0181	2.1195	2.4185	2.6981	3.296	3.538
43	1.302	1.681	2.0167	2.1179	2.4163	2.6951	3.291	3.532
44	1.301	1.680	2.0154	2.1164	2.4142	2.6923	3.286	3.526
45	1.301	1.679	2.0141	2.1150	2.4121	2.6896	3.281	3.520
46	1.300	1.679	2.0129	2.1136	2.4102	2.6870	3.277	3.515
47	1.300	1.678	2.0117	2.1123	2.4083	2.6846	3.273	3.510
48	1.299	1.677	2.0106	2.1111	2.4066	2.6822	3.269	3.505
49	1.299	1.677	2.0096	2.1099	2.4049	2.6800	3.265	3.500
50	1.299	1.676	2.0086	2.1087	2.4033	2.6778	3.261	3.496
51	1.298	1.675	2.0076	2.1076	2.4017	2.6757	3.258	3.492
52	1.298	1.675	2.0067	2.1066	2.4002	2.6737	3.255	3.488
53	1.298	1.674	2.0058	2.1055	2.3988	2.6718	3.251	3.484
54	1.297	1.674	2.0049	2.1046	2.3974	2.6700	3.248	3.480
55	1.297	1.673	2.0041	2.1036	2.3961	2.6682	3.245	3.476
56	1.297	1.673	2.0033	2.1027	2.3948	2.6665	3.242	3.473
57	1.297	1.672	2.0025	2.1018	2.3936	2.6649	3.239	3.470
58	1.296	1.672	2.0017	2.1010	2.3924	2.6633	3.237	3.466
59	1.296	1.671	2.0010	2.1002	2.3912	2.6618	3.234	3.463
60	1.296	1.671	2.0003	2.0994	2.3902	2.6604	3.232	3.460
70	1.294	1.667	1.9944	2.0927	2.3808	2.6480	3.211	3.435
80	1.292	1.664	1.9901	2.0878	2.3739	2.6387	3.195	3.416
90	1.291	1.662	1.9867	2.0840	2.3685	2.6316	3.183	3.402
100	1.290	1.660	1.9840	2.0809	2.3642	2.6259	3.174	3.391
110	1.289	1.659	1.9818	2.0784	2.3607	2.6213	3.166	3.381
120	1.289	1.658	1.9799	2.0763	2.3578	2.6174	3.160	3.373
150	1.287	1.655	1.9759	2.0718	2.3515	2.6090	3.145	3.357
200	1.286	1.653	1.9719	2.0672	2.3451	2.6006	3.131	3.340
250	1.285	1.651	1.9695	2.0645	2.3414	2.5956	3.123	3.330
300	1.284	1.650	1.9679	2.0627	2.3388	2.5923	3.118	3.323
350	1.284	1.649	1.9668	2.0614	2.3371	2.5900	3.114	3.319
400	1.284	1.649	1.9659	2.0605	2.3357	2.5882	3.111	3.315
∞	1.282	1.645	1.9600	2.0537	2.3263	2.5758	3.090	3.291

B.2   Tables of the χ^2 distribution

Table B.3: *Percentage points of the χ^2 distribution.*

				α levels				
Two-sided	0.002	0.01	0.02	0.04	0.05	0.10	0.20	0.40
One-sided	0.001	0.005	0.01	0.02	0.025	0.05	0.10	0.20
					Percentiles			
df	0.001	0.005	0.01	0.02	0.025	0.05	0.10	0.20
1	0.000	0.000	0.000	0.001	0.001	0.004	0.016	0.064
2	0.002	0.010	0.020	0.040	0.051	0.103	0.211	0.446
3	0.024	0.072	0.115	0.185	0.216	0.352	0.584	1.005
4	0.091	0.207	0.297	0.429	0.484	0.711	1.064	1.649
5	0.210	0.412	0.554	0.752	0.831	1.145	1.610	2.343
6	0.381	0.676	0.872	1.134	1.237	1.635	2.204	3.070
7	0.598	0.989	1.239	1.564	1.690	2.167	2.833	3.822
8	0.857	1.344	1.646	2.032	2.180	2.733	3.490	4.594
9	1.152	1.735	2.088	2.532	2.700	3.325	4.168	5.380
10	1.479	2.156	2.558	3.059	3.247	3.940	4.865	6.179
11	1.834	2.603	3.053	3.609	3.816	4.575	5.578	6.989
12	2.214	3.074	3.571	4.178	4.404	5.226	6.304	7.807
13	2.617	3.565	4.107	4.765	5.009	5.892	7.042	8.634
14	3.041	4.075	4.660	5.368	5.629	6.571	7.790	9.467
15	3.483	4.601	5.229	5.985	6.262	7.261	8.547	10.307
16	3.942	5.142	5.812	6.614	6.908	7.962	9.312	11.152
17	4.416	5.697	6.408	7.255	7.564	8.672	10.085	12.002
18	4.905	6.265	7.015	7.906	8.231	9.390	10.865	12.857
19	5.407	6.844	7.633	8.567	8.907	10.117	11.651	13.716
20	5.921	7.434	8.260	9.237	9.591	10.851	12.443	14.578
21	6.447	8.034	8.897	9.915	10.283	11.591	13.240	15.445
22	6.983	8.643	9.542	10.600	10.982	12.338	14.041	16.314
23	7.529	9.260	10.196	11.293	11.689	13.091	14.848	17.187
24	8.085	9.886	10.856	11.992	12.401	13.848	15.659	18.062
25	8.649	10.520	11.524	12.697	13.120	14.611	16.473	18.940
26	9.222	11.160	12.198	13.409	13.844	15.379	17.292	19.820
27	9.803	11.808	12.879	14.125	14.573	16.151	18.114	20.703
28	10.391	12.461	13.565	14.847	15.308	16.928	18.939	21.588
29	10.986	13.121	14.256	15.574	16.047	17.708	19.768	22.475
30	11.588	13.787	14.953	16.306	16.791	18.493	20.599	23.364
31	12.196	14.458	15.655	17.042	17.539	19.281	21.434	24.255
32	12.811	15.134	16.362	17.783	18.291	20.072	22.271	25.148
33	13.431	15.815	17.074	18.527	19.047	20.867	23.110	26.042
34	14.057	16.501	17.789	19.275	19.806	21.664	23.952	26.938
35	14.688	17.192	18.509	20.027	20.569	22.465	24.797	27.836

Table B.4: *Percentage points of the χ^2 distribution.*

				α levels				
Two-sided	.40	.20	.10	.05	.04	.02	.01	.002
One-sided	.20	.10	.05	.025	.02	.01	.005	.001
					Percentiles			
df	0.80	0.90	0.95	0.975	0.98	0.99	0.995	0.999
1	1.642	2.706	3.841	5.024	5.412	6.635	7.879	10.828
2	3.219	4.605	5.991	7.378	7.824	9.210	10.597	13.816
3	4.642	6.251	7.815	9.348	9.837	11.345	12.838	16.266
4	5.989	7.779	9.488	11.143	11.668	13.277	14.860	18.467
5	7.289	9.236	11.070	12.833	13.388	15.086	16.750	20.515
6	8.558	10.645	12.592	14.449	15.033	16.812	18.548	22.458
7	9.803	12.017	14.067	16.013	16.622	18.475	20.278	24.322
8	11.030	13.362	15.507	17.535	18.168	20.090	21.955	26.125
9	12.242	14.684	16.919	19.023	19.679	21.666	23.589	27.877
10	13.442	15.987	18.307	20.483	21.161	23.209	25.188	29.588
11	14.631	17.275	19.675	21.920	22.618	24.725	26.757	31.264
12	15.812	18.549	21.026	23.337	24.054	26.217	28.300	32.910
13	16.985	19.812	22.362	24.736	25.471	27.688	29.819	34.528
14	18.151	21.064	23.685	26.119	26.873	29.141	31.319	36.124
15	19.311	22.307	24.996	27.488	28.259	30.578	32.801	37.697
16	20.465	23.542	26.296	28.845	29.633	32.000	34.267	39.254
17	21.615	24.769	27.587	30.191	30.995	33.409	35.718	40.789
18	22.760	25.989	28.869	31.526	32.346	34.805	37.156	42.312
19	23.900	27.204	30.143	32.852	33.687	36.191	38.582	43.819
20	25.038	28.412	31.410	34.170	35.020	37.566	39.997	45.315
21	26.171	29.615	32.671	35.479	36.343	38.932	41.401	46.797
22	27.301	30.813	33.924	36.781	37.660	40.290	42.796	48.270
23	28.429	32.007	35.172	38.076	38.968	41.638	44.181	49.726
24	29.553	33.196	36.415	39.364	40.270	42.980	45.559	51.179
25	30.675	34.382	37.653	40.647	41.566	44.314	46.928	52.622
26	31.795	35.563	38.885	41.923	42.856	45.642	48.290	54.054
27	32.912	36.741	40.113	43.195	44.140	46.963	49.645	55.477
28	34.027	37.916	41.337	44.461	45.419	48.278	50.994	56.893
29	35.139	39.087	42.557	45.722	46.693	49.588	52.336	58.303
30	36.250	40.256	43.773	46.979	47.962	50.892	53.672	59.703
31	37.359	41.422	44.985	48.232	49.226	52.192	55.003	61.100
32	38.466	42.585	46.194	49.480	50.487	53.486	56.328	62.486
33	39.572	43.745	47.400	50.725	51.743	54.775	57.648	63.868
34	40.676	44.903	48.602	51.966	52.995	56.061	58.964	65.246
35	41.778	46.059	49.802	53.204	54.244	57.342	60.275	66.622

Table B.5: *Percentage points of the χ^2 distribution.*

	α levels							
Two-sided	0.002	0.01	0.02	0.04	0.05	0.10	0.20	0.40
One-sided	0.001	0.005	0.01	0.02	0.025	0.05	0.10	0.20
				Percentiles				
df	0.001	0.005	0.01	0.02	0.025	0.05	0.10	0.20
36	15.32	17.887	19.233	20.783	21.336	23.269	25.64	28.74
37	15.96	18.586	19.960	21.542	22.106	24.075	26.49	29.64
38	16.61	19.289	20.691	22.304	22.878	24.884	27.34	30.54
39	17.26	19.996	21.426	23.069	23.654	25.695	28.20	31.44
40	17.92	20.707	22.164	23.838	24.433	26.509	29.05	32.34
41	18.58	21.421	22.906	24.609	25.215	27.326	29.91	33.25
42	19.24	22.138	23.650	25.383	25.999	28.144	30.76	34.16
43	19.91	22.859	24.398	26.159	26.785	28.965	31.62	35.06
44	20.58	23.584	25.148	26.939	27.575	29.787	32.49	35.97
45	21.25	24.311	25.901	27.720	28.366	30.612	33.35	36.88
46	21.93	25.041	26.657	28.504	29.160	31.439	34.22	37.80
47	22.61	25.775	27.416	29.291	29.956	32.268	35.08	38.71
48	23.30	26.511	28.177	30.080	30.755	33.098	35.95	39.62
49	23.98	27.249	28.941	30.871	31.555	33.930	36.82	40.53
50	24.67	27.991	29.707	31.664	32.357	34.764	37.69	41.45
51	25.37	28.735	30.475	32.459	33.162	35.600	38.56	42.36
52	26.06	29.481	31.246	33.256	33.968	36.437	39.43	43.28
53	26.76	30.230	32.018	34.055	34.776	37.276	40.31	44.20
54	27.47	30.981	32.793	34.856	35.586	38.116	41.18	45.12
55	28.17	31.735	33.570	35.659	36.398	38.958	42.06	46.04
56	28.88	32.490	34.350	36.464	37.212	39.801	42.94	46.96
57	29.59	33.248	35.131	37.270	38.027	40.646	43.82	47.88
58	30.30	34.008	35.913	38.078	38.843	41.492	44.70	48.80
59	31.02	34.770	36.698	38.888	39.662	42.339	45.58	49.72
60	31.74	35.535	37.485	39.699	40.482	43.188	46.46	50.64
70	39.04	43.275	45.442	47.893	48.758	51.739	55.33	59.90
80	46.52	51.172	53.540	56.213	57.153	60.391	64.28	69.21
90	54.16	59.196	61.754	64.635	65.647	69.126	73.29	78.56
100	61.92	67.328	70.065	73.142	74.222	77.930	82.36	87.94
110	69.79	75.550	78.458	81.723	82.867	86.792	91.47	97.36
120	77.76	83.852	86.923	90.367	91.573	95.705	100.62	106.81
150	102.11	109.142	112.668	116.608	117.984	122.692	128.28	135.26
200	143.84	152.241	156.432	161.100	162.728	168.279	174.84	183.00
250	186.55	196.161	200.939	206.249	208.098	214.392	221.81	231.01
300	229.96	240.663	245.972	251.864	253.912	260.878	269.07	279.21
350	273.90	285.608	291.406	297.831	300.064	307.648	316.55	327.56
400	318.26	330.903	337.155	344.078	346.482	354.641	364.21	376.02

Table B.6: *Percentage points of the χ^2 distribution.*

	α levels							
Two-sided	.40	.20	.10	.05	.04	.02	.01	.002
One-sided	.20	.10	.05	.025	.02	.01	.005	.001
	Percentiles							
df	0.80	0.90	0.95	0.975	0.98	0.99	0.995	0.999
36	42.88	47.212	50.998	54.437	55.489	58.619	61.58	67.99
37	43.98	48.363	52.192	55.668	56.731	59.893	62.89	69.35
38	45.08	49.513	53.384	56.896	57.969	61.163	64.18	70.71
39	46.17	50.660	54.572	58.120	59.204	62.429	65.48	72.06
40	47.27	51.805	55.759	59.342	60.437	63.691	66.77	73.41
41	48.36	52.949	56.942	60.561	61.665	64.950	68.05	74.75
42	49.46	54.090	58.124	61.777	62.892	66.207	69.34	76.09
43	50.55	55.230	59.303	62.990	64.115	67.459	70.62	77.42
44	51.64	56.369	60.481	64.201	65.337	68.709	71.89	78.75
45	52.73	57.505	61.656	65.410	66.555	69.957	73.17	80.08
46	53.82	58.640	62.829	66.616	67.771	71.201	74.44	81.40
47	54.91	59.774	64.001	67.820	68.985	72.443	75.70	82.72
48	55.99	60.907	65.171	69.022	70.196	73.682	76.97	84.03
49	57.08	62.038	66.339	70.222	71.406	74.919	78.23	85.35
50	58.16	63.167	67.505	71.420	72.613	76.154	79.49	86.66
51	59.25	64.295	68.669	72.616	73.818	77.386	80.75	87.97
52	60.33	65.422	69.832	73.810	75.021	78.616	82.00	89.27
53	61.41	66.548	70.993	75.002	76.222	79.843	83.25	90.57
54	62.50	67.673	72.153	76.192	77.422	81.070	84.50	91.88
55	63.58	68.796	73.312	77.381	78.619	82.292	85.75	93.17
56	64.66	69.919	74.469	78.568	79.815	83.515	87.00	94.47
57	65.74	71.040	75.624	79.752	81.009	84.733	88.24	95.75
58	66.82	72.160	76.777	80.935	82.200	85.949	89.47	97.03
59	67.90	73.279	77.931	82.118	83.392	87.167	90.72	98.34
60	68.97	74.397	79.082	83.298	84.581	88.381	91.96	99.62
70	79.72	85.527	90.531	95.023	96.387	100.424	104.21	112.31
80	90.41	96.578	101.879	106.628	108.069	112.328	116.32	124.84
90	101.05	107.565	113.145	118.135	119.648	124.115	128.30	137.19
100	111.67	118.499	124.343	129.563	131.144	135.811	140.18	149.48
110	122.25	129.385	135.480	140.917	142.562	147.416	151.95	161.59
120	132.81	140.233	146.567	152.211	153.918	158.950	163.65	173.62
150	164.35	172.580	179.579	185.798	187.675	193.202	198.35	209.22
200	216.61	226.022	233.997	241.062	243.192	249.455	255.28	267.62
250	268.60	279.052	287.884	295.694	298.045	304.951	311.37	324.93
300	320.40	331.787	341.393	349.870	352.419	359.896	366.83	381.34
350	372.05	384.305	394.624	403.720	406.454	414.466	421.89	437.43
400	423.59	436.647	447.628	457.298	460.201	468.707	476.57	492.99

B.3 Tables of the W' statistic

This table was obtained by taking the mean of ten estimates of the percentile, each based on a sample of 500 observations. Estimates with standard errors of about .002 or less are reported to three decimal places. The estimates reported with two decimal places have standard errors between about .002 and .008.

Table B.7: *Percentiles of the W' statistic.*

n	.01	.05	n	.01	.05
5	0.69	0.77	36	0.91	0.940
6	0.70	0.79	38	0.915	0.942
7	0.72	0.81	40	0.918	0.946
8	0.75	0.82	45	0.928	0.951
9	0.75	0.83	50	0.931	0.952
10	0.78	0.83	55	0.938	0.958
11	0.79	0.85	60	0.943	0.961
12	0.79	0.86	65	0.945	0.961
13	0.81	0.870	70	0.953	0.966
14	0.82	0.877	75	0.954	0.968
15	0.82	0.883	80	0.957	0.970
16	0.83	0.886	85	0.958	0.970
17	0.84	0.896	90	0.960	0.972
18	0.85	0.896	95	0.961	0.972
19	0.86	0.902	100	0.962	0.974
20	0.86	0.902	120	0.970	0.978
22	0.87	0.910	140	0.973	0.981
24	0.88	0.915	160	0.976	0.983
26	0.89	0.923	180	0.978	0.985
28	0.89	0.924	200	0.981	0.986
30	0.89	0.928	250	0.984	0.988
32	0.90	0.933	300	0.987	0.991
34	0.91	0.936			

B.4 Tables of the Studentized range

These tables are largely those from May (1952) and are presented with the permission of the Trustees of *Biometrika*. Comparisons with several other tables have been made and the values that appear to be most accurate have been used. In doubtful cases, values have been rounded up.

Table B.8: $Q(.95, r, dfE)$.

dfE	2	3	4	5	6	7	8	9	10	11
1	18.0	27.0	32.8	37.1	40.4	43.1	45.4	47.4	49.1	50.6
2	6.09	8.33	9.80	10.88	11.74	12.44	13.03	13.54	13.99	14.39
3	4.50	5.91	6.83	7.50	8.04	8.48	8.85	9.18	9.46	9.72
4	3.93	5.04	5.76	6.29	6.71	7.05	7.35	7.60	7.83	8.03
5	3.64	4.60	5.22	5.67	6.03	6.33	6.58	6.80	7.00	7.17
6	3.46	4.34	4.90	5.31	5.63	5.90	6.12	6.32	6.49	6.65
7	3.34	4.17	4.68	5.06	5.36	5.61	5.82	6.00	6.16	6.30
8	3.26	4.04	4.53	4.89	5.17	5.40	5.60	5.77	5.92	6.05
9	3.20	3.95	4.42	4.76	5.02	5.24	5.43	5.60	5.74	5.87
10	3.15	3.88	4.33	4.65	4.91	5.12	5.31	5.46	5.60	5.72
11	3.11	3.82	4.26	4.57	4.82	5.03	5.20	5.35	5.49	5.61
12	3.08	3.77	4.20	4.51	4.75	4.95	5.12	5.27	5.40	5.51
13	3.06	3.74	4.15	4.45	4.69	4.89	5.05	5.19	5.32	5.43
14	3.03	3.70	4.11	4.41	4.64	4.83	4.99	5.13	5.25	5.36
15	3.01	3.67	4.08	4.37	4.59	4.78	4.94	5.08	5.20	5.31
16	3.00	3.65	4.05	4.33	4.56	4.74	4.90	5.03	5.15	5.26
17	2.98	3.63	4.02	4.30	4.52	4.71	4.86	4.99	5.11	5.21
18	2.97	3.61	4.00	4.28	4.50	4.67	4.82	4.96	5.07	5.17
19	2.96	3.59	3.98	4.25	4.47	4.65	4.79	4.92	5.04	5.14
20	2.95	3.58	3.96	4.23	4.45	4.62	4.77	4.90	5.01	5.11
24	2.92	3.53	3.90	4.17	4.37	4.54	4.68	4.81	4.92	5.01
30	2.89	3.49	3.85	4.10	4.30	4.46	4.60	4.72	4.82	4.92
40	2.86	3.44	3.79	4.04	4.23	4.39	4.52	4.64	4.74	4.82
60	2.83	3.40	3.74	3.98	4.16	4.31	4.44	4.55	4.65	4.73
120	2.80	3.36	3.69	3.92	4.10	4.24	4.36	4.47	4.56	4.64
∞	2.77	3.31	3.63	3.86	4.03	4.17	4.29	4.39	4.47	4.55

Table B.9: $Q(.95, r, dfE)$.

dfE	12	13	14	15	16	17	18	19	20
1	52.0	53.2	54.3	55.4	56.3	57.2	58.0	58.8	59.6
2	14.75	15.08	15.38	15.65	15.91	16.14	16.37	16.57	16.77
3	9.95	10.15	10.35	10.53	10.69	10.84	10.98	11.11	11.24
4	8.21	8.37	8.53	8.66	8.79	8.91	9.03	9.13	9.23
5	7.32	7.47	7.60	7.72	7.83	7.93	8.03	8.12	8.21
6	6.79	6.92	7.03	7.14	7.24	7.34	7.43	7.51	7.59
7	6.43	6.55	6.66	6.76	6.85	6.94	7.02	7.10	7.17
8	6.18	6.29	6.39	6.48	6.57	6.65	6.73	6.80	6.87
9	5.98	6.09	6.19	6.28	6.36	6.44	6.51	6.58	6.64
10	5.83	5.94	6.03	6.11	6.19	6.27	6.34	6.41	6.47
11	5.71	5.81	5.90	5.98	6.06	6.13	6.20	6.27	6.33
12	5.62	5.71	5.80	5.88	5.95	6.02	6.09	6.15	6.21
13	5.53	5.63	5.71	5.79	5.86	5.93	6.00	6.06	6.11
14	5.46	5.55	5.64	5.71	5.79	5.85	5.92	5.97	6.03
15	5.40	5.49	5.57	5.65	5.72	5.79	5.85	5.90	5.96
16	5.35	5.44	5.52	5.59	5.66	5.73	5.79	5.84	5.90
17	5.31	5.39	5.47	5.54	5.61	5.68	5.73	5.79	5.84
18	5.27	5.35	5.43	5.50	5.57	5.63	5.69	5.74	5.79
19	5.23	5.32	5.39	5.46	5.53	5.59	5.65	5.70	5.75
20	5.20	5.28	5.36	5.43	5.49	5.55	5.61	5.66	5.71
24	5.10	5.18	5.25	5.32	5.38	5.44	5.49	5.55	5.59
30	5.00	5.08	5.15	5.21	5.27	5.33	5.38	5.43	5.48
40	4.90	4.98	5.04	5.11	5.16	5.22	5.27	5.31	5.36
60	4.81	4.88	4.94	5.00	5.06	5.11	5.15	5.20	5.24
120	4.71	4.78	4.84	4.90	4.95	5.00	5.04	5.09	5.13
∞	4.62	4.69	4.74	4.80	4.85	4.89	4.93	4.97	5.01

Table B.10: $Q(.99, r, dfE)$.

dfE	2	3	4	5	6	7	8	9	10	11
1	90.0	135	164	186	202	216	227	237	246	253
2	14.0	19.0	22.3	24.7	26.6	28.2	29.5	30.7	31.7	32.6
3	8.26	10.6	12.2	13.3	14.2	15.0	15.6	16.2	16.7	17.1
4	6.51	8.12	9.17	9.96	10.6	11.1	11.6	11.9	12.3	12.6
5	5.70	6.98	7.80	8.42	8.91	9.32	9.67	9.97	10.24	10.48
6	5.24	6.33	7.03	7.56	7.97	8.32	8.61	8.87	9.10	9.30
7	4.95	5.92	6.54	7.01	7.37	7.68	7.94	8.17	8.37	8.55
8	4.75	5.64	6.20	6.63	6.96	7.24	7.47	7.68	7.86	8.03
9	4.60	5.43	5.96	6.35	6.66	6.92	7.13	7.33	7.50	7.65
10	4.48	5.27	5.77	6.14	6.43	6.67	6.88	7.06	7.21	7.36
11	4.39	5.15	5.62	5.97	6.25	6.48	6.67	6.84	6.99	7.13
12	4.32	5.05	5.50	5.84	6.10	6.32	6.51	6.67	6.81	6.94
13	4.26	4.96	5.40	5.73	5.98	6.19	6.37	6.53	6.67	6.79
14	4.21	4.90	5.32	5.63	5.88	6.09	6.26	6.41	6.54	6.66
15	4.17	4.84	5.25	5.56	5.80	5.99	6.16	6.31	6.44	6.56
16	4.13	4.79	5.19	5.49	5.72	5.92	6.08	6.22	6.35	6.46
17	4.10	4.74	5.14	5.43	5.66	5.85	6.01	6.15	6.27	6.38
18	4.07	4.70	5.09	5.38	5.60	5.79	5.94	6.08	6.20	6.31
19	4.05	4.67	5.05	5.33	5.55	5.74	5.89	6.02	6.14	6.25
20	4.02	4.64	5.02	5.29	5.51	5.69	5.84	5.97	6.09	6.19
24	3.96	4.55	4.91	5.17	5.37	5.54	5.69	5.81	5.92	6.02
30	3.89	4.46	4.80	5.05	5.24	5.40	5.54	5.65	5.76	5.85
40	3.83	4.37	4.70	4.93	5.11	5.27	5.39	5.50	5.60	5.69
60	3.76	4.28	4.60	4.82	4.99	5.13	5.25	5.36	5.45	5.53
120	3.70	4.20	4.50	4.71	4.87	5.01	5.12	5.21	5.30	5.38
∞	3.64	4.12	4.40	4.60	4.76	4.88	4.99	5.08	5.16	5.23

Table B.11: $Q(.99, r, dfE)$.

dfE	12	13	14	15	16	17	18	19	20
1	260	266	272	277	282	286	290	294	298
2	33.4	34.1	34.8	35.4	36.0	36.5	37.0	37.5	38.0
3	17.5	17.9	18.2	18.5	18.8	19.1	19.3	19.6	19.8
4	12.8	13.1	13.3	13.5	13.7	13.9	14.1	14.2	14.4
5	10.70	10.89	11.08	11.24	11.40	11.55	11.68	11.81	11.9
6	9.49	9.65	9.81	9.95	10.08	10.21	10.32	10.43	10.5
7	8.71	8.86	9.00	9.12	9.24	9.35	9.46	9.55	9.65
8	8.18	8.31	8.44	8.55	8.66	8.76	8.85	8.94	9.03
9	7.78	7.91	8.03	8.13	8.23	8.33	8.41	8.50	8.57
10	7.49	7.60	7.71	7.81	7.91	7.99	8.08	8.15	8.23
11	7.25	7.36	7.47	7.56	7.65	7.73	7.81	7.88	7.95
12	7.06	7.17	7.27	7.36	7.44	7.52	7.59	7.67	7.73
13	6.90	7.01	7.10	7.19	7.27	7.35	7.42	7.49	7.55
14	6.77	6.87	6.96	7.05	7.13	7.20	7.27	7.33	7.40
15	6.66	6.76	6.85	6.93	7.00	7.07	7.14	7.20	7.26
16	6.56	6.66	6.74	6.82	6.90	6.97	7.03	7.09	7.15
17	6.48	6.57	6.66	6.73	6.81	6.87	6.94	7.00	7.05
18	6.41	6.50	6.58	6.66	6.73	6.79	6.85	6.91	6.97
19	6.34	6.43	6.51	6.59	6.65	6.72	6.78	6.84	6.89
20	6.29	6.37	6.45	6.52	6.59	6.65	6.71	6.77	6.82
24	6.11	6.19	6.26	6.33	6.39	6.45	6.51	6.56	6.61
30	5.93	6.01	6.08	6.14	6.20	6.26	6.31	6.36	6.41
40	5.76	5.84	5.90	5.96	6.02	6.07	6.12	6.17	6.21
60	5.60	6.67	5.73	5.79	5.84	5.89	5.93	5.97	6.02
120	5.44	5.51	5.56	5.61	5.66	5.71	5.75	5.79	5.83
∞	5.29	5.35	5.40	5.45	5.49	5.54	5.57	5.61	5.65

B.5   The Greek alphabet

Table B.12: *The Greek alphabet.*

[b] Capital	Small	Name	Capital	Small	Name
A	α	alpha	N	ν	nu
B	β	beta	Ξ	ξ	xi
Γ	γ	gamma	O	o	omicron
Δ	δ, ∂	delta	Π	π	pi
E	ε, ϵ	epsilon	P	ρ	rho
Z	ζ	zeta	Σ	σ	sigma
H	η	eta	T	τ	tau
Θ	θ	theta	Υ	υ	upsilon
I	ι	iota	Φ	ϕ	phi
K	κ	kappa	X	χ	chi
Λ	λ	lambda	Ψ	ψ	psi
M	μ	mu	Ω	ω	omega

B.6   Tables of the F distribution

Table B.13: *90th percentiles of the F distribution.*

Den.	Numerator degrees of freedom							
df	1	2	3	4	5	6	7	8
1	39.862	49.500	53.593	55.833	57.240	58.204	58.906	59.439
2	8.5263	9.0000	9.1618	9.2434	9.2926	9.3255	9.3491	9.3668
3	5.5383	5.4625	5.3908	5.3427	5.3092	5.2847	5.2662	5.2517
4	4.5449	4.3246	4.1909	4.1072	4.0506	4.0098	3.9790	3.9549
5	4.0604	3.7797	3.6195	3.5202	3.4530	3.4045	3.3679	3.3393
6	3.7760	3.4633	3.2888	3.1808	3.1075	3.0546	3.0145	2.9830
7	3.5895	3.2574	3.0741	2.9605	2.8833	2.8274	2.7849	2.7516
8	3.4579	3.1131	2.9238	2.8065	2.7265	2.6683	2.6241	2.5894
9	3.3603	3.0065	2.8129	2.6927	2.6106	2.5509	2.5053	2.4694
10	3.2850	2.9245	2.7277	2.6054	2.5216	2.4606	2.4140	2.3772
11	3.2252	2.8595	2.6602	2.5362	2.4512	2.3891	2.3416	2.3040
12	3.1765	2.8068	2.6055	2.4801	2.3941	2.3310	2.2828	2.2446
13	3.1362	2.7632	2.5603	2.4337	2.3467	2.2830	2.2341	2.1954
14	3.1022	2.7265	2.5222	2.3947	2.3069	2.2426	2.1931	2.1539
15	3.0732	2.6952	2.4898	2.3614	2.2730	2.2081	2.1582	2.1185
16	3.0481	2.6682	2.4618	2.3328	2.2438	2.1783	2.1280	2.0880
17	3.0263	2.6446	2.4374	2.3078	2.2183	2.1524	2.1017	2.0613
18	3.0070	2.6240	2.4160	2.2858	2.1958	2.1296	2.0785	2.0379
19	2.9899	2.6056	2.3970	2.2663	2.1760	2.1094	2.0580	2.0171
20	2.9747	2.5893	2.3801	2.2489	2.1583	2.0913	2.0397	1.9985
21	2.9610	2.5746	2.3649	2.2334	2.1423	2.0751	2.0233	1.9819
22	2.9486	2.5613	2.3512	2.2193	2.1279	2.0605	2.0084	1.9668
23	2.9374	2.5493	2.3387	2.2065	2.1149	2.0472	1.9949	1.9531
24	2.9271	2.5384	2.3274	2.1949	2.1030	2.0351	1.9826	1.9407
25	2.9177	2.5283	2.3170	2.1842	2.0922	2.0241	1.9714	1.9293
26	2.9091	2.5191	2.3075	2.1745	2.0822	2.0139	1.9610	1.9188
28	2.8939	2.5028	2.2906	2.1571	2.0645	1.9959	1.9427	1.9002
30	2.8807	2.4887	2.2761	2.1422	2.0492	1.9803	1.9269	1.8841
32	2.8693	2.4765	2.2635	2.1293	2.0360	1.9669	1.9132	1.8702
34	2.8592	2.4658	2.2524	2.1179	2.0244	1.9550	1.9012	1.8580
36	2.8504	2.4563	2.2426	2.1079	2.0141	1.9446	1.8905	1.8471
38	2.8424	2.4479	2.2339	2.0990	2.0050	1.9352	1.8810	1.8375
40	2.8354	2.4404	2.2261	2.0909	1.9968	1.9269	1.8725	1.8289
60	2.7911	2.3932	2.1774	2.0410	1.9457	1.8747	1.8194	1.7748
80	2.7693	2.3702	2.1536	2.0165	1.9206	1.8491	1.7933	1.7483
100	2.7564	2.3564	2.1394	2.0019	1.9057	1.8339	1.7778	1.7324
150	2.7393	2.3383	2.1207	1.9827	1.8861	1.8138	1.7572	1.7115
200	2.7308	2.3293	2.1114	1.9732	1.8763	1.8038	1.7470	1.7011
300	2.7224	2.3203	2.1021	1.9637	1.8666	1.7939	1.7369	1.6908
400	2.7182	2.3159	2.0975	1.9590	1.8617	1.7889	1.7319	1.6856
∞	2.7055	2.3026	2.0838	1.9449	1.8473	1.7741	1.7167	1.6702

Table B.14: *90th percentiles of the F distribution.*

Den.	Numerator degrees of freedom							
df	9	10	11	12	13	14	15	16
1	59.858	60.195	60.473	60.705	60.903	61.072	61.220	61.350
2	9.3805	9.3915	9.4005	9.4080	9.4144	9.4198	9.4245	9.4286
3	5.2401	5.2305	5.2226	5.2158	5.2098	5.2047	5.2003	5.1964
4	3.9357	3.9199	3.9067	3.8956	3.8859	3.8776	3.8704	3.8639
5	3.3163	3.2974	3.2816	3.2682	3.2568	3.2468	3.2380	3.2303
6	2.9577	2.9369	2.9195	2.9047	2.8920	2.8809	2.8712	2.8626
7	2.7247	2.7025	2.6839	2.6681	2.6545	2.6426	2.6322	2.6230
8	2.5613	2.5381	2.5186	2.5020	2.4876	2.4752	2.4642	2.4545
9	2.4404	2.4164	2.3961	2.3789	2.3640	2.3511	2.3396	2.3295
10	2.3473	2.3226	2.3018	2.2841	2.2687	2.2553	2.2435	2.2331
11	2.2735	2.2482	2.2269	2.2087	2.1930	2.1792	2.1671	2.1563
12	2.2135	2.1878	2.1660	2.1474	2.1313	2.1173	2.1049	2.0938
13	2.1638	2.1376	2.1155	2.0966	2.0802	2.0659	2.0532	2.0419
14	2.1220	2.0954	2.0730	2.0537	2.0370	2.0224	2.0095	1.9981
15	2.0862	2.0593	2.0366	2.0171	2.0001	1.9853	1.9722	1.9605
16	2.0553	2.0282	2.0051	1.9854	1.9682	1.9532	1.9399	1.9281
17	2.0284	2.0010	1.9777	1.9577	1.9404	1.9252	1.9117	1.8997
18	2.0047	1.9770	1.9535	1.9334	1.9158	1.9004	1.8868	1.8747
19	1.9836	1.9557	1.9321	1.9117	1.8940	1.8785	1.8647	1.8524
20	1.9649	1.9367	1.9129	1.8924	1.8745	1.8588	1.8450	1.8325
21	1.9480	1.9197	1.8957	1.8750	1.8570	1.8412	1.8271	1.8147
22	1.9328	1.9043	1.8801	1.8593	1.8411	1.8252	1.8111	1.7984
23	1.9189	1.8903	1.8659	1.8450	1.8267	1.8107	1.7964	1.7837
24	1.9063	1.8775	1.8530	1.8319	1.8136	1.7974	1.7831	1.7703
25	1.8947	1.8658	1.8412	1.8200	1.8015	1.7853	1.7708	1.7579
26	1.8841	1.8550	1.8303	1.8090	1.7904	1.7741	1.7596	1.7466
28	1.8652	1.8359	1.8110	1.7895	1.7708	1.7542	1.7395	1.7264
30	1.8490	1.8195	1.7944	1.7727	1.7538	1.7371	1.7223	1.7090
32	1.8348	1.8052	1.7799	1.7581	1.7390	1.7222	1.7072	1.6938
34	1.8224	1.7926	1.7672	1.7452	1.7260	1.7091	1.6940	1.6805
36	1.8115	1.7815	1.7559	1.7338	1.7145	1.6974	1.6823	1.6687
38	1.8017	1.7716	1.7459	1.7237	1.7042	1.6871	1.6718	1.6581
40	1.7929	1.7627	1.7369	1.7146	1.6950	1.6778	1.6624	1.6486
60	1.7380	1.7070	1.6805	1.6574	1.6372	1.6193	1.6034	1.5890
80	1.7110	1.6796	1.6526	1.6292	1.6087	1.5904	1.5741	1.5594
100	1.6949	1.6632	1.6360	1.6124	1.5916	1.5731	1.5566	1.5417
150	1.6736	1.6416	1.6140	1.5901	1.5690	1.5502	1.5334	1.5182
200	1.6630	1.6308	1.6031	1.5789	1.5577	1.5388	1.5218	1.5065
300	1.6525	1.6201	1.5922	1.5679	1.5464	1.5273	1.5102	1.4948
400	1.6472	1.6147	1.5868	1.5623	1.5408	1.5217	1.5045	1.4889
∞	1.6315	1.5987	1.5705	1.5458	1.5240	1.5046	1.4871	1.4714

Table B.15: *90th percentiles of the F distribution.*

Den.	Numerator degrees of freedom							
df	18	20	25	30	40	50	100	200
1	61.566	61.740	62.054	62.265	62.528	62.688	63.006	63.163
2	9.4354	9.4408	9.4513	9.4579	9.4662	9.4712	9.4812	9.4861
3	5.1900	5.1846	5.1747	5.1681	5.1597	5.1546	5.1442	5.1389
4	3.8531	3.8444	3.8283	3.8175	3.8037	3.7952	3.7781	3.7695
5	3.2172	3.2067	3.1873	3.1741	3.1573	3.1471	3.1263	3.1157
6	2.8481	2.8363	2.8147	2.8000	2.7812	2.7697	2.7463	2.7343
7	2.6074	2.5947	2.5714	2.5555	2.5351	2.5226	2.4971	2.4841
8	2.4381	2.4247	2.3999	2.3830	2.3614	2.3481	2.3208	2.3068
9	2.3123	2.2983	2.2725	2.2547	2.2320	2.2180	2.1892	2.1743
10	2.2153	2.2008	2.1739	2.1554	2.1317	2.1171	2.0869	2.0713
11	2.1380	2.1231	2.0953	2.0762	2.0516	2.0364	2.0050	1.9888
12	2.0750	2.0597	2.0312	2.0115	1.9861	1.9704	1.9379	1.9210
13	2.0227	2.0070	1.9778	1.9576	1.9315	1.9153	1.8817	1.8642
14	1.9785	1.9625	1.9326	1.9119	1.8852	1.8686	1.8340	1.8159
15	1.9407	1.9243	1.8939	1.8728	1.8454	1.8284	1.7928	1.7743
16	1.9079	1.8913	1.8603	1.8388	1.8108	1.7935	1.7570	1.7380
17	1.8792	1.8624	1.8309	1.8090	1.7805	1.7628	1.7255	1.7059
18	1.8539	1.8368	1.8049	1.7827	1.7537	1.7356	1.6976	1.6775
19	1.8314	1.8142	1.7818	1.7592	1.7298	1.7114	1.6726	1.6521
20	1.8113	1.7938	1.7611	1.7382	1.7083	1.6896	1.6501	1.6292
21	1.7932	1.7756	1.7424	1.7193	1.6890	1.6700	1.6298	1.6085
22	1.7768	1.7590	1.7255	1.7021	1.6714	1.6521	1.6113	1.5896
23	1.7619	1.7439	1.7101	1.6864	1.6554	1.6358	1.5944	1.5723
24	1.7483	1.7302	1.6960	1.6721	1.6407	1.6209	1.5788	1.5564
25	1.7358	1.7175	1.6831	1.6589	1.6272	1.6072	1.5645	1.5417
26	1.7243	1.7059	1.6712	1.6468	1.6147	1.5945	1.5513	1.5281
28	1.7039	1.6852	1.6500	1.6252	1.5925	1.5718	1.5276	1.5037
30	1.6862	1.6673	1.6316	1.6065	1.5732	1.5522	1.5069	1.4824
32	1.6708	1.6517	1.6156	1.5901	1.5564	1.5349	1.4888	1.4637
34	1.6573	1.6380	1.6015	1.5757	1.5415	1.5197	1.4727	1.4470
36	1.6453	1.6258	1.5890	1.5629	1.5282	1.5061	1.4583	1.4321
38	1.6345	1.6149	1.5778	1.5514	1.5163	1.4939	1.4453	1.4186
40	1.6249	1.6052	1.5677	1.5411	1.5056	1.4830	1.4336	1.4064
60	1.5642	1.5435	1.5039	1.4755	1.4373	1.4126	1.3576	1.3264
80	1.5340	1.5128	1.4720	1.4426	1.4027	1.3767	1.3180	1.2839
100	1.5160	1.4944	1.4527	1.4227	1.3817	1.3548	1.2934	1.2571
150	1.4919	1.4698	1.4271	1.3960	1.3534	1.3251	1.2595	1.2193
200	1.4799	1.4575	1.4142	1.3826	1.3390	1.3100	1.2418	1.1991
300	1.4679	1.4452	1.4013	1.3691	1.3246	1.2947	1.2236	1.1779
400	1.4619	1.4391	1.3948	1.3623	1.3173	1.2870	1.2143	1.1667
∞	1.4439	1.4206	1.3753	1.3419	1.2951	1.2633	1.1850	1.1301

Table B.16: *95th percentiles of the F distribution.*

Den.	Numerator degrees of freedom							
df	1	2	3	4	5	6	7	8
1	161.45	199.50	215.71	224.58	230.16	233.99	236.77	238.88
2	18.513	19.000	19.164	19.247	19.296	19.329	19.353	19.371
3	10.128	9.552	9.277	9.117	9.013	8.941	8.887	8.845
4	7.709	6.944	6.591	6.388	6.256	6.163	6.094	6.041
5	6.608	5.786	5.409	5.192	5.050	4.950	4.876	4.818
6	5.987	5.143	4.757	4.534	4.387	4.284	4.207	4.147
7	5.591	4.737	4.347	4.120	3.972	3.866	3.787	3.726
8	5.318	4.459	4.066	3.838	3.687	3.581	3.500	3.438
9	5.117	4.256	3.863	3.633	3.482	3.374	3.293	3.230
10	4.965	4.103	3.708	3.478	3.326	3.217	3.135	3.072
11	4.844	3.982	3.587	3.357	3.204	3.095	3.012	2.948
12	4.747	3.885	3.490	3.259	3.106	2.996	2.913	2.849
13	4.667	3.806	3.411	3.179	3.025	2.915	2.832	2.767
14	4.600	3.739	3.344	3.112	2.958	2.848	2.764	2.699
15	4.543	3.682	3.287	3.056	2.901	2.790	2.707	2.641
16	4.494	3.634	3.239	3.007	2.852	2.741	2.657	2.591
17	4.451	3.592	3.197	2.965	2.810	2.699	2.614	2.548
18	4.414	3.555	3.160	2.928	2.773	2.661	2.577	2.510
19	4.381	3.522	3.127	2.895	2.740	2.628	2.544	2.477
20	4.351	3.493	3.098	2.866	2.711	2.599	2.514	2.447
21	4.325	3.467	3.072	2.840	2.685	2.573	2.488	2.420
22	4.301	3.443	3.049	2.817	2.661	2.549	2.464	2.397
23	4.279	3.422	3.028	2.796	2.640	2.528	2.442	2.375
24	4.260	3.403	3.009	2.776	2.621	2.508	2.423	2.355
25	4.242	3.385	2.991	2.759	2.603	2.490	2.405	2.337
26	4.225	3.369	2.975	2.743	2.587	2.474	2.388	2.321
28	4.196	3.340	2.947	2.714	2.558	2.445	2.359	2.291
30	4.171	3.316	2.922	2.690	2.534	2.421	2.334	2.266
32	4.149	3.295	2.901	2.668	2.512	2.399	2.313	2.244
34	4.130	3.276	2.883	2.650	2.494	2.380	2.294	2.225
36	4.113	3.259	2.866	2.634	2.477	2.364	2.277	2.209
38	4.098	3.245	2.852	2.619	2.463	2.349	2.262	2.194
40	4.085	3.232	2.839	2.606	2.449	2.336	2.249	2.180
60	4.001	3.150	2.758	2.525	2.368	2.254	2.167	2.097
80	3.960	3.111	2.719	2.486	2.329	2.214	2.126	2.056
100	3.936	3.087	2.696	2.463	2.305	2.191	2.103	2.032
150	3.904	3.056	2.665	2.432	2.274	2.160	2.071	2.001
200	3.888	3.041	2.650	2.417	2.259	2.144	2.056	1.985
300	3.873	3.026	2.635	2.402	2.244	2.129	2.040	1.969
400	3.865	3.018	2.627	2.394	2.237	2.121	2.032	1.962
∞	3.841	2.996	2.605	2.372	2.214	2.099	2.010	1.938

Table B.17: *95th percentiles of the F distribution.*

Den.	Numerator degrees of freedom							
df	9	10	11	12	13	14	15	16
1	240.54	241.88	242.98	243.91	244.69	245.36	245.95	246.46
2	19.385	19.396	19.405	19.412	19.419	19.424	19.429	19.433
3	8.812	8.786	8.763	8.745	8.729	8.715	8.703	8.692
4	5.999	5.964	5.936	5.912	5.891	5.873	5.858	5.844
5	4.772	4.735	4.704	4.678	4.655	4.636	4.619	4.604
6	4.099	4.060	4.027	4.000	3.976	3.956	3.938	3.922
7	3.677	3.637	3.603	3.575	3.550	3.529	3.511	3.494
8	3.388	3.347	3.313	3.284	3.259	3.237	3.218	3.202
9	3.179	3.137	3.102	3.073	3.048	3.025	3.006	2.989
10	3.020	2.978	2.943	2.913	2.887	2.865	2.845	2.828
11	2.896	2.854	2.818	2.788	2.761	2.739	2.719	2.701
12	2.796	2.753	2.717	2.687	2.660	2.637	2.617	2.599
13	2.714	2.671	2.635	2.604	2.577	2.554	2.533	2.515
14	2.646	2.602	2.566	2.534	2.507	2.484	2.463	2.445
15	2.588	2.544	2.507	2.475	2.448	2.424	2.403	2.385
16	2.538	2.494	2.456	2.425	2.397	2.373	2.352	2.333
17	2.494	2.450	2.413	2.381	2.353	2.329	2.308	2.289
18	2.456	2.412	2.374	2.342	2.314	2.290	2.269	2.250
19	2.423	2.378	2.340	2.308	2.280	2.256	2.234	2.215
20	2.393	2.348	2.310	2.278	2.250	2.225	2.203	2.184
21	2.366	2.321	2.283	2.250	2.222	2.197	2.176	2.156
22	2.342	2.297	2.259	2.226	2.198	2.173	2.151	2.131
23	2.320	2.275	2.236	2.204	2.175	2.150	2.128	2.109
24	2.300	2.255	2.216	2.183	2.155	2.130	2.108	2.088
25	2.282	2.236	2.198	2.165	2.136	2.111	2.089	2.069
26	2.265	2.220	2.181	2.148	2.119	2.094	2.072	2.052
28	2.236	2.190	2.151	2.118	2.089	2.064	2.041	2.021
30	2.211	2.165	2.126	2.092	2.063	2.037	2.015	1.995
32	2.189	2.142	2.103	2.070	2.040	2.015	1.992	1.972
34	2.170	2.123	2.084	2.050	2.021	1.995	1.972	1.952
36	2.153	2.106	2.067	2.033	2.003	1.977	1.954	1.934
38	2.138	2.091	2.051	2.017	1.988	1.962	1.939	1.918
40	2.124	2.077	2.038	2.003	1.974	1.948	1.924	1.904
60	2.040	1.993	1.952	1.917	1.887	1.860	1.836	1.815
80	1.999	1.951	1.910	1.875	1.845	1.817	1.793	1.772
100	1.975	1.927	1.886	1.850	1.819	1.792	1.768	1.746
150	1.943	1.894	1.853	1.817	1.786	1.758	1.734	1.711
200	1.927	1.878	1.837	1.801	1.769	1.742	1.717	1.694
300	1.911	1.862	1.821	1.785	1.753	1.725	1.700	1.677
400	1.903	1.854	1.813	1.776	1.745	1.717	1.691	1.669
∞	1.880	1.831	1.789	1.752	1.720	1.692	1.666	1.644

Table B.18: *95th percentiles of the F distribution.*

Den.	Numerator degrees of freedom							
df	18	20	25	30	40	50	100	200
1	247.32	248.01	249.26	250.09	251.14	251.77	253.04	253.68
2	19.440	19.446	19.456	19.462	19.470	19.475	19.486	19.491
3	8.675	8.660	8.634	8.617	8.594	8.581	8.554	8.540
4	5.821	5.803	5.769	5.746	5.717	5.699	5.664	5.646
5	4.579	4.558	4.521	4.496	4.464	4.444	4.405	4.385
6	3.896	3.874	3.835	3.808	3.774	3.754	3.712	3.690
7	3.467	3.445	3.404	3.376	3.340	3.319	3.275	3.252
8	3.173	3.150	3.108	3.079	3.043	3.020	2.975	2.951
9	2.960	2.936	2.893	2.864	2.826	2.803	2.756	2.731
10	2.798	2.774	2.730	2.700	2.661	2.637	2.588	2.563
11	2.671	2.646	2.601	2.570	2.531	2.507	2.457	2.431
12	2.568	2.544	2.498	2.466	2.426	2.401	2.350	2.323
13	2.484	2.459	2.412	2.380	2.339	2.314	2.261	2.234
14	2.413	2.388	2.341	2.308	2.266	2.241	2.187	2.159
15	2.353	2.328	2.280	2.247	2.204	2.178	2.123	2.095
16	2.302	2.276	2.227	2.194	2.151	2.124	2.068	2.039
17	2.257	2.230	2.181	2.148	2.104	2.077	2.020	1.991
18	2.217	2.191	2.141	2.107	2.063	2.035	1.978	1.948
19	2.182	2.156	2.106	2.071	2.026	1.999	1.940	1.910
20	2.151	2.124	2.074	2.039	1.994	1.966	1.907	1.875
21	2.123	2.096	2.045	2.010	1.965	1.936	1.876	1.845
22	2.098	2.071	2.020	1.984	1.938	1.909	1.849	1.817
23	2.075	2.048	1.996	1.961	1.914	1.885	1.823	1.791
24	2.054	2.027	1.975	1.939	1.892	1.863	1.800	1.768
25	2.035	2.007	1.955	1.919	1.872	1.842	1.779	1.746
26	2.018	1.990	1.938	1.901	1.853	1.823	1.760	1.726
28	1.987	1.959	1.906	1.869	1.820	1.790	1.725	1.691
30	1.960	1.932	1.878	1.841	1.792	1.761	1.695	1.660
32	1.937	1.908	1.854	1.817	1.767	1.736	1.669	1.633
34	1.917	1.888	1.833	1.795	1.745	1.713	1.645	1.609
36	1.899	1.870	1.815	1.776	1.726	1.694	1.625	1.587
38	1.883	1.853	1.798	1.760	1.708	1.676	1.606	1.568
40	1.868	1.839	1.783	1.744	1.693	1.660	1.589	1.551
60	1.778	1.748	1.690	1.649	1.594	1.559	1.481	1.438
80	1.734	1.703	1.644	1.602	1.545	1.508	1.426	1.379
100	1.708	1.676	1.616	1.573	1.515	1.477	1.392	1.342
150	1.673	1.641	1.580	1.535	1.475	1.436	1.345	1.290
200	1.656	1.623	1.561	1.516	1.455	1.415	1.321	1.263
300	1.638	1.606	1.543	1.497	1.435	1.393	1.296	1.234
400	1.630	1.597	1.534	1.488	1.425	1.383	1.283	1.219
∞	1.604	1.571	1.506	1.459	1.394	1.350	1.243	1.170

Table B.19: *99th percentiles of the F distribution.*

Den.	Numerator degrees of freedom							
df	1	2	3	4	5	6	7	8
1	4052	5000	5403	5625	5764	5859	5928	5981
2	98.50	99.00	99.17	99.25	99.30	99.33	99.36	99.37
3	34.12	30.82	29.46	28.71	28.24	27.91	27.67	27.49
4	21.20	18.00	16.69	15.98	15.52	15.21	14.98	14.80
5	16.26	13.27	12.06	11.39	10.97	10.67	10.46	10.29
6	13.75	10.92	9.78	9.15	8.75	8.47	8.26	8.10
7	12.25	9.55	8.45	7.85	7.46	7.19	6.99	6.84
8	11.26	8.65	7.59	7.01	6.63	6.37	6.18	6.03
9	10.56	8.02	6.99	6.42	6.06	5.80	5.61	5.47
10	10.04	7.56	6.55	5.99	5.64	5.39	5.20	5.06
11	9.65	7.21	6.22	5.67	5.32	5.07	4.89	4.74
12	9.33	6.93	5.95	5.41	5.06	4.82	4.64	4.50
13	9.07	6.70	5.74	5.21	4.86	4.62	4.44	4.30
14	8.86	6.51	5.56	5.04	4.69	4.46	4.28	4.14
15	8.68	6.36	5.42	4.89	4.56	4.32	4.14	4.00
16	8.53	6.23	5.29	4.77	4.44	4.20	4.03	3.89
17	8.40	6.11	5.19	4.67	4.34	4.10	3.93	3.79
18	8.29	6.01	5.09	4.58	4.25	4.01	3.84	3.71
19	8.19	5.93	5.01	4.50	4.17	3.94	3.77	3.63
20	8.10	5.85	4.94	4.43	4.10	3.87	3.70	3.56
21	8.02	5.78	4.87	4.37	4.04	3.81	3.64	3.51
22	7.95	5.72	4.82	4.31	3.99	3.76	3.59	3.45
23	7.88	5.66	4.76	4.26	3.94	3.71	3.54	3.41
24	7.82	5.61	4.72	4.22	3.90	3.67	3.50	3.36
25	7.77	5.57	4.68	4.18	3.85	3.63	3.46	3.32
26	7.72	5.53	4.64	4.14	3.82	3.59	3.42	3.29
28	7.64	5.45	4.57	4.07	3.75	3.53	3.36	3.23
30	7.56	5.39	4.51	4.02	3.70	3.47	3.30	3.17
32	7.50	5.34	4.46	3.97	3.65	3.43	3.26	3.13
34	7.44	5.29	4.42	3.93	3.61	3.39	3.22	3.09
36	7.40	5.25	4.38	3.89	3.57	3.35	3.18	3.05
38	7.35	5.21	4.34	3.86	3.54	3.32	3.15	3.02
40	7.31	5.18	4.31	3.83	3.51	3.29	3.12	2.99
60	7.08	4.98	4.13	3.65	3.34	3.12	2.95	2.82
80	6.96	4.88	4.04	3.56	3.26	3.04	2.87	2.74
100	6.90	4.82	3.98	3.51	3.21	2.99	2.82	2.69
150	6.81	4.75	3.91	3.45	3.14	2.92	2.76	2.63
200	6.76	4.71	3.88	3.41	3.11	2.89	2.73	2.60
300	6.72	4.68	3.85	3.38	3.08	2.86	2.70	2.57
400	6.70	4.66	3.83	3.37	3.06	2.85	2.68	2.56
∞	6.63	4.61	3.78	3.32	3.02	2.80	2.64	2.51

Table B.20: *99th percentiles of the F distribution.*

Den.	Numerator degrees of freedom							
df	9	10	11	12	13	14	15	16
1	6022	6056	6083	6106	6126	6143	6157	6170
2	99.39	99.40	99.41	99.42	99.42	99.43	99.43	99.44
3	27.35	27.23	27.13	27.05	26.98	26.92	26.87	26.83
4	14.66	14.55	14.45	14.37	14.31	14.25	14.20	14.15
5	10.16	10.05	9.96	9.89	9.82	9.77	9.72	9.68
6	7.98	7.87	7.79	7.72	7.66	7.60	7.56	7.52
7	6.72	6.62	6.54	6.47	6.41	6.36	6.31	6.28
8	5.91	5.81	5.73	5.67	5.61	5.56	5.52	5.48
9	5.35	5.26	5.18	5.11	5.05	5.01	4.96	4.92
10	4.94	4.85	4.77	4.71	4.65	4.60	4.56	4.52
11	4.63	4.54	4.46	4.40	4.34	4.29	4.25	4.21
12	4.39	4.30	4.22	4.16	4.10	4.05	4.01	3.97
13	4.19	4.10	4.02	3.96	3.91	3.86	3.82	3.78
14	4.03	3.94	3.86	3.80	3.75	3.70	3.66	3.62
15	3.89	3.80	3.73	3.67	3.61	3.56	3.52	3.49
16	3.78	3.69	3.62	3.55	3.50	3.45	3.41	3.37
17	3.68	3.59	3.52	3.46	3.40	3.35	3.31	3.27
18	3.60	3.51	3.43	3.37	3.32	3.27	3.23	3.19
19	3.52	3.43	3.36	3.30	3.24	3.19	3.15	3.12
20	3.46	3.37	3.29	3.23	3.18	3.13	3.09	3.05
21	3.40	3.31	3.24	3.17	3.12	3.07	3.03	2.99
22	3.35	3.26	3.18	3.12	3.07	3.02	2.98	2.94
23	3.30	3.21	3.14	3.07	3.02	2.97	2.93	2.89
24	3.26	3.17	3.09	3.03	2.98	2.93	2.89	2.85
25	3.22	3.13	3.06	2.99	2.94	2.89	2.85	2.81
26	3.18	3.09	3.02	2.96	2.90	2.86	2.81	2.78
28	3.12	3.03	2.96	2.90	2.84	2.79	2.75	2.72
30	3.07	2.98	2.91	2.84	2.79	2.74	2.70	2.66
32	3.02	2.93	2.86	2.80	2.74	2.70	2.65	2.62
34	2.98	2.89	2.82	2.76	2.70	2.66	2.61	2.58
36	2.95	2.86	2.79	2.72	2.67	2.62	2.58	2.54
38	2.92	2.83	2.75	2.69	2.64	2.59	2.55	2.51
40	2.89	2.80	2.73	2.66	2.61	2.56	2.52	2.48
60	2.72	2.63	2.56	2.50	2.44	2.39	2.35	2.31
80	2.64	2.55	2.48	2.42	2.36	2.31	2.27	2.23
100	2.59	2.50	2.43	2.37	2.31	2.27	2.22	2.19
150	2.53	2.44	2.37	2.31	2.25	2.20	2.16	2.12
200	2.50	2.41	2.34	2.27	2.22	2.17	2.13	2.09
300	2.47	2.38	2.31	2.24	2.19	2.14	2.10	2.06
400	2.45	2.37	2.29	2.23	2.17	2.13	2.08	2.05
∞	2.41	2.32	2.25	2.18	2.13	2.08	2.04	2.00

Table B.21: *99th percentiles of the F distribution.*

Den.	Numerator degrees of freedom							
df	18	20	25	30	40	50	100	200
1	6191	6209	6240	6261	6287	6302	6334	6350
2	99.44	99.45	99.46	99.46	99.47	99.48	99.49	99.49
3	26.75	26.69	26.58	26.50	26.41	26.35	26.24	26.18
4	14.08	14.02	13.91	13.84	13.75	13.69	13.58	13.52
5	9.61	9.55	9.45	9.38	9.29	9.24	9.13	9.08
6	7.45	7.40	7.30	7.23	7.14	7.09	6.99	6.93
7	6.21	6.16	6.06	5.99	5.91	5.86	5.75	5.70
8	5.41	5.36	5.26	5.20	5.12	5.07	4.96	4.91
9	4.86	4.81	4.71	4.65	4.57	4.52	4.41	4.36
10	4.46	4.41	4.31	4.25	4.17	4.12	4.01	3.96
11	4.15	4.10	4.01	3.94	3.86	3.81	3.71	3.66
12	3.91	3.86	3.76	3.70	3.62	3.57	3.47	3.41
13	3.72	3.66	3.57	3.51	3.43	3.38	3.27	3.22
14	3.56	3.51	3.41	3.35	3.27	3.22	3.11	3.06
15	3.42	3.37	3.28	3.21	3.13	3.08	2.98	2.92
16	3.31	3.26	3.16	3.10	3.02	2.97	2.86	2.81
17	3.21	3.16	3.07	3.00	2.92	2.87	2.76	2.71
18	3.13	3.08	2.98	2.92	2.84	2.78	2.68	2.62
19	3.05	3.00	2.91	2.84	2.76	2.71	2.60	2.55
20	2.99	2.94	2.84	2.78	2.69	2.64	2.54	2.48
21	2.93	2.88	2.79	2.72	2.64	2.58	2.48	2.42
22	2.88	2.83	2.73	2.67	2.58	2.53	2.42	2.36
23	2.83	2.78	2.69	2.62	2.54	2.48	2.37	2.32
24	2.79	2.74	2.64	2.58	2.49	2.44	2.33	2.27
25	2.75	2.70	2.60	2.54	2.45	2.40	2.29	2.23
26	2.72	2.66	2.57	2.50	2.42	2.36	2.25	2.19
28	2.65	2.60	2.51	2.44	2.35	2.30	2.19	2.13
30	2.60	2.55	2.45	2.39	2.30	2.25	2.13	2.07
32	2.55	2.50	2.41	2.34	2.25	2.20	2.08	2.02
34	2.51	2.46	2.37	2.30	2.21	2.16	2.04	1.98
36	2.48	2.43	2.33	2.26	2.18	2.12	2.00	1.94
38	2.45	2.40	2.30	2.23	2.14	2.09	1.97	1.90
40	2.42	2.37	2.27	2.20	2.11	2.06	1.94	1.87
60	2.25	2.20	2.10	2.03	1.94	1.88	1.75	1.68
80	2.17	2.12	2.01	1.94	1.85	1.79	1.65	1.58
100	2.12	2.07	1.97	1.89	1.80	1.74	1.60	1.52
150	2.06	2.00	1.90	1.83	1.73	1.66	1.52	1.43
200	2.03	1.97	1.87	1.79	1.69	1.63	1.48	1.39
300	1.99	1.94	1.84	1.76	1.66	1.59	1.44	1.35
400	1.98	1.92	1.82	1.75	1.64	1.58	1.42	1.32
∞	1.93	1.88	1.77	1.70	1.59	1.52	1.36	1.25

Table B.22: *99.9th percentiles of the F distribution.*

Den.	Numerator degrees of freedom							
df	1	2	3	4	5	6	7	8
1	405292	500009	540389	562510	576416	585949	592885	598156
2	998.54	999.01	999.18	999.26	999.31	999.35	999.37	999.39
3	167.03	148.50	141.11	137.10	134.58	132.85	131.59	130.62
4	74.138	61.246	56.178	53.436	51.712	50.526	49.658	48.997
5	47.181	37.123	33.203	31.085	29.753	28.835	28.163	27.650
6	35.508	27.000	23.703	21.924	20.803	20.030	19.463	19.030
7	29.245	21.689	18.772	17.198	16.206	15.521	15.019	14.634
8	25.415	18.494	15.830	14.392	13.485	12.858	12.398	12.046
9	22.857	16.387	13.902	12.560	11.714	11.128	10.698	10.368
10	21.040	14.905	12.553	11.283	10.481	9.926	9.517	9.204
11	19.687	13.812	11.561	10.346	9.578	9.047	8.655	8.355
12	18.643	12.974	10.804	9.633	8.892	8.379	8.001	7.710
13	17.816	12.313	10.209	9.073	8.354	7.856	7.489	7.206
14	17.143	11.779	9.729	8.622	7.922	7.436	7.077	6.802
15	16.587	11.339	9.335	8.253	7.567	7.092	6.741	6.471
16	16.120	10.971	9.006	7.944	7.272	6.805	6.460	6.195
17	15.722	10.658	8.727	7.683	7.022	6.563	6.223	5.962
18	15.379	10.390	8.488	7.459	6.808	6.355	6.021	5.763
19	15.081	10.157	8.280	7.265	6.623	6.175	5.845	5.590
20	14.819	9.953	8.098	7.096	6.461	6.019	5.692	5.440
21	14.587	9.772	7.938	6.947	6.318	5.881	5.557	5.308
22	14.380	9.612	7.796	6.814	6.191	5.758	5.438	5.190
23	14.195	9.469	7.669	6.696	6.078	5.649	5.331	5.085
24	14.028	9.339	7.554	6.589	5.977	5.550	5.235	4.991
25	13.877	9.223	7.451	6.493	5.885	5.462	5.148	4.906
26	13.739	9.116	7.357	6.406	5.802	5.381	5.070	4.829
28	13.498	8.931	7.193	6.253	5.657	5.241	4.933	4.695
30	13.293	8.773	7.054	6.125	5.534	5.122	4.817	4.581
32	13.118	8.639	6.936	6.014	5.429	5.021	4.719	4.485
34	12.965	8.522	6.833	5.919	5.339	4.934	4.633	4.401
36	12.832	8.420	6.744	5.836	5.260	4.857	4.559	4.328
38	12.714	8.331	6.665	5.763	5.190	4.790	4.494	4.264
40	12.609	8.251	6.595	5.698	5.128	4.731	4.436	4.207
60	11.973	7.768	6.171	5.307	4.757	4.372	4.086	3.865
80	11.671	7.540	5.972	5.123	4.582	4.204	3.923	3.705
100	11.495	7.408	5.857	5.017	4.482	4.107	3.829	3.612
150	11.267	7.236	5.707	4.879	4.351	3.981	3.706	3.493
200	11.155	7.152	5.634	4.812	4.287	3.920	3.647	3.434
300	11.044	7.069	5.562	4.746	4.225	3.860	3.588	3.377
400	10.989	7.028	5.527	4.713	4.194	3.830	3.560	3.349
∞	10.828	6.908	5.422	4.617	4.103	3.743	3.475	3.266

Table B.23: *99.9th percentiles of the F distribution.*

Den.	Numerator degrees of freedom							
df	9	10	11	12	13	14	15	16
1	602296	605634	608381	610681	612636	614316	615778	617058
2	999.40	999.41	999.42	999.43	999.44	999.44	999.45	999.45
3	129.86	129.25	128.74	128.32	127.96	127.65	127.38	127.14
4	48.475	48.053	47.705	47.412	47.163	46.948	46.761	46.597
5	27.245	26.917	26.646	26.418	26.224	26.057	25.911	25.783
6	18.688	18.411	18.182	17.989	17.825	17.683	17.559	17.450
7	14.330	14.083	13.879	13.707	13.561	13.434	13.324	13.227
8	11.767	11.540	11.353	11.195	11.060	10.943	10.841	10.752
9	10.107	9.894	9.718	9.570	9.443	9.334	9.238	9.154
10	8.956	8.754	8.587	8.445	8.325	8.220	8.129	8.048
11	8.116	7.922	7.761	7.626	7.510	7.409	7.321	7.244
12	7.480	7.292	7.136	7.005	6.892	6.794	6.709	6.634
13	6.982	6.799	6.647	6.519	6.409	6.314	6.231	6.158
14	6.583	6.404	6.256	6.130	6.023	5.930	5.848	5.776
15	6.256	6.081	5.935	5.812	5.707	5.615	5.535	5.464
16	5.984	5.812	5.668	5.547	5.443	5.353	5.274	5.205
17	5.754	5.584	5.443	5.324	5.221	5.132	5.054	4.986
18	5.558	5.390	5.251	5.132	5.031	4.943	4.866	4.798
19	5.388	5.222	5.084	4.967	4.867	4.780	4.704	4.636
20	5.239	5.075	4.939	4.823	4.724	4.637	4.562	4.495
21	5.109	4.946	4.811	4.696	4.598	4.512	4.437	4.371
22	4.993	4.832	4.697	4.583	4.486	4.401	4.326	4.260
23	4.890	4.730	4.596	4.483	4.386	4.301	4.227	4.162
24	4.797	4.638	4.505	4.393	4.296	4.212	4.139	4.074
25	4.713	4.555	4.423	4.312	4.216	4.132	4.059	3.994
26	4.637	4.480	4.349	4.238	4.142	4.059	3.986	3.921
28	4.505	4.349	4.219	4.109	4.014	3.932	3.859	3.795
30	4.393	4.239	4.110	4.001	3.907	3.825	3.753	3.689
32	4.298	4.145	4.017	3.908	3.815	3.733	3.662	3.598
34	4.215	4.063	3.936	3.828	3.735	3.654	3.583	3.520
36	4.144	3.992	3.866	3.758	3.666	3.585	3.514	3.451
38	4.080	3.930	3.804	3.697	3.605	3.524	3.454	3.391
40	4.024	3.874	3.749	3.642	3.551	3.471	3.400	3.338
60	3.687	3.541	3.419	3.315	3.226	3.147	3.078	3.017
80	3.530	3.386	3.265	3.162	3.074	2.996	2.927	2.867
100	3.439	3.296	3.176	3.074	2.986	2.908	2.840	2.780
150	3.321	3.180	3.061	2.959	2.872	2.795	2.727	2.667
200	3.264	3.123	3.005	2.904	2.816	2.740	2.672	2.612
300	3.207	3.067	2.950	2.849	2.762	2.686	2.618	2.558
400	3.179	3.040	2.922	2.822	2.735	2.659	2.592	2.532
∞	3.097	2.959	2.842	2.742	2.656	2.580	2.513	2.453

Table B.24: *99.9th percentiles of the F distribution.*

Den.	Numerator degrees of freedom							
df	18	20	25	30	40	50	100	200
1	619201	620922	624031	626114	628725	630301	633455	635033
2	999.46	999.46	999.47	999.48	999.49	999.49	999.50	999.50
3	126.74	126.42	125.84	125.45	124.96	124.67	124.07	123.77
4	46.322	46.101	45.699	45.429	45.089	44.884	44.470	44.261
5	25.568	25.395	25.080	24.869	24.602	24.441	24.115	23.951
6	17.267	17.120	16.853	16.673	16.445	16.307	16.028	15.887
7	13.063	12.932	12.692	12.530	12.326	12.202	11.951	11.824
8	10.601	10.480	10.258	10.109	9.919	9.804	9.571	9.453
9	9.012	8.898	8.689	8.548	8.369	8.260	8.039	7.926
10	7.913	7.804	7.604	7.469	7.297	7.193	6.980	6.872
11	7.113	7.008	6.815	6.684	6.518	6.417	6.210	6.105
12	6.507	6.405	6.217	6.090	5.928	5.829	5.627	5.524
13	6.034	5.934	5.751	5.626	5.467	5.370	5.172	5.070
14	5.655	5.557	5.377	5.254	5.098	5.002	4.807	4.706
15	5.345	5.248	5.071	4.950	4.796	4.702	4.508	4.408
16	5.087	4.992	4.817	4.697	4.545	4.451	4.259	4.160
17	4.869	4.775	4.602	4.484	4.332	4.239	4.049	3.950
18	4.683	4.590	4.418	4.301	4.151	4.058	3.868	3.770
19	4.522	4.430	4.259	4.143	3.994	3.902	3.713	3.615
20	4.382	4.290	4.121	4.005	3.856	3.765	3.576	3.478
21	4.258	4.167	3.999	3.884	3.736	3.645	3.456	3.358
22	4.149	4.058	3.891	3.776	3.629	3.538	3.349	3.251
23	4.051	3.961	3.794	3.680	3.533	3.442	3.254	3.156
24	3.963	3.873	3.707	3.593	3.447	3.356	3.168	3.070
25	3.884	3.794	3.629	3.516	3.369	3.279	3.091	2.992
26	3.812	3.723	3.558	3.445	3.299	3.208	3.020	2.921
28	3.687	3.598	3.434	3.321	3.176	3.085	2.897	2.798
30	3.581	3.493	3.330	3.217	3.072	2.981	2.792	2.693
32	3.491	3.403	3.240	3.128	2.983	2.892	2.703	2.603
34	3.413	3.325	3.163	3.051	2.906	2.815	2.625	2.524
36	3.345	3.258	3.096	2.984	2.839	2.748	2.557	2.456
38	3.285	3.198	3.036	2.925	2.779	2.689	2.497	2.395
40	3.232	3.145	2.984	2.872	2.727	2.636	2.444	2.341
60	2.912	2.827	2.667	2.555	2.409	2.316	2.118	2.009
80	2.763	2.677	2.518	2.406	2.258	2.164	1.960	1.846
100	2.676	2.591	2.431	2.319	2.170	2.076	1.867	1.749
150	2.564	2.479	2.319	2.206	2.056	1.959	1.744	1.618
200	2.509	2.424	2.264	2.151	2.000	1.902	1.682	1.552
300	2.456	2.371	2.210	2.097	1.944	1.846	1.620	1.483
400	2.429	2.344	2.184	2.070	1.917	1.817	1.589	1.448
∞	2.351	2.266	2.105	1.990	1.835	1.733	1.495	1.338

References

Agresti, A. and Coull, B.A. (1998), Approximate is Better than Exact for Interval Estimation of Binomial Proportions. *The American Statistician*, **52**, 119–126.

Aitchison, J. and Dunsmore, I. R. (1975). *Statistical Prediction Analysis*. Cambridge University Press, Cambridge.

Atkinson, A. C. (1973). Testing transformations to normality. *Journal of the Royal Statistical Society, Series B*, **35**, 473–479.

Atkinson, A. C. (1985). *Plots, Transformations, and Regression: An Introduction to Graphical Methods of Diagnostic Regression Analysis*. Oxford University Press, Oxford.

Bailey, D. W. (1953). *The Inheritance of Maternal Influences on the Growth of the Rat*. Ph.D. Thesis, University of California.

Baten, W. D. (1956). An analysis of variance applied to screw machines. *Industrial Quality Control*, **10**, 8–9.

Beineke, L. A. and Suddarth, S. K. (1979). Modeling joints made with light-gauge metal connector plates. *Forest Products Journal*, **29**, 39–44.

Berry, D. A. (1996). *Statistics: A Bayesian Perspective*. Wadsworth, Belmont, CA.

Bethea, R. M., Duran, B. S., and Boullion, T. L. (1985). *Statistical Methods for Engineers and Scientists*, Second Edition. Marcel Dekker, New York.

Bickel, P. J., Hammel, E. A., and O'Conner, J. W. (1975). Sex bias in graduate admissions: Data from Berkeley. *Science*, **187**, 398–404.

Bisgaard, S. and Fuller, H. T. (1995). Reducing variation with two-level factorial experiments. *Quality Engineering*, **8**, 373–377.

Bissell, A. F. (1972). A negative binomial model with varying element sizes. *Biometrika*, **59**, 435–441.

Bliss, C. I. (1947). 2×2 factorial experiments in incomplete groups for use in biological assays. *Biometrics*, **3**, 69–88.

Bliss, C. I. and James, A. T. (1966). Fitting the rectangular hyperbola. *Biometrics*, **22**, 573–602.

Box, G. E. P. (1950). Problems in the analysis of growth and wear curves. *Biometrics*, **6**, 362–389.

Box, G. E. P. and Cox, D. R. (1964). An analysis of transformations. *Journal of the Royal Statistical Society, Series B*, **26**, 211–246.

Box, G. E. P. and Draper, N. R. (1987). *Empirical Model-Building and Response Surfaces*. John Wiley and Sons, New York.

Box, G. E. P. and Tidwell, P. W. (1962). Transformations of the independent variables. *Technometrics*, **4**, 531–550.

Brownlee, K. A. (1960). *Statistical Theory and Methodology in Science and Engineering*. John Wiley and Sons, New York.

Burt, C. (1966). The genetic determination of differences in intelligence: A study of monozygotic twins reared together and apart. *Br. J. Psych.*, **57**, 137–153.

Carroll, R. J. and Ruppert, D. (1984). Power transformations when fitting theoretical models to data. *Journal of the American Statistical Association*, **79**, 321–328.

Casella, G. (2008). *Statistical Design*. Springer-Verlag, New York.

Cassini, J. (1740). *Eléments d'astronomie*. Imprimerie Royale, Paris.

Chapman, R. E., Masinda, K. and Strong, A. (1995). Designing simple reliability experiments. *Quality Engineering*, **7**, 567–582.

Christensen, R. (1987). *Plane Answers to Complex Questions: The Theory of Linear Models*. Springer-Verlag, New York.

Christensen, R. (1996). *Analysis of Variance, Design, and Regression: Applied Statistical Methods*. Chapman and Hall/CRC, Boca Raton, FL.

Christensen, R. (1997). *Log-Linear Models and Logistic Regression*, Second Edition. Springer-Verlag, New York.

Christensen, R. (2001). *Advanced Linear Modeling* (previously *Linear Models for Multivariate, Time Series, and Spatial Data*). Springer-Verlag, New York.

Christensen, Ronald (2000). Linear and log-linear models. *Journal of the American Statistical Association*, **95**, 1290-1293.

Christensen, R. (2011). *Plane Answers to Complex Questions: The Theory of Linear Models*, Fourth Edition. Springer-Verlag, New York.

Christensen, R. and Bedrick, E. J. (1997). Testing the independence assumption in linear models. *Journal of the American Statistical Association*, **92**, 1006–1016.

Christensen, R., Johnson, W., Branscum, A. and Hanson, T. E. (2010). *Bayesian Ideas and Data Analysis*, CRC Press, Boca Raton.

Cochran, W. G. and Cox, G. M. (1957). *Experimental Designs*, Second Edition. John Wiley and Sons, New York.

Coleman, D. E. and Montgomery, D. C. (1993). A systematic approach to planning for a designed industrial experiment (with discussion). *Technometrics*, **35**, 1–27.

Conover, W. J. (1971). *Practical Nonparametric Statistics*. John Wiley and Sons, New York.

Cook, R. D. and Tsai, C.-L. (1985). Residuals in nonlinear regression. *Biometrika*, **72**, 23–29.

Cook, R. D. and Tsai, C.-L. (1990). Diagnostics for assessing the accuracy of normal approximations in exponential family nonlinear models. *Journal of the American Statistical Association*, **85**, 770–777.

Cook, R. D. and Weisberg, S. (1982). *Residuals and Influence in Regression*. Chapman and Hall, New York.

Cornell, J. A. (1988). Analyzing mixture experiments containing process variables. A split plot approach. *Journal of Quality Technology*, **20**, 2–23.

Cox, D. R. (1958). *Planning of Experiments*. John Wiley and Sons, New York.

Cox, D. R. and Reid, N. (2000). The Theory of the Design of Experiments. Chapman and Hall/CRC, Boca Raton, FL.

Cramér, H. (1946). *Mathematical Methods of Statistics*. Princeton University Press, Princeton.

Dalal, S.R., Fowlkes, E.B., and Hoadley, B. (1989). Risk analysis of the space shuttle: Pre-*Challenger* prediction of failure. *Journal of the American Statistical Association*, **84**, 945–957.

David, H. A. (1988). *The Method of Paired Comparisons*. Methuen, New York.

Day, B. B. and del Priore, F. R. (1953). The statistics in a gear-test program. *Industrial Quality Control*, **7**, 16–20.

Day, N. E. (1966). Fitting curves to longitudinal data. *Biometrics*, **22**, 276–291.

Deming, W. E. (1986). *Out of the Crisis*. MIT Center for Advanced Engineering Study, Cambridge, MA.

Devore, Jay L. (1991). *Probability and Statistics for Engineering and the Sciences*, Third Edition. Brooks/Cole, Pacific Grove, CA.

Dixon, W. J. and Massey, F. J., Jr. (1969). *Introduction to Statistical Analysis*, Third Edition. McGraw-Hill, New York.

Dixon, W. J. and Massey, F. J., Jr. (1983). *Introduction to Statistical Analysis*, Fourth Edition. McGraw-Hill, New York.

Draper, N. and Smith, H. (1966). *Applied Regression Analysis*. John Wiley and Sons, New York.

Emerson, J. D. (1983). Mathematical aspects of transformation. In *Understanding Robust and Exploratory Data Analysis*, edited by D.C. Hoaglin, F. Mosteller, and J.W. Tukey. John Wiley and Sons, New York.

Everitt, B. J. (1977). *The Analysis of Contingency Tables*. Chapman and Hall, London.

Feigl, P. and Zelen, M. (1965). Estimation of exponential probabilities with concomitant information. *Biometrics*, **21**, 826–838.

Fienberg, S. E. (1980). *The Analysis of Cross-Classified Categorical Data*, Second Edition. MIT Press, Cambridge, MA. Reprinted in 2007 by Springer-Verlag.

Finney, D. J. (1964). *Statistical Method in Biological Assay*, Second Edition. Hafner Press, New York.

Fisher, R. A. (1925). *Statistical Methods for Research Workers*, 14th Edition, 1970. Hafner Press, New York.

Fisher, R. A. (1935). *The Design of Experiments*, Ninth Edition, 1971. Hafner Press, New York.

Fisher, R. A. (1947). The analysis of covariance method for the relation between a part and the whole. *Biometrics*, **3**, 65–68.

Forbes, J. D. (1857). Further experiments and remarks on the measurement of heights by the boiling point of water. *Transactions of the Royal Society of Edinburgh*, **21**, 135– 143.

Fuchs, C. and Kenett, R. S. (1987). Multivariate tolerance regions on F-tests. *Journal of Quality Technology*, **19**, 122–131.

Garner, N. R. (1956). Studies in textile testing. *Industrial Quality Control*, **10**, 44–46.

Hader, R. J. and Grandage, A. H. E. (1958). Simple and multiple regression analyses. In *Experimental Designs in Industry*, edited by V. Chew, pp. 108–137. John Wiley and Sons, New York.

Hahn, G. J. and Meeker, W. Q. (1993). Assumptions for statistical inference. *The American Statistician*, **47**, 1–11.

Heyl, P. R. (1930). A determination of the constant of gravitation. *Journal of Research of the National Bureau of Standards*, **5**, 1243–1250.

Hinkelmann, K. and Kempthorne, O. (2005). *Design and Analysis of Experiments: Volume 2, Advanced Experimental Design*. John Wiley and Sons, Hoboken, NJ.

Hinkelmann, K. and Kempthorne, O. (2008). *Design and Analysis of Experiments: Volume 1, Introduction to Experimental Design*, Second Edition. John Wiley and Sons, Hoboken, NJ.

Hochberg, Y. and Tamhane, A. (1987). *Multiple Comparison Procedures*. John Wiley and Sons, New York.

Hosmer, D. W., Hosmer, T., le Cessie, S. and Lemeshow, S. (1997). A comparison of goodness-of-fit tests for the logistic regression model. *Statistics in Medicine*, **16(9)**, 965–980.

Hsu, J.C. (1996). *Multiple Comparisons, Theory and methods*. Chapman & Hall, Boca Raton.

Inman, J., Ledolter, J., Lenth, R. V., and Niemi, L. (1992). Two case studies involving an optical emission spectrometer. *Journal of Quality Technology*, **24**, 27–36.

Jaswal, I. S., Mann, R. F., Juusola, J. A., and Downie, J. (1969). The vapour-phase oxidation of benzene over a vanadium pentoxide catalyst. *Canadian Journal of Chemical Engineering*, **47**, No. 3, 284–287.

Jensen, R. J. (1977). Evinrude's computerized quality control productivity. *Quality Progress*, **X, 9**, 12–16.

John, P. W. M. (1961). An application of a balanced incomplete block design. *Technometrics*, **3**, 51–54.

John, P. W. M. (1971). *Statistical Design and Analysis of Experiments*. Macmillan, New York.

Johnson, F. J. (1978). Automated determination of phosphorus in fertilizers: Collaborative study. *Journal of the Association of Official Analytical Chemists*, **61**, 533–536.

Johnson, D. E. (1998). *Applied multivariate methods for data analysts*. Duxbury Press, Belmont, CA.

Johnson, R. A. and Wichern, D. W. (2007). *Applied Multivariate Statistical Analysis*, Sixth Edition. Prentice-Hall, Englewood Cliffs, NJ.

Jolicoeur, P. and Mosimann, J. E. (1960). Size and shape variation on the painted turtle: A principal component analysis. *Growth*, **24**, 339–354.

Kempthorne, O. (1952). *Design and Analysis of Experiments*. Krieger, Huntington, NY.

Kihlberg, J. K., Narragon, E. A., and Campbell, B. J. (1964). Automobile crash injury in relation to car size. Cornell Aero. Lab. Report No. VJ-1823-Rll.

Koopmans, L. H. (1987). *Introduction to Contemporary Statistical Methods*, Second Edition. Duxbury Press, Boston.

Lazerwitz, B. (1961). A comparison of major United States religious groups. *Journal of the American Statistical Association*, **56**, 568–579.

Lehmann, E. L. (1975). *Nonparametrics: Statistical Methods Based on Ranks*. Holden-Day, San Francisco.

Lindgren, B.W. (1968). *Statistical Theory*, Second Edition. Macmillan, New York.

McCullagh, P. (2000). Invariance and factorial models, with discussion. *Journal of the Royal Statistical Society, Series B*, **62**, 209-238.

McCullagh, P. and Nelder, J. A. (1989). *Generalized Linear Models*, Second Edition. Chapman and Hall, London.

McDonald, G. C. and Schwing, R. C. (1973). Instabilities in regression estimates relating air pollution to mortality. *Technometrics*, **15**, 463–481.

Mandel, J. (1972). Repeatability and reproducibility. *Journal of Quality Technology*, **4**, 74–85.

Mandel, J. (1989a). Some thoughts on variable-selection in multiple regression. *Journal of Quality Technology*, **21**, 2–6.

Mandel, J. (1989b). The nature of collinearity. *Journal of Quality Technology*, **21**, 268–276.

Mandel, J. and Lashof, T. W. (1987). The nature of repeatability and reproducibility. *Journal of Quality Technology*, **19**, 29–36.

Marquardt, D. W. (1963). An algorithm for least-squares estimation of nonlinear parameters. *SIAM Journal of Applied Mathematics*, **11**, 431–441.

May, J. M. (1952). Extended and corrected tables of the upper percentage points of the studentized range. *Biometrika*, **39**, 192–193.

Mercer. W. B. and Hall, A. D. (1911). The experimental error of field trials. *Journal of Agricultural Science*, **iv**, 107–132.

Miller, R. G., Jr. (1981). *Simultaneous Statistical Inference*, Second Edition. Springer-Verlag, New York.

Milliken, G. A. and Graybill, F. A. (1970). Extensions of the general linear hypothesis model. *Journal of the American Statistical Association*, **65**, 797–807.

Mosteller, F. and Tukey, J. W. (1977). *Data Analysis and Regression*. Addison-Wesley, Reading, MA.

Mulrow, J. M., Vecchia, D. F., Buonaccorsi, J. P., and Iyer, H. K. (1988). Problems with interval estimation when data are adjusted via calibration. *Journal of Quality Technology*, **20**, 233–247.

Nelson, P. R. (1993). Additional uses for the analysis of means and extended tables of critical values. *Technometrics*, **35**, 61–71.

Ott, E. R. (1949). Variables control charts in production research. *Industrial Quality Control*, **3**, 30–31.

Ott, E. R. and Schilling, E. G. (1990). *Process Quality Control: Trouble Shooting and Interpretation of Data*, Second Edition. McGraw-Hill, New York.

Patterson, H. D. (1950). The analysis of change-over trials. *Journal of Agricultural Science*, **40**, 375–380.

Pauling, L. (1971). The significance of the evidence about ascorbic acid and the common cold. *Proceedings of the National Academy of Science*, **68**, 2678–2681.

Pritchard, D. J., Downie, J., and Bacon, D. W. (1977). Further consideration of heteroscedasticity in fitting kinetic models. *Technometrics*, **19**, 227–236.

Quetelet, A. (1842). *A Treatise on Man and the Development of His Faculties*. Chambers, Edinburgh.

Rao, C. R. (1965). *Linear Statistical Inference and Its Applications*. John Wiley and Sons, New York.

Rao, C. R. (1973). *Linear Statistical Inference and Its Applications*, Second Edition. John Wiley and Sons, New York.

Reiss, I. L., Banward, A., and Foreman, H. (1975). Premarital contraceptive usage: A study and some theoretical explorations. *Journal of Marriage and the Family*, **37**, 619–630.

Ryan, T. P. (1989). *Statistical Methods for Quality Improvement*. John Wiley and Sons, New York.

St. Laurent, R. T. and Cook, R. D. (1992). Leverage and superleverage in nonlinear regression. *Journal of the American Statistical Association*, **87**, 985–990.

Satterthwaite, F. E. (1946). An approximate distribution of estimates of variance components. *Biometrics*, **2**, 110–114.

Scheffé, H. (1959). *The Analysis of Variance*. John Wiley and Sons, New York.

Schneider, H. and Pruett, J. M. (1994). Control charting issues in the process industries. *Quality Engineering*, **6**, 345–373.

Seber, G. A. F. and Wild, C. J. (1989). *Nonlinear Regression*. John Wiley and Sons, New York. (The 2003 version appears to be just a reprint of this.)

Shapiro, S. S. and Francia, R. S. (1972). An approximate analysis of variance test for normality. *Journal of the American Statistical Association*, **67**, 215–216.

Shewhart, W. A. (1931). *Economic Control of Quality*. Van Nostrand, New York.

Shewhart, W. A. (1939). *Statistical Method from the Viewpoint of Quality Control*. Graduate School of the Department of Agriculture, Washington. Reprint (1986), Dover, New York.

Shumway, R. H. and Stoffer, D. S. (2000). *Time Series Analysis and Its Applications*. Springer-Verlag, New York.

Smith, H., Gnanadesikan, R., and Hughes, J. B. (1962). Multivariate analysis of variance (MANOVA). *Biometrics*, **18**, 22–41.

Snedecor, G. W. (1945a). Query. *Biometrics*, **1**, 25.

Snedecor, G. W. (1945b). Query. *Biometrics*, **1**, 85.

Snedecor, G. W. and Cochran, W. G. (1967). *Statistical Methods*, Sixth Edition. Iowa State University Press, Ames, IA.

Snedecor, G. W. and Cochran, W. G. (1980). *Statistical Methods*, Seventh Edition. Iowa State University Press, Ames, IA.

Snedecor, G. W. and Haber, E. S. (1946). Statistical methods for an incomplete experiment on a perennial crop. *Biometrics*, **2**, 61–69.

Stigler, S. M. (1986). *The History of Statistics*. Harvard University Press, Cambridge, MA.

Sulzberger, P. H. (1953). The effects of temperature on the strength of wood, plywood and glued joints. Aeronautical Research Consultative Committee, Australia, Department of Supply, Report ACA–46.

Tibshirani, R. (1996). Regression shrinkage and selection via the lasso. *Journal of the Royal Statistical Society, Series B*, **58**, 267–288.

Tukey, J. W. (1949). One degree of freedom for nonadditivity. *Biometrics*, **5**, 232–242.

Utts, J. (1982). The rainbow test for lack of fit in regression. *Communications in Statistics—Theory and Methods*, **11**, 2801–2815.

Wahba, G. (1990). *Spline Models for Observational Data*. (Vol. 59, CBMS-NSF Regional Conference Series in Applied Mathematics.) SIAM, Philadelphia.

Watkins, D., Bergman, A., and Horton, R. (1994). Optimization of tool life on the shop floor using design of experiments. *Quality Engineering*, **6**, 609–620.

Weisberg, S. (1985). *Applied Linear Regression*. Second Edition. John Wiley and Sons, New York.

Williams, E. J. (1959). *Regression Analysis*. John Wiley and Sons, New York.

Wilm, H. G. (1945). Notes on analysis of experiments replicated in time. *Biometrics*, **1**, 16–20.

Woodward, G., Lange, S. W., Nelson, K. W., and Calvert, H. O. (1941). The acute oral toxicity of acetic, chloracetic, dichloracetic and trichloracetic acids. *Journal of Industrial Hygiene and Toxicology*, **23**, 78–81.

Younger, M. S. (1979). *A Handbook for Linear Regression*, Duxbury Press, Belmont, CA.

Author Index

Subject Index

Printed in the United States
by Baker & Taylor Publisher Services